Elements of Spacecraft Design

Elements of Spacecraft Design

Charles D. Brown
Wren Software, Inc.
Castle Rock, Colorado

EDUCATION SERIES
J. S. Przemieniecki
Series Editor-in-Chief

Published by
American Institute of Aeronautics and Astronautics, Inc.
1801 Alexander Bell Drive, Reston, VA 20191-4344

American Institute of Aeronautics and Astronautics, Inc., Reston, Virginia

2 3 4 5

Library of Congress Cataloging-in-Publication Data

Brown, Charles D., 1930–
Elements of spacecraft design / Charles D. Brown.
p. cm.
Includes bibliographical references and index.
1. Space vehicles—Design and construction. I. Title.
TL875 .B76 2002 629.47—dc21 2002010232
ISBN 1-56347-524-3 (alk. paper)

Foreword

The latest text by Charles Brown, *Elements of Spacecraft Design*, complements his other two texts, *Spacecraft Mission Design* and *Spacecraft Propulsion*, previously published in this series. This new text starts first with a comprehensive discussion of the conceptual stages of mission design, systems engineering, and orbital mechanics, all of which provide the basis for the design process for different components and functions. Included are propulsion and power systems, structures, attitude control, thermal control, command and data systems, and telecommunications. This text evolved from the spacecraft design course taught by the author for many years at the University of Colorado.

The author is eminently qualified to write on the subject of spacecraft design, having been involved in directing various design teams at Martin Marietta, for the Mariner 9, the first spacecraft to orbit another planet in 1971, the Viking orbiter spacecraft, and the Magellan spacecraft, which produced the first high-resolution imaging of the planet Venus and was the first planetary spacecraft to fly on the Shuttle. In 1992, Charles Brown received the Goddard Memorial Trophy for his Magellan project leadership and the NASA Public Service Medal, just to mention a few of his accomplishments and awards.

The AIAA Education Series of textbooks and monographs, inaugurated in 1984, embraces a broad spectrum of theory and application of different disciplines in aeronautics and astronautics, including aerospace design practice. The series also includes texts on defense science, engineering, and management. The books serve as both teaching texts for students and reference materials for practicing engineers, scientists, and managers. The complete list of textbooks published in the series can be found on the end pages of this volume.

J. S. Przemieniecki
Editor-in-Chief
AIAA Education Series

Table of Contents

1
Introduction

1.1 First Spacecraft

The first spacecraft was a publicity stunt. At the conclusion of World War II, the United States and the USSR eyed each other across a destroyed Europe, neither trusting the other, both holding the recipe for the atomic bomb. It was clear, from captured German work, that it was feasible to design a ballistic missile capable of delivering the bomb anywhere in the world. It was also clear that if the USSR were first to have the capability, they could destroy the United States in a few hours without any possibility of retaliation during the "missile gap" before we finished our development. Both nations strained to develop the first intercontinental missile capability, and the USSR was first to have it. They could not just call the president and say "We've got it." They decided instead to make the announcement in a very dramatic way. It turns out that if you can put a given-size bomb on an intercontinental trajectory, you can put almost that much mass into low Earth orbit.

The USSR announced their intercontinental missile capability to the world on 4 October 1957 by launching the world's first spacecraft, Sputnik I, into low Earth orbit. The impact of Sputnik (Fig. 1.1), on public opinion was immense. The airspace over the continental United States was never penetrated during World War II despite the efforts of the two largest air forces ever assembled. Now Sputnik cut through our airspace with impunity every 90 min.

It had been our national outlook that technical superiority counterbalanced the huge human resources of the USSR. Suddenly and dramatically our technical superiority was undermined. President Eisenhower spoke on television, then new, to calm national fears. An outright race ensued between the two countries, which came to be called the "space race." Clearly, the USSR won the first inning.

The spacecraft that changed world opinion was not impressive by today's standards. It was an 84-kg, 58-cm-diam ball that contained a battery, transmitter, and whip antenna. The transmitter produced a monotonous beep that could be readily received anywhere in the industrialized world. The beep continued incessantly until the spacecraft reentered in January 1958.

After several embarrassing failures, the United States responded with the launch of our first spacecraft, Explorer I, on 31 January 1958. The 14-kg Explorer I is shown in Fig. 1.2; the lower half is actually a solid motor, which provided the final velocity increment. The spacecraft is the upper half, which consisted of a particles and fields experiment, micrometeorite experiment, cosmic ray detector, and a low power transmitter. Explorer I was in orbit for two months during which it discovered the Van Allen belt.

By the late 1960s the United States spacecraft launch capability had grown from 14 to 56,000 kg, and we had sent manned spacecraft to the moon.

Fig. 1.1 Replica of Sputnik I—the first spacecraft. (Courtesy of National Air and Space Museum, Smithsonian Institution; photo by Mark Avino, SI Neg. No. 87-14645.)

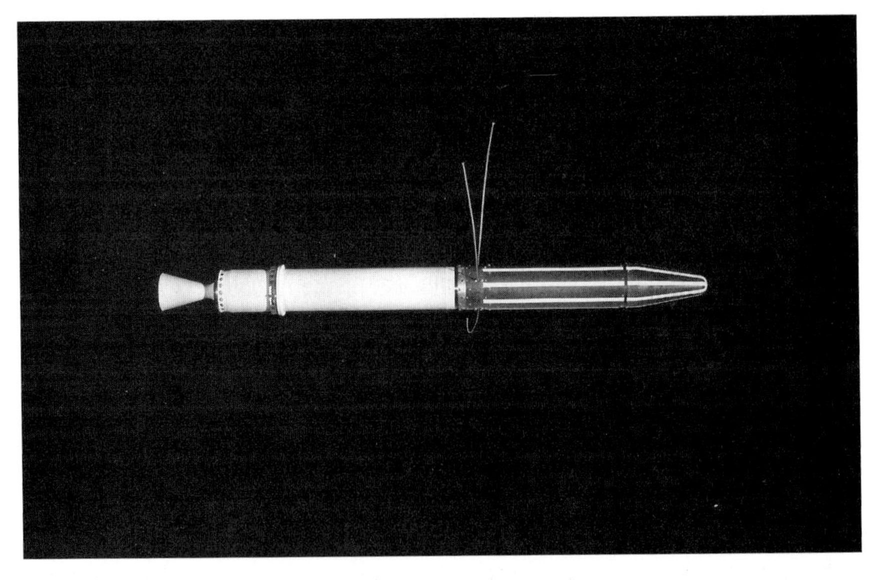

Fig. 1.2 Replica of Explorer I—first U.S. spacecraft. (Courtesy of National Air and Space Museum, Smithsonian Institution; photo by Dane Penland, SI Neg. No. 80-4976.)

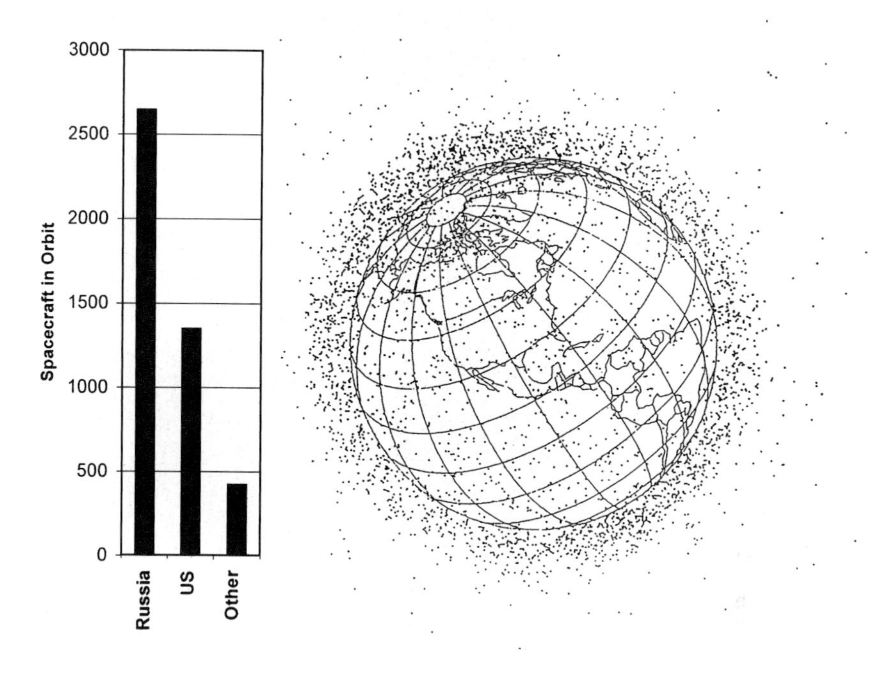

Fig. 1.3 Crowd in Earth orbit.

Since Sputnik, there have been 4400 successful spacecraft launches (through 2000) by all nations.[2,3] About 60% of these were launched by the USSR/CIS, about half that many by the United States, and about 10% by the rest of the world (see Fig. 1.3). Spacecraft launches have changed in recent years in three major ways:

1) The launch rate of the USSR/CIS has declined as economic problems increased. In 1996 the launches by Russia fell below those of the United States for the first time in 30 years.

2) In the early years spacecraft were all built and launched by governments. In recent years U.S. commercial launches almost equal government launches. Commercial launches, particularly Iridium and Globalstar, are also the cause of the recent increases in the U.S. launch rate.

3) The national character of spacecraft launches has diminished. It is now common for a U.S. commercial spacecraft to be launched on a Russian launch vehicle and contain equipment from many countries. In 1960 that would have been unheard of; a launch contained equipment of a single nation from launch bolts to nose cone.

1.2 Spacecraft Missions

You might well ask, "What are all of these spacecraft doing?" They are doing more things than anyone could possibly have imagined in 1957. These myriad missions can be sorted into three primary classes: 1) Earth orbiters, 2) planetary and lunar explorers, and 3) manned missions, as shown in Fig. 1.4. The first spacecraft to

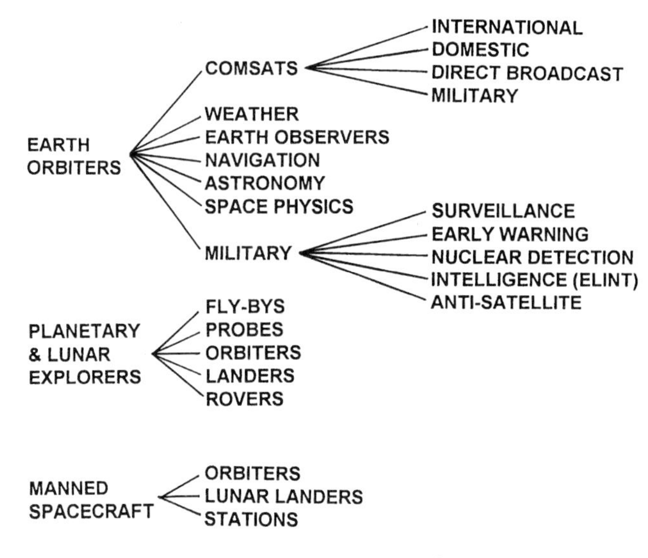

Fig. 1.4 Spacecraft missions.

leave the Earth's sphere of influence was Luna I in January 1959. It was intended for the moon but missed its target by 37,600 miles. The first U.S. spacecraft to attempt the same mission was Pioneer 4. Both spacecraft are still in orbit about the sun.

Since the advent of spacecraft in 1957, these odd devices have provided seven classes of service to mankind that cannot be provided in any other way, at least not well in any other way.

1.2.1 Communication

Few activities have provided the tremendous benefit that communication spacecraft have. As late as the 1960s, calls went overseas by cable. Six-hundred sixty of them connected the United States with the rest of the world. Circuits were busy and expensive. The phrase "like trying to call Paris" was a common expression for any task that was time consuming, frustrating, and failure prone.

By 1998 we watched the Olympics in Japan brought to us by communication spacecraft in near real time and thought nothing of it. Today we can go to an automated teller machine in rural Spain, withdraw from our home-town bank, and get our money in local currency. The entire transaction takes place by communication satellite in less time than it takes to tell about it. The phenomenal growth of satellite communication is shown in Fig. 1.5.

Current advanced communications spacecraft designs are striving to provide personal television, voice, and Internet services to any point on the face of the Earth. In theory, a customer could sit in the middle of the Gobi Desert and watch a baseball game on television, or phone a friend in New York, or conduct research on the Internet. The global coverage is of particular interest to developing nations because an enormous investment in Earth-based cabling would be avoided.

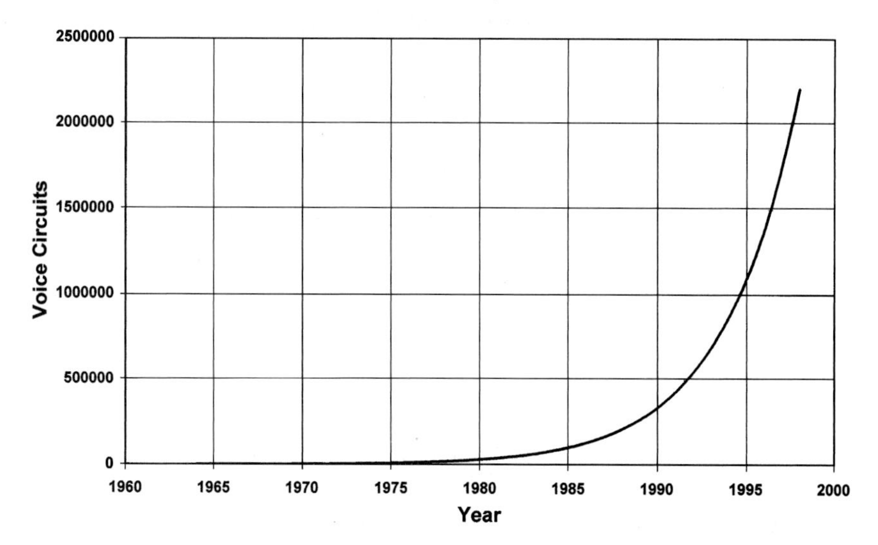

Fig. 1.5 Transoceanic telephone service.[4,5]

1.2.2 Weather

Early in the 19th century weather reports were gathered by mail. It was noted that the great storms were rotary in nature and moved from place to place. Construction of the first telegraph line in 1884 made it possible to gather reports on the weather from many places simultaneously and also to track its changes and movement. For the first time only a little over a century ago it was possible to make relatively scientific weather forecasts, and some of the larger countries established storm-warning services. The U.S. weather service was first established under the U.S. Army Signal Corps in 1870 and became the U.S. Weather Bureau in 1891 (Ref. 5).

Meanwhile investigators had been sending weather instruments up on kites and balloons to learn more about the upper atmosphere; these first atmospheric "soundings" were made in the mid-18th century. However, there was little progress in this direction until the 20th century, when the growth in aviation made it essential to have information about winds, clouds, icing, and turbulence aloft. By 1938 daily soundings were being made with radiosondes—balloons with weather instruments and radio transmitters to return the data. Directional tracking capabilities were added after 1940. These instruments, coupled with reports from ordinary weather stations, from ships at sea, and from pilots of aircraft, provided the raw data for meteorologists to use in making their science more exact. The rapid development of electronic computers in the decades following 1940 made it possible for the meteorologists to use all of these data without being inundated.

Today the NOAA spacecraft (Fig. 1.6) observe world-wide motion of weather systems and provide these observations four times a day. Forecast weather based on these observations are available at least twice per day and are available continuously on the Internet. All nations can access the National Oceanic and Atmospheric Administration (NOAA), and for many NOAA is the only weather forecasting reference. The instrument suite meets the data requirements of 140 nations.

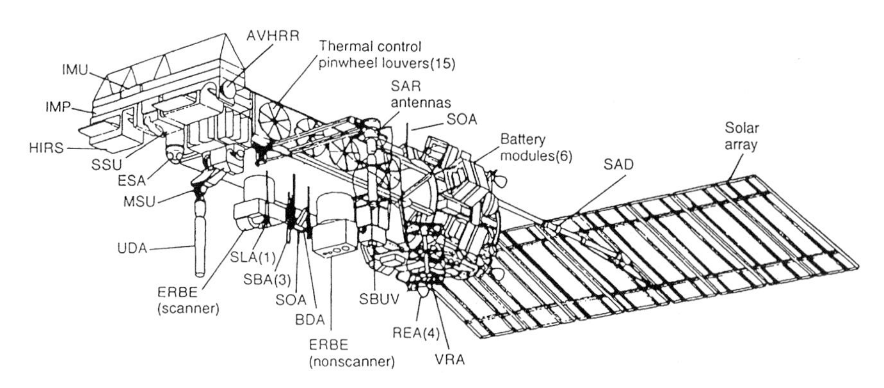

Fig. 1.6 NOAA 9 meteorology spacecraft. (Courtesy of Lockheed Martin.)

1.2.3 Navigation

The use of heavenly bodies—i.e., the sun, the moon, stars, and planets—for purposes of navigation was started centuries ago and was utilized extensively by the early Portuguese navigators to explore this planet. These early explorers made position fixes by combining the known positions of the heavenly bodies with onboard position measurements made with an instrument called an *astrolabe*. The results of this simple technique were sufficiently accurate to allow navigators to find their approximate positions even when far from land. In the 18th century the sextant, compass, and star and sun tables were integrated with the clock to improve navigation performance significantly. By the 1940s the use of radio signal direction finding led to Loran and Omega, which were much more accurate systems. In general, navigators thus equipped were well satisfied to find a position within a mile of the correct one.

Today a hand-held unit can use global positioning satellites to determine position anywhere in the world with an accuracy of less than 30 m (Ref. 6). The hand-held unit is available for less than $300.

1.2.4 Astronomy

The importance of spacecraft as a tool for astronomy was recognized immediately. Spacecraft avoid the optical distortion of the Earth's atmosphere and provide unique positions from which to observe. The current series of astronomy spacecraft are doing nothing less than revolutionizing our understanding of astronomy. The current great observatories of the United States and the European Space Agency (ESA) are as follows: 1) International Ultraviolet Explorer (IUE)[7,8] launched in 1978; 2) Hubble Space Telescope (HST), at 11.6 tons, is the first of NASA's great telescopes, launched in 1989; 3) Compton Gamma Ray Observatory (GRO), the second of NASA's great observatories, was launched in April 1991; 4) Chandra X-ray Observatory was launched 23 July 1999; 5) Hipparchus is an 1130-kg astronomy spacecraft launched by ESA in August 1989; and 6) SOHO, solar observatory mission, is a large, cooperative venture between ESA and NASA. A single example of the many accomplishments of these machines is the HST photograph of the three pillars of creation in the Eagle Nebula (Fig. 1.7), each pillar being more than a light year (6 trillion miles) long. Gaseous globules at the tips of each

Fig. 1.7 Pillars of creation in M16. (Courtesy of J. J. Hester, Arizona State University.)

pillar, larger than our solar system, are embryonic stars. This photograph would have been impossible to the point of absurdity, without the Hubble spacecraft.

1.2.5 Earth Resources

The first photo of the Earth taken from a satellite was made in 1959 by a camera aboard the Mark II Reentry Vehicle, which was launched by an early Thor rocket. The Gemini astronauts, using a hand-held Hasselblad camera, showed the potential of the view from space.

Remote sensing is now an established technology. Every 18 days a LANDSAT spacecraft crosses over every point on Earth, including your home. It can tell what type of crops are growing, it they are healthy or diseased, and if they need water. It can determine if ponds or lakes are clear, brackish, or salty. It makes these determinations on a global basis to a resolution of one-quarter acre.[9]

Table 1.1 Major achievements in manned space flights through STS-1

Spacecraft	Flag	Astronauts	Launch	Achievements
Vostok 1	USSR	Yuri A. Gagarin	12 April 1961	**First manned space flight.**
Mercury 3	USA	Alan B. Shepherd, Jr.	5 May 1961	**First American in space,** suborbital flight
Mercury 6	USA	John H. Glenn, Jr.	20 Feb. 1962	**First American in orbit,** 3 orbits
Vostok 6	USSR	Valentina V. Tereshkova	16 June 1963	**First woman in space.**
Voskhod 2	USSR	Aleksei A. Leonov Pavel I. Belyayev	18 March 1965	**First spacewalk outside spacecraft** (Leonov).
Gemini 8	USA	Neil Armstrong David R. Scott	16 March 1966	**First orbital docking**
Apollo 8	USA	Frank Borman James A. Lovell, Jr. William A. Anders	21 Dec. 1969	**First manned flight around the Moon.** 10 orbits.
Apollo 11	USA	Neil A. Armstrong Edwin E. Aldrin, Jr. Michael Collins	16 July 1969	**First lunar landing.** **First lunar EVA** (Armstrong, Aldrin).
Apollo 13	USA	James A. Lovell, Jr. Fred W. Haise, Jr. John L. Swigert	11 April 1970	**First rescue.** Failure of spacecraft oxygen tank about 56 h into the flight. Monumental effort and ingenuity resulted in safe return to Earth.
Salyut 1	USSR		19 April 1971	**First manned space station.**
Space Shuttle, STS-1	USA	Robert Crippen John Young	12 April 1981	**First reusable manned space vehicle.**

1.2.6 Manned Spacecraft

Manned spacecraft are the most dramatic space machines; the bulk of the NASA budget is spent on them and much has been written about them. Although manned spacecraft are not the focus of this book, Table 1.1 commemorates some of the major achievements in this field. In subsequent years these early achievements were followed by the Space Shuttle, Mir, Soyuz, and the Space Station. Through the end of 1996 there have been 192 manned flights of all nations, manned by a total of 354 different astronauts.

1.2.7 Planetary Exploration

One of the most interesting things we can do with spacecraft is the exploration of the solar system. Since 1957, spacecraft of the United States have

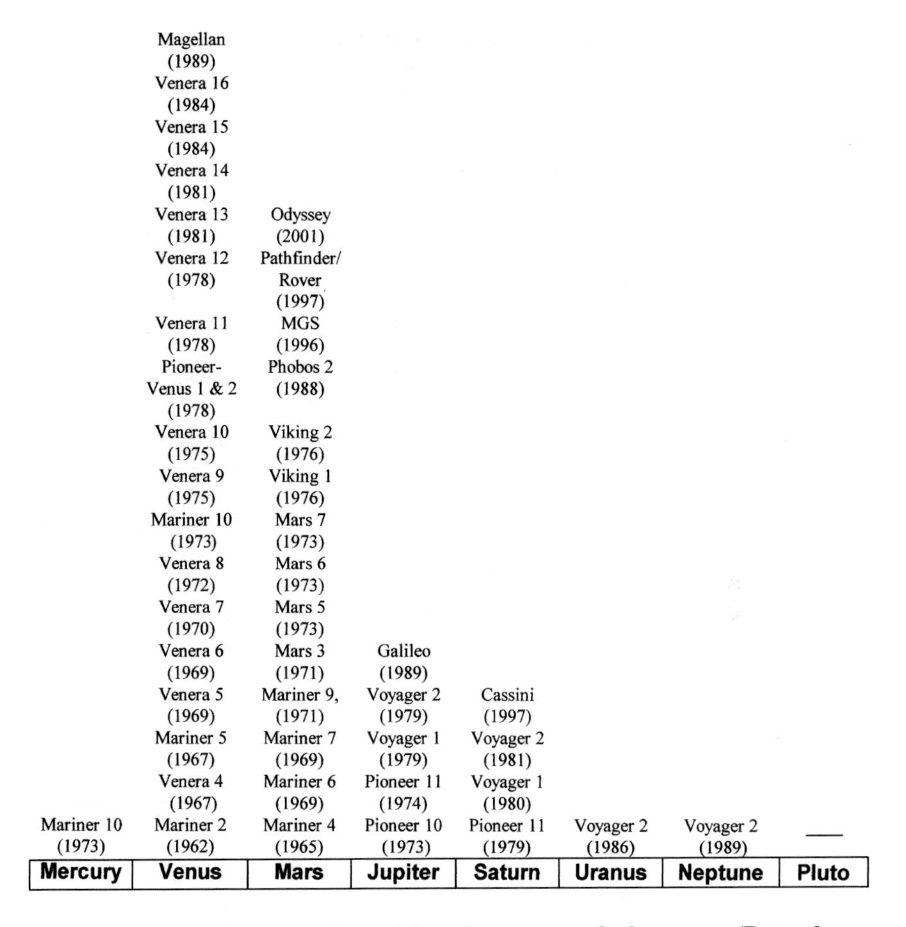

Fig. 1.8 Spacecraft investigations of the solar system—the box score. (Dates shown are launch dates.)

visited and gathered data from eight of the nine planets, with Pluto being the only exception. The first 40 years of successful planetary spacecraft and their missions are shown in Fig. 1.8. These missions provide photographic coverage, atmospheric data, planetary weather, comparative geology, and an opportunity to study the variation in the ways planets form and evolve. These spacecraft have also provided an opportunity to search for life on other planets, so far unsuccessful. The moons of Jupiter and Saturn have been studied and are equally interesting.

Figure 1.9 is just one small sample of what these machines can do. It shows the Magellan images of the Eistla Regio on Venus. These images were made, by synthetic aperture radar techniques, through the perpetual thick clouds of Venus.

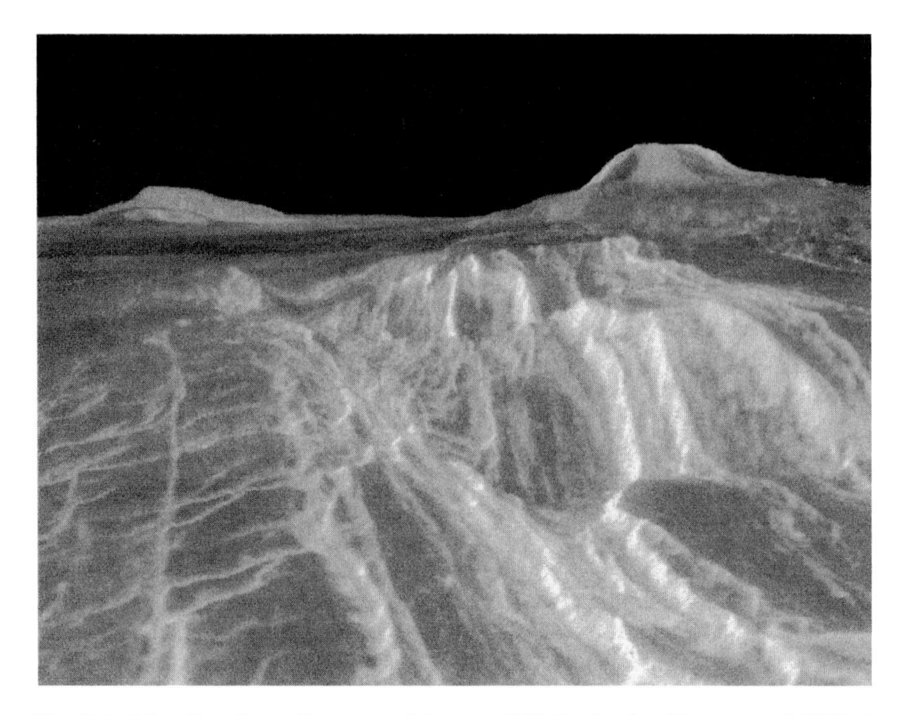

Fig. 1.9 Magellan three-dimensional image of Eistla Regio. (Courtesy of NASA/JPL/Caltech.)

The digital data from Magellan were adjusted by computer to bring the observers' point of view down from orbital altitudes to a position at the foot of the mountains. Without spacecraft our knowledge of Venus geology would have been virtually nonexistent. (The author of this book was the Magellan spacecraft program director.)

In our day it is impossible to imagine life without the benefits spacecraft bring. But the purpose of this book is a different one: to show how rewarding it is to design them.

References

[1]Bryan, C. D. B., Larkin, D., et al., *The National Air and Space Museum*, Vol. 2, Peacock Press/Bantam Books, Smithsonian Inst., New York, 1979, 1982.

[2]Thompson, T. D. (ed.), *Space Log, 1996*, TRW Space and Electronics Group, Redondo Beach, CA, 1997.

[3]Curtis, A. R. (ed.), *Space Satellite Handbook*, 3rd ed., Gulf Publishing Co., Houston, TX, 1994.

[4]Alper, J., and Pelton, J. N. (eds.), *The INTELSAT Global Satellite System*, AIAA, Washington, DC, 1993.

[5]Downey, P., and Davis, B., *Space, A Resource for Earth,* AIAA, Washington, DC, 1977.

[6]*1992 Federal Radionavigation Plan*, National Technical Information Service, Springfield, VA, May 1993.

[7]Freeman, H. R., and Longanecker, G. W., *The International Ultraviolet Explorer (IUE) Case Study in Spacecraft Design*, AIAA Professional Study Series, AIAA, Washington, DC, 1979.

[8]Kondo, Y. K., Wamsteker, W., and Stickland, D., "IUE: 15 Years and Counting," *Sky & Telescope*, Sky Publishing Corp., Cambridge, MA, Sept. 1993.

[9]"The Future of Remote Sensing from Space: Civilian Satellite Systems and Applications," U.S. Congress, Office of Technical Assessment, OTA-ISC-558, U.S. Government Printing Office, Washington, DC, July 1993.

2
System Engineering

System engineering is the comprehensive engineering task in spacecraft design. It makes the subsystems work together and makes the whole greater than the parts. System engineering includes the following major functions: 1) requirements definition, 2) resource allocation and management, especially power and mass, 3) conduct of system level trade studies, for example comparing navigation by ground operations and by onboard navigation, and 4) technical integration of the subsystems into a spacecraft system and integration of the spacecraft with other project elements: launch vehicle, ground systems, etc. In this chapter we will take a top view of a spacecraft and a design project. We will discuss the anatomy of a spacecraft, the resources and margins needed, and the launch vehicle interface. This chapter will follow NASA practice and terminology[1]; however, all government agencies have similar practices.

2.1 Anatomy of a Spacecraft

In this section we will discuss the life cycle of a spacecraft project, the composition of a spacecraft project, and the subsystems of which a spacecraft is made.

2.1.1 Spacecraft Life Cycle

A spacecraft is born by stages, as shown in Fig. 2.1. The design activities on a spacecraft vary substantially during the life cycle. The first step is an idea: "I wonder if we couldn't get surface pictures of Venus through the clouds using synthetic aperture techniques?". An idea can originate anywhere in the aerospace community, from a customer, from a university, or from a contractor. Most corporations and customers maintain small organizations with the purpose of generating and studying mission ideas.

The second step is a feasibility check. There was a time when this step meant a feasibility study to make sure the idea could be implemented. Feasibility is not much of a step these days; almost everything is feasible: a phone call or two will likely clear up the feasibility. "Yes, we could get SAR (synthetic aperture radar) images from an altitude of 1000 km with a resolution of about 100 m." After feasibility checks it is time to find out if the customer needs the mission bad enough to justify it. In a planetary or scientific mission, a selected, high level, science group will answer this question. If the answer is favorable it is time to do concept trades; this is called phase A.

2.1.1.1 Phase A: Preliminary analysis. Preliminary studies are usually done by a number of companies in industry with customer oversight, a customer mission statement, and some customer money. The studies are awarded

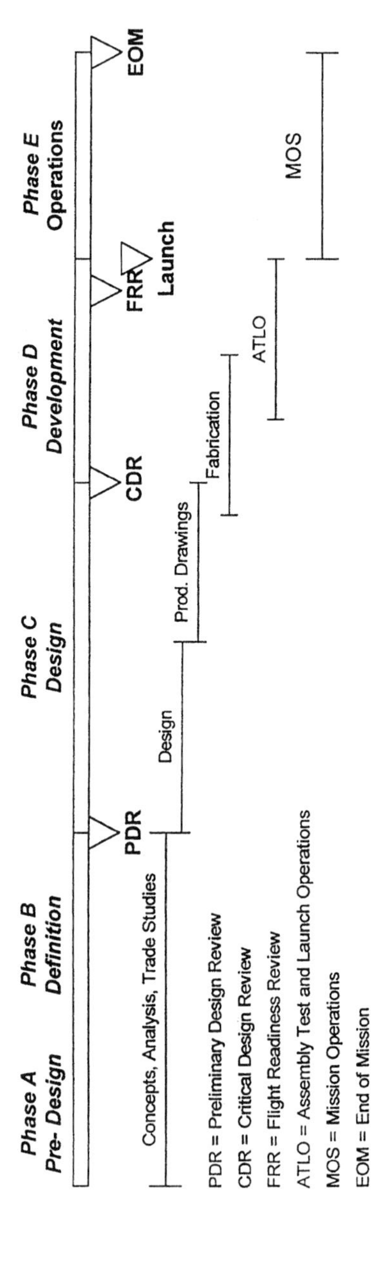

Fig. 2.1 Life cycle of a spacecraft project.

competitively. There are usually customer studies as well. The primary questions that are answered during phase A are the following:

1) What is a reasonable spacecraft configuration that will do the mission? (Not the best; one that will work reasonably well.)

2) Are there any "tall poles" (significant cost, schedule or technical risks) in the development?

3) What major trade studies should be made?

4) About what is it going to cost?

5) About how long will it take?

A phase A design is a conceptual design, professionally done, that will meet the mission statement, has no technical flaws, and is internally consistent. It is not yet necessary to find the best design; that happens in phase B.

2.1.1.2 Phase B: Definition.

Phase B studies are usually awarded competitively to at least two contractors; they are not necessarily the phase A contractors. The questions to be answered by the winning phase B bidders are the following:

1) What is the best spacecraft design for the mission? (and why?)

2) What are the risks involved?

3) What is your implementation plan?

4) What is your company's cost estimate?

5) How much time would it take your company?

6) Are any long-lead actions necessary to protect schedule?

During this phase the technical and business baselines for the project are defined.

2.1.1.3 Phase C/D: Full-scale development.

After phase B studies are completed, a winner-take-all competition is conducted. The winner is awarded a full-scale development contract. Theoretically the winning design is implemented; however, that almost never happens. The competitive phases take time (five years in the case of Magellan). During that time, changes have probably occurred in the customer's mission statement and certainly have occurred in his funding plans. Thus, in the real world, the winning contractor normally conducts a short, delta-phase-B study on revised design requirements; this step took a year in the Magellan case.

A full-scale development program progresses through recognized phases of its own. In the *preliminary design phase*, requirements and performance are defined to the point that detailed design drawings can be made. In preliminary design, the engineering emphasis is on 1) functional performance, 2) requirements definition, and 3) interface definition.

The phase ends with the preliminary design review (PDR). The PDR is a formal customer review to evaluate the adequacy of the preliminary design and compliance with the customer requirements. After PDR the design is partially frozen and specifications are put under change control.

In the *design phase* the build drawings are made and the software is coded. Subcontracts are started for components (reaction wheels, computers, valves and the like) or subsystems that are buy items. (In real life, it is often necessary to start some subcontracts prior to PDR.) Subsystem-level build and test is started in this phase. The design phase ends with the critical design review (CDR), which is a formal customer review that evaluates the adequacy of the design and the interface definitions.

2.1.1.4 Assembly, test, and launch operations (ATLO). ATLO starts when flight-qualified subassemblies are available for assembly into a flight spacecraft. Usually spacecraft assembly starts with units identical to the qualification units that are qualified in parallel with spacecraft assembly. Even if there are qualification failures, time is saved. The system level tests include the following:

1) System level functional tests of each mission phase. These are repeated between environmental tests.

2) Thermal testing simulating space, either solar thermal vacuum (technically best) or infrared vacuum (cheapest).

3) Acoustics, simulating the environment in the launch vehicle shroud.

4) End-to-end communications check using a ground station, for example the deep space network station at Kennedy Space Center.

5) Mission simulations and environmental tests (acoustics, thermal vacuum). This phase ends with the pre-ship review which considers the acceptability of the spacecraft in the light of system test data.

2.1.1.5 Launch phase. This phase normally includes reassembly after shipment to the launch site and retesting. Just prior to launch there is one final review to asses the readiness of the spacecraft.

2.1.1.6 Phase E: Mission operations. Phase E starts immediately after launch. The mission operations team is typically trained for some months during assembly and test, to control and command the spacecraft remotely and to be proficient at understanding the spacecraft condition from a flight-like data stream. The mission operations (MOS) team can be as large as several hundred people or as small as a several people, depending on the complexity of the spacecraft and of the mission.

2.1.1.7 Reviews. The most common formal customer reviews that are conducted in the course of a program are listed in Table 2.1. These reviews have evolved over the last four decades and follow released guidelines that are essentially the same for all government agencies.

It is normal practice to conduct subsystem PDRs and CDRs in the same general time frame as the spacecraft PDR and CDR. These reviews are also used in industry to denote a rough measure of program maturity: for example, "The program just completed PDR."

Table 2.1 Formal customer reviews

Abbrev.	Review name
CoDR	Conceptual design review
PDR	Preliminary design review
CDR	Critical design review
PRR	Preshipment readiness review
FRR	Flight readiness review

2.1.1.8 Commercial procurement. The phases, reviews, and steps just described are for a government procurement. A commercial procurement can vary considerably because of the following factors:

1) Design is driven by financial concerns (return on investment, market share).
2) The approval chain extends to the board of directors.
3) There is usually a single customer engineering point of contact.
4) There is usually no set of established procedures for guidance.
5) Once the contract is signed there is minimal customer oversight.
6) There is legal and regulatory impact on the design, especially in communication spacecraft.

2.1.2 Project Elements

If you observe the myriad activities on a spacecraft project you will discover that they can be divided into five reasonably self-contained project elements, as shown in Fig. 2.2.

2.1.2.1 Science payload. The science payload is the set of instruments that perform the mission. On Magellan the science payload was a single instrument, the synthetic aperture radar. Magellan was unusual in that it had only one instrument; the Cassini spacecraft will explore Saturn and its moons with 12 different instruments on the Orbiter and six more on the Probe. The instrument payload usually comes from a different organization than the spacecraft, often a university or the customer. The primary payload interfaces with the spacecraft are power, data management, command, thermal, mechanical, and field of view. The instruments sometimes come with a built-in data collection and formatting capability.

2.1.2.2 Launch system. The vehicle and support elements that put the spacecraft in orbit about Earth or on an escape trajectory to a planet comprise the launch system. Sometimes the spacecraft supplies some of this energy (e.g., Voyager), but the major energy source is the launch vehicle. The launch vehicle is

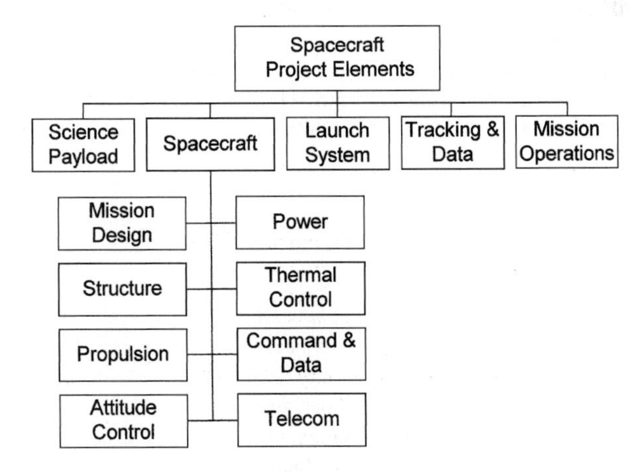

Fig. 2.2 Spacecraft project elements.

a primary interface; the design loads for the structure are from the launch vehicle. The launch vehicle shroud determines the maximum spacecraft dimensions during launch. This constraint results in a collection of mechanisms that change the spacecraft from launch configuration to mission configuration. There is also a strong launch vehicle functional interface with power, command, telecommunication, and command and data systems.

For Space Shuttle launches there is an astronaut safety interface requiring significant spacecraft resources and approximately three years of coordination culminating in full-scale spacecraft/shuttle emergency exercises. In addition, there are two countdowns to prepare for. The first countdown is to the launch of the shuttle, with the added constraint of good weather at all shuttle emergency landing sites. The second countdown is with shuttle in orbit. For a planetary mission there is a point in space on each orbit from which the mission can be initiated. A countdown is required culminating in separation from shuttle at that point. If there is a spacecraft failure in this on-orbit countdown and the spacecraft is too heavy for a shuttle landing, the spacecraft will be jettisoned as space junk.

Detailed engineering integration with the launch vehicle is an involved process and should start about 36 months before launch, up to 48 months for a shuttle launch. The structural loads analysis and integration requires a special emphasis and a series of formal meetings.

There are now more launch systems to choose from than ever before. The premier source of launch vehicle information is the *International Reference Guide to Space Launch Systems*, by Isakowitz et al.[2]

2.1.2.3 Tracking and data systems. After liftoff, the link to a spacecraft is through tracking and data stations that 1) receive the spacecraft downlink and relay it to mission operations and 2) uplink commands to the spacecraft. These stations also use the radio link to provide the range, azimuth, and elevation to the spacecraft. These systems are discussed in more detail in Chapter 9, "Telecommunication."

2.1.2.4 Mission operations (MOS). At liftoff a hand-off from the launch team to the MOS team is made as the vehicle leaves the launch pad. In the case of a planetary launch, the launch team is at Cape Canaveral and the MOS team is typically located at the Jet Propulsion Laboratory (JPL) in Pasadena, California. The MOS team provides analysis of spacecraft performance from the downlink and provides spacecraft commands for uplink. The downlink data are decoded and delivered to the consoles of the MOS team. Normal performance parameters are stored in a database. Any anomalous behavior is dealt with by the team. This may be a minor issue or a spacecraft life-threatening event. (To paraphrase an old pilot saying, MOS can be hours and hours of boredom punctuated by moments of sheer terror.)

The MOS team also prepares and tests the commands that are uplinked to the spacecraft. A command error can, and has, caused total spacecraft failure, so commands are heavily reviewed and tested. (Work on the Magellan Venus orbit insertion commands started eight months before encounter.) It is helpful for the MOS team to have a spacecraft simulator on which to test commands. These simulators are very expensive and vary widely in definition from program to program.

Planetary MOS work is complicated by the time delay resulting from planetary distance. For Mars and Venus the data stream is telling you what went on 15–30 min ago. For a Jupiter mission the delay is about 1 h. So if data from Jupiter show a failure and it takes 1 h to decide what to do, prepare and test the commands, it will be 3 h after the failure when a command reaches the spacecraft. Modern planetary spacecraft have an onboard fault protection system because of this delay. In many fault protection system designs, failures too difficult for the fault protection system to solve cause the vehicle to go into a safe mode and call for help from MOS. There is a major trade-off here between how much the spacecraft can do for itself and how much MOS support to provide.

2.1.3 *Spacecraft Subsystems*

The spacecraft is traditionally subdivided into eight subsystems as shown in Fig. 2.2 and Table 2.2. These subsystems align with the general academic specialties of an engineering staff and provide reasonably manageable interfaces among subsystems. The spacecraft subsystems will be covered briefly in this section and in more depth in the remainder of the book, the pertinent chapter is shown in the right column of Table 2.2.

The ACS and the CDS subsystems are heavy users of onboard computer resource. The two subsystems may share a single flight computer or, more often, two separate computers custom designed to the tasks. The flight and ground software is sometimes organized as a separate spacecraft subsystem; in this book software and computers are discussed with the CDS in Chapter 8.

2.1.4 *Functional Block Diagram*

The functional block diagram is a useful, concise way to visualize the relationships between the subsystems. A highly simplified example is shown in Fig. 2.3. These diagrams are prepared and maintained by the system engineering team; in some cases their distribution may be a customer requirement.

2.2 Mass Properties

There are three types of spacecraft mass properties that must be controlled: 1) spacecraft moments of inertia, 2) spacecraft mass, and 3) center of mass. Each of these must be calculated, controlled, and kept current for each of the spacecraft mission modes (launch mode, cruise mode, mission mode, etc.). For a spinning spacecraft the ratios of axial moments of inertia are also critical; a flat spin can result if the ratio favors the wrong axis. Center of mass is important for determining the torque resulting from motor firings and for thrust vector control during motor firings.

Although there are several mass property functions, spacecraft weight is always the major noisy issue on a spacecraft project. Experience shows that spacecraft mass always increases as understanding of the design increases. Very early in the process, systems engineering should provide mass and power reserves and allocate the remainder to each of the subsystems. Current best estimate (CBE) of power and mass compared with the allocations provides a running measure of how the design team is doing.

Table 2.2 Spacecraft subsystems

Subsystem	Function	Chapter
Orbital mechanics (ORB)	Develop the optimum path, or trajectory, through space for the performance of the mission. Determine the velocity changes that will be needed to send the spacecraft on the desired trajectory. Determine the impulse that will be needed from the launch vehicle and from the spacecraft propulsion system.	3
Propulsion (PROP)	Provide the translational velocity changes necessary to establish and maintain the required orbit. Control thrust direction, duration, and magnitude. Provide vehicle rotation about all axes at the command of the attitude control system.	4
Attitude control (ACS)	Determine vehicle attitude and correct it to desired attitude. Implement and control commanded changes in velocity or attitude. Control the articulation of appendages (antennas, solar panels, cameras).	5
Power (EPS)	Generate, store, regulate, and distribute electrical power for all equipment in all mission modes. Provide grounding and fusing for all equipment. Switch all spacecraft equipment on and off as required by the command system.	6
Thermal control (TC)	Maintain temperature of all spacecraft equipment within allowable limits for all mission modes.	7
Command and data handling (CDS or C&DH)	Receive commands from the communication system. Decode, store, distribute, and initiate all commands. Collect, process, format, store, and deliver data to the communication system. Data from the payload may be handled by CDS or handled directly by the payload.	8
Telecommunication (COM)	Receive commands from the ground communication facility and transmit to the CDS system (uplink). Transmit payload and engineering data to the ground facility (downlink). Receive and re-transmit signals for navigation and tracking.	9
Structure and mechanisms (STR)	Establish the general arrangement for all spacecraft equipment 1) in the launch configuration and 2) in the mission configuration. Provide mechanisms to convert the spacecraft from one configuration to the other. Provide structure to support all spacecraft equipment during launch and during the mission. Provide a launch vehicle adapter to join the spacecraft and launch vehicle structurally. Provide mechanisms to articulate antennas, cameras, and solar panels, etc.	10

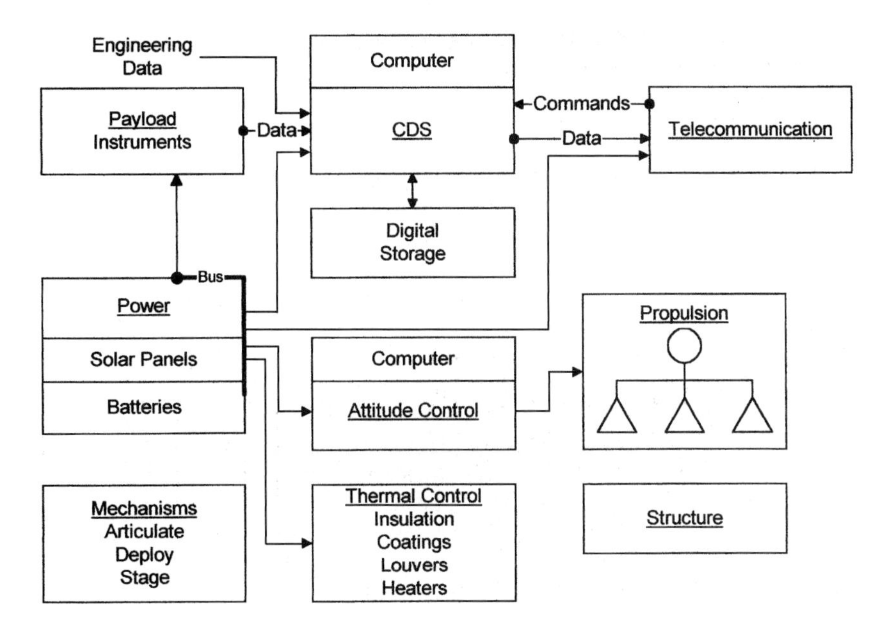

Fig. 2.3 Simplified spacecraft block diagram.

2.2.1 Mass Definitions

The following mass definitions are generally used; however, there is no standard, and detailed definitions vary. Figure 2.4 illustrates the definitions and shows actual mass breakdown of the Viking Orbiter in kilograms.

Science (or payload) mass is the mass of the science instruments plus all equipment used in direct support of the instruments. The support equipment usually includes mounting structure, power cabling, engineering instrumentation, thermal control heaters, blankets, and radiators. It may also include a dedicated data system. It is the U.S. Air Force custom to call this equipment the *payload*. (This use of payload is not to be confused with payload as used by launch vehicle teams. In the launch vehicle world, payload means the spacecraft launch mass.)

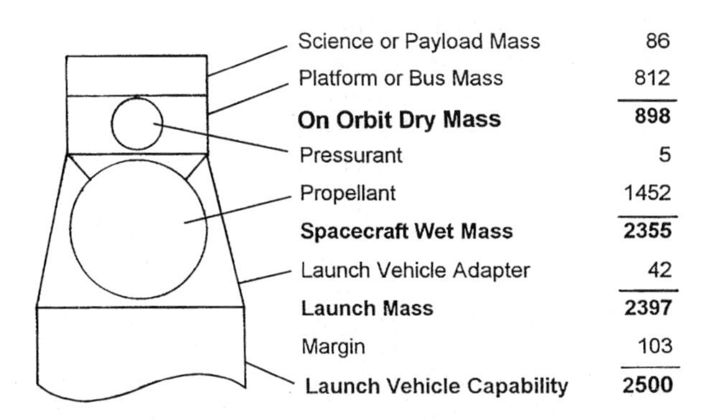

Fig. 2.4 Spacecraft mass definitions.

Bus or platform mass is the total, on-orbit, dry mass of the spacecraft subsystems, structure, propulsion, thermal control, power, attitude control, telecommunication, and data systems. The masses of the science, propellants, and gasses are not included. Dry mass means mass without propellant or gasses loaded.

Launch vehicle adapter (LVA) mass is the mass of the structure, separation devices, cabling, and thermal control equipment necessary to adapt the spacecraft to the launch vehicle. (The launch vehicle adapter is provided by the spacecraft team, but it is left with the launch vehicle at staging.)

Injected mass is the mass of a planetary spacecraft that is accelerated to the Earth departure velocity. Normally this is the cruise mass plus the launch vehicle adapter plus the burn-out mass of the last stage.

Launch mass is the total mass of the spacecraft as it rests on the launch vehicle in launch configuration. It includes propellants, gases, and the launch vehicle adapter.

Cruise mass is the wet or dry mass in interplanetary cruise configuration. It is the launch mass minus the launch vehicle adapter mass. (All spacecraft with multiple configurations have a mass statement for each).

On-orbit dry mass consists of two major parts, the mass of the science instruments plus the platform or bus mass. Propellant and gas masses are not included (hence dry mass). When spacecraft mass is specified, discussed, and controlled it is at this level.

Burn-out mass is the mass of the spacecraft just after shut-down from any particular propulsion event. It is the spacecraft dry mass plus the pressurizing gas plus propellant remaining.

Mass uncertainty is an estimate of the growth in mass from a given time to the launch day. Uncertainty is calculated by various algorithms based on the maturity of the mass estimate.

Mass maturity is reflected by the commonly used degrees of maturity, which are *estimated*, *calculated*, and *actual*. Actual mass is a weight measurement of actual flight hardware. Calculated mass is an engineering calculation of mass based on the dimensions on the flight equipment drawings. Estimated mass is an estimate based on history or analogous units.

Mass margin or contingency is the difference between the current spacecraft launch mass estimate and the launch vehicle capability.

2.2.2 Preliminary Estimates

This section contains weight-estimating relationships that can be used for very early estimates of on-orbit dry mass. These relationships are based on statistics from prior spacecraft. Note that estimating relationships based on history, by definition, do not reflect the latest state of the art. Figure 2.5 shows the on-orbit dry mass of seven geosynchronous communication satellites as a function of payload mass. In Fig. 2.5, the on-orbit dry mass can be estimated to be about 3.6 times the payload mass. Recall that the on-orbit dry weight contains the payload weight; therefore, the data in Fig. 2.5 also show that the supporting subsystems will weigh about 2.6 times the weight of the payload supported.

A JPL study[3] of 46 Earth-orbiting navigation, communications, meteorology, Earth resources, and astronomy spacecraft, showed that the on-orbit dry mass was

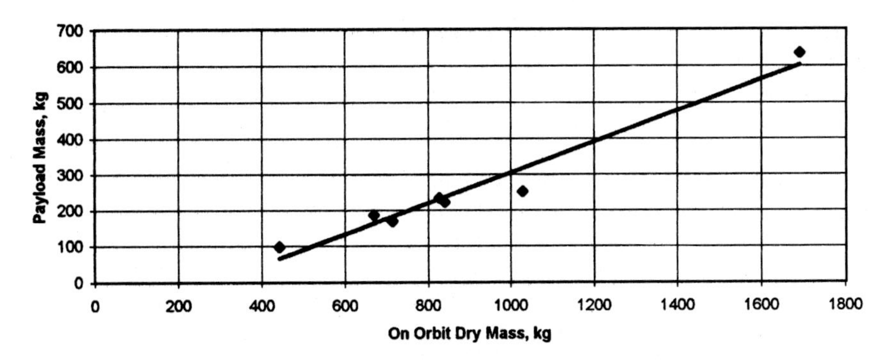

Fig. 2.5 Communication spacecraft mass.

bounded by the limits of 3 and 7 times the payload mass with the average ratio being 4.8; see Fig. 2.6.

Figure 2.7 shows planetary spacecraft on-orbit dry mass vs payload mass for 11 planetary spacecraft including Magellan, Mars Orbiter, Mars Global Surveyor, Galileo, Voyager, and four Mariners. From these historical data, the on-orbit dry mass of planetary spacecraft can be estimated as about 7.5 times payload mass.

It should be emphasized that spacecraft level estimates of this type are for preliminary work only. Such estimates should be replaced by more detailed subsystem level (bottoms up) estimates as soon as possible in a study.

2.2.3 Mass Growth

Spacecraft mass always increases, and it grows more than you think it will. Magellan mass history shows that even after flight equipment was weighed the growth was still 1.2% because of thermal paint, lockwire, QC putty, brackets, pigtails, attach nuts, etc.

The spacecraft on-orbit dry mass growth from program start to launch is summarized in Table 2.3 for seven spacecraft programs. Table 2.3 shows that a margin of 27% between the estimated spacecraft mass and the launch vehicle capability has a 50/50 chance of being adequate. Five of these seven spacecraft were

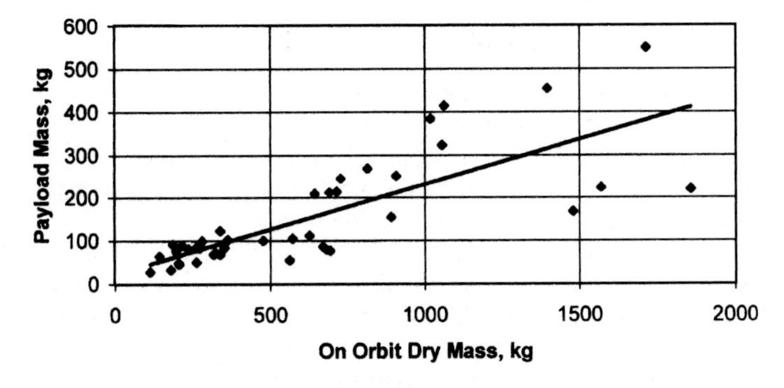

Fig. 2.6 Earth-orbiting spacecraft mass.

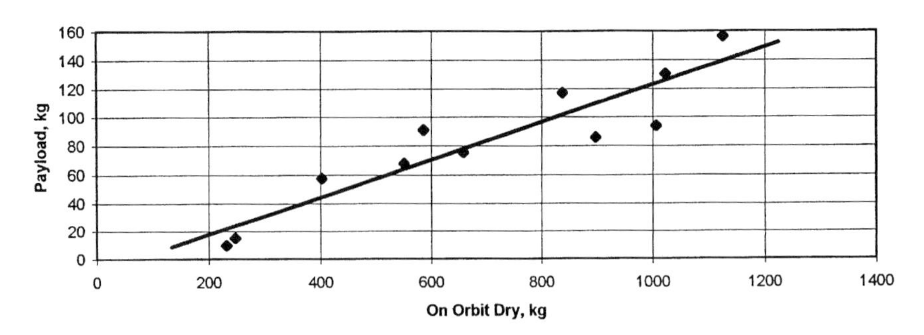

Fig. 2.7 Planetary spacecraft mass.

designed under intense pressure from the launch vehicle mass constraint. Inadequate mass margin causes programs to be more expensive than they would be otherwise. To further appreciate the mass growth shown in Table 2.3, you must realize that the ATP weights were results of substantial preliminary design study. Magellan was in preliminary design for five years prior to ATP. The Magellan ATP mass also contained actual weights for a substantial amount of equipment. As a design matures the mass *will* grow; it is essential to have adequate margin.

The U.S. Air Force conducted a study of the mass growth of 11 spacecraft programs. The mass growth of spacecraft subsystems is shown in Table 2.4. The command and data subsystem shows the most erratic mass history. This may have been caused by the rapid improvement in computer state of the art. All subsystems reflect the growth trend exhibited by the spacecraft as a whole. The study showed an unusual number of subsystems with mass losses (mass losses are usually accounting changes or configuration changes).

Mass histories are plentiful; it can be shown that mass increases regardless of carefully preliminary work. Why does this happen?

First, improved understanding. For example, you may have made an initial estimate of the launch vehicle adapter using historical data similar to Eq. (2.1).

Table 2.3 Mass growth from ATP to launch, kg

Program	Span, months	ATP mass, kg	Launch mass, kg	% Growth	Ref.
Pioneer Venus	52	292	374	28	4, 5
Scatha	25	360	396	10	
FLTSATCOM	50	645	840	30	6
Magellan	72	830	1032	25	7
HEAO-2	60	2223	3016	36	8
HEAO-3	60	2313	2722	18	8
Mars Observer	71	827	1125	36	9
Average				27	

Table 2.4 Mass growth of spacecraft from proposal to launch[a]

Program	Structure, %	Power, %	CDS, %	ACS, %	COM, %
1	8	3	12	42	16
2	11	27	−50	−3	82
3	24	4	−11	16	44
4	10	15	−28	43	25
5	57	3	8	−7	82
6	29	14	−1	8	NA
7	28	9	10	−23	NA
8	29	7	−4	43	4
9	105	66	4	46	NA
10	65	8	58	9	47
11	50	−27	42	62	69
Average	37.5	11.7	3.6	21.5	46.1

[a]Courtesy U.S. Air Force Space Division (Ref. 10).

After the first stress analysis the weight might increase substantially. After final loads data are available, it would increase again. When the build drawings are released it would increase again. When the actual truss was weighed it would increase again. When the thermal coatings were put on the truss, the *actual* weight would increase.

The electrical cables are notorious for this type of weight increase. In preliminary design the cable weights will be estimated by a historical relation similar to Eq. (2.1). Accurate cable weight estimates cannot be made until detailed drawings of the spacecraft structure, electrical equipment installation drawings, and detailed knowledge of each connector are available. Even armed with this much detail, it requires an elaborate CAD/CAM type of analysis to produce a reasonable cable weight estimate. For this reason, cable weights are suspect until actual weights are available; these will be higher than estimates (sometimes substantially higher). Several spacecraft have gotten in serious weight trouble because of this problem alone.

Second, make-play changes. Changes of this type are myriad as the equipment is tested. An easy example is a structural test failure. Almost by definition, the corrective action for these failures adds weight. Another example is a supplier going out of business. The substitute equipment is, of course, heavier. Almost all adversity encountered during the definition of the design requires weight to solve.

Third, improvement changes. As the design progresses, better ideas will occur. These ideas are often cost issues but they can have weight impact as well.

Of these three causes of weight increase, only the improvement changes can be managed. Make-play changes and improved understanding cannot be avoided. As a result, weight will increase with design maturity. The best defense is intelligent provision of weight margin.

2.2.4 AIAA Recommended Mass Margin

Mass contingency, or margin, is the allowance for growth resulting from design definition and development. Mass margin is defined as

$$\text{Margin} = \text{Total capability} - \text{Current best estimate} \tag{2.1}$$

$$\%\text{Margin} = \frac{\text{margin}}{\text{capability}} \times 100 \tag{2.2}$$

A committee of the AIAA, in cooperation with the American National Standards Institute (ANSI), has reviewed industry-wide historical data from numerous DOD and NASA manned and unmanned programs. Using these data, ANSI and AIAA recommended the dry mass contingencies shown in Table 2.5 as a function of design maturity.

Design maturity is measured in Table 2.5 by the major reviews conducted during a design: the bid (proposal or bid stage), PRR, FRR, CoDR, PDR, and CDR (see Table 2.1). The classes of design listed in Table 2.5 are class 1: a new spacecraft (one of a kind, first generation); class 2: the next-generation spacecraft based on a previously developed family, which expands in complexity or capability within an established design envelope; and class 3: a production-level development based on an existing design for which multiple units are planned and where a significant amount of standardization exists.

Table 2.5 AIAA recommended weight contingencies[a]

Description/ categories	Minimum standard weight contingencies, %														
	Proposal stage			Design development stage											
	Bid Class			*CoDR* Class			*PDR* Class			*CDR* Class			*PRR* Class		
	1	2	3	1	2	3	1	2	3	1	2	3	1	2	3
Category AW, 0–50 kg 0–110 lb	50	30	4	35	25	3	25	20	2	15	12	1	0	0	0
Category BW, 50–500 kg 110–1102 lb	35	25	4	30	20	3	20	15	2	10	10	1	0	0	0
Category CW, 500–2500 kg 1102–5511 lb	30	20	2	25	15	1	20	10	0.8	10	5	0.5	0	0	0
Category DW, 2500 kg and up	28	18	1	22	12	0.8	15	10	0.6	10	5	0.5	0	0	0

[a]Data copyright AIAA, reproduced with permission; Ref. 11.

2.2.5 *Allocating Subsystem Dry Mass*

One of the first actions in initiating a spacecraft design is the allocation of available mass to the subsystems. The steps in the process are as follows:

1) Determine the maximum spacecraft launch mass for the mission.

2) Deduct launch vehicle adapter mass from launch mass.

3) Determine the propellants and pressurants required for the mission.

4) Determine the total allowable on-orbit dry mass.

5) Establish the total allowable payload weight (usually specified by the customer).

6) Evaluate the mass margin to be set aside. The AIAA guidelines[11] are very useful in this step.

7) Allocate mass budgets to each subsystem.

The less obvious of these steps are discussed in the following sections.

2.2.5.1 Maximum spacecraft launch mass. The maximum spacecraft launch mass is derived directly from the launch vehicle capability, which in turn is a function of the mission design. This key requirement is usually specified by the customer, although data are readily available from Isakowitz's *International Reference Guide to Space Launch Systems*[2] or from individual manufacturers.

2.2.5.2 Launch vehicle adapter. This assembly adapts the launch vehicle structure to the spacecraft structure and provides for spacecraft separation. It is designed by the spacecraft team but is left with the launch vehicle at separation. The mass of the adapter comes out of the spacecraft mass budget. As you would expect, the adapter mass is a strong function of the spacecraft mass, as shown in Fig. 2.8.

The equation in Fig. 2.8 can be rounded to

$$\text{LVA} = 0.0755\text{LM} + 50 \tag{2.3}$$

where

LVA = launch vehicle adapter mass, kg
LM = launch mass, kg

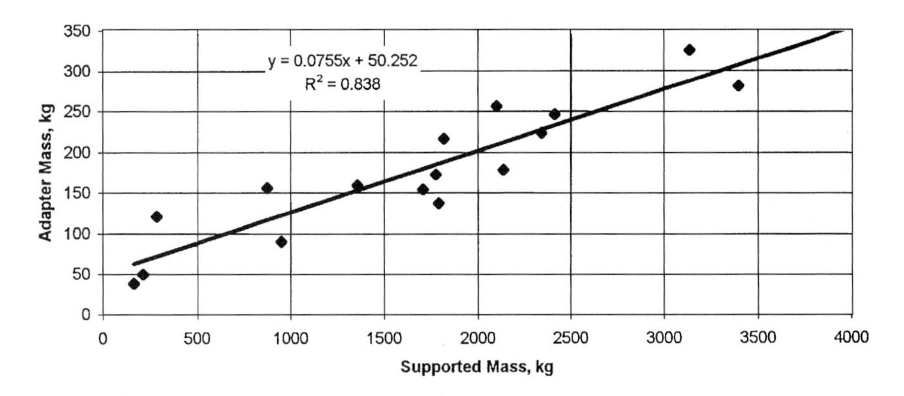

Fig. 2.8 Launch vehicle adapter mass.

Table 2.6 Subsystem on-orbit dry mass allocation guide

	Comsats[a]		Metsats[b]		Planetary		Other	
Subsystem	with P/L[c]	GFE P/L	with P/L	GFE P/L	with P/L	GFE P/L	with P/L	GFE P/L
Structure, %	21	29	20	29	26	29	21	30
Thermal, %	4	6	3	4	3	3	3	4
ACS, %	7	10	9	13	9	10	8	11
Power, %	26	35	16	23	19	21	21	29
Cabling, %	3	4	8	12	7	8	5	7
Propulsion, %	7	10	5	7	13	15	5	7
Telecom, %	—	—	4	6	6	7	4	6
CDS, %	4	6	4	6	6	7	4	6
Payload, %	28	—	31	—	11	—	29	—

[a]Comsat = communication satellite. [b]Metsat = meteorology or weather satellite. [c]P/L = payload.

2.2.5.3 Propellant and pressurant mass. If the spacecraft will perform translational velocity change maneuvers, such as orbit insertion, the propellant load may be substantial and can be estimated using the Tsiolkowski equations discussed in Chapter 4, Section 4.3.

2.2.5.4 Total allowable subsystem on-orbit dry mass. This mass can be computed as follows:

Total allowable dry weight for the spacecraft subsystems = LM − LVA
− Margin − Payload mass − Propellant mass − Pressurant mass

The preceding equation assumes that the payload is customer furnished and not part of the spacecraft development task (the most common case). If the payload is being developed along with the other spacecraft subsystems, it should be treated as another subsystem and not be deducted from the equation.

2.2.5.5 Allocating subsystem mass budgets. The final step is the intelligent division of the total allowable dry weight into subsystem allocations. For this process the best guide is history. Table 2.6 provides a subsystem mass allocation guide for initial mass budgets. This table was developed by analysis of several spacecraft of each type. Two conditions are shown: 1) payload supplied by the spacecraft team and 2) customer-supplied payload (or government-furnished equipment, GFE).

In Table 2.6, for communication satellites, telecommunication is not shown separately; it is included in the CDS mass. The subsystem weights are shown as percentages of spacecraft dry weight.

Example 2.1 Subsystem Mass Allocation

A new design communication spacecraft in bid phase is being designed for the following conditions:

Launch weight capability = 1868 kg
Payload weight (supplied by customer = 222 kg
Orbit insertion Delta V propellant required = 1027 kg

Launch vehicle adapter mass:

$$\text{LVA} = 0.076\,(1868) + 50 = 192\,\text{kg}$$

Propellants required: Calculation of required propellant is discussed in Chapter 4. For now, assume that the propellant required is 1027 kg.

On-orbit dry mass: Spacecraft maximum on-orbit dry mass is equal to

$$\text{LM capability} - \text{Adapter} - \text{Propellant} = 1868 - 192 - 1027$$
$$= 649\ \text{kg}\ (1431\ \text{lb})$$

Allowable subsystem mass: The maximum allowable subsystem or bus mass is equal to

$$\text{On-orbit dry mass} - \text{Payload mass} = 649 - 222 = 427\,\text{kg}$$

Mass margin: Taken from Table 2.5, Category CW, bid stage, class 1 = 30%. The total subsystem mass for allocation = 427/1.30 = 328 kg; mass margin = 427 − 328 = 99 kg. (It is a common error to calculate margin as a percentage of maximum allowable (427 × 0.30); however, the approach shown is preferred.)

Allocation: 328 kg may now be allocated to subsystems using the percentages given in Table 2.6. The result is shown in Table 2.7.

2.2.5.6 Launch mass prediction algorithm. Once the design process starts, the mass properties activity changes from budget allocation to monitoring progress and estimating what the actual launch weight will be. For mass monitoring, it is customary to tabulate detailed weights along with the maturity of each element.

Table 2.7 Subsystem mass allocation

Subsystem	Percent of total	Allocated mass, kg
Structure	29	95
Thermal	6	20
ACS	10	33
Power	35	114
Cabling	4	13
Propulsion	10	33
Telecom (Comsat)	—	—
CDS	6	20
Budget total		**328**
Mass margin		**99**
Max bus mass		**427**
Max payload mass		**222**
Max on-orbit dry		**649**

Table 2.8 Mass growth algorithm

Mass category	Estimated	Calculated	Actual
Structure	25	4.6	2.6
Propulsion	5	4.6	2.6
Electronics	15	3.2	1.2
Cabling	50	5.0	1.2

The classes of maturity are *estimated, calculated,* and *actual.* For example, the launch vehicle adapter weight initially is estimated using an estimating relationship similar to Fig. 2.8. The adapter mass is entered into the project weight tables as an *estimated* weight. After the adapter structure is designed, the adapter weight is calculated from the drawing dimensions and material densities. This weight is entered into the project weight tables as *calculated.* Finally, after the adapter is built, it is weighed and the weight entered as *actual.*

Periodically, usually monthly, the total set of weight tables is summarized to produce 1) the current spacecraft weight estimate, 2) an estimate of the weight at launch, 3) the percentage of the total weight in each maturity category, and 4) a report card on the weight status of each subsystem. Item 2, the estimate of launch weight, is the difficult issue. Analysis of spacecraft history suggests the launch weight prediction algorithm shown in Table 2.8. The weight growth remaining to launch is estimated by multiplying the current best estimate by the factors shown in Table 2.8. These factors are a function of the maturity of the weight item (estimated, calculated, or actual), and also a function of the equipment type, structure, propulsion, electronics, or cabling.

To arrive at a projected launch weight, the sum of all structure mass in the *estimated* category should be increased by 25%, the sum of all structure mass in the *calculated* category increased by 4.6%, and the *actual* category increased by 2.6%. The sum of these adjusted masses is the predicted launch mass for structure. The process is repeated for propulsion mass, electronics mass, and cabling mass to produce the projected launch mass of the spacecraft:

$$\text{Structure LM} = 1.25(\text{Est Stru}) + 1.046(\text{Calc Stru}) + 1.026(\text{Actual Stru})$$

$$\text{Propulsion LM} = 1.05(\text{Est Prop}) + 1.046(\text{Calc Prop}) + 1.026(\text{Actual Prop})$$

$$\text{Electronics LM} = 1.15(\text{Est Elect}) + 1.032(\text{Calc Elect}) + 1.012(\text{Actual Elect})$$

$$\text{Cabling LM} = 1.50(\text{Est Cable}) + 1.05(\text{Calc Cable}) + 1.012(\text{Actual Cable})$$

The projected dry launch mass is the sum of structure, propulsion, electronics, and cable estimates just shown. Figure 2.9 is a report card on the algorithm using Magellan weights. It shows a comparison of the projection of the algorithm, as a function of time, with the actual launch mass. The algorithm served the program well. Its worst projection was 4% (40 kg) high.

2.2.6 Moments of Inertia

Spacecraft moments of inertia are important for two reasons: 1) The propellant required to maintain stability or to make rotational maneuvers is proportional to

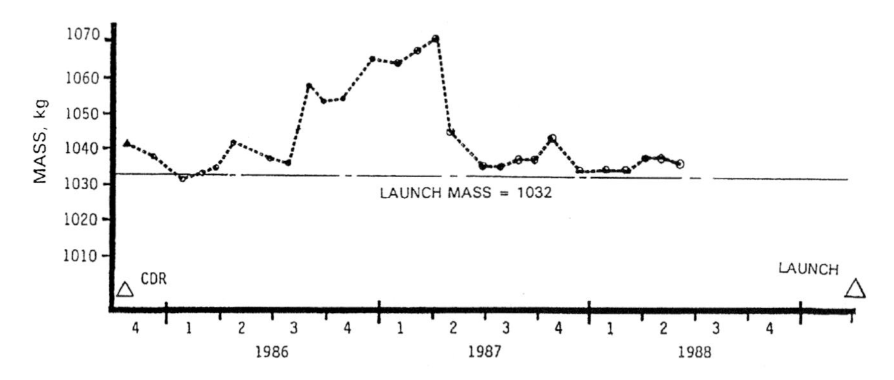

Fig. 2.9 Predicted launch mass vs time.

moment of inertia. 2) The spin axis of a spinning spacecraft must be the axis of highest moment of inertia or the spacecraft is unstable. The moments of inertia must be actively controlled to maintain spin stability.

The moment of inertia about any axis is the sum of the elemental masses times the distances from the axis; see Fig. 2.10:

$$I_x = \Sigma m r^2 \tag{2.4}$$

where r is the distance from the center of mass of element m to the axis x. One of the bookkeeping chores of a project is to record m and r for each part or box to be installed on the spacecraft. In the preliminary design phases of a project, the moments of inertia are estimated from the larger elements of the spacecraft (for example the equipment module). In the later phases, the moments will be calculated at the box level (for example a battery).

The moments of inertia of some useful shell shapes are summarized in Fig. 2.11. In the figure, moment of inertia is represented by I, surface area by S, radius by R, and mass by W. Uniform density is assumed. Figure 2.11 is only a sample of the data available for shells and solids, see *AIAA Aerospace Design Engineer's Guide*[12] and other similar handbooks.

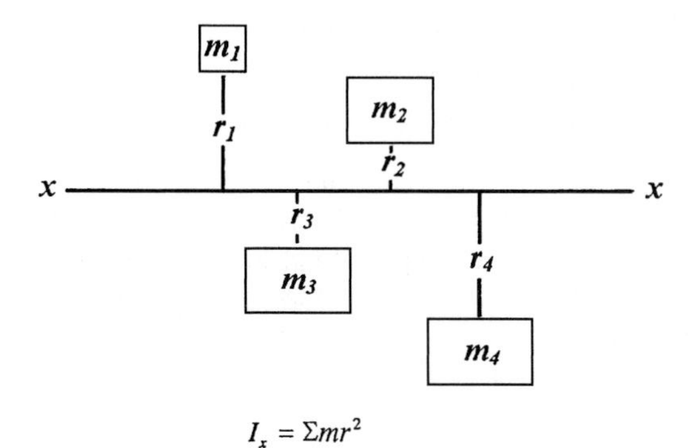

Fig. 2.10 Moment of inertia.

Spherical Shell	Properties	Moment of Inertia
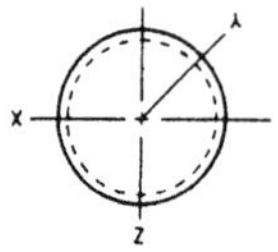	**Surface Area** $S = 4\pi R^2$	$I_x = I_y = I_z = \frac{2}{3} W R^2$
Cube-Solid		
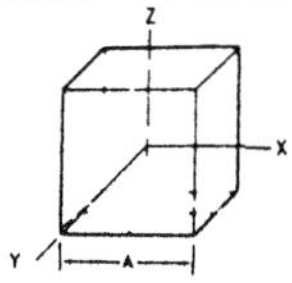	**Center of Mass** $\bar{x} = \bar{y} = \bar{z} = \frac{A}{2}$	$I_x = I_y = I_z = \frac{WA^2}{6}$
Hollow Box-Shell		
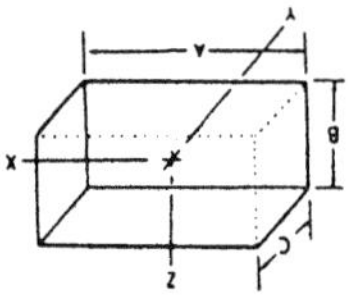	**Surface Area** $S = AB + BC + A$	$I_x = \frac{W}{12}(B^2 + C^2) + \frac{W}{6}\left(\frac{ABC(B+C)}{AB+BC+AC}\right)$ $I_y = \frac{W}{12}(A^2 + B^2) + \frac{W}{6}\left(\frac{ABC(A+B)}{AB+BC+AC}\right)$ $I_z = \frac{W}{12}(A^2 + C^2) + \frac{W}{6}\left(\frac{ABC(A+B)}{AB+BC+AC}\right)$
Cylindrical Can -Shell		
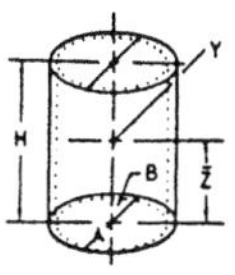	**Surface Area** $S = 2\pi R(R+H)$ **Center of Mass** $\bar{z} = \frac{H}{2}$	$I_x = \frac{W}{12(R+H)}\left[3R^2(R+2H) + H^2(3R+H)\right]$ $I_Y = I_X$ $I_Z = \frac{WR^2}{2}\left(\frac{R+2H}{R+H}\right)$
Frustrum of Cone-Shell		
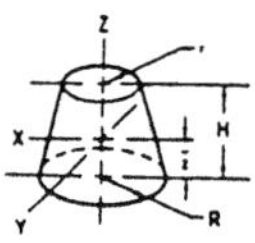	**Surface Area** $(R+r)\sqrt{H^2+(R-$ **Center of Mass** $\bar{z} = \frac{H}{3}\left(\frac{2r+R}{r+R}\right)$	$I_X = \frac{W}{4}\left(R^2 + r^2\right) + \frac{WH^2}{18}\left[1 + \frac{2Rr}{(R+r)^2}\right]$ $I_Y = I_X$ $I_Z = \frac{W}{2}\left(R^2 + r^2\right)$

Fig. 2.11 Moment of inertia for some shells and solids.

Table 2.9 Total spacecraft power estimating relationships

Spacecraft mission	Power estimating relationship
Communications	$P_t = 1.1568P_{pl} + 55.497$
Meteorology	$P_t = 602.18\ln(P_{pl}) - 2761.4$
Planetary	$P_t = 332.93\ln(P_{pl}) - 1046.6$
Other missions	$P_t = 210 + 1.3P_{pl}$

2.3 Power

Power is the second critical resource, along with mass, in the design of a spacecraft. In this section the process of estimating, allocating, and controlling this expendable resource will be described.

2.3.1 Estimating Total Power Requirement

A spacecraft design starts with a working definition (not necessarily a final definition) of the payload requirements. There is usually little or no knowledge of the spacecraft power requirement at the start. However, total spacecraft power is a strong function of the payload power; initial power requirements can be estimated on this basis. Table 2.9 summarizes estimating relationships of this kind based on a statistical analysis of past spacecraft designs.

As a spacecraft design progresses, the early power estimates made with the relations in Table 2.9 will be replaced with better estimates. Additional information on the formulation of the table can be found in Chapter 6, "Power." Note that the total spacecraft power computed from the relations in the table includes payload power but no contingency. Contingency should be added to the total power calculated from the relations.

In setting the payload power for use in a Table 2.9 estimate, it is necessary to be careful of the heaters used in the payload instruments. There are two common types of payload heaters: *replacement heaters* and *bake-out heaters*.

Replacement heaters are used for thermal control reasons. When the equipment is turned off, a replacement heater is turned on. The power level of the replacement heater may be the same as that of the equipment it replaces or, more often, slightly lower. High enough for thermal protection but as low as possible to relieve the power load. In building the payload power requirement for such equipment, you clearly do not want to add the replacement heater power to the equipment maximum power.

Bake-out heaters are used with optical instruments to bake out the volatiles in the elastomers, coatings, insulations, and other organics used in the construction of the instrument. This vacuum bake-out is to make certain the volatiles do not later condense on the lenses. The bake-out is done early in the flight and can take many days. It is done in early cruise so that the instrument is ready for use when the mission starts. Bake-out is free to the power system because it occurs when the solar panels are new. When you design the panels in Chapter 6, you will note that degradation is significant with time. Beginning-of-life (BOL) power is substantially greater than end-of-life (EOL) power. (EOL power is the design point).

Table 2.10 Subsystem power allocation guide

	Percentage of subsystem total			
Subsystem	Comsats	Metsats	Planetary	Other
Thermal control	30	48	28	33
Attitude control	28	19	20	11
Power	16	5	10	2
CDS	19	13	17	15
Communications	0	15	23	30
Propulsion	7	0	1	4
Mechanisms	0	0	1	5

On planetary missions, early cruise takes place in full sun, no battery charging is required, and the payload is turned off. Power is so plentiful in early cruise that the design problem is to make sure the shunt radiators are big enough to dump the excess. Clearly, bake-out heaters should not be included in establishing the power requirement or in estimating total spacecraft power.

2.3.2 Power Allocation

Table 2.10 is a guide for the initial allocation of power for each subsystem; the recommendation is expressed as a percentage of total power available to subsystem loads after deducting margin and power specified for the payload. Table 2.10 was developed by statistical analysis of several spacecraft of each type.

In Table 2.10, cable losses are included in the power system allotment. For communication satellites the engineering communication is included in the command and data subsystem (CDS); there is no engineering communication system separate from the payload. In the communication satellite industry the command and data system is generally called the telemetry, tracking, and command (TT&C) subsystem.

2.3.3 Power Margin

Power margin or contingency is defined as

$$\text{Margin} = \text{Total capability} - \text{Current best estimate} \tag{2.5}$$

$$\%\text{Margin} = \frac{\text{Margin}}{\text{Total capability}} \times 100 \tag{2.6}$$

Total capability is the total power capability of the power system, the maximum output of the power source. In a solar powered planetary mission, the total capability at Earth and at the target planet will be different because of the difference in solar distance. In any solar panel powered mission, total capability decreases with age of the panel. In radioisotope thermoelectric generator (RTG) powered missions the total capability decreases with time as the isotope decays.

AIAA conducted a review of historical data from prior NASA and DOD programs to establish industrial guidelines for the power margin shown in Table 2.11. Design maturity is expressed in Table 2.11 as a function of the major

Table 2.11 AIAA recommended power contingencies[a]

Description/ categories	Minimum standard power contingencies, %														
	Proposal stage			Design development stage											
	Bid Class			*CoDR* Class			*PDR* Class			*CDR* Class			*PRR* Class		
	1	2	3	1	2	3	1	2	3	1	2	3	1	2	3
Category AP, 0–500 W	90	40	13	75	25	12	45	20	9	20	15	7	5	5	5
Category BP, 500–1500 W	80	35	13	65	22	12	40	15	9	15	10	7	5	5	5
Category CP, 1500–5000 W	70	30	13	60	20	12	30	15	9	15	10	7	5	5	5
Category DP, 5000 W and up	40	25	13	35	20	11	20	15	9	10	7	7	5	5	5

[a]Copyright AIAA, data reproduced with permission; Ref. 11.

reviews conducted during a design: the bid, CoDR, PDR, CDR, PRR, and FRR (see Table 2.1).

The classes of design listed in Table 2.11 are class 1: a new spacecraft (one of a kind, first generation); class 2: the next-generation spacecraft based on a previously developed family, which expands in complexity or capability within an established design envelope; and class 3: a production-level development based on an existing design for which multiple units are planned and where a significant amount of standardization exists. Note that there should be at least a 10% power margin at liftoff to ensure adequate power for changes in the power profile during the flight mission.

Example 2.2 Power Allocation

Estimate spacecraft power consumption for a communications spacecraft with a payload power requirement of 1500 W (from Table 2.9):

$$P_t = 1.1568\,(1500) + 55.497 = 1791\ \text{W}$$

The subsystem power allocation is 1791 − 1500 = 291 W.

Apply AIAA recommended contingency for proposal stage, class 1, new design. From Table 2.11, the recommended contingency is 70%. Assuming that contingency for the payload is held by the payload provider, a (291)(0.70) = 204 W contingency should be provided for the subsystem design. The total spacecraft power allocation is shown in Table 2.12.

Power now can be allocated to the subsystems using the percentages in Table 2.13. Note that power contingency is handled differently than mass contingency. The purpose of power margin is to make certain that the initial estimate

Table 2.12 Spacecraft power summary

	Allocation, W
Payload	1500
Subsystems	291
Margin	204
Total	1995

of solar panel size is adequate. The solar panels are normally long-lead items. If the panel size is initially underestimated and later needs to be increased, a serious panel delivery schedule problem will likely result. It may become necessary to comb back through the design, decreasing the power requirement. Either approach will result in substantial unplanned cost.

2.4 Other Margins

Mass margins and power margins are normally the most troublesome and were covered in earlier sections. Other design margins are also required; guidelines for these are discussed in this section and exhaustively covered in *Design, Verification and Operations Principles for Flight Systems*, from JPL.[13]

2.4.1 Propulsion

The required propellant load should be based on the maximum possible spacecraft mass, including unallocated reserves (the launch vehicle limit), minimum engine specific impulse, and maximum possible mission ΔV requirement. For a liquid propellant system, propellant margin above requirements should be provided if at all possible. The more margin you can place here, the better the chances for an extended mission or desirable, unplanned maneuvers. Propellant margin can be reduced late in the program if necessary.

Propellant temperatures, and hardware in contact with propellant, should be maintained >10°C above propellant freezing temperature.

Propellant upper temperature must be controlled with substantial margin. Serious failures have occurred by overheating propellant. Note that most elevated

Table 2.13 Spacecraft subsystem power budget

Subsystem	Allocation, %	Power, W
Thermal control	28	82
Attitude control	20	58
Power	10	29
C & DH	17	49
Communications	23	67
Propulsion	1	3
Mechanisms	1	3
Total		291

temperature data on propellants are obtained in laboratory conditions (sterile quartz glass containers and the like); real conditions aboard a spacecraft can lead to propellant decomposition at temperatures lower than laboratory data.

2.4.2 Computer Resource

JPL recommends[13] the following for all computer resource parameters, including memory, CPU speed, and throughput:

At computer selection	400%
At start of phase C/D	60%
At launch	20%(to accommodate post-launch fixes, new sequences, and adequate operating margin)

Margin percentage is computed as shown in Eq. (2.2).

2.4.3 Processing Time and Data Bus Usage

The computer processing maximum throughput requirement should not exceed 50% of computer capacity at the time of computer selection.

2.4.4 Thermal Margin

The thermal margin should be kept at the component level. The calculated temperatures for components need to be kept well inside the component design, qualification, and flight acceptance limits. Figure 2.12 shows the recommended thermal margin pyramid. Note that the temperature calculations are for the stacked worst case. What Fig. 2.12 shows is that a component which the thermal analysis predicts would see 85°C in the hot case and 20°C in the cold case would be designed to the following temperatures.

Hot case:

$$85 + 10 + 10 + 15 = 120^{\circ}\mathrm{C}$$

Cold case:

$$20 - 10 - 10 - 15 = -15^{\circ}\mathrm{C}$$

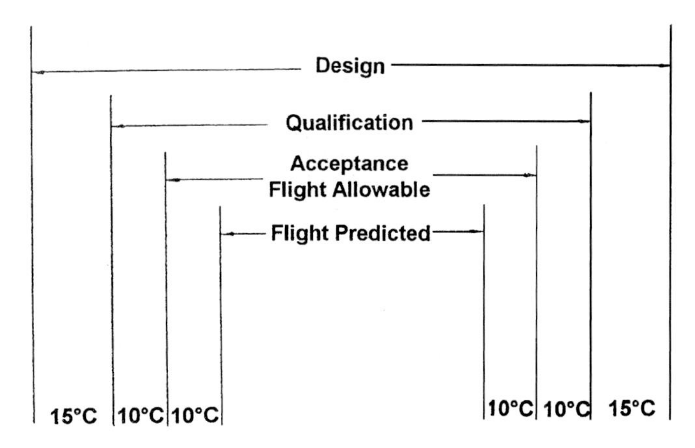

Fig. 2.12 Recommended thermal margin system.

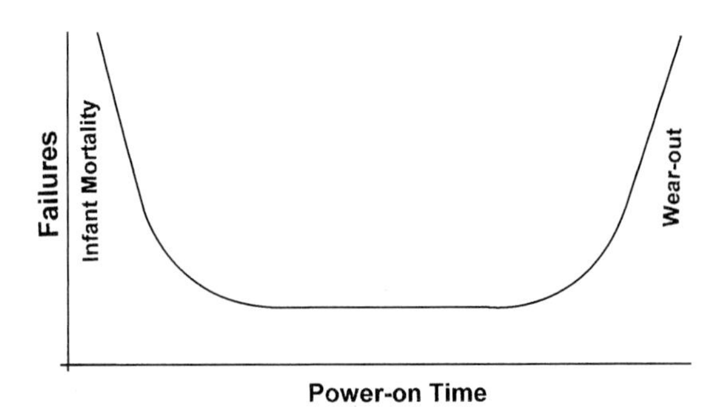

Fig. 2.13 Bathtub curve.

2.4.5 Battery Margin

The recommended[13] battery energy margin is 40%, minimum, at phase C/D start.

2.4.6 Force/Torque Margin

Mission-critical deployments and separations, e.g., solar panel deployment, antenna deployment, and launch vehicle separation require a force/torque margin of at least 100% under worst-case conditions including cold, stiff cables, and vacuum coefficients of friction.

2.4.7 Electronics Minimum Operating Time

The flight spacecraft electronic systems should have about 1000 h of operating time, at system level prior to launch. This requirement is based on the long-held belief that electronic failures as a function of time follow a bathtub curve (Fig. 2.13), with infant mortality failures occurring in the first 1000 or so hours and wear-out failures occurring after many thousand hours.

The Space Shuttle system is an almost ideal test-bed for this theory because of the large number of systems and the long duration of use. Space Shuttle data, compiled by T. Weber, Space System Division of Rockwell International (Ref. 1, p. 93) tends to confirm this long-held belief.

2.4.8 Schedule Margin

Schedule margin is a period of time in which there are no scheduled activities. Recommended schedule margin is shown in Table 2.14.

Table 2.14 Recommended schedule margin (Ref. 13)

Phase	Margin, months
Phase C/D start to ATLO Start	1 month/year
ATLO start to ship to launch site	2 months/year
Arrival at launch site to launch	1 week/month

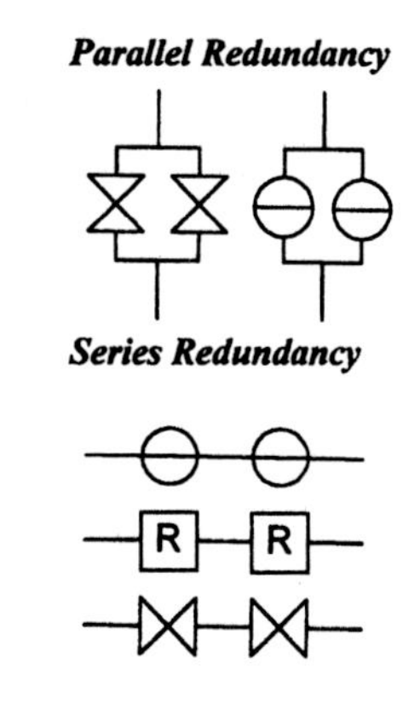

Parallel, normally closed, ordnance valves essentially eliminate failure to open

Parallel solenoid valves essentially eliminate failure to open; however, leakage probability and quantity are increased, as is the probability of failure to close.

Normally open ordinance valves or solenoid valves in series essentially eliminate failure to close.

Series redundant regulators significantly reduce the quantity and probability of leakage. Probability of failure to close is essentially eliminated. (These are the most serious regulator failures in most designs.) Probability of failure to open is increased.

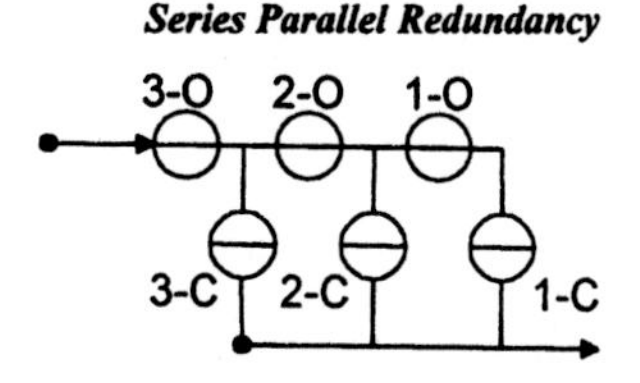

Ordnance valves in sets of series-parallel pairs are a common design in propulsion systems in cases where there are long coast periods between firings. This arrangement allows protection from failure to open and failure to close as well a providing a sealed system during coast periods. The penalty is weight and cost.

Fig. 2.14 Redundancy techniques for propulsion and fluid systems.

Note that going to two shifts in ATLO yields one week of margin per week. There are numerous advantages to conducting a 7×24 ATLO phase: Building time-powered-up is one advantage, and building schedule margin is another.

2.5 Redundancy Techniques

Redundancy is the most common method of increasing the failure tolerance of a spacecraft. Redundancy is used at the piece part level (transistors, diodes, etc.), at the circuit level, and the box level in electronics. In fluid systems it is used at the component level (valves, regulators). In this section the types of redundancy in common use will be described briefly with the characteristics of each.

Figure 2.14 shows the most commonly used redundancy techniques for propulsion and other fluid systems. Figure 2.15 shows the most commonly used redundancy techniques for electronic systems. It should be noted that all redundancy schemes require additional weight and cost. The cost is more than the cost of the additional units; the analysis and testing of a redundant system is more time consuming and expensive than single-string systems.

2.6 Launch Vehicle Interface

The launch vehicle represents the most critical interface with the spacecraft. The selection of the launch vehicle is a major issue that is normally decided by the customer very early in the design process. The critical technical interfaces in the early phases of design are 1) launch mass capability, which determines

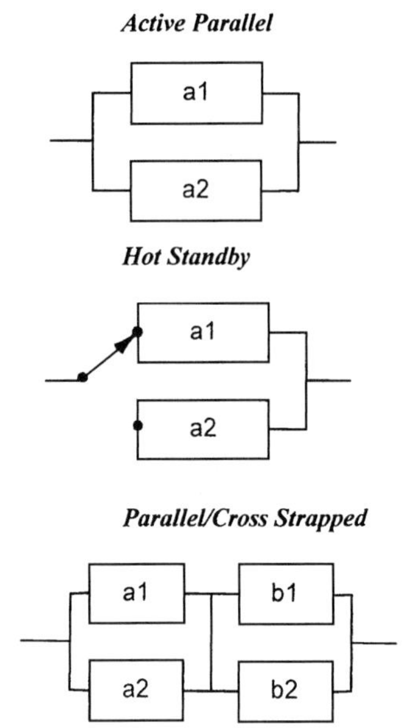

Two active units in parallel redundancy reduce the probability of failure of the function if:
1) additional equipment is not required to sense the failure and switch the units.
2) the failure cause is very improbable or not common to both units.

Redundant units with one unit in hot standby will reduce the probability of failure of the function if:

1) the switching function has a vanishingly small probability of failure.
2) the failure cause is either very improbable or not common to both units

This form of redundancy can withstand any failure of unit *a* or unit *b* and some combinations of *a* and *b* failures, unit a_1 and b_2 for example. However, cross strapping adds significant complexity, provides possible "sneak path" failure modes (which in turn, requires extensive fault analysis). In addition, test time is increased in order to cover all of the cross strapped configurations.

Fig. 2.15 Redundancy techniques for electronic systems.

the maximum acceptable mass of the spacecraft, and 2) fairing dynamic envelope, which determines the maximum dimensions of the spacecraft in the launch configuration.

As the design progresses the interface issues with the launch vehicle become myriad. The integration process with the launch vehicle should start at least 36 months before launch for an expendable launch vehicle and 48 months for a shuttle launch. In the detail design, the launch environment, loads, power, and other issues become important.

Selected elements of launch vehicle integration are contained in this section. A wealth of additional information can be obtained from the AIAA publication, *International Reference Guide to Space Launch Systems*, Third Edition.[2]

2.6.1 Launch Mass

In recent years the number of available launch vehicles has increased enormously. The launch mass capability of a small sample of the available vehicles is shown in Fig. 2.16.

2.6.2 Upper Stages

There are a number of upper stages (sometimes called orbital transfer vehicles) available to provide additional capability for launch vehicles. A given upper stage may be used on several launch vehicles; see Isakowitz[2] for additional detail. Note that the upper stage weight, as loaded onto a launch vehicle, includes the pad weight plus the airborne support equipment.

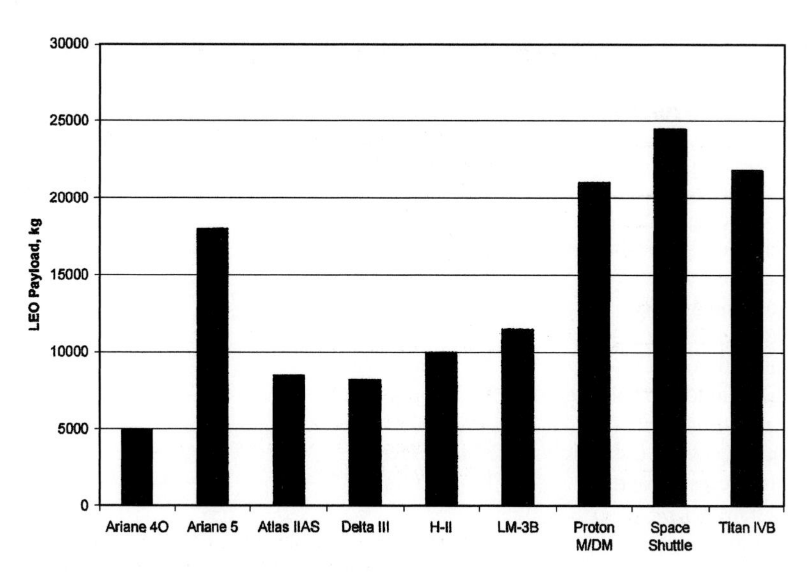

Fig. 2.16 LEO performance capability of selected launch vehicles. Copyright AIAA, reproduced with permission; Ref. 2.

Table 2.15 Fairing dimensions[a]

Launch vehicle	Upper stage	Diameter, m	Length, m
Atlas I	Centaur–1	2.9	7.7
Atlas II	Centaur–2	4.2	9.7
Delta II, 6920/25	PAM–D	2.5	4.8, 6.8
Delta II, 7920/25	PAM–D	2.8	4.2, 5.7
Pegasus		1.2	1.9
Scout 1		0.76	1.2, 1.6
		0.97	1.6
Shuttle	IUS, TOS, PAM–D	4.5	18.0
Titan II		2.8	6.0, 7.6, 9.0
Titan III		3.6	12.4
	PAM–D2	3.6	15.5
	Transtage	3.6	16.0
	TOS	3.6	16.0
Titan IV	Centaur, IUS	4.5	17.0, 20.0, 23.0, 26.0
Ariane 40	H–10	3.6	8.6 to 12.4
H-1		2.2	4.6, 6.5
H-2		3.7	10
MV		2.2	4.5
Long March CZ1D		1.9	2.8
Long March CZ3		2.3	3.1
Long March CZ2E		3.8	7.5
Proton	D1	3.3	4.2, 7.5
Proton	D1e	4.1	4.2, 7.5
Energia	EUS, RCS	5.5	19, 37
Zenit 2		3.3	5.8

[a]Data copyright AIAA, reproduced with permission; Ref. 2.

Table 2.16 Launch vehicle integration issues[a]

Integration issue	Titan IVB accommodation	STS accommodation
Payload compartment		
Maximum payload diameter	4.57 m	4.570 m (180.0 in)
Maximum length	9.7, 12.75, 15.6, 18.8 m	18.3 m (720 in)
Payload integration		
Payload adapter diameter	?	Mission unique
Integration start	T − 33 months	T−36 to 45 months
Last countdown hold w/o recycling	T − 5 min	T − 31 s
On-pad storage capability	30 days fueled	8 h fueled
Last access to payload	T − 15 days	T − 48 to 17 h
Environment		
Max axial load	+5.0 g	3.2 g axial
Max lateral load	±1.5 g	±2.5 g lateral, 4.2 g landing
Min lateral/longitudinal frequency	>2.5 Hz, avoid 6–10 Hz, 17–24 Hz	15 Hz/35 Hz
Max acoustic level	129 dB at 100–600 Hz	
Overall sound pressure	139.3 dB (full octave)	140 dB (one third octave)
Max flight shock	2000 g at 5000 Hz	5500 g at 4000 Hz
Max dynamic pressure on fairing	47 kPa (975 lb/ft^2)	39.2 kPa (819 psi)
Max aeroheating rate at fairing separation	Centaur & IUS: 315 W/m^2 IUS: 473 W/m^2	
Max pressure change rate in fairing	3.5 kPa/s (0.5 psi/s)	3.45 kPa/s (0.5 psi/s)
Cleanliness level in fairing	Class 5000	Class 10,000
Payload delivery		
Orbit injection accuracy (3 sigma)	Perigee: 111 ±2 km Apogee: 328 ±8.1 km Inclination 28.6 ±0.01 deg	Circular orbit: ±18 km Inclination ±0.5 deg
Attitude accuracy (3 sigma)	?	±0.1 deg, ±0.01 deg/s
Nominal payload separation rate	0.6 m/s (2 ft/s)	0.3 m/s (1 ft/s)
Deployment rotation rate available	0–2 rpm	Special spin table req'd
Loiter duration in orbit	No	7–16 days
Maneuvers available (thermal, collision avoidance)	Yes	Yes
Multiple payloads possible	Yes	Yes

[a]Data copyright AIAA, reproduced with permission; Ref. 2. Payload accommodation data of this type are available for essentially any launch vehicle in Ref. 2.

2.6.3 Fairing Dimensions

The launch vehicle aerodynamic fairing dimensions control the size of the spacecraft in the launch configuration; these dimensions are summarized in Table 2.15. A spacecraft may exceed these dimensions in the cruise or on-orbit configurations only through the use of deployment mechanisms.

The spacecraft cannot take full advantage of the fairing dimensions in Table 2.15; provision must be made for the dynamic deflection of the fairing. The reduced dimensions available to the spacecraft are called the *dynamic envelope.*

2.6.4 Spacecraft/Launch Vehicle Interface

Table 2.16 lists the primary areas in which technical integration with a launch vehicle occurs and the accommodations available from two launch vehicles, Titan IVB and the Space Shuttle.

References

[1]Shishko, R., *NASA Systems Engineering Handbook*, NASA, SP-6105, Washington, DC, 1995.

[2]Isakowitz, S. J., Hopkins, J. P., and Hopkins, J. B., *International Reference Guide to Space Launch Systems*, 3rd ed., AIAA, Washington, DC, 1991.

[3]Nagler, R. G., and Schule, J. W., *Satellite Capabilities Handbook and Data Sheets*, Jet Propulsion Lab., JPL 624-3, California Inst. of Technology, Pasadena, CA, July 1976.

[4]*Pioneer Venus Case Study in Spacecraft Design*, Hughes Aircraft Co., AIAA, Washington, DC, 1979.

[5]Fimmel, R. O., Colin, L., and Burgess, E., *Pioneer Venus*, NASA SP-461, NASA, Washington, DC, 1983.

[6]Reeves, E. I., *FLTSATCOM Case Study in Spacecraft Design*, TRW, AIAA, Washington, DC, 1979.

[7]*Magellan Final Report, Vol 1, Pre Launch*, Martin Marietta, Denver, CO, May 1989.

[8]Wihilden, R. D. C., *HEAO: Case Study in Spacecraft Design*, TRW, AIAA, Washington, DC, 1981.

[9]Albaugh, D. H. (ed.), *Mars Observer Mission, Technical Articles*, Jet Propulsion Lab., California Inst. of Technology, Pasadena, CA, 1993.

[10]*Unmanned Spacecraft Cost Model*, U.S. Air Force Space Division, Los Angeles, CA, Aug. 1981.

[11]*Guide for Estimating and Budgeting Weight and Power Contingencies*, ANSI/AIAA-G-020, April 1992.

[12]Whitting, K., and Kovalcik, E. S., *AIAA Aerospace Design Engineers Guide*, 3rd ed., AIAA, Washington, DC, 1993.

[13]*Design, Verification/Validation and Operations Principles for Flight Systems*, Rev. 1, Jet Propulsion Lab., D-17868, California Inst. of Technology, Pasadena CA, 2001.

Problems

2.1 Consider a spacecraft with the following payload instruments and requirements:

	Mass, kg	Operating power, W	Replacement heaters, W	Bake-out heaters, W
MOLA	25.894	33.1	10	
MO Camera	20.955	12.5	11	52.4
MBR	8.829	9.3		
TES	14.139	17.6	8.4	

The maximum spacecraft launch mass is 3230 kg. The required propellant load is 1788 kg and the pressurant mass is 8 kg. Make the following calculations for this spacecraft.

(a) Establish a mass contingency using the AIAA recommendations; use category CW, new spacecraft in the bid phase. Estimate launch vehicle adapter mass. Establish an on-orbit dry mass budget for each subsystem.

Partial solution: Structure dry mass budget = 239 kg.

(b) Prepare a spreadsheet the project might use for a running comparison of subsystem on-orbit dry mass budget to the current estimate.

(c) Estimate the total power requirement for the spacecraft. Set aside power contingency using AIAA recommendations and allocate subsystem power budgets (power to the loads). (In this case, add margin on top of the estimated subsystem power. The total power estimate will be used to design the power system; the object is to make sure the power system is big enough. The mass limit is absolute.)

Partial solution: Attitude control power budget is 61 W.

2.2 You plan to use an existing, qualified computer in a spacecraft design. It has been qualified for a temperature range of 5 to 62°C. What are the thermal control requirements for the computer compartment using the recommended margin requirements (Fig. 2.12)?

2.3 Your project has just completed phase A. Your current best estimate of launch weight is:

	Mass in each category		
	Estimated	Calculated	Actual
Structure	190 kg	25 kg	
Propulsion, dry	62 kg	23	5
Electronics	128 kg	34	23
Cabling	80 kg		

Using the growth algorithm from Section 2.2.5.6, what would the predicted launch mass be?

2.4 Assuming your spacecraft can be represented by a solid cube 1.6 m on a side and it weighs 1320 kg, what would the moment of inertia be about the center of gravity?

2.5 You are planning to launch a spacecraft using the Space Shuttle to reach LEO and the IUS upper stage to provide the additional velocity needed for your mission. If the shuttle will put 24,900 kg into LEO, what is the maximum launch weight for your spacecraft?

3
Orbital Mechanics

Much of the history of mathematical and physical thought was inspired by curiosity about the motion of the planets—the very same laws that govern the motion of spacecraft. The first observations of the celestial bodies predate recorded history. The inertial position of the vernal equinox vector was observed and recorded in stone constructions—Stonehenge, for example—as early as 1800 B.C. Written evidence of stellar observations was left by the Egyptians and the Babylonians from about 3500 years ago. (The Babylonians of this era divided time into 60 even units, a tradition that survives to this day.[1])

In about 350 B.C. Aristotle explained the wandering motion of the planets by proposing that the universe was composed of 55 concentric rotating spheres centered in the Earth. The outermost sphere contained the fixed stars; its rotation is a very adequate explanation of the observed motion of stars in the night sky and the irresistible image of a celestial sphere. The rotation of the inner sphere containing the moon was also a simple, descriptive idea. The motion of the planets, however, was much more difficult. Not only were the observing instruments crude and the mathematical tools nonexistent, the Earth is a singularly poor observation post for heliocentric motion. Usually the planets move slowly eastward across a background of fixed stars; however, at times they reverse direction and move westward. The retrograde loop of Mars is renowned. To explain this motion, Aristotle invented the remaining 53 concentric spheres. Each planet was located in one of the spheres, and its motion was influenced by the rotation of several other spheres.[2]

At about the same time a Greek named Aristarchus proposed a much simpler theory in which the sun and stars were fixed and the planets rotated about the sun, but this theory was not accepted. Aristotle's theory dominated scientific thought for 1800 years.

In about 150 A.D. the Greek astronomer Ptolemy presented a more elaborate Earth-centered theory, which held that the planets moved around the Earth in small circles called epicycles, whose centers moved around in large circles called deferents. The tables of planetary motion computed by Ptolemy based on this theory were used for 1400 years.

In 1543 Nicholas Copernicus broke with Aristotle's theory and advocated sun-centered rotation. His theory neatly explained the retrograde motion of the planets as observed from Earth; however, measured positions were so crude at the time that they fit Ptolemy's conception as well as that of Copernicus.

In about 1610 the Italian scientist Galileo Galilei made two observations that reinforced the theory of Copernicus. First, he observed the motion of moons orbiting Jupiter; thus at least some bodies must not orbit Earth. Second, he observed moonlike phases of the sunlight on Venus that could not be explained by Ptolemy's theory. Galileo attracted the wrath of the Catholic Church and was forced to recant his observations.

In the late 1500s Tycho Brahe made the first accurate measurements of the positions of the planets as a function of time. His achievement is all the more remarkable because the telescope had not yet been invented. Brahe himself believed in Ptolemy's theory of the universe, but his careful observations allowed Johannes Kepler to describe mathematically the heliocentric motion of the planets and to lay to rest the ancient theories of Aristotle and Ptolemy.

In early 1600, Kepler presented his three laws of planetary motion, which are the basis of our understanding of planetary (and spacecraft) motion.

First Law: The orbit of each planet is an ellipse, with the sun at one focus.

Second Law: The line joining the planet to the sun sweeps out equal areas in equal times.

Third Law: The square of the period of a planetary orbit is proportional to the cube of its mean radius.

In addition, Kepler contributed Kepler's equation, which relates position and time. Kepler's equation is the most famous transcendental equation ever discovered. Solving it for the time elapsed since periapsis when one is given orbital elements is trivial; solving it for orbital elements when one is given the time elapsed since periapsis was the "Mount Everest" of mathematics for three centuries.

With Kepler's laws, the world had a description of planetary motion, but the underlying cause of the motion was yet to come, and that would require the genius of Isaac Newton. Newton was 23 and a student at Cambridge in 1665 when an epidemic of the plague broke out. He moved to the relative safety of the countryside, where he spent the most productive two years of his life. Just one of his achievements during that period was to develop the physics of planetary motion. He postulated that all masses are attracted to one another with a force proportional to the product of their masses and inversely proportional to the square of the distance between them. He further postulated that the mass of symmetrical bodies could be concentrated at their centers. To test these assumptions, he formulated the differential equations of motion for the planets and invented calculus to solve them. The result confirmed Kepler's laws. By 1666 one man understood planetary motion, but it would be another 20 years before the world had the news.

Newton put this incredible piece of work aside and neither published it nor discussed it until 20 years later, when he was questioned by his friend Edmund Halley about planetary motion. Newton casually replied that he had already worked it out and had it somewhere among his papers. At the urging of an astonished Halley, Newton published his work in 1687 in *Principia.*[2]

3.1 Two-Body Motion

All of the celestial bodies, from a fleck of dust to a supernova, are attracted to each other in accordance with Newton's law of universal gravitation:

$$F_g = \frac{MmG}{r^2} \tag{3.1}$$

where

F_g = universal gravitational force between bodies
M, m = mass of the two bodies
G = universal gravitational constant
r = distance between the center of masses of the two bodies

The motion of a spacecraft in the universe is governed by an infinite network of attractions to all celestial bodies. A rigorous analysis of this network would be impossible; fortunately, the motion of a spacecraft in the solar system is dominated by one central body at a time. This observation leads to the very useful two-body assumptions:

1) The motion of a spacecraft is governed by attraction to a single central body.

2) The mass of the spacecraft is negligible compared to that of the central body.

3) The bodies are spherically symmetric with the masses concentrated at the centers.

4) No forces act on the bodies except for gravitational forces and centrifugal forces acting along the line of centers.

If the two-body assumptions hold, it can be shown that conic sections are the only possible paths for orbiting bodies and that the central body must be at a focus of the conic.[4]

The two-body assumptions are very nearly true. Table 3.1 shows the most significant relative accelerations on a low Earth orbiter. The influence of Earth on the spacecraft is more significant than any other influence by more than a factor of 1000. The oblateness of the Earth also leads to errors in two-body solutions; however, these errors are small and can be accurately predicted.

No explicit solution has been found for the N-body problem except for $N = 2$; however, numerical solutions are available for the N-body situation. These solutions require a large computing capacity and are used only when the two-body solution is suspect (e.g., a Mercury orbiter) or when high accuracy is required (e.g., navigation calculations).

3.1.1 Circular Orbits

Figure 3.1 shows the forces on a spacecraft in a circular orbit under two-body conditions. The gravitational force on the spacecraft is defined by Eq. (3.1); the

Table 3.1 Accelerations on a low Earth orbiter[a]

Body	Acceleration, g
Earth	0.9
Sun	6×10^{-4}
Moon	3×10^{-6}
Jupiter	3×10^{-8}
Venus	2×10^{-8}

[a]From Ref. 3, p. 11; reproduced through the courtesy of Dover Publications, Inc.

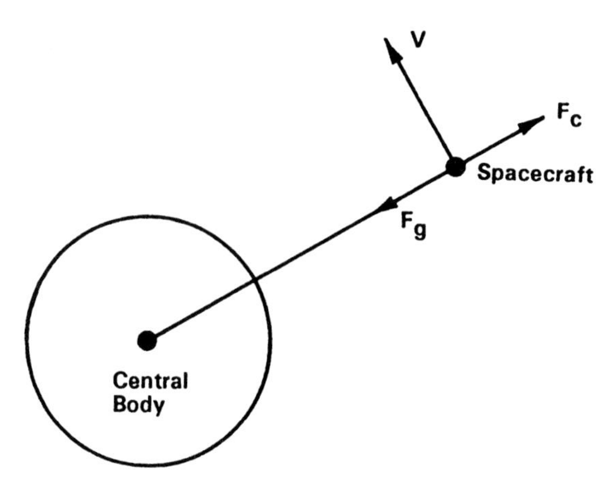

Fig. 3.1 Two-body motion—circular orbit.

centrifugal force on the spacecraft is

$$F_c = \frac{mV^2}{r} \tag{3.2}$$

where

F_c = centrifugal force on the spacecraft
V = velocity of the spacecraft
m = mass of the spacecraft
r = radius from the spacecraft center of mass to the central body center of mass

For circular, steady-state motion to occur, the gravitational and centrifugal forces must be equal; therefore,

$$\frac{mV^2}{r} = \frac{MmG}{r^2} \tag{3.3}$$

$$V = \sqrt{MG/r} \tag{3.4}$$

It is convenient to assign a gravitational parameter μ, which is the product of the central body mass and the universal gravitational constant. In other words,

$$\mu = MG \tag{3.5}$$

allows the simplification

$$V = \sqrt{\mu/r} \tag{3.6}$$

for circular orbits.

The gravitational parameter is a property of the central body; a table that lists values for each of the major bodies in the solar system is given in Appendix B (Substantial improvement in the accuracy of planetary constants is one of the by-products of planetary exploration.)

The period of a circular orbit, derived with equal simplicity, is given by

$$P = \text{circumference/velocity} = 2\pi\sqrt{r^3/\mu} \tag{3.7}$$

Example 3.1 Circular Orbit Velocity and Period

What is the velocity of the Space Shuttle in a 150-n mile circular orbit? For Earth,

$$R_0 = 6378.14\,\text{km}$$

$$\mu = 398{,}600\,\text{km}^3/\text{s}^2$$

Spacecraft altitude h is specified more frequently than radius r in practical applications. It is understood that altitude, used as an orbital element, is given with respect to the mean equatorial radius R_0.

Calculate r:

$$r = R_0 + h = 6378.14 + (150)(1.852) = 6655.94\text{ km}$$

Calculate shuttle velocity for a circular orbit by using Eq. (3.6):

$$V = \sqrt{398{,}600/6655.94} = 7.739\text{ km/s}$$

Calculate orbit period by using Eq. (3.7):

$$P = 2\pi\sqrt{r^3/\mu} = 2\pi\sqrt{(6655.94)^3/398{,}600} = 5404\text{ s} \approx 90\,\text{min}$$

3.1.2 General Solution

Circular motion is a special case of two-body motion. Solving the general case requires integration of the equations of motion; this solution is summarized in the work of Koelle[4] and elsewhere. The conclusions that can be drawn from the general solution are more interesting than the solution itself:

1) Kepler's laws of planetary motion are confirmed and generalized to allow orbits of any conic section, not just elliptical orbits. (Two-body motion is often called Keplerian motion.)

2) The sum of the potential energy and kinetic energy of the orbiting body, per unit mass, is a constant at all points in the orbit and is

$$\varepsilon = \frac{V^2}{2} - \frac{\mu}{r} \tag{3.8}$$

where ε is the total mechanical energy per unit mass, or specific energy, of an object in any orbit about a central body. The kinetic energy term in Eq. (3.8) is $V^2/2$ and the potential energy term is $-\mu/r$. Potential energy is considered to be zero at infinity and negative at radii less than infinity. Equation (3.8) can be reduced to

$$\varepsilon = -\frac{\mu}{2a} \tag{3.9}$$

where a is the semimajor axis (see Fig. 3.3). The total energy of any orbit depends on the semimajor axis of the orbit only. For a circular orbit, $a = r$ and specific

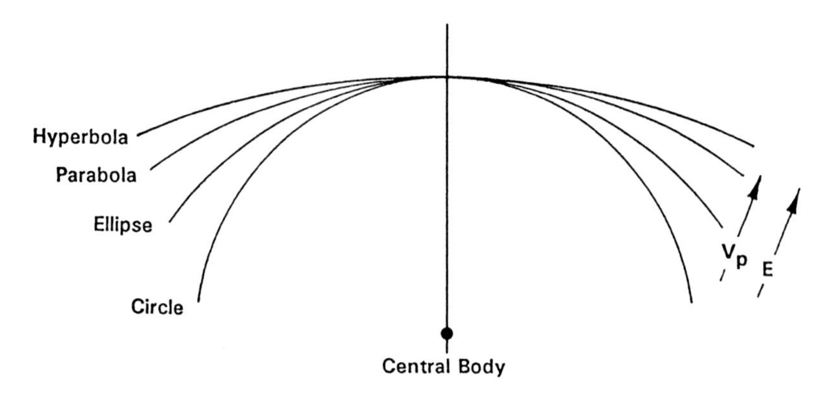

Fig. 3.2 Relative energy of orbit types.

energy is negative. For an elliptical orbit, a is positive and specific energy is negative. Thus, for all closed orbits specific energy is negative. For parabolic orbits, $a = \infty$ and specific energy is zero; as we will see, a parabolic orbit is a boundary condition between hyperbolas and ellipses. For hyperbolic orbits, a is negative and specific energy is positive. Figure 3.2 shows the relative energy for orbit types.

At a given radius, velocity and specific energy increase in the following order: circular, elliptical, parabolic, hyperbolic; total spacecraft energy increases in the same order. Additional energy must be added to a spacecraft to change an orbit from circular to elliptical. Energy must be removed to change from an elliptical to a circular orbit. Both adding and removing energy requires a force on the vehicle and in general that means consumption of propellant.

A particularly useful form of Eq. (3.9) is

$$a = -\frac{\mu}{2\varepsilon} \tag{3.10}$$

3) Total angular momentum of the orbiting body is a constant, equal to the cross product of the radius and the velocity vectors:

$$\boldsymbol{H} = \boldsymbol{r} \times \boldsymbol{V} \tag{3.11}$$

where $\boldsymbol{H}$ is the angular momentum per unit mass (or specific momentum) and is a vector quantity. From vector mechanics, the magnitude of $\boldsymbol{H}$ can be determined by

$$H = rV\cos\gamma \tag{3.12}$$

where

H = magnitude of the specific momentum, km^2/s
r = magnitude of the radius vector (the distance from the spacecraft to the center of mass of the central body), km
V = magnitude of the velocity vector, km/s
γ = flight path angle (the angle between the local horizontal and the velocity vector; see Fig. 3.3), deg

Eccentricity e defines the shape of a conic orbit and is equal to c/a in Fig. 3.3. Eccentricity equals zero for a circular orbit, is less than one for an elliptical orbit,

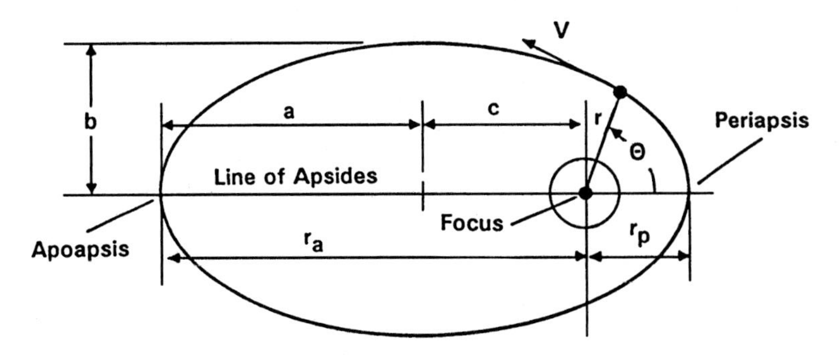

Fig. 3.3 Elliptical orbit

equal to one for a parabolic orbit, and is greater than one for a hyperbolic orbit. Specific energy and eccentricity are related as follows:

$$e = \sqrt{1 - \frac{H^2}{\mu a}} \tag{3.13}$$

The most useful relation resulting from the general two-body solution is the energy integral (also called the vis-viva integral), which yields the general relation for the velocity of an orbiting body:

$$V = \sqrt{(2\mu/r) - (\mu/a)} \tag{3.14}$$

Equation (3.14) yields spacecraft velocity at any point on any conic orbit. For each conic it can be reduced, if desired, to a specific relation.

For a circle, $a = r$, and

$$V = \sqrt{\mu/r} \tag{3.6}$$

as was derived for circular orbits in the previous section. For an ellipse, $a > 0$, and

$$V = \sqrt{(2\mu/r) - (\mu/a)} \tag{3.14}$$

For a parabola, $a = \infty$, and

$$V = \sqrt{2\mu/r} \tag{3.15}$$

For a hyperbola, $a < 0$, and

$$V = \sqrt{(2\mu/r) + (\mu/a)} \tag{3.16}$$

Table 3.2 summarizes the distinguishing characteristics of the four conic orbits. It is important to note that Eqs. (3.8–3.16) are general equations. They are valid at any point on any orbit and can be used in cases where the orbit type is not known. These relations also are summarized in Table 3.3.

Equations (3.8–3.14) can be used to define an orbit and discover its type given only r, V, and γ at a point. The steps required are as follows:

Table 3.2 Characteristics of conic orbits

Element	Circle	Ellipse	Parabola	Hyperbola
Eccentricity e	Zero	<1	1	>1
Semimajor axis a	$=r$	Positive	∞	Negative
Specific energy ε	Negative	Negative	Zero	Positive

1) Given r and V, the specific energy can be calculated from Eq. (3.8).
2) With specific energy, the semimajor axis can be obtained from Eq. (3.10).
3) Given r, V, and γ, the magnitude of specific momentum can be obtained from Eq. (3.12).
4) With specific momentum and the semimajor axis, eccentricity can be obtained from Eq. (3.13).
5) From the characteristics of the eccentricities, the orbit type can be determined from inspection, Table 3.2.

Eccentricity and the semimajor axis define a conic orbit. Knowing these two elements and the orbit type, any other element can be obtained using the relations derived in subsequent sections.

Example 3.2 Defining an Orbit Given *r*, *V*, and γ

An Earth-orbiting spacecraft has been observed to have a velocity of 10.7654 km/s at an altitude of 1500 km and a flight path angle of 23.174 deg; determine the orbit elements e and a and the orbit type.

The orbital radius is $1500 + 6378.14 = 7878.14$ km. Determine the specific energy from Eq. (3.8):

$$\varepsilon = \frac{(10.7654)^2}{2} - \frac{398600.4}{7878.14} = 7.351169\,\text{km}^2/\text{s}^2$$

[One of the problems with using Eq. (3.8) is the subtraction of two large numbers, which reduces accuracy. Intermediate steps must be taken to four or five places.]

With ε set, a can be calculated from Eq. (3.10):

$$a = -\frac{398600.4}{2(7.351169)} = -27111.36\text{ km}$$

The negative semimajor axis indicates that the orbit is a hyperbolic departure. Calculating specific momentum from Eq. (3.12),

$$H = (7878.14)(10.7654)\cos(23.174) = 77968.2\text{ km}^3/\text{s}$$

With the semimajor axis and specific momentum, eccentricity can be calculated from Eq. (3.13)

$$e = \sqrt{1 - \frac{(77968.2)^2}{(398600.4)(-27111.36)}} = 1.250$$

An eccentricity larger than 1 confirms that the orbit is a hyperbolic departure.

Table 3.3 Relations defining an elliptical orbit

Eccentricity e

$$e = \frac{c}{a} \quad (3.19) \qquad e = \frac{r_a}{a} - 1 \quad (3.39)$$

$$e = \frac{(r_a - r_p)}{(r_a + r_p)} \quad (3.20) \qquad e = 1 - \frac{r_p}{a} \quad (3.40)$$

$$e = \frac{r_2 - r_1}{r_1 \cos\theta_1 - r_2 \cos\theta_2} \quad (3.28)$$

Flight path angle γ

$$\tan\gamma = \frac{e \sin\theta}{1 + e\cos\theta} \quad (3.31)$$

Mean motion n

$$n = \sqrt{\mu/a^3} \quad (3.36)$$

Period P

$$P = 2\pi/n \quad (3.37)$$

$$P = 2\pi\sqrt{a^3/\mu} \quad (3.38)$$

Radius (general) r

$$r = \frac{a(1 - e^2)}{1 + e\cos\theta} \quad (3.22)$$

$$r = \frac{r_p(1 + e)}{1 + e\cos\theta} \quad (3.23)$$

Radius of apoapsis r_a

$$r_a = a(1 + e) \quad (3.41)$$

$$r_a = 2a - r_p \quad (3.42)$$

$$r_a = r_p\frac{(1 + e)}{(1 - e)} \quad (3.43)$$

Radius of periapsis r_p

$$r_p = a(1 - e) \quad (3.44)$$

$$r_p = r_a\frac{(1 - e)}{(1 + e)} \quad (3.45)$$

$$r_p = 2a - r_a \quad (3.46)$$

$$r_p = \frac{r_1(1 + e\cos\theta_1)}{1 + e} \quad (3.29)$$

Semimajor axis a

$$a = \frac{(r_a + r_p)}{2} \quad (3.17)$$

$$a = \frac{\mu r}{2\mu - V^2 r} \quad (3.47)$$

$$a = \frac{r_p}{(1 - e)} \quad (3.48)$$

$$a = \frac{r_a}{(1 + e)} \quad (3.49)$$

(continued)

Table 3.3 Relations defining an elliptical orbit (continued)

Time since periapsis t		
	$t = (E - e \sin E)/n$	(3.34)
	$\cos E = \dfrac{e + \cos\theta}{1 + e\cos\theta}$	(3.35)
True anomaly θ		
	$\cos\theta = \dfrac{r_p(1+e)}{re} - \dfrac{1}{e}$	(3.24)
	$\cos\theta = \dfrac{a(1-e^2)}{re} - \dfrac{1}{e}$	(3.25)
Velocity V		
	$V = \sqrt{\dfrac{2\mu}{r} - \dfrac{\mu}{a}}$	(3.14)
	$r_p V_p = r_a V_a$	(3.33)

3.1.3 Elliptical Orbits

Elliptical orbits are by far the most common orbits. All planets and most spacecraft move in elliptical orbits. The geometry of an elliptical orbit is shown in Fig. 3.3.

3.1.3.1 Defining an elliptical orbit. An elliptical orbit is most frequently defined in terms of these orbital elements:

$$a = \text{semimajor axis}$$
$$e = \text{eccentricity}$$
$$r_a = \text{apoapsis radius}$$
$$r_p = \text{periapsis radius}$$

The periapsis of an orbit is the point of closest approach to the central body or the point of minimum radius. The apoapsis is the point of maximum radius. The apoapsis, periapsis, and center of mass of the central body are joined by the line of apsides.

Periapsis and apoapsis are general terms for orbits about any central body; there are also body-specific terms:

General:	Periapsis	Apoapsis
Sun:	Perihelion	Aphelion
Earth:	Perigee	Apogee
Moon:	Perilune	Apolune

By inspection, the long axis of an elliptical orbit is the sum of the apoapsis radius and the periapsis radius. It is useful to define the semimajor axis a as one-half of the long axis. Therefore,

$$a = (r_a + r_p)/2 \tag{3.17}$$

The semimajor axis is one of the classical orbital elements. It defines the size of the orbit and indicates the energy of the orbit. In astronomical work the semimajor axis is often called the mean distance; this term is misleading, however, because the semimajor axis is not equal to the time-average radius.

Similarly, the distance between elliptical foci is $2c$, and

$$c = (r_a - r_p)/2 \tag{3.18}$$

Eccentricity e is one of the classical orbital elements. As previously stated, eccentricity defines the shape of an orbit, and it is defined as

$$e = c/a \tag{3.19}$$

Thus,

$$e = \frac{r_a - r_p}{r_a + r_p} \tag{3.20}$$

The semiminor axis b of an ellipse is related to a and c as follows:

$$a^2 = b^2 + c^2 \tag{3.21}$$

As shown in Fig. 3.4, a spacecraft position in orbit is defined by the radius r and the position angle θ, called the true anomaly, which is measured from the periapsis to the spacecraft in the direction of motion. Given an orbit defined by e and a, the radius to a position can be calculated using the true anomaly as follows:

$$r = \frac{a(1 - e^2)}{(1 + e\cos\theta)} \tag{3.22}$$

$$r = \frac{r_p(1 + e)}{(1 + e\cos\theta)} \tag{3.23}$$

Given a defined orbit, the true anomaly can be calculated from the radius as follows:

$$\cos\theta = \frac{r_p(1 + e)}{re} - \frac{1}{e} \tag{3.24}$$

$$\cos\theta = \frac{a(1 - e^2)}{re} - \frac{1}{e} \tag{3.25}$$

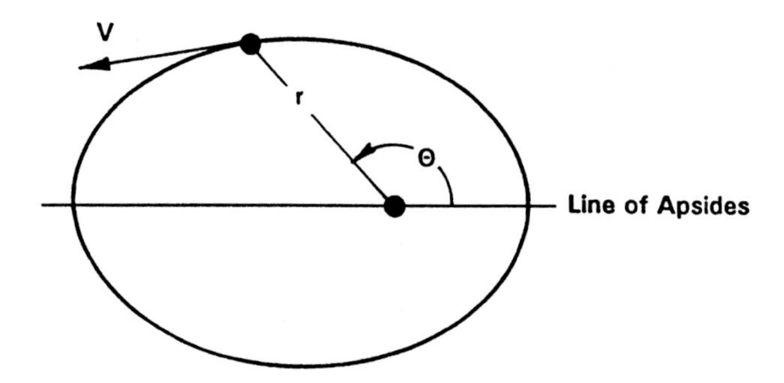

Fig. 3.4 Spacecraft position in orbit.

Example 3.3 True Anomaly at a Point

Given an elliptical Earth orbit with a perigee radius of 6500 km and apogee radius of 60,000 km, find the true anomaly of the spacecraft position as it enters the Van Allen belt at an altitude of about 500 km.

Find the eccentricity by using Eq. (3.20):

$$e = \frac{60{,}000 - 6500}{60{,}000 + 6500} = 0.8045$$

Find true anomaly at an altitude of 500 km (Earth radius is from Appendix B):

$$r = 6378.14 + 500 = 6878.14 \text{ km}$$

From Eq. (3.24)

$$\cos\theta = \frac{(6500)(1 + 0.8045)}{(6878.14)(0.8045)} - \frac{1}{0.8045}$$

$$\theta = 28.755 \text{ deg}$$

Note that the altitude (or radius) defines two positions on an orbit; therefore, the radius in this example will be 6878.14 km when the true anomaly is either 28.755 deg or 331.245 deg.

It is sometimes necessary to design an elliptical orbit to pass through two given points, as shown in Fig. 3.5; two points are sufficient to design a unique elliptical orbit. Intercept trajectories and interplanetary orbits are designed in this way.

From Eq. (3.23), the relations for r_1 and r_2 can be given as

$$r_1 = \frac{r_p(1 + e)}{1 + e\cos\theta_1} \tag{3.26}$$

$$r_2 = \frac{r_p(1 + e)}{1 + e\cos\theta_2} \tag{3.27}$$

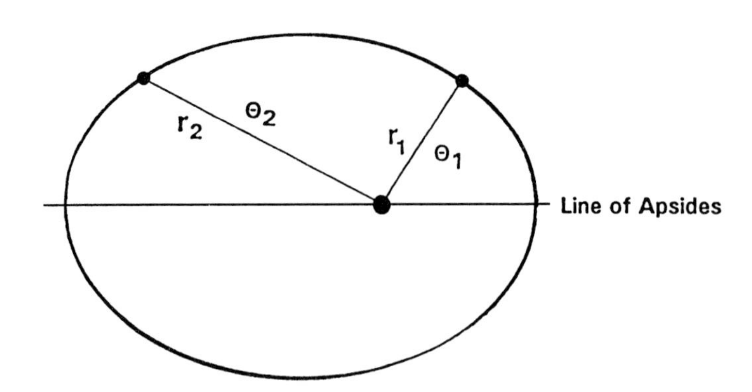

Fig. 3.5 Elliptical orbit defined by two points.

Equations (3.26) and (3.27) are two equations with two unknowns; they can be solved to produce the equations defining an orbit from two points:

$$e = \frac{r_2 - r_1}{r_1 \cos\theta_1 - r_2 \cos\theta_2} \tag{3.28}$$

$$r_p = r_1 \frac{(1 + e\cos\theta_1)}{1 + e} \tag{3.29}$$

Example 3.4 Defining an Ellipse from Two Points

Design a transfer ellipse from Earth at a heliocentric position of $r = 1.00$ AU and a longitude of 41.26° to Pluto at $r = 39.5574$ AU and a longitude of 194.66°. Place the line of apsides at a longitude of 25°.

The true anomaly of a spacecraft at Earth's position is 41.26 deg − 25 = 16.26 deg. Similarly, at Pluto's position, the true anomaly is 194.66 deg − 25 = 169.66 deg.

The radii of date of the planets are

$$r_1(\text{Earth}) = 1.49598 \times 10^8 \text{ km}$$

$$r_2(\text{Pluto}) = 5.9177 \times 10^9 \text{ km}$$

Find the eccentricity by using Eq. (3.28):

$$e = \frac{5.9177 \times 10^9 - 1.49598 \times 10^8}{(1.49598 \times 10^8)\cos 16.26 - (5.9177 \times 10^9)\cos 169.66}$$

$$= 0.9670$$

Find r_p by using Eq. (3.29):

$$r_p = (1.49598 \times 10^8)\frac{(1 + 0.9670\cos 16.26)}{(1 + 0.9670)} = 1.4666 \times 10^8 \text{ km}$$

Any of the remaining elements of this colossal transfer ellipse can be found from e and r_p.

3.1.3.2 Defining parameters at a point. Having defined the orbit, it is now possible to define the parameters at any point on the orbit. The radius and true anomaly define the orbit point. The parameters of interest at a point are flight path angle γ, velocity V, and time since periapsis t.

Flight path angle. Flight path angle is defined as the angle between the local horizontal and the velocity vector, as shown in Fig. 3.6. It might seem strange to consider local horizontal and vertical in a 0-g situation; however, the horizontal at any point can be defined as perpendicular to the radius vector.

The relation between radius and flight path angle can be readily derived by noting that

$$\tan\gamma = \frac{\mathrm{d}r}{r}\mathrm{d}\theta \tag{3.30}$$

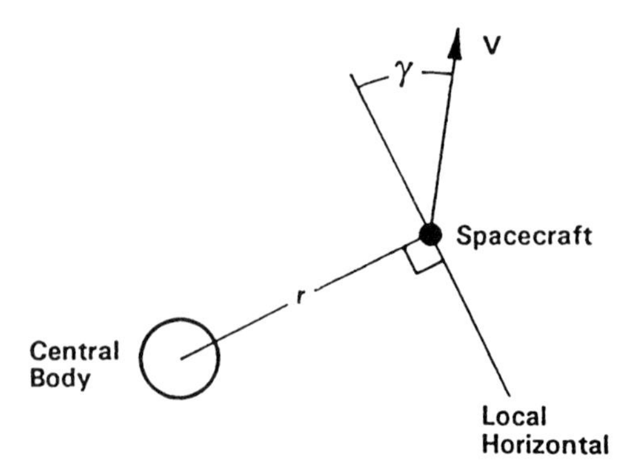

Fig. 3.6 Flight path angle.

Differentiating Eq. (3.30) and rearranging yields

$$\tan \gamma = \frac{e \sin \theta}{1 + e \cos \theta} \tag{3.31}$$

Flight path angle varies with orbital position as shown in Fig. 3.7. As a spacecraft flies around an orbit, its flight path angle is zero at periapsis; it is positive as the spacecraft rises to apoapsis. It is zero again at apoapsis and negative as the spacecraft descends to periapsis.

Velocity. Velocity at any point is calculated from the general velocity equation as previously discussed:

$$V = \sqrt{(2\mu/r) - (\mu/a)} \tag{3.14}$$

Another useful velocity relationship can be obtained from the equation for angular momentum. From Eq. (3.11), the angular momentum vector is

$$\boldsymbol{H} = \boldsymbol{r} \times \boldsymbol{V} \tag{3.11}$$

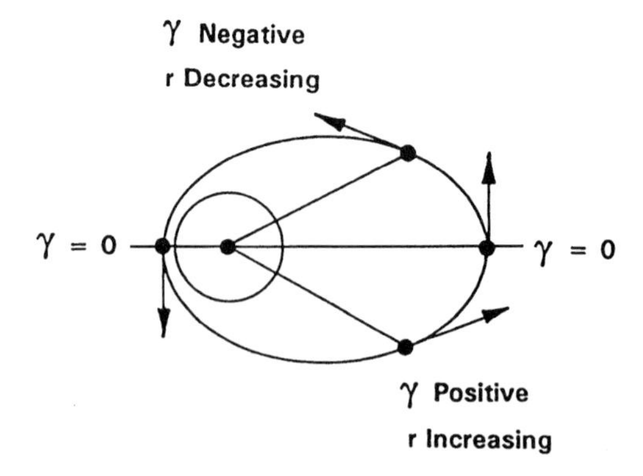

Fig. 3.7 Flight path angle as a function of position.

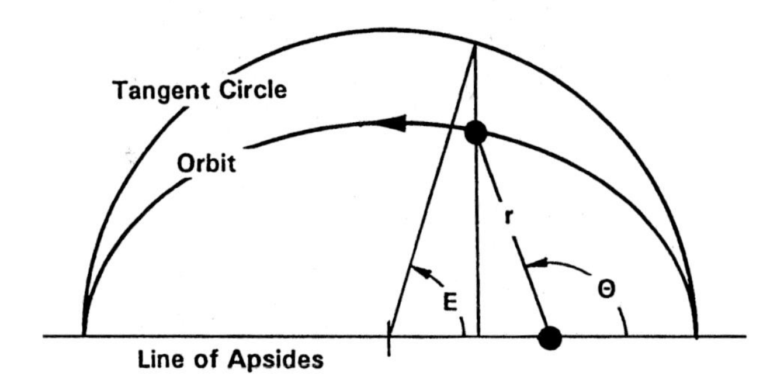

Fig. 3.8 Eccentric anomaly and true anomaly.

Angular momentum is constant for any point on an orbit; therefore,

$$r_1 V_{t1} = r_2 V_{t2} \tag{3.32}$$

where V_t is tangential velocity component.

Since flight path angle is zero at periapsis and apoapsis,

$$r_p V_p = r_a V_a \tag{3.33}$$

Time since periapsis. The time taken by a spacecraft to move from periapsis to a given true anomaly (time since periapsis) is computed using the famous Kepler equation:

$$t = (E - e \sin E)/n \tag{3.34}$$

where

t = time since periapsis
E = eccentric anomaly, rad
e = eccentricity of orbit
n = mean motion

Figure 3.8 shows the geometric relationship between the eccentric anomaly and the true anomaly.

The eccentric anomaly traces a point on a circle, with radius equal to a, that circumscribes the elliptical orbit. As eccentricity goes to zero, the eccentric anomaly and true anomaly merge. The relation between eccentric and true anomaly is

$$\cos E = \frac{e + \cos\theta}{1 + e\cos\theta} \tag{3.35}$$

If the spacecraft were traveling on the circumscribing circle shown in Fig. 3.8, rather than the elliptical orbit, it would have an angular velocity equal to the mean motion,

$$n = \sqrt{\mu/a^3} \tag{3.36}$$

Equation (3.34) does not yield time values greater than one half of the orbit period. For true anomalies greater than π, the result obtained from Eq. (3.34) must be subtracted from the orbit period to obtain the correct time since periapsis. Note also that in using Eqs. (3.34) and (3.35), all angles must be expressed in radians.

Kepler's equation has attracted the attention of mathematicians for centuries. It is tractable when being used to calculate the time since periapsis given the orbit parameters. The historic interest, however, stems from attempts to deduce orbit parameters knowing the time since periapsis. Obtaining a solution in this direction is very difficult indeed. Many of the great mathematical minds of all time attempted to solve Kepler's equation—Newton, Euler, Gauss, Laplace, and Lagrange, to name a few. It is interesting to speculate as to how mathematics would have developed if Kepler's equation were trivial.

Orbital period. When $E = 2\pi$, Kepler's equation reduces to Kepler's third law, the relation for elliptical orbit period:

$$P = 2\pi / n \tag{3.37}$$

$$P = 2\pi \sqrt{a^3/\mu} \tag{3.38}$$

where P equals the orbital period in seconds. The orbital period for a circular orbit, given by Eq. (3.7), is a special case of Eq. (3.38), with $a = r$.

Example 3.5 Parameters at a Point

The elements of the Magellan mapping orbit about Venus are as follows:

$$a = 10{,}424.1 \text{ km}$$

$$e = 0.39433$$

The mapping pass is started at a true anomaly of 280 deg. What are the altitude, flight path angle, velocity, and time since periapsis at this point?

Calculate the radius by using Eq. (3.22):

$$r = \frac{10424.1[1 - (0.39433)^2]}{1 + 0.39433 \cos 280 \text{ deg}} = 8239 \text{ km}$$

$$h = 8239 - 6052 = 2187 \text{ km}$$

Calculate the flight path angle by using Eq. (3.31):

$$\tan \gamma = \frac{0.39433 \sin 280 \text{ deg}}{1 + 0.39433 \cos 280 \text{ deg}}$$

$$\gamma = -19.97 \text{ deg}$$

Calculate the velocity by using Eq. (3.14):

$$V = \sqrt{\frac{2(324858.81)}{8239} - \frac{(324858.81)}{10424.1}} = 6.906 \text{ km/s}$$

Calculate the eccentric anomaly by using Eq. (3.35) (in preparation for calculation of time since periapsis):

$$\theta = 280 \text{ deg} = 4.8869 \text{ rad}$$

$$\cos E = \frac{0.39433 + \cos 4.8869}{1 + 0.39433 \cos 4.8869}$$

$$E = 1.01035 \text{ rad}$$

Calculate the mean motion by using Eq. (3.36):

$$n = \sqrt{\frac{324858.81}{(10424.1)^3}} = 0.0005355\ \text{s}^{-1}$$

Calculate the time since periapsis by using Eq. (3.34):

$$t = \frac{1.01035 - 0.39433 \sin 1.01035}{0.0005355} = 1263\ \text{s}$$

Recall that Eq. (3.34) gives the time since periapsis in the shortest direction. Since the true anomaly is greater than 180 deg, the result of Eq. (3.34) must be subtracted from orbit period. Calculate the orbit period by using Eq. (3.37),

$$P = 2\pi/0.0005355 = 11{,}733\ \text{s}$$

The time at which the mapping starts, measured in the direction of flight, is

$$t = 11{,}733 - 1263 = 10{,}470\ \text{s or } 174.5\ \text{min}$$

3.1.3.3 Summary of relations for elliptical orbits. There are myriad relations between the principle elements of elliptical orbits that can be derived algebraically from the foregoing definitions; many of these are tabulated in the work of Wood[5] and elsewhere. The relations in Table 3.3 are an adequate working set.

3.1.4 Parabolic Orbits

A parabolic orbit would be achieved by an object falling from an infinite distance toward a central body. Such a fall essentially describes the motion of comets, and as a result, comets approach parabolic orbits. The process is reversible. If an object were propelled to the velocity for a parabolic orbit, it would just reach infinity.

A parabolic orbit, shown in Fig. 3.9, represents the boundary condition between an elliptic orbit and a hyperbolic orbit. A parabola can be considered an ellipse with an infinite semimajor axis. The arms become parallel as r approaches infinity and when $e = 1$ and $a = \infty$. The velocity along a parabolic orbit is

$$V = \sqrt{2\mu/r} \tag{3.15}$$

Parabolic orbits are the least energetic open orbits. The velocity on a parabolic orbit is the minimum velocity needed for a spacecraft to escape the central body; i.e., parabolic velocity is the escape velocity. Note that the escape velocity is an inverse function of the square root of the radius; the greater the spacecraft altitude, the lower the escape velocity.

Example 3.6 Escape Velocity

What is the escape velocity from the surface of the moon? For the moon,

$$\mu = 4902.8\ \text{km}^3/\text{s}^2$$

$$R_0 = 1738\ \text{km}$$

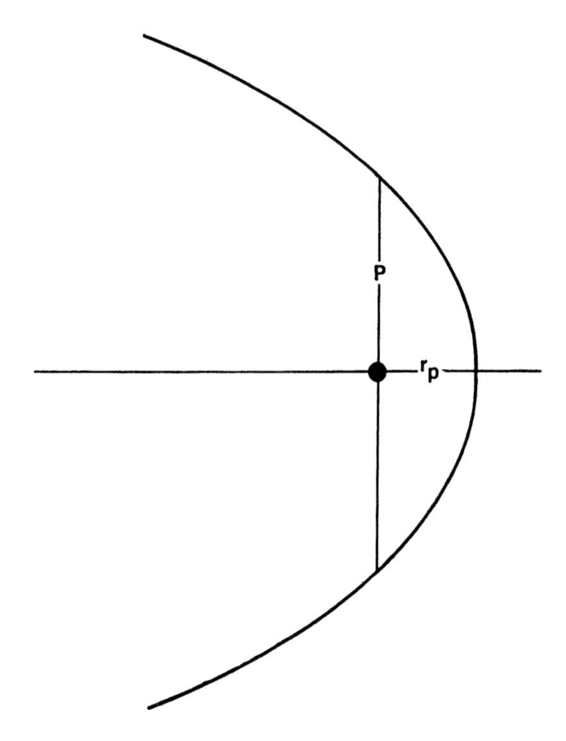

Fig. 3.9 Parabolic orbit.

From Eq. (3.15) the lunar escape velocity is

$$V = \sqrt{(2)(4902.8)/1738} = 2.375 \text{ km/s}$$

Parabolic orbits are an interesting boundary condition but not a useful spacecraft trajectory.

3.1.5 Hyperbolic Orbits

Hyperbolic orbits are used for Earth departure on planetary flights and for planetary arrival and targeting. Hyperbolic planetary flyby orbits are used for energetic gravity-assist maneuvers that change the direction and magnitude of spacecraft velocity without expending spacecraft resources. At any radius, a spacecraft on a hyperbolic orbit has a greater velocity than it would on a parabolic orbit; thus all hyperbolas are escape trajectories. Figure 3.10 shows the geometry of a hyperbolic trajectory. The orbital parameters are similar to those of an ellipse:

r_p = periapsis radius
a = semimajor axis, the distance from the center to the periapsis
b = semiminor axis, the distance from the asymptote to a parallel passing through the central body
e = eccentricity, c/a (greater than 1)
β = angle of the asymptote
θ_a = true anomaly of the asymptote

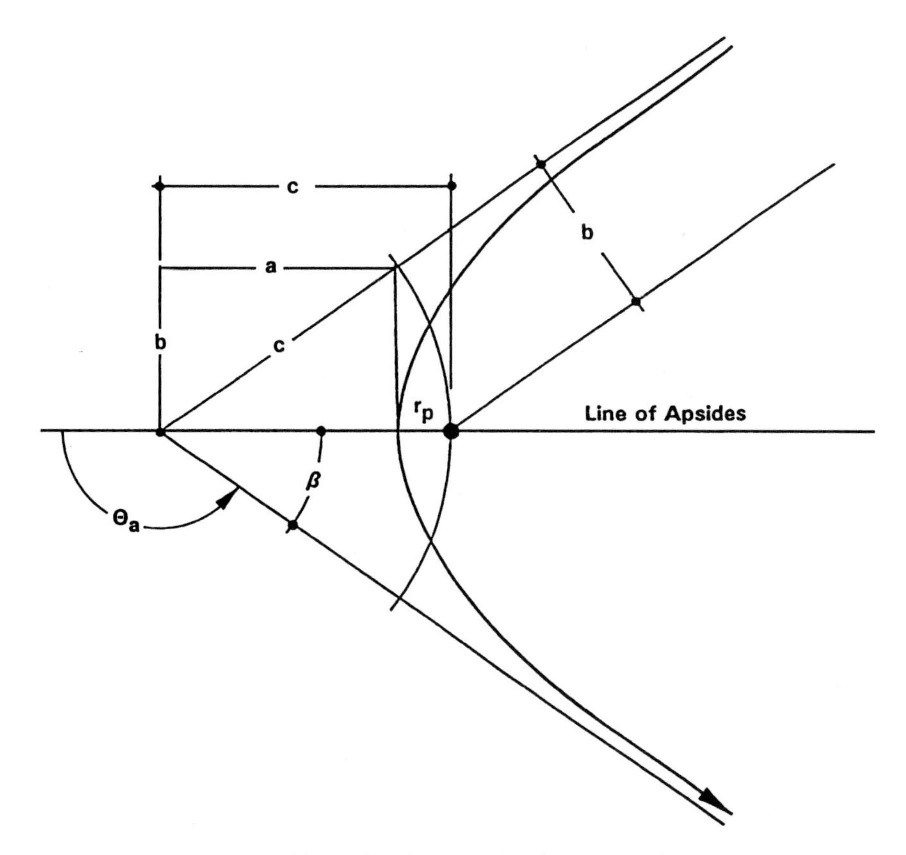

Fig. 3.10 Elements of hyperbola.

It is important to note that the semimajor axis of a hyperbola is considered negative; however, the equations in this book are adjusted to accept a positive semimajor axis.

As with an ellipse,

$$c^2 = a^2 + b^2 \tag{3.21}$$

$$e = c/a \tag{3.19}$$

The angle of the asymptote is

$$\cos\beta = 1/e \tag{3.50}$$

The position of a spacecraft on a hyperbolic orbit is defined by radius and true anomaly; in a manner similar to that for ellipses,

$$r = \frac{a(e^2 - 1)}{1 + e\cos\theta} \tag{3.51}$$

$$\cos\theta = \frac{a(e^2 - 1)}{re} - \frac{1}{e} \tag{3.52}$$

The true anomaly of the asymptote θ_a is

$$\theta_a = 180\,\text{deg} \pm \beta \tag{3.53}$$

In the region between the minimum θ_a and the maximum θ_a, the hyperbolic radius is infinite. From Eq. (3.52), the true anomaly of the asymptote can also be expressed as

$$\cos\theta_a = -1/e \tag{3.54}$$

The flight path angle relation is the same as for an ellipse and is derived in the same way:

$$\tan\gamma = \frac{e\sin\theta}{1 + e\cos\theta} \tag{3.31}$$

The velocity at any point on a hyperbola is

$$V = \sqrt{(2\mu/r) + (\mu/a)} \tag{3.16}$$

The velocity on a hyperbolic trajectory is greater than the velocity on a parabola at any radius. The parabolic velocity goes to zero for an infinite radius; on a hyperbolic trajectory, the velocity at infinity is finite and equal to

$$V_\infty = \sqrt{\mu/a} \tag{3.55}$$

V_∞ is the velocity in excess of the escape velocity and is called the hyperbolic excess velocity (V_{HE}) when Earth escape is intended. For all situations other than Earth escape, V_∞ is the preferable term. For the equations in this text, V_{HE} and V_∞ may be used interchangeably.

$$V_\infty = V_{\text{HE}} = \sqrt{\mu/a} \tag{3.56}$$

$$V = \sqrt{(2\mu/r) + V_{\text{HE}}^2} \tag{3.57}$$

V_{HE} is the velocity that must be added to the Earth's velocity to achieve departure on a planetary mission. It is traditional to express the energy required of a launch vehicle for a planetary mission as C3, which is the square of V_{HE}:

$$\text{C3} = V_{\text{HE}}^2 \tag{3.58}$$

C3 is used to describe hyperbolic departure from Earth; it is not used to describe an arrival at a planet.

Example 3.7 Hyperbolic Earth Departure

The elements of the departure hyperbola of the Viking I Mars Lander were[6]

$$a = 18{,}849.7\,\text{km}$$

$$e = 1.3482$$

What C3 value was provided by the lander's Titan IIIE launch vehicle? From Eqs. (3.56) and (3.58),

$$\text{C3} = V_{\text{HE}}^2 = \frac{398{,}600.4}{18{,}849.7} = 21.146\,\text{km}^2/\text{s}^2$$

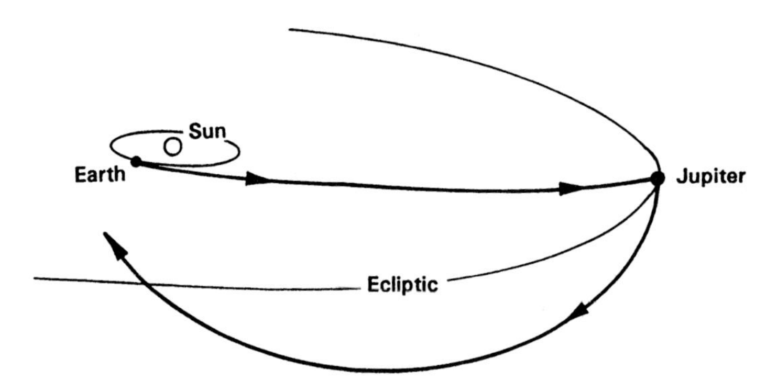

Fig. 3.11 Ulysses mission gravity-assist maneuver.

The angle of the asymptote is given by Eq. (3.50) as

$$\cos\beta = 1/1.3482 = 0.7417$$

$$\beta = 42.12\,\text{deg}$$

3.1.5.1 Gravity-assist maneuvers. The angle through which a spacecraft velocity vector is turned by an encounter with a planet is 180 deg $-\ 2\beta$. This type of encounter is called a gravity-assist maneuver; it is a very energetic maneuver that can be accomplished without expending spacecraft resources.

The 1989 Galileo mission would not have been possible without multiple gravity-assist turns at Venus and Earth. Gravity-assist trajectories were also used by Voyager to target from one outer planet to the next at a substantial reduction in time of flight. The Ulysses mission, to take scientific data over the polar region of the sun, would not be possible in any year without a gravity turn out of the ecliptic. As shown in Fig. 3.11, Ulysses uses the gravitational attraction of Jupiter to bend its trajectory out of the ecliptic plane and send it on its way over the polar region of the sun. The design of this important maneuver will be discussed in Section 3.5.

3.1.5.2 Time of flight. The time since periapsis can be determined in a manner analogous to that for elliptical orbits with the aid of the hyperbolic eccentric anomaly F:

$$t = (e\sinh F - F)/n \tag{3.59}$$

$$\cosh F = \frac{e + \cos\theta}{1 + e\cos\theta} \tag{3.60}$$

where

t = time since periapsis passage, s
F = hyperbolic eccentric anomaly, rad
e = eccentricity
a = semimajor axis, km
θ = true anomaly, deg
n = mean motion, $1/s$

This expression for F yields inaccurate results for e values near 1. See Ref. 3, page 191, for a universal variable solution for near-parabolic orbits.

The following hyperbolic relations are useful in solving for time:

$$F = \ln(\cosh F + \sqrt{\cosh^2 F - 1}) \tag{3.61}$$

$$\sinh F = \tfrac{1}{2}[\exp(F) - \exp(-F)] \tag{3.62}$$

Example 3.8 Time Since Periapsis—Hyperbola

On 24 August 1989, Voyager 2 flew past the north pole of Neptune.[7] The elements of the Voyager 2 encounter hyperbola were

$$a = 19{,}985\ \text{km}$$

$$e = 2.45859$$

During departure, Voyager 2 passed Triton, one of the moons of Neptune, at a radius of 354,600 km. What was the time since periapsis for the encounter with Triton?

Calculate the mean motion by using Eq. (3.36):

$$n = \sqrt{\frac{6{,}871{,}307.8}{(19{,}985)^3}} = 0.0009278\ \text{s}^{-1}$$

Calculate the cosine of the true anomaly by using Eq. (3.52):

$$\cos\theta = \frac{19{,}985[(2.45859)^2 - 1]}{(354{,}600)(2.45859)} - \frac{1}{2.45859} = -0.2911$$

From Eq. (3.60)

$$\cosh F = \frac{2.45859 - 0.2911}{1 + (2.45859)(-0.291096)} = 7.6236$$

From Eq. (3.61)

$$F = \ln[(7.6236) + \sqrt{(7.6236)^2 - 1}] = 2.720$$

From Eq. (3.62),

$$\sinh F = \tfrac{1}{2}[\exp(2.720) - \exp(-2.720)] = 7.5577 \tag{3.63}$$

Finally, calculate time since periapsis by using Eq. (3.59):

$$t = \frac{(2.45859)(7.5577) - 2.720}{0.0009278} = 17095\ \text{s or } 4.75\ \text{h}$$

3.1.5.3 Summary of relations defining a hyperbolic orbit. Additional relations for hyperbolic orbits can be found by algebraic manipulation or on page 201 of Ref. 5 and elsewhere. Table 3.4 summarizes the most frequently used equations.

The equations in Table 3.4 have been arranged to accept semimajor axis as a positive number. V_{HE} denotes an Earth-centered hyperbola; V_∞ is the general case. V_∞ and V_{HE} are used interchangeably in Table 3.4.

3.1.6 Time Systems

Mission design calculations, especially ephemeris calculations, require a more precise definition of how time is measured and of the relationship between time and planetary position. Five different time measurement systems must be understood.

3.1.6.1 Apparent solar time. The most ancient measure of time is the apparent solar day. It is the time interval between two successive solar transits across a local meridian, i.e., two successive high noons. Two motions are involved in this definition: the rotation of the Earth about its axis and the revolution of the Earth about the sun, as shown in Fig. 3.12. The apparent solar day can be measured by a sundial and was an adequate standard for thousands of years.

3.1.6.2 Mean solar time. One of the problems with apparent solar time is that the days are all of different lengths. This variation occurs because the Earth's axis is not perpendicular to the ecliptic plane, the Earth's orbit is not circular, and the Earth's axis wobbles slightly with respect to the ecliptic plane. These effects are all small, regular, and predictable; therefore, it is possible to establish a mean solar day that has an invariant length. A mean solar day is defined based on the assumptions that the Earth is in a circular orbit with period exactly equal to the actual period of the Earth's orbit and that the Earth's axis is perpendicular to the ecliptic plane. The mean solar day is the common time standard; it is the time you read from your watch. Seconds, minutes, and hours are defined in duration by dividing a mean solar day into equal parts. A mean solar day is equal to exactly 24 h, or 1440 min, or 86,400 s.

3.1.6.3 Sidereal time. For some purposes, notably astronomy, it is necessary to measure time with respect to the fixed stars rather than the solar zenith. A mean sidereal day is the mean time required for the Earth to rotate once on its axis with respect to fixed stars, or with respect to inertial space. A mean sidereal day is slightly shorter than a mean solar day, as shown in Fig. 3.13. A sidereal

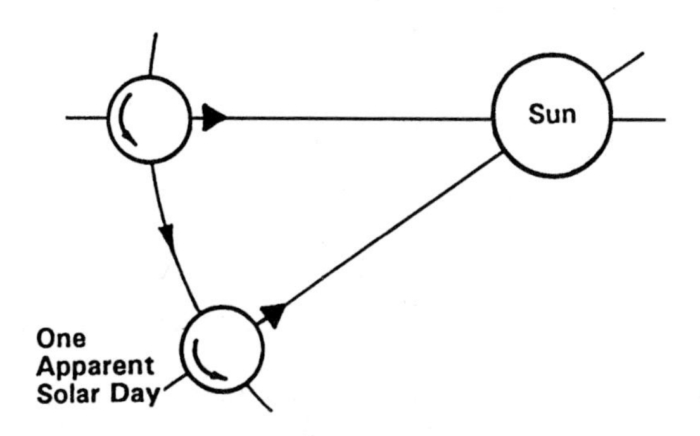

Fig. 3.12 Apparent solar day.

Table 3.4 Relations defining a hyperbolic orbit

Angle of asymptote β

$\tan\beta = b/a$ (3.63) $\tan\beta = bV_{HE}^2/\mu$ (3.64)

$\tan\beta = \dfrac{2br_p}{b^2 - r_p^2}$ (3.65) $\cos\beta = 1/e$ (3.50)

Eccentricity e

$e = 1/\cos\beta$ (3.66) $e = 1 + (r_p/a)$ (3.67)

$e = \sqrt{1 + (b^2/a^2)}$ (3.68)

Flight path angle γ

$$\tan\ \gamma = \frac{e\sin\theta}{1 + e\cos\theta} \quad (3.31)$$

Mean motion n

$$n = \sqrt{\mu/a^3} \quad (3.36)$$

Radius (general) r

$$r = \frac{a(e^2 - 1)}{1 + e\cos\theta} \quad (3.51)$$

Radius of periapsis r_p

$r_p = b\sqrt{(e-1)/(e+1)}$ (3.69) $r_p = a(e - 1)$ (3.70)

$r_p = c - a$ (3.71) $r_p = b\tan(\beta/2)$ (3.72)

$r_p = \dfrac{2\mu + \mu(e-1)}{V_p^2}$ (3.73) $r_p = -\dfrac{\mu}{V_{HE}^2} + \sqrt{\left(\dfrac{\mu}{V_{HE}^2}\right)^2 + b^2}$ (3.74)

$r_p = -a + \sqrt{a^2 + b^2}$ (3.75)

Semimajor axis a

$a = b/\sqrt{e^2 - 1}$ (3.76) $a = \dfrac{r_p}{e - 1}$ (3.77)

$a = \mu/V_{HE}^2$ (3.78) $a = \left(b^2 - r_p^2\right)/2r_p$ (3.79)

$a = \dfrac{\mu r_p}{r_p V_p^2 - 2\mu}$ (3.80)

Semiminor axis b

$b = r_p\sqrt{(e+1)/(e-1)}$ (3.81) $b = a\sqrt{e^2 - 1}$ (3.82)

$b = r_p\sqrt{\dfrac{2\mu}{r_p V_{HE}^2} + 1}$ (3.83)

(**continued**)

Table 3.4 Relations defining a hyperbolic orbit (continued)

Time since periapsis t

$$t = (e \sinh F - F)/n \tag{3.59}$$

$$\cosh F = \frac{(e + \cos\theta)}{(1 + e\cos\theta)} \tag{3.60}$$

$$F = \ln(\cosh F + \sqrt{\cosh^2 F - 1}) \tag{3.61}$$

$$\sinh F = \tfrac{1}{2}[\exp(F) - \exp(-F)] \tag{3.62}$$

True anomaly θ

$$\cos\theta = \frac{a(e^2 - 1)}{re} - \frac{1}{e} \tag{3.52}$$

True anomaly of asymptote θ_a

$$\theta_a = 180\,\text{deg} \pm \beta \tag{3.53}$$

$$\cos\theta_a = -\frac{1}{e} \tag{3.54}$$

Velocity V

$V_\infty = V_{HE} = \sqrt{\mu/a}$	(3.56)	$V = \sqrt{(2\mu/r) + V_{HE}^2}$	(3.57)
$V = \sqrt{(2\mu/r) + (\mu/a)}$	(3.16)	$C3 = V_{HE}^2$	(3.58)

day is subdivided into sidereal hours, minutes, and seconds just as a solar day is; however, the lengths are slightly different:

1 mean solar day = 1.0027379093 mean sidereal days[8]
= 24 h, 3 min, 56.5536 s of sidereal time
= 86,636.55536 sidereal seconds
= 86,400.00000 mean solar seconds

1 mean sidereal day = 86,164.091 mean solar seconds

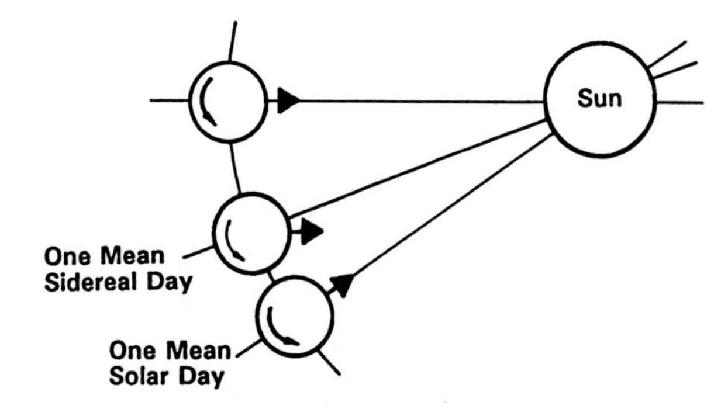

Fig. 3.13 Mean sidereal day. (Courtesy of Dover Publications, Inc.; Ref. 3, p. 102.)

Table 3.5 Conversion of UT to local time

Eastern Standard Time	(EST)	+	5 h	=	UT
Eastern Daylight Time	(EDT)	+	4 h	=	UT
Central Standard Time	(CST)	+	6 h	=	UT
Central Daylight Time	(CDT)	+	5 h	=	UT
Mountain Standard Time	(MST)	+	7 h	=	UT
Mountain Daylight Time	(MDT)	+	6 h	=	UT
Pacific Standard Time	(PST)	+	8 h	=	UT
Pacific Daylight Time	(PDT)	+	7 h	=	UT

3.1.6.4 Time zones and universal time. Another problem with solar time is that the time of day is different at every longitude on Earth. Up until the mid-1800s every town in the United States set its clock by local high noon. The advent of railroads changed all that. To operate a train schedule it was necessary to standardize time. The Earth is now subdivided into 24 standard time zones, each encompassing approximately 15° of longitude. Greenwich, England is the index mark for time zones. The mean solar time at Greenwich is called universal time (UT). (Greenwich mean time, an early standard used prior to 1925, was similar to universal time except that a new day was started at noon rather than at midnight.) Table 3.5 shows the conversion of universal time to local mean solar times for the United States. Interestingly enough, universal time is computed from solar motion in mean sidereal time and then converted to mean solar time. Universal time is expressed by the 24-h clock method; i.e., 4 p.m. is stated as 16:00.

3.1.6.5 Julian days. The Julian day system is a means of providing a unique number to all days that have elapsed since a standard reference day in the distant past. The day selected for the starting point of the system is 1 January 4713 B.C. The days are in mean solar measure. The Julian day (JD) numbers are never repeated and are not partitioned into weeks or months. As a result, the number of days between two dates may be obtained by subtracting Julian day numbers.

There are 36,525 mean solar days in a Julian century and 86,400 s in a day. The Julian century does not refer to some time system; it is merely a count of a fixed number of days. Ephemeris calculations are done in Julian days and Julian centuries.

This curious system was devised by Joseph Scaliger in 1582 to provide a calendar suitable for recording astronomical observations.[9] The starting date was selected because it is the starting point of three cycles: the 28-year solar cycle, the 19-year lunar cycle, and the 15-year tax cycle in use at the time. In spite of the general inconvenience of the system, it is still, four centuries later, the only generally recognized system of unique day numbers.

A Julian day starts at noon UT rather than at midnight, an astronomical custom; astronomers find it disconcerting for the day number to change in the middle of a night's observations. This custom has a curious effect on the conversion of Julian days to equivalent Gregorian calendar (the common calendar) days, as shown in Fig. 3.14. Julian dates may be calculated from calendar dates using Eq. (3.84),

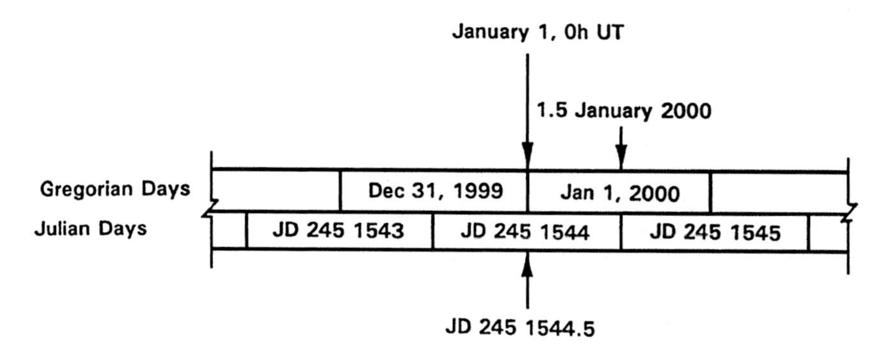

Fig. 3.14 Conversion of Gregorian days to Julian days.

adapted from a remarkably compact relation devised by Thomas C. Van Flandern (see Ref. 10). The result is accurate for dates between 1901 A.D. and 2099 A.D.

$$J = 367Y - \frac{7[Y + (M + 9)/12]}{4} + \frac{275M}{9} + D + 1{,}721{,}013.5 \qquad (3.84)$$

where

J = Julian day number
Y = calendar year
M = calendar month number (e.g., July = 7)
D = calendar day and fraction

All divisions must be integer divisions. Only the integer is kept; the fraction is discarded.

Example 3.9 Conversion to Julian Days

What is the number of the Julian day that started at noon UT on 1 January 2000? That is, find the Julian day for

$$M = 1$$
$$Y = 2000$$
$$D = 1.5$$

From Eq. (3.84),

$$J = 367(2000) - 7\{[2000 + (1 + 9)/12]/4\} + 275(1)/9 + 1.5 + 1{,}721{,}013.5$$
$$J = 734{,}000 - 3500 + 30 + 1.5 + 1{,}721{,}013.5 = 2{,}451{,}545$$

3.1.7 Coordinate Systems

Four types of coordinate systems are common in mission design work: 1) the geocentric–inertial system, 2) the heliocentric–inertial system, 3) the geographic–body-fixed system, and 4) the International Astronomical Union (IAU) cartographic system (for the planets). The systems are designed to make various types of motion easy to visualize; selection of the proper coordinate system has a profound effect on the difficulty of a given type of problem. Each system is defined by the selection of the origin, selection of axes, and the determination of what is fixed.

A body-fixed coordinate system measures all motion relative to that body with the assumption that the body is stationary. Our daily experience is in body-fixed coordinates where the sun appears to rise in the east and set in the west. An inertially fixed coordinate system is one which is referenced to stellar positions. The vernal equinox vector is the primary reference in such systems.

3.1.7.1 Vernal equinox. There are two equinoxes in a year, one in the spring and one in the fall. On these days the Earth is located at the intersection line of the equatorial and ecliptic planes. The axis of rotation of the Earth is in a plane perpendicular to the sun's rays; as a result the length of the day and night is the same everywhere on Earth. As shown in Fig. 3.15, the vernal equinox vector is the vector from the center of mass of the Earth to the center of mass of the sun on the spring (northern hemisphere) equinox day, which occurs around 21 March.

The vernal equinox was first observed more than 5000 years ago; at that time the vector passed through Aries (a constellation in the zodiac also called the Ram). The sign of the Ram, ♈, is used to this day to indicate the vernal equinox; however, over the years the vector has moved through Aries and into Pisces (the Fish).[11] This small precession of the equinoxes, about 0.014 deg per year, does not prevent consideration of the vector as fixed for most purposes; however, this precession

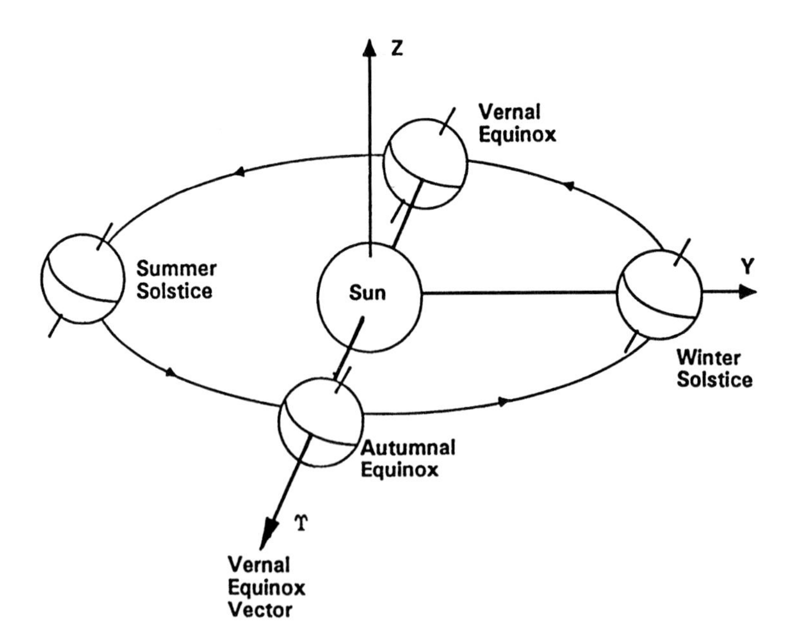

Fig. 3.15 Vernal equinox. (Courtesy of Dover Publications, Inc.; Ref. 3, p. 54.)

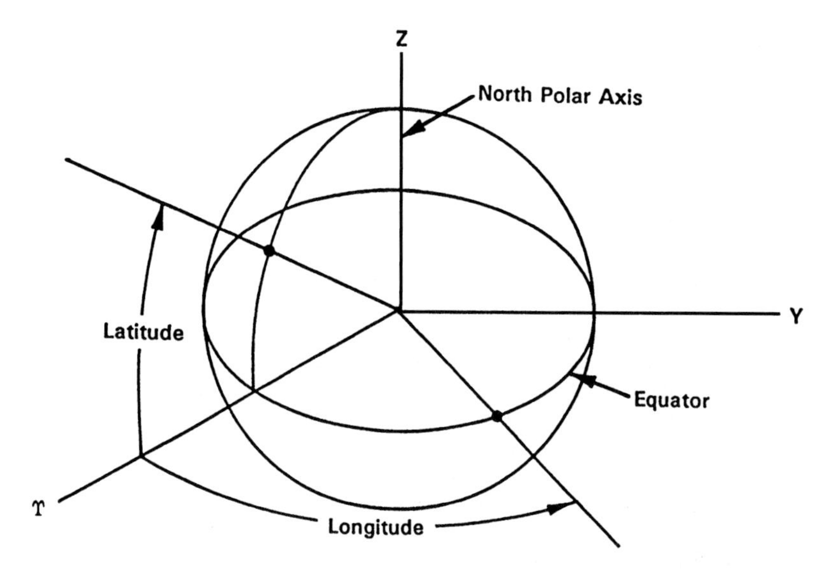

Fig. 3.16 Geocentric–inertial coordinate system.

and the motion of the ecliptic plane are important for ephemeris calculations. Ephemeris tabulations are noted to indicate the instant of time that defines the exact position of the reference axes.

3.1.7.2 Geocentric–inertial coordinate system. For most orbital calculations the coordinate system of choice is the geocentric–inertial system, shown in Fig. 3.16. The origin for the geocentric system is the center of mass of the central body, which is usually, but not necessarily, the Earth. The equatorial plane (the plane of the Earth's equator) is the reference plane. The X axis is the vernal equinox vector, and the Z axis is the spin axis of the Earth; north is positive. The axes are fixed in inertial space or fixed with respect to the stars.

Positions are measured by latitude and longitude; longitudes are measured eastward (i.e., counterclockwise as viewed from celestial north) from the vernal equinox vector and centered in the Earth. North latitudes are measured in the positive Z direction from the equatorial plane, and south latitudes are measured in a negative Z direction.

3.1.7.3 Heliocentric–inertial system. The heliocentric–inertial system is used for interplanetary mission design. The ecliptic plane, the plane that contains the center of mass of the sun and the orbit of the Earth, is the reference plane for the heliocentric system. The system is identical to the geocentric-inertial system shown in Fig. 3.16 except for the reference plane and the central body. The origin for the system is centered in the sun, and the system is fixed with respect to the stars. The equatorial plane is inclined at an angle of approximately 23.5 deg with respect to the ecliptic. The X axis for the heliocentric system is the vernal equinox vector, which is common to the ecliptic plane and the equatorial plane. The Z axis is

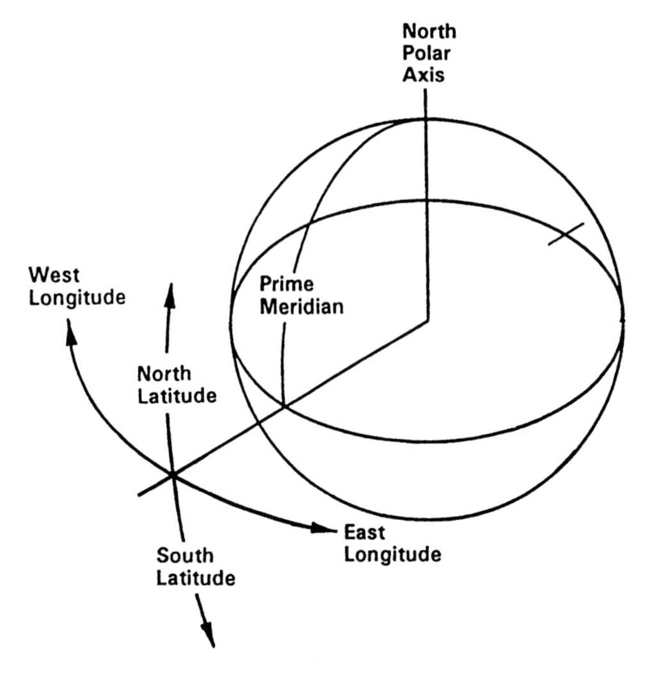

Fig. 3.17 Geographic–body-fixed coordinate system.

perpendicular to the ecliptic; positive is north. Latitude and longitude are measured as in the geocentric system.

3.1.7.4 Geographic–body-fixed coordinate system. The geographic–body-fixed coordinate system, shown in Fig. 3.17, has been used for centuries to locate and map positions on the Earth. The system is Earth-centered and body-fixed; the surface of the Earth is divided into a grid of latitude and longitude measured in degrees. Spacecraft ground track is commonly plotted in this coordinate system.

Longitude is a spherical angle measured around the polar axis, starting at the prime meridian. The prime meridian is the great circle passing through Greenwich, England, and the Earth's poles. Longitude is measured in degrees east or west from the prime meridian (the highest longitude being 180°). The longitude of New York is approximately 75°W. Both east and west longitudes are positive.

Latitude is a spherical angle measured around the center of the Earth starting from 0° at the equator. Latitude is measured north or south of the equator with the highest latitude being 90°. Both north and south latitudes are positive. Philadelphia and Denver are at approximately 40°N.

The geographic system is most frequently mapped as a Mercator projection, which projects an Earth map on a cylinder wrapped around the equator.

3.1.7.5 International Astronomical Union cartographic coordinates. The International Astronomical Union (IAU) cartographic system is a body-centered, body-fixed system for mapping solar system bodies. International agreement has been established by the IAU on the placement of the north polar axis, the equatorial plane, and the prime meridian for the planets and their satellites.[12] The

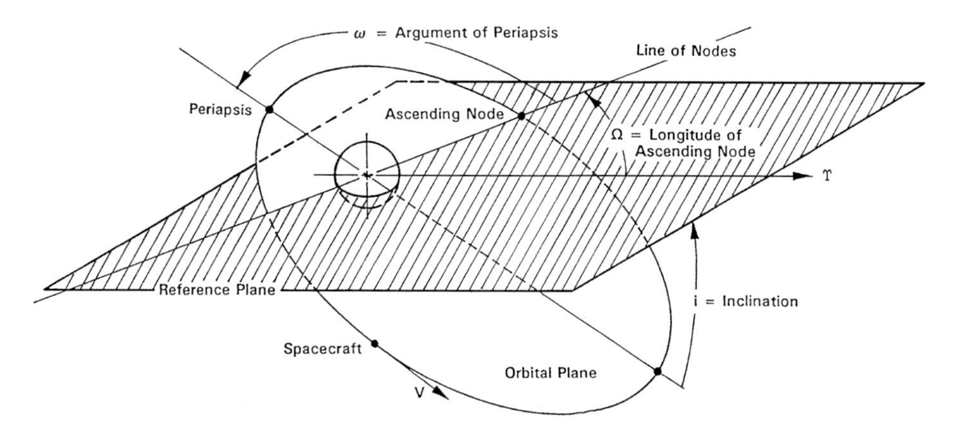

Fig. 3.18 Classical orbital elements.

north pole is placed in the northern celestial hemisphere regardless of the direction of rotation of the body. Parameters specifying the orientation of the north pole and the location of the prime meridian vary slowly with time and can be obtained from Ref. 12.

Longitudes are reckoned in an eastward direction from the prime meridian, i.e., in a counterclockwise direction as viewed from the north pole. Unlike the geographic system, longitudes increase from 0° to 360°; latitudes north of the equator are positive, and southern latitudes are negative.

3.1.8 Classical Orbital Elements

There are a number of independent parameters describing the size, shape, and spatial position of an orbit. Six of these have become the parameters of choice to define and describe an orbit. These six parameters (see Fig. 3.18) are called the classical orbital elements:

$e =$ Eccentricity: the ratio of minor to major dimensions of an orbit defines the shape.

$a =$ Semimajor axis: the orbit size is defined by one half of the major axis dimension. (Circular orbits are defined by radius.)

$i =$ Inclination: the angle between the orbit plane and the reference plane or the angle between the normals to the two planes.

$\omega =$ Argument of periapsis: the angle from the ascending node to the periapsis, measured in the orbital plane in the direction of spacecraft motion. The *ascending node* is the point where the spacecraft crosses the reference plane headed from south to north. The *line of nodes* is the line formed by the intersection of the orbit plane and the reference plane. The ascending node and the descending node are on this line.

$\Omega =$ Longitude of the ascending node: the angle between the vernal equinox vector and the ascending node measured in the reference plane in a counterclockwise direction as viewed from the northern hemisphere.

$\Theta =$ True anomaly: the sixth element locates the spacecraft position on the orbit. (Time since periapsis is also used as this orbital element.[3])

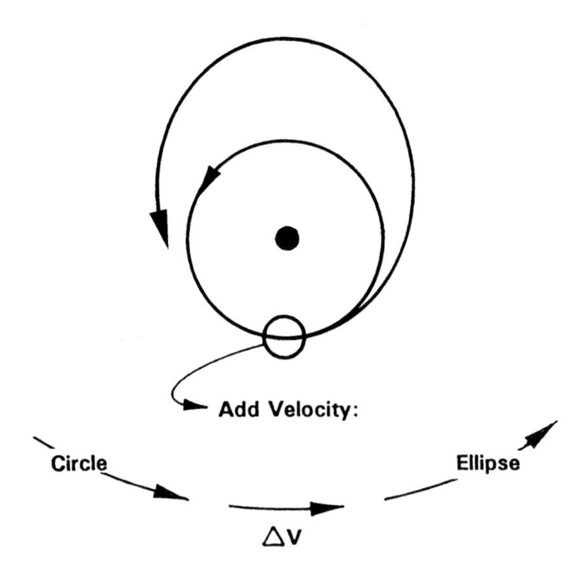

Fig. 3.19 Changing a circular orbit to an elliptical orbit.

For orbits about the Earth or planets, the elements are located with respect to the geocentric system. For interplanetary orbits the elements are given with respect to the heliocentric system. The coordinate system, the orbital elements, and the orbit itself are fixed in inertial space and do not rotate with the central body.

3.2 Orbital Maneuvers

Orbital maneuvering is based on the fundamental principle that an orbit is uniquely determined by the position and velocity at any point. Conversely, changing the velocity vector at any point instantly transforms the trajectory to correspond to the new velocity vector. For example, to change from a circular orbit to an elliptical orbit, the spacecraft velocity must be increased to that of an elliptic orbit, as shown in Fig. 3.19.

In Fig. 3.19,

$$V(\text{circle}) + \Delta V = V(\text{ellipse}) \tag{3.85}$$

3.2.1 In-Plane Orbit Changes

Any conic orbit can be converted to any other conic orbit by adjusting velocity; a spacecraft travels on the trajectory defined by its velocity at a point. Circular trajectories can be converted to ellipses, ellipses can be changed in eccentricity, and circles or ellipses can be changed to hyperbolas—all by adjusting velocity.

Example 3.10 Simple Coplanar Orbit Change

Consider an initial circular, low Earth orbit at a 300-km altitude. What velocity increase would be required to produce an elliptical orbit with a 300-km altitude at periapsis and a 3000-km altitude at apoapsis?

Table 3.6 Effect of velocity change

ΔV, km/s	Impulse location	Resulting trajectory
0.624	Periapsis	Ellipse: $h_p = 300$ km, $h_a = 3000$ km
−0.029	Apoapsis	Ellipse: $h_p = 200$ km, $h_a = 300$ km
3.200	Periapsis	Parabola: $h_p = 300$ km, $e = 1.0$
4.490	Periapsis	Hyperbola: $h_p = 300$ km, $e = 1.5$

The velocity on the initial circular orbit is, from Eq. (3.6),

$$V = \sqrt{\frac{398{,}600.4}{(300 + 6378.14)}} = 7.726 \text{ km/s}$$

The semimajor axis on the final orbit is, from Eq. (3.17),

$$a = \frac{(300 + 6378.14) + (3000 + 6378.14)}{2} = 8028.14 \text{ km}$$

The velocity at periapsis is, from Eq. (3.14),

$$V = \sqrt{\frac{2(398{,}600.4)}{6678.14} - \frac{398{,}600.4}{8028.14}} = 8.350 \text{ km/s}$$

The velocity increase required to convert the initial circular orbit to the final elliptical orbit is $8.350 - 7.726 = 0.624$ km/s. The velocity should be increased at the point of desired periapsis placement. Table 3.6 shows the results of various velocity changes from an initial 300-km Earth orbit. In Table 3.6 it is assumed that the magnitude of velocity is changed without changing the direction. The radius at the point at which the velocity is changed remains unchanged. Velocity changes made at the periapsis change the apoapsis radius but not the periapsis radius, and vice versa. As would be expected, the plane of orbit in inertial space does not change as velocity along the orbit is changed. Orbital changes are a reversible process.

Figure 3.20 shows the general coplanar maneuver, which changes the initial orbit velocity V_i to an intersecting coplanar orbit with velocity V_f. The velocity on the final orbit is equal to the vector sum of the initial velocity and the velocity change vector. Applying the cosine law,

$$\Delta V^2 = V_i^2 + V_f^2 - 2V_i V_f \cos\alpha \tag{3.86}$$

where

ΔV = the velocity added to spacecraft initial velocity to change from the initial orbit to the final orbit; ΔV can be either positive or negative

V_i = the velocity on the initial orbit at the point of intersection of the two orbits

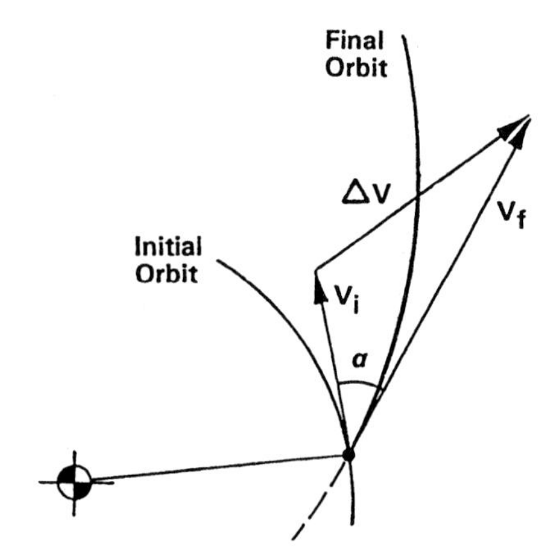

Fig. 3.20 Generalized coplanar maneuver.

V_f = the velocity on the final orbit at the point of intersection of the two orbits
α = the angle between vectors $\boldsymbol{V}_i$ and $\boldsymbol{V}_f$

The transfer can be made at any intersection of two orbits. Note that the necessary velocity change is lowest when the orbits are tangent and $\alpha = 0$.

Example 3.11 General Coplanar Maneuver

Referring to Fig. 3.20, consider an initial direct, circular Earth orbit of radius 9100 km and a final direct, coplanar, elliptical orbit with $e = 0.1$ and $r_p = 9000$ km. What velocity change is required to make the transfer?

The velocity on the initial orbit is, from Eq. (3.6),

$$V = \sqrt{\frac{398{,}600.4}{9100}} = 6.618 \text{ km/s}$$

For the final orbit, the semimajor axis is, from Eq. (3.48),

$$a = 9000/(1 - 0.1) = 10{,}000 \text{ km}$$

The velocity at the point of intersection is, from Eq. (3.14),

$$V = \sqrt{\frac{2(398{,}600.4)}{9100} - \frac{398{,}600.4}{10{,}000}} = 6.910 \text{ km/s}$$

The true anomaly on the final orbit (at the intersection points) is, from Eq. (3.24),

$$\cos\Theta = \frac{9000(1.1)}{9100(0.1)} - \frac{1}{0.1}$$

$$\Theta = 28.464 \text{ deg and } 331.536 \text{ deg}$$

The flight path angle on the final orbit (at the intersection points) is, from Eq. (3.31),

$$\tan\gamma = \frac{(0.1)\sin(28.464)}{1 + (0.1)\cos(28.464)}$$

$$\gamma = 2.508 \text{ deg} \quad \text{and} \quad -2.508 \text{ deg}$$

At either of the intercept points, the velocity change necessary to convert from the initial to the final orbit is, from Eq. (3.86),

$$\Delta V^2 = (6.618)^2 + (6.910)^2 - 2(6.618)(6.910)\cos(2.508)$$

$$\Delta V = 0.4158 \text{ km/s}$$

3.2.2 Hohmann Transfer

Suppose you need to transfer between two nonintersecting orbits; how can you do it? The Hohmann transfer, shown in Fig. 3.21, answers this need in a most efficient way. The Hohmann transfer, devised by Walter Hohmann in 1925, employs an elliptical transfer orbit that is tangent to the initial and final orbits at the apsides.[13]

To design a Hohmann transfer between two circular, coplanar orbits, set the periapsis radius of the transfer ellipse equal to the radius of the initial orbit, and set the apoapsis radius equal to the radius of the final orbit:

$$r_{pt} = r_i \tag{3.87}$$

$$r_{at} = r_f \tag{3.88}$$

With these two radii set, the transfer ellipse is defined.

There are two velocity increments required to accomplish the transfer. One increment changes the initial velocity of the spacecraft to the velocity needed on

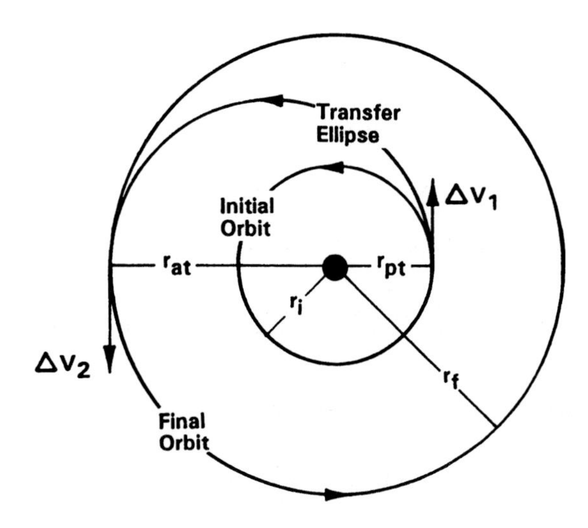

Fig. 3.21 Hohmann transfer.

the transfer ellipse, and a second increment changes from the velocity needed on the transfer ellipse to the velocity needed on the final orbit:

$$\Delta V_1 = V_{pt} - V_i \tag{3.89}$$

$$\Delta V_2 = V_{at} - V_f \tag{3.90}$$

where

V_{pt} = the periapsis velocity on the transfer ellipse
V_{at} = the apoapsis velocity on the transfer ellipse
V_i = the spacecraft velocity on the initial orbit
V_f = the spacecraft velocity on the final orbit

A transfer between two circular orbits is shown in Fig. 3.21, but the transfer could as well have been between elliptical orbits. Similarly, the transfer could have been from a high orbit to a low orbit.

Example 3.12 Hohmann Transfer

Design a Hohmann transfer from a circular Mars orbit of radius 8000 km to a circular Mars orbit of radius 15,000 km.

For Mars,

$$\mu = 42{,}828.3 \text{ km}^3/\text{s}^2$$

The velocity on the initial orbit is, from Eq. (3.6),

$$V = \sqrt{\frac{42{,}828.3}{8000}} = 2.314 \text{ km/s}$$

Similarly, the velocity on the final orbit is found to be 1.690 km/s.

The semimajor axis of the transfer ellipse is, from Eq. (3.17),

$$a = (8000 + 15{,}000)/2 = 11{,}500 \text{ km}$$

The velocity at periapsis of the transfer ellipse is, from Eq. (3.14),

$$V_p = \sqrt{\frac{2(42{,}828.3)}{8000} - \frac{42{,}828.3}{11{,}500}} = 2.642 \text{ km/s}$$

Similarly the velocity at apoapsis is

$$V_a = \sqrt{\frac{2(42{,}828.3)}{15{,}000} - \frac{42{,}828.3}{11{,}500}} = 1.409 \text{ km/s}$$

There is another, possibly easier way to calculate the velocity at apoapsis by using Eq. (3.32):

$$V_a = \frac{(8000)(2.642)}{15{,}000} = 1.409 \text{ km/s}$$

The velocity change required to enter the transfer orbit is found by using Eq. (3.89):

$$\Delta V_1 = 2.642 - 2.314 = 0.328 \text{ km/s}$$

Similarly from Eq. (3.90), the velocity change to "circularize" is 0.281 km/s, and the total velocity change for the transfer is 0.609 km/s.

The time required to make the transfer is one half the period of the transfer ellipse, or from Eq. (3.38),

$$P = 2\pi\sqrt{\frac{(11{,}500)^3}{42{,}828.3}} = 37{,}442 \text{ s}$$

The time for transfer is 18,271 s, or about 5.2 h.

The efficiency of the Hohmann transfer stems from the fact that the two velocity changes are made at points of tangency between the trajectories; therefore, only the magnitude of the velocity is changed without the energy losses associated with a change in direction.

3.2.3 Plane Changes

So far only in-plane changes to orbits have been considered. What happens when a spacecraft must change orbit planes to accomplish its mission? Such a change is accomplished by applying an out-of-plane impulse at the intersection of the initial and final orbit planes, as shown in Fig. 3.22. By inspection of Fig. 3.22, or from the law of cosines,

$$\Delta V = 2V_i \sin(\alpha/2) \tag{3.91}$$

where

ΔV = the velocity change required to produce a plane change
V_i = the velocity of the spacecraft on the initial orbit at the intersection of the initial and final orbit planes
α = the angle of the plane change

Spacecraft speed is unaltered by the plane change; thus,

$$|V_i| = |V_f| \tag{3.92}$$

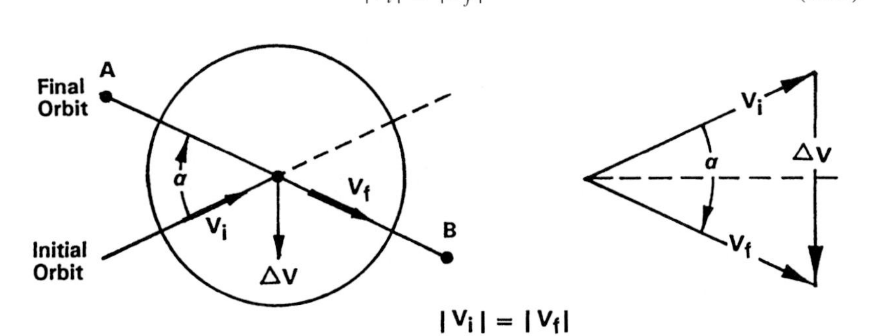

Fig. 3.22 Plane change maneuver.

Orbit shape is unaffected; therefore, eccentricity, semimajor axis, and radii are unchanged.

Plane changes are expensive as far as energy is concerned. A 10-deg plane change in low Earth orbit requires a velocity change of 1.4 km/s. Equation (3.91) shows that it is important to change planes through the smallest possible angle and at the lowest possible velocity. The lowest possible velocity occurs at the greatest radius, i.e., at the apoapsis.

The maximum displacement between positions on the initial and final orbits occurs at 90 deg from the point of plane change. There are two such points, A and B, in Fig. 3.22.

A change in orbit plane can result in a change in inclination, the longitude of the ascending node, or both, as shown in Fig. 3.23. A spherical triangle, with one side and two angles known, is formed by i_i, $180 - i_f$, and $\Delta\Omega$. To design a plane change with the desired effect on inclination and longitude of ascending node, the necessary angle of plane change can be found from the cosine law of spherical trigonometry:

$$\cos\alpha = \cos i_i \cos i_f + \sin i_i \sin i_f \cos(\Delta\Omega) \tag{3.93}$$

The velocity change ΔV required to provide a plane change α can be obtained from Eq. (3.91). The argument of latitude on the initial orbit, at which the maneuver must be performed, is given by the sine law of spherical trigonometry:

$$\sin A_{La} = \frac{\sin i_f \sin(\Delta\Omega)}{\sin\alpha} \tag{3.94}$$

and

$$A_{La} = \omega_i + \Theta_i \tag{3.95}$$

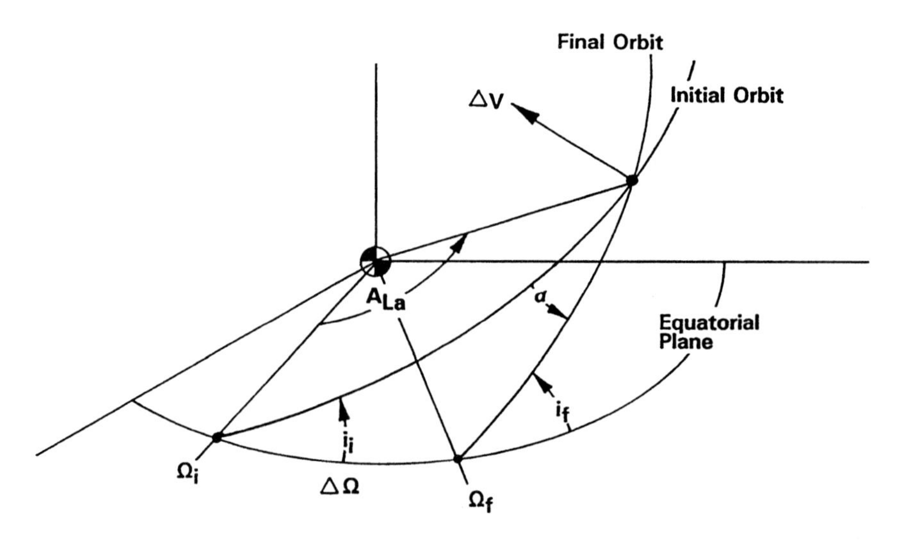

Fig. 3.23 General plane change.

where

A_{La} = the argument of latitude, which is the angle from the reference plane to the spacecraft position, measured in the direction of motion; equal to the true anomaly plus the argument of periapsis
ω_i = the argument of periapsis on the initial orbit
Θ_i = the true anomaly on the initial orbit at which the plane change must occur
$\Delta\Omega$ = the change in longitude of the ascending node caused by the plane change $(\Delta\Omega = \Omega_f - \Omega_i)$
α = the angle between initial and final planes

Example 3.13 General Plane Change

Consider an initial circular Earth orbit with the following characteristics:

$$h = 275 \text{ km}$$
$$i = 28.5 \text{ deg}$$
$$\Omega = 60^\circ \text{ west}$$

It is desired to make a plane change to a circular orbit with the following final characteristics:

$$h = 275 \text{ km}$$
$$i = 10 \text{ deg}$$
$$\Omega = 100^\circ \text{ west}$$

Design the plane change. The angle of the plane change, from Eq. (3.93), is

$$\cos\alpha = \cos(28.5)\cos(10) + \sin(28.5)\sin(10)\cos(40)$$
$$\alpha = 21.730 \text{ deg}$$

The argument of latitude on the initial orbit at which the impulse must be applied, from Eq. (3.94), is

$$\sin A_{La} = \frac{\sin(10)\sin(40)}{\sin(21.730)}$$
$$A_{La} = 17.547 \text{ deg}$$

The velocity on the initial and final orbit, found by using Eq. (3.6), is 7.740 km/s. The velocity change required for the plane change, from Eq. (3.91), is

$$\Delta V = 2(7.740)\sin(21.730/2) = 2.918 \text{ km/s}$$

If spacecraft velocity is changed at the reference plane, i.e., at one of the nodes, the inclination of the orbit will be changed, and the longitude of the ascending node will not be changed. The change in inclination will be

$$\Delta i = \alpha \tag{3.96}$$

where α is the angle between initial and final velocity vectors. Inclination is the only orbital parameter changed by adding out-of-plane velocity at one of the nodes.

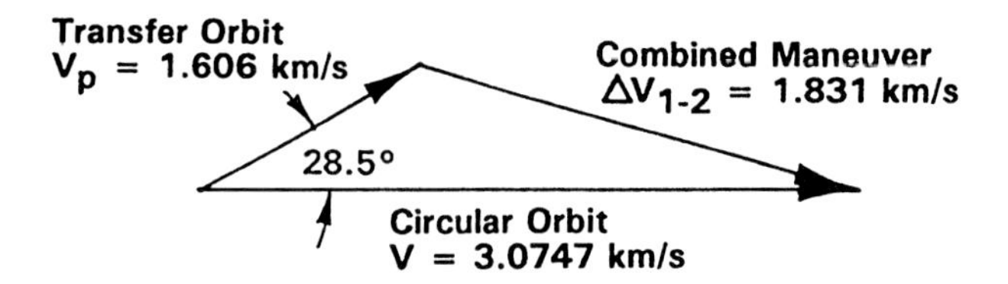

Separate Maneuvers:

1) Plane Change Maneuver ΔV = 0.791 km/s

2) Circularization Maneuver ΔV = 1.469 km/s

Total ΔV = 2.260 km/s

Fig. 3.24 Combined maneuver.

3.2.4 Combined Maneuvers

Significant energy savings can be made if in-plane and out-of-plane maneuvers can be combined. The savings result from the fact that any side of a vector triangle is always smaller than the sum of the other two sides. The maneuvers necessary to establish a geosynchronous orbit provide an excellent example of combined maneuver savings. The final two maneuvers, shown in Fig. 3.24, are an in-plane orbit circularization and a 28.5-deg plane change. The plane change maneuver at a velocity of 1.606 km/s would require a ΔV of 0.791 km/s. The circularization maneuver would change the velocity at apoapsis on the transfer ellipse to the circular orbit velocity of 3.0747 km/s; this maneuver would require a ΔV of 1.4687 km/s. The total for velocity change for the two maneuvers, conducted separately, would be 2.260 km/s. Combining the two maneuvers, however, requires a velocity change of only 1.831 km/s, a savings of 0.429 km/s. (See Section 3.4 for a complete description of the geosynchronous mission design.) Watch for opportunities to combine maneuvers. Every plane change offers an opportunity of this kind.

3.2.5 Propulsion for Maneuvers

The ultimate result of a velocity change requirement is reflected in the propellant mass required to accomplish it. The equations for converting a velocity change to an equivalent propellant mass are derived and discussed in Chapter 4, "Propulsion." For here, suffice it to say that velocity changed is very expensive in mass terms. Any orbital mechanics devices, such as combined maneuvers, which reduce velocity change requirements, have a very beneficial effect on the overall spacecraft design.

3.3 Observing the Central Body

One of the principle uses of spacecraft is for observation of a central body from the unique position of space. The position of a spacecraft over the central body is determined by the orbital parameters already discussed and by the properties of the launch, launch site, and perturbations to the orbit. This section discusses each of these effects, and also discusses the view from a spacecraft given its position. Observations of the Earth are charted from space in the geographic coordinate system; observations of other planets are recorded in the similar IAU cartographic system.

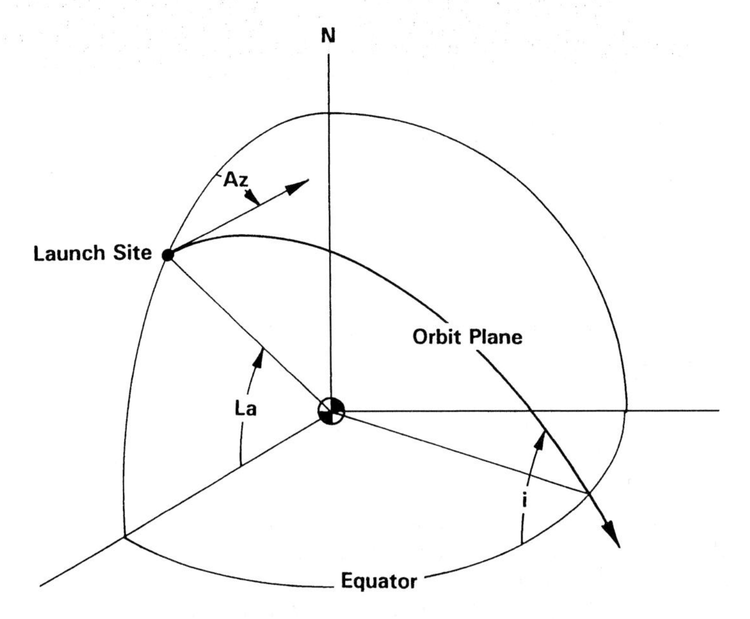

Fig. 3.25 Launch azimuth and altitude.

3.3.1 *Effect of the Launch Site*

Both the latitude of the launch location and the allowable launch azimuth range have a profound effect on orbital parameters. In general, southerly launches allow polar orbits, and easterly launches take advantage of the Earth's rotation and produce low-inclination orbits. The more nearly equatorial the launch site is, the greater the range of orbit inclinations that can be achieved.

Figure 3.25 shows a launch site at latitude La and a launch azimuth Az. Latitude is measured on a great circle through the north pole, perpendicular to the equatorial plane. Azimuth is measured from true north, which is a vector in the latitude circle, to the launch direction vector. The launch direction vector is in the orbit plane. On the first orbit after launch, the spacecraft crosses the equatorial plane at the orbit inclination angle i.

A right spherical triangle is formed by the side La and by the two angles Az and i. From spherical trigonometry,

$$\cos i = \cos La \sin Az \tag{3.106}$$

where

i = the orbit inclination
Az = the launch azimuth, the angle from true north to the departure trajectory
La = the latitude of the launch platform

Equation (3.106) assumes a nonrotating Earth. To adjust for a rotating Earth, adjust the azimuth and orbital velocity by adding the Earth's eastward rotational vector, as shown in Fig. 3.26. In Fig. 3.26,

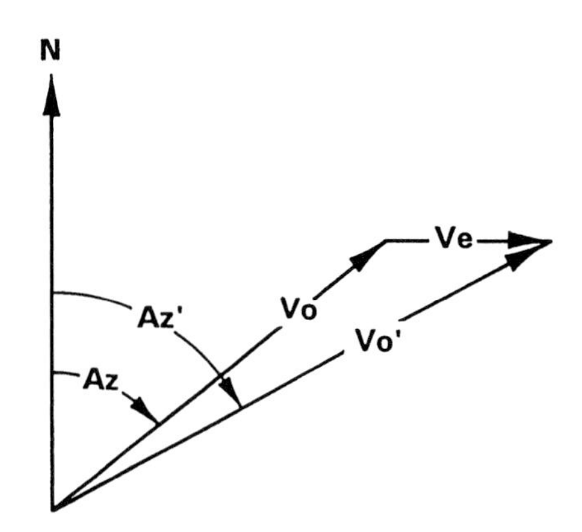

Fig. 3.26 Adjusting for Earth rotation.

V_e = the eastward rotational velocity of Earth
V_o = the orbital velocity over a nonrotating Earth
V_o' = the orbital velocity corrected for Earth's rotation
Az' = the launch azimuth corrected for Earth's rotation

The rotating Earth correction of the azimuth is usually small enough to be neglected; however, the velocity addition is important. Missions requiring high launch energy (e.g., geosynchronous or planetary missions) usually launch due east to obtain the maximum benefits of Earth rotation.

Equation (3.106) produces a minimum inclination when the launch azimuth is due east (90°) or west (270°). The latitude of the launch site is numerically equal to the minimum orbital inclination that can be achieved from that site.

3.3.1.1 Launch latitude. The latitudes of selected launch sites of the world are listed in Table 3.7. The United States operates two major launch sites. The eastern test range at Cape Canaveral, Florida, is called the Kennedy Space Flight Center (KSC) by NASA and Cape Canaveral or eastern test range (ETR) by the Air Force; in this chapter we will use ETR. The western test range, near Lompoc, California, is called either Vandenberg Air Force Base or WTR; we will use WTR. These two sites were chosen for their advantageous latitude and coastal positions.

The minimum orbit inclination that a U.S. spacecraft can have without a plane change is 28.5°, the latitude of ETR. The European Space Agency has a launch site in South America at a latitude of 5.2° N. The former USSR, with launch sites at latitudes of 45.6° and 62.9° has difficulty achieving low orbit inclinations. The new mobile Sea Launch Odyssey can achieve any inclination from 0 to 100°.

3.3.1.2 Launch azimuth. Given a launch site, the desired orbit inclination is obtained by selection of a launch azimuth in accordance with Eq. (3.106). However, the usable range of launch azimuths is restricted by safety considerations. The area underneath a departing launch vehicle is clearly unsafe if the vehicle malfunctions, and even a normal launch sheds spent stages. The location of downrange populated areas limits the acceptable launch azimuth range for a launch complex. Figure 3.27 shows the acceptable azimuth range for ETR and WTR.

Table 3.7 Selected worldwide launch sites[a]

Launch site	Country	Latitude, deg	Longitude, deg	Achievable inclinations, deg
ETR	United States	28.5 N	81.0 W	28.5–57
WTR	United States	34.7 N	120.6 W	63.4–110
Baikonur	Russia	45.6 N	63.4 E	50.5–99.0[b]
Plesetsk	Russia	62.9 N	40.3 E	63–86[b]
Sea Launch Odyssey	Consortium	0 N Mobile[c]	154 W Mobile[c]	0–100
Kourou	France, ESA	5.2 N	52.8 W	5.2–100
Talyuan	China	37.8 N	130.6 E	87, 96–98
Xichang	China	28.2 N	102.0 E	27.5–31.1
Woomera	Australia	31.1 S	136.6 E	45–60, 84–99

[a]Data copyright AIAA, reproduced with permission; Ref. 44.
[b]Not a continuum; certain forbidden inclinations.
[c]Odyssey Sea Launch platform is mobile, normal launch location given.

The restrictions on launch azimuth and the latitudes of ETR and WTR control the inclination capabilities of the United States as shown in Table 3.7. Polar orbits, with $i \approx 90$ deg, can be achieved by launching south out of WTR. Polar orbits cannot be achieved from ETR without a plane change (dog leg).

Direct orbit. Orbits are classified as direct or retrograde depending on their direction of rotation about the central body. In a direct orbit, a spacecraft rotates counterclockwise around the central body as viewed from the north pole; orbit inclination is between 0 and 90 deg. (All planets are in direct orbits around the sun.) The launch azimuth for direct orbits is 0 to 180 deg. Direct orbits are usually achieved by launches from ETR.

Retrograde orbit. In a retrograde orbit, a spacecraft rotates clockwise around the central body as viewed from the north pole; orbit inclination is between 90 and 180 deg. Launch azimuths for retrograde orbits are between 180 and 360 deg; the United States can achieve retrograde orbits only by launching from WTR.

Fig. 3.27 Acceptable launch azimuth range for the United States.

Example 3.14 Effect of Launch Azimuth

The Solar Mesosphere Explorer spacecraft was launched on a two-stage Delta into a circular orbit with the following characteristics:

$$\text{Altitude} = 500 \text{ km}$$
$$\text{Inclination} = 97.4 \text{ deg}$$
$$\text{Period} = 94.6 \text{ min}$$

What was the launch azimuth, and where was the launch site?

WTR is the only U.S. launch site that can achieve an inclination of 97.4 deg. (Technically, an inclination of 97.4 deg could be achieved from ETR with a plane change maneuver called a dog leg, but this maneuver is expensive and highly unlikely.)

The latitude of WTR is 34.5°. The launch azimuth can be obtained by rearranging Eq. (3.106) as follows:

$$\sin Az = \frac{\cos i}{\cos La} = \frac{\cos 97.4}{\cos 34.5} = -0.1563$$
$$Az = 188.99 \text{ deg}$$

Since the inclination is greater than 90 deg the orbit is retrograde. In fact, the orbit is sun-synchronous, a type of orbit that will be discussed in Section 3.4.

3.3.2 Orbit Perturbations

In our analysis of two-body motion, we have assumed that the mass of the central body is spherically symmetrical and can be considered concentrated at the geometric center. In addition, we have assumed that gravitational attraction is the only force acting on the spacecraft. Both of these assumptions are very nearly true, but for refined calculations it is necessary to consider the orbit perturbations caused by forces due to 1) oblateness of the Earth, 2) drag, 3) attraction of the sun, 4) attraction of the moon, and 5) solar radiation pressure. The relative importance of these forces is shown in Fig. 3.28. For each effect the logarithm of the disturbing acceleration, normalized to 1 g, is plotted as a function of altitude. Earth's gravity is the dominant acceleration at altitudes above 100 km (below 100 km, reentry conditions prevail and atmospheric drag dominates). At altitudes below about 200 km, atmospheric drag must be considered in projecting the life of a spacecraft. For orbits above 200 km the effects of Earth's oblateness J_2 is the major force, second only to Earth's gravity. In this section the effects of central body oblateness on spacecraft orbits will be considered.

The Earth is not spherically symmetric. The equatorial radius is 6378.14 km and the polar radius 6356.77 km.[12] This oblateness is caused by the axial rotation rate of the Earth. All planets share this oblateness to one degree or another. Oblateness of the planets causes two orbital perturbations: 1) regression of nodes and 2) rotation of apsides.

3.3.2.1 Regression of nodes. An equatorial bulge causes a component of the gravitational force to be out of the orbit plane, as shown in Fig. 3.29. The out-of-plane force causes the orbit plane to precess gyroscopically; the resulting orbital

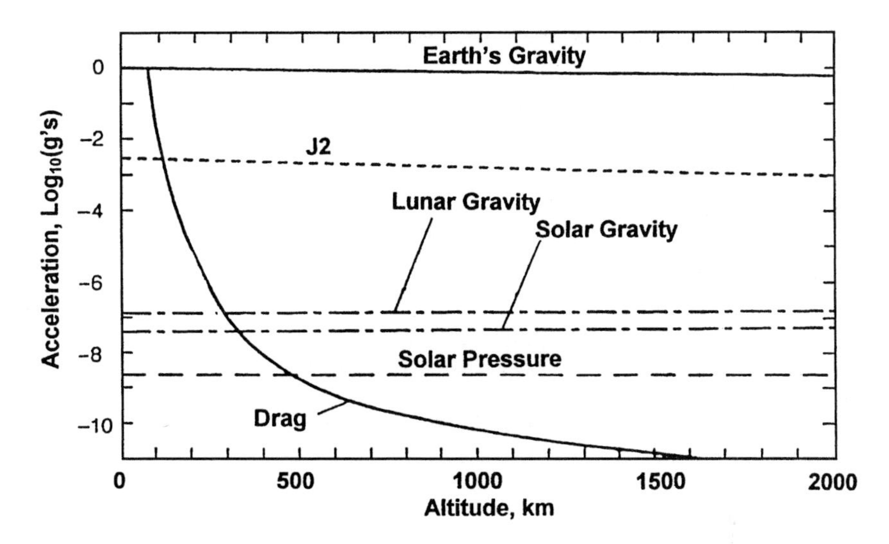

Fig. 3.28 Relative importance of orbit perturbations. (Reproduced with permission of Wiley; Ref. 16, p. 99.)

rotation is called regression of nodes. Regression of nodes can be approximated by an expression of the following form[11,15]:

$$\frac{\mathrm{d}\Omega}{\mathrm{d}t} = \frac{-3nJ_2R_0^2(\cos i)}{2a^2(1-e^2)^2} \tag{3.107}$$

where

$\mathrm{d}\Omega/\mathrm{d}t$ = rate of change of the longitude of the ascending node, rad/s
R_0 = mean equatorial radius of the central body
a = semimajor axis of the orbit
i = inclination of the orbit
e = eccentricity of the orbit
n = mean motion
J_2 = zonal coefficient, a constant peculiar to each celestial body

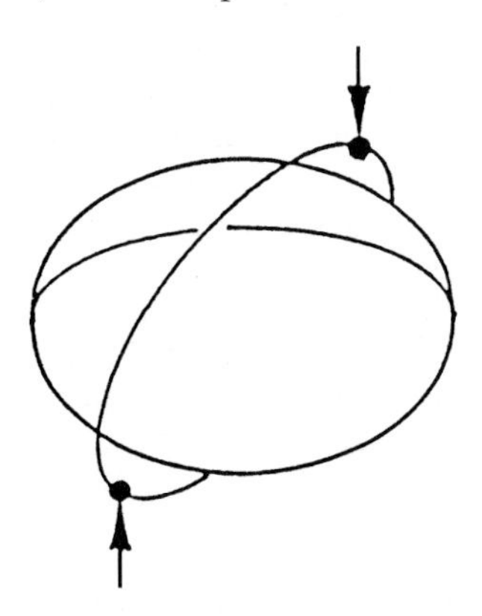

Fig. 3.29 Regression of nodes.

Table 3.8 Zonal coefficients[a]

Planet	J_2
Venus	0.000027
Earth	0.00108263
Mars	0.001964
Jupiter	0.01475
Saturn	0.01645
Uranus	0.012
Neptune	0.004
(Moon)	0.0002027

[a]Data from Ref. 38, p. E 88.

This equation is derived by expressing the gravitational force as a polynomial, by assuming that longitudinal variations in force can be ignored, and by observing that J_2 dominates all terms.

Table 3.8 shows J_2 for the planets.[12] Oblateness is large for the gaseous outer planets, which have high rotation rates; it is small for Venus, which is almost stationary.

For Earth orbits Eq. (3.107) can be reduced to

$$\frac{\mathrm{d}\Omega}{\mathrm{d}t} = -2.06474 \times 10^{14} \frac{\cos i}{a^{3.5}(1 - e^2)^2} \tag{3.108}$$

Equation (3.108) yields nodal regression in degrees per mean solar day.

Figure 3.30 (arrangement of Fig. 3.30 based on the work of Bate[3]) shows the nodal regression rate for some circular Earth orbits. Elliptical orbits of low

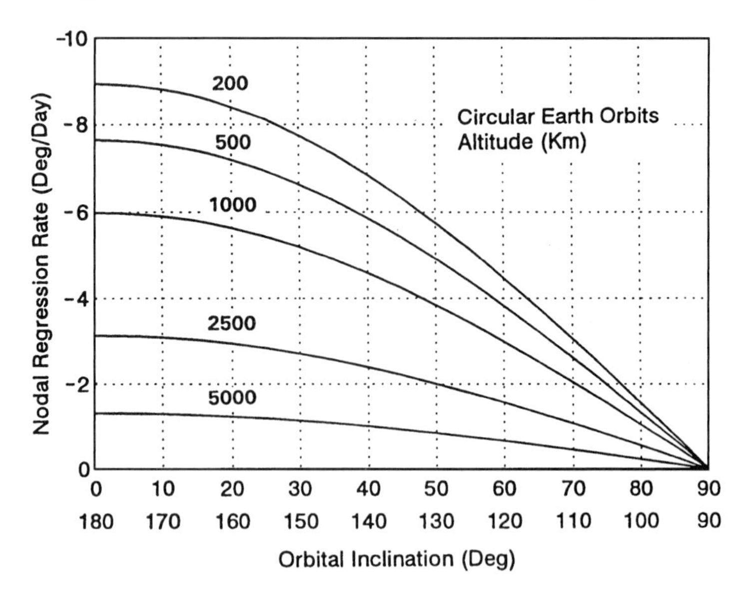

Fig. 3.30 Nodal regression rates for various orbital inclinations. (Reproduced courtesy of Dover Publications, Inc.; Ref. 3, p. 157.)

eccentricity exhibit nodal regression very nearly equal to that of a circular orbit of the same semimajor axis. As shown in Fig. 3.30, regression of nodes is greater at low inclinations and low altitudes, where it can reach 9 deg per mean solar day. For orbits above 10,000 km in altitude, the regression of nodes is less than 0.5 deg per mean solar day and can be ignored for most purposes. Care must be taken, however, in the analysis of geosynchronous orbits. Even though the regression is small at that altitude, it is cumulative and important in planning for station keeping.

The orbit plane rotates clockwise for direct orbits and counterclockwise for retrograde orbits. For direct orbits, regression of nodes adds to the apparent westward motion caused by the rotation of the Earth.

Taking advantage of the regression of nodes makes a sun-synchronous orbit possible. In such an orbit, regression of nodes is used to hold the orbit plane at a constant angle with respect to the sun vector; see Section 3.4 for details.

Example 3.15 Regression of Nodes

What is the regression of nodes for the STS30, which achieved a near-circular orbit with the following characteristics:

$$\text{Periapsis altitude} = 270\ \text{km}$$
$$\text{Apoapsis altitude} = 279\ \text{km}$$
$$\text{Inclination} = 28.5\ \text{deg}$$

The semimajor axis is, from Eq. (3.17),

$$a = \frac{(270 + 6378.14) + (279 + 6378.14)}{2} = 6652.6\ \text{km}$$

The mean motion is, from Eq. (3.36),

$$n = \sqrt{398{,}600.4/(6652.6)^3} = 0.001164$$

From Table 3.8, $J_2 = 1.08263 \times 10^{-3}$. Eccentricity, 0.00068, can be considered to be zero. From Eq. (3.107) the regression of nodes is

$$\frac{\mathrm{d}\Omega}{\mathrm{d}t} = -1.5(0.001164)(1.08263 \times 10^{-3})\frac{(6378.14)^2}{(6652.64)^2}\cos 28.5$$
$$\frac{\mathrm{d}\Omega}{\mathrm{d}t} = -1.5265 \times 10^{-6}\ \text{rad/s} = -7.556\ \text{deg/day}$$

(Correction for eccentricity would produce a regression of nodes of −7.5559 deg/day.)

3.3.2.2 Rotation of apsides. Rotation of apsides is an orbit perturbation due to the Earth's bulge and is similar to regression of nodes (see Fig. 3.31). Rotation of apsides is caused by a greater than normal acceleration near the equator and subsequent overshoot at periapsis. A rotation of periapsis results. This motion occurs only in elliptical orbits. The polynomial expansion described for Eq. (3.107) produces the following description of the motion[11,15]:

$$\frac{\mathrm{d}\omega}{\mathrm{d}t} = \frac{3nJ_2R_0^2(4 - 5\sin^2 i)}{4a^2(1 - e^2)^2} \tag{3.109}$$

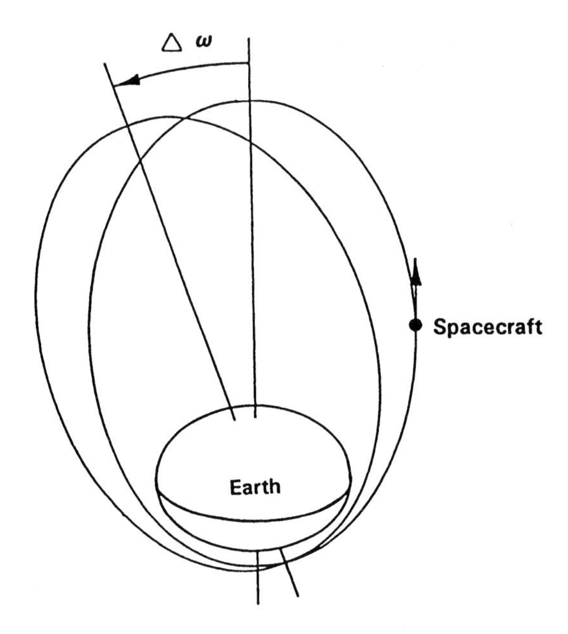

Fig. 3.31 Rotation of apsides.

where

$d\omega/dt$ = apsidal rotation rate, rad/s
R_0 = mean equatorial radius of central body
e = orbit eccentricity
i = orbit inclination
n = mean motion

For an Earth orbit Eq. (3.109) can be reduced to

$$\frac{d\omega}{dt} = 1.0324 \times 10^{14} \frac{(4 - 5\sin^2 i)}{a^{3.5}(1 - e^2)^2} \tag{3.110}$$

Equation (3.110) produces apsidal rotation rate in degrees per mean solar day.

Figure 3.32 (arrangement of Fig. 3.32 based on the work of Bate[3]) shows the apsidal rotation rate, in degrees per mean solar day, for orbits with perigee altitudes of 200 km and various apogee altitudes and inclinations. Figure 3.32 shows that apsidal rotation can be either positive or negative and is greater for low inclination or low altitude orbits.

Example 3.16 Rotation of Apsides

What is the rotation of apsides for an Earth orbit with the following characteristics:

Periapsis altitude = 185 km
Apoapsis altitude = 555 km
Inclination = 30 deg

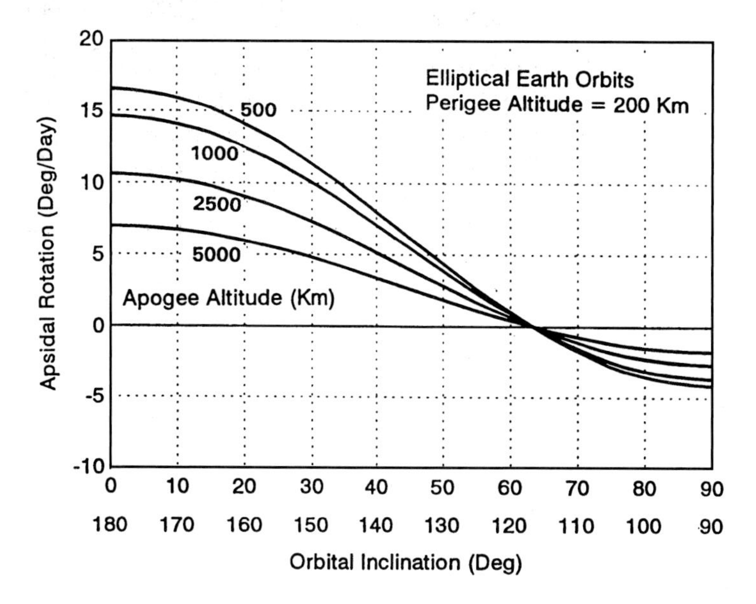

Fig. 3.32 Apsidal rotation for various orbital inclinations. (Reproduced courtesy of Dover Publications, Inc.; Ref. 3, p. 158.)

The semimajor axis is, from Eq. (3.17),

$$a = (6563.14 + 6933.14)/2 = 6748.14 \text{ km}$$

The eccentricity is, from Eq. (3.20),

$$e = \frac{6933.14 - 6563.14}{6933.14 + 6563.14} = 0.027415$$

The mean motion is, from Eq. (3.36),

$$n = \sqrt{\frac{398{,}600.4}{(6748.14)^3}} = 1.1389 \times 10^{-3}$$

From Table 3.8, $J_2 = 1.08263 \times 10^{-3}$. From Eq. (3.108), the rotation of apsides is

$$\frac{\mathrm{d}\omega}{\mathrm{d}t} = \frac{0.75(1.1389 \times 10^{-3})(1.08263 \times 10^{-3})(6378.14)^2[4 - 5\sin^2(30)]}{(6748.14)^2[1 - (0.027415)^2]^2}$$

$$\frac{\mathrm{d}\omega}{\mathrm{d}t} = 2.2753 \times 10^{-6} \text{ rad/s} = 11.26 \text{ deg/day}$$

Because this is an Earth orbit, the same result could be obtained from Eq. (3.110).

It can be seen from Eq. (3.109) that when $\sin^2 i = 4/5$, apsidal rotation is zero regardless of eccentricity. This inclination, 63.435 or 116.565 deg, is called the *critical inclination*. The Molniya orbit is at the critical inclination (see Section 3.4). Note that critical inclination is independent of J_2 and therefore is the same for all celestial bodies.

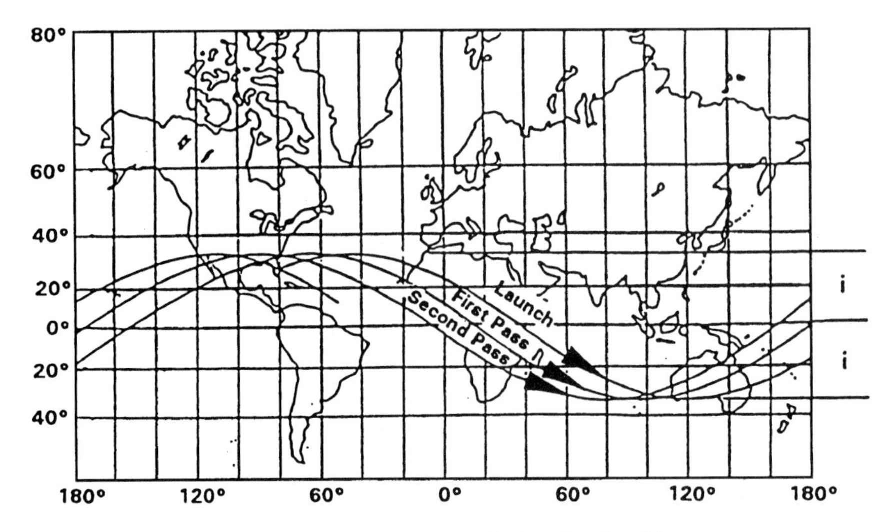

Fig. 3.33 Ground track of a low Earth orbit.

3.3.3 Ground Track

The ground track (or ground trace) of a spacecraft is the locus of nadir positions traced on the surface of the central body by a spacecraft as a function of time. Figure 3.33 shows the ground track of a spacecraft in a direct, circular low Earth orbit for the first three orbits after launch from ETR.

Spacecraft ground track is the result of three motions: 1) the motion of the spacecraft in orbit, 2) the rotation of the central body, and 3) the perturbation of the orbit caused by the equatorial bulge of the central body. Consider a spacecraft in orbit over the surface of a central body, as shown in Fig. 3.34.

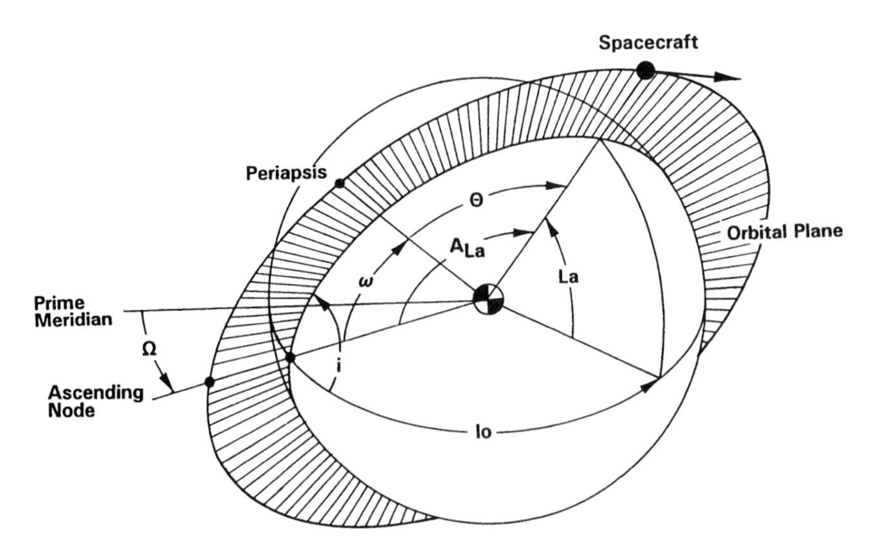

Fig. 3.34 Spacecraft position.

In the figure,

La = latitude of spacecraft at time t
Lo = longitude of spacecraft at time t
Ω = longitude of ascending node at the last crossing (measured from the prime meridian)
i = orbit inclination
A_{La} = the argument of latitude (the angle in the orbital plane from the ascending node to the spacecraft position; $A_{La} = \omega + \Theta$)
lo = longitudinal angle to spacecraft position measured in the equatorial plane from the ascending node
t = time to travel from the ascending node to the current position, s
ω = argument of periapsis
Θ = true anomaly of the current position

For a right spherical triangle,

$$\sin La = \sin i \sin A_{La} \tag{3.111}$$

$$\sin lo = \tan La / \tan i \tag{3.112}$$

The argument of latitude is measured in the direction of motion and is equal to the true anomaly plus the argument of periapsis. The maximum latitude occurs when A_{La} is 90 deg and numerically equal to the orbit inclination. When A_{La} is 0 or 180 deg, the spacecraft position is on the equator, and the latitude is 0°. When A_{La} is between 0 and 180 deg, the spacecraft is in the northern hemisphere, and latitude is noted as north for Earth and as positive for the other planets. When A_{La} is between 180 and 360 deg, latitude is noted as south for Earth and as negative for the other planets.

The longitude of the spacecraft for a spherical nonrotating Earth would be

$$Lo = \Omega + lo \tag{3.113}$$

For a rotating Earth, the spacecraft longitude must be adjusted to account for the Earth's rotation rate of 360 deg per mean sidereal day (86,164 s). This adjustment Re is added to western longitudes and subtracted from eastern longitudes and is expressed as

$$Re = 360t/86{,}164 = 0.0041781t \tag{3.114}$$

3.3.3.1 Regression of nodes. The longitude of the spacecraft ground track must be further adjusted to account for regression of nodes $\Delta\Omega$. The orbit plane rotates westward for direct orbits and eastward for retrograde orbits. For direct orbits, regression of nodes adds to the apparent westward motion caused by the rotation of the Earth.

3.3.3.2 Signs of adjustment. In calculating the longitude of the ground track, it is important to keep track of the sign of each of the three adjustments made to the longitude of the ascending node due to regression of nodes, Earth's rotation (or apparent westward progression), and orbital motion. The signs vary with orbit direction and hemisphere as shown in Fig. 3.35.

In Fig. 3.35,

Lo = longitude of the ground track
Ω = longitude of the ascending node at the last equatorial crossing

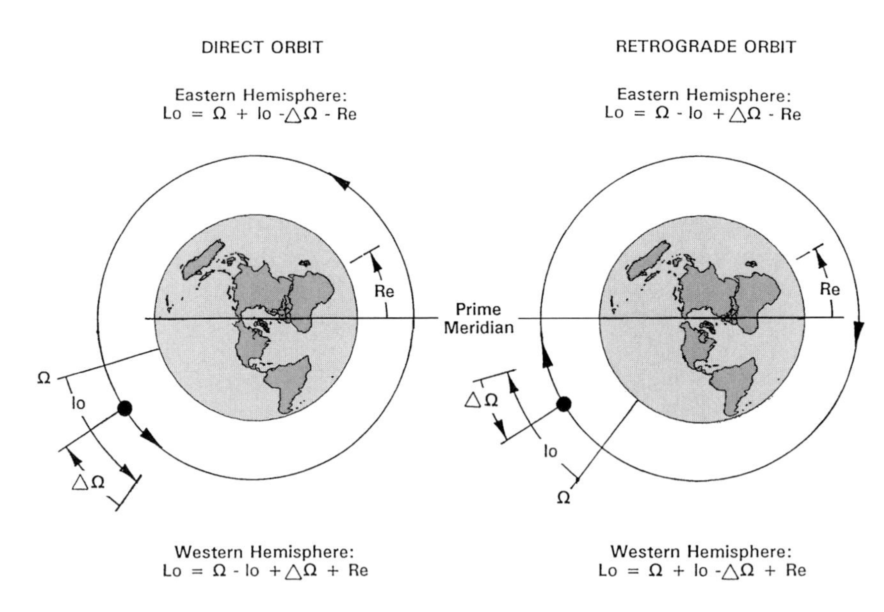

Fig. 3.35 Adjustments to longitude.

$\Delta\Omega$ = adjustment for regression of nodes since passing the ascending node
lo = longitudinal angle to spacecraft position measured in the equatorial plane from the ascending node
Re = longitudinal adjustment for rotation of Earth since passing the ascending node

When using the geocentric coordinate system, the calculation of longitude can be simplified by ignoring the east/west longitude convention and making all calculations in east (or west) longitude measure in accordance with Eqs. (3.115) and (3.116) for east and Eqs. (3.117) and (3.118) for west:

$$Lo = \Omega + lo - \Delta\Omega - Re \tag{3.115}$$

$$Lo = \Omega - lo + \Delta\Omega - Re \tag{3.116}$$

$$Lo = \Omega - lo + \Delta\Omega + Re \tag{3.117}$$

$$Lo = \Omega + lo - \Delta\Omega + Re \tag{3.118}$$

As a final step, convert the resulting east longitude to the east/west measure. To convert, reduce the east longitude to less than 360° by subtracting all full revolutions. If the residual east longitude Lo' is greater than 180°, the west longitude is

$$Lo = 360^\circ - Lo' \tag{3.119}$$

For example, 1020° east longitude would become 60° west longitude. The IAU cartographic system used for planets other than Earth has been simplified by eliminating the east/west convention.

Example 3.17 Ground Track of Space Shuttle *Atlantis*

The elements of the May 1989 flight of the Space Shuttle *Atlantis* were as follows[41]:

Perigee altitude = 270 km
Apogee altitude = 279 km
Eccentricity = 0.000676
Semimajor axis = 6652.64 km
Period = 90.00 min
Argument of periapsis = 25 deg
Inclination = 28.5 deg
Longitude of ascending node = 167° east (third orbit)

Find the latitude and longitude of *Atlantis* for a true anomaly of 20 deg.

The true anomaly at the ascending node was 360 deg − 25 deg = 335 deg.

Calculate the eccentric anomaly by using Eq. (3.35) in preparation for calculating the time since passing the ascending node. (Note that this orbit is so nearly circular that a circular period could be used.)

$$\cos E = \frac{(0.000676 + \cos 335)}{1 + (0.000676)(\cos 335)} = 0.9064$$

$$E = 0.4360 \text{ rad}$$

Calculate mean motion by using Eq. (3.36):

$$n = \sqrt{\frac{398{,}600.4}{(6652.64)^3}} = 0.0011635$$

Calculate the time since passing the ascending node by using Eq. (3.34):

$$t = \frac{(0.4360) - (0.000676)(\sin 0.4360)}{0.0011635} = 374.5 \text{ s}$$

Recall that Eq. (3.34) has the curious property of calculating the shortest time since periapsis; that is, the calculated time is always shorter than one half of the orbit period. This property is convenient in this case, because Eq. (3.34) produces the time being sought.

Repeating the preceding steps yields a time of 299.6 s from periapsis to a true anomaly of 20 deg; therefore, the time from the ascending node to the current position is 674.1 s.

The argument of latitude A_{La} for the current position is 20 deg + 25 deg = 45 deg. From Eq. (3.111) the current latitude of the *Atlantis* is

$$\sin La = (\sin 28.5)(\sin 45) = 0.3374$$

$$La = 19.72^\circ \text{ North}$$

The latitude is north because A_{La} is between 0 and 180 deg.

In preparation for calculating longitude, we need 1) the longitudinal angle, 2) the rotation of the Earth since passing the ascending node, and 3) the regression of nodes since passing the ascending node.

From Eq. (3.112) the longitudinal angle is

$$\sin lo = \tan(19.72)/\tan(28.5) = 0.660$$
$$lo = 41.31^\circ$$

The longitudinal adjustment for the rotation of the Earth since passing the ascending node is, from Eq. (3.114),

$$Re = 0.0041781(674.1) = 2.82 \text{ deg}$$

The regression of nodes can be calculated most simply from Eq. (3.108) since the central body is Earth:

$$\frac{d\Omega}{dt} = \frac{-2.06474 \times 10^{14}(\cos 28.5)}{(6652.64)^{3.5}[1 - (0.000676)^2]^2} = -7.556 \text{ deg/day}$$
$$= -8.745 \times 10^{-5} \text{ deg/s}$$

The regression since passing the ascending node is

$$\Delta\Omega = (-8.745 \times 10^{-5})(674.1) = -0.059 \text{ deg}$$

We now have everything we need to calculate longitude. For the sake of simplicity, we will calculate longitude in east longitude measure and convert later. From Eq. (3.115) for a direct orbit, the longitude of *Atlantis* is

$$Lo = 167 + 41.31 - 0.059 - 2.82 = 205.55^\circ \text{ east}$$

Convert east longitude to the east–west convention used for Earth by using Eq. (3.119):

$$Lo = 360 - 205.43 = 154.6^\circ \text{ west}$$

If we were proceeding to calculate additional positions, we would now calculate the rotation of apsides to reposition the perigee for the next point.

3.3.4 Spacecraft Horizon

The spacecraft horizon, shown in Fig. 3.36, forms a circle on the spherical surface of the central body. The spacecraft horizon circle encloses the area in which the following conditions exist:

1) The spacecraft can be seen from the central body.

2) Two-way microwave communication can be established with the spacecraft from the central body.

3) The spacecraft can observe the central body.

The line from the spacecraft to the local horizon is, at all points, perpendicular to the line from the center of the central body to the horizon, as shown in Fig. 3.37.

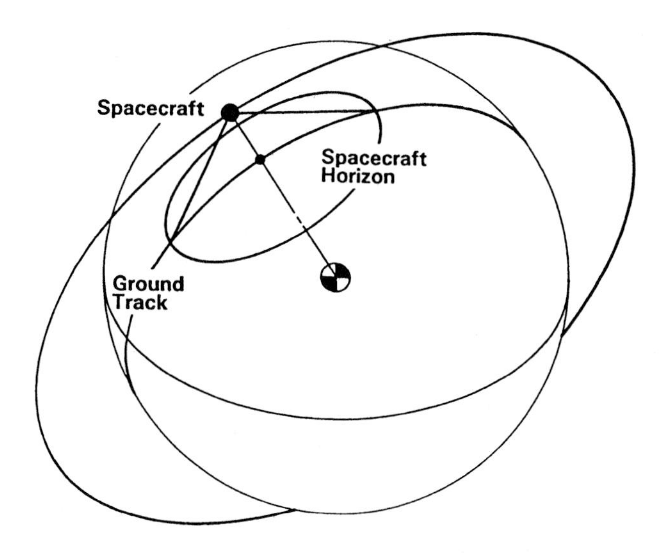

Fig. 3.36 Spacecraft horizon.

In Fig. 3.37,

R_s = the radius to the surface of the central body at the spacecraft position; for most calculations it can be assumed that R_s is equal to the mean equatorial radius R_0

h_s = the instantaneous altitude of the spacecraft above the local terrain

s_w = the swath width, which is an arc on the surface of the central body running from spacecraft horizon to horizon

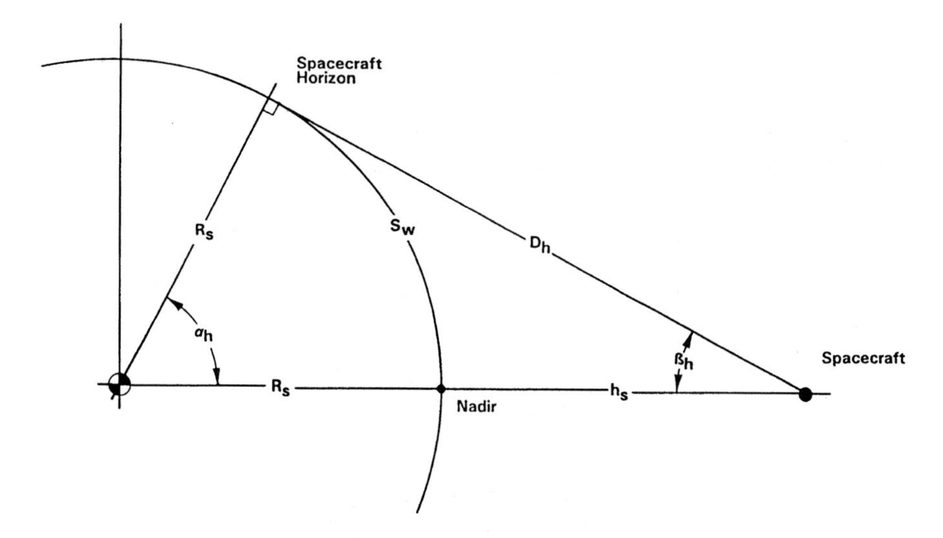

Fig. 3.37 Horizon angle and distance.

β_h = the horizon angle, which is centered at the spacecraft center of mass and measured from the nadir to the horizon
α_h = the central angle, which is centered at the central body center of mass and measured from the spacecraft horizon to the nadir vector
D_h = the horizon distance, which is measured from the spacecraft to the horizon

Figure 3.37 shows that the spacecraft position, the horizon point, and the central body center of mass form a plane right triangle with one side R_s and hypotenuse $R_s + h_s$. Note that $R_s + h_s$ is equal to the orbital radius r regardless of surface irregularities.

The central half angle to the horizon α_h can be found from

$$\cos \alpha_h = R_s/(R_s + h_s) = R_s/r \tag{3.120}$$

Similarly, the angle from the nadir to the spacecraft horizon β_h can be determined from

$$\sin \beta_h = R_s/r \tag{3.121}$$

The distance from the spacecraft to the horizon D_h is, therefore,

$$D_h = r \cos \beta_h \tag{3.122}$$

As the spacecraft moves over the surface, the locus of horizon circles forms a ribbon, or swath, of land centered on the ground track, as shown in Fig. 3.38. The swath width S_w is a linear measurement on the spherical surface of the central body from horizon to horizon:

$$S_w = 2\alpha_h R_s \tag{3.123}$$

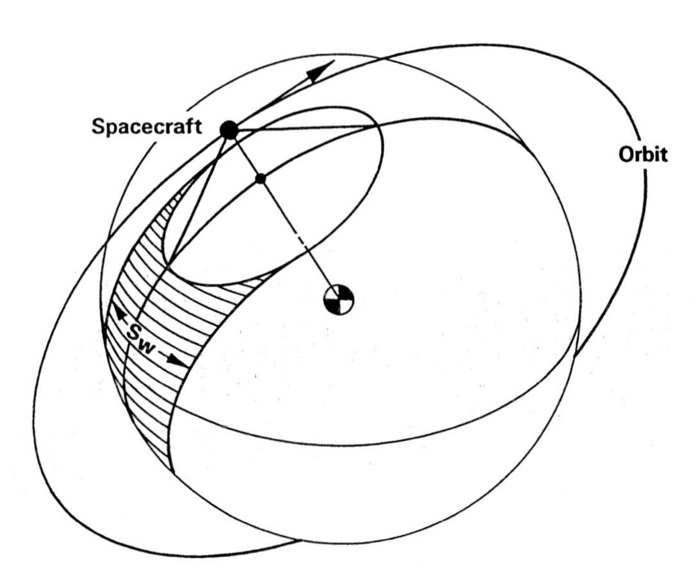

Fig. 3.38 Spacecraft swath.

Note that Eqs. (3.120), (3.121), and (3.123) use the instantaneous radius to the surface of the central body R_s, which is a function of position. For most purposes it is adequate to substitute the mean equatorial radius R_0; however, for very precise work (e.g., cartography) it is necessary to make adjustments for the oblateness of the central body and for terrain roughness. Mapping spacecraft often carry radar altimeters for this reason.

Example 3.18 Swath Width

The LANDSAT D is in a circular, near-polar Earth orbit at an altitude of 709 km. What is the swath width when it passes over mile-high Denver, Colorado?

Spacecraft altitude, when used as an orbital element, refers to the mean equatorial radius; therefore,

$$r = 6378.14 + 709 = 7087.14 \text{ km}$$

The radius of the surface at Denver is

$$R_s = 1.609 + 6378.14 = 6379.75 \text{ km}$$

From Eq. (3.120) the central angle to the spacecraft horizon is

$$\cos\alpha_h = 6379.75/7087.14 = 0.9002$$

$$\alpha_h = 0.4506 \text{ rad} = 25.82 \text{ deg}$$

From Eq. (3.123), the swath width is

$$S_w = 2(0.4506)(6379.75) = 5749 \text{ km}$$

How much difference would it have made if we had ignored the altitude of Denver and used the mean equatorial radius as R_s? The central angle would become

$$\cos\alpha_h = 6387.14/7087.14 = 0.8999$$

$$\alpha_h = 0.4511 \text{ rad}$$

The swath width would be

$$S_w = 2(0.4511)(6378.14) = 5754 \text{ km}$$

Assuming that $R_s = R_0$ causes an error in swath width of about 0.1% at a surface altitude of 5280 ft. The difference is negligible for most purposes.

3.4 Special Earth Orbits

In this section the mission designs of four important orbit types—geosynchronous, sun-synchronous, Molniya, and low Earth—are analyzed.

3.4.1 Geosynchronous Orbit

A spacecraft in a circular, equatorial orbit with a period equal to one sidereal day has a rotation rate exactly equal to the rotation rate of the Earth's surface directly below. The spacecraft appears to be stationary as observed from Earth. This is called a geosynchronous orbit. The north polar view of a geosynchronous orbit is shown to scale in Fig. 3.39. Positions on this orbit are so important commercially that

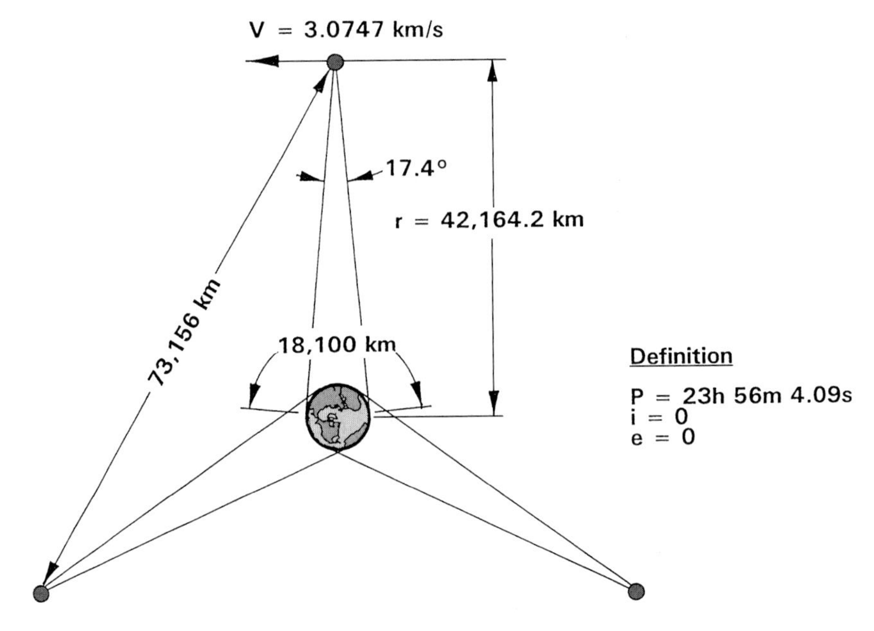

Fig. 3.39 Geosynchronous orbit.

they are rationed by international agreement. Most of the world's communications satellites and meteorological satellites are positioned on this ring.

The importance of this orbit was well understood long before it could be reached. Arthur C. Clarke emphasized what could be accomplished from such an orbit in 1945 in an essay in *Wireless World*.[21] It was 1963, however, before launch vehicle and electronics technology would enable the world's first geosynchronous satellite, Syncom II. In 1964 Syncom III brought the Olympic games directly from Japan to the United States in real time, a feat that had never before been possible.

3.4.1.1 Geosynchronous mission design. By definition, the period of a geosynchronous orbit must be equal to the time required for the Earth to rotate once with respect to the stars. This period is one sidereal day and is 23 h 56 min 4.09 s, or 86,164.09 s, long.

Equation (3.7) can be rearranged to yield the radius of a circular orbit given the period

$$r = (P^2 \mu / 4\pi^2)^{\frac{1}{3}} \tag{3.124}$$

From Eq. (3.124) the radius of a geosynchronous orbit can be obtained as follows:

$$r = [(86{,}164.09)^2\ 398{,}600.4/4\pi^2]^{\frac{1}{3}} = 42{,}164.17 \text{ km}$$

$$h = r - R_0$$

$$h = 42{,}164.17 - 6378.14 = 35{,}786 \text{ km}$$

Surprisingly, placing a spacecraft in geosynchronous orbit requires as much energy as launching a planetary flight. As shown in Fig. 3.40, four distinct velocity

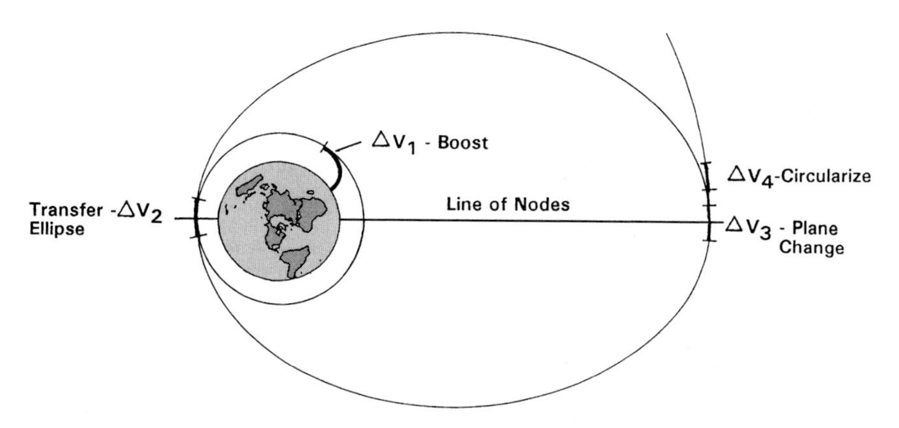

Fig. 3.40 Velocity increments needed to achieve geosynchronous orbit.

changes are required: 1) launch to low Earth orbit, 2) Hohmann transfer from the parking orbit to $r = 42{,}164.2$ km, 3) plane change from $i = 28.5$ deg to the equatorial plane, and 4) circularization of the transfer ellipse to form the geosynchronous orbit.

3.4.1.2 Launch to low Earth orbit. A launch to geosynchronous orbit would be made due east out of ETR (for a U.S. spacecraft) to 1) take advantage of the Earth's rotational velocity and 2) provide the minimum orbit inclination. In a typical launch, the Space Shuttle accelerates the spacecraft to a 280-km circular parking orbit. The velocity of the spacecraft in the parking orbit, from Eq. (3.6), is

$$V = \sqrt{\mu/r} = \sqrt{398{,}600.4/6658.14} = 7.737 \text{ km/s}$$

The launch vehicle contributes 7.329 km/s, and the rotation of the Earth contributes 408 m/s. In addition, the launch vehicle contributes the potential energy of the spacecraft and the substantial energy that goes to gravity and drag losses.

3.4.1.3 Parking orbit to transfer ellipse. The perigee radius of the Hohmann transfer ellipse must be tangent to the parking orbit. The apogee radius must be 42,164.2 km to match the geosynchronous orbit; thus, the semimajor axis is 24,411.2 km. The line of apsides must be in the equatorial plane, so that the apogee will be a point in the geosynchronous orbit. Therefore, the velocity increase needed to change the parking orbit to the transfer ellipse must take place on the equator. An upper stage usually performs this burn at the first equatorial crossing. Note that two impulses are required of this stage. A single stage with restart capability or two stages are required.

The velocity at the perigee of the transfer ellipse is, from Eq. (3.14),

$$V_p = \sqrt{\frac{2(398{,}600.4)}{6658.14} - \frac{398{,}600.4}{24{,}411.2}}$$

$$V_p = 10.169 \text{ km/s}$$

The required velocity increase is

$$\Delta V_2 = 10.169 - 7.737 = 2.432 \text{ km/s}$$

3.4.1.4 Plane change and circularization. A plane change is required to rotate the transfer ellipse from an inclination of 28.5 to 0 deg. Equation (3.91) shows that to conserve energy it is important to make the plane change at the lowest velocity that will occur on the entire mission. The lowest velocity occurs at the apogee of the transfer ellipse; that velocity is

$$V_a = \sqrt{\frac{2(398{,}600.4)}{42{,}164.2} - \frac{398{,}600.4}{24{,}411.2}}$$
$$V_a = 1.606 \text{ km/s}$$

Apogee velocity could also be obtained from Eq. (3.33):

$$r_p V_p = r_a V_a$$
$$V_a = r_p V_p / r_a = 6658.14(10.169)/42{,}164.2$$
$$V_a = 1.606 \text{ km/s}$$

The velocity change that would be required to make a 28.5-deg plane change at a velocity of 1.606 km/s is 0.791 km/s. Note that if the plane change were made after circularization, the velocity change required would be 1.514 km/s. In practice, the plane change is combined with the circularization burn because 1) less ΔV is required with a combined burn and 2) a solid motor can be used for the burn if restart is not required. The solid motor that performs this burn is often called a kick stage.

The velocity on a geosynchronous orbit is

$$V = \sqrt{398{,}600.4/42{,}164.2} = 3.0747 \text{ km/s}$$

The vector diagram for the combined maneuver is shown in Fig. 3.41. From the law of cosines [see Eq. (3.86)], the velocity change required for the combined maneuver is 1.831 km/s. Note that if two individual maneuvers were used, the velocity change required would have been 2.260 km/s. It is always advantageous to combine individual burns in this fashion; watch for plane changes that can be combined with in-plane maneuvers.

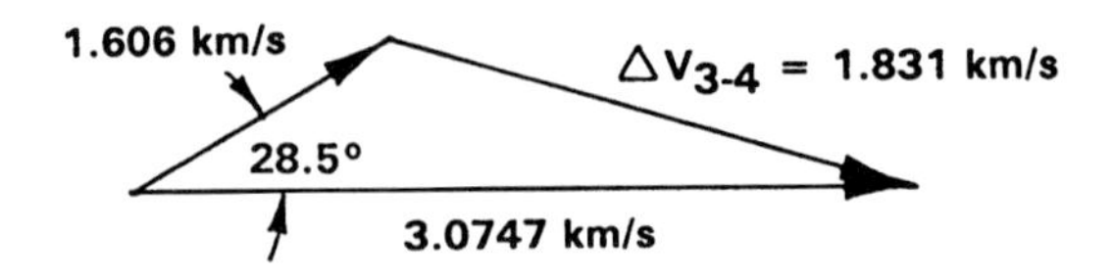

Fig. 3.41 Vector diagram for the final impulse.

The total velocity change required to place a spacecraft in geosynchronous orbit is 12.0 km/s as shown here:

ΔV_1 boost to parking orbit	$= 7.737$
ΔV_2 for transfer ellipse	$= 2.432$
ΔV_3 and V_4 plane change and circularization	$= \underline{1.831}$
Total ΔV	12.000 km/s

The required total ΔV will vary slightly with the selection of parking orbit altitude. If we had chosen a 180-km parking orbit altitude, the total ΔV would have been 12.09 km/s. The parking orbit altitude is usually chosen to accommodate launch vehicle constraints rather than ΔV considerations.

3.4.1.5 View from geosynchronous orbit. As shown in Fig. 3.42, three spacecraft positioned on this orbit can observe the Earth's entire surface except for the extreme polar regions. The horizon angle from a geosynchronous spacecraft can be obtained from Eq. (3.121):

$$\sin\beta = 6378.14/42{,}164.17 = 0.15127$$

$$\beta = 8.70\,\text{deg} = 0.1518\text{ rad}$$

The broad expanse shown in Fig. 3.42 represents a field of view of only 17.4 deg to the spacecraft. Figure 3.39 shows why this is true.

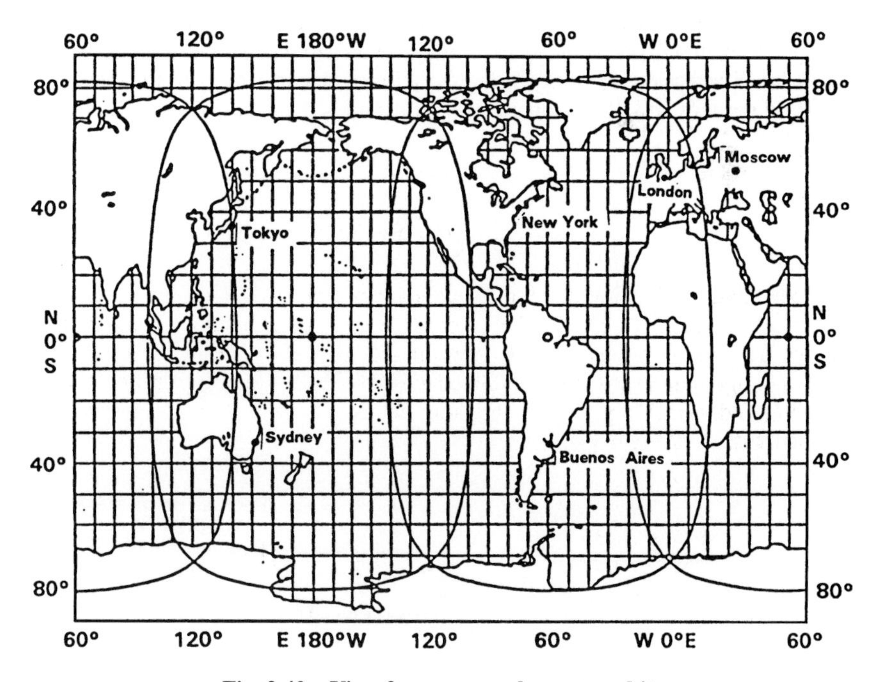

Fig. 3.42 View from geosynchronous orbit.

The central half angle is, from Eq. (3.120),

$$\cos\alpha = 0.15127$$

$$\alpha = 81.30 \text{ deg} = 1.4189 \text{ rad}$$

The swath width is, from Eq. (3.128),

$$S_w = (2)(1.4189)(6378.14) = 18{,}100 \text{ km}$$

The horizon circles shown in Fig. 3.42 are 18,100 km across, a vast field of view.

3.4.2 Sun-Synchronous Orbit

The sun-synchronous orbit is an Earth orbit with the curious property of providing a constant sun angle for the observation of Earth. A constant sun angle is very desirable for cameras and other instruments that observe reflected light.

A sun-synchronous orbit is designed by matching the regression of nodes to the rotation of the Earth around the sun. For example, assume that a polar spacecraft is launched with the orbit plane on an Earth–sun line, as shown in Fig. 3.43. As the Earth travels around the sun, the orbit plane would depart from the sun line by 360 deg during the 365 days required to make a complete orbit. The orbit plane would depart from the sun line at a clockwise rate of about 1 deg per day. In 90 days the orbit plane would be perpendicular to the sun line. However, suppose we design the orbit such that its regression of nodes is equal and opposite the mean daily motion of

$$360 \text{ deg}/365.242 \text{ days} = 0.9856 \text{ deg/day}$$

In such a case the orbit plane would always be on a sun line. So how do we get a regression of nodes of -0.9856 deg/day? From Eq. (3.108) it can be seen that there is a range of orbital inclinations and altitudes that will produce the proper

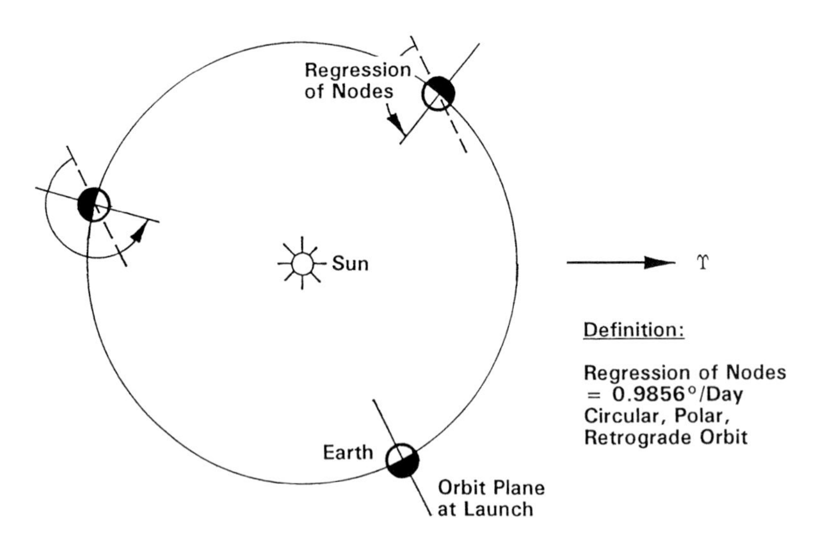

Fig. 3.43 Sun-synchronous orbit.

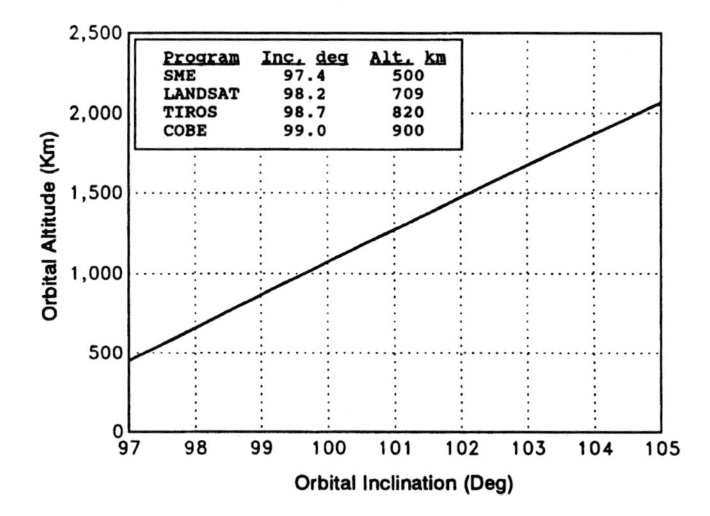

Fig. 3.44 Orbital inclinations and altitudes for sun-synchronous orbits.

regression of nodes. This range is shown in Fig. 3.44. Note that a sun-synchronous orbit must be retrograde.

LANDSAT is probably the best-known sun-synchronous spacecraft. LANDSAT D was placed in a 709-km-altitude orbit with a 98.2-deg inclination.[17] Circling the globe every 103 min, its sensors view a 185-km strip of the surface running nearly north and south. It covers the entire surface of the Earth every 20 days. The spacecraft crosses the equator at 9:30 a.m. local time every orbit. The spacing of the swath is 138 km at the equator. This orbit produces a consistent and constant lighting of the Earth, the best condition for an imaging system.

3.4.3 Molniya Orbit

The Molniya orbit, shown in Fig. 3.45, was devised by the USSR to provide features of a geosynchronous orbit with better coverage of the northern latitudes and without the large plane change that would be required from their far northern launch sites. The approximate orbital elements are as follows:

$$P = 43{,}082 \text{ s (one half of a sidereal day)}$$
$$a = 26562 \text{ km}$$
$$i = 63.4 \text{ deg}$$

Viewed in Earth-fixed coordinates, the orbit rises alternately above the North American continent and the Eurasian continent. As shown in Fig. 3.46, a Molniya spacecraft alternates 12-h periods above each continent.

The time ticks in Fig. 3.46 show that the spacecraft spends most of its time in the high-altitude portion of the orbit. There is an 8-h period over the North American continent each day when Eurasia is also in view. During that period, a single spacecraft can serve as the communication link between continents. A constellation of three spacecraft would provide a continuous direct link. The Molniya ground

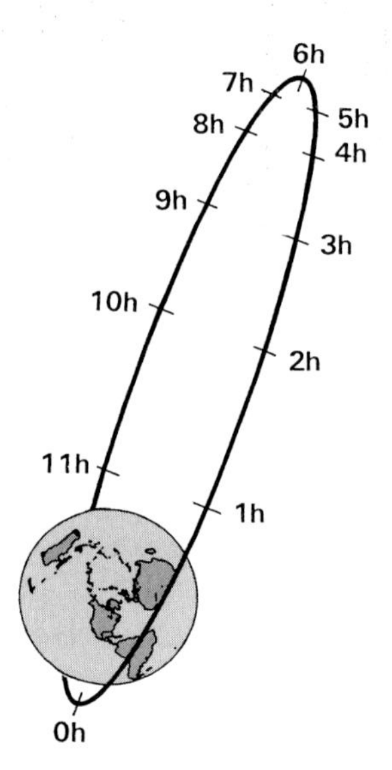

Fig. 3.45 Molniya orbit.

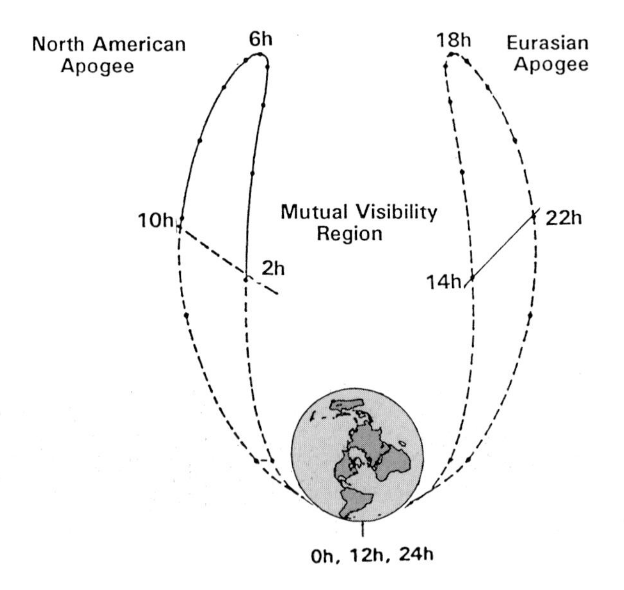

Fig. 3.46 Molniya orbit in Earth-fixed coordinates.

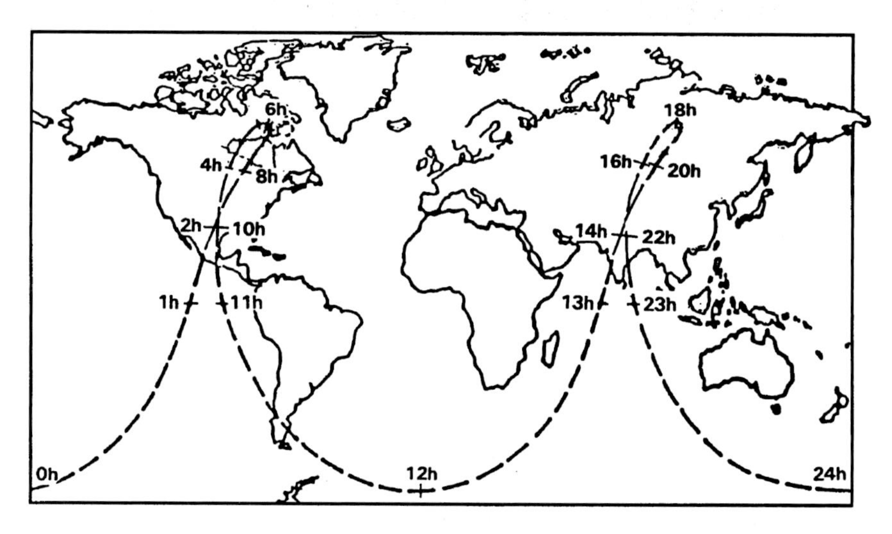

Fig. 3.47 Molniya ground track.

track is shown in Fig. 3.47. The orbit retraces this ground track each day. Although it does not hover in a fixed position as a geosynchronous spacecraft would, it holds a regional position for 8 h a day in two important regions.

A Molniya orbit requires only two impulses to become established: 1) launch to parking orbit and 2) velocity increase at perigee to establish the ellipse.

3.4.3.1 Launch to parking orbit. A launch due east from Plesetsk, the busiest launch site on Earth, would produce a parking orbit with an inclination of 62.8 deg. A small adjustment in azimuth will produce an inclination of 63.4 deg and thereby eliminate rotation of apsides. The Earth's rotational assist at Plesetsk is 200 m/s compared with 408 m/s for a launch from ETR.

The parking orbit altitude is important because, unlike the geosynchronous orbit, the parking orbit radius becomes the perigee radius of the final orbit. The parking orbit must be high enough to minimize drag; 300 km is an adequate altitude. Several different perigee altitudes have been used; consider the 504-km altitude used by Molniya 1–73.[23] The velocity change for a 504-km parking orbit would be 7.610 km/s.

3.4.3.2 Parking orbit to final orbit. The orbit period must be half of a sidereal day, or 43,082 s. The semimajor axis, computed directly from the period, must be 26,562 km. Any combination of apogee and perigee radii with a semimajor axis of 26,562 km will have the correct period. The simplest design is one with a perigee radius the same as, and tangent to, the parking orbit radius, which we have assumed to be 6882 km. Therefore, the apogee radius must be 46,241 km. The perigee velocity is 10.04 km/s for this orbit. The velocity increase to place a spacecraft on this orbit is 2.43 km/s from a 504-km parking orbit.

There is no plane change or circularization required. The Molniya orbit requires a ΔV of 2.43 km/s, above the parking orbit velocity, compared with 5.175 km/s

Table 3.9 Some Molniya orbits[a]

Spacecraft	i, deg	h_p, km	h_a, km	P, s
Molniya 1–74	62.8	623	39,721	43,044
Molniya 1–73	63.1	504	39,834	43,068
Example	63.4	504	39,863	43,082

[a]From Ref. 23, pp. 59 and 62.

for a geosynchronous orbit from Plesetsk. This energy difference translates into twice as much spacecraft mass in a Molniya orbit than in a geosynchronous orbit (for a given launch vehicle—the normal situation).

Table 3.9 compares the example design with two actual Molniya orbits.

3.5 Interplanetary Trajectories

Humans have been on Earth for at least 500,000 years. For most of that period they have puzzled over the planets; for the last 40 years we have had the ability to explore these worlds with spacecraft. Understanding interplanetary trajectories was one of the most interesting achievements underlying that capability. This chapter discusses how an interplanetary mission is analyzed and designed.

A brief review of the pertinent geometry of the solar system is in order. Figure 3.48 shows an isometric view of the ecliptic plane with the nine planets in their positions as of 1 January 1992, rotating in direct, slightly elliptical, slightly inclined orbits around the sun. The distances involved are so enormous that two different scales are required to accommodate all planets on one page. The semi-major axis of the Earth's orbit is 149.597870×10^6 km,[12] a distance that is called an astronomical unit (AU). The length of an astronomical unit is shown for both scales used in Fig. 3.48.

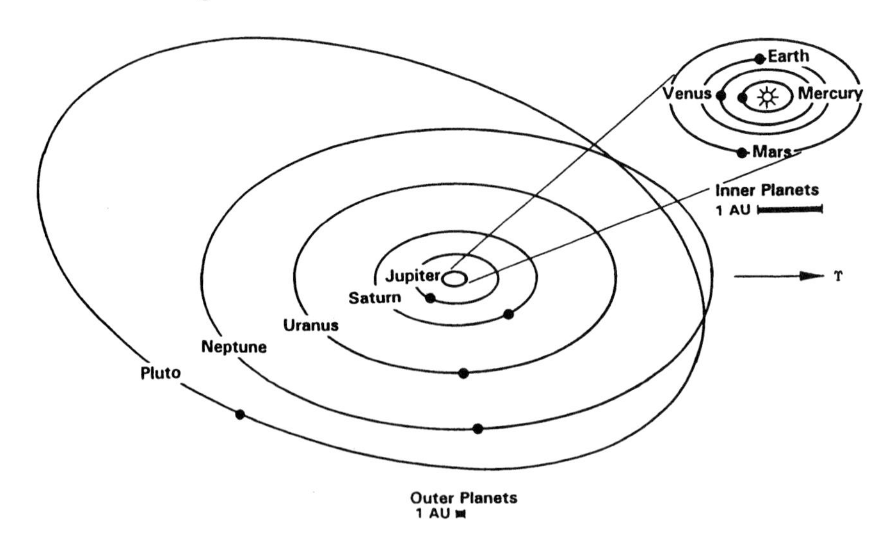

Fig. 3.48 Solar system.

Table 3.10 Mean orbital parameters of the planets[a]

Planet	a, AU	r_p, $\times 10^6$ km	e	i, deg	Velocity, km/s
Mercury	0.387	45.99	0.2056	7.005	47.89
Venus	0.723	107.437	0.0068	3.395	35.05
Earth	1.000	147.10	0.0167	0.001	29.77
Mars	1.524	206.72	0.0933	1.850	24.13
Jupiter	5.203	740.84	0.0482	1.305	13.05
Saturn	9.516	1345.02	0.0552	2.487	9.64
Uranus	19.166	2729.29	0.0481	0.772	6.80
Neptune	30.011	4447.85	0.0093	1.772	5.43
Pluto	39.557	4436.42	0.2503	17.150	4.73

[a]From Ref. 12, p. E3; Ref. 25, pp. 14, 15, and Ref. 26, p. 17.

As shown in Table 3.10, the perihelion distances vary from 46 million km for tiny Mercury to over 4 billion km for Neptune and Pluto. Pluto is the maverick planet; the inclination of its orbit, 17 deg, is two and a half times greater than that of its nearest rival, and its orbit is so eccentric ($e = 0.25$) that at times Neptune is the outer planet.

The slight eccentricities of these orbits means that not only are the distances large, but they are time variant. These eccentricities also imply that planetary arrivals and departures must take into account flight path angles.

The planets are not quite in the ecliptic plane. The significance of the inclination of these orbits is that a plane change at high speed must be accounted for in mission design.

The velocities shown in Table 3.10 are the mean velocities of the planets around the sun; since the orbits are elliptical, the velocities are time variant. The velocities of the planets are large compared to the velocities that launch vehicles can provide. It would not be possible to send spacecraft to the planets without taking advantage of the orbital velocity of the Earth; the contribution of the launch vehicle is minor in comparison.

In Table 3.10, semimajor axis, eccentricity, and inclination are given as of 24 January 1991 (JD 244 8280.5) and refer to the mean ecliptic and equinox of J2000.0. The orbital elements of the planets change slowly with time. For precise work *The Astronomical Almanac* should be consulted.[12]

3.5.1 Patched Conic Approximation

A planetary trajectory is a four-body motion involving the Earth, the target planet, the sun, and the spacecraft. It would be possible and proper to use an N-body simulation to study the trajectory; however, there are numerous disadvantages in this approach. The patched conic technique is a brilliant approximation of this four-body motion; it provides adequate accuracy for almost all purposes.

The patched conic approximation subdivides the planetary mission into three parts (see Fig. 3.49):

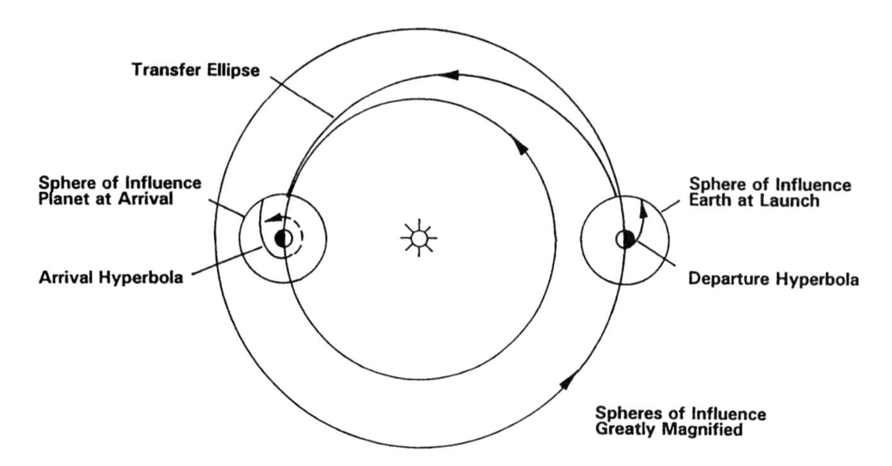

Fig. 3.49 Patched conic orbit.

1) *The departure phase*, in which the two relevant bodies are Earth and the spacecraft. The trajectory is a departure hyperbola with Earth at the focus. The influences of the sun and target planet are neglected.

2) *The cruise phase*, in which the two bodies are the sun and the spacecraft. The trajectory is a transfer ellipse with the sun at the focus. The influences of the planets are neglected.

3) *The arrival phase*, in which the two bodies are the target planet and the spacecraft. The trajectory is an arrival hyperbola with the planet at the focus. The influences of the sun and Earth are neglected.

An arbitrary sphere of influence is defined about the Earth and the target planet. The transfer ellipse is "patched" to the hyperbolas at the boundaries of the spheres of influence. The radius of the sphere of influence suggested by Laplace is[3,15]

$$R_s \approx R\left(\frac{M_{\text{planet}}}{M_{\text{sun}}}\right)^{2/5} \tag{3.125}$$

where

R_s = radius of the sphere of influence of a planet
R = mean orbital radius of the planet
M_{planet} = mass of the planet
M_{sun} = mass of the sun

At radius R_s the gravitational attractions of the planet and the sun are approximately equal. Other definitions of R_s are possible; specifically, an exponent of $1/3$ rather than $2/5$ is sometimes used in Eq. (3.125). Fortunately, a precise definition of sphere of influence is not critical to planetary trajectory design.

Table 3.11 shows the sphere of influence values found by using the Laplace method. Inside the sphere of influence, spacecraft times and positions are calculated on the departure or arrival hyperbola. Outside the sphere, times and positions are calculated on the transfer ellipse.

Table 3.11 Sphere of influence

Planet	R_s, $\times 10^6$ km
Mercury	0.111
Venus	0.616
Earth	0.924
Mars	0.577
Jupiter	48.157
Saturn	54.796
Uranus	51.954
Neptune	80.196
Pluto	3.400
Moon	0.0662

Technically, the transfer ellipse should be designed to terminate at the boundary of the sphere of influence. However, the sphere of influence is so small compared to the transfer ellipse that this small correction is negligible. For example, the true anomaly on the transfer ellipse at Earth should be reduced by 0.15 deg to account for Earth's sphere of influence.

It is important to remember that the actual trajectory converts smoothly from the departure hyperbola to the transfer ellipse and back to the arrival hyperbola. Nonetheless, the patched conic technique is a surprisingly accurate approximation—accurate enough for all but the most demanding work (e.g., navigation).

3.5.2 Highly Simplified Example

To get an overview of a planetary mission, consider the simplest possible case. Take a mission to Venus as an example, and assume the following:

1) The planetary orbits are circular. This assumption will allow the use of mean orbital velocities without the complication of velocity variation and flight path angles.

2) The orbits are coplanar. This assumption will eliminate plane changes.

Venus can be arbitrarily placed at arrival such that it is diametrically opposed to the Earth position at departure, thereby eliminating the ephemeris calculations and making the transfer ellipse tangent to the planetary orbits. The resulting trajectory is a simple Hohmann transfer on a grand scale, as shown in Fig. 3.50. The velocities, radii, and positions shown in Fig. 3.50 are given with respect to the sun and are mean values assuming circular orbits.

The periapsis of the transfer ellipse is equal to the radius of the Venus orbit, and the apoapsis radius is equal to the radius of the Earth orbit. The transfer orbit elements can be calculated readily:

$$r_a = 149.59 \times 10^6 \text{ km}, \qquad V_a = 27.29 \text{ km/s}$$
$$r_p = 108.21 \times 10^6 \text{ km}, \qquad V_p = 37.73 \text{ km/s}$$
$$a = 128.90 \times 10^6 \text{ km}, \qquad P = 292 \text{ days}$$
$$e = 0.1605$$

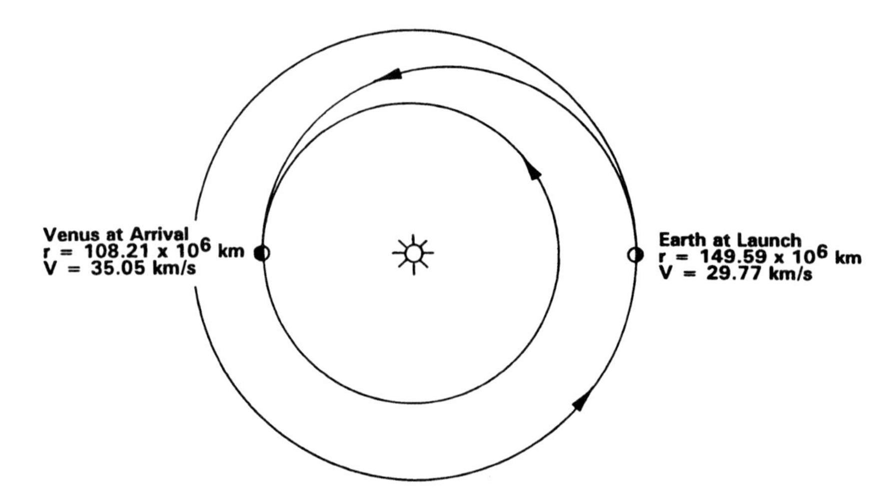

Fig. 3.50 Simplified Venus trajectory.

The spacecraft velocity at departure must be equal to the apoapsis velocity on the transfer ellipse, or 27.29 km/s. The arrival velocity will be the periapsis velocity, 37.73 km/s. The time of flight will be half the transfer orbit period, or 146 days.

3.5.2.1 Hyperbolic excess velocity and C3. The vector difference between the velocity of the Earth with respect to the sun and the velocity required on the transfer ellipse is called the hyperbolic excess velocity V_{HE} (see Fig. 3.51). The hyperbolic excess velocity is V_∞ on the departure hyperbola. Recall that the velocity at infinity along a hyperbolic trajectory is the excess amount above the escape velocity (hence the name). In Fig. 3.51,

$V_{s/s}$ = required velocity of the spacecraft with respect to the sun on the transfer ellipse
$V_{e/s}$ = velocity of the Earth with respect to the sun
V_{HE} = hyperbolic excess velocity, which is the required spacecraft velocity with respect to the Earth
α = included angle between $\boldsymbol{V}_{s/s}$ and $\boldsymbol{V}_{e/s}$

In this example it has been assumed that the transfer ellipse and the Earth's orbit are tangent and coplanar; therefore, the included angle α is zero and

$$V_{\mathrm{HE}} = V_{s/s} - V_{e/s}$$

$$V_{\mathrm{HE}} = 27.29 - 29.77 = -2.48 \text{ km/s} \tag{3.126}$$

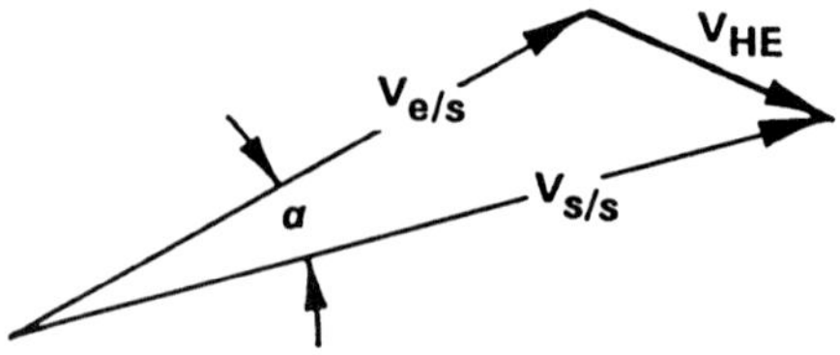

Fig. 3.51 Definition of hyperbolic excess velocity.

Table 3.12 C3 required for various missions

Mission	C3, km^2/s^2
Venus, 1986	11
Mars, 1990	17
Jupiter, 1987	80
Ulysses, 1986	123

V_{HE} is negative for a Venus mission or for any mission to an inner planet, indicating that the Earth's orbital velocity must be reduced to enter the transfer ellipse.

The hyperbolic excess velocity is important because it is a measure of the energy required from the launch vehicle system. It is traditional to use C3, which is V_{HE}^2, as the major performance parameter agreed on between the launch vehicle system and a planetary spacecraft.

From the point of view of the spacecraft, C3 comes from a mission design calculation as previously shown and represents the minimum energy requirement needed to accomplish the mission. Table 3.12 shows some typical C3 requirements. From the point of view of the launch vehicle, C3 is computed as the maximum energy the launch vehicle can deliver carrying a spacecraft of a given weight. Hopefully the launch vehicle C3 capability at the expected spacecraft weight will be above the C3 required for the mission.

3.5.2.2 V_∞ at the planet. When the spacecraft arrives at the target planet, a velocity condition analogous to departure occurs. The hyperbolic excess velocity on arrival at the planet is called V_∞ or V_{HP}; we will use V_∞. V_∞ is the vector difference between the arrival velocity on the transfer ellipse and the orbital velocity of the planet. In this example, the included angle α would be zero at arrival also; therefore,

$$V_\infty = V_{s/s} - V_{p/s}$$

$$V_\infty = 37.73 - 35.05 = 2.68 \text{ km/s} \tag{3.127}$$

where

$V_{s/s}$ = arrival velocity of the spacecraft on the transfer ellipse with respect to the sun
$V_{p/s}$ = velocity of the target planet with respect to the sun
V_∞ = velocity at infinity along the arrival hyperbola

V_∞ at the target planet is positive in this example, indicating that velocity must be reduced for capture.

3.5.3 Patched Conic Procedure

Consider next a realistic planetary mission with actual planetary positions and elliptical, inclined planetary orbits. This example will use the patched conic procedure, which consists of four steps:

1) Pick a launch date and an arrival date during the launch opportunity (the period in which the mission is possible). Accurately determine the position of the Earth and the target planet on the chosen dates.

2) Design a transfer ellipse from Earth to the target planet. The transfer ellipse must contain the Earth's position at launch and the planet's position at arrival. The time of flight on the transfer arc must be equal to the time between the launch and arrival dates. This is a trial and error process. Each trial transfer ellipse is defined by an arbitrary selection of the longitude of the line of apsides. Trials are made until the transfer conditions are met.

3) Design the departure hyperbola such that it will deliver the spacecraft to the transfer ellipse.

4) Design the approach hyperbola and the arrival mission.

Each step will be considered in terms of theory and by example in subsequent sections. The example that will be used is the type I Venus mission in the 1988 launch opportunity.

3.5.4 Locating the Planets

In the prior simplified design, the planets were placed arbitrarily at locations that were convenient to the calculation. Ephemeris calculations allow the determination of actual locations for the planets at the times needed.

3.5.4.1 Launch opportunity. It is desirable for the Earth at launch and target planet at arrival to be directly opposed, as in a Hohmann transfer, to minimize launch energy. The years for which an approximation of this position occurs are said to offer a launch opportunity. To see how this concept works, consider the launch opportunities between the Earth and an imaginary planet with a period exactly twice that of the Earth, as shown in Fig. 3.52. (This situation is similar to that between Earth and Mars.) Assume that Earth and the target planet are perfectly aligned for a Hohmann transfer on 1 January of year 1. In one year the Earth will make one revolution and will be back to its original position. The period of the target planet is two years, so that it will be halfway around the sun on arrival day. Its arrival position will be adjacent to the Earth departure position. Initiating an interplanetary transfer is impossible in this situation, regardless of launch vehicle capability.

At the end of the second year (1 January, year 3), the Earth has made its second orbit of the sun. The target planet will complete its orbit of the sun on arrival day. The planets are again in ideal positions for a mission.

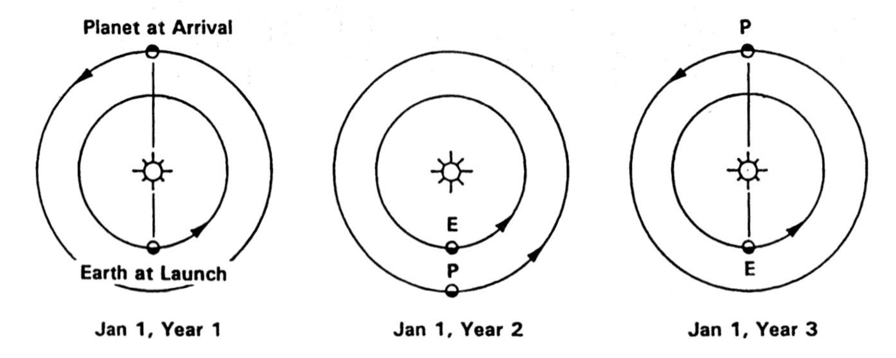

Fig. 3.52 Relative position of Earth and a planet with twice the Earth's period.

Table 3.13 Synodic periods

Planet	S, days
Mercury	116
Venus	584
Mars	780
Jupiter	399
Saturn	378
Uranus	370
Neptune	367
Pluto	367
Moon	30

3.5.4.2 Synodic period. For this imaginary planet the synodic period is two years. The synodic period is the time interval between launch opportunities and is a characteristic of each planet. The synodic period of a planet is

$$S = \frac{2\pi}{\omega_e - \omega_p} = \frac{1}{1 - 1/P_p} \tag{3.128}$$

where

S = synodic period, year
ω_e, ω_p = angular velocity of Earth and planet, rad/year
P_p = period of the planet, year

Table 3.13 shows the synodic period of the planets. Note that the derivation of synodic period is based on an assumption of circular orbits. The synodic period of the outer planets is essentially one year because their orbital motion is very slow in comparison to that of the Earth. In one year the Earth is back in a favorable launch position, and the outer planet has not moved much.

3.5.4.3 Trajectory type and class. Planetary trajectories are classified based on the length of the transfer ellipse (see Fig. 3.53). If the spacecraft will

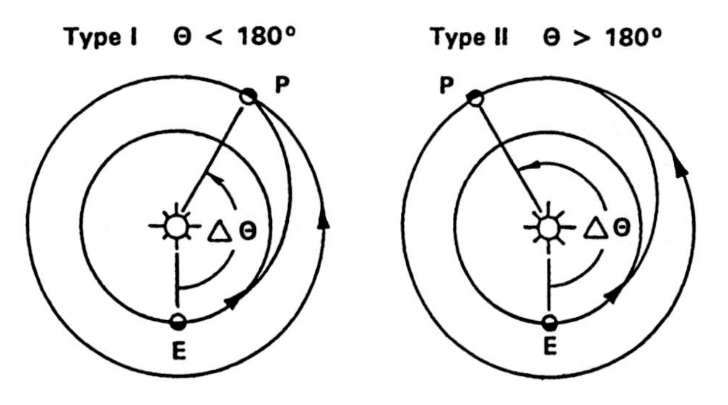

Fig. 3.53 Trajectory types.

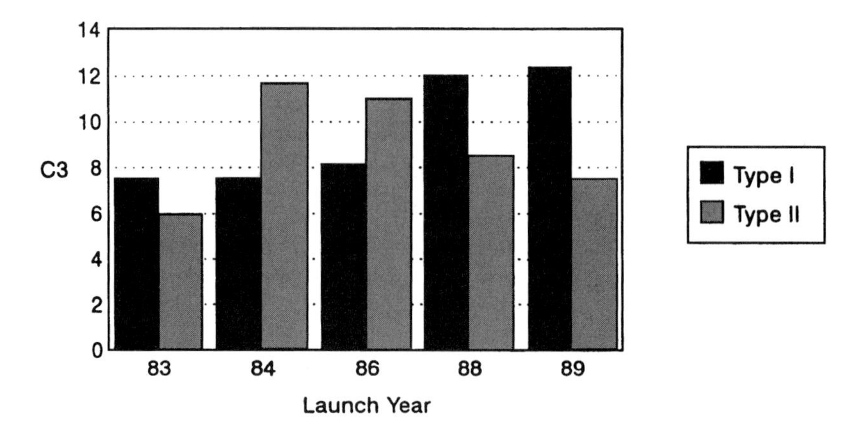

Fig. 3.54 Minimum C3 for trajectories to Venus.[27]

travel less than a 180-deg true anomaly, the trajectory is called type I. If the spacecraft will travel more than 180 deg and less than 360 deg, the trajectory is called a type II. Types III and IV exist but are seldom used. Trajectories are also organized into classes. A class I transfer trajectory reaches the target planet before apoapsis (or before periapsis for inbound missions). Class II trajectories reach the target planet after apoapsis (or after periapsis for inbound missions). The Viking mission to Mars, for example, used a type II, class II trajectory.

A good launch opportunity depends on more than the relative position of the planets. The location of the line of nodes for the target planet's orbit is important in sizing the plane change and therefore is important in setting the energy required of the launch vehicle. Launch opportunities also vary in quality because of the eccentricity of the orbits. Figure 3.54 shows the variation in C3 for type I and type II trajectories to Venus over five launch opportunities. The variation shown is due to all causes: the relative positions, the plane change, the velocities of the planets, and the eccentricity of the orbits.

3.5.4.4 Planet locations. Strangely enough, for a planetary trajectory, the launch and arrival dates may be arbitrarily picked; launch energy is the dependent variable. The Voyager 1 trajectory to Jupiter was designed to launch after Voyager 2 and arrive at Jupiter four months ahead of Voyager 2.

In 1988 the launch opportunity to Venus ran from 1 March to about 10 April. The example trajectory to Venus may be arbitrarily selected as follows:

$$\text{Launch} = 8 \text{ April } 1988 \text{ (JD } 244\ 7259.5)$$

$$\text{Arrival} = 26 \text{ July } 1988 \text{ (JD } 244\ 7368.5)$$

3.5.4.5 Ephemeris calculations. The sizes, shapes, and locations of the planetary orbits change slowly with time; these changes are caused by perturbations to the orbits. The changes are small but important for ephemeris calculations. To make the situation even more interesting, the ecliptic plane moves slightly with time, as does the location of the vernal equinox vector; thus, the axis system is moving. To accurately locate the two planets we must 1) arbitrarily fix the reference system, 2) determine the size, shape, and location of planetary orbits in the period of interest, and 3) hold these parameters fixed and determine the latitude and longitude of Earth on launch day and of the target planet on arrival day.

There are two methods currently in use for fixing the axis system. One approach is to pick a date, locate the ecliptic plane and the vernal equinox vector, and assume that the axes are fixed for a period of time (usually 50 years). A calculation of this kind is labeled "of epoch J2000.0" or "of epoch J1950.0." These expressions mean that the measurements are made from the locations of the ecliptic and the vernal equinox vector on 1 January 2000 or 1 January 1950, respectively.

The second method of handling the moving axis system is to measure from the instantaneous position of the ecliptic and the equinox vector. Calculations of this type are labeled "of date."

The standard reference for ephemeris data is *The Astronomical Almanac*,[12] available annually from the U.S. Government Printing Office. Prior to 1981 the publication was referred to as *The American Ephemeris and Nautical Almanac*.

The *Astronomical Almanac* provides tabulations of the heliocentric latitude, longitude, and radius of the planets. These tabulations were given daily for the inner planets and at longer intervals for the outer planets. Interestingly, the Earth's heliocentric coordinates must be calculated from the geocentric coordinates of the sun by adding (or subtracting) 180° to the longitude and reversing the sign of the latitude. In addition, the almanac gives the instantaneous planetary orbital elements at 11 times spaced equally over a year. The instantaneous two-body elements are known as *osculating orbital elements*.

Another method of obtaining ephemeris information employs the polynomial fit technique, using short numerical series of the form

$$E_{\mathrm{FUTURE}} = E_{\mathrm{REF}} + C_1 T + C_2 T^2 \tag{3.129}$$

where

E_{FUTURE} = an orbital element (parameter) at some future date
E_{REF} = the same orbital element at a reference date in the past
C_1, C_2 = polynomial coefficients
T = Julian centuries from the reference date to the future date

Given the future Julian date $\mathrm{JD_{FUTURE}}$ and the reference Julian date $\mathrm{JD_{REF}}$,

$$T = (\mathrm{JD_{FUTURE}} - \mathrm{JD_{REF}})/36{,}525 \tag{3.130}$$

Polynomial coefficients of this form are given in Refs. 28 and 29.

Ephemeris data are also calculated by an alternate relation of the following form:

$$E = a_0 + a_1 p + a_2 p^2 \tag{3.131}$$

The value of p is determined by using the following equation:

$$p = (\mathrm{JD_{FUTURE}} - \mathrm{JD_{REF}})/2280 \tag{3.132}$$

Polynomial coefficients of this type are given for all planets in Ref. 30.

The geometry of the 1988 Venus mission and the corresponding ephemeris data are shown in Fig. 3.55, which shows a view of the ecliptic plane from the northern

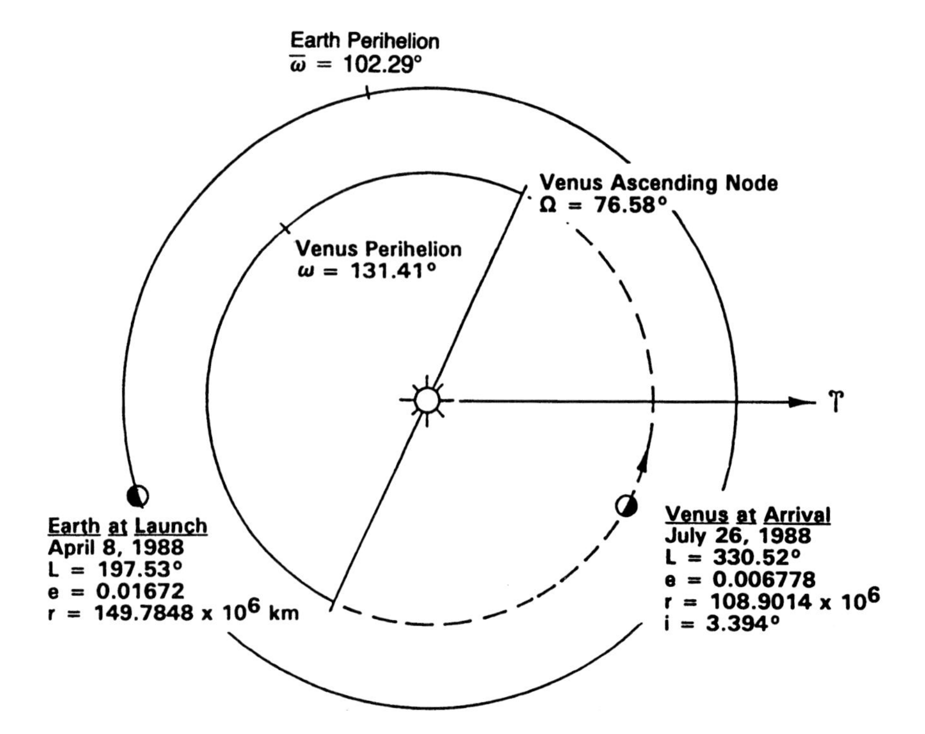

Fig. 3.55 1988 type I mission to Venus.

celestial sphere. The orbit of Venus is slightly out of plane; Venus is below the ecliptic at arrival. Longitudes are measured in the ecliptic plane counterclockwise in the direction of motion, from the vernal equinox vector. The Earth at launch is less than 180 deg from Venus at arrival; therefore, this is a type I trajectory.

From Fig. 3.55, the true anomaly of Earth at launch is

$$\Theta_e = 197.53 - 102.29 = 95.24 \text{ deg}$$

At this position the velocity of the Earth with respect to the sun is, from Eq. (3.14),

$$V_{e/s} = 29.75 \text{ km/s}$$

and the flight path angle is, from Eq. (3.31),

$$\gamma_e = 0.9554 \text{ deg}$$

From Fig. 3.55, for Venus at arrival,

$$\Theta_p = 330.52 - 131.41 = 199.11 \text{ deg}$$
$$V_{p/s} = 34.80 \text{ km/s}$$
$$\gamma_p = -0.1280 \text{ deg}$$

3.5.5 Design of the Transfer Ellipse

The transfer ellipse of this example is defined by the two following requirements: 1) It must pass through the Earth position at launch and through the Venus position at arrival. 2) The time of flight between these positions must be exactly equal to the number of days between 8 April 1988 and 26 July 1988. These requirements constitute the rendezvous conditions.

The transfer ellipse can be found by trial and error. The two planet positions define two points on a family of transfer ellipses, each defined by the radial position of the line of apsides. Selecting the position of the line of apsides defines the true anomaly for each planet position and the time of flight between points. The trial and error process then becomes a series of selections of line of apsides positions and subsequent calculations of times of flight on the resulting transfer ellipse.

The required time of flight on the transfer ellipse is the number of days between the Julian dates of launch and arrival:

Arrival:	26 July 1988	JD 244 7368.5
Launch:	8 April 1988	JD 244 7259.5
Time of Flight:		109.0 days

For the first attempt, try placing the line of apsides through the Earth position at launch (see Fig. 3.56). The true anomaly of the Earth's position becomes 180 deg. Consulting Fig. 3.55, Venus is 132.99 deg ahead of the Earth position (330.52 − 197.53), indicating a true anomaly of 180 + 132.99 = 312.99 deg. The planet radii and true anomalies define the trial transfer ellipse shown in Fig. 3.56.

The eccentricity of the transfer ellipse, from Eq. (3.28), is

$$e = \frac{(108.9014 \times 10^6) - (149.7848 \times 10^6)}{(149.7848 \times 10^6)(\cos 180) - (108.9014 \times 10^6)(\cos 312.99)}$$

$$e = 0.1825$$

The periapsis radius, from Eq. (3.29), is

$$r_p = \frac{149.7848 \times 10^6[1 + (0.1825)(-1)]}{(1.1825)} = 103.5550 \times 10^6 \text{ km}$$

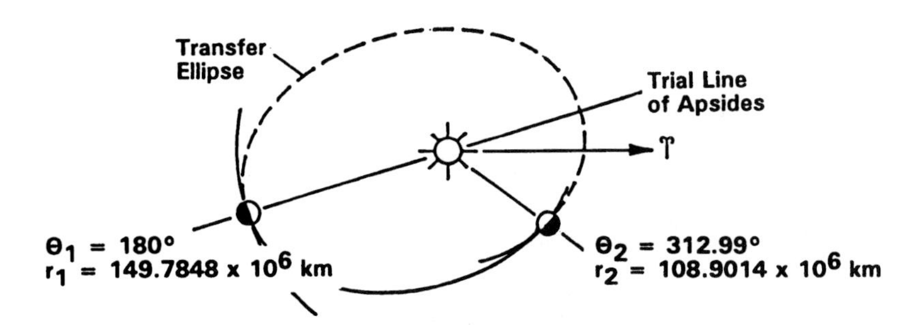

Fig. 3.56 Trial transfer ellipse.

Table 3.14 Trial and error properties of transfer ellipse

	Θ at Earth	Θ at Venus	e	$a \times 10^6$	TOF, days
1	180	312.99	0.1825	126.669	116.15
2	190	322.99	0.1744	127.954	112.16
3	**199.53**	**332.52**	**0.17194**	**129.336**	**109.02**

The semimajor axis is, from Eq. (3.48),

$$a = \frac{103.555 \times 10^6}{(1 - 0.1825)} = 126.673 \times 10^6 \text{ km}$$

From the Kepler equation comes the time of flight:

	θ, deg	T, days
Periapsis to Earth	180	142.30
Venus to periapsis	312.99	26.15
Transfer time		116.15 (109.0 required)

The first trial transfer ellipse yields a time of flight of 116.15 days vs a requirement of 109.0 days. Subsequent trials must be made holding the following properties on the transfer ellipse constant and varying the true anomaly at Earth and Venus:

1) radius at Earth $= 149.7848 \times 10^6$ km
2) radius at Venus $= 108.9014 \times 10^6$ km
3) longitudinal angle from Earth to Venus $= 132.99^\circ$

Table 3.14 shows the results of three trials. Trial 3 is the transfer ellipse.

The spacecraft reaches Venus before periapsis; therefore, this is a class I trajectory. The velocities required on the transfer ellipse and the associated flight path angles can now be calculated.

$$\text{Near Earth:} \quad V = 27.312 \text{ km/s}, \quad \gamma = -3.924 \text{ deg}$$

$$\text{Near Venus:} \quad V = 37.566 \text{ km/s}, \quad \gamma = -3.938 \text{ deg}$$

3.5.6 Design of the Departure Trajectory

On a planetary flight, the launch vehicle sends the spacecraft off on a hyperbolic escape trajectory. As the spacecraft speeds along, the Earth's influence diminishes with distance and the sun's influence increases. The hyperbolic orbit gradually becomes an elliptic orbit about the sun. To send a spacecraft to a planet, send it away on the right hyperbolic orbit (right V_∞) and in the right plane (the transfer plane). We can then sit back and watch physics do the rest. Like a cannon, if we point it correctly, it will hit the target. The key results of the patched conic calculation are the definition of the transfer plane and V_∞ (i.e., how to point the cannon). Calculation of these key results will be discussed in this section.

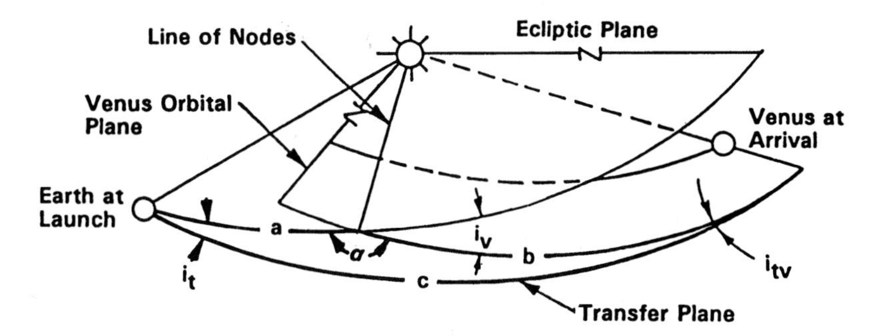

Fig. 3.57 Transfer plane.

3.5.6.1 Plane change. The plane change is made at departure to take advantage of the energy economy of combining the plane change with injection. In addition, launch vehicle energy is used rather than spacecraft energy. As a result, the transfer ellipse is not in the ecliptic plane; it is in an intersecting plane that contains the center of mass of the sun, the Earth at launch, and Venus at arrival, as shown in Fig. 3.57. In Fig 3.57,

i_t = the inclination of the transfer plane
i_v = the inclination of the Venus orbital plane
α = $180 - i_v = 180 - 3.394 = 176.61$ deg
a, b, c = spherical angles measured on the surface of a sphere of radius r_e centered at the sun

To get the inclination of the transfer plane i_t requires solution of the spherical triangle a, b, c in Fig. 3.57. The side a is in the ecliptic plane and can be obtained directly from longitudes (see Fig. 3.57):

$$a = (\Omega + 180) - L_e$$

$$a = 76.58 + 180 - 197.53 = 59.05 \text{ deg} \tag{3.133}$$

The arc b', measured from the line of nodes to the Venus longitude can be obtained by

$$b' = L_v - (\Omega + 180)$$

$$b' = 330.52 - 76.58 - 180 = 73.94 \text{ deg} \tag{3.134}$$

Because longitudes are measured in the ecliptic plane, a small adjustment could be made in b' to yield b. If the inclination of the target planet orbit is small and preliminary design work is being done, this step can be ignored. In this example, solving the right spherical triangle to obtain b yields $b = 73.967$ deg.

The arc c can be obtained using the law of cosines from spherical trigonometry:

$$\cos c = \cos a \cos b + \sin a \sin b \cos \alpha$$

$$\cos c = -0.6807 \tag{3.135}$$

$$c = 132.90 \text{ deg}$$

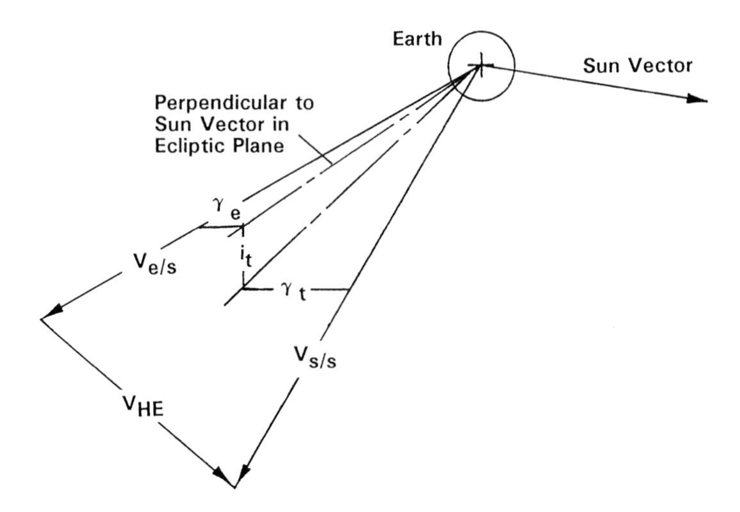

Fig. 3.58 Vector diagram at departure.

The inclination of the transfer plane can be obtained by using the law of sines from spherical trigonometry:

$$\sin i_t = \frac{\sin \alpha \sin b}{\sin c} = 0.07768 \qquad (3.136)$$
$$i_t = 4.455 \text{ deg}$$

3.5.6.2 Calculating V_{HE} and C3. The $\boldsymbol{V}_{\text{HE}}$ vector is designed so that when the spacecraft reaches "infinity" on the departure hyperbola it will have the proper speed and direction to establish the transfer ellipse in the transfer plane. The size of the transfer ellipse is so large compared to the hyperbolic departure orbit that, for preliminary calculations, we do not have to find an X, Y position where the departure (or arrival) hyperbola and the transfer ellipse are patched. As shown in Fig. 3.58, the direction established for V_{HE} accommodates the flight path angle of Earth, the flight path angle of the spacecraft on the transfer ellipse, and the plane change to the transfer plane.

In Fig. 3.58,

$V_{e/s}$ = velocity of Earth with respect to the sun
$\boldsymbol{V}_{s/s}$ = velocity of the spacecraft with respect to the sun on the transfer ellipse
$\boldsymbol{V}_{\text{HE}}$ = hyperbolic excess velocity, which is the velocity with respect to Earth at infinity on the departure hyperbola
γ_e = flight path angle of Earth
γ_t = flight path angle of the spacecraft on the transfer ellipse
i_t = inclination of the transfer plane with respect to the ecliptic

The angles in Fig. 3.58 are greatly exaggerated. The vector geometry shown is for the example problem. The geometry will vary widely depending on the sign and magnitude of each angle.

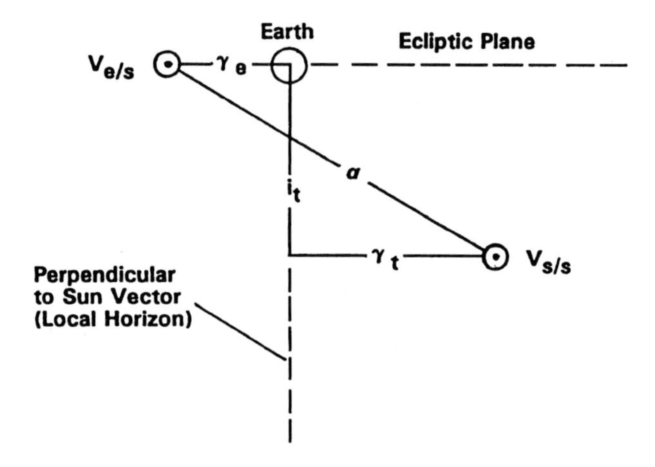

Fig. 3.59 Spherical triangles containing departure vectors.

Figure 3.59 shows the geometry as seen when looking down the departure vector. It is necessary to know α because it is the angle between the two vectors of interest and is needed for a cosine law solution. For a right spherical triangle,

$$\cos\alpha = \cos i_t \cos(\gamma_e + \gamma_t)$$

$$\cos\alpha = \cos 4.455 \cos(0.9554 + 3.924) = 0.99337 \tag{3.137}$$

$$\alpha = 6.604 \text{ deg}$$

Note: The above cosine solution step (Eq. 3.137) is a frequent source of error in patched conic solutions. You can see from the vector diagrams, Fig. 3.58 and Fig. 3.59, that angles γe and γ_c should add. This is not always the case. Sometimes the two angles must be subtracted, as you will see at Venus arrival in this example (Fig. 3.67). You can tell from the vector diagram if γe and γ_c should be added or subtracted (the technically correct way) or you can remember the following rule of thumb: If γe and γ_c have the same sign they should be subtracted; if the signs of γe and γ_c differ they should be added. You can see from the vector diagrams, Figs. 3.58 and 3.59, why this is so.

From the law of cosines,

$$C3 = V_{e/s}^2 + V_{s/s}^2 - 2V_{e/s}V_{s/s}\cos\alpha \tag{3.138}$$

Since

$$V_{e/s} = 29.75 \text{ km/s}$$

$$V_{s/s} = 27.312 \text{ km/s}$$

C3 for the 1988 type I Venus mission is

$$C3 = 16.73 \text{ km}^2/\text{s}^2$$

$$V_{\text{HE}} = 4.090 \text{ km/s}$$

for launch on 8 April and arrival on 26 July.

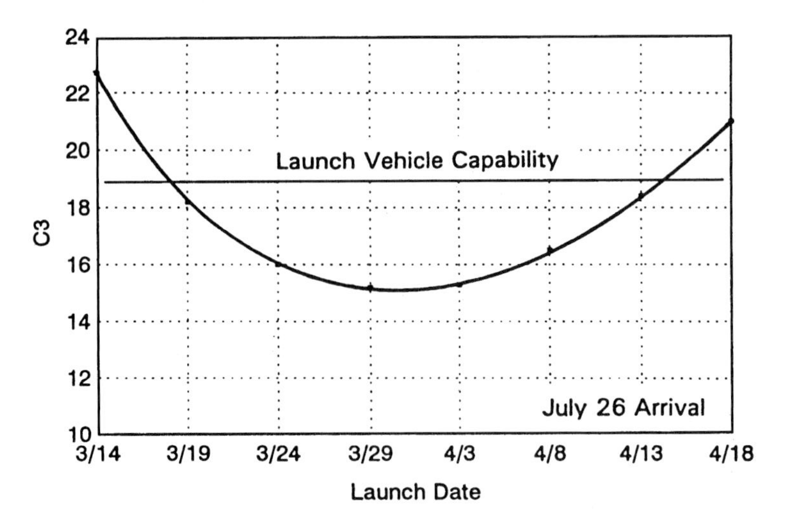

Fig. 3.60 C3 vs launch date for 1988 Venus mission.

In any year that contains a launch opportunity, there are a number of acceptable launch dates and a number of acceptable arrival dates. Holding the arrival date of 26 July fixed, and varying the launch date, yields a series of patched conic solutions with various times of flight and various C3 values. The result is that only certain days in a launch opportunity are feasible for any given launch vehicle, as Fig. 3.60 shows.

In a year with a launch opportunity there will be a best day that has relative planetary positions requiring the least launch vehicle energy. There will be a period of a few days on either side of the best day when a mission is possible with a given launch vehicle. For Magellan (Venus, type IV, 1989) the launch opportunity was 28 days long.

Note how rapidly C3 rises at the edges of the opportunity; substantially more launch energy would not greatly extend the window.

It is common practice to calculate a large number of trajectories to obtain contours of C3 for a range of launch dates and arrival dates. Figure 3.61 shows such a plot for the Venus 1988 launch opportunity. Figure 3.61 shows that there are two regions of minimum C3 in the 1988 opportunity. The upper region contains the type II missions, and the lower region contains the type I missions. Lower C3 values could be obtained with type II missions in 1988. A C3 of 8.5 is attainable, and time of flight (TOF) is about 200 days. The type I missions have a minimum C3 of 12.0 and a TOF of about 100 days. The type I and II launch periods overlap; it would be possible to launch two spacecraft at close intervals and have widely spaced arrivals (as the Voyager program did at Jupiter). Figure 3.61 also shows how quickly C3 increases on either side of the minimum. In just a matter of days, a mission on any conceivable launch vehicle becomes impossible.

The launch date selected for a project is usually a compromise between C3 and other concerns, usually arrival conditions. Thus a launch seldom takes place at the minimum C3.

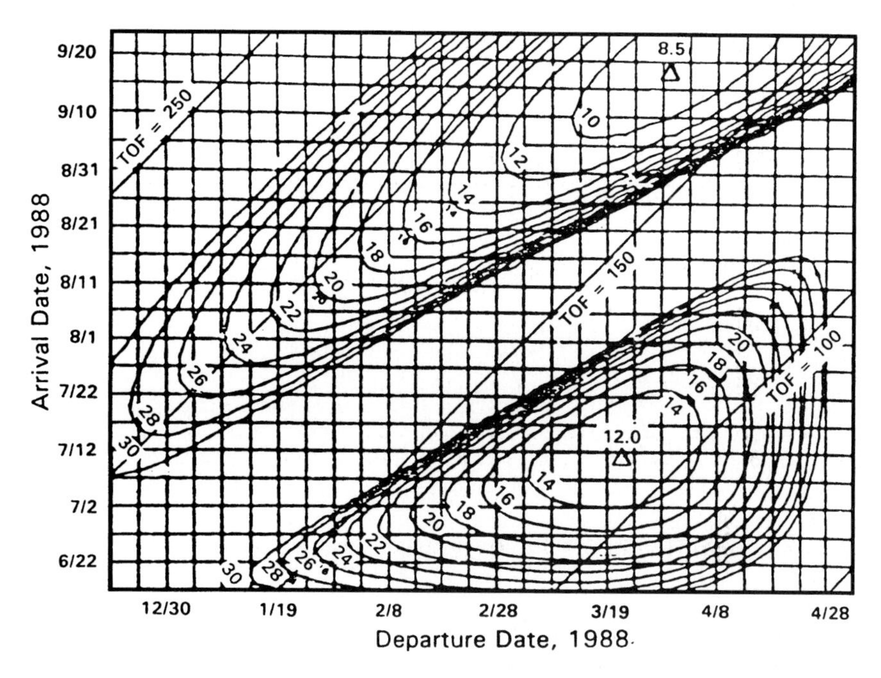

Fig. 3.61 Contours of C3: Earth to Venus 1988. (Provided through the courtesy of NASA/JPL/Caltech; Ref. 27, pp. 2–113.)

3.5.6.3 Departure hyperbola. The departure hyperbola, shown in Fig. 3.62, has a periapsis radius equal to the radius of the parking orbit and a V_{HE} magnitude and direction that are calculated to place the spacecraft on the proper transfer ellipse as previously discussed. It is convenient to visualize the V_{HE} vector passing through the center of the Earth. The acceptable departure hyperbolas form a body of revolution about the V_{HE} vector, as shown in Fig. 3.63. Departure on any hyperbola in the body of revolution is acceptable. The *point of*

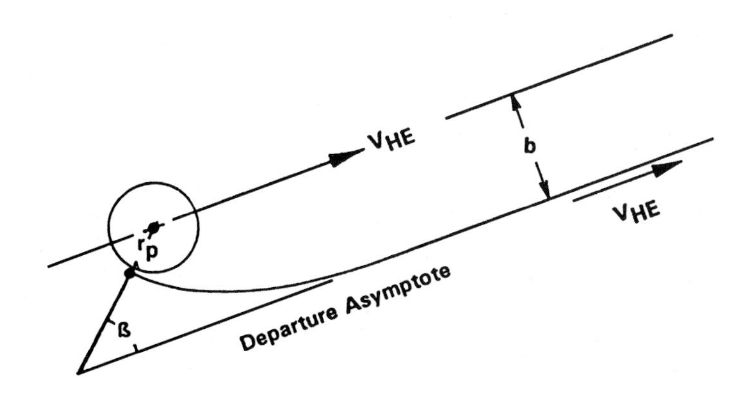

Fig. 3.62 Departure hyperbola.

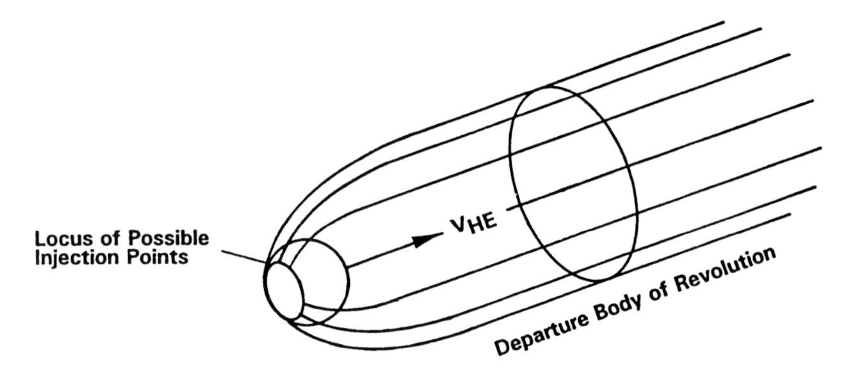

Fig. 3.63 Departure body of revolution.

injection is the location at which the spacecraft reaches the velocity needed to enter the proper hyperbolic departure trajectory; this point is normally at the periapsis of the hyperbola. The velocity required at the injection point is

$$V = \sqrt{(2\mu/r_p) + V_{\rm HE}^2} \tag{3.57}$$

where

V = the spacecraft velocity at the point of injection into the departure hyperbola
$V_{\rm HE}$ = hyperbolic excess velocity
r_p = radius at the point of injection (assumed to be at periapsis)

Although the launch vehicle interface is specified in terms of C3, the correct velocity, as defined in Eq. (3.57), is what the launch vehicle must actually achieve.

$\boldsymbol{V}_{\rm HE}$ and r_p define the departure hyperbola; the remaining elements can be obtained using the relations provided in Section 3.1. For the example mission, assuming $h_p = 330$ km, the periapsis velocity is, from Eq. (3.57),

$$V = \sqrt{\frac{(2)398{,}600}{6708} + 16.73} = 11.64\,\text{km/s}$$

The semiminor axis of the departure hyperbola, from Eq. (3.83), is

$$b = r_p\sqrt{\frac{2\mu}{r_p V_{\rm HE}^2} + 1} = 6708\sqrt{\frac{2(398{,}600)}{(6708)(16.73)} + 1} = 19{,}097\,\text{km}$$

The angle of asymptote, from Eq. (3.64), is

$$\tan\beta = \frac{bV_{\rm HE}^2}{\mu} = \frac{(19{,}097)(16.73)}{(398{,}600)} = 0.8015$$

$$\beta = 38.71\,\text{deg}$$

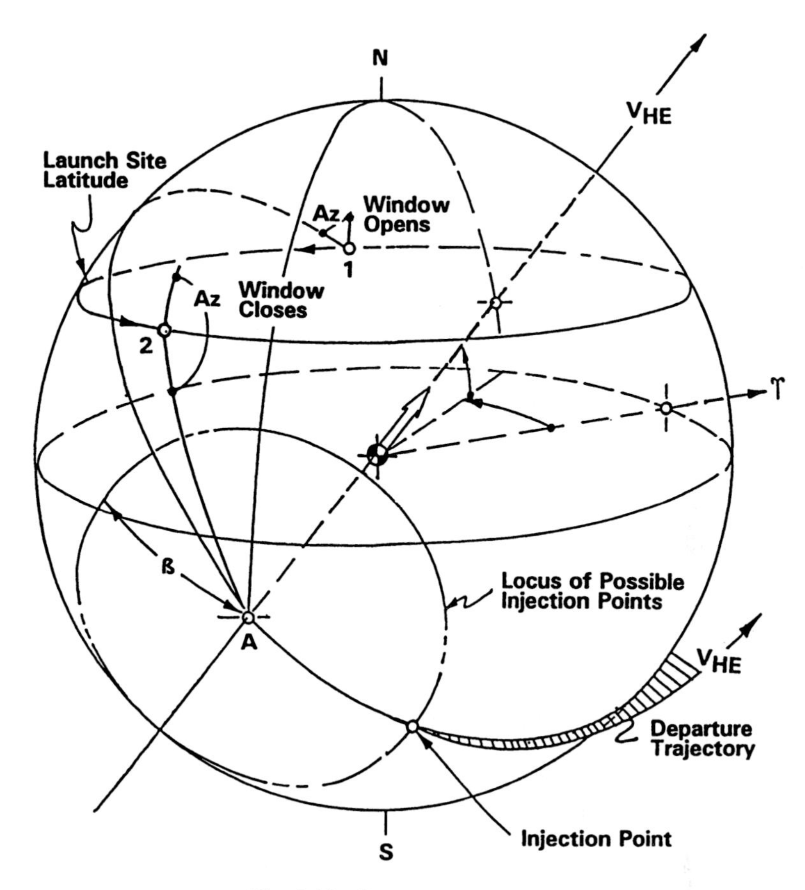

Fig. 3.64 Departure geometry.

3.5.6.4 Launch window. The acceptable launch azimuth range at the launch site constrains the allowable launch times to a daily launch window. Figure 3.64 shows how this works. The Earth rotates about its axis with the V_{HE} vector passing through its center of mass. The locus of possible injection points is a circle of radius β centered about the V_{HE} vector at point A. To inject the spacecraft onto an acceptable departure hyperbola, the launch trajectory must follow a great circle from the launch site to point A.

Consider all of the features of the figure as fixed in space while the launch site rotates counterclockwise around the Earth's axis on a latitude line. The daily launch window opens at point 1 when the required launch azimuth reaches the minimum acceptable angle for the launch site (about 35 deg for ETR; see Section 3.3).

As time passes the launch site rotates from point 1 to point 2. The required launch azimuth increases with time until the maximum acceptable angle is exceeded at point 2 and the window closes. The departure path shown in Fig. 3.64 assumes that the launch is made at the instant the window opens.

Note from Fig. 3.64 that there are two opportunities to launch on some of the departure trajectories, one early in the window and another late in the window where the great circle crosses the launch site latitude. Also note that many of the acceptable hyperbolas are not achievable.

The position of the V_{HE} vector changes slowly, and as a result the length of the launch window changes with time. The Magellan launch window varied from 28 min on the first day (28 April 1989) to 126 min on the last, or thirtieth, day.

To recap, a planetary launch can occur only in certain years, only for a few days in those years, and only for a few minutes in those days. If the launch opportunity is missed there will be a significant delay (one synodic period) before another opportunity occurs. These facts are key scheduling and planning factors in any planetary program.

3.5.7 *Design of the Arrival Trajectory*

There are three basic arrival mission types:

Mission	Example
Direct impact	Ranger
Gravity-assist	Voyager
Planetary orbiter	Magellan

Arrival design for gravity-assist missions is dominated by targeting considerations; arrival for orbiting missions is dominated by arrival energy considerations.

3.5.7.1 Plane change. The design of the approach hyperbola is similar to the design of the departure hyperbola. The first step is the determination of the inclination of the transfer plane with respect to the orbital plane of the target planet. This calculation requires considering again the spherical triangle formed by the transfer ellipse, the ecliptic plane, and the Venus orbital plane, as shown in Fig. 3.65.

The angle of interest is i_{tp}, the inclination of the transfer plane with respect to the target planet Venus orbital plane. By the sine law of spherical trigonometry, the following equation may be obtained:

$$\sin i_{tp} = \sin\alpha \sin a / \sin c \tag{3.139}$$

$$i_{tp} = 3.975 \text{ deg}$$

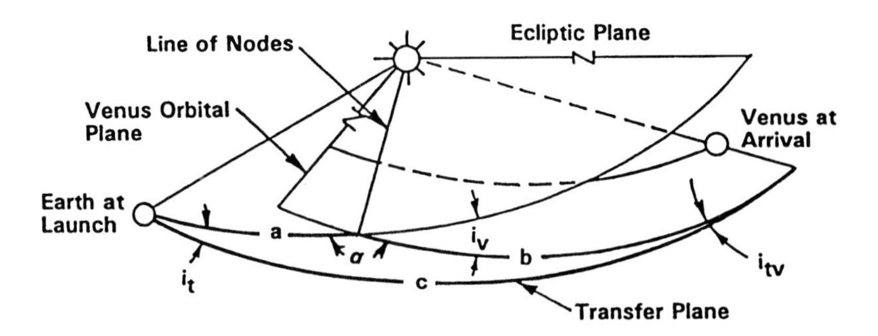

Fig. 3.65 Transfer plane.

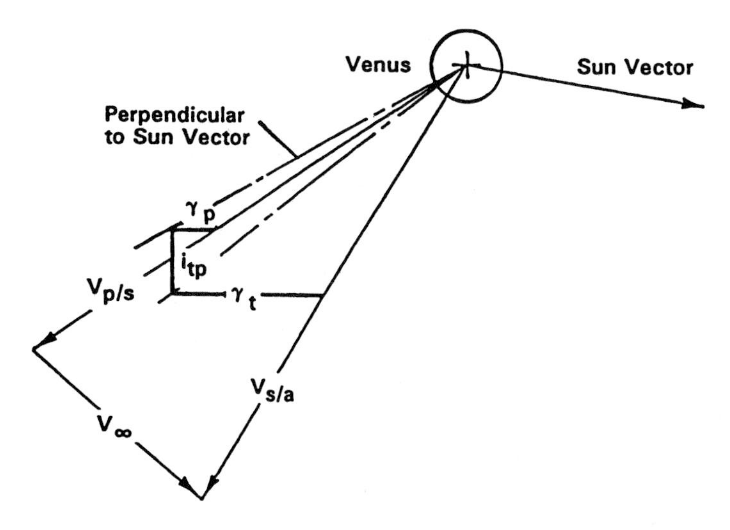

Fig. 3.66 Vector diagram at arrival.

3.5.7.2 Calculating V_∞. V_∞ at the target planet, analogous to $\boldsymbol{V}_{\mathrm{HE}}$ at departure, is a dependent variable determined by the departure and transfer trajectories. The arrival vector diagram is conceptually the same as the departure diagram, as shown in Fig. 3.66. In Fig. 3.66,

$\boldsymbol{V}_{p/s}$ = velocity of the target planet with respect to the sun
$\boldsymbol{V}_{sa}$ = velocity of the spacecraft on the transfer ellipse, with respect to the sun, at arrival
$\boldsymbol{V}_\infty$ = spacecraft velocity at infinity on the arrival hyperbola, with respect to the target planet
γ_p = flight path angle of target planet
γ_t = flight path angle of the spacecraft on the transfer ellipse
i_{tp} = inclination of the transfer plane, with respect to the target planet orbital plane

Figure 3.67 shows a spherical triangle, with sides α_a, i_{tp}, and $(\gamma_t - \gamma_p)$, centered on the target planet, looking into the arrival vectors $V_{p/s}$ and $V_{s/a}$. The angle between the two vectors of interest is α_a. For a right spherical triangle,

$$\cos\alpha_a = \cos i_{tp}\cos(\gamma_t - \gamma_p) \tag{3.140}$$

$$\cos\alpha_a = \cos 3.975\cos(3.938 - 0.1280)$$

$$\alpha_a = 5.5039\,\text{deg}$$

Since $\boldsymbol{V}_{p/s} = 34.80$ km/s and $\boldsymbol{V}_{s/a} = 37.57$ km/s, using the law of cosines yields the following $\boldsymbol{V}_\infty$ for the 1988 type I Venus mission:

$$\boldsymbol{V}_\infty = 4.442\ \text{km/s}$$

Figure 3.67 is appropriate only for the example problem and is not a general solution. An actual arrival geometry will vary widely depending on the signs and magnitudes of the angles involved.

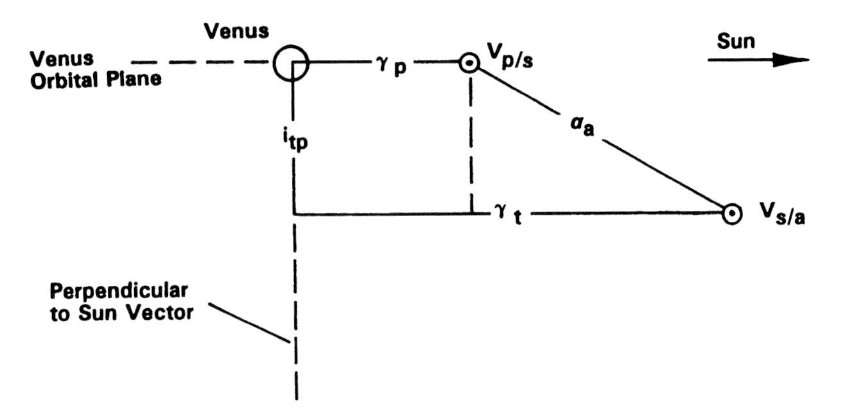

Fig. 3.67 Spherical triangle containing the arrival vectors.

3.5.8 Establishing Planetary Orbit

Frequently it is desired to place a spacecraft in orbit about the target planet. Mariner 9 in 1971 was the first spacecraft to orbit another planet, followed by Viking, Pioneer 10, and Magellan. Establishing a planetary orbit requires a simple orbit change maneuver as shown in Fig. 3.68.

The velocity at periapsis of the approach hyperbola is

$$V_p = \sqrt{V_\infty^2 + 2\mu/r_p} \tag{3.141}$$

The velocity at periapsis of the desired orbit is

$$V_p' = \sqrt{2\mu/r_p - \mu/a} \tag{3.142}$$

To put a spacecraft into a planetary orbit, the velocity at periapsis must be reduced from V_p to V_p'. Substantial spacecraft energy and weight are usually required. For

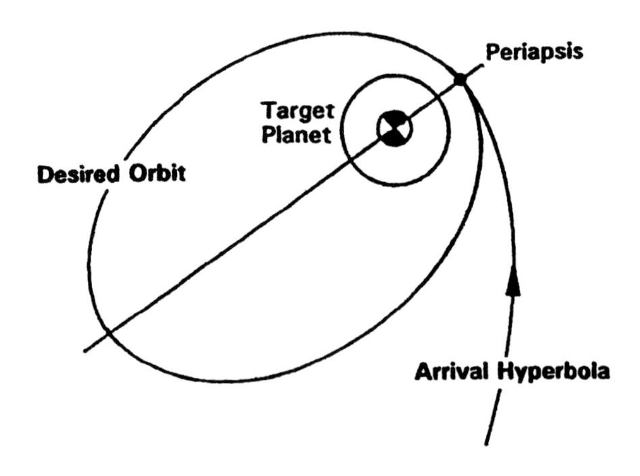

Fig. 3.68 Establishing a planetary orbit.

example, 60% of the Magellan cruise weight was dedicated to putting the spacecraft into Venus orbit. The velocity required to establish an orbit is proportional to V_∞. Therefore, for an orbital mission, the optimum launch date is a compromise between the minimum C3 and the minimum V_∞.

Note that for capture the velocity of the spacecraft must be reduced to a value below $\sqrt{2\mu/r}$.

3.5.9 Gravity-Assist Maneuver

A gravity-assist maneuver is a planetary encounter designed to produce a specific effect on the departure velocity vector of the spacecraft. When a spacecraft passes near a planet, the spacecraft velocity vector is rotated, and the magnitude of the velocity with respect to the sun is changed. This is a very useful maneuver because significant changes can be made to the velocity vector without expending spacecraft resources. After 20 years of analysis, speculation, and debate, the technique was successfully demonstrated by Mariner 10 in 1974.[31] After the flight of Mariner 10, the maneuver became an indispensable part of planetary mission design; three of the last four planetary missions—Voyager, Galileo, and Ulysses—would not have been possible without it.

Figure 3.69 shows the maneuver, in planet-centered coordinates, designed for an inside pass by the target planet (i.e., between the planet and the sun). In Fig. 3.69,

δ = angle through which the spacecraft velocity vector will be $(180 - 2\beta)$ turned
β = the hyperbolic asymptote angle

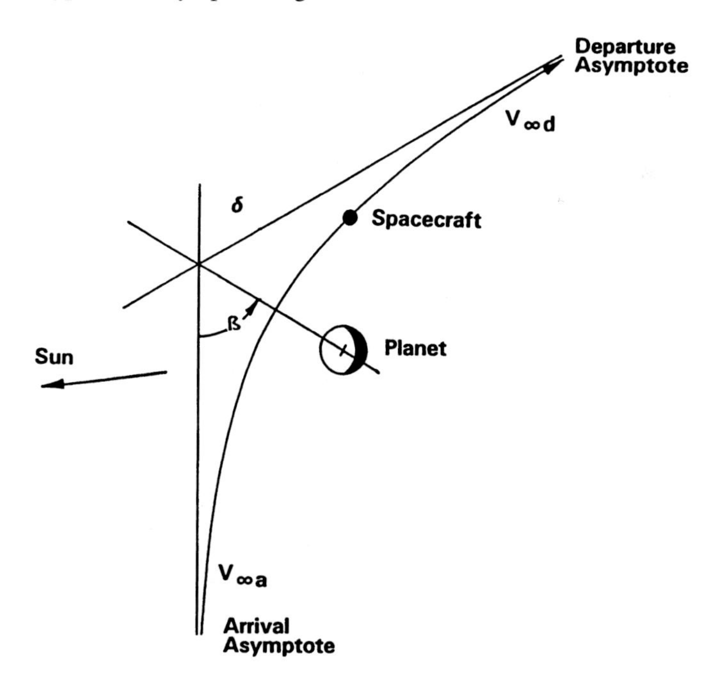

Fig. 3.69 Planetary encounter in geocentric coordinates.

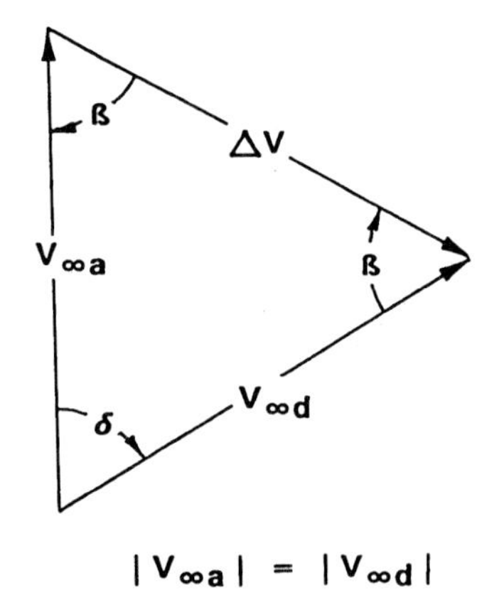

Fig. 3.70 Effect of a flyby on V_∞.

$V_{\infty a}$ = velocity at infinity on the arrival asymptote
$V_{\infty d}$ = velocity at infinity on the departure asymptote

As shown in Fig. 3.70, the effect of a planetary encounter is to add the vector $\Delta \boldsymbol{V}$ to the arrival velocity $\boldsymbol{V}_{\infty a}$. The magnitude of the $\Delta \boldsymbol{V}$ vector is

$$\Delta V = 2V_\infty \cos \beta \tag{3.143}$$

$$\Delta V = 2V_\infty / e \tag{3.144}$$

The spacecraft $\boldsymbol{V}_\infty$ vector is rotated through the angle δ by the gravitational effect of the planet. The magnitude of the $\boldsymbol{V}_\infty$ vector is not changed; however, the velocity of the spacecraft with respect to the sun is changed. Recall that $\boldsymbol{V}_\infty$ is the velocity of the spacecraft with respect to the planet. To obtain the velocity of the spacecraft with respect to the sun, the velocity of the planet with respect to the sun must be added to $\boldsymbol{V}_\infty$.

The velocity of the spacecraft with respect to the sun before and after the encounter is shown in Fig. 3.71. In Fig. 3.71,

$\boldsymbol{V}_{sa}$ = spacecraft velocity with respect to the sun on arrival at the sphere of influence of the target planet
$\boldsymbol{V}_{sd}$ = spacecraft velocity with respect to the sun at departure from the sphere of influence of the planet
$\boldsymbol{V}_{p/s}$ = velocity of the target planet with respect to the sun

It can be seen from Fig. 3.71 that the velocity of the spacecraft with respect to the sun has been increased by the encounter. (In Fig. 3.71, the central portions of vectors $\boldsymbol{V}_{sa}$, $\boldsymbol{V}_{sd}$, and $\boldsymbol{V}_{p/s}$ have been deleted, so that the area of interest may be seen more clearly.)

The arrival vectors are fixed by the transfer trajectory design; δ, β, $\Delta \boldsymbol{V}$, and $\boldsymbol{V}_{sd}$ can be changed. (The arrival parameters change slightly each day of the launch opportunity but can be considered fixed for any single day.) Selection of any one orbital parameter defines the encounter trajectory. During design, targeting is specified by periapsis radius or altitude; during the mission it is specified by the target point in the B plane.

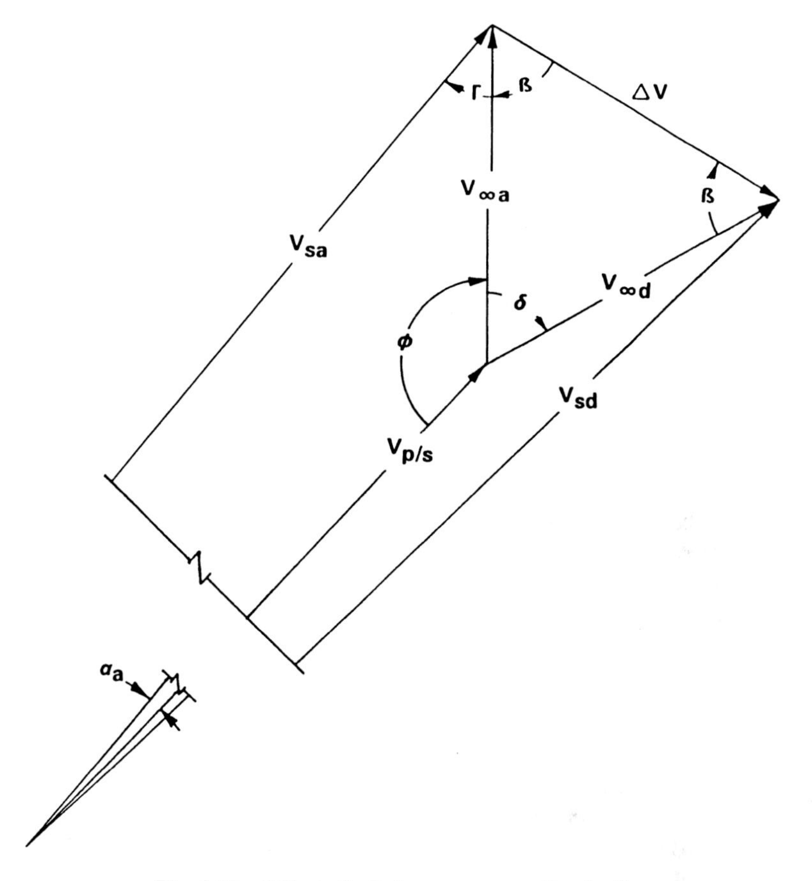

Fig. 3.71 Effect of a flyby on spacecraft velocity.

The angle α_a between the velocity of the planet and the velocity of the spacecraft is determined by the design of the transfer orbit. The angles δ and β are defined by the design of the flyby hyperbola. The angle Γ can be calculated from the law of sines as follows:

$$\sin \Gamma = \frac{V_{p/s} \sin \alpha_a}{V_{\infty a}} \tag{3.145}$$

and angle $\phi = 180 - \Gamma - \alpha_a$.

The cosine law can now be used to compute V_{sd}:

$$V_{sd}^2 = V_{sa}^2 + \Delta V^2 - V_{sa} \Delta V \cos(\beta \pm \Gamma) \tag{3.146}$$

The sign of gamma may be determined by inspection of the vector diagram.

The departure angle α_d can be obtained from the law of cosines as follows:

$$\cos \alpha_d = \frac{V_{p/s}^2 + V_{sd}^2 - V_\infty^2}{2 V_{p/s} V_{sd}} \tag{3.147}$$

For the Venus mission example ($V_\infty = 4.442$ km/s), targeting to achieve a periapsis altitude of 5000 km defines an encounter hyperbola with the following elements:

Asymptote angle:	$\beta = 53.25$ deg
Semiminor axis:	$b = 22{,}047$ km
Semimajor axis:	$a = 16{,}464$ km
Eccentricity:	$e = 1.6713$
Periapsis velocity:	$V_p = 8.861$ km/s

To achieve a periapsis altitude of 5000 km requires a target point in the B plane, which is 22,047 km from the center of the planet.

From prior calculations,

$$V_{sa} = 37.57\ \text{km/s} \qquad V_{p/s} = 34.80\ \text{km/s}$$

$$\alpha_a = 5.5039\ \text{deg} \qquad V_\infty = 4.442\ \text{km/s}$$

The value of ΔV from Eq. (3.144) is

$$\Delta V = 2(4.442)/1.6713 = 5.316\ \text{km/s}$$

Equation (3.145) yields

$$\sin \Gamma = \frac{34.80}{4.442} \sin 5.5039 = 0.7514$$

$$\Gamma = 48.713\ \text{deg}$$

$$\delta = 180 - 48.713 - 5.5039 = 125.783\ \text{deg}$$

Using the law of cosines to calculate V_{sd} yields

$$V_{sd}^2 = (37.57)^2 + (5.316)^2 - 2(199.72)\cos(48.713 + 53.249)$$

$$V_{sd} = 39.02\ \text{km/s}$$

The spacecraft gained 1.45 km/s in the encounter.

The departure angle, from Eq. (3.147), is

$$\cos \alpha_d = \frac{(34.80)^2 + (39.02)^2 - (4.442)^2}{(2)(34.80)(39.02)}$$

$$\alpha_d = 2.156\ \text{deg}$$

3.5.9.1 Maximizing spacecraft velocity increase. The departure velocity of the spacecraft with respect to the sun is at a maximum when $V_{p/s}$ and $V_{\infty d}$ are collinear, as shown in Fig. 3.72. In this case, the velocity of the spacecraft at departure is the arithmetic sum of the velocity of the planet and V_∞.

The turning angle to maximize V_{sd} can be determined as follows:

$$\delta' = 180 - \phi \tag{3.148}$$

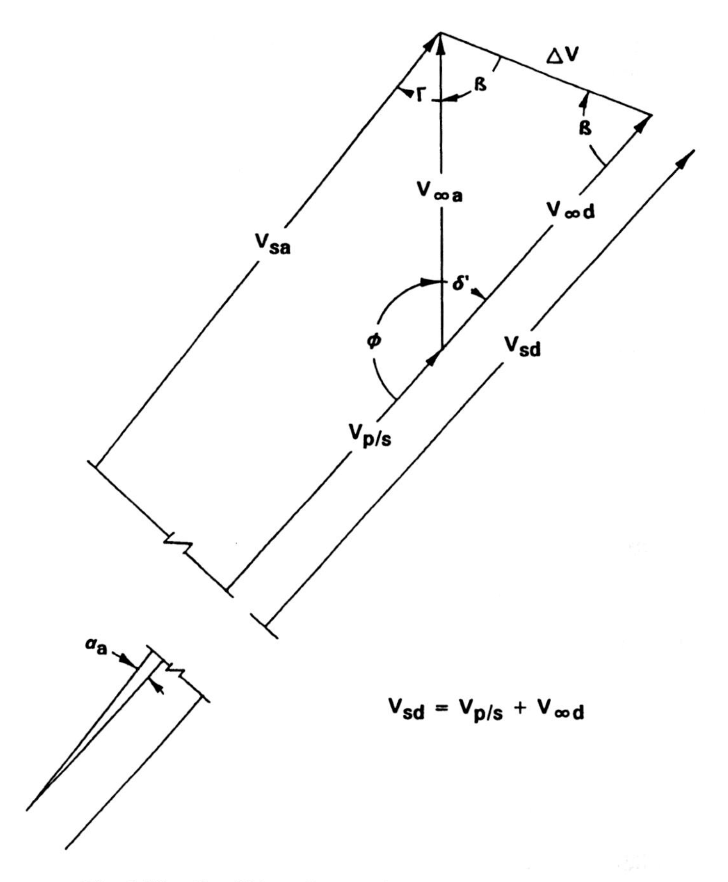

Fig. 3.72 Conditions for maximum departure velocity.

or

$$\beta' = \phi/2 \tag{3.149}$$

where δ' = the angle of turn that produces maximum spacecraft velocity at departure V_{sd}, and β' = the asymptote angle that produces maximum velocity at departure.

The maximum V_{sd} may not be achievable; the encounter hyperbola for β' may require a periapsis radius lower than the surface of the target planet.

3.5.9.2 Maximum angle of turn. It is sometimes desirable to achieve a given angle of turn rather than a velocity increase; the Ulysses mission offers an example of this situation. The maximum theoretical angle of turn is 180 deg, which occurs when $\beta = 0$ and ΔV is a maximum. From Eq. (3.143),

$$\Delta V_{\max} = 2V_{\infty} \tag{3.150}$$

The theoretical maximum angle of turn represents an elastic collision with the target planet. The largest practical angle of turn occurs with the closest acceptable approach to the target planet. For the Venus example, selecting a periapsis altitude of 400 km as the closest approach yields the following encounter hyperbola:

Asymptote angle:	$\beta = 44.07$ deg
Semiminor axis:	$b = 15,940$ km
Semimajor axis:	$a = 16,464$ km
Eccentricity:	$e = 1.392$
Periapsis velocity:	$V_p = 10.974$ km/s

The maximum practical turning angle for the Venus example is

$$\alpha_{\max} = 180 - (2)(44.05) = 91.9 \text{ deg}$$

The examples discussed so far show a velocity increase during the encounter; however, the maneuver can be designed to provide a velocity increase or decrease.

3.6 Lunar Trajectories

Lunar trajectories were the premier problem in mission design in the 1960s. In this section the characteristics of this trajectory and the techniques used will be summarized. A patched conic method will be discussed; however, the method is not as accurate for lunar trajectories as planetary ones because of the influence of both the Earth and sun. Accurate solutions must be accomplished by numerical analysis.

3.6.1 Motion of the Earth–Moon System

The Earth–moon system is unique; the two bodies are so close to the same mass that, had the moon been slightly larger, they would be the only known binary planet system. It is a common misconception that the moon revolves around the Earth. In fact, the Earth and moon revolve around a common center of mass that is 4671 km from the center of the Earth and 379,729 km from the center of the moon; see Fig. 3.73. One sidereal rotation about the common center takes 27.32 days.

Solar perturbations change the rotation period by as much as 7 h. The orbit is slightly elliptical with an eccentricity of 0.0549 and semimajor axis of 384,400 km. As a result the Earth–moon distance changes slightly with true anomaly. In addition the semimajor axis is increasing with time as the tides about Earth take energy from the moon orbit and slow its orbital velocity. Small changes in eccentricity occur with a period of 31.8 days; this effect is called *evection* and was observed by the Greeks 2000 years ago.[3]

The average orbit inclination, with respect to the ecliptic, is 5.145 ± 0.15 deg varying with a period of 173 days.[36] The inclination of the equatorial plane with the ecliptic is 23.45 deg and the equatorial plane is relatively stable with a period of 26,000 years. When the ascending node of the Earth–moon orbit is aligned with the vernal equinox, the inclination of the moon orbit with the equator is at a maximum of $23.45 + 5.145$ or 28.6 deg. Conversely, when the descending node is

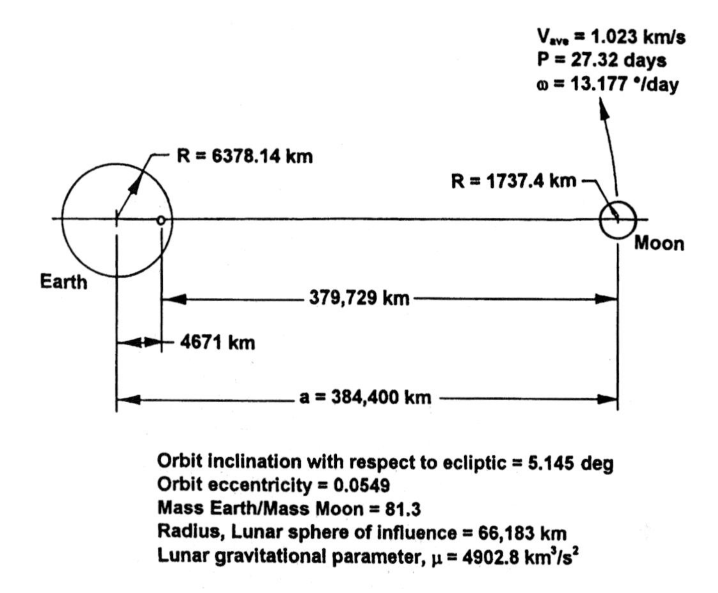

Fig. 3.73 Characteristics of the Earth–moon orbit.

at the equinox, the inclination of the moon orbit with the equator is 18.3 deg. The period of this variation is 18.6 years. Recall that the minimum inclination that can be achieved from ETR without a plane change is 28.5 deg; therefore, good launch years are on 18.6 year centers. It is not a coincidence that the moon launches in 1969 were during a good year.

3.6.2 Time of Flight and Injection Velocity

Time of flight was a very serious consideration in the Apollo manned missions because of the mass of provisions required to sustain life. Injection velocity is a serious consideration in any mission because the chosen launch vehicle imposes an absolute limit. Lunar time of flight and injection velocity can be bounded using a simplified case with the following assumptions:

1) The lunar orbit has a circular radius of 384,400 km.
2) The transfer ellipse is in the lunar orbit plane.
3) The gravitational effect of the moon is negligible.
4) The injection point is at the perigee of the transfer ellipse.

Unlike planetary launches, a lunar departure orbit is elliptical rather than hyperbolic. The minimum energy trajectory is an ellipse just tangent to the lunar orbit, orbit 1 in Fig. 3.74. Any less energetic orbit would not reach the lunar radius. The minimum energy trajectory has the longest possible transfer time and the lowest injection velocity. Assuming a transfer ellipse with perigee of 275 km, the nominal shuttle orbit, produces a minimum energy transfer ellipse with a time of flight of 119.5 h and an injection speed of 10.853 km/s.

Shorter flight times can be obtained by increasing the injection speed, orbits 2 and 3 in Fig. 3.74. Figure 3.75 shows the relation between time of flight and

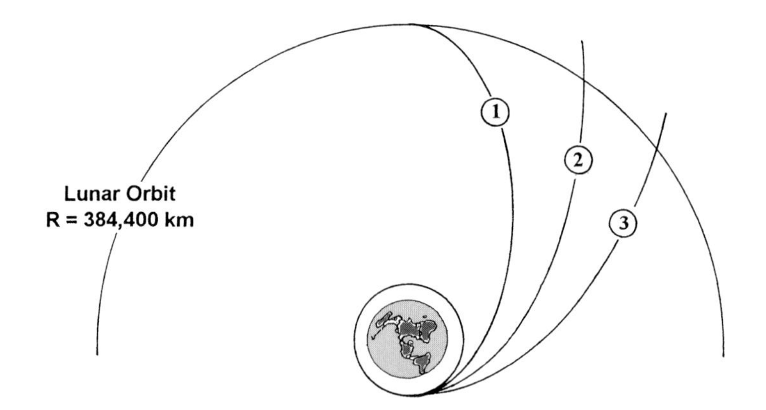

Fig. 3.74 Lunar trajectories.

injection speed for injection altitudes of 275 km. You can see that the 72-h time of flight used in the Apollo program is a reasonable compromise between time and launch vehicle energy. A curve similar to Fig. 3.76 can be constructed for any chosen injection altitude.

3.6.3 Sphere of Influence

In the patched conic analysis of a planetary mission, it was possible and accurate to assume that the sphere of influence was negligibly small compared with the transfer ellipse, and essentially to ignore it. That assumption is not accurate for a lunar trajectory; it is necessary to acknowledge the sphere of influence and make a trajectory patch at its boundary.

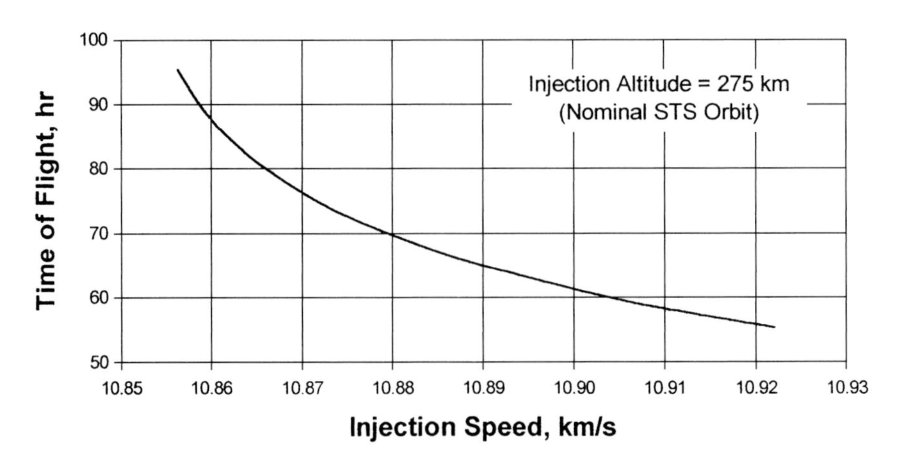

Fig. 3.75 Time of flight vs injection speed.

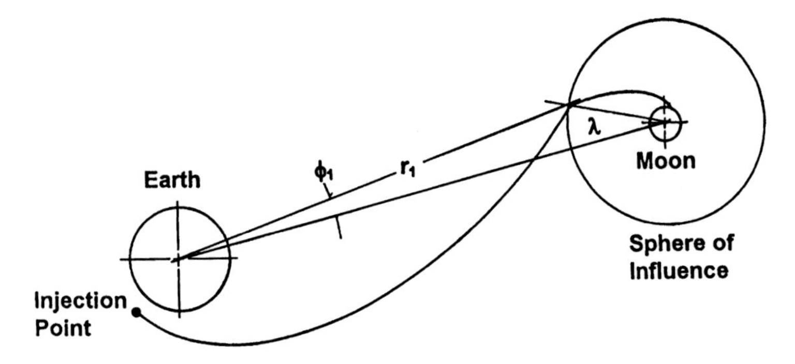

Fig. 3.76 Lunar patched conic.

The radius of the sphere of influence may be calculated by the Laplace method. From Eq. (3.125),

$$r_s = r_1 \left(\frac{M_{\text{moon}}}{M_{\text{Earth}}} \right)^{2/5} \tag{3.151}$$

where

r_s = radius of the lunar sphere of influence
r_1 = distance between centers of mass for Earth and moon, 384,400 km
$M_{\text{moon}}/M_{\text{Earth}}$ = ratio of mass for moon and Earth, 1/81.3

$$r_s = (384{,}400)(1/81.3)^{2/5} \tag{3.152}$$

$$r_s = 66{,}183\,\text{km} \tag{3.153}$$

3.6.4 Lunar Patched Conic

Using the patched conic method for a moon mission is not as accurate as it is for a planetary mission, primarily because of the influence of the sun on both bodies and the short distance between the two, compared with the sphere of influence. In the following analysis, we will assume that the lunar transfer orbit is coplanar with the lunar orbit. The mission is shown schematically in Fig. 3.76.

3.6.4.1 Designing a lunar mission. The procedure for designing a lunar mission is as follows:

1) Set initial conditions. To define the transfer ellipse, it is necessary to pick injection altitude (or radius), velocity, and flight path angle. (If injection is made at perigee the flight path angle is zero.) In addition, it is necessary to define the location of the arrival point at the sphere of influence; the most convenient method is to set the angle λ, as shown in Fig. 3.76.

2) Define the transfer ellipse given r, V, and γ at the point of injection using the energy/momentum technique described in Section 3.1.2 and Example 3.2. If the initial velocity is not high enough, the departure ellipse will not intersect the moon sphere of influence and a second set of initial conditions must be chosen.

Note that the departure trajectory is an ellipse rather than a hyperbola as it was in a planetary mission. A lunar mission can be done without reaching the escape velocity.

3) Find the radius to the sphere of influence, r_1 in Fig. 3.76, from trigonometry.

4) Given r_1, find time of flight to the sphere of influence boundary.

5) Define V_2, and γ_2 inside the sphere of influence at the arrival point. The radius is the radius of the sphere of influence, 66,183 km.

6) Given r_2, V_2, and γ_2 inside the sphere of influence, define the arrival orbit. (It is not reasonable to assume that the sphere of influence will be pierced at V_∞ on the arrival hyperbola as it is with a planetary mission.)

7) If the arrival orbit is satisfactory, find the launch day using the time of flight calculated in step 3 and average orbital velocity.

8) If the arrival orbit is not satisfactory (e.g., if the arrival hyperbola impacts the surface when a lunar orbit was desired), adjust initial conditions and start over at step 1.

Example 3.19 Lunar Patched Conic

Assume the lunar orbit is circular with radius 384,400 km and is coplanar with the transfer ellipse. Define a lunar trajectory with the following initial conditions:

Injection at perigee $\gamma_0 = 0$
Injection radius $r_0 = 6700$ km
Injection velocity $V_0 = 10.88$ km/s
Arrival angle $\lambda = 60$ deg

Using Eq. (3.8) for the stated initial conditions, the specific energy on the transfer ellipse is

$$\varepsilon = \frac{V_0^2}{2} - \frac{\mu}{r_0}$$

$$\varepsilon = \frac{(10.88)^2}{2} - \frac{398600.4}{6700} \tag{3.8}$$

$$\varepsilon = -0.305397 \text{ km}^2/\text{s}^2$$

The specific momentum is

$$\boldsymbol{H} = r_0 V_0 \cos \gamma_0$$

$$\boldsymbol{H} = (6700)(10.88) \cos (0) \tag{3.12}$$

$$\boldsymbol{H} = 72{,}896 \text{ km}^3/\text{s}$$

and from Eq. (3.10)

$$a = -\frac{\mu}{2\varepsilon}$$

$$a = -\frac{398600.4}{(2)(-0.305397)} = 652{,}594 \text{ km} \tag{3.10}$$

From Eq. (3.13),

$$e = \sqrt{1 - \frac{H^2}{\mu a}} \tag{3.13}$$

$$e = \sqrt{1 - \frac{(72896)^2}{(398600)(652594)}} = 0.98973$$

You can use Eq. (3.14) to assure yourself that the calculated semimajor axis and a periapsis velocity of 10.88 km/s are consistent.

Arrival conditions. Defining arrival as the point on the transfer ellipse at the intersection with the sphere of influence, the radius of the arrival point is r_1 in Fig. 3.76. The radius r_1 can be obtained from trigonometry, specifically the cosine law. The full solution of the triangle containing r_1 is shown in Fig. 3.77. The phase angle ϕ_1 is 9.2662 deg and r_1 is 355,953 km. Given r_1, the arrival velocity V_1, the flight path angle γ_1, the true anomaly θ_1, and the time of flight can be determined by evaluating parameters at a point. The technique is described in Section 3.1.3 and Example 3.5.

The parameters at a point evaluation yields $V_1 = 1.276$ km/s, $\gamma_1 = 80.766$ deg, $\theta_1 = 166.54$ deg, and time of flight = 49.752 h.

Defining the lunar orbit. The lunar orbit, inside the sphere of influence, is defined by the radius, velocity, and flight path angle. The average velocity of the moon about the Earth–moon center of mass is $V_m = 1.023$ km/s in a counterclockwise direction perpendicular to the Earth–moon radius. The arrival geometry is shown in Fig. 3.78; note that the angle between the known velocity vectors V_m and V_1 is $\gamma_1 - \phi_1$.

The spacecraft velocity, with respect to the moon V_2, can be obtained from the cosine law; the full solution of the vector diagram is shown in Fig. 3.79. You can see from Fig. 3.78, the arrival geometry, that the flight path angle associated with V_2 is

$$\gamma_2 = 180 - \lambda - \beta$$

$$\gamma_2 = 57.05 \text{ deg}$$

The orbital elements inside the sphere of influence are determined:

$$r_2 = 66{,}183 \text{ km}$$
$$V_2 = 1.359 \text{ km/s}$$
$$\gamma_2 = 57.05 \text{ deg}$$

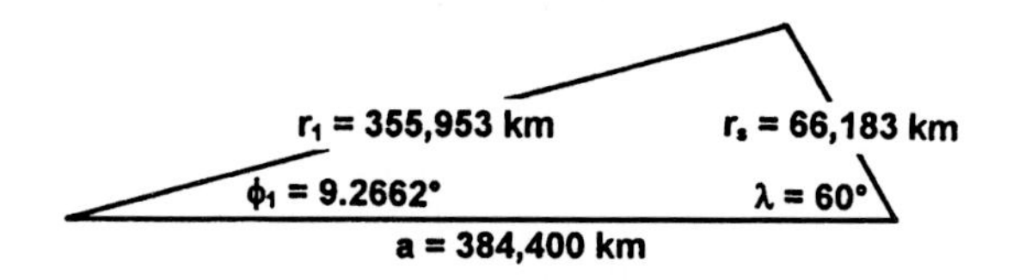

Fig. 3.77 Triangle solution for r_1.

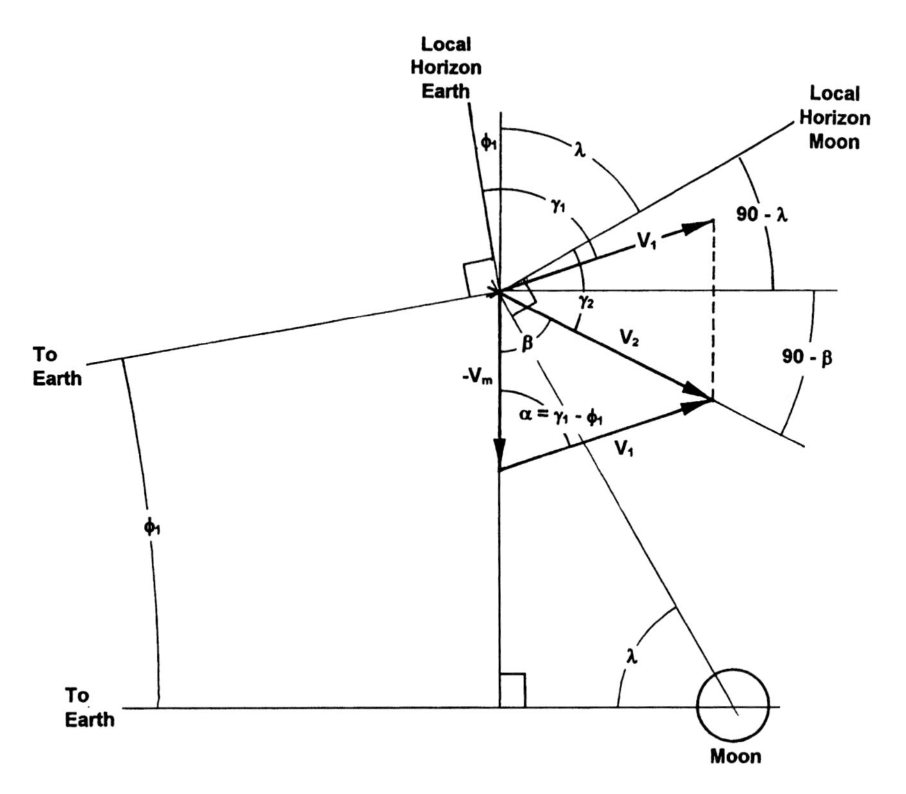

Fig. 3.78 Lunar arrival geometry.

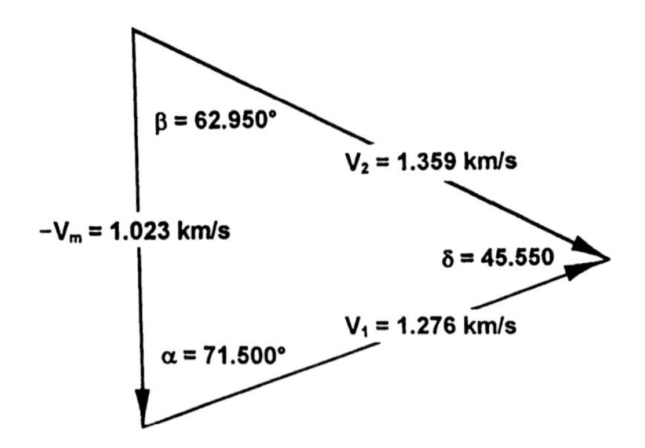

Fig. 3.79 Arrival vector diagram.

The lunar orbit can now be defined using the energy/momentum technique. (It is not adequate to assume that the spacecraft arrives at V_∞ on a hyperbolic orbit, as is done in planetary trajectories, because the sphere of influence is relatively small.) From Eq. (3.8), the specific energy of the lunar orbit is

$$\varepsilon = \frac{(1.359)^2}{2} - \frac{4902.8}{66183}$$

$$\varepsilon = 0.84936\,\text{km}^2/\text{s}^2$$

(You will recognize that μ for the moon is 4902.8.) A positive specific energy signals a hyperbolic orbit. Calculating specific momentum,

$$H = (66183)(1.359)\cos(57.05)$$

$$H = 48{,}920.5\,\text{km}^3/\text{s}$$

and from Eq. (3.10)

$$a = -\mu/2\varepsilon$$

$$a = -\frac{4902.8}{(2)(0.84936)} = -2886.2\,\text{km} \tag{3.10}$$

From Eq. (3.15),

$$e = \sqrt{1 - \frac{H^2}{\mu a}} \tag{3.15}$$

$$e = \sqrt{1 - \frac{(48920.5)^2}{(4902.8)(-2886.2)}} = 13.0432$$

A negative semimajor axis and an eccentricity larger than one both confirm the orbit as hyperbolic.

The resulting lunar orbit is the relatively flat hyperbola shown to scale in Fig. 3.80. The periapsis radius is 34,759 km and the time of flight from sphere of influence to periapsis is 11.59 h, making the total time of flight from injection to periapsis 61.34 h. Note that V_∞ is 1.3033 km/s, while velocity at the patch point is 1.39 km/s. The common planetary trajectory assumption that velocity at the patch point is V_∞ would have led to a serious error.

Recall the initial assumption that the trajectory was in the lunar orbit plane. If a noncoplanar trajectory is desired, the inclination of the transfer plane can be incorporated into the calculations using the methods described in Section 3.5.

3.6.4.2 Evaluation of the orbit. With the lunar trajectory elements in hand, evaluate the orbit against what is needed for the mission. For example, if a lunar

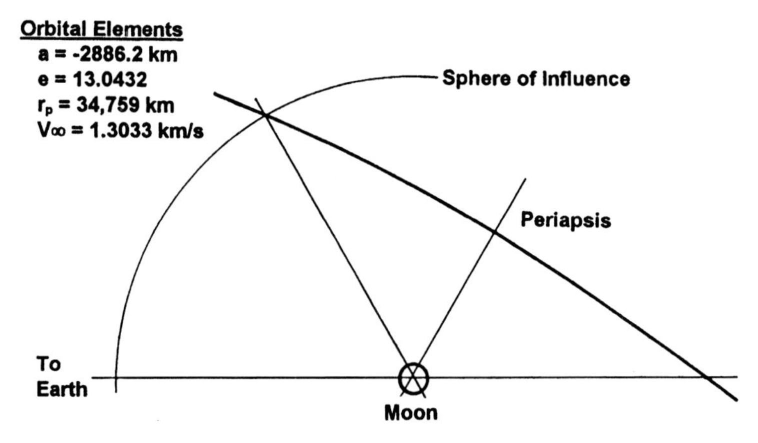

Fig. 3.80 Lunar orbit.

landing were desired, this is clearly the wrong orbit. If the orbit is not satisfactory, change the initial conditions, particularly λ, and recalculate.

3.6.4.3 Phasing. After a satisfactory lunar orbit is found, the phasing of the lunar position at injection can be determined. The time of flight from injection to the sphere of influence is 49.752 h. The average lunar angular speed is 13.177 deg per day; therefore, the moon, at the time of injection, must be 27.3 deg before its position at spacecraft arrival (see Fig. 3.81).

It is worth repeating that a patched conic analysis of a lunar trajectory is for preliminary design only. For accurate work, a numerical analysis is required.

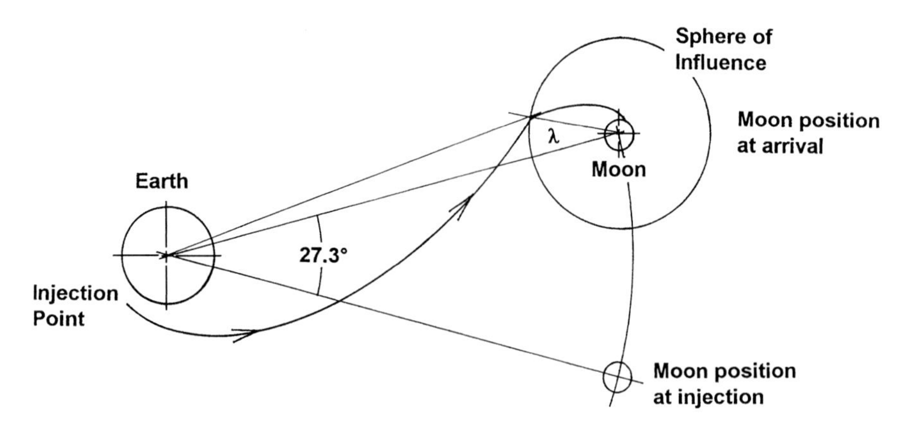

Fig. 3.81 Moon position at injection.

References

[1]Pasachoff, J. M., *Contemporary Astronomy*, Saunders, Philadelphia, PA, 1977.

[2]Newton, I., *Sir Isaac Newton's Mathematical Principles of Natural Philosophy and His System of the World*, translated by Andrew Motte, 1729. Revised translation by F. Cajori, Univ. of California Press, Berkeley, CA, 1934.

[3]Bate, R. R., Mueller, D. D., and White, J. E., *Fundamentals of Astrodynamics*, Dover, New York, 1971.

[4]Koelle, H. H. (ed.), *Handbook of Astronautical Engineering*, McGraw-Hill, New York, 1961.

[5]Wood, K. D., *Aerospace Vehicle Design*, Vol. II, Johnson, Boulder, CO, 1964.

[6]O'Neil, W. J., Rudd, R. P., Farless, D. L., Hildebrand, C. E., Mitchell, R. T., Rourke, K. H., and Euler, E. A., *Viking Navigation*, Jet Propulsion Lab., Pub. 78-38, California Inst. of Technology, Pasadena, CA, 1979.

[7]*Voyager at Neptune: 1989*, Jet Propulsion Lab., JPL 400-353, California Inst. of Technology, Pasadena, CA, 1989.

[8]Taff, L. G., *Computational Spherical Astronomy*, Wiley, New York, 1981.

[9]Wertz, J. R. (ed.), *Spacecraft Attitude Determination and Control*, D. Reidel, The Netherlands, 1978.

[10]Van Flandern, T. C., "Bits and Bytes," *Sky and Telescope*, Sky Publishing, Cambridge, MA, Aug. 1991, pp. 1–183.

[11]Wertz, J. R., and Larson, W. J. (eds.), *Space Mission Analysis and Design*, Kluwer Academic Press, Dordrecht, The Netherlands, 1991.

[12]*The Astronomical Almanac For The Year 1991*, U.S. Naval Observatory and The Royal Greenwich Observatory, U.S. Government Printing Office, Washington, DC, 1991.

[13]Hohmann, W., *Erreichbarkeit der Himmelskörper*, Munich, Germany, 1925.

[14]Chobotov, V. A. (ed.), *Orbital Mechanics*, AIAA Education Series, AIAA, Washington, DC, 1991, pp. 1–365.

[15]Griffin, M. D., and French, J. R., *Space Vehicle Design*, AIAA Education Series, AIAA, Washington, DC, 1991.

[16]Fortescue, P., and Stark, J., *Spacecraft Systems Engineering*, 2nd ed., Wiley, New York, 1995.

[17]Bachofer, B. T., *Landsat D Case Study in Spacecraft Design*, AIAA, Washington, DC, 1979.

[18]Easton, R. L., and Brescia, Continuously Visible Satellite Constellations, Naval Research Lab. Rept. 6896, 1969.

[19]Walker, J. G., "Some Circular Orbit Patterns Providing Continuous Whole Earth Coverage," *Journal of the British Interplanetary Society*, Vol. 24, 1971, pp. 369–384.

[20]Draim, J., "Three- and Four-Satellite Continuous-Coverage Constellations," *Journal of Guidance, Control, and Dynamics*, Vol. 8, No. 6, 1985, pp. 725–730.

[21]Clarke, A. C., "Extra-Terrestrial Relays," *Wireless World*, Vol. 51, No. 10, 1945, pp. 305–308.

[22]Agrawal, B. N., *Design of Geosynchronous Spacecraft*, Prentice–Hall, Englewood Cliffs, NJ, 1986.

[23]Thompson, T. D. (ed.), *1988 TRW Space Log*, TRW Space and Technology Group, Redondo Beach, CA, 1989.

[24]*Air & Space*, National Air and Space Museum, Washington, DC, April/May 1987, pp. 1–49.

[25]Kendrick, J. B. (ed.), *TRW Space Data*, TRW Systems Group, Redondo Beach, CA, 1967.

[26]Hunten, D. M., Colin, L., Donahue, T. M., and Moroz, V. I. (eds.), *Venus*, Univ. of Arizona Press, Tucson, AZ, 1983.

[27]Sergeyevsky, A. B., *Mission Design Data for Venus, Mars and Jupiter Through 1990*, Jet Propulsion Lab., TM 33-736, California Inst. of Technology, Pasadena, CA, 1975.

[28]*American Ephemeris and Nautical Almanac*, U.S. Government Printing Office, Washington, DC, published annually prior to 1981.

[29]Michaux, C. M., *Handbook of the Physical Properties of Mars*, NASA SP-3030, U.S. Government Printing Office, Washington, DC, 1967.

[30]*Planetary and Lunar Coordinates for the Years 1984–2000*, U.S. Naval Observatory and The Royal Greenwich Observatory, U.S. Government Printing Office, Washington, DC, 1983.

[31]Dunne, J. A., and Burgess, E., *The Voyage of Mariner 10*, NASA SP-424, U.S. Government Printing Office, Washington, DC, 1978.

[32]Neihoff, J., "Pathways to Mars: New Trajectory Opportunities," American Academy of Science, Paper 86-1782, July 1986.

[33]Aldren, E. E., "Cyclic Trajectory Concepts," Science Applications International Corp., Aerospace Systems Group, Hermosa Beach, CA, Oct. 1985.

[34]Hollister, W. M., "Periodic Orbits for Interplanetary Flight," *Journal of Spacecraft and Rockets*, Vol. 6, No. 4, 1969, pp. 366–369.

[35]Byrnes, D. V., Longuski, J. M., and Aldrin, E. E., "Cycler Orbit Between Earth and Mars," *Journal of Spacecraft and Rockets*, Vol. 30, No. 3, 1993, pp. 334–336.

[36]Lang, K. R., *Astrophysical Data: Planets and Stars*, Springer-Verlag, New York, 1992.

[37]Lyons, D. T., and Dallas, S. S., *Magellan Planetary Constants and Models Document*, Jet Propulsion Lab., PD 630-79, Rev. C, California Inst. of Technology, Pasadena, CA, 1988.

[38]*The Astronomical Almanac For The Year 1998*, U.S. Naval Observatory and The Royal Greenwich Observatory, U.S. Government Printing Office, Washington, DC, 1998.

[39]Boyce, J. M., and Maxwell, T., *Our Solar System—A Geological Snapshot*, NASA, Washington, DC, 1992.

[40]Fordyce, J., Kwok, J. H., Cutting, E., and Dallas, S. S., *Magellan Trajectory Characteristics Document*, Jet Propulsion Lab., PD 630-76, California Inst. of Technology, Pasadena, CA, 1988.

[41]"STS 30 Mission Chart," Defense Mapping Agency, St. Louis, MO, 1988.

[42]*The Magellan Final Mission Design*, Jet Propulsion Lab., JPL-D-2331, California Inst. of Technology, Pasadena, CA, 1986.

[43]Meeus, J., *Astronomical Algorithms*, Willmann–Bell, Richmond, VA, 1991.

[44]Isakowitz, S. J., Hopkins, J. P., Jr., and Hopkins, J. B., *International Reference Guide to Space Launch Systems*, Third Edition, AIAA, Washington, DC, 1999.

Problems

3.1 An Earth satellite is in an orbit with a perigee altitude of 400 km and an eccentricity of 0.6. Find the following:

(a) the perigee velocity

(b) the apogee radius

(c) the apogee velocity
(d) the orbit period
(e) the satellite velocity when its altitude is 3622 km
(f) the true anomaly at altitude 3622 km
(g) the flight path angle at altitude 3622 km

3.2 The LANDSAT C Earth resources satellite is in a near-polar, near-circular orbit with a perigee altitude of 917 km, an eccentricity of 0.00132, and an inclination of 89.1 deg. What are the apogee altitude, the orbit period, and the perigee velocity?
Solution: 936.3 km, 1.726 h, 7.397 km/s.

3.3 Two radar fixes on an unidentified Earth orbiter yield the following positions:

$$\text{Altitude} = 1545 \text{ km at a true anomaly of } 126 \text{ deg}$$
$$\text{Altitude} = 852 \text{ km at a true anomaly of } 58 \text{ deg}$$

What are the eccentricity, altitude of perigee, and semimajor axis of the spacecraft orbit?

3.4 The Magellan spacecraft was placed in an elliptical orbit around Venus with a periapsis altitude of 250 km and a period of 3.1 h. What is the apoapsis altitude?
Solution: h = 7810 km.

3.5 Consider an elliptical Earth orbit with a semimajor axis of 12,500 km and an eccentricity of 0.472. What is the time from periapsis passage to a position with a true anomaly of 198 deg?

3.6 A spacecraft is approaching Venus with $V_\infty = 10$ km/s and $b = 10{,}000$ km. What will be the periapsis radius at Venus?
Solution: 7266 km.

3.7 A hyperbolic Earth departure trajectory has a periapsis velocity of 15 km/s at an altitude of 300 km. Find the following:
(a) the hyperbolic excess velocity
(b) the radius when the true anomaly is 100 deg
(c) the velocity when the true anomaly is 100 deg
(d) the time from periapsis to a true anomaly of 100 deg

3.8 The Magellan approach hyperbola at Venus had the following elements:

$$a = 17{,}110 \text{ km}$$
$$e = 1.3690$$

The spacecraft was placed in a nearly polar, elliptical mapping orbit with the following elements:

$$a = 10{,}424.1 \text{ km}$$
$$e = 0.39433$$

If the two orbits were tangent at periapsis, what velocity change was required to establish the mapping orbit? Was it a velocity increase or decrease?

Solution: $\Delta V = -2.571$ km/s.

3.9 The Thor-Delta placed GEOS-A in a 500-km-altitude circular parking orbit. Design a Hohmann transfer to lift it to a 36,200-km-radius circular orbit. Define each velocity change, and find the time required for the transfer.

3.10 Determine the velocity change required to convert a direct, circular, Earth orbit with a radius of 15,000 km into a coplanar, direct, elliptical orbit with the following elements:

$$\text{Periapsis altitude} = 500 \text{ km}$$
$$\text{Apoapsis radius} = 22{,}000 \text{ km}$$

Would the velocity change have been different if the orbits had been about Mars?

Solution: $\Delta V = 1.361$ km/s.

3.11 Design a plane change for the following circular Earth orbit:

$$\text{Altitude} = 1000 \text{ km}$$
$$\text{Inclination} = 37 \text{ deg}$$
$$\text{Longitude of ascending node} = 30^\circ$$

which results in an inclination of 63 deg and a longitude of the ascending node of 90° west. What are the angle of the plane change and the change in velocity?

3.12 The orbit of the GOES-B spacecraft has the following elements:

$$\text{Inclination} = 28.8 \text{ deg}$$
$$\text{Eccentricity} = 0.732$$
$$\text{Period} = 10.6 \text{ h}$$

What are the rotation of apsides and regression of nodes?

Solution: Rotation of apsides = 0.5912 deg/day.

3.13 The encounter hyperbola of the Voyager spacecraft at Neptune had an eccentricity of 2.4586. What was the width of the horizon (swath width) as the spacecraft streaked over the north pole at a periapsis altitude of 4850 km?

3.14 Design a mission to place a spacecraft in a geosynchronous equatorial orbit from the Russian launch site at Plesetsk. How much velocity change is required? Compare the result with an ETR launch.

3.15 The mean daily motion of Mars is 0.5240 deg/day. Design a sun-synchronous orbit for Mars. Give altitude, inclination, and regression of nodes. The requirement is to match the mean daily motion within 0.01 deg.

3.16 What is the maximum radius for a circular sun-synchronous Earth orbit? Why?

3.17 The Voyager 2 grand tour of the outer planets started with launch on 20 August 1997, from ETR and ended with the encounter of Neptune on 24 August 1989. In transit there were gravity-assist maneuvers at Jupiter, Saturn, and Uranus. (The planetary alignment that allowed this mission occurs once every 176 years.) The mission time from Earth to Neptune was 12 years. How long would the mission have taken by way of a simple Hohmann transfer without the gravity-assist maneuvers?

Solution: 30.7 years.

3.18 The Voyager 2 encounter hyperbola at Neptune had the following elements:

$$\text{Eccentricity} = 2.4586$$

$$\text{Semimajor axis} = 20{,}649\ \text{km}$$

The periapsis was placed near the north pole.

What was the maximum relative velocity that the camera system had to deal with while the strikingly clear surface pictures were being taken?

3.19 Magellan approached Venus with a velocity at infinity of 4.357 km/s and a semiminor axis b of 16,061.4 km. What was the periapsis radius at Venus? What was the minimum velocity change required for capture by Venus?

Solution: Periapsis radius = 6356.7 km.

3.20 Design a type I mission to Mars for a launch on 22 March 2001 and arrival at Mars on 8 October 2001. The time of flight is the difference between these dates or 200 days. The ephemeris data of Earth at launch and Mars at arrival are shown in the following table:

Element	Earth	Mars
Longitude, deg	181.444	333.221
Eccentricity	0.0167084	0.0934025
Semimajor Axis, km	149,598,020	227,939,133
Inclination, deg	0	1.8497
Long. of Perihelion, deg	102.958	336.093
Long. of Ascending Node, deg	0	49.572
Radius, km	149,9059097	206,671,197
Velocity, km/s	29.892	26.4964
True Anomaly, deg	78.4854	357.128
Flight Path Angle, deg	0.93486	−0.24524

(a) Design the transfer ellipse; determine the longitude of the line of apsides, the eccentricity, and the semimajor axis.

(b) Calculate the inclination of the transfer ellipse with respect to the ecliptic plane.

(c) Calculate C3.

(d) Calculate the inclination of the transfer ellipse with respect to the Mars orbit plane.

(e) Calculate velocity at infinity on the arrival hyperbola.

(f) Calculate the elements of the arrival hyperbola for a periapsis altitude of 378 km.

(g) Calculate the velocity change required at periapsis of the approach hyperbola to establish a circular Mars orbit of radius 378 km.

3.21 Define a lunar trajectory using the patched conic technique assuming circular coplanar transfer. Calculate the elements of the transfer ellipse and the arrival hyperbola given the following:

$$\text{Injection velocity at perigee of transfer ellipse} = 10.738 \text{ km/s}$$
$$\text{Injection altitude} = 500 \text{ km}$$
$$\text{Arrival angle } \lambda = 30 \text{ deg}$$

What is the time of flight from injection to the arrival at the sphere of influence? Is this a lunar landing trajectory?

4
Propulsion

4.1 Introduction

Konastantin Eduardovich Tsiolkowski, a Russian mathematics professor, was the first to observe that rocket propulsion was a prerequisite for space exploration. As early as 1883, Tsiolkowski noted that gas expulsion could create thrust; thus, a rocket could operate in a vacuum. In 1903 he published a milestone paper describing how space flight could be accomplished with rockets. He advocated the use of liquid hydrogen and liquid oxygen as propellants. He described staged rockets and showed mathematically that space exploration would require staging. Every time you use Eq. (4.19), think of Tsiolkowski.

Robert H. Goddard, professor of physics at Clark University, also observed that space flight would require liquid rocket propulsion. With this goal he was the first to design, build, and fly a liquid rocket, a feat that required over 200 patented inventions and his entire life. He flew the first liquid rocket in 1926. Using liquid oxygen and gasoline, it flew for 2.5 s, reaching an altitude of 41 ft and a speed of 63 mph. He developed pump-fed engines, clustered stages, quick disconnects, pressurization systems, and gyro stabilization. By the time he died in 1945, he had developed every type of equipment that would be required for a vehicle like the Saturn V. His rockets had reached a size of 2200 N (500 lb) and had reached altitudes of 1.9 km (2 miles).

Goddard's work was largely forgotten in the United States. Germany however followed his work closely and vigorously developed propulsion systems as weapons. After World War II the German technology was brought to the United States (ironically) and Russia, where it was developed to its current state.

4.1.1 Units

A word about units: All of the development work done in the United States was done in the English system of units. The conversion to the SI units is most particularly trying in propulsion. As you probably know, this conversion difficulty contributed to the loss of a planetary spacecraft in 1999. Table 4.1 summarizes the most frequently used conversion factors. In this section SI units will be shown first and English units following in parentheses.

4.1.2 Arrangement of Chapter 4

The fundamentals of propulsion are described in Section 4.2. Section 4.3 describes how the performance requirements of a propulsion system are derived. Sections 4.4–4.7 describe the considerations peculiar to each of the most common types of spacecraft propulsion systems. Monopropellant systems are described first. Equipment common to each system (for example, tankage) are discussed in

Table 4.1 Unit conversions and constants

Parameter or constant	English unit	Multiply by	To get SI unit
Acceleration	feet/second^2	0.3048	meters/second^2
Density	pounds mass/cubic feet	16.01846	kilograms/cubic meter
Distance	statute miles	1.609347	kilometers
Mass, m	pounds mass	0.4535924	kilograms
Pressure, P	pounds/square inch	6894.757	Pascal (N/m^2)
Specific impulse, I_{sp}	seconds or lb_f-s/lb_m	9.80665	Newton-second/ kilogram
Thrust, F	pound, force	4.448222	Newton (kg-m/s^2)
Total impulse, I	pound-seconds	4.448222	Newton-second
Velocity	feet/second	0.3048	meters/second
Gravitational acceleration	32.174 ft/s	9.80665	meters/second
Universal gas constant, R_u	1545 ft-lb_f/lb_m-mole °R	—	8314 joule/kg-mole °K
Mechanical equivalent heat, J	778 ft-lb/Btu	—	4190 calories/joule
Standard atmosphere	14.696 psia	—	1.01325E5 Pascal

Section 4.4 and referenced thereafter. Propulsion reference data, propellant properties for example, are included in Appendix B.

4.2 Theoretical Rocket Performance

A rocket generates thrust by accelerating a high-pressure gas to supersonic velocities in a converging-diverging nozzle. In most cases the high-pressure gas is generated by high-temperature combustion of propellants. As shown in Fig. 4.1, a rocket, of any type, consists of a combustion chamber, throat, and nozzle.

In a bipropellant rocket engine the gases are generated by the rapid combustion of a liquid oxidizer and liquid fuel in the combustion chamber, for example, liquid hydrogen, and liquid oxygen. In a monopropellant system only one propellant is used. High-pressure, high-temperature gases are generated by decomposition of a single propellant. Hydrazine is the most common monopropellant. In a solid system a solid fuel and oxidizer are mechanically mixed and cast as a solid-propellant grain. The grain occupies most of the volume of the combustion chamber. In a cold-gas system there is no combustion involved. A gas, like helium, is stored at high pressure and injected into the chamber without combustion.

4.2.1 Thrust

Rocket thrust is generated by momentum exchange between the exhaust and the vehicle and by the pressure imbalance at the nozzle exit. The thrust caused by

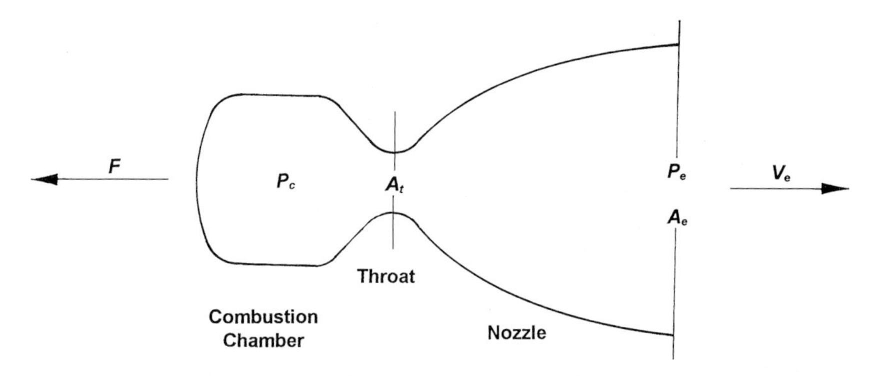

Fig. 4.1 Rocket nozzle.

momentum exchange can be derived from Newton's second law:

$$F_m = ma \tag{4.1}$$

$$F_m = \dot{m}_p(V_e - V_0) \tag{4.2}$$

where F_m is the thrust generated as a result of momentum exchange, N, (lb); a the acceleration, m/s^2 (ft/s^2); $\dot{m}_p$ the mass flow rate of propellants flowing into the chamber or mass of the exhaust gas, kg/s (lbm/s); V_e the average velocity of the exhaust gas, m/s (fps); and V_0 the initial velocity of the gases, m/s (fps). Because the initial velocity of the gases is essentially zero. Equations (4.1) and (4.2) assume that the velocity of the exhaust gas acts along the nozzle centerline.

Example 4.1 Thrust from Momentum

A man is standing in a row boat throwing bricks out the back. The bricks weigh 1 kg each, and he can throw them at a rate of six bricks per minute with a consistent speed of 10 m/s. If this process is frictionless, how much thrust does he generate? From Eq. (4.2)

$$F = \frac{6}{60}10 = 1\,\text{kg-m/s}^2 = 1\,\text{N}$$

In addition to the thrust caused by momentum, thrust is generated by a pressure-area term at the nozzle exit. If the nozzle were exhausting into a vacuum, the pressure-area thrust would be

$$F_p = P_e A_e \tag{4.3}$$

If the ambient pressure is not zero,

$$F_p = P_e A_e - P_a A_e \tag{4.4}$$

$$F_p = (P_e - P_a)A_e \tag{4.5}$$

where F_p is the thrust from exit plane pressure, N (lbf); P_e the static pressure in the exhaust gas, Pa (psia); A_e the area of the nozzle exit when hot, m^2 (in.^{2}); and

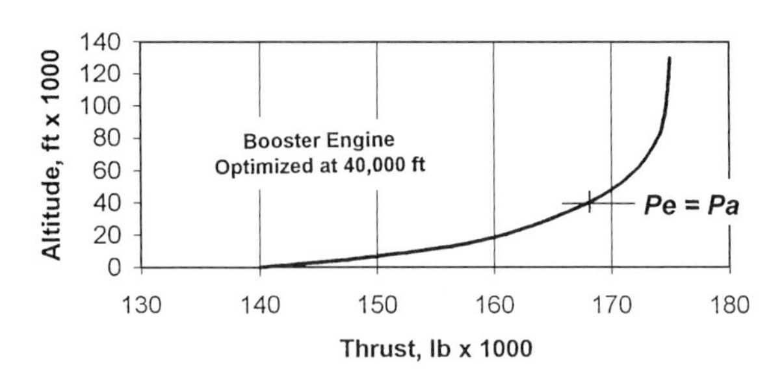

Fig. 4.2 Thrust vs altitude.

P_a the ambient static pressure, Pa (psia). Because the total thrust on the vehicle is the sum of the thrust caused by momentum exchange and thrust caused by exit plane pressure, the fundamental relationship for thrust becomes

$$F = \dot{m}V_e + (P_e - P_a)A_e \tag{4.6}$$

For any given engine, thrust is greater in a vacuum than at sea level by an amount equal to $P_e a_e$. A rocket is the only engine that will operate in a vacuum, and it is also substantially more efficient in a vacuum, as shown in Fig. 4.2.

For engines that must operate in atmosphere, for example, launch vehicle engines, the diverging section of the nozzle is designed so that $Pe = Pa$ at design point. This is called *optimum expansion* and provides best performance for launch vehicle engines. For optimum expansion to occur in a vacuum engine would require an infinite area ratio and an infinitely long nozzle.

4.2.2 Ideal Rocket Thermodynamics

By somewhat idealizing the flow in a rocket engine, thermodynamics can be used to predict rocket performance parameters to within a few percent of measured values. The following seven assumptions define what is known as theoretical performance:

1) The propellant gases are homogeneous and invariant in composition throughout the nozzle. This condition requires good mixing and rapid completion of combustion and, in the case of solids, homogeneous grain. Good design will provide these conditions.

2) The propellant gases follow the perfect gas laws. The high temperature of rocket exhaust is above vapor conditions, and these gases approach perfect gas behavior.

3) There is no friction at the nozzle walls, therefore, no boundary layer.

4) There is no heat transfer across the nozzle wall.

5) Flow is steady and constant.

6) All gases leave the engine with an axial velocity.

7) The gas velocity is uniform across any section normal to the nozzle axis.

Assumptions 3, 4, 6, and 7 permit the use of one-dimensional isentropic expansion relations. Assumption 1 defines what is called *frozen equilibrium conditions*. The gas composition can be allowed to vary from section to section in what is called a *shifting equilibrium calculation*. The shifting equilibrium calculation accounts for exothermic recombinations, which occur in the exhaust stream; higher propellant performance results. With these assumptions theoretical rocket engine performance parameters can be calculated, as well as measured, and the results compared.

4.2.2.1 Specific impulse I_{sp}. Specific impulse describes how much thrust is delivered by an engine per unit propellant mass flow rate. It is the premier measurement of rocket performance. The utility of the parameter was recognized at least as early as Goddard's work. During the postwar development of launch vehicles in the United States, it was common to assume that an extra second of specific impulse was worth a million dollars in engine development. I_{sp} is the conventional method of comparing propellants, propellant combinations, and the efficiency of rocket engines. In the English system of units it is

$$I_{sp} = \frac{F}{\dot{m}} \tag{4.7}$$

where I_{sp} is the specific impulse, s; F the thrust, lbf; and $\dot{m}$ the propellant mass flow rate, lbm/s.

In the English system, pounds mass and pounds force are allowed to cancel, and I_{sp} is measured, conveniently, in seconds. In very careful work you will sometimes see lbf-s/lbm. In the SI system mass is in kilograms, and force is in newtons or kilograms-meters/second. In equations with SI units, it is necessary to multiply I_{sp} in seconds times the SI gravitational constant, 9.80665. Almost all published material, even British material, still reports I_{sp} in seconds, and this chapter will follow that custom. However, the student needs to be careful in the use of I_{sp} and note when it should be multiplied by g_c. With force and flow rate in SI units and I_{sp} in seconds, Eq. (4.7) becomes

$$I_{sp} = \frac{F}{\dot{m} g_c} \tag{4.7a}$$

The thermodynamic expression for theoretical vacuum specific impulse in English units is[1,2]

$$I_{sp} = \sqrt{\frac{2kRT_c}{g_c(k-1)}\left[1-\left(\frac{P_e}{P_c}\right)^{\frac{k-1}{k}}\right]} + \frac{P_e A_e}{P_c A_t}\sqrt{\frac{RT_c}{kg_c(2/k+1)^{\frac{k+1}{k-1}}}} \tag{4.8}$$

where

A_{ge} = exit area of the nozzle, ft^2
A_t = throat area of the nozzle, ft^2
k = ratio of specific heats of the gas, C_p/C_v
Pe = absolute pressure of the gas at the exit plane (psia)
Pa = ambient absolute pressure, Pa (psia)
T = absolute, total temperature of the gas, °R
R = specific gas constant 1545 ft-lb_f/lb_m mole °R
M = molecular weight of gas, lb

Adjusting Eq. (4.8) for SI units produces

$$I_{sp}g_c = \sqrt{\frac{2kRT_c}{(k-1)}\left[1-\left(\frac{P_e}{P_c}\right)^{\frac{k-1}{k}}\right]} + \frac{P_eA_e}{P_cA_t}\sqrt{\frac{RT_c}{kg_c(2/k+1)^{\frac{k+1}{k-1}}}} \tag{4.9}$$

Equations (4.8) and (4.9) show that I_{sp} is a thermodynamic property of the propellants, approximately proportional to $\sqrt{T_c/M}$. The left term in Eq. (4.9) is dominant. The effect of chamber pressure is second order; therefore, spacecraft engines can and do have low chamber pressures compared to their sea-level counterparts.

To fully define specific impulse, it is necessary to state 1) ambient pressure, 2) chamber pressure, 3) area ratio, 4) shifting or frozen equilibrium conditions, and 5) real or theoretical. Specific impulse calculated from Eqs. (4.8) and (4.9) is theoretical, frozen equilibrium and vacuum conditions; chamber pressure and area ratio can be anything you wish. If gas composition is allowed to vary in the nozzle, shifting equilibrium conditions prevail, and a slightly higher I_{sp} results. A real engine can be expected to have a specific impulse about 93% of theoretical.

Figure 4.3 compares the theoretical, vacuum, frozen equilibrium I_{sp} of nine propellant combinations at a chamber pressure of 500 psia and an area ratio of 50. Fluorine is the most energetic oxidizer ever tested; however, it is very difficult to contain because it reacts vigorously with tank materials. The use of fluorine never progressed beyond research. Liquid oxygen and liquid hydrogen are the Space Shuttle propellants. They are difficult to handle because they are *cryogenics*, boiling point 90°K and 20°K respectively, but the I_{sp} obtainable makes them the best performing practical propellants. The nitrogen tetroxide propellant combinations shown are called *storable* because they are liquids at room temperatures. These propellants are used on the Titan family of launch vehicles and on spacecraft. Monopropellant hydrazine is another popular spacecraft propellant because it is a monopropellant; it can be spontaneously decomposed, without an oxidizer, to yield hot propulsion gases. Thermodynamic properties of combustion gases from

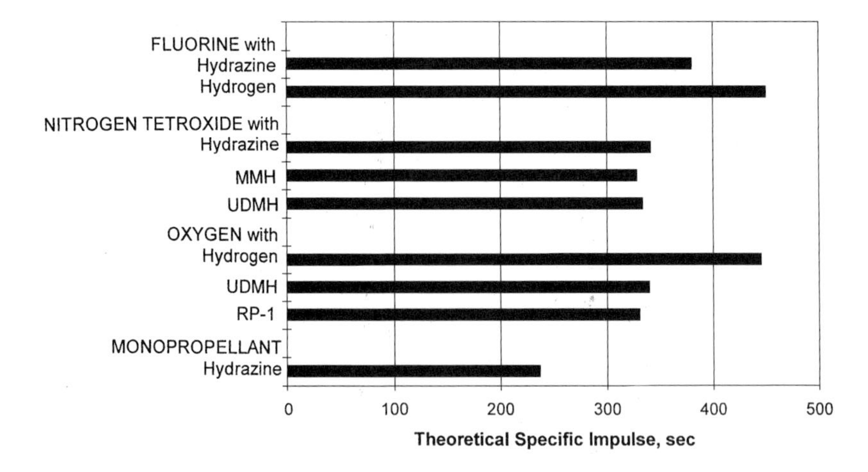

Fig. 4.3 Theoretical specific impulse of some propellants.

Table 4.2 Thermodynamic properties for some propellants[a]

Oxidizer	Fuel	T_c °K	T_c °R	M	k
Fluorine	Hydrazine	4660	8390	19.4	1.33
	Hydrogen	3960	7130	11.8	1.33
Nitrogen tetroxide	Hydrazine	3280	5910	18.9	1.26
	MON	3145	6145	21.5	1.25
	UDMH	3435	6185	21.0	1.25
Oxygen	Hydrogen	2985	5375	10.0	1.26
	UDMH	3615	6505	21.3	1.25
	RP-1	3670	6605	23.3	1.24
Monoprop	Hydrazine	1750	2210	13	1.27

[a]Data in part from Ref. 3, pp. 20–21.

some propellant combinations are shown in Table 4.2. Note in Table 4.2 the high-combustion gas temperatures for bipropellant combinations. It is clear why cooling is a key problem in rocket engine design from Goddard's time to today. One of the advantages of monopropellants is that radiative cooling techniques can be used.

4.2.2.2 Area ratio. The area ratio of a rocket engine is the ratio of the exit area to the throat area,

$$e = A_e/A_t \tag{4.10}$$

where e is the area ratio; A_e the exit area measured hot; and A_t the throat area measured hot. Area ratio is a measure of the gas expansion provided by an engine. Optimum area ratio provides an exit-plane pressure equal to local ambient pressure. For a sea-level or first-stage engine the optimum area ratio is selected near midpoint of the flight; an area ratio around 12 is common. For a spacecraft engine (or an upper-stage engine) the optimum area ratio is infinite; the largest area ratio allowed by space and weight is used. An area ratio in the range 50 to 100 is common for spacecraft engines.

4.2.2.3 Mass flow rate. The mass flow rate of gas through a supersonic, isentropic nozzle is[1,2]

$$\dot{m}_p = \frac{P_c A_t k}{\sqrt{kRT_c}} \sqrt{\left(\frac{2}{k+1}\right)^{\frac{k+1}{k-1}}} \tag{4.11}$$

Equation (4.11) shows that for a given propellant and stagnation temperature the flow rate through a nozzle is proportional to chamber pressure and throat area and those parameters only.

4.2.2.4 Thrust coefficient C_f. Thrust coefficient is a useful term which first arose during rocket engine testing. It is the proportionality constant between thrust and the product of chamber pressure and throat area:

$$F = P_c A_t C_f \tag{4.12}$$

where P_c is the chamber pressure, Pa (psia); A_t the area of the nozzle throat when hot, m^2 (in.^{2}); and C_f the thrust coefficient. An engine without a diverging section would have a thrust coefficient of 1; thus, the coefficient is always larger than 1. The improvement in thrust provided by the nozzle is characterized by thrust coefficient. From thermodynamics theoretical C_f is

$$C_f = \sqrt{\frac{2k^2}{k-1}\left(\frac{2}{k+1}\right)^{\frac{k+1}{k-1}}\left[1-\left(\frac{P_e}{P_c}\right)^{\frac{k-1}{k}}\right]} + \left(\frac{P_e - P_a}{P_c}\right)\frac{A_e}{A_t} \tag{4.13}$$

Figure 4.5 shows theoretical C_f as a function of k and area ratio for vacuum conditions. In practice, about 98% of theoretical C_f is achieved in steady-state performance.

4.2.2.5 Total impulse I. Impulse is defined as a change in momentum caused by a force acting over time, for constant thrust,

$$I = Ft \tag{4.14}$$

where I is the total impulse delivered to the spacecraft, N-s (lbf-s); F the thrust, N (lbf); and t the burn time (time of application of F), s.

Equation (4.14) assumes that thrust is a constant with time, which is usually true for liquid engines. For cases where thrust is not constant, as in solid rockets, total impulse can be computed by integrating for the area under the thrust-time curve or from Eq. (4.15), which is independent of thrust.

$$I = w_p g_c I_{\text{sp}} \tag{4.15}$$

where w_p is the total weight of propellant consumed in kilograms and g_c is the gravitational constant. Propulsion system size is rated based on total impulse; this is a particularly useful measure of solid rocket motor systems, although it is used for all types of propulsion systems.

4.2.2.6 Mixture ratio (MR). Mixture ratio is an important parameter for bipropellant systems. It is the ratio of oxidizer to fuel flow rate, on a weight basis,

$$MR = \frac{\dot{w}_o}{\dot{w}_f} \tag{4.16}$$

where $\dot{w}_o$ is the oxidizer mass flow rate, kg/s (lb/s) and $\dot{w}_f$ the fuel mass flow rate, kg/s (lb/s). Volumetric mixture ratio is sometimes used in conjunction with tank sizing; mixture ratio can be converted to volumetric mixture ratio as follows:

$$VMR = MR\frac{\rho_f}{\rho_o} \tag{4.17}$$

where ρ_f is the density of the fuel, kg/m^3 (lb/ft^3) and ρ_o the density of the oxidizer, kg/m^3 (lb/ft^3).

The volumetric mixture ratio, consumed by the engine, has a major effect on system design because it determines the relative sizes of the propellant tanks. In a spacecraft design the loaded mixture ratio is sometimes designed off optimum in order to have both oxidizer and fuel tanks the same volume. The I_{sp} loss is slight, and two identical tanks are cheaper than two individually sized tanks. (By convention, mixture ratio is always by mass unless volume is specified.)

4.2.2.7 Bulk density. It is often convenient to use the bulk density of a propellant combination to expedite approximate calculation. Bulk density is the mass of a unit volume of a propellant combination "mixed" at the appropriate mixture ratio. Bulk density can be computed as follows:

$$\rho_b = \frac{MR + 1}{MR/\rho_o + 1/\rho_f} \tag{4.18}$$

where ρ_b is the bulk density of a propellant combination in the same units as ρ_o and ρ_f, usually in kilograms/cubic meters.

4.3 Propulsion Requirements

There are three major activities involved in setting the requirements for a spacecraft propulsion system. The first decision, and the one which has the most far reaching impact, is the selection of propulsion system type. This choice is described in Section 4.3.1. Secondarily, the translational and rotational thrust levels and propellant allocations are determined. These activities are discussed in Sections 4.3.2 and 4.3.3.

4.3.1 Propulsion System Types

Current propulsion technology provides five basically different, widely used, propulsion choices: cold-gas systems, solid motor systems, monopropellant systems, bipropellant systems, and dual-mode systems; each has its niche in spacecraft design. The selection of propulsion system type has substantial impact on the total spacecraft and is a key selection in early design.

There are two other system types, which are feasible and are beginning to be used. These are the following: ion propulsion, which received an exhaustive flight demonstration on Deep Space 1, and Hall effect thrusters, which were extensively used by the former USSR and are now being used on U.S. spacecraft. Both of these new systems are for low thrust applications.

4.3.1.1 Cold-gas systems. Almost all of the spacecraft of the 1960s used this system (see Fig. 4.4). It is the simplest choice and the least expensive. Cold-gas systems can provide multiple restarts and pulsing. The major disadvantage of the system is low specific impulse (about 40 s) and low thrust levels (less than a newton) with resultant high weight for all but low total impulse missions.

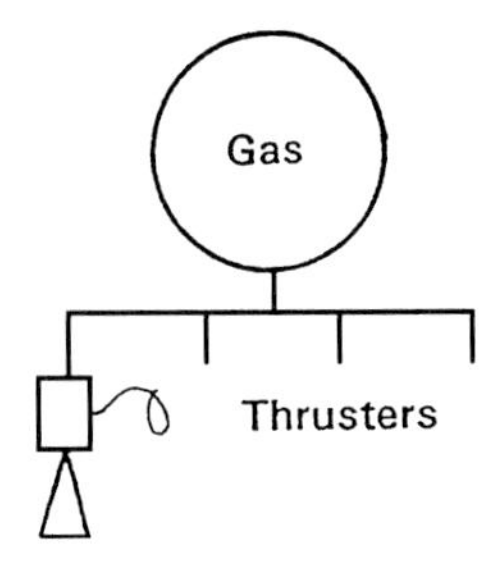

Fig. 4.4 Cold-gas system.

4.3.1.2 Monopropellant systems. Figure 4.5 is the next step up in complexity and cost. Monopropellant systems can supply pulsing or steady-state thrust. The specific impulse is about 225 s, and thrusts from 0.5 to several hundred newtons are available. The system is a common choice for attitude control and midrange impulse requirements. The only monopropellant in flight use is hydrazine (N_2H_4).

4.3.1.3 Bipropellant systems. Figure 4.6 is the top of the complexity and expense scale; however, it is a very versatile and high performance system. Bipropellants provide an I_{sp} of about 310 s and a wide range of thrust capability (from a few pounds to many thousands of pounds). They can be used in pulsing or steady-state modes. The most common propellants for spacecraft use are nitrogen tetroxide and monomethyl hydrazine. Bipropellant systems are not as common as monopropellants because high total impulse with restart is not usually required.

4.3.1.4 Dual-mode systems. Figure 4.7 shows the dual-mode system, a new arrangement specifically designed for situations that require high-impulse burns in addition to low-impulse attitude control pulses. High-performance high-impulse burns are provided by a bipropellant nitrogen tetroxide/hydrazine system. The bipropellant fuel, hydrazine, is also used as a monopropellant for low-impulse attitude control pulses. This system type is ideal for geosynchronous satellites, which require a final high-impulse burn to achieve geosynchronous orbit and require low-impulse attitude control and stationkeeping. Planetary spacecraft,

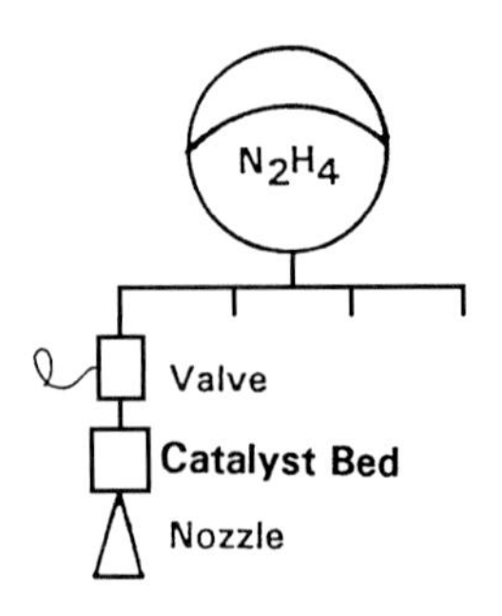

Fig. 4.5 Monopropellant system.

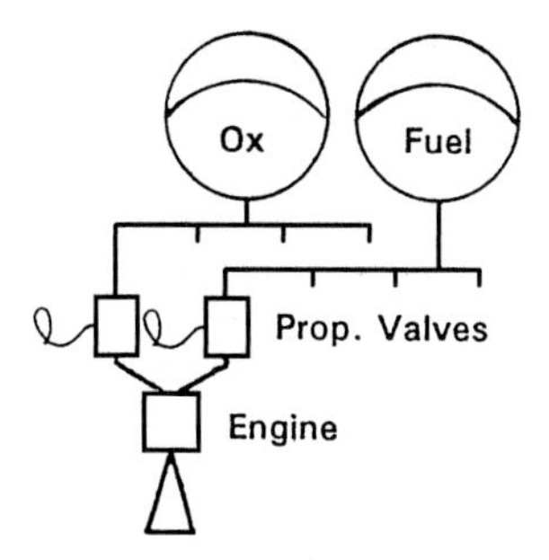

Fig. 4.6 Bipropellant system.

which require an orbit insertion burn as well as attitude control, is another mission for which this system has advantages.

4.3.1.5 Solid motor systems. Solid motor systems (see Fig. 4.8) are candidates when all of the impulse is to be delivered in a single burn and the impulse can be accurately calculated in advance (shutdown/restart is not current state of the art for solids). Examples of situations of this type are planetary orbit insertion and geosynchronous apogee burns (apogee kick motors). If the single burn criteria are met, solids provide simplicity and reasonable performance (specific impulse about 290 s).

Table 4.3 compares the capabilities of each type of propulsion system and the missions for which they are suited. The impulse ranges and specific impulses listed in Table 4.3 are approximate. Solid shutdown-on-command has been demonstrated as has bipropellant pulsing. Neither process is common in spacecraft design.

As you can see from Table 4.3, monopropellants offer wide versatility at medium performance, which explains their wide use. Bipropellant systems offer the same

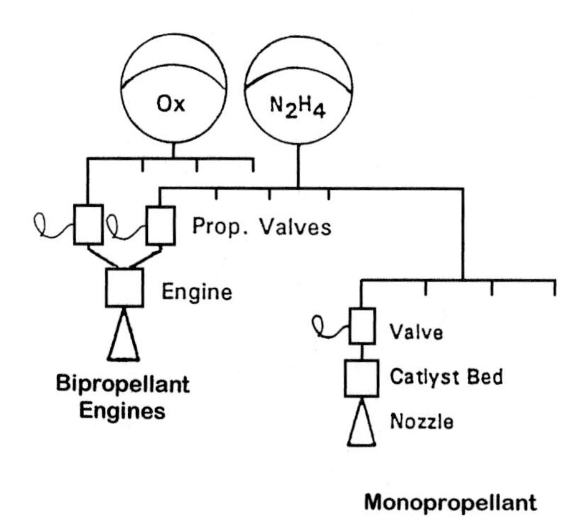

Fig. 4.7 Dual-mode system.

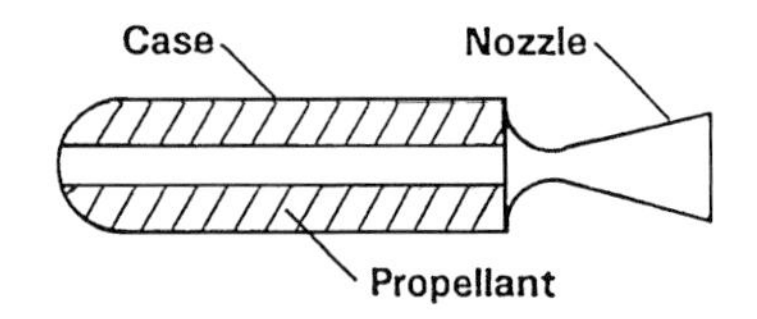

Fig. 4.8 Solid motor.

flexibility at high performance and highest complexity and cost. Planetary missions are now using dual-mode systems to provide high-performance bipropellant translation burns for tasks like orbit insertion and monopropellants for rotational attitude control. These systems use hydrazine as a monopropellant and as a bipropellant fuel with nitrogen tetroxide oxidizer.

Solid systems are relatively inflexible; their greatest utility is for situations in which the required impulse is accurately known years before launch, at the time the motor is poured. They give good performance at reduced complexity. The motor itself is simple; however, in addition to the motor you must provide thrust vector control, ignition, and safe/arm.

Cold-gas systems are a clear choice for low performance, low total impulse, and inexpensive simplicity.

The propulsion system also includes any pyrotechnic devices used by the spacecraft. A typical list of pyrotechnic functions is as follows: 1) upper-stage separation; 2) release of deployable equipment such as solar panels, antennas, scan platforms; 3) solid rocket motor ignition; 4) solid motor jettison; and 5) pyro valve operation.

Propulsion is the only method of applying an external force to a spacecraft. Aside from the launch of a spacecraft, which requires a great deal of propulsion, there are two functions that require propulsion. These are as follows: 1) translational velocity changes, typically to change orbits (the source of these requirements is from "Orbital Mechanics," Chapter 3); and 2) rotational velocity changes for rotational maneuvers, typically to change rotation rate, attitude, or pointing direction. (The source of these requirements is the attitude control system; see Chapter 5.)

Table 4.3 Mission suitability matrix

Requirement	Cold gas	Monopropellant	Bipropellant	Dual mode	Solid
Specific impulse, s	<150	230	310	310 (SS) 230 (pulse)	<300
Impulse range, N-s (lb-s)	<2500 (<500)	<45,000 (<10,000)	>45,000 (>10,000)	>45,000 (>10,000) +pulsing	>45,000 (>10,000)
Restart	Yes	Yes	Yes	Yes	No
Pulsing	Yes	Yes	No	Yes	No
Command shutdown	Yes	Yes	Yes	Yes	No

Table 4.4 Spacecraft propulsion functions

Task	Description
Translational velocity change	(Usually for orbit changes)
Orbit changes	Convert one orbit to another
Plane changes	Rotate the orbit plane
Orbit trim	Remove launch vehicle errors
Stationkeeping	Maintain constellation position
Repositioning	Change spacecraft position
Rotational velocity change	
Thrust vector control	Remove vector errors
Attitude control	Maintain an attitude
Attitude changes	Change attitudes
Reaction wheel unloading	Remove stored momentum
Maneuvering	Repositioning the spacecraft axes

The character of these two situations is quite different. Orbit changes are relatively high total impulse and relatively high thrust maneuvers. Attitude control maneuvers are low total impulse, low thrust maneuvers called pulsing.

Table 4.4 shows the two basic types of tasks performed by spacecraft propulsion and maneuvers of each type.

4.3.2 Translational Velocity Change

One of the main products of the mission design is a statement of the ΔV required for the maneuvers required for the mission. To convert these velocity change requirements to propellant requirements, we need the Tsiolkowski equation and its corollaries. Tsiolkowski published these equations in 1895, before the invention of the automobile or the airplane, much less the invention of spacecraft:

$$\Delta V = g_c I_{\text{sp}} \ln\left(\frac{m_i}{m_f}\right) \tag{4.19}$$

$$m_p = m_i\left[1 - \exp\left(-\frac{\Delta V}{g_c I_{\text{sp}}}\right)\right] \tag{4.20}$$

$$m_p = m_f\left[\exp\left(\frac{\Delta V}{g_c I_{\text{sp}}}\right) - 1\right] \tag{4.21}$$

where

m_i = initial vehicle mass, kg (lb)
m_f = final vehicle mass, kg (lb)
$m_p = m_i - m_f$ = propellant mass consumed to produce the given ΔV, kg (lb)
ΔV = velocity increase of the vehicle, mps (fps)
g_c = gravitational constant, 9.80665 m/s^2 (32.1740 ft/s^2)
I_{sp} = Specific impulse at which propellant m_p was burned, s

The assumptions implicit in Eq. (4.19–4.21) are as follows:

1) Thrust is the only unbalanced force on the vehicle, that is, force of gravity balanced by centrifugal force and drag is zero. (The equations are not valid for launch vehicle first-stage operation in atmosphere.).

2) Thrust is tangential to vehicle trajectory (aligned with instantaneous vehicle velocity).

3) The exhaust gas velocity is constant, which implies a fixed nozzle throat area and sonic velocity at the throat. The throat area of a solid motor may vary with ablation; however, this error is usually small.

All rocket engines are sonic at the throat; this condition occurs at a relatively low pressure ratio. Equations (4.19–4.21) are particularly important because they yield the propellant mass required for a spacecraft of a given mass to perform a given ΔV maneuver. You should add Eqs. (4.19–4.21) to your life list of equations you remember.

4.3.2.1 Derivation of Tsiolkowski equation [Eq. (4.19)]. It is worth a moment to see how Tsiolkowski derived his important equations. From the preceding assumptions thrust on the spacecraft is exactly equal to the force generated by the exiting exhaust gas and opposite in sign. Therefore, the force on the vehicle (thrust F) is equal to the mass of the vehicle m times the acceleration of the vehicle

$$F = m\frac{\mathrm{d}V}{\mathrm{d}t}$$

where V is the velocity of the spacecraft and m the mass of the spacecraft. Thrust is also equal to the time rate of change of momentum, or the exhaust gas velocity $-Ve$, times the time rate of change of mass $\mathrm{d}m/\mathrm{d}t$,

$$F = -V_e\frac{\mathrm{d}m}{\mathrm{d}t}$$

Equating the two expressions for thrust,

$$m\frac{\mathrm{d}V}{\mathrm{d}t} = -V_e\frac{\mathrm{d}m}{\mathrm{d}t}$$

$$\mathrm{d}V = -V_e\frac{\mathrm{d}m}{m}$$

where i and f indicate initial and final conditions and integrating yields

$$\int_i^f \mathrm{d}V = -V_e\int_i^f \frac{\mathrm{d}m}{m}$$

$$V_f - V_i = V_e(\ln m_i - \ln m_f)$$

For $Pe = Pa$, which is conservative in this case,

$$V_e = g_c I_{\mathrm{sp}}$$

and

$$\Delta V = g_c I_{\mathrm{sp}} \ln\left(\frac{m_i}{m_f}\right) \tag{4.19}$$

where g_c is the gravity constant, 32.1740 ft/s^2 or 9.8066 m/s^2. Mass ratio is independent of unit system. Equations (4.20) and (4.21) are algebraic manipulations of Eq. (4.19).

Example 4.2 Translational Velocity Requirement

The geosynchronous orbit provides a very common example of translational velocity requirements derived from orbital mechanics. The geosynchronous orbit is obtained by first placing a spacecraft on a low Earth parking orbit. Typically, an upper stage of the launch vehicle then places the spacecraft in an elliptical transfer orbit tangent to the parking orbit with apogee at geosynchronous altitude, 35,786 km. It is a spacecraft task to circularize the transfer orbit and to change the orbit plane from the transfer inclination, 28.5 deg for a U.S. launch, to zero inclination. Plane change and circularization requires a velocity increase of 1.831 km/s for a U.S. launch (see Ref. 5, pp. 81–87 for details).

A solid rocket motor, called an *apogee kick motor* (AKM), is a common choice for this task. Such a motor would typically deliver an I_{sp} of 290 s. How much propellant is required for this maneuver if the spacecraft weighs 2000 lb at burn out, including AKM inert weight? Equation (4.20) provides for this situation (solving first in English units).

Converting kilometers/second to feet/second,

$$\Delta V = 1.831\ \text{km/s}\,(3280.84) = 6007.2\ \text{ft/s}$$

$$m_p = m_f\left[\exp\left(\frac{\Delta V}{g_c I_{sp}}\right) - 1\right]$$

$$m_p = 2000\left[\exp\left(\frac{(6007.2)}{(32.1740)(290)}\right) - 1\right]$$

$$m_p = 2000\,[0.90375] = 1807.5\ \text{lb propellant}$$

The propellant weight required for this maneuver is 1807.5 lb. Therefore, the spacecraft launch weight is 3807.5 lb. The importance of high specific impulse is evident.

Repeating the same calculation in SI units, $m_f = 907$ kg:

$$m_p = 907\left[\exp\left(\frac{(1.831)}{(9.8066)(290)}\right) - 1\right]$$

$$m_p = 819.88\ \text{kg} = 1807.5\ \text{lb propellant}$$

4.3.2.2 Finite burn losses. Mission design calculations assume that velocity is changed at a point on the trajectory, that is, that the velocity change is instantaneous. If this assumption is not valid, serious energy losses can occur. These losses, called *finite burn losses*,[5] are caused by the rotation of the spacecraft velocity vector during the burn, as shown in Fig. 4.9.

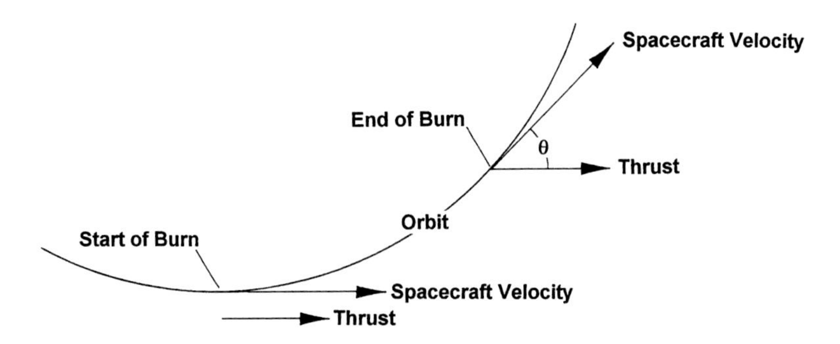

Fig. 4.9 Velocity change during finite burn.

In Fig. 4.9 it is assumed that the thrust vector is held inertially fixed during the burn and oversimplifies the case in two respects:

1) The orbit at the end of burn is not the same orbit as the start of the burn. The orbit may not even be the same type; the initial orbit might be a circle and the final orbit an ellipse. The orbital elements are continually changing during the burn.

2) If the final orbit in Fig. 4.9 were an ellipse, θ would be increased by the amount of the final flight-path angle. In reality, there will be a continuous change in flight-path angle as a function of time.

Also note that the change in spacecraft velocity is not linear with time [see Eq. (4.19)]; acceleration is much larger at the end of the burn because the spacecraft is lighter.

The finite burn losses can be a significant percentage of the maneuver energy. Figure 4.10 shows the finite burn losses for a Venus orbit insertion as a function of thrust-to-weight ratio.

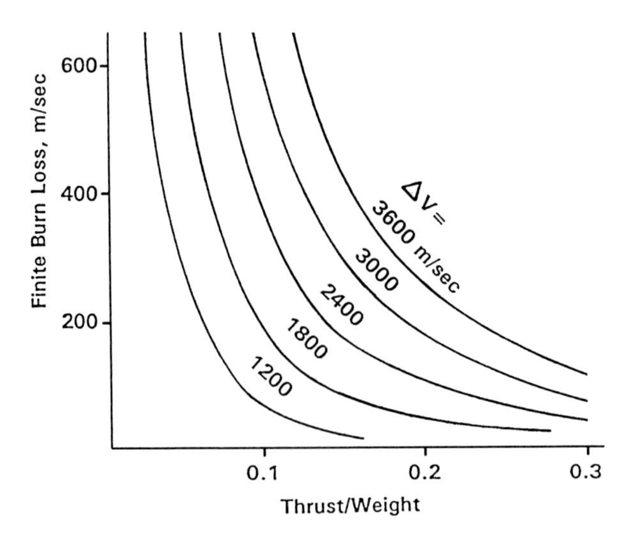

Fig. 4.10 Finite burn loss (Ref. 6, p. 27).

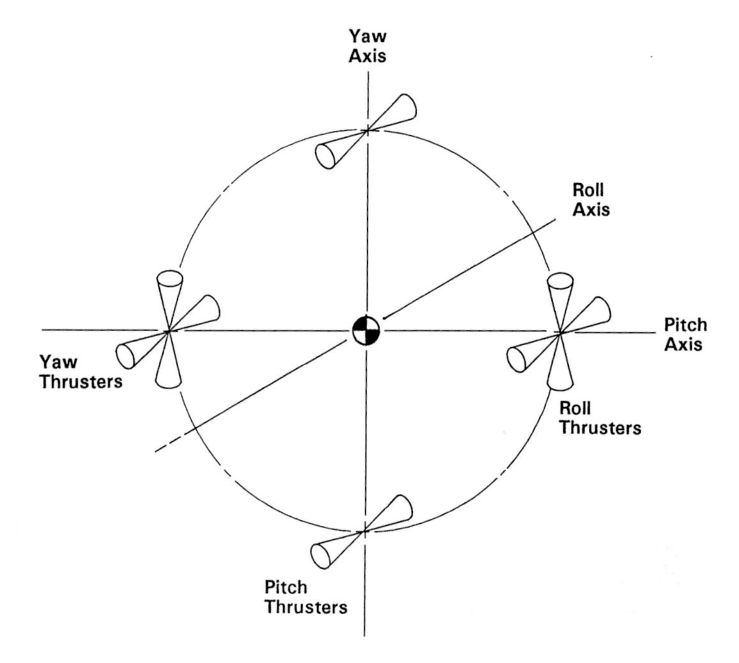

Fig. 4.11 Typical thruster installation for three-axis spacecraft.

Figure 4.10 is not general; each suspect situation requires a numerical integration to evaluate accurately. Low thrust-to-weight ratios are to be avoided; situations where thrust to weight is less than about 0.5 should be analyzed. Finite burn losses can be avoided by steering to hold the thrust vector on the spacecraft velocity vector; losses are reduced to the error in steering. The trade is a saving of propellant mass vs increased complexity in the attitude control system.

4.3.3 Attitude Maneuvers

Some, if not all, of the thrusters on a spacecraft are devoted to attitude control. Because these motors must restart frequently, only the fluid motors (bipropellant, monopropellant, cold gas) are candidates.

In a three-axis stabilized system an attitude maneuver consists of rotations about each of the spacecraft axes. Figure 4.11 is a typical thruster installation for a three-axis stabilized spacecraft. The thrusters are arranged so that the torques applied to the spacecraft are pure couples (no translational component).

4.3.3.1 Applying torque to a spacecraft. The elemental action in any maneuver is applying a torque to the spacecraft about an axis. To apply torque, spacecraft thrusters are fired in pairs producing a torque of

$$T = nFL \tag{4.22}$$

where

T = torque on the spacecraft, N-m (ft-lb)
F = thrust of a single motor, N (lb)
n = number of motors firing usually two; must be a multiple of two for pure rotation
L = radius from the vehicle center of mass to the thrust vector, m (ft)

From kinetics,

$$\theta = \tfrac{1}{2}\alpha t_b^2 \tag{4.23}$$

$$\alpha = \frac{T}{I_v} \tag{4.24}$$

$$\omega = \alpha t_b \tag{4.25}$$

$$H = I_v \omega \tag{4.26}$$

$$H = T t_b \tag{4.27}$$

where

θ = angle of rotation of the spacecraft, rad
ω = angular velocity of the spacecraft, rad/s
α = angular acceleration of the spacecraft during a firing, rad/s^2
I_v = mass moment of inertia of the vehicle, kg-m^2 (slug-ft^2)
t_b = duration of the burn, s
H = change of spacecraft angular momentum during the firing, kg-m^2/s (slug-ft^2/s)

During the burn, the angular acceleration of the spacecraft will be [from Eq. (4.24)]

$$\alpha = \frac{nFL}{I_v} \tag{4.28}$$

When the thrusters are shut down, the vehicle will have turned. From Eqs. (4.23) and (4.28),

$$\theta = \frac{nFLt_b^2}{2I_v} \tag{4.29}$$

At shutdown, acceleration goes to zero, and the spacecraft is left rotating at a velocity of ω. From Eqs. (4.25) and (4.28),

$$\omega = \frac{nFL}{I_v} t_b \tag{4.30}$$

From Eq. (4.22) the angular momentum produced by a single firing is

$$H = nFLt_b \tag{4.31}$$

The propellant consumed during a single burn is

$$m_p = \frac{nFt_b}{I_{sp}} \tag{4.32}$$

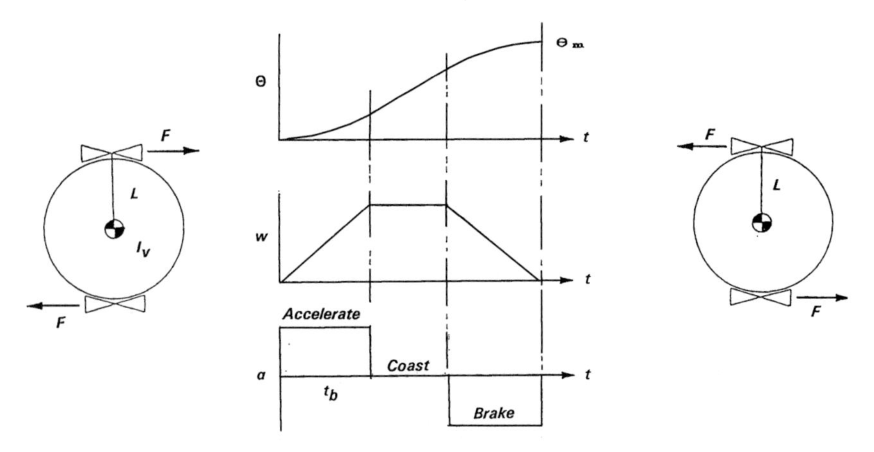

Fig. 4.12 One-axis maneuver.

or

$$m_p = \frac{H}{LI_{\text{sp}}} \tag{4.33}$$

Equation (4.33) shows the advantage of a long moment arm. Expendable propellant can be saved by increasing moment arm. The maximum moment arm is constrained in a surprising way—by the inside diameter of the launch vehicle payload fairing. The fairing dimensions of various launch vehicles are listed in Chapter 2. It is not uncommon to design the thruster support structures to take the full diameter of the dynamic envelope.

4.3.3.2 One-axis maneuver. A maneuver about one axis consists of three parts: 1) angular acceleration, 2) coasting, and 3) braking. Angular acceleration is produced by a thruster pair firing; braking is caused by a firing of the opposite pair. Figure 4.12 shows a one-axis maneuver.

The total angle of rotation is

$$\theta_m = \theta(\text{accelerating}) + \theta(\text{coasting}) + \theta(\text{braking})$$

The rotation during coasting is

$$\theta = \omega t_c \tag{4.34}$$

where t_c is the duration of the coasting, s. Using Eqs. (4.30) and (4.34), the coasting rotation angle is

$$\theta = \frac{nFL}{I_v} t_b t_c \tag{4.35}$$

The rotation during acceleration or braking is given by Eq. (4.29); therefore, the total rotation during the maneuver is

$$\theta_m = \frac{nFL}{I_v} t_b^2 + \frac{nFL}{I_v} t_b t_c \tag{4.36}$$

or

$$\theta_m = \frac{nFL}{I_v}\left(t_b^2 + t_b t_c\right) \tag{4.37}$$

Note that t_b is the burn time for either of the two burns, and the maneuver time is

$$t_m = t_c + 2t_b \tag{4.38}$$

If there is no coast period, then

$$t_m = 2t_b = 2\sqrt{\frac{\theta_m I_v}{nFL}} \tag{4.39}$$

The propellant required for a one-axis maneuver is twice the single burn consumption given by Eq. (4.32),

$$m_p = 2\frac{nFt_b}{I_{sp}} \tag{4.40}$$

Virtually all maneuvers are three-axis maneuvers. Equation (4.40) must be applied to each axis to determine total propellant required. Most maneuvers come in pairs, a maneuver to commanded attitude and a maneuver back to normal attitude. Therefore, the propellant required for a complete maneuver is usually twice that obtained from a three-axis application of Eq. (4.40).

Example 4.3 One-Axis Maneuver

Find the minimum time required for a spacecraft to perform a 90-deg turn about the z axis with two thrusters if the spacecraft has the following characteristics: moment of inertia about the z axis $= 500$ kg-m^2; moment arm $= 1.8$ m; thrust of each engine $= 3.5$ N; and

$$\theta_m = \frac{\pi}{2} = 1.5708 \text{ rad}$$

For minimum maneuver time the coast time must be zero, and from Eq. (4.39)

$$t_m = 2\sqrt{\frac{(1.5708)(500)}{(2)(3.5)(1.8)}}$$

$$t_m = 15.79 \text{ s}, \quad t_b = 7.90 \text{ s}$$

How much propellant was consumed by the maneuver if the $I_{sp} = 190$ s? From Eq. (4.40),

$$m_p = 2\frac{(2)(3.5)(7.9)}{(190)(9.8067)} = 0.0594 \text{ kg}$$

4.3.4 Limit Cycles

Propulsion is frequently used to control spacecraft attitude in a limit-cycle mode (see Chapter 5). In this mode attitude is allowed to drift until a limit is reached, then thrusters are used for correction.

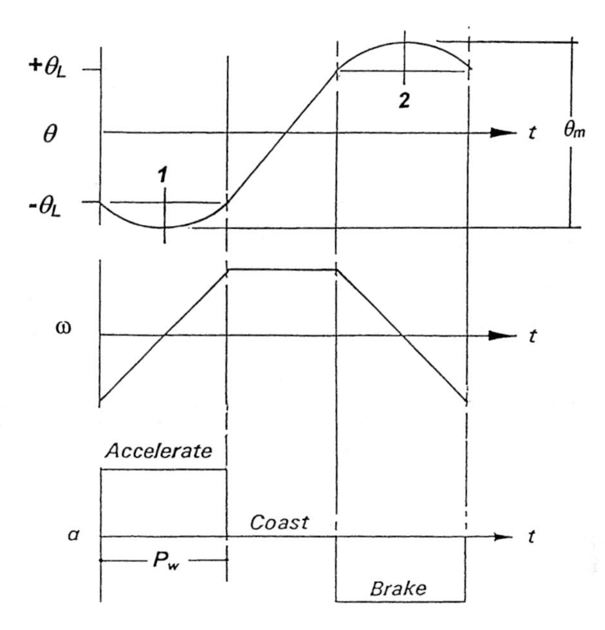

Fig. 4.13 Limit-cycle motion without external torque.

4.3.4.1 Without external torque. A limit cycle without external torque swings the spacecraft back and forth between preset angular limits as shown in Fig. 4.13 for one axis. When the spacecraft drifts across one of the angular limits θ_L, the attitude control system will fire a thruster pair for correction. The spacecraft rotation will reverse and continue until the opposite angular limit is reached at which time the opposite thruster pair will be fired. It is important that the smallest possible impulse be used for the corrections because the impulse must be removed by the opposite thruster pair.

Note from Fig. 4.13 that this limit cycle is an extension of the one-axis maneuver described mathematically in the preceding section and shown in Fig. 4.12. The one-axis maneuver equations describe the motion from point 1 to point 2 in Fig. 4.13. The pulse width P_w, shown in Fig. 4.13, is twice the duration of t_b, as discussed in conjunction with one-axis maneuvers. The total angle of rotation θ_m overshoots the control limits during the time P_w. The angle of vehicle rotation can be found by adapting Eq. (4.37).

$$\theta_m = \frac{nFL}{I_v}\left(\frac{P_w^2}{4} + \frac{t_c P_w}{2}\right) \tag{4.41}$$

The limit settings $\pm\theta_L$ are one-half of the coasting angle, from Eq. (4.36)

$$\theta_L = \frac{nFL}{4I_v} P_w t_c \tag{4.42}$$

Table 4.5 Representative pulsing performance

Propulsion system type	Min thrust, N	Min impulse bit, N-s	Pulsing I_{sp}, s
Cold gas–helium	0.05	0.0005	80
Cold gas–nitrogen	0.05	0.0005	50
Monopropellant–N_2H_4	0.5	0.005	120
Bipropellant–N_2O_4/MMH	8	0.025	120

Each cycle includes two pulses; the propellant consumed per cycle is

$$m_p = 2\frac{nFP_w}{I_{sp}} \tag{4.43}$$

Equation (4.43) shows the importance of low thrust, short burn time, and high specific impulse in pulsing operation. Pulsing engines are characterized by minimum impulse bit I_{min}, where

$$I_{min} = F(P_w)_{min} \tag{4.44}$$

The minimum pulse width $(P_w)_{min}$ is a characteristic of a given thruster/valve combination. Table 4.5 shows typical pulsing performance of the propulsion system types. During pulsing, the engine cools between pulses. Some of the energy from the next pulse goes into reheating the engine. The steady-state specific impulse can be approached if the pulses are very close together and are of long duration. The pulsing I_{sp} shown in Table 4.5 is for cold engines with infrequent pulses, also called a minimum duty cycle.

In a limit cycle the propellant consumed per unit time is a key issue. The length of a cycle is

$$t_{cy} = 2P_w + 2t_c \tag{4.45}$$

The coast time is

$$t_c = \frac{4I_v\theta_L}{nFLP_w} \tag{4.46}$$

Thus, if minimum impulse bits are used, the length of a cycle is

$$t_{cy} = \frac{8I_v\theta_L}{nLI_{min}} + 2P_w \tag{4.47}$$

The burn time is negligible (milliseconds) compared to the coast time (seconds) and can be neglected (This is also a conservative assumption.). The propellant consumption per unit time is, from Eqs. (4.43) and (4.47),

$$\frac{m_p}{t_{cy}} = \frac{n^2 I_{min}^2 L}{4I_{sp}I_v\theta_L} \tag{4.48}$$

where m_p/t is the propellant consumption per unit mission time, kg/s. Equation (4.48) shows that a wide control band is desirable as is a low minimum impulse bit and a high I_{sp}.

Example 4.4 Limit-Cycle Operation

A spacecraft with 2000 kg-m^2 moment of inertia uses 1.5-N thruster pairs mounted at a radius of 1.6 m from the center of mass. For limit-cycle control to $\theta_L = 0.5$ deg (0.008727 rad), what is the propellant consumption rate if the I_{sp} is 170 s? The pulse duration is 30 ms, and there are no external torques.

The propellant consumed per cycle is, from Eq. (4.43),

$$m_p = 2\frac{(2)(1.5)(0.03)}{(170)(9.8067)} = 0.000107969 \text{ kg/cycle}$$

The minimum impulse bit I_{min} is

$$I_{min} = Ft_{min} = (1.5)(0.30) = 0.045 \text{ N-s}$$

The duration of a cycle is, from Eq. (4.47),

$$t_{cy} = \frac{(8)(2000)(0.008727)}{(2)(1.6)(1.5)(0.045)} + 2(0.030) = 646.5 \text{ s}$$

The propellant consumption rate is

$$\frac{m_p}{t_{cy}} = \frac{0.000107969}{646.5} = 1.67 \times 10^{-7} \text{ kg/s}$$

4.3.4.2 External torque. Spacecraft are subject to a number of external torques: 1) gravity gradient, 2) solar pressure, 3) magnetic field, and 4) aerodynamic. Aerodynamic torques are proportional to atmospheric density and hence are important for low Earth orbits and rapidly become negligible in higher orbits. Solar pressure is the dominant disturbance above geosynchronous altitudes. Gravitygradient torques are dependent on spacecraft mass and mass distribution. Gravity gradients have been used to stabilize spacecraft. Methods for calculating the magnitude of these torques are discussed in "Attitude Control," Chapter 5. Propulsion requirements evolve from the necessity to correct for these torques.

4.3.4.3 One-sided limit cycle. With an external torque on the spacecraft, rotation occurs until a limit line is reached, and a thruster pair is fired for correction. If the limit lines are wider than a certain value, a one-sided limit cycle occurs. Figure 4.14 shows a one-sided limit cycle.

In Fig. 4.14 when the rotation reaches one of the limit lines, a thruster pair is pulsed. The unbalanced torque reverses the rotation, and the thrusters are fired to start another cycle. In effect the unbalanced torque has replaced one of the thruster pairs.

For this cycle the momentum supplied by the propulsion system exactly equals the momentum induced by the external torque

$$H = T_x t \tag{4.49}$$

where T_x is the external torque on the spacecraft, kg-m and t the mission duration, s. The propellant weight required to compensate for the external torque is

$$m_p = \frac{T_x t}{LI_{sp}} \tag{4.50}$$

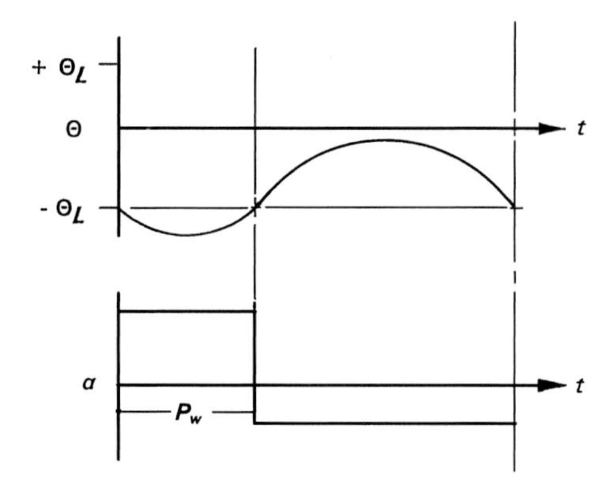

Fig. 4.14 One-sided limit cycle.

A one-sided limit cycle is clearly an efficient way to deal with an external torque. For the cycle to occur, the rotational limits must be wide enough to prevent a thruster firing from assisting the external torque.

4.3.5 Reaction Wheels

A reaction wheel is a flywheel driven by a reversible dc motor. To perform a maneuver with a reaction wheel, the flywheel is accelerated. The spacecraft accelerates in the opposite direction. To do a maneuver as shown in Fig. 4.12, with reaction wheels and with a zero coast period,

$$\theta_v = \frac{T_v t_m^2}{4I_v} \tag{4.51}$$

and torque is

$$T_v = \alpha_w I_w \tag{4.52}$$

where subscript w refers to the properties of the reaction wheel and subscript v refers to properties of the vehicles. The angle of spacecraft rotation is

$$\theta_v = \frac{\alpha_w I_w t_m^2}{4I_v} \tag{4.53}$$

The net change in wheel speed is zero, which is the major advantage of reaction wheels. However, eventually unbalanced torques will load the wheel (bring it to maximum speed). The wheel can be unloaded by reversing the motor and holding the spacecraft in position with reaction jets. The momentum of the wheel H is

$$H = Tt = nFLt \tag{4.54}$$

and

$$H = I_w \omega_w \tag{4.55}$$

The total impulse required to unload the reaction wheel is

$$I = nFt = \frac{I_w \omega_w}{L} \tag{4.56}$$

During unloading, the momentum will be transferred from the propulsion system. Assuming wheel speed is brought to zero, the propellant required to unload is

$$m_p = \frac{I_w \omega_w}{L I_{\text{sp}}} \tag{4.57}$$

The time required to unload is

$$t = \frac{I_w \omega_w}{nFL} \tag{4.58}$$

When reaction wheels are controlling external torques, the momentum required for wheel unloading is equal to the momentum caused by the external torque; Eq. (4.56) applies.

Example 4.5 Reaction Wheel Unloading

How much propellant does it take to unload one of the Magellan wheels, and how long does it take? The Magellan wheel characteristics are maximum momentum = 27 N-m-s and maximum wheel speed = 4000 rpm = 418.879 rad/s.

The thruster pair to be used has the following characteristics: thrust = 0.889 N and moment arm = 2.134 m.

Assuming that the wheel is fully loaded, the engine burn time required to unload the wheel is, from Eq. (4.58),

$$t = \frac{27}{(2.134)(2)(0.889)} = 7.116\,\text{s}$$

If the I_{sp} of the engines is 150 s for short burns, the propellant required to unload is from Eq. (4.57)

$$m_p = \frac{27}{(150)(9.80665)(2.134)} = 0.0086\,\text{kg}$$

4.3.6 Spinning Spacecraft Maneuvers

To this point we have been considering three-axis oriented spacecraft. In a spin-stabilized spacecraft attitude maneuvers consist of translations for stationkeeping and repositioning, reorienting the spin axis, and adjusting spin velocity. Figure 4.15 shows a typical thruster installation for a spinner.

Spin-stabilized spacecraft is maneuvered by precession of the spin axis. Reorientation of the spin axis constitutes a change in the direction of the momentum vector, as shown in Fig. 4.16.

Putting a torque on the spacecraft rotates the momentum vector through angle ϕ,

$$\phi \approx \frac{H_a}{H_i} = \frac{nFLt}{I_y \omega} \tag{4.59}$$

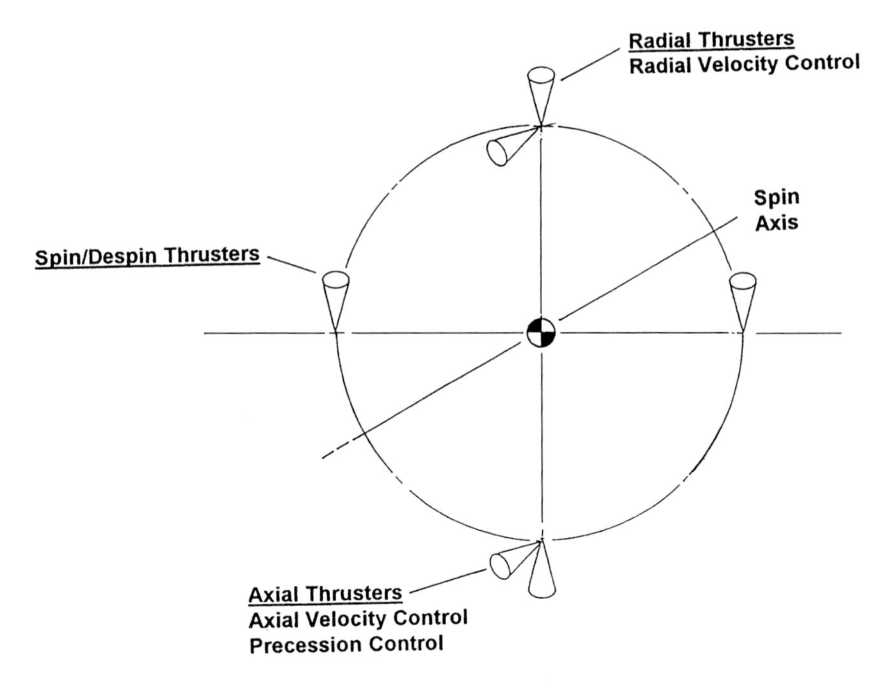

Fig. 4.15 Typical thruster installation for a spinning spacecraft.

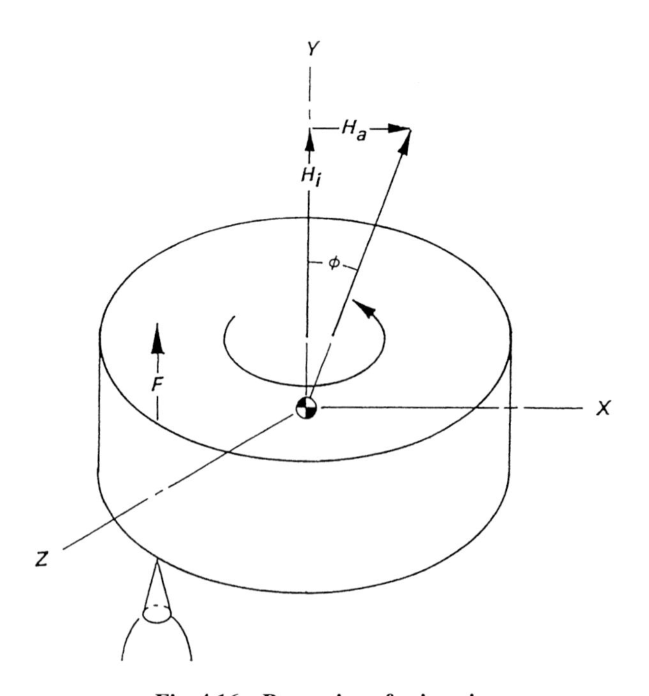

Fig. 4.16 Precession of spin axis.

where

H_a = momentum vector added, slug-ft/s
H_i = initial spin momentum, slug-ft/s
I_y = moment of inertia about the spin axis, slug-ft^2
ϕ = precession angle, rad
ω = angular velocity of the spacecraft, rad/s

Note that the pulse width t must be short compared to the period of spin. A pulse during an entire revolution would result in $H_a = 0$.

Example 4.6 Precession of Spin Axis

What burn time, or pulse width, is required to precess a spacecraft spin axis by 3 deg (0.05236 rad) under the following conditions:

Thrust = 10 N
Moment arm = 1.5 m
Spacecraft spin rate = 2 rpm (0.2094 rad/s)
Moment of inertia = 1100 N-m^2
Specific impulse = 185 s

From Eq. (4.59) the burn time is

$$t_b = \frac{\phi I_v \omega}{nFL}$$

$$t_b = \frac{(0.05236)(1100)(0.2094)}{(1)(10)(1.5)} = 0.8040 \text{ s}$$

Compare burn time to rotation time: Period = 30 s, given (large compared with burn time). The propellant consumed is, from Eq. (4.32),

$$m_p = \frac{nFt_b}{I_{\text{sp}}}$$

$$m_p = \frac{(1)(10)(0.8040)}{(185)(9.80665)} = 0.00443 \text{ kg}$$

The spin axis will continue to precess until a second pulse of equal magnitude and opposite direction is fired. The spin axis can be repositioned by this method.

4.4 Monopropellant Systems

A monopropellant system generates hot, high-velocity gas by triggering decomposition of a single chemical—a monopropellant. The concept is shown in Fig. 4.17.

The monopropellant is injected into a catalyst bed, where it decomposes; the resulting hot gases are expelled through a converging/diverging nozzle generating thrust. A monopropellant must be a slightly unstable chemical, which will decompose exothermally to produce a hot gas. There are a number of chemicals, which will do this. Table 4.6 lists some of these.

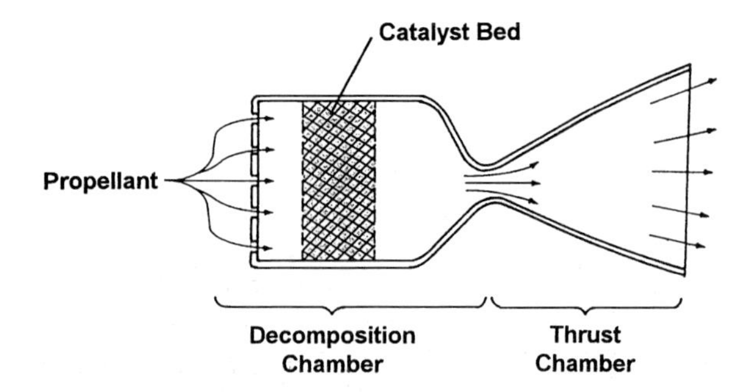

Fig. 4.17 Monopropellant thruster concept.

There are a number of practical considerations, notably stability, that thin the list in Table 4.6. Only three monopropellants have ever been used on flight vehicles: hydrazine, hydrogen peroxide, and n-propyl nitrate. Shock sensitivity eliminated n-propyl nitrate after limited use for jet engine starters. Hydrogen peroxide saw considerable service as a monopropellant in the 1950s and 1960s (starting with the V-2). The persistent problem with hydrogen peroxide is slow decomposition during storage. The decomposition products cause a continuous increase in pressure in the storage vessel and water dilution of the propellant. The pressure rise complicates flight and ground tank design; the water dilution reduces performance.

Table 4.6 shows that hydrazine has desirable properties across the board, including the highest specific impulse of the stable chemical group. A serious limitation on the early use of hydrazine was ignition; hydrazine is relatively difficult to ignite. In the early years hydrazine was ignited by injecting a start slug of nitrogen tetroxide. Once lit, the combustion was continued by a catalyst. Ranger and Mariner II, as well as Mariner IV, used a hydrazine start slug system. Using start slugs limited the number of burns to the number of slugs, two in the case of Mariner IV. Start slugs also complicate the system. The great advantage of monopropellants is the elimination of the oxidizer system. With the start slug the oxidizer system was still necessary.

Table 4.6 Characteristics of some monopropellants[a]

Chemical	Density	Flame temperature, °K	I_{sp}, s	Sensitivity
Methyl nitrate	1.21	3716	259	Yes
Nitromethane	1.13	2479	244	Yes
Nitroglycerine	1.60	3309	244	Yes
Ethyl nitrate	1.10	1944	224	Yes
Hydrazine	1.01	1394	230	No
Tetronitromethane	1.65	2170	180	Yes
Hydrogen peroxide	1.45	1277	165	No
Ethylene oxide	0.87	1233	189	No
n-Propyl nitrate	1.06	1693	201	Yes

[a]Courtesy of McGraw–Hill; Ref. 3, pp. 20–41.

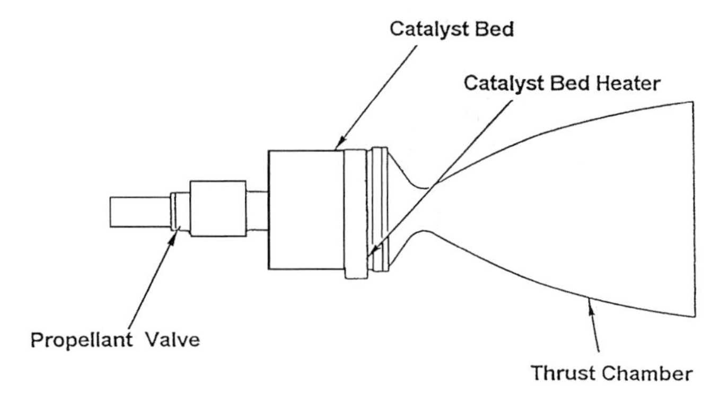

Fig. 4.18 Typical monopropellant thruster.

The need for a spontaneous catalyst led the Shell Development Company and the Jet Propulsion Laboratory to develop the Shell 405 iridium pellet catalyst bed in 1962. Almost unlimited spontaneous restart capability resulted. The spontaneous restart capability along with relative stability, high performance, clean exhaust, and low flame temperature have made hydrazine the only monopropellant in use today.

4.4.1 Monopropellant Hydrazine Thrusters

A monopropellant hydrazine thruster is shown in Fig. 4.18. The propellant flow into the chamber is controlled by a propellant valve, which is usually an integral part of the thruster. The propellant is injected into a catalyst bed, where it decomposes into hydrogen, nitrogen, and ammonia. The gases are expelled through a converging/diverging nozzle producing thrust. The fundamental rocket engine equation applies; the thrust generated is

$$F = \dot{m} V_e + (P_e - P_a) A_e \tag{4.6}$$

The gas temperature is in the 1200°C (2200°F) range. High-temperature alloys can be used for the converging/diverging nozzle, and radiation cooling is adequate. Catalyst bed heaters are often used for pulsing engines to improve first pulse performance and improve bed cycle life. The valve performance is an integral of the thruster performance during pulsing, and the two are considered as a unit. Monopropellant thrusters of this configuration have flown in the sizes ranging from 0.5 to 2600 N and blowdown ratios up to 6. Pulse widths as low as 7 ms have been demonstrated,[7] as have thrust levels in the 4500 N class.

The decomposition of hydrazine leads first to hydrogen and ammonia. The reaction is exothermic, and the adiabatic flame temperature is about 1700°K; however, the ammonia further decomposes into hydrogen and nitrogen. This reaction is endothermic and leads to a reduction in flame temperature and I_{sp}. Figure 4.19 shows the theoretical vacuum specific impulse of anhydrous hydrazine, area ratio = 50, as a function of percent ammonia dissociation. It is desirable to limit the dissociation of ammonia as much as possible. Ammonia dissociation can be held to about 50% in current engine designs.

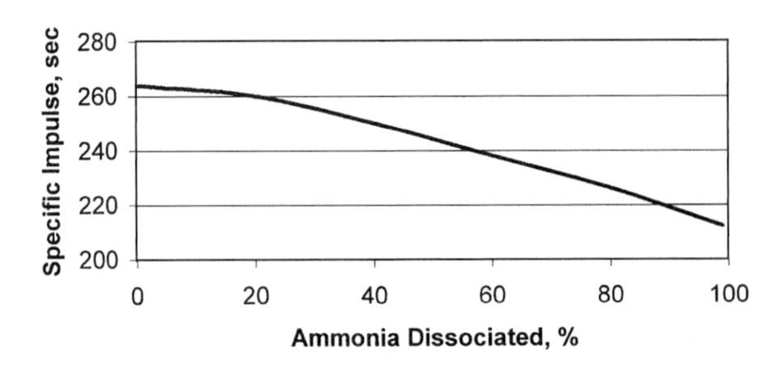

Fig. 4.19 Hydrazine monopropellant performance vs ammonia dissociation. (Data from Ref. 7.)

4.4.1.1 Steady-state performance. A steady-state, theoretical, vacuum specific impulse of about 240 s can be expected at an area ratio of 50:1. Real system specific impulse will be about 93% of theoretical. Specific impulse at other area ratios can be estimated from the ratio of thrust coefficients (see Section 4.2). The ratio of specific heats for the exhaust is about 1.27.

4.4.1.2 Pulsing performance. Attitude control applications of monopropellant hydrazine engines require operation over wide ranges of duty cycles and pulse widths. Such use makes the pulsing specific impulse and minimum impulse bit very important. Pulsing involves the performance of the propellant valve and feed tubing as well as the chamber itself. A typical pulse is shown in Fig. 4.20.

Pressure response time is the time (measured from the propellant valve actuation signal) required to reach an arbitrary percentage of steady-state chamber pressure. Response time is affected by 1) the valve response characteristics, 2) time for the propellant to flow from the valve to the injector (hydraulic delay), 3) the ignition delay (time to wet the catalyst bed and for initial decomposition heat release to bring the catalyst to a temperature at which ignition becomes rapid), and 4) the pressure rise time (time to decompose enough hydrazine to fill the void space in the catalyst bed and heat the whole bed).

Valve response characteristics and feed line hydraulic delay vary for each specific design. Valves are available with response times better than 10 ms. The distance between the valve and injector must be minimized; however, the control of heat soak back to the valve sets a minimum length for the injector tube.

With proper injector design, ignition delay will be approximately 10 to 20 ms for catalyst and propellant temperatures in the range of 5–20°C. With a catalyst bed temperature in the range of 260°C, ignition delay will be approximately 1–2 ms. Pressure rise time is a much larger fraction of the response time than ignition delay for most thrust chambers. Response times from valve signal to 90% of steady-state chamber pressure of 15 ms have been demonstrated with tail-off time (signal to 10% thrust) of 20 ms.[7]

In Fig. 4.20 the response from signal to 90% thrust is 15 ms; the tail-off time is 20 ms. If the duty cycle is long, the engine will cool between pulses, and thrust will not reach rated value because of energy losses to engine heating. If pulses are frequent, thrust will reach rated value.

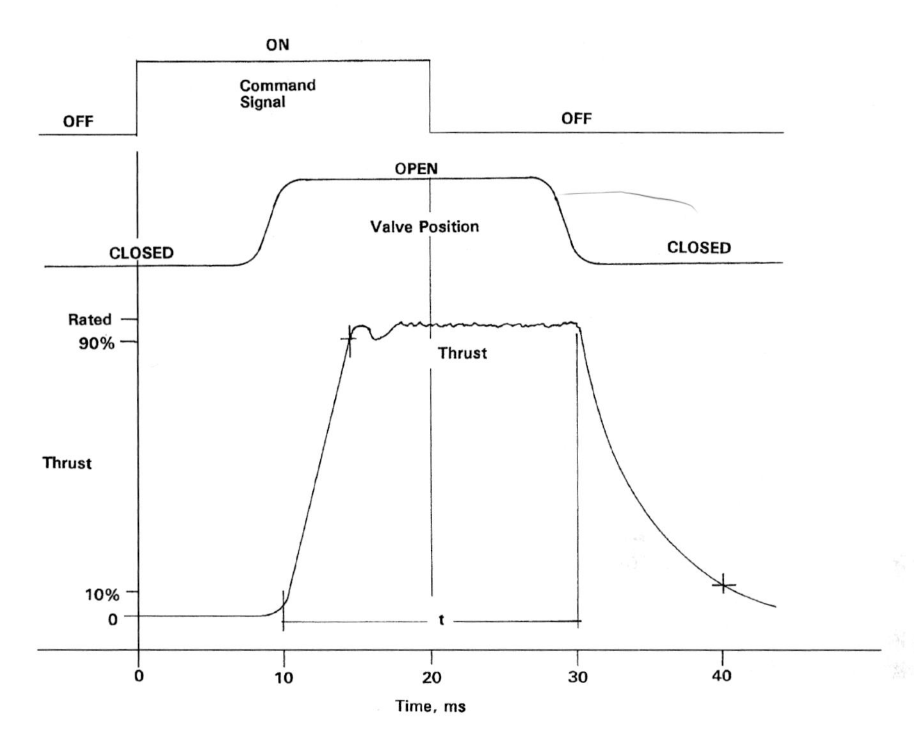

Fig. 4.20 Typical 20-ms pulse.

In the absence of test data, the minimum impulse bit can be estimated by assuming typical system response and that time t in Fig. 4.20, from ignition to first loss of thrust, is equal to the pulse width. Because impulse is the area under the thrust time curve,

$$I_{\min} = I(\text{Startup}) + I(\text{Steady state}) + I(\text{Shutdown}) \tag{4.60}$$

By trapezoidal approximation minimum impulse bit is

$$I_{\min} = \frac{0.005}{2}F + (t - 0.005)F + \frac{0.010}{2}F \tag{4.61}$$

$$I_{\min} \approx Ft \tag{4.62}$$

where F is the steady-state thrust reached, N; and t is the pulse width, time from valve on to valve off, s.

For all practical purposes impulse bit can be estimated as Ft. For infrequent pulses the thruster will be cold, and full rated thrust will not be reached. For such cases the thrust level corresponding to the expected gas temperature should be used.

Pulsing specific impulse is low at low duty cycles because energy is lost reheating the motor. Short pulses deliver low specific impulse for the same reason. Low duty

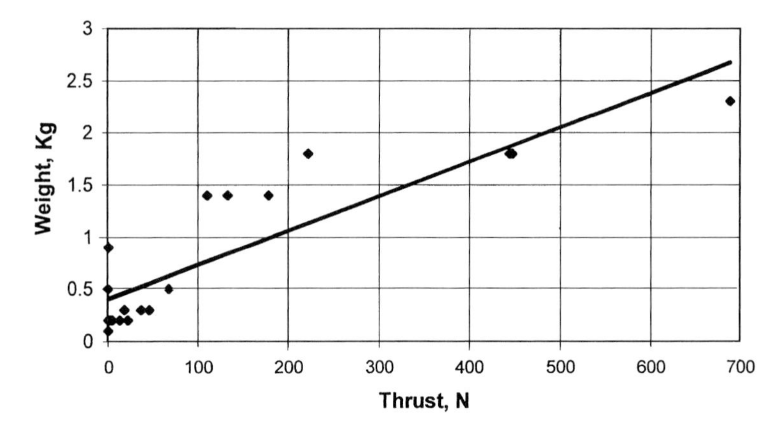

Fig. 4.21 Monopropellant thruster weight.

cycles and short pulses in combination deliver specific impulse as low as 115 s. Specific impulse is limited at low duty cycles by the performance of an ambient temperature bed. The performance of a thruster operating with a cold bed can be estimated by assuming the exhaust gas exits at the bed temperature.

All of the hydrazine that reaches the catalyst bed is decomposed; there are no liquid losses on start or shutdown.

4.4.1.3 Negative pulses. In the course of a steady-state burn, it is often desirable to pulse off to create attitude control torques. The Magellan and Viking spacecraft have used the pulse-off technique. In the Magellan case the 445-N thrusters were fired in steady state during the solid motor burn for Venus orbit insertion. (The 445-N engines are aligned with the motor thrust vector.) Pulsing off was used for thrust vector control during the burn. A similar strategy was used by Viking Lander during the deorbit phase.

4.4.1.4 Thruster weight. A least-squares curve fit of the weight of 19 actual monopropellant thruster/valve designs (Fig. 4.21) produces the following relation for estimating thruster/valve weight,

$$W_t = 0.4 + 0.0033F \tag{4.63}$$

For low thrust levels the thruster weight approaches the valve weight as shown by Fig. 4.21. Use 0.3 kg as a minimum thruster/valve weight for low thrust levels. Figure 4.21 is for a thruster and a single valve; the estimated weight must be increased if redundant valves are used.

4.4.2 Propellant Systems

The purpose of the propellant system is to contain the propellants and to serve them to the engine on demand, at the proper pressure, quality, and cleanliness. Anhydrous hydrazine, the only monopropellant in use today, is a clear, colorless hygroscopic liquid with a distinct, ammonia-like odor. It is a stable chemical that

Table 4.7 Propellant inventory

Propellant use	Load, kg
Usable propellant	1000
Trapped propellant, 3%	30
Loading error, 0.5%	5
Loaded propellant	1035

can be stored for long periods without loss of purity. It is relatively (compared to other monopropellants) insensitive to shock.

Hydrazine is a strong reducing agent, and it is toxic. Special preparation, special equipment, and special procedures are required for handling it.

4.4.2.1 Propellant inventory. The propellant inventory is a subdivided tabulation of the loaded propellant weight. Table 4.7 shows an example for 1000 lb of usable monopropellant.

The usable propellant is that portion of the propellant loaded, which is actually burned. This is the quantity required for all maneuvers and all attitude control functions. It is important that the usable propellant quantity be calculated under worst-case conditions (see Section 4.3).

Not all of the propellant loaded can be used; a certain amount is trapped or loading uncertainty. Trapped propellant is the propellant remaining in the feed lines, tanks and valves, hold-up in expulsion devices, and retained vapor left in the system with the pressurizing gas. Trapped propellant is about 3% of the usable propellant.

The measurement uncertainty in propellant loading is about 0.5%. The uncertainty is added to the load to ensure that the usable propellant can be no less than worst-case requirement.

Propellant reserves are a very valuable commodity. The more of the project reserves placed in propellant the better. There are three primary reasons for this: 1) propellant is usually the life-limiting expendable on a spacecraft; 2) it is often desirable to make unplanned maneuvers in response to unexpected results or emergency conditions; 3) there are usually extended mission objectives that can be achieved after the primary mission—if there is propellant. Voyager, for example, visited Uranus and Neptune after the primary mission was over. Propellant reserve can be used in many useful ways during the mission, unlike weight margin, power margin, and most other margin types.

4.4.2.2 Zero-g propellant control. The location of the ullage bubble, or bubbles, becomes indeterminate in zero-g conditions unless the liquid and gas are positively separated. Figure 4.22 shows zero-g bubble shapes.

Unless the gas location is controlled, there is no way to guarantee gas-free liquid to the engine. Although an engine can tolerate small amounts of gas, serious failures have occurred as a result of gas ingestion. It is good design practice to prevent any gas ingestion. Spin-stabilized spacecraft can take advantage of centrifugal force. Three-axis spacecraft or spacecraft with a low spin rate must use one of the following schemes: 1) capillary devices, which use surface tension to

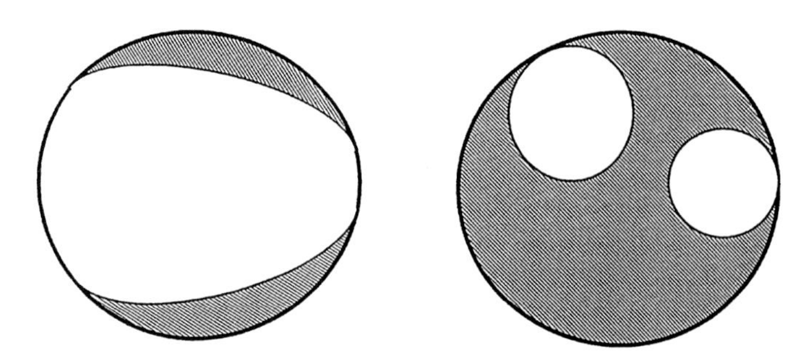

Fig. 4.22 Zero-*g* bubble shapes.

keep gas and liquid separated (particularly useful for bipropellant systems, like the Space Shuttle and Viking Orbiter, because they are compatible with strong oxidizers); 2) diaphragms and bladders, which are physical separation devices made of elastomer or Teflon® (Voyager, Mariner 9, and Magellan used these; elastomers are not compatible with oxidizers); 3) bellows, a metal separation device, used by Minuteman; and 4) traps, a check valve protected compartment, used by Transtage. Figure 4.23 shows the concepts in schematic.

4.4.2.3 Capillary devices. Capillary devices take advantage of the small pressure differences between a wetting liquid and a gas. For a simple screen device (Fig. 4.24) the ΔP capability is inversely proportional to radius of the screen openings.

The Young–Laplace equation[1] states that the maximum differential pressure across a spherical surface is

$$\Delta P = \frac{2\sigma}{r} \tag{4.64}$$

where ΔP is the pressure difference sustained by a screen, *Pa* (lb/ft^2); σ is the surface tension of the liquid, N/m^2 (lb/ft); and r is the radius of the screen pores, m (ft). Extremely fine screen will support approximately 0.5 m (hydrostatic head) of hydrazine against an adverse 1-g acceleration.

Screen devices are shaped to maintain liquid over the tank outlet for any acceleration direction the spacecraft will experience. The capillary device used to control the Space Shuttle reaction control system (RCS) propellants (N_2O_4 and MMH) is shown in Fig. 4.25. In the Space Shuttle, a system with complicated maneuvering, the capillary system also becomes complex. Both capillary devices and bladders have demonstrated an expulsion efficiency of 99%.

4.4.2.4 Propellant control: spin-stabilized spacecraft. A spinning spacecraft can use centrifugal force to control propellant position. Spin rates above about 6 rpm are required for centrifugal force to dominate surface tension and slosh forces; it takes an analysis of the expected spacecraft loads to verify the adequacy of centrifugal force. An expulsion efficiency of over 97% can be expected.[8]

The tank outlet is placed perpendicular to the spin axis pointing outboard for the liquid to be held over the outlet by centrifugal force. Unfortunately the tank

System Type	Features
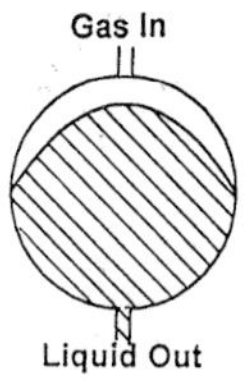 Gas In Liquid Out	**Diaphragm Tank** **Example of Flight Use:** Cassini RCS, Magellan RCS (both monopropellant) **Operation:** Hemispherical diaphragm separates the ullage gas from propellant **Compatibility:** Can be used with N2O4: Material = Teflon Can be used with Hydrazines: Material = Elastomer **Cycles:** Can be used for numerous cycles **One g Operation:** Can be tested at 1g
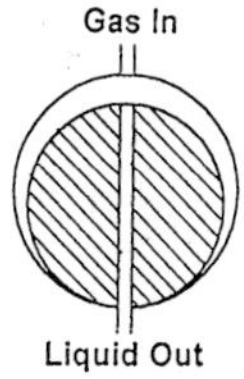 Gas In Liquid Out	**Bladder Tank** **Example of Flight Use:** Mariner 9 (Bipropellant) **Operation:** Spherical bladder separates the ullage gas from propellant **Compatibility:** Can be used with N2O4: Material = Teflon Can be used with Hydrazines: Material = Elastomer **Cycle Life:** Can be used for numerous cycles **One g Operation:** Can be tested at 1g
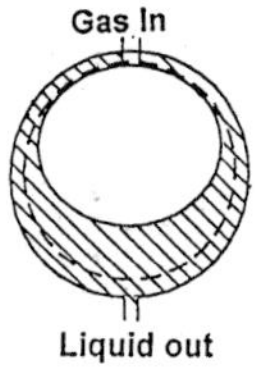 Gas In Liquid out	**Capillary Screen Tank** **Example of Flight Use:** Shuttle RCS (Bipropellant) **Operation:** Very fine screen mesh separates propellant and ullage gas using surface tension forces. **Compatibility:** Compatible with N2O4 and hydrazines for long term storage Material for N2O4 = fine mesh CRES screen Material for hydrazines = fine mesh CRES screen **Cycle Life:** Can be used for numerous cycles **One g Operation:** Can not be performance tested at 1 g
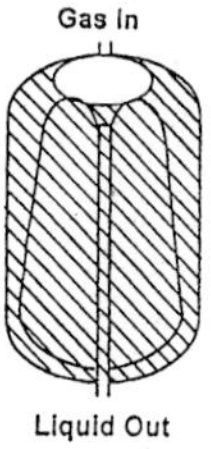 Gas In Liquid Out	**Capillary Vane Tank** **Example of Flight Use:** Cassini (Bipropellant) MGS (Dual Mode) **Operation:** Vanes designed such that minimum energy position of ullage bubble is at the top and sides of the tank. **Compatibility:** Compatible with N2O4 and hydrazines for long term storage Material for N2O4 = Titanium Material for hydrazines = Titanium **Cycle Life:** Can be used for numerous cycles **One g Operation:** Can not be performance tested at 1 g
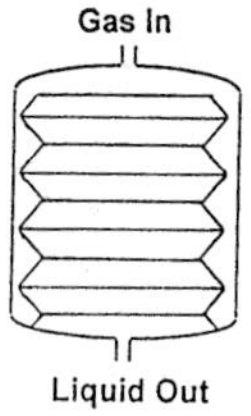 Gas In Liquid Out	**Compressible Bellows** **Example of Flight Use:** Minuteman RCS (Bipropellant) **Operation:** Compressible CRES bellows Propellant and ullage gas **Compatibility:** Compatible with N2O4 and hydrazines for long term storage Material for N2O4 = CRES Material for hydrazines = CRES **Cycle Life:** Multiple cycles **One g Operation:** Can be performance tested at 1 g

Fig. 4.23 Zero-*g* propellant control devices.

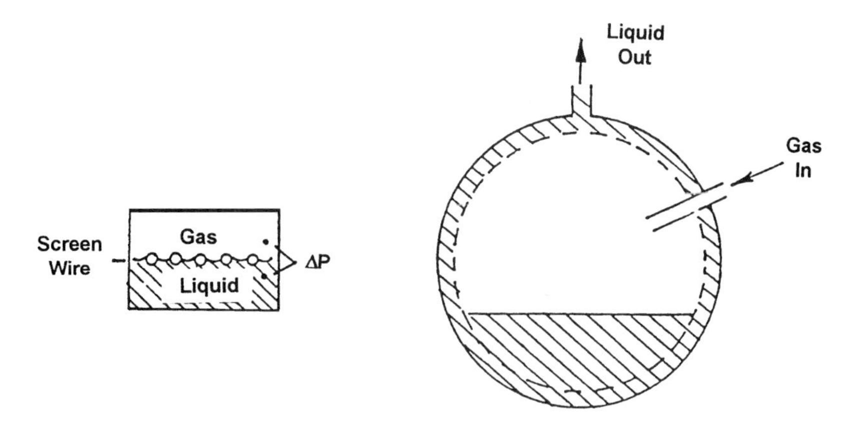

Fig. 4.24 Simple capillary device.

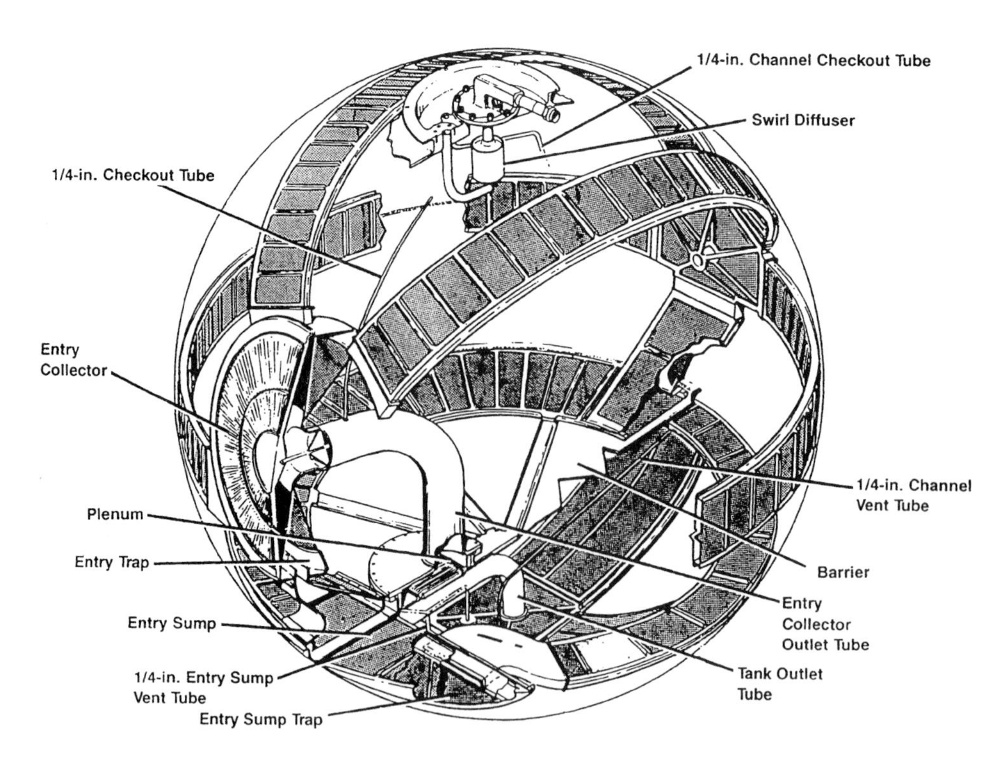

Fig. 4.25 Space Shuttle RCS tank capillary device. (Reproduced with permission of Lockheed Martin; Ref. 6.)

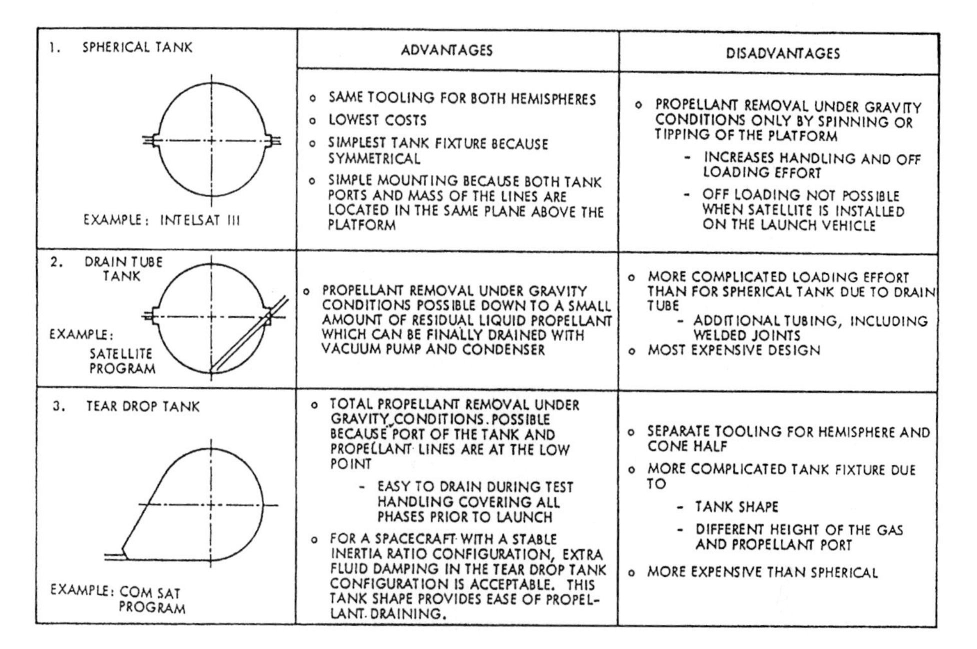

	ADVANTAGES	DISADVANTAGES
1. SPHERICAL TANK EXAMPLE: INTELSAT III	o SAME TOOLING FOR BOTH HEMISPHERES o LOWEST COSTS o SIMPLEST TANK FIXTURE BECAUSE SYMMETRICAL o SIMPLE MOUNTING BECAUSE BOTH TANK PORTS AND MASS OF THE LINES ARE LOCATED IN THE SAME PLANE ABOVE THE PLATFORM	o PROPELLANT REMOVAL UNDER GRAVITY CONDITIONS ONLY BY SPINNING OR TIPPING OF THE PLATFORM - INCREASES HANDLING AND OFF LOADING EFFORT - OFF LOADING NOT POSSIBLE WHEN SATELLITE IS INSTALLED ON THE LAUNCH VEHICLE
2. DRAIN TUBE TANK EXAMPLE: SATELLITE PROGRAM	o PROPELLANT REMOVAL UNDER GRAVITY CONDITIONS POSSIBLE DOWN TO A SMALL AMOUNT OF RESIDUAL LIQUID PROPELLANT WHICH CAN BE FINALLY DRAINED WITH VACUUM PUMP AND CONDENSER	o MORE COMPLICATED LOADING EFFORT THAN FOR SPHERICAL TANK DUE TO DRAIN TUBE - ADDITIONAL TUBING, INCLUDING WELDED JOINTS o MOST EXPENSIVE DESIGN
3. TEAR DROP TANK EXAMPLE: COM SAT PROGRAM	o TOTAL PROPELLANT REMOVAL UNDER GRAVITY CONDITIONS. POSSIBLE BECAUSE PORT OF THE TANK AND PROPELLANT LINES ARE AT THE LOW POINT - EASY TO DRAIN DURING TEST HANDLING COVERING ALL PHASES PRIOR TO LAUNCH o FOR A SPACECRAFT WITH A STABLE INERTIA RATIO CONFIGURATION, EXTRA FLUID DAMPING IN THE TEAR DROP TANK CONFIGURATION IS ACCEPTABLE. THIS TANK SHAPE PROVIDES EASE OF PROPELLANT DRAINING.	o SEPARATE TOOLING FOR HEMISPHERE AND CONE HALF o MORE COMPLICATED TANK FIXTURE DUE TO - TANK SHAPE - DIFFERENT HEIGHT OF THE GAS AND PROPELLANT PORT o MORE EXPENSIVE THAN SPHERICAL

Fig. 4.26 Tank configurations for spin-stabilized spacecraft. (Reproduced with permission of R. L. Sackheim; Ref. 9.)

outlet is then horizontal when the spacecraft is being assembled and tested; in this position the tank is difficult to drain. There are three solutions to this problem shown in Fig. 4.26 along with the pros and cons of each:

1) The tank can be kept simple and spherical, and the entire spacecraft (or propulsion module) can be rotated to drain the tank. Alternatively, an additional port can be installed in the tank for drainage.

2) An internal drain tube can be installed sloped so that either acceleration vector will drain the liquid.

3) Or the tank can be shaped so that it will drain with either acceleration vector. Such a tank has a teardrop or conosphere shape. This design has been used successfully on a number of recent spacecraft. One disadvantage of this design is that energy dissipation from fuel slosh reduces the stability margin of a dual-spin spacecraft; this reduction is worse than a spherical tank and can be as much as a factor of 30 (Ref. 9).

4.4.2.5 Propellant loading. Spacecraft systems use the weight method to measure propellant loaded. The system is moved to a remote area and weighed empty. The propellant is then loaded using special clothing and procedures to protect the personnel. Loading is a hazardous event; one or more dry runs of the procedure will precede the actual event. After loading the system is weighed again. The weight change is the propellant load.

Because of the loading process, it is highly desirable to design any liquid propulsion system to be readily removable from the spacecraft as a module, because

1) the propulsion system can be taken to the remote area for loading, independent of the spacecraft operations; 2) work can proceed on the spacecraft in parallel to the loading process; and 3) if a spill should occur during loading, the equipment at risk is limited. The loading takes place at the launch site close to the launch date. This is a time and place when these advantages are very important.

4.4.3 Pressurization Subsystems

The purpose of a pressurization system is to control the gas pressure in the propellant tanks. For spacecraft systems the tank pressure must be higher than the engine chamber pressure by an amount equal to the system losses; in addition a significant delta pressure must be maintained across the injector for combustion stability.

4.4.3.1 Pressurants. The pressurant, or pressurizing gas, must be inert in the presence of the propellants, and a low molecular weight is desirable. There are two pressurants in spacecraft use: nitrogen and helium. Helium provides the lightest system, but it is difficult to prevent helium leakage; therefore, nitrogen is used if the weight situation will allow it.

4.4.3.2 Ullage. The volume that the pressurant occupies above the propellant is called the *ullage*. This curious term was borrowed from the wine industry.

4.4.3.3 Blowdown system. There are two basic types of pressurization systems in use today. They are regulated and blowdown. A blowdown system is shown in Fig. 4.27. In a blowdown system the tank is pressurized to an initial level, and the pressure is allowed to decay as propellant is used.

The advantages of a blowdown system are as follows: 1) it is the simplest method and hence more reliable, and 2) it is less expensive because of fewer components. The disadvantages are as follows: 1) Tank pressure, thrust, and propellant flow rate vary as a function of time, and 2) I_{sp} is a second-order function of chamber pressure and drops as a function of time. Variability of flow rate and engine inlet

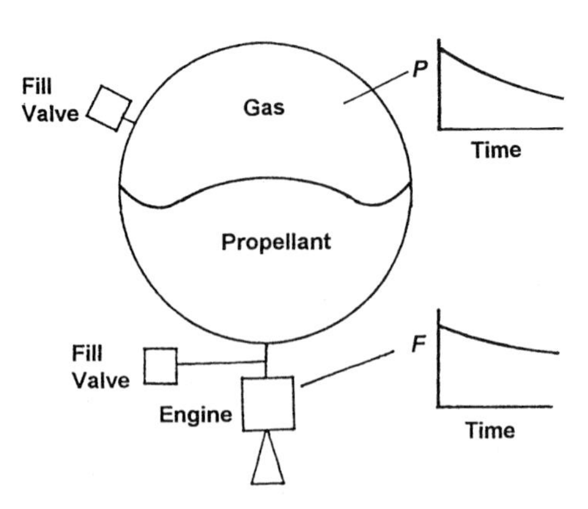

Fig. 4.27 Blowdown pressurization.

pressure make the blowdown system difficult to use with bipropellant systems. However, the disadvantages are slight for monopropellant systems; modern spacecraft monopropellant systems use blowdown exclusively.

4.4.3.4 Tank gas thermodynamics

Blowdown ratio. The ratio of initial pressure to final pressure is called the blowdown ratio B. The blowdown ratio is

$$B = \frac{P_{\text{gi}}}{P_{\text{gf}}} = \frac{V_{\text{gf}}}{V_{\text{gi}}} \tag{4.65}$$

and

$$V_{\text{gi}} = \frac{V_u}{B - 1} \tag{4.66}$$

$$V_{\text{gi}} = \frac{W_u}{\rho(B - 1)} \tag{4.67}$$

where

V_{gi} = initial ullage volume, m^3 (ft^3)
P_{gi} = initial gas pressure, Pa (psia)
V_{gf} = final ullage volume, m^3 (ft^3)
P_{gf} = final gas pressure, Pa (psia)
V_u = volume of useable propellant, m^3 (ft^3)
W_u = weight of useable propellant, kg (lbm)
B = blowdown ratio unitless
ρ = propellant density, kg/m^3 (lb/ft^3)

The maximum blowdown ratio is determined by the inlet pressure range the engines can accept. Ratios of three to four are in common use today; a blowdown ratio of six has been flown.[9]

Equation of state. From thermodynamics we know that for an ideal gas at any state point, the product of the tank pressure and the ullage volume is

$$PV = WRT \tag{4.68}$$

where

W = gas weight, kg (lb)
P = gas pressure (psia)
V = ullage volume, m^3 (ft^3)
R = specific gas constant = universal gas constant/molecular weight; universal gas constant is 8314.14 joule/kg-mole °K (1545 ft-lbf/lb-mole °R)
T = temperature of the gas, °K (°R)

Equation (4.68) is the equation of state; it relates the thermodynamic parameters for a perfect gas and steady-state conditions. (The departure of real gases from the behavior of perfect gases is normally negligible.) The equation of state is particularly useful for calculating pressurant weight in the form

$$W = \frac{PV}{RT} \tag{4.69}$$

Isothermal expansion. For spacecraft the outflow of propellant is usually slow, and heat transfer will keep the gas temperature fixed at or near the propellant temperature, and the process is isothermal. For an isothermal process

$$P_2 = P_1 \frac{V_1}{V_2} \tag{4.70}$$

During the blowdown process, the tank pressure at any time t is

$$P(t) = P_{\text{gi}} \left(\frac{V_{\text{gi}}}{m_p(t)/\rho + V_{\text{gi}}} \right) \tag{4.71}$$

where $P(t)$ is the tank gas pressure at any time t, P_{gi} is initial gas pressure; $m_p(t)$ is the total propellant mass removed from the tank in the time interval ending at time t, kg (lb); and ρ is the propellant density, kg/m^3 (lb/ft^3).

Isentropic expansion. If propellant is withdrawn rapidly as it would be in a translation burn, the gas expansion in the ullage will be isentropic, and the gas temperature will drop during the process. For isentropic expansion

$$P_1 V_1^k = P_2 V_2^k \tag{4.72}$$

and the tank pressure at any time t is

$$P(t) = P_{\text{gi}} \left(\frac{V_{\text{gi}}}{m_p(t)/\rho + V_{\text{gi}}} \right)^k \tag{4.73}$$

The difference between isothermal expansion [Eq. (4.71)] and isentropic expansion [Eq. (4.73)] is the exponent of the volumetric expansion term, k.

Figure 4.28 shows the difference in tank pressure with rapid, isentropic expansion and slow isothermal expansion for the same volume change. With isentropic

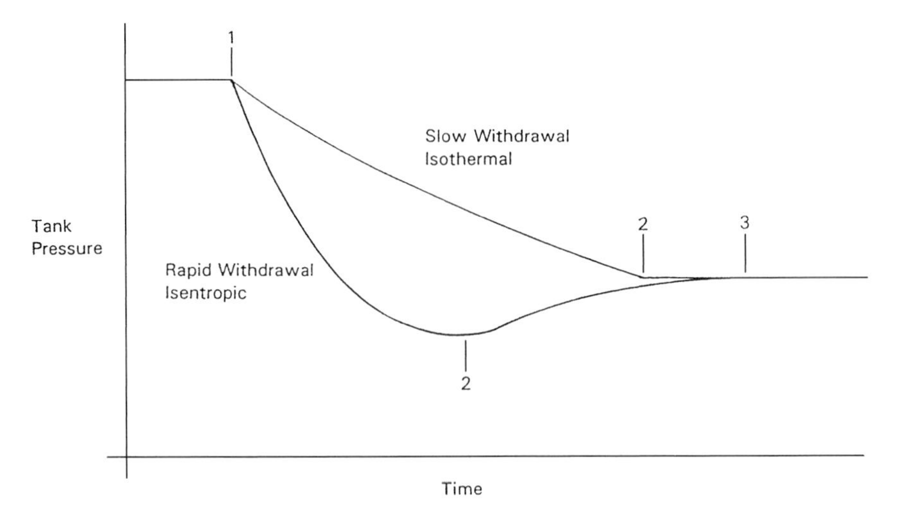

Fig. 4.28 Isothermal and isentropic blowdown.

expansion the temperature drops more rapidly and drops below propellant temperature. After engine shutdown (point 2 in Fig. 4.28), the gas temperature warms up to the propellant temperature in time. With isentropic expansion the tank pressure and thrust will be lower during a portion of the burn than isothermal calculations would predict.

Blowdown system thrust. With a blowdown system the tank pressure, propellant flow rate, and thrust are a function of time, dropping as the burn progresses. Thus a blowdown system cannot be used in a situation that requires a constant thrust.

Bipropellant systems have not yet been flown with a blowdown system because of the difficulty in holding the two propellant tanks at the same pressure and the concern about operating bipropellant engines over a wide pressure range; however, there is serious development work going on in this area.[10] Regulated pressurization systems will be discussed in Section 4.5.

4.4.4 Tankage Weight

(Note: The information in this section is intended for tank weight estimation only. The structural design of a propulsion system tank requires analysis beyond what is described in this section.)

The major dry weight component in a liquid-propellant propulsion system is the tankage. In this section we will show how tank weight can be estimated. First, the volume of the tank must be established and maximum tank pressure set. Then the tank weight can be estimated. For a propellant tank there are four components to the volume: 1) the initial ullage; 2) the useable propellant volume; 3) the unusable propellant volume, which is 3–4% of usable; and 4) the volume occupied by the zero-g device.

Example 4.7 Blowdown Propellant Tank Volume

What is the propellant tank size for a blowdown hydrazine monopropellant system with the following specifications?

$$I_{sp} = 225 \text{ s}$$

$$I = 50{,}000 \text{ N-s}$$

$$B = 4$$

Assume an elastomeric bladder, a spherical tank, and a design temperature range of 5–37.8°C.

The useable propellant weight

$$W_u = \frac{50{,}000}{(225)(9.80665)} = 22.66 \text{ kg}$$

The density of hydrazine at 37.8°C (100°F) = 992.15 kg/m^3 and volume of the usable propellant is $Vu = 0.022839$ m^3 or 22,839 cc.

The volume of unusable propellant is about 3% of useable for a monopropellant system; thus, the total volume of propellant is $Vp = 23{,}525$ cc.

The initial ullage volume is, from Eq. (4.66),

$$V_{gi} = \frac{22839}{3} = 7613 \text{ cc}$$

The tank volume before provisions for the bladder is

$$Vp + V_{gi} = 23{,}525 + 7613 = 31138 \text{ cc}$$

A bladder is essentially a hemispherical shell that nests within the tank shell. The internal radius of the bladder is approximately

$$r = \sqrt[3]{\frac{(0.75)(31138)}{\pi}} = 19.516 \text{ cm}$$

The area of the bladder is half the surface of a sphere

$$A_b = 2\pi r^{2..} = 2\pi(19.516)^{2..} = 2393.21 \text{ cm}^2$$

The bladder thickness can be expected to be about 0.20 cm; therefore, the volume of the bladder can be approximated as

$$V_b = 0.02(2393.21) = 479 \text{ cc}$$

and the required volume of the tank is 31,617 cc.

4.4.4.1 Spherical tanks. After determining the volume required, the tank weight can be estimated. The common tank configurations are spherical tanks and barrel tanks with hemispherical domes. Tank weights are a byproduct of the structural design of the tanks. For spheres the load in the walls is pressure times area as shown in Fig. 4.29. The force *PA* tending to separate the tank is

$$PA = P\pi r^2 \tag{4.74}$$

where P is the maximum pressure in the tank, Pa (pressure above atmospheric); r the internal radius of the tank, m; and A the area, m^2.

Because stress is load divided by the area carrying the load,

$$\sigma = \frac{P\pi r^2}{2\pi r t} = \frac{Pr}{2t} \tag{4.75}$$

and

$$t = \frac{Pr}{2\sigma} \tag{4.76}$$

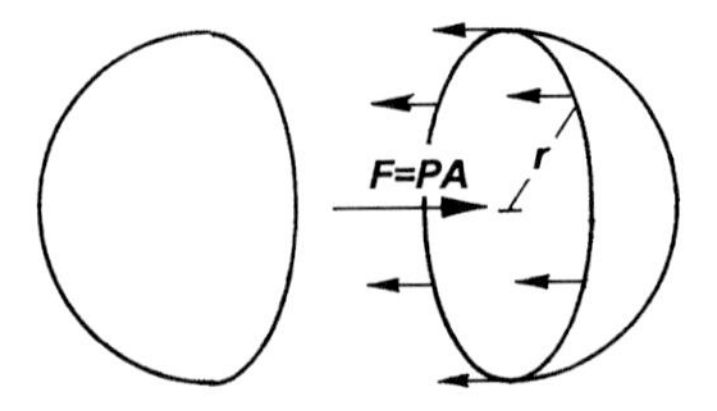

Fig. 4.29 Spherical tank stresses.

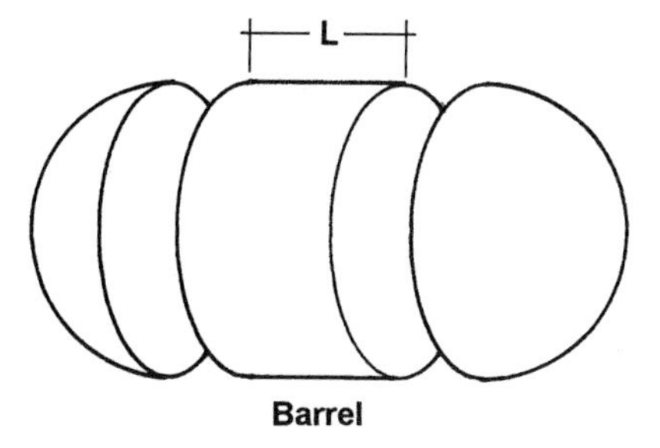

Fig. 4.30 Cylindrical tank.

where σ is the allowable stress, N/m^2 (after all structural margins are accounted for, including man rating if required) and t is the thickness of the tank wall, m.

Knowing the thickness of the walls, the weight of the tank membrane can be calculated as

$$R = r + t \tag{4.77}$$

$$W = 4/3\pi\rho(R^3 - r^3) \tag{4.78}$$

where W is the tank membrane weight, kg; ρ the density of the tank material, kg/m^3; and R the outside radius of the sphere, m.

4.4.4.2 Cylindrical barrels. For cylindrical barrels the hoop stress is twice that in a spherical shell, as shown in Fig. 4.30:

$$t = \frac{Pr}{\sigma} \tag{4.79}$$

The weight of the barrel is

$$W = \pi L\rho(R^2 - r^2) \tag{4.80}$$

where L is the length of the barrel section, m.

4.4.4.3 Penetrations and girth reinforcements. The membrane weight is for a perfect pressure shell. Areas of reinforcement, called *land areas*, are needed for 1) girth welds (the weld between two hemispheres), 2) penetration welds (welds for inlet and outlets), 3) bladder attachment (attachment between a bladder and the tank wall), and 4) structural mounting of the tank.

Weight must be added to the membrane weight to account for these reinforcements. Weld areas must be reinforced because of the reduction in material properties near welds. For the girth weld two reinforcing bands with a thickness of t, a width of about 5 cm each are added, as shown in Fig. 4.31.

Penetrations for attachments, inlet and outlet tubes weaken the shell and require reinforcement rings. These are about 15 cm in diameter centered on the penetration, with a thickness of t. Mounting pads on the tank add about 2% of the supported weight. Penetrations, girth land areas, and mounting pads normally add about 25% to the shell weight.

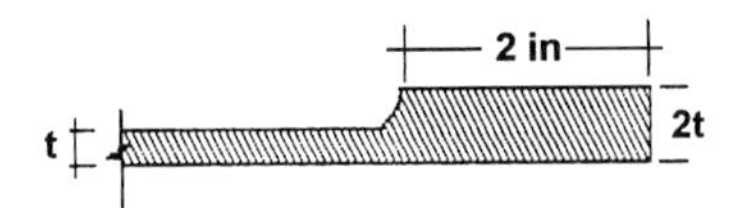

Fig. 4.31 Land area for welds.

It is important to design the tank mounting so that the tank can expand and contract without being constrained. Otherwise expansion will put unexpected loads into the walls.

Example 4.8 Spherical Propellant Tank Weight Estimate

Estimate the weight of a spherical titanium propellant tank with the following specifications: maximum working pressure = 4653.96 kPa, allowable stress = 690,000 kPa, internal volume = 0.02812 m^3, and elastomeric diaphragm.

Provide land areas for penetrations at each pole, a girth weld, and a structural mount pad at the lower pole.

Calculate the membrane thickness first:

$$r = \sqrt[3]{\frac{0.75(0.02812)}{\pi}} = 0.18865\,\text{m (inside, minimum)}$$

From Eq. (4.76), using allowable stress of 690,000 kPa (for example only, the allowable stress is a matter for considerable analysis and debate),

$$t = \frac{(4653.96)(0.18865)}{2(689{,}476)} = 0.0006367\text{ m} = 0.0637\text{ cm (0.025 in.)}$$

The thickness, 0.0647 cm, is the minimum acceptable; therefore, tolerances must add to the thickness

$$t = 0.0657 \pm 0.002\text{ cm}$$

Weight estimates should be made for the maximum thickness. With thickness determined, the tank outside radius is

$$R = 18.865 + 0.0677 = 18.932\text{ cm (outside)}$$

Calculate membrane weight using a density of 4429.89 kg/m^3 or 0.004430 kg/cc:

$$W = \frac{4}{3}\pi(0.00443)[(18.932)^3 - (18.865)^3] = 1.332\,\text{kg}$$

Now calculate the reinforced areas starting with the girth weld land:

$$W = 2\pi Rtw\rho$$

$$W = 2\pi(18.9)(0.0637)(5)(2)(0.004430) = 0.335\,\text{kg}$$

The penetration land weight, two 15-cm-diam disks,

$$W = 2\pi r^2 t\rho = 2\pi(7.5)^2(0.0657)(0.004430) = 0.103\,\text{kg}$$

The structural attachment weight equals 2% of supported weight.

1.332 kg (membrane) + 0.335 kg (girth land) + 0.103 kg (penetrations)
= 1.77 kg
+ 0.04 kg (structural attachment, 2%)
= 1.81 kg (tank shell weight)

Calculating the diaphragm volume and weight assuming a thickness of 0.2 cm,

$$V_d = \frac{4}{3}\pi[(18.865)^3 - (18.665)^3] = 885\,\text{cc}$$

$$\text{Diaphragm weight} = (0.0009965)(885) = 0.087\,\text{kg}$$

The tank assembly weight is as follows: tank shell, 1.81 kg; diaphragm, 0.087; and total, 1.897 kg.

Example 4.9 Spherical Tank Weight Estimate: Short Form

The weight estimate for a spherical tank can be approximated with less computation by noting that the tank shell weight is proportional to the product of pressure and volume. The proportionality constant can be developed by a series of calculations similar to the preceding example. The resulting approximation is

$$W_t = 0.0116PV \tag{4.81}$$

where

W_t = weight of a titanium spherical tank shell, land areas and attachments, kg
P = maximum tank pressure, kPa
V = tank volume, m^3

For the preceding problem

$$W_t = (0.0116)(4653.96)(0.02812) = 1.52\,\text{kg}$$

4.4.5 Monopropellant System Design Example

This section illustrates how the information in the preceding sections can be used to produce a complete monopropellant system design to the phase A level. The spacecraft under consideration is a geosynchronous communication satellite, which is placed in orbit by Atlas Centaur and an apogee kick motor. The spacecraft is three-axis stabilized except during the solid motor burn for which the spacecraft is spun and despun. The steps in a monopropellant propulsion system design are summarized here.

1) Define the requirements: steady-state impulse required from orbital mechanics (set maneuvering thrust level); pulsing impulse required from attitude control system (ACS); maneuvering thrust level required from ACS; wheel unloading if any from ACS; minimum revolutions per minute if spinner from ACS; maximum moment arm from launch vehicle shroud; fault protection from customer; temperature limits from thermal control.

2) Calculate the propellant required; add margin.
3) Select the propellant control device.
4) Decide dual-vs single-propellant tanks.
5) Decide propellant tank type, sphere, barrel, conosphere.
6) Select the pressurant: helium if mass is critical, otherwise nitrogen.
7) Select pressurization system type, and set the performance parameters, max tank pressure, and blowdown ratio.
8) Design the propellant tank.
9) Design the engine modules and general arrangement.
10) Design the system schematic; plan redundancy.
11) Calculate system mass.
12) Conduct trade studies of system alternatives; repeat the process.

4.4.5.1 Requirements. The following requirements have been established for the system: steady-state impulse required N-S stationkeeping, 234866 N-s; E-W stationkeeping, 25443 N-s; reaction wheel unloading, 37854 N-s; and pulsing impulse required spin/despin/nutation damping, 9340 N-s; initial orientation, 15570 N-s; propellant system temperature range 5°–50°C; maneuver 180 deg about any axis in 20 s; spacecraft max moment of inertia = 1000 kg-m^2; launch vehicle = Titan 4/Centaur. No single, nonstructural failure can cause loss of more than 50% of the mission objectives or mission duration.

4.4.5.2 Propellant inventory. All of the burns are pulsing or short steady-state burns. Select catalyst bed heating to improve I_{sp} and to improve catalyst bed life. Estimate the average I_{sp} for short burns to be 215 s and for pulsing 110 s. With these assumptions the propellant inventory, Table 4.8, can be constructed. Providing propellant reserves is always an area for lively project discussion; alternatively, it may be specified by your customer. A 50% reserve in not excessive for early stages of a project.

4.4.5.3 Initial selections. Now let us make some arbitrary choices so that the design can proceed; these choices can be revisited by later trade studies. Initial choices are blowdown pressurization (blowdown is almost universally used for monopropellant systems); blowdown ratio is 4; helium pressurant; initial tank pressure is 3619.75 kPa; diaphragm propellant control device; dual spherical tanks (you can readily show that this propellant load is too big for a single tank); and allowable stress 690,000 kPa.

Table 4.8 Propellant inventory

Propellant use	Propellant weight
1) Propellant for short burns	298150/(215)(9.8065) = 141.4 kg
2) Propellant for pulsing	24910/(110)(9.8065) = 23.1 kg
3) Reserves	50% = 82 kg
4) Subtotal—Usable propellant	246.5 kg
5) Trapped propellant	3% of usable = 7.4
6) Loading uncertainty	0.5% of usable = 1.2
7) Loaded propellant	255.1 kg

4.4.5.4 Propellant tank weight. The propellant tank volume should be calculated based on hydrazine density at maximum temperature. Hydrazine density is[7] in kilograms/cubic meter

$$\rho = 1025.817 - 0.8742(^{\circ}\mathrm{C}) - 0.0005(^{\circ}\mathrm{C})^2 \tag{4.82}$$

At 50 deg

$$\rho = 980.98\ \mathrm{kg/m^3}$$

The hydrazine volume is

$$V_p = \frac{255.1}{980.98} = 0.2600\ \mathrm{m^3}$$

The volume of the usable propellant is

$$V_u = \frac{246.5}{980.98} = 0.2512\ \mathrm{m^3}$$

The initial ullage volume is

$$V_{\mathrm{gi}} = \frac{0.2512}{3} = 0.08376\ \mathrm{m^3}$$

The tank volume, before provisions for a diaphragm, is

$$V_t = 0.2600 + 0.08376 = 0.3437\ \mathrm{m^3}\ (0.1719\ \mathrm{m^3 each})$$

Estimating the diaphragm volume at 1% of the tank volume or 0.0018 m^3 produces a volume of 0.1737 m^3 for each tank. The tank can now be designed using the procedure demonstrated in Example 4.8, or by the short method, with the following result. For the tank design the inside diameter is 0.6916 m, the wall thickness is 0.0931 cm, and the shell weight is 7.595 kg. For the diaphragm design the diaphragm volume is 1524 cc, and the weight is 1.52 kg. The tank assembly weight is 9.115 kg each or 18.23 kg for the pair. The allowance made for the diaphragm is 0.00021 m^3 larger than required. The tank design could be iterated, but the gain would be small.

4.4.5.5 Pressurant weight. The adjusted initial ullage volume for each tank is

$$V_{\mathrm{gi}} = 0.08376 + 0.00021 = 0.06397\ \mathrm{m^3}$$

The helium mass loaded in each tank from the equation of state (specific gas constant for helium = 2078.5) is

$$W_g = \frac{PV}{RT} = \frac{(3619750)(0.06397)}{(2078.5)(290)} = 0.384\ \mathrm{kg}$$

4.4.5.6 Thruster arrangement. This spacecraft has a spin mode and a three-axis stabilized mode; the latter mode governs thruster arrangement. It takes 12 thrusters to apply pure couples to a three-axis stabilized vehicle.

4.4.5.7 Thrust level. This vehicle is required to make a 180-deg turn about any axis in 20 s. The maximum moment of inertia about any axis is 1000 kg/m. The launch vehicle is Titan 4 Centaur; therefore, from Table 2.15 the maximum shroud diameter is 4.5 m; allowing clearance, 4.4 m, and $l = 2.2$ m. (This choice of radius to the thrusters will have to be reviewed to determine if it is clear of the dynamic envelope of the shroud, which is smaller than the physical dimensions.) Rearranging Eq. (4.39) yields

$$F = \frac{4\theta_m I_v}{n\, l\, t_m^2} \tag{4.83}$$

and minimum thrust is

$$F = \frac{4\pi(1000)}{2(2.2)(400)} = 7.14\,\text{N}$$

At this point an industry search would be made for a thruster with a thrust level equal to or greater than 7.14 N. Let us assume that an acceptable engine was found with a 7.5-N thrust level.

4.4.5.8 System schematic. Figure 4.32 shows a simplified system schematic for the propulsion system as it has evolved so far.

Table 4.9 summarizes the logic used to establish the system schematic in a failure modes and effects (FMEA) table. (Some form of FMEA is normally a contractual requirement.) Valves 1A and 1B are opened at the start of the flight; therefore, they should be considered open as you read Table 4.9. Failure types are considered only once; symmetrical failures are not tabulated.

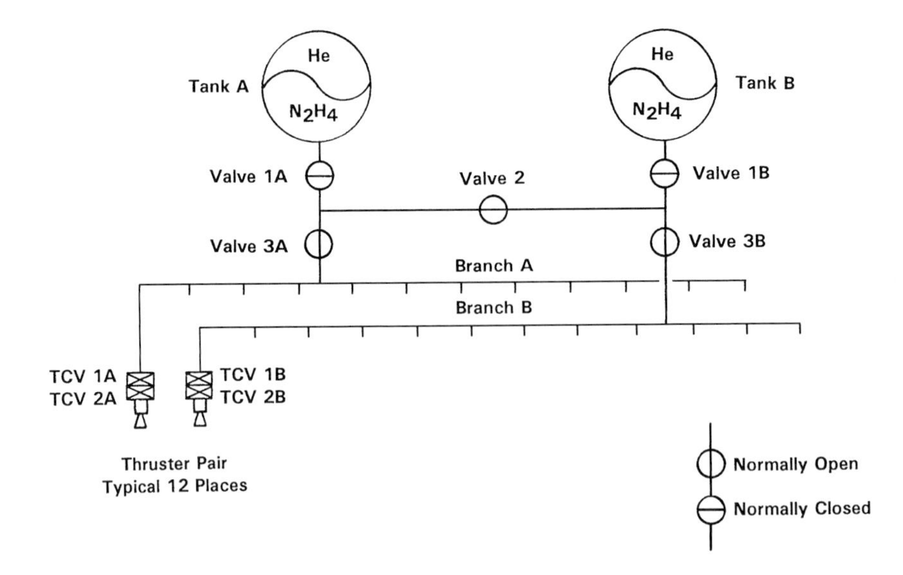

Fig. 4.32 System functional schematic—partial.

Table 4.9 Failure modes and effects

Failure	Corrective action	Mission effect
Thruster A failure off	Operate with thruster B	None
Thruster A failure on	Close TCV 1A and 2A	None
TCV 1A fail open	Operate with TCV 2A	None
TCV 1A fail closed	Operate with thruster B	None
Leakage branch A below valve 3A	Close valve 3A	None
Leakage branch A above valve 3A	Close valve 2	Reduced duration
Diaphragm rupture	Close valve 2 and 3A	Reduced duration
Valve 1A fail closed	None; use branch B	Reduced duration

Fault tolerance requirements dictate redundant branches of 12 engines each to accommodate an engine failure. If the thrust-chamber valves are placed in series, their dominant failure mode (leakage or failure open) can be tolerated. Thrust-chamber valve failure closed can be tolerated because the engines are redundant. (Series-parallel valves are an alternative solution.) If the dual tanks are cross strapped, a failure of either propellant system can be accommodated. If ordnance valves are used for cross strapping, the system can be sealed from propellant loading to flight use.

Table 4.9 shows that the system in Fig. 4.32 is satisfactorily failure tolerant. The schematic can now be completed with the addition of 1) fill and drain provisions, 2) heaters and thermostats, 3) instrumentation, and 4) filters. The schematic with these additions is shown in Fig. 4.33.

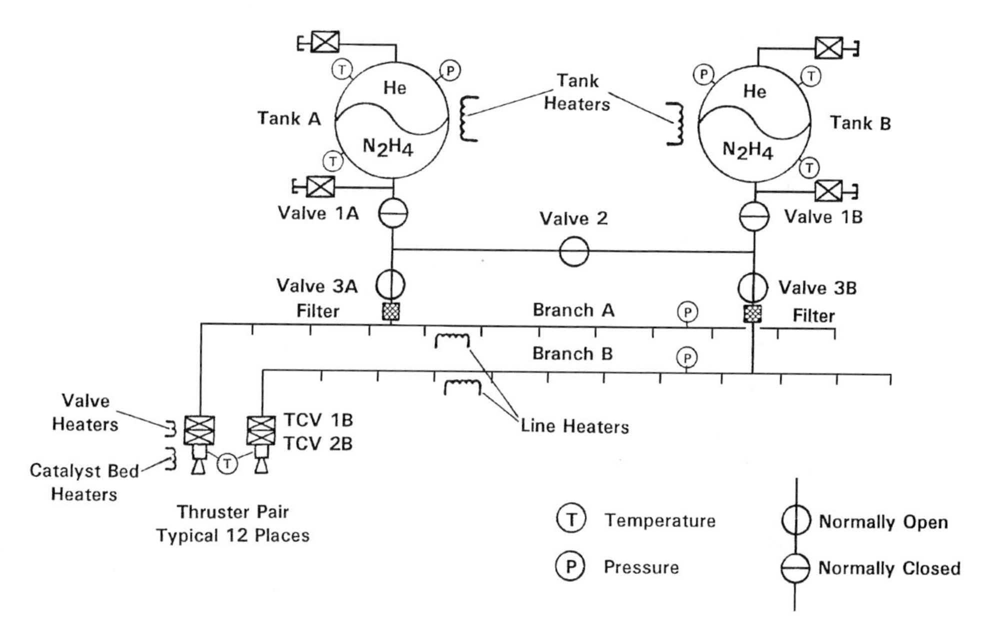

Fig. 4.33 System functional schematic.

Fill and drain connections are needed to load helium and hydrazine. The fill valves must be capped to provide a double seal against leakage. Filters are required on the ground equipment side of each fill and drain connection, thus avoiding additional flight filters. Filters are required downstream of all pyrovalves and upstream of all valve seats. For the schematic shown in Fig. 4.33, two filters are adequate. There are several filters manufactured for this use that have a 15–20 μ absolute rating.

Hydrazine must be prevented from freezing; the freezing point is about 2°C. If freezing should occur, the hydrazine shrinks. Line rupture will occur during thaw if liquid fills the volume behind the frozen hydrazine and is trapped. Satellite failures have occurred as result of this process. The power required for propulsion heating is a product of the thermal design of the vehicle. The most common technique is to provide heaters and thermostats on the lines, tanks, and thrust chamber valves. Catalyst beds are also heated to increase performance and bed life.

Instrumentation is selected so that any failure can be detected from analysis of the data. This requires temperature and pressure in each segment of the system. The FMEA should be reviewed with the instrument list in mind. The instrumentation shown in Fig. 4.33 is adequate if ground testing with more complete instrumentation has been performed. Chamber pressure is a valuable measurement; however, the measurement is not usually attempted in very small thrusters. Chamber pressure can be calculated from upstream pressure given adequate ground-test data.

4.4.5.9 First pulse considerations. The first pulses from a flight system are usually required to stabilize the residual rotation rates just after separation from the upper stage. This process is very important; the first pulses must have the full impulse that the attitude control system expects. For good first pulses the liquid system must be 1) gas free, 2) filled with hydrazine from the tank down to the thrust chamber valves, and 3) the catalyst bed must be hot.

Turning the catalyst bed heaters on before separation is easy; filling the hydrazine system is not. At launch the hydrazine system is full of dry nitrogen from valve 1A and 1B down to the thrust-chamber valves. If nothing were done, the first pulses would be nitrogen not hot gas. One solution to this problem is to open valve 1A and 1B during the upper-stage flight and then open all thrust-chamber valves until each chamber temperature rises. This approach might not be satisfactory to the launch vehicle because of the upsetting torques produced and because of exhaust impingement on the upper stage. The alternative approach is to open the thrust-chamber valves during the upper-stage flight with valve 1A and 1B closed. This step bleeds the trapped dry nitrogen into the hard vacuum of space. If valves 1A and 1B are now opened, hydrazine would fill the lines at very high velocity. When the hydrazine reaches the thrust-chamber valves, the deceleration shock can be energetic enough to decompose the hydrazine. To avoid this problem, orifices are required in the propellant lines to limit the initial hydrazine flow rate. If these orifices can be made small enough to force orderly initial filling and large enough not to obstruct the maximum thruster demand, the problem is solved. If not, separate bleed lines are required for the orifice flow; this is additional hardware not shown in Fig. 4.33.

Table 4.10 Propulsion mass estimate

Components	Unit mass, kg	Number	Total mass, kg
Propellant			255.1
Usable	246.5	1	246.5
Unusable	8.6	1	8.6
Pressurant	0.5	2.0	1.0
Feed system			
Tanks	7.6	2	15.2
Diaphragms	1.52	2	3.1
Valves	0.76	9	6.9
Filters	0.3	2	0.6
Lines and fittings	5.0	—	5.0
Temperature transducers	0.1	28	2.8
Pressure transducers	0.3	4	1.2
Heaters	0.5	—	0.5
Thrusters and valves			
7.5 N	0.82	16	13.1
Wet mass	—	—	304.5
Burn-out mass	—	—	58.0
Dry mass	—	—	49.4

4.4.5.10 System mass estimate. The system is now well enough defined for the first mass estimate shown in Table 4.10. The thruster weights, from Eq. (4.63), are

$$Wt = 0.4 + 0.0033(7.5) = 0.42\,\text{kg}$$

However, Eq. (4.63) was compiled for a thruster and one valve. The mass was adjusted upward to 0.82 kg to account for the additional valve. The remaining masses in Table 4.10 were estimated by analogy or calculated in the preceding paragraphs.

4.4.5.11 Trade studies. The first cycle of propulsion system design has been completed and a baseline established. The next step is to find ways to improve it. Potential improvements should be evaluated by comparison to the baseline by way of trade studies. Some of the trade studies that could be made against the preceding baseline are as follows: 1) evaluate the use of a single-propellant tank, and 2) evaluate the use of a dual-mode system.

4.4.6 Flight Monopropellant Systems

In this section several monopropellant systems with a successful flight history are described. They are arranged in the order of their launch date and hence in the order of increasing design sophistication. Each made a significant contribution to monopropellant system technology. The characteristics of the systems are summarized in Table 4.11.

Table 4.11 Flight monopropellant systems

Characteristic	Mariner 4	LANDSAT	Viking	HEAO	Voyager	Pioneer Venus	INTELSAT V	IUS	Magellan
Launch date	1964	1972	1976	1977	1977	1978	1980	1982	1989
Attitude control	3 axis	3 axis	3 axis	3 axis	3 axis	Spin	3 axis	3 axis	3 axis
No. thrusters	1	3	4, 3	12	16, 4, 4	7	20	12	12, 4, 8
Initial thrust, lb	50	1.0	10, 600	1.1	0.2, 5, 100	1.5	0.1, 0.6, 5	—	0.2, 5, 100
Pressurization	Regulated	Blowdown	Blowdown	Blowdown	Blowdown	Blowdown	Blowdown	Blowdown	Blowdown
Pressurant	N_2	N_2	N_2	N_2	N_2	He	N_2	N_2	He
No. propellant tanks	1	1	2	2	1	2	2	1, 2, or 3	1
Initial pressure, psia	—	—	530	350	450	350	270	—	450
Blowdown ratio	—	3.3	—	3.5	—	1.8	1.8	—	4
Repressurization	—	No	No	No	No	No	Yes	No	Yes
Propellant control	Bladder	Diaphragm	Deceleration	Diaphragm	Diaphragm	5-rpm spin	Capillary	Diaphragm	Diaphragm
Tank shape	Spherical	Spherical	Spherical	Spherical	Spherical	Conosphere	Barrel	Spherical	Spherical
Crossover	—	—	—	Yes	—	Yes	Yes	Yes	No
Dry mass, lb	26.7	—	—	56.2	—	—	78	—	135
Propellant mass	21.5	67	185	300	230	86.2	410		293.2
Features	Slug start jet vanes	Simplicity	Throttlable	—	400,000 cycle pulsing	—	Electrothermal thrusters	Removable tanks	—
Primary reference	11	12	13	14	15	16	17	18	19

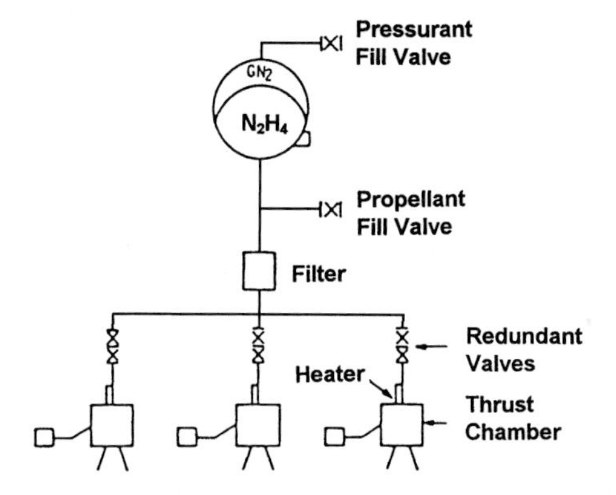

Fig. 4.34 LANDSAT 3 propulsion system. (Reproduced with permission of Lockheed Martin; Ref. 12, pp. 3–19.)

4.4.6.1 LANDSAT 3. The LANDSAT 3 offers an illustration of a flight monopropellant system at its simplest. Ground control uses the system to establish precise orbital parameters after orbit insertion and to make orbit adjustments throughout the mission in order to maintain overlapping coverage in the imagery.

The system, shown in Fig. 4.34, is constructed as a single module consisting of three rocket engines, a propellant tank, and feed system.

Each of the engine assemblies consists of a series redundant propellant valve, a catalyst bed, and a nozzle. Operation of the solenoid valves by electrical command produces thrust. Thirty kilograms of anhydrous hydrazine are loaded through a propellant fill valve. The single spherical titanium propellant tank contains an elastomeric diaphragm for propellant position control. The nitrogen pressurant is loaded through a fill valve. The pressurization is by a 3.3 to 1 blowdown system. The thrust range operating in a blowdown mode is from 4.4 N initially to 1.3 N finally. The system total impulse is 67,700 N-s. The system is mounted to the spacecraft sensory ring with the three thrusters located along the pitch and roll axes. The thrusters are aligned such that each thrust vector passes approximately through the spacecraft center of mass. With these thrust vectors the system is capable of imparting incremental velocity to the spacecraft to correct orbital errors and perturbations.

More complicated monopropellant systems would provide ACS pulsing, thrust vector control, and maneuvering. These capabilities would be provided by additional thrusters. A planetary system would require additional redundancy, and the system would be sealed by ordnance valves to prevent leakage during coast periods.

4.4.6.2 Voyager. The Voyager spacecraft, developed by the Jet Propulsion Laboratory, is one of the most successful scientific spacecraft ever to fly. It was launched in 1979; it gathered scientific data at each of the outer planets arriving at Neptune in August of 1989. It used monopropellant hydrazine system for orbit trims and attitude control. It used a solid rocket motor to supplement the launch vehicle energy at launch. The monopropellant system supplied thrust vector control during the solid motor burn. The system schematic is shown in Fig. 4.35 (Ref. 15).

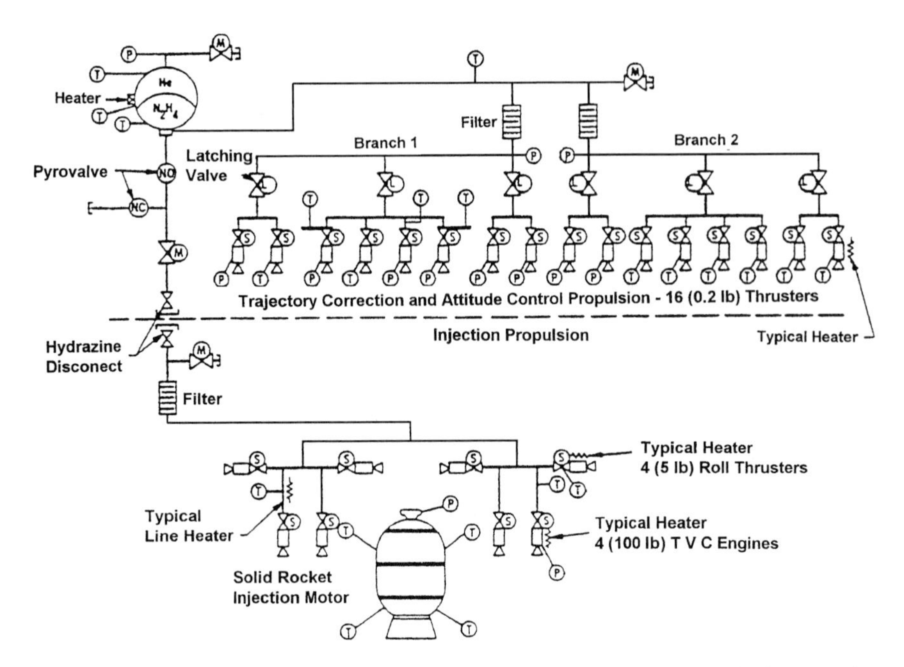

Fig. 4.35 Voyager propulsion system. (Copyright AIAA, reproduced with permission; Ref. 15.)

During solid motor burn, pitch and yaw were provided by four 445-N thrusters; roll control was provided by four 22-N thrusters. At the completion of the solid motor burn, the spent motor and the eight hydrazine thrusters associated with it were jettisoned. Note the hydrazine disconnect in Fig. 4.35. It is unusual to break a hydrazine system in flight; however, this disconnect operated satisfactorily.

The trajectory correction and attitude control system contained 104 kg of hydrazine in a 71-cm-diam titanium tank, which incorporated an elastomeric bladder. The tank was pressurized initially to 3103 kPa at 40.6°C. Pressurization was by blowdown.

Two branches of thrusters were provided, along with ability to switch pitch-yaw thrusters and roll thrusters independently. Switching was done with latch valves; latch valve position was telemetered. There were 16 0.89-N thrusters and valves. Heaters (1.4 W) were provided for each catalyst bed to maintain a minimum temperature of 115 ± 8°F.

During development of the system, a degradation in peak Pc and an increase of ignition delay were observed in vacuum testing the 0.89-N (0.2-lb) engines at 10-ms pulse widths. It was determined that the degradation was caused by trace amounts (<0.5%) of aniline in the hydrazine. The use of super pure (Viking Grade) hydrazine cured the problem.[15] The purification technique was developed by Lockheed Martin for the Viking Lander. The unusually long mission provided interesting flight data on the propulsion system. The cycles accumulated on the thrusters in the first 18 months of the mission are shown in Table 4.12 (Ref. 15).

Table 4.12 0.89-N (0.2-lb) thruster operating cycles

	Voyager 1		Voyager 2[a]	
Thruster	Br 1	Br 2	Br 1	Br 2
+ Pitch	29,510	521	16,416	2,986
− Pitch	44,235	332	22,612	6,162
+ Yaw	44,935	547	22,511	19,027
− Yaw	42,698	485	21,639	9,074
+ Roll	50,428	5,777	42,851	0
− Roll	45,236	6,188	44,840	0

[a]Voyager 2 data for 569 days after launch; Ref. 15.

The total mission time for Voyager 2 from launch to Neptune encounter was 4388 mission days. Linear extrapolation would place the total number of cycles on the Voyager 2 thrusters between 175,000 and 345,000 depending on the sharing between branch 1 and 2. The thrusters were qualified for 400,000 cycles.

4.4.6.3 INTELSAT V system. The INTELSAT V spacecraft is the fifth high-capacity communications spacecraft to be developed for INTELSAT corporation. At the time of its launch, it was the largest communications spacecraft ever built.[17] The INTELSAT V monopropellant system consists of two titanium propellant tanks, redundant sets of 10 thrusters each, and associated valves. Figure 4.36 shows the system arrangement.[20]

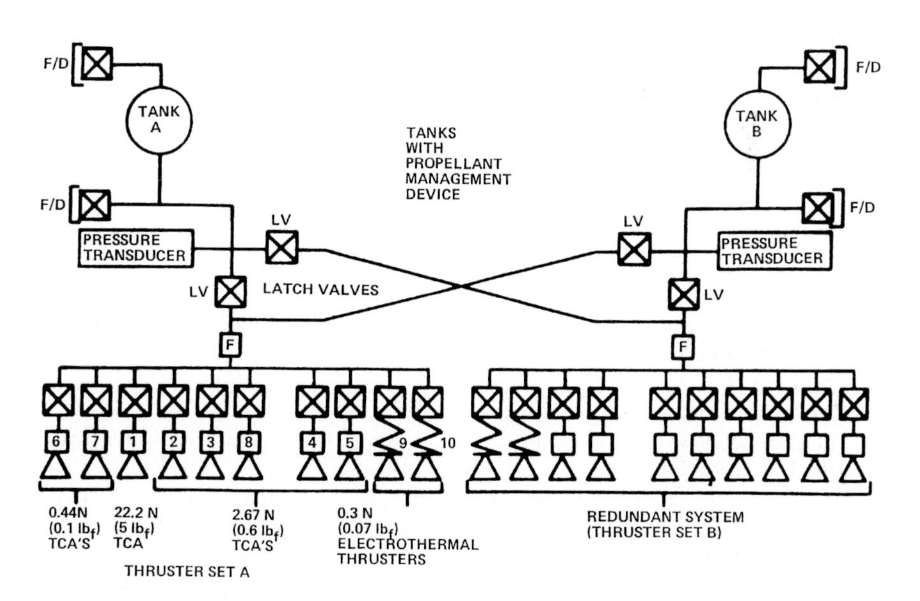

Fig. 4.36 INTELSAT V monopropellant system. (Reproduced courtesy of Space Systems/Loral; Ref. 20.)

Twenty-two N thrusters are used for orientation and orbit trim. Spacecraft spin/despin, east/west stationkeeping, and pitch/yaw control are performed by 2.7-N thrusters. Electrothermal thrusters are used for north/south stationkeeping, and these 0.4-N thrusters are backed up by the 3.7-N thrusters. Roll maneuvers are performed by 0.4-N thrusters.

The plumbing is arranged so that using isolation valves either tank can feed either or both sets of thrusters. Capillary propellant management devices control the hydrazine under 0-*g* or 1-*g* conditions in all tank positions. The tank internal volume is 140.7 liters and allows a loading of 213 kg of propellant. Nominally 185 kg are required for the mission.

A blowdown pressurization system is used. The tanks are high-strength titanium. All other components are stainless steel; joints are all welded. Electrochemical 0.4-N thrusters were chosen for the high I_{sp} (average 304 s) derived from operating at 2470 K.

4.4.6.4 Magellan system.[19] The mission of the Magellan spacecraft was to provide a first full surface map of Venus. The map was obtained from an elliptical Venus orbit using a synthetic aperture radar. The Magellan monopropellant blowdown system provided tip-off control during vehicle separations, reaction wheel desaturation, orbit trim maneuvers (for both the sun orbit and the Venus orbit), rate damping after any burn, and thrust vector control and attitude control during the solid rocket motor firing. The system was designed so that no single malfunction could prevent mission completion.

Super pure hydrazine is fed from the titanium diaphragm tank through the isolation valves to the four engine modules. Each module contains two 0.9-N, one 22-N, and two 445-N thrusters. The primary purpose of the aft-facing 445-N engines was thrust vector control during solid rocket motor firing. They were also used for orbit trims. The 22-N engines and 0.9-N engines were used for attitude control functions. The thrusters were redundant and were fed by dual-propellant feed systems. Thrusters also served as a backup in the event of a reaction wheel failure. The system schematic is shown in Fig. 4.37.

The pressurization system was a blowdown type using helium pressurant. The maximum operating propellant tank pressure was 3103 kPa (450 psig), and the blowdown ratio was 4. The pressurization system is unusual in that a single recharge was provided by an auxiliary tank containing helium at 22,060 kPa (3200 psig).

The propellant lines were evacuated during the launch process by opening the thrust chamber valves. During the IUS burn, after deployment from the Space Shuttle, the isolation valves (P0 and P1 in Fig. 4.37) were opened to allow propellant to fill the lines. Flow rate during line filling was controlled by orifices to prevent shock to the propellant. The propellant and pressurization systems were all welded stainless steel (except for the titanium tanks). There were no mechanical joints in the system.

The liquid propulsion module was readily separable from the spacecraft so that propellant loading could be done at a remote area apart from the spacecraft and off the critical path.

The spacecraft was developed by Lockheed Martin (then Martin Marietta) under contract to the Jet Propulsion Laboratory. The spacecraft was launched on Space Shuttle *Atlantis* on 9 May 1989; the mission was completely successful.

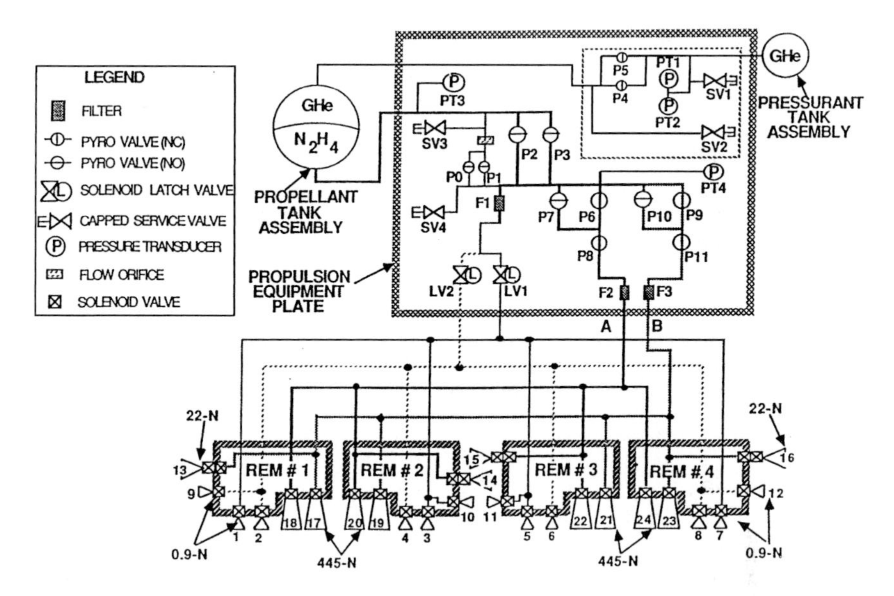

Fig. 4.37 Magellan monopropellant system. (Reproduced with permission of Lockheed Martin; Ref. 6.)

4.5 Bipropellant Systems

Bipropellant systems offer the most performance (I_{sp} as high as 450 s) and the most versatility (pulsing, restart, variable thrust). They also offer the most failure modes and the highest price tags.

4.5.1 Bipropellant Rocket Engines

The major parts of a pressure-fed engine are the injector, the nozzle and the cooling system, and thrust-chamber valves, as shown in Fig. 4.38. (Launch vehicle systems are pump fed and considerably more complicated.)

4.5.1.1 Injector. The injector introduces the oxidizer and fuel into the combustion chamber in such a way as to promote stable, efficient combustion without

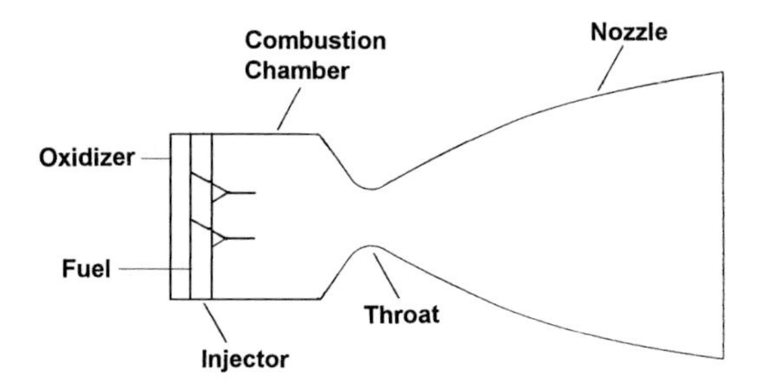

Fig. 4.38 Bipropellant rocket engine.

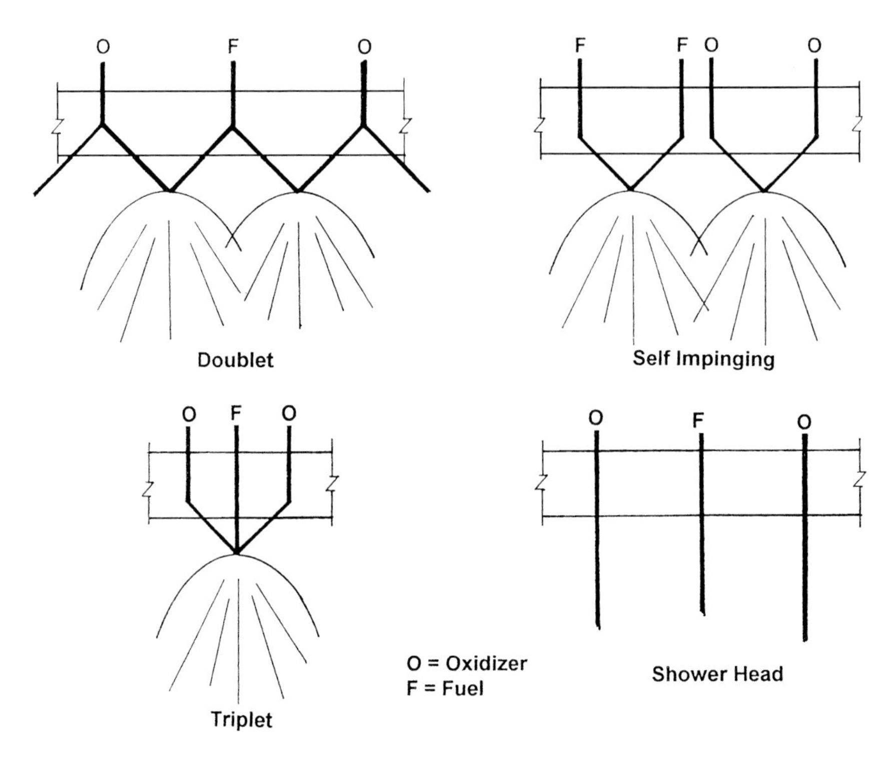

Fig. 4.39 Injector configurations in use.

overheating the injector face or the chamber walls. The injector design is the single most important contributor to engine performance. It also determines to a large measure whether combustion will be stable. Figure 4.39 shows injector configurations in use.

The most sensitive area in a bipropellant engine is the combustion zone of the thrust chamber just downstream of the injector. Propellants enter this high-pressure zone as room-temperature liquids and leave as supersonic gases at velocities greater than 600 m/s and temperatures as high as 3500°C. Energy release rates may be as great as 200 kW/cc. Mixing, ignition, and combustion occur in milliseconds. Propellants are introduced through the injector plate at high velocity so that the fuel and oxidizer are shattered into droplets small enough to complete combustion upstream of the throat. Continuous and rapid combustion must occur near the injector face so that pockets of mixed but unreacted propellants do not form and explode. The thrust developed by a single pair of injector orifices can be as high as 2000 N.

4.5.1.2 Cooling. The adiabatic flame temperature for nitrogen tetroxide and monomethylhydrazine is 3414°K; the heat-transfer rates at the throat can be as high as 3200 J/s-cm^2 during operation. Clearly, throat and nozzle cooling is an important design problem. The cooling methods used are 1) regenerative cooling, 2) film cooling, 3) ablative cooling, and 4) radiation cooling.

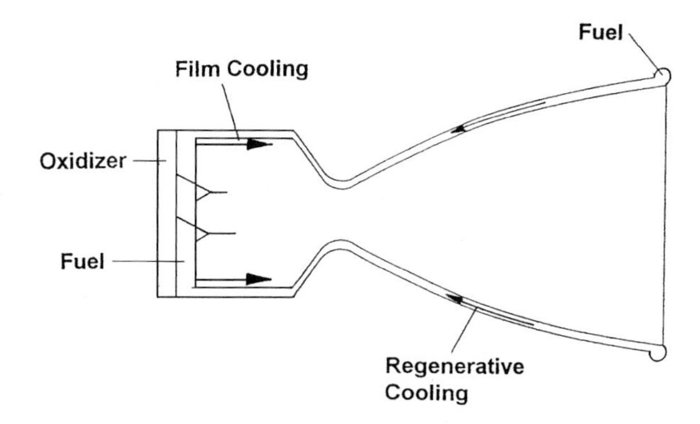

Fig. 4.40 Regenerative coolant flow.

Regenerative cooling is used on all large, launch vehicle engines. As shown in Fig. 4.40, the nozzle construction is a bundle of shaped cooling tubes brazed together and radially reinforced. The fuel is routed through the tubes from the exit plane up to the injector. The energy collected from the gas stream in cooling is returned to the combustion process by the hot fuel as it enters the chamber; hence, the process is regenerative.

Film cooling is used in all engines as a supplement to the primary cooling method. A cylindrical film of fuel is injected into the chamber such that it washes down the wall. The flow is designed so that some liquid will reach the throat. Cooling comes from the heat of vaporization of the fuel. Combustion near the wall is fuel rich and inefficient; therefore, specific impulse is reduced, and the amount of fuel used in film cooling is minimized in engine designs.

In *ablative cooling* the chamber wall is constructed of material that will be evaporated and eroded away during the firing. Typically the throat will be a machined carbon block to prevent throat erosion. Ablative cooling takes place primarily in the combustion chamber and diverging section, which are usually a fiberglass epoxy material. The material thickness is sized so that the expected burn time will remove an allowable amount of material.

For *radiation cooling* the chamber is made of a refractory material and radiative heat rejection to space plus enhanced film cooling keeps the chamber walls within acceptable limits.

4.5.2 Bipropellants

In Goddard's early work with rockets, he chose liquid oxygen and kerosene for their availability and performance. When the Germans started their work with rockets, they chose oxygen but elected to use alcohol as the fuel because of a scarcity of petroleum. After World War II, when the work returned to the United States, kerosene came back as the fuel and was renamed RP-1 (Rocket Propellant 1). Liquid oxygen was not a good choice for strategic weapons because of the difficulty storing it under field conditions. As a result "storable propellants" were developed by the U.S. Air Force. Nitric acid was used as an oxidizer in very early work; however, in the 1960s the U.S. Air Force selected less corrosive nitrogen tetroxide for the oxidizer. The fuels most seriously studied were hydrazine

Table 4.13 Properties of some propellants

Propellant	Symbol	Mole weight	Freezing point, °K	Boiling point, °K	Density at 20°C	Vapor pressure	
						kPa	°C
Chlorine triflouride	ClF_3	92.46	191	284.9	1.825	143	43.3
Fluorine	F_2	38	53.7	84.8	1.51	34	−197
Hydrazine	N_2H_4	32.05	274.7	386.4	1.008	1.4	20
Hydrogen	H_2	2.02	13.7	20.4	0.071	7.0	−259
MMH	$CH_3N_2H_3$	46.08	220.9	359.9	0.8765	4.8	20
Nitric Acid	HNO_3	63.02	231.9	358.4	1.513	6.41	20
Nitrogen tetroxide	N_2O_4	92.02	261.9	294.3	1.447	103	20
Oxygen	O_2	32	54.4	90.0	1.14	50.7	−189
RP-1	$CH_{1.9-2.0}$	175	228.7	455–533	0.806	0.14	20
UDMH	$(CH_3)_2N_2H_2$	60.10	215.9	337	0.793	16.4	20

and unsymmetrical dimethylhydrazine (UDMH). Hydrazine was less stable but it delivered higher I_{sp}. The final selection for the U.S. Air Force launch vehicle program was a 50/50 mixture of hydrazine and UDMH. This combination, named Aerozine 50, was first used in Titan II and is the Titan propellant combination to this day. In later developments the storable fuel of choice became monomethylhydrazine (MMH), which has the properties of Aerozine 50 without the problems of mixing.

Meanwhile NASA, with its need for heavy launch vehicles, developed Saturn series (Saturn 1, 1B, 5) with oxygen/hydrogen. The Space Shuttle, in turn, was developed with this combination. Research persisted with the more energetic fluorine oxidizer and derivatives like chlorine triflouride; these were eventually dropped because of the difficulty of containing fluorine in metal tanks. The properties of some propellants are shown in Table 4.13.

Fluorine, oxygen, and hydrogen are called *cryogenics* because of the extremely low temperatures at which they are liquids. There is a body of technology associated with the use of cryogenics that will not be covered in this text. As one example, note that a liquid hydrogen tank must be insulated in such a way that air cannot penetrate the insulation. If air penetrates to the tank wall and approaches liquid hydrogen temperatures, it will liquefy with resulting very high heat transfer to the hydrogen. As another example, if a cryogenic is allowed to warm in a long vertical feed line it will form sudden large bubbles as it reaches the boiling point. These bubbles will coalesce and rise through the liquid with destructive force; this series of events is called *geysering*. One cure for geysering is to inject gaseous helium into the cryogenic, then the cryogenic rapidly evaporates into the helium bubble seeking to reach a saturated vapor state; the result is rapid cooling of the cryogenic. Studies of the use of cryogenics in spacecraft have been conducted with the hope of using the cold temperatures achievable in space for cryogenic storage; however, cryogenics have never been used for anything other than launch vehicles.

4.5.2.1 Nitrogen tetroxide (N_2O_4). Nitrogen tetroxide is an equilibrium solution of nitrogen dioxide NO_2 and nitrogen tetroxide N_2O_4. The equilibrium state varies with temperature. It is a relative of nitric acid, which it smells like. It is reddish brown and very toxic. It is hypergolic (ignites spontaneously on contact) with hydrazine, Aerozine 50, and MMH; therefore, igniters are not required. This property makes pulsing performance practical with storable propellants. It is compatible with stainless steel, aluminum, and Teflon®; it is incompatible with virtually all elastomers.

4.5.2.2 Monomethylhydrazine (MMH). MMH is a clear water-white toxic liquid. It has a sharp decaying fish smell typical of amines. It is not sensitive to impact or friction; it is more stable than hydrazine when heated, but decomposes with catalytic oxidation. Nitrogen tetroxide and MMH is the dominant spacecraft propellant combination for bipropellant spacecraft propulsion. However recent planetary spacecraft have used hydrazine fuel in dual-mode systems, which will be discussed later.

4.5.2.3 Hydrazine (N_2H_4). The advent of dual-mode propulsion systems in the 1990s brought hydrazine to the forefront as the spacecraft fuel of choice. It provides slightly higher I_{sp} with nitrogen tetroxide than MMH and can be used as a monopropellant for pulsing. The properties of hydrazine were covered in detail in Section 4.4.

4.5.3 Bipropellant Fluid Systems

Pressurization, propellant systems, and tankage were discussed in Section 4.4.; these discussions are equally applicable to bipropellant systems. Only the peculiarities of bipropellant systems will be discussed here.

4.5.3.1 Propellant inventory. The propellant inventory is a subdivided tabulation of the loaded propellant weight. In preparation of a bipropellant inventory, the usable propellant is first subdivided into oxidizer and fuel. By definition, mixture ratio is

$$MR = \frac{w_o}{w_f} \tag{4.84}$$

where w_o is the oxidizer weight flow rate and w_f the fuel weight flow rate. Nitrogen tetroxide and MMH are a commonly used mixture ratio of about 1.6. Rearranging Eq. (4.84) produces an expression for separating usable propellant into oxidizer and fuel:

$$W_f = \frac{W_u}{1 + MR} \tag{4.85}$$

and

$$W_o = W_u - W_f \tag{4.86}$$

Table 4.14 Propellant inventory, kg

Propellant use	Fuel	Oxidizer	Total
Usable propellant	384.6	615.4	1000
Trapped propellant, 3%	11.5	18.5	30
Outage, 1%	3.8	6.2	10
Loading error, 0.5%	1.9	3.1	5
Loaded propellant	401.8	643.2	1045

where

W_u = total (oxidizer and fuel) usable propellant weight, kg
W_f = total usable fuel load, kg
W_o = total usable oxidizer load, kg

Table 4.14 shows an example propellant inventory for 1000 kg of usable propellant at a mixture ratio of 1.6.

The usable propellant is that portion of the propellant loaded which is actually burned. Not all of the propellant loaded can be used; a certain amount is trapped, drop out, or loading uncertainty. Trapped propellant is the propellant remaining in the feed lines, tanks and valves, hold up in expulsion devices and retained vapor left in the system with the pressurizing gas. Trapped propellant is about 3% of the usable propellant in small bipropellant systems.

The two primary activities that dictate the required usable propellant are 1) the propellant required for velocity change maneuvers as dictated by the mission design and 2) the propellant required to control attitude. The Tsiolkowski equations give usable propellant for velocity change maneuvers. For example, it takes a velocity change of about 2.25 km/s to move a spacecraft from a parking orbit to a geosynchronous orbit; the propellant for this maneuver is velocity change propellant Attitude control propellant is required for duty-cycle operation, wheel unloading, or spacecraft pointing. A given project can separate usable propellant into these functions in the propellant inventory.

Outage is caused by the difference between the mixture ratio loaded and the mixture ratio at which the engine actually burned. There is always some of one propellant left when the other propellant is depleted; the remaining propellant is called *outage*. Outage depends on loading accuracy and on the burn-to-burn repeatability of engine mixture ratio. One percent outage can be used for initial estimates; however, it takes statistical engine data and loading data to make an informed estimate.

Some design groups calculate the propellant requirements under nominal conditions. If this is done, it is necessary to calculate the additional propellant necessary to do the mission under worst-case conditions and show this in the propellant inventory. The most common practice is to calculate the propellant requirements using worst-case conditions and show that in the propellant inventory as just stated.

Propellant reserves can be shown as a separate item in the propellant inventory. Propellant reserves are a very valuable commodity. The more of the project reserves placed in propellant the better; there are three primary reasons for this: 1) propellant is usually the life-limiting expendable; 2) it is often desirable to make unplanned maneuvers in emergency conditions; and 3) there are usually extended mission objectives that can be achieved after the primary mission—if there is propellant. If reserves are held, these are itemized as a subdivision of usable propellant.

The measurement uncertainty in propellant loading is about 0.5%. The uncertainty is added to the load to ensure that the usable propellant can be no less than worst-case requirement.

Drop out (launch vehicle systems only) is the propellant left in the feed line when the propellant surface in the tank drops into the line.

4.5.3.2 Propellant control and tankage. Titanium, aluminum, and stainless steel are compatible with both nitrogen tetroxide and the hydrazines, and are the tank materials used. Titanium is the lightest and most common. The maximum performance mixture ratio and the equal volume mixture ratio are so close that it is sometimes possible to use identical tank shells for oxidizer and fuel as a cost-saving measure.

Nitrogen tetroxide is not compatible with elastomers; as a result, the propellant control devices are limited to metals and Teflon®. Bellows, capillaries, and Teflon® bladders have been used; see Fig. 4.23.

The Space Shuttle RCS chose titanium tanks with complex stainless-steel capillary systems. These were discussed in Section 4.4 and shown in Fig. 4.24. Mariner 9 and Transtage chose Teflon® bladders. In each case the propellant control device choice was driven by the oxidizer but used in both oxidizer and fuel tanks.

The Minuteman postboost propulsion system used a stainless-steel bellows assembly to control propellant position and to accommodate long-term propellant storage. Minuteman, like most other projects, decided to make both oxidizer and fuel tanks essentially identical.

4.5.3.3 Pressurization. Blowdown pressurization has not been used with bipropellants because of 1) the difficulty in keeping both tanks at the same pressure and 2) difficulties with varying inlet pressure to bipropellant engines. Either helium or nitrogen systems are used, although helium is the most frequent choice. A typical, simplified, bipropellant pressurization system is shown in Fig. 4.41.

A regulated system controls the pressure in the propellant tanks at a preset pressure. The pressurant (pressurizing gas) is stored at a high pressure—21,000–34,500 kPa (3,000–5,000 psi). The engine is supplied propellant at a tightly controlled lower pressure and flow rate. Thrust does not vary during propellant consumption. For bipropellant systems regulation is essential in order to keep the flow rate of each propellant constant and at the correct mixture ratio.

The pressurant tank weight is essentially constant for any initial pressure for a given gas weight, for example, it takes about 3.2 kg (7 lb) of tank to contain 1 lb of helium. Thus, the initial storage pressure selected is a second-order trade. A regulator requires an inlet pressure around 100 psi higher than the outlet pressure in order to operate properly. Therefore, the unusable pressurant is trapped in the pressurant tank at or above the minimum regulator inlet pressure. A typical pressure profile for a regulated system is as follows:

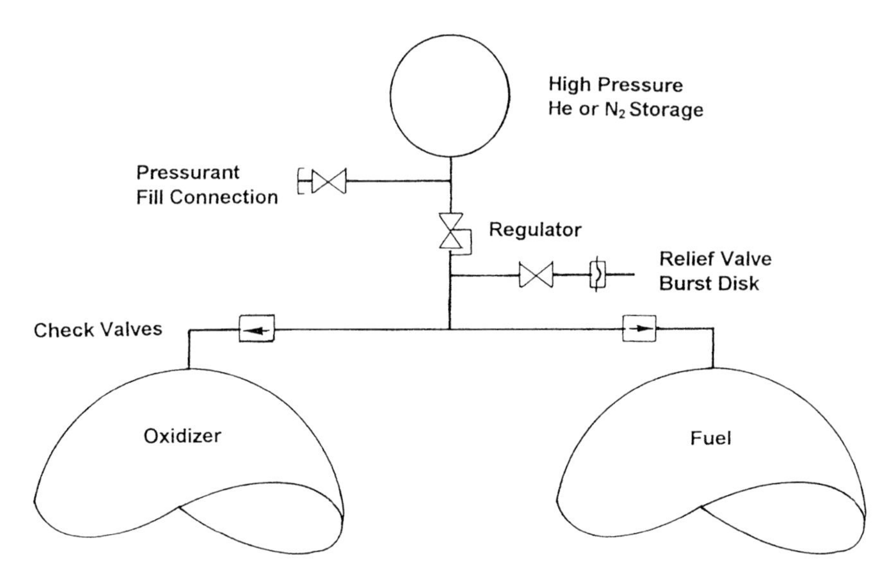

Fig. 4.41 Simplified bipropellant pressurization system.

Maximum pressurant tank pressure	31,000 kPa	(4500 psia)
Minimum pressurant tank pressure	4,500 kPa	(650 psia)
Minimum regulator inlet pressure	4,100 kPa	(600 psia)
Nominal propellant tank pressure	3,450 kPa	(500 psia)
Minimum engine valve inlet pressure	3,100 kPa	(450 psia)
Nominal chamber pressure	2,400 kPa	(350 psia)

The minimum ullage volume must be about 3% of the propellant tank volume in order for the regulator to have a stable response when outflow starts. In addition to the pressure regulator, relief valves are necessary to protect the tank in case the regulator fails open. It is good practice to place a burst disk upstream of the relief valve so that pressurant is not lost overboard at the valve seat leakage rate. This precaution is particularly important if helium is the pressurant. For bipropellant systems each propellant tank ullage must be isolated to prevent vapor mixing.

For constant temperature conditions the volume of the pressurant sphere is

$$V_s = \frac{P_r V_u}{P_1 - P_2} \tag{4.87}$$

where V_s is the volume of pressurant sphere, P_r the regulated propellant tank pressure, V_u the volume of the usable propellant (both), P_1 the initial pressurant sphere pressure, and P_2 the final pressurant sphere pressure.

Once the volume of the pressurant sphere is calculated, the initial weight of the stored pressurant can be calculated from the equation of state. The total weight of the pressurant on board is the initial weight of pressurant in the sphere plus the initial weight of pressurant in each tank ullage. (The initial pressurant in the ullage is normally supplied from ground sources.)

Note from Eq. (4.87) that the sphere volume, and hence weight, is independent of the pressurizing gas chosen.

Example 4.10 Pressurant Sphere Sizing: Regulated System

Calculate the pressurant sphere size and loaded gas weight for a regulated system under the following conditions:

Helium pressurant	$R = 8314.41/4 = 2078.6$	
Regulated propellant tank pressure	3551 kPa	(515 psia)
Initial pressurant sphere pressure	33095 kPa	(4800 psia)
Final pressurant sphere pressure	3965 kPa	(575 psia)
Volume of oxidizer consumed	0.66828 m^3	(23.6 ft^3)
Volume of fuel consumed	0.52103 m^3	(18.4 ft^3)
Initial oxidizer ullage	0.04021 m^3	(1.42 ft^3)
Initial fuel ullage	0.031149 m^3	(1.10 ft^3)
Temperature	21°C	(70°F)

Calculate the pressurant sphere volume from Eq. (4.87):

$$V_s = \frac{(3551)(0.66828 + 0.521003)}{(33095 - 3965)} = 0.14498\,\text{m}^3$$

The weight of the initial gas load from the equation of state, Eq. (4.69), is

$$w_s = \frac{(33095000)(0.14498)}{(2078.6)(294.3)} = 7.843\,\text{kg}$$

The weight of pressurant initially loaded into the propellant tanks, from the equation of state (4.68), is

$$w_u = \frac{(3551000)(0.04021 + 0.031149)}{(2078.6)(294.3)} = 0.414\,\text{kg}$$

The total weight of the helium on board at launch is 8.257 kg.

Careful consideration should be given to propellant vapor mixing between the check valves after permeation of the check valve seals. Regulator leakage occurred as Viking 1 neared Mars, caused by an obstruction on the valve seat. The spacecraft orbit insertion was successfully accomplished in spite of the leak as was the remainder of the multiyear mission. The leakage was exhaustively investigated. The probable cause was propellant vapor mixing between the check valves. The same phenomena was the most probable cause of the recent total failure of the Mars Observer. Positive mixing prevention is particularly important on long duration missions.

4.5.3.4 Unusable propellant. The unusable propellant is greater with bipropellant systems than monopropellant systems. The difference between loaded mixture ratio and burned mixture ratio results in residual of one or the other commodities. This residual is called *outage*. The unusable propellant is about 4% of the total load by weight.

4.5.3.5 Pulsing performance. Sutton[2] states that an I_{sp} of 50% of theoretical can be expected with a bipropellant engine producing 0.01-s pulses (as compared to steady state of 92%). With 0.1-s pulses 75 to 85% can be excepted. Reference 9 gives a pulsing I_{sp} for bipropellant engines of 170 s. Transtage motor testing with 0.02-s pulses indicates a delivered I_{sp} of 128 s or 45% of steady state for a 25-lb engine and a delivered I_{sp} of 191 s or 65% of steady state for a 45-lb engine. Transtage engine firings produced an impulse repeatability of $+/-10\%$ pulse to pulse and $+/-30\%$ engine to engine.

4.5.4 Flight Bipropellant Systems

Five spacecraft bipropellant systems with a successful flight history are summarized in Table 4.15. The systems are arranged in the order of first flight to show the progression of technology. The three most recent systems—Galileo, INTELSAT VI, and Mars Global Surveyor—are described in this section.

4.5.4.1 Galileo. Galileo is the next step in the exploration of Jupiter following the flyby missions of Pioneer 10 and 11 and Voyager 1 and 2. The Galileo mission is to find out more about the chemical composition and physical state of the atmosphere and to study the satellites of Jupiter. The spacecraft is a dual-spin orbiter and an attached probe. Galileo was the second interplanetary spacecraft to be launched on the shuttle, following the Magellan flight in 1989.

The Galileo propulsion system is a highly redundant bipropellant MMH/nitrogen tetroxide system shown in Fig. 4.42.

The system provides midcourse corrections, orbit adjustments, and Jupiter orbit insertion, which occurs after release of the probe. Maneuvering propulsion is accomplished with a single 400-N main engine; attitude control is provided with two clusters of six 10-N thrusters. The two clusters are on opposing booms. The 400-N engine can be fired for up to 70 min; the 10-N thrusters can be used in pulse mode or steady state. The minimum impulse bit is 0.09 N-s with an on time of 22 ms. The 10-N thrusters can back up the 400-N engine in the event of a failure. Three parallel supply branches feed propellant to the two redundant thruster branches and to the 400-N engine. Dual helium tanks provide helium for the regulated pressurization of the four propellant tanks. Special low leakage check valves were designed to prevent propellant vapor mixing between the two check valves.

The Galileo propulsion system was designed by MBB and was provided to Jet Propulsion Laboratory/NASA by Germany.

4.5.4.2 INTELSAT VI. The INTELSAT VI spacecraft are the latest in a long line of INTELSAT communication satellites; the first satellite of the series was launched in October 1989. The propulsion system, shown in Fig. 4.43, supplies thrust for apogee injection (replacing a kick motor) and orbital adjustments.

Table 4.15 Spacecraft bipropellant systems

Characteristic	Viking Orbiter	Shuttle RCS	Galileo	INTELSAT VI	Mars Global Surveyor	Cassini
First launch	1975	1981	1989	1989	1996	1997
No. thrusters	1 (ACS by cold gas)	44	13	8	13	2
Thrust, N	1335	111, 3870	10, 400	22.2, 490		445
Engine cooling	Beryllium	Radiation cooled and insulated	Radiation	Radiation	Radiation	Radiation
Fuel	MMH	MMH	MMH	MMH	Hydrazine	MMH
Oxidizer	NTO	NTO	NTO	NTO	NTO	NTO
Mixture ratio	1.50		1.6	1.6		1.6
Propellant control	Capillary vane devices	Capillary screens	Centrifugal (10 rpm)	Centrifugal	Capillary vane	Capillary vane
Propellant tanks	Titanium equal volume barrel	Titanium, equal volume, spherical	Four equal volume, titanium, spherical	Eight equal volume, titanium, spherical	Three equal volume	
Pressurization	Regulated helium	—	Regulated helium	Regulated helium	Titanium barrel	Titanium barrel
Dry mass, kg	200				139	
Propellants, kg	1423		925	2313 to 2717	836	3100
Features	Beryllium cooling	Large size, multi-use	Spinner, flushing burns	Spinner, redundant half systems	Dual mode operation	Large size long mission

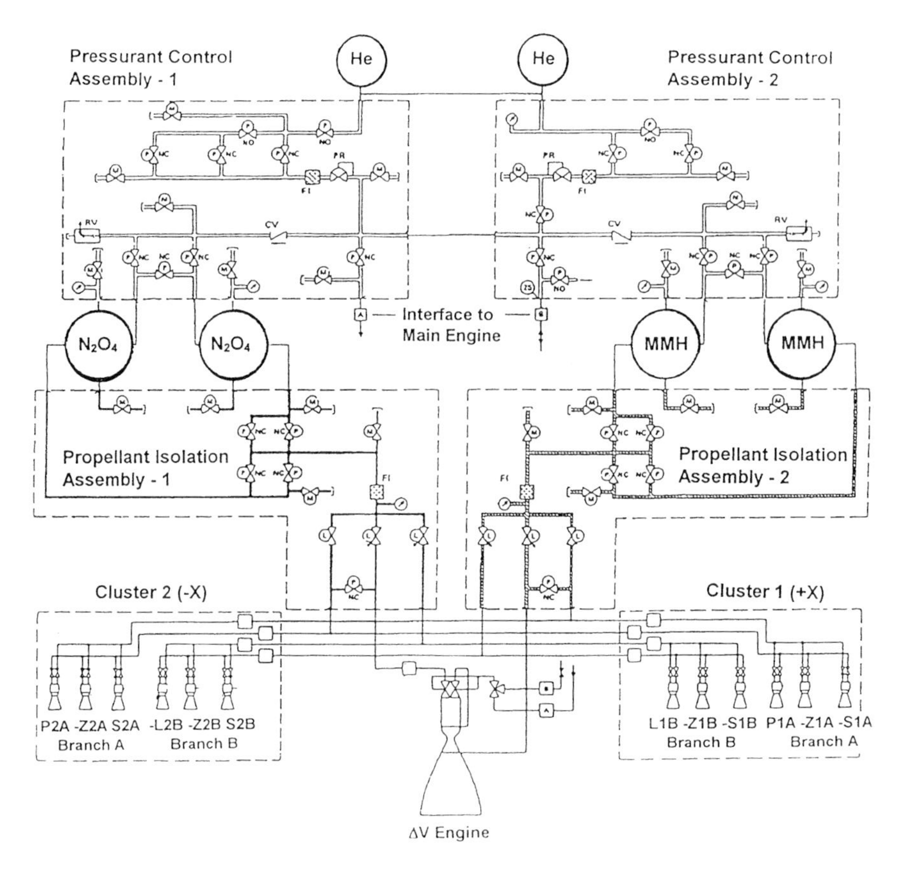

Fig. 4.42 Galileo propulsion system. (Copyright AIAA, reproduced with permission; Ref. 23.)

The system consists of two functionally independent half-systems connected by latch valves on the liquid side and normally open squib valves on the pressurant side. Each half-system can perform all of the propulsion functions required for spacecraft operation. Individual fuel and oxidizer tanks can be isolated on the pressurant side to prevent propellant migration. Two 490-N liquid apogee motors (LAMs) provide the impulse for geosynchronous orbit insertion and spacecraft reorientation during the orbit-transfer portion of the mission. The LAMs are located in the aft end of the spin-stabilized spacecraft and are used in steady state or pulse mode. Six 22.2-N reaction control thrusters are used for control functions. Two are used for axial impulse and are located at the aft end of the spacecraft near the LAMs. Four thrusters are located radially on the spun shelf and are used for spin up and spin down. The thrusters also provide impulse precession, orbit trim, nutation control, north–south stationkeeping and east–west station change, and deorbit control. The thrusters are separated into two redundant groups.

Helium pressurization is supplied through series redundant regulators. Eight propellant tanks are symmetrically located around the center of mass. Both the

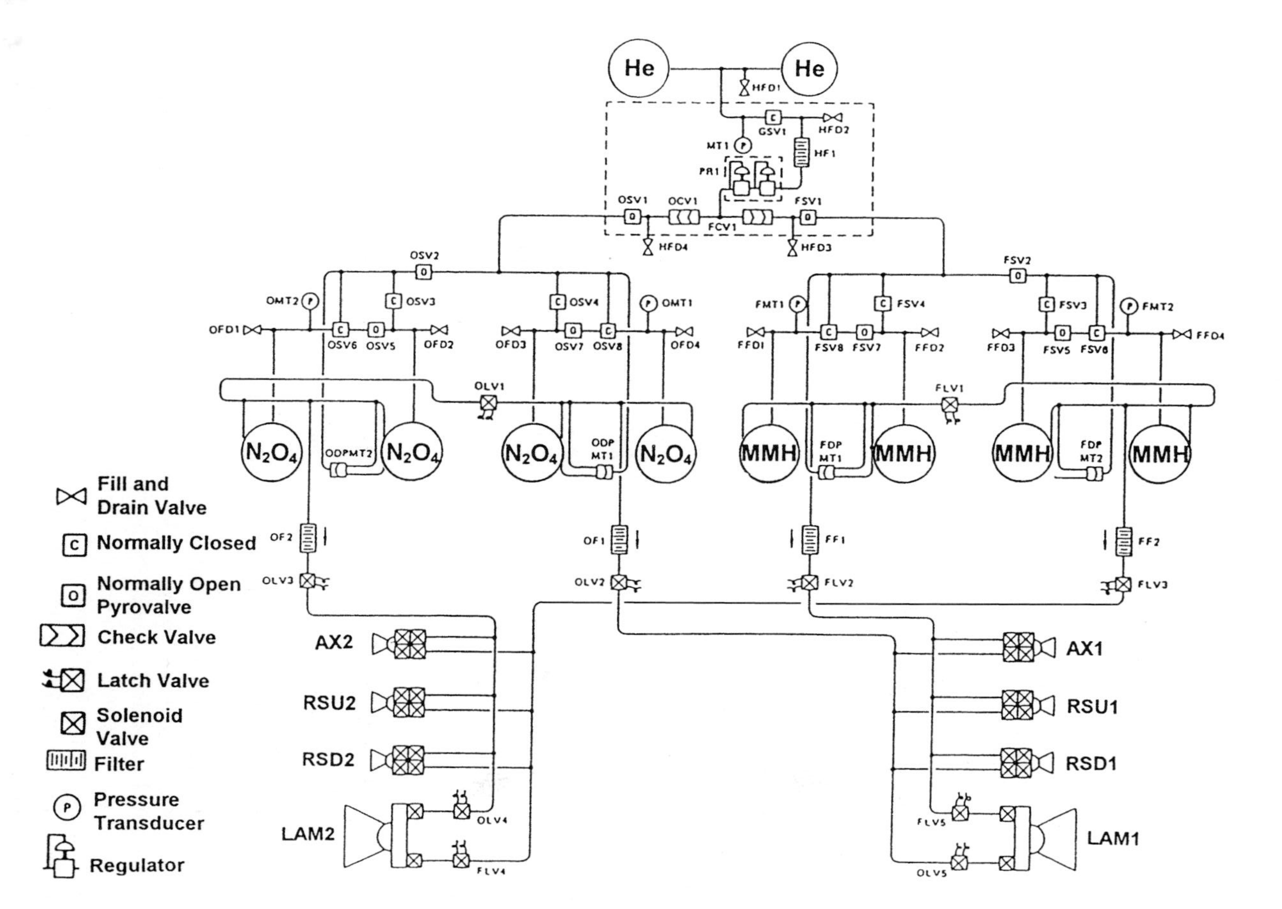

Fig. 4.43 INTELSAT VI propulsion system. (Copyright AIAA, reproduced with permission; Ref. 24.)

oxidizer and fuel tanks are manifolded to maintain spacecraft balance. Centrifugal forces from spinning provide propellant position control; no propellant management device is necessary.

The propellants and pressurant are launched isolated by latching valves and pyrovalves. Initial orientation and control is provided by the 22.2-N thrusters in the blowdown mode. The pressurization system is opened for the apogee burn. After the apogee burn and trim maneuvers the LAMs are isolated from the propellant supply by closing the isolation latch valves (FLV4 and 5 and OLV4 and 5) to preclude leakage failure. Some time after LAM isolation, the pressurization system is sealed, and oxidizer and fuel systems are isolated to prevent propellant vapor mixing or propellant migration. After isolation the system operates in blowdown mode for the remainder of the mission. Mixture ratio can be controlled during blowdown by operating FSV1 and OSV1.[24]

The INTELSAT VI propulsion system was designed by Hughes Space and Communications Company under contract to International Telecommunication Satellite Organization (INTELSAT). Five INTELSAT VI spacecraft were launched; each propulsion system performed satisfactorily.

4.6 Dual-Mode Systems

Many missions require a high-impulse velocity change burn as well as pulse mode attitude control operation. For example, a planetary orbiter requires a high-impulse single burn for orbit insertion and pulse mode for attitude control. Geosynchronous spacecraft also require a high-impulse burn to establish the required orbit as well as pulsing performance for ACS. Magellan used a solid motor for orbit insertion and a monopropellant system for pulse mode and small translation burns. Galileo used a bipropellant system for high-impulse burns and for pulse mode operation. Viking Orbiter used a MMH/N_2O_4 bipropellant system for orbit insertion and trajectory correction maneuvers and a cold-gas system for attitude control. Geosynchronous orbiters commonly use a solid motor for orbit insertion and a monopropellant system for ACS. There is a new technology that promises a better solution by using hydrazine as bipropellant fuel, replacing monomethyl hydrazine, and also as a monopropellant for pulsing. This is called the *dual-mode system*. The system is shown schematically in Fig. 4.44.

The advantages of the dual-mode system are 1) the ability to use the hydrazine as a monopropellant in attitude control thrusters and as the fuel in bipropellant main engines with resulting system simplification, and 2) a significant increase in I_{sp} for the orbit insertion burn. A typical solid motor delivers an I_{sp} of 290 s; the dual-mode bipropellant system delivers about 317 s. The dual-mode system is used on the Mars Global Surveyor spacecraft launched in 1996. It will also fly in upcoming communication satellites and planetary missions.

As you can see from Fig. 4.44, the detailed design aspects of the bipropellant half of the system are as discussed in Section 4.5; the details of the monopropellant half of the system are as described in Section 4.4. In this section the features peculiar to dual-mode operation are shown by a flight system example.

The Mars Global Surveyor, launched in 1996, is a significant element of the NASA overall strategy for the exploration of Mars over the next decade. It is also one of the first spacecraft to use the dual-mode system. The propulsion system,

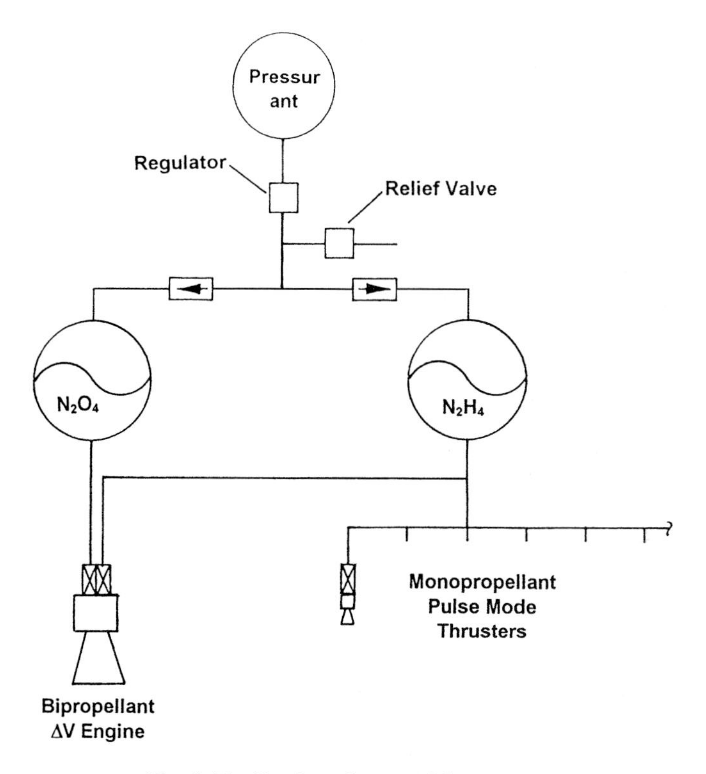

Fig. 4.44 Dual-mode propulsion system.

shown in Fig. 4.45, provides Mars orbit insertion, attitude control pulsing, and a number of orbit trim maneuvers.

The orbit insertion burn was accomplished by the single 490-N bipropellant velocity control thruster. Thrust vector control, orbit trim, and attitude control maneuvers are accomplished by 12 4.45-N thrusters. The system uses three identical titanium barrel tanks (two fuel, one oxidizer) with capillary vane propellant positioning systems similar to those used on Viking Orbiter. The total propellant load is 386 kg. First pulse problems for attitude control are prevented by launching with propellant loaded to the thruster valves.

The system contains 1.36 kg of helium stored at 25,510 kPa. Vapor migration is prevented by a ladder of pyrovalves in the oxidizer pressurization line, which allow the oxidizer tank to be isolated between bipropellant burns. Specially designed, low vapor leak, quad-redundant, check valves are used in the pressurization system to reduce vapor migration during periods when the pyrovalves are open.

The Mars Global Surveyor spacecraft was designed by Lockheed Martin under contract with the Jet Propulsion Laboratory.

4.7 Solid Rocket Systems

In a solid motor the oxidizer and fuel are stored in the combustion chamber as a mechanical mixture in solid form. When the propellants are ignited, they

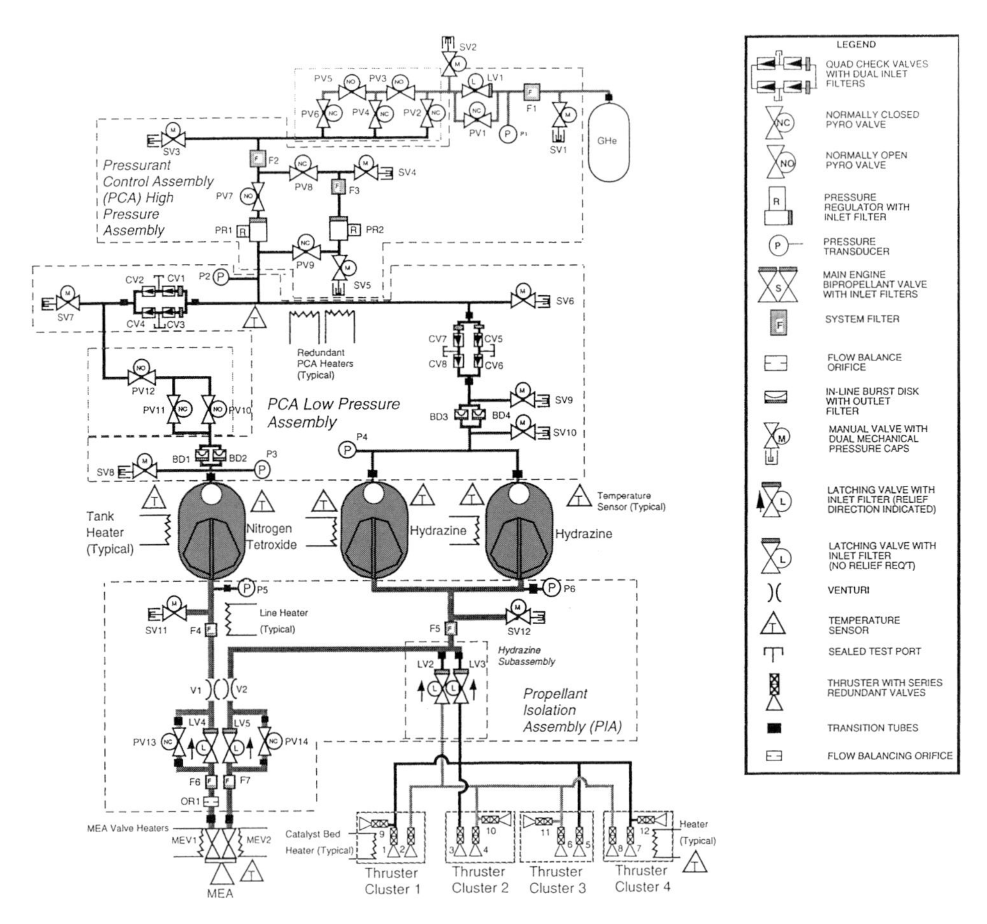

Fig. 4.45 Mars Global Surveyor dual-mode propulsion system. (Reproduced with permission of Lockheed Martin; Ref. 6, p. 133.)

burn in place. Solid rocket systems are used extensively in situations where 1) the total impulse requirement is known accurately in advance and 2) restart is not required. Boosters for Titan IV and the Space Shuttle are good examples of such a situation. Spacecraft examples are the kick stage for a geosynchronous orbiter and the orbit insertion motor for a planetary orbiter. Solid motors are used in applications requiring impulse as small as 196,000 N-s (Explorer I motor) up to the 1.45 billion N-s (one shuttle booster motor). Sizes range from 15 cm diam to 6.6 m.

The elements of a solid rocket motor are shown in Fig. 4.46 and discussed in subsequent sections of this chapter.

4.7.1 Propellants

Solid rockets are over 700 years old, having been used by the Chinese in 1232. The basis of Chinese rockets was ordinary black powder, the gun powder of muzzle loaders. Smokeless powder was invented by Alfred Nobel in 1879, but it was

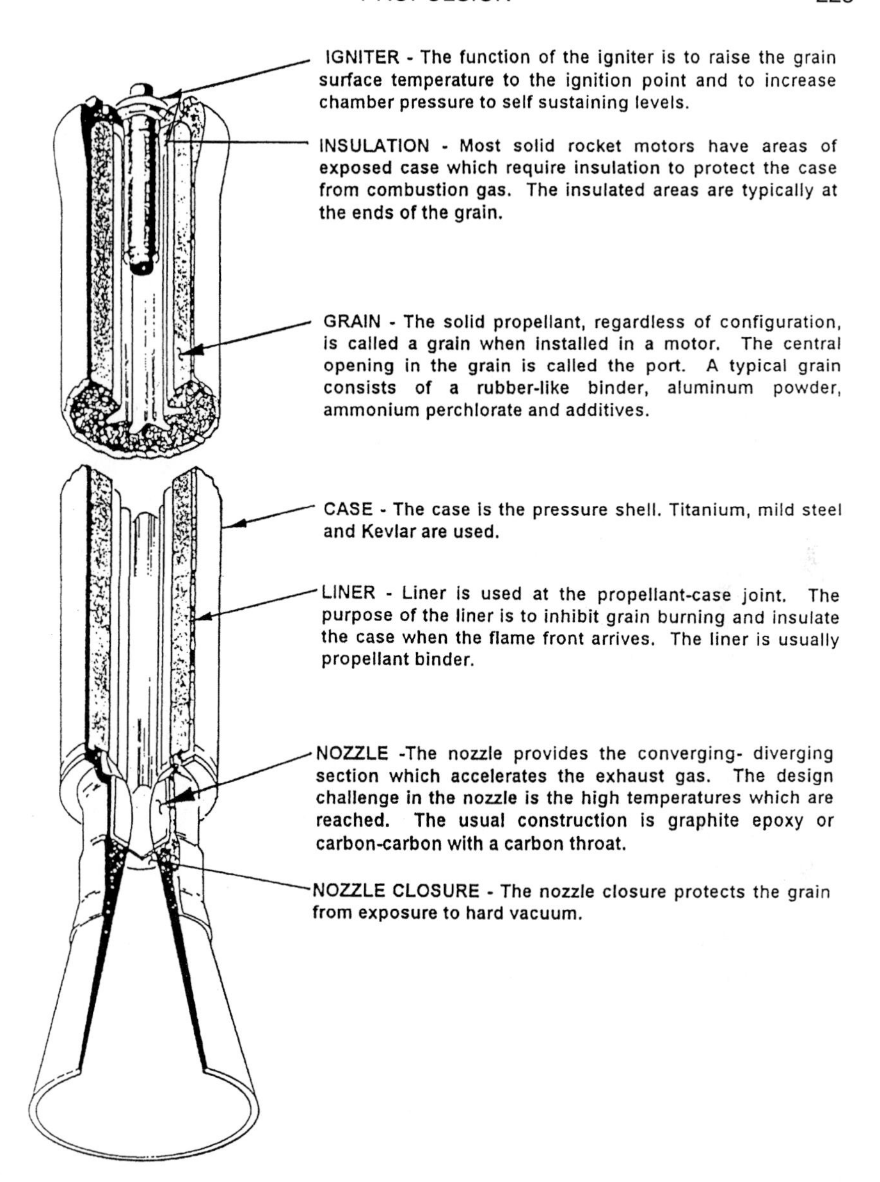

Fig. 4.46 Elements of a solid rocket motor.

not used as solid rocket fuel until Goddard tried it in 1918. He concluded that liquid propellants would be the best development path for space work, and his conclusion held true for 50 years. *Double-base* propellant, which is the interesting combination of nitrocellulose dissolved in nitroglycerin, was used for a time, and combinations of composite and double-based propellants have been used. Modern, high-performance, solid rockets were made possible in 1942 when the Jet Propulsion Laboratory introduced the first *composite propellant*. By 1960 composite propellants were the industry standard.

Composite propellants are composed of one of several organic binders, aluminum powder and an oxidizer, which is nearly always ammonium perchlorate (NH_4ClO_4). The binders are rubber-like polymers, which serve a dual purpose as fuel and, in addition, bind the ammonium perchlorate and aluminum powders into a solid capable of being shaped into a grain. Two common binders are hydroxy terminated polybutadiene (HTPB), which is used in the Thiokol Star series of motors, and polybutadiene acrylonitrile (PBAN), which is used in the Titan solid rocket motors.

Composites also contain small amounts of chemical additives to improve various physical properties, for example, improve burn rate, promote smooth burning (flash depressor), improve casting characteristics, improve structural properties (plasticizer), and absorb moisture during storage (stabilizer).

In general, solid propellants produce vacuum specific impulse of 300 s or lower, somewhat lower than bipropellants. Solid-propellant densities, however, are higher, resulting in smaller systems for a given impulse. As higher specific impulse is sought, less stable ingredients, such as boron and nitronium perchlorate, are necessary. The characteristics of some propellants are shown in Table 4.16.

4.7.1.1 Hazard classification. Propellants are divided into two explosive hazard designations by DOD and DOT: 1) class 2 (or DOT class B)—catastrophic failure produces burning or explosion (most motors) and 2) DOD class 7 (or DOT class A)—catastrophic failure produces detonation (uncommon). The motor classification dictates the safety requirements for using and handling the motor. A class 2 motor is obviously the least expensive motor to transport and handle.

4.7.1.2 Burning rate. Control of the exhaust gas flow rate is achieved, not by precise metering of propellant flows as in a liquid system, but by precise control of the exposed grain surface area and the burning rate of the propellant mixture. The solid is transformed into combustion gases at the grain surface. The surface regresses normal to itself in parallel layers; the rate of regression is called the *burning rate*. Burning rate can be measured by measuring flame front velocity on a strand (Fig. 4.47).

$$\dot{w}_p = \rho A_p r \tag{4.88}$$

where $\dot{w}_p$ is the combustion gas flow rate, kg/s; ρ the gas density, kg/cc; A_p the surface area of burning propellant, cm^2; and r the burning rate, speed at which the flame front is progressing, cm/s.

Pressure effect. Because grain density is essentially constant during a burn, gas flow rate is controlled by grain area and burning rate. Gas flow rate, thrust, and chamber pressure go up with an increase in burning area. Also burning rate goes up exponentially with pressure:

$$r = cP^n \tag{4.89}$$

where n and c are empirical constants.

Temperature effect. Chamber pressure increases nonlinearly with grain temperature because burning rate increases with temperature and pressure. Figure 4.48 shows the effect of grain initial temperature on chamber pressure and hence thrust.

Table 4.16 Typical solid-propellant characteristics[a]

Propellant type	Approximate, I_{sp}, s[b]	Flame temperature, °K	Density, kg/m^3	Metal content, wt%	Burning rate, cm/s	Pressure exponent, n	Hazard Hazard class, DOT/ DOD	Processing method
Double base	250–260	2500	1605	0	1.1	0.30	A/7	Extruded
DB/AP/Al[c]	285–295	3900	1800	20–21	2.0	0.40	B/2	Extruded
PVC[d]/AP	260–260	2800	1690	0	1.1	0.38	B/2	Solvent cast
PVC/AP/Al	290–295	3400	1770	21	1.1	0.35	B/2	Cast or extruded
PS[e]/AP	260–265	2900	1720	0	0.89	0.43	B/2	Cast or extruded
PS/AP/Al	270–2750	3000	1720	3	0.79	0.33	B/2	Cast
PU[f]/AP/Al	290–295	3300–3600	1770	16–20	0.69	0.15	B/2	Cast
PBAN[g]/AP/Al	290–295	3500	1770	16	1.4	0.33	B/2	Cast
CTPB[h]/AP/Al	290–295	3400–3500	1770	15–17	1.1	0.40	B/2	Cast
HTPB[i]/AP/Al	290–295	3400–3500	1850	4–17	1.0	0.40	B/2	Cast
PBAA[j]/AP/Al	290–295	3300–3600	1770	14	0.81	0.35	B/2	Cast

[a]*Rocket Propulsion Elements*, Fifth Ed., George P. Sutton, copyright John Wiley and Sons, Inc., 1986. This material is used by permission of John Wiley and Sons, Inc.
[b]I_{sp} conditions: $e = 40$, vacuum; $Pc = 4137$ kPa; Faverage = 67,000 N.
[c] DB, double base; AP, ammonium perchlorate; Al, aluminum.
[d]PVC, polyvinyl chloride.
[e]PS, polysulfide.
[f]PU, polyurethane.
[g]PBAN, polybutadiene-acrylic acid-acrylonitrile terpolymer.
[h]CTPB, carboxy-terminated polybutadiene.
[i]HTPB, hydroxy-terminated polybutadiene.
[j]PBAA, polybutadiene-acrylic acid polymer.

Fig. 4.47 Burning rate.

The sensitivity of chamber pressure to temperature is typically 0.12 to 0.50% per °C.

Spacecraft motor temperature is carefully controlled for two reasons: 1) to limit variation in performance and 2) to minimize thermal stress in the grain and hence minimize the possibility of grain cracks. A typical temperature control range is 15° to 38°C; the control range can be tightened before the motor firing. Magellan used electrical heaters, thermostats, and insulation to control the motor temperature.

Acceleration effect. Burning rate is increased by acceleration perpendicular to the burning surface, a phenomena of particular interest in the design of spinning spacecraft.[24] Acceleration in the range of 30 g can double the burn rate. The enhancement of burning rate has been observed for acceleration vectors at any angle of 60 to 90 deg. with the burning surface. The increase in burning rate is attributed to the presence of molten metal and metal oxide particles that are retained against the grain surface by the radial acceleration with an attendant increase in heat transfer at the grain surface. The effect is enhanced by increased aluminum concentration and particle size.[25]

4.7.1.3 Space storage. There is some controversy and little data regarding the ability of a solid motor grain to withstand long periods of space exposure with an open port. The concern is that evaporation could affect the chemical composition of the propellant or bond line and affect performance. There are two options available: 1) conduct adequate testing to establish postvacuum performance, or 2) design a sealed nozzle closure.

Most missions, kick stages and upper stages, entail only short-duration exposure. These missions have been flown for years with no closure. A planetary mission, like Magellan, is another story. Thiokol Corporation space aging data[26] for

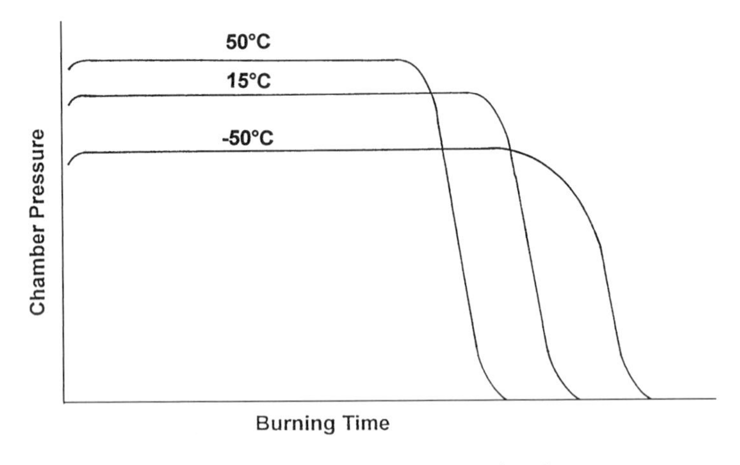

Fig. 4.48 Effect of grain temperature on chamber pressure.

TPH-3135 shows low risk for 10-month exposure. LDEF results[26] indicate that 5.5-year exposures may be acceptable. The Magellan mission itself shows (one data point) that 15-month space vacuum exposure, without a nozzle plug, is acceptable.

4.7.2 Grains

In a solid motor the propellant tank and the combustion chamber are the same vessel. The viscous propellant mixture is cast and cured in a mold to achieve the desired shape and structural strength. After casting the propellant is referred to as the grain. The most frequent method of casting is done in place in the motor case using a mandrill to form the central port. When cast in this way, the grain is called case bonded.

Grains can also be cast separately and loaded into the case at a later time; such grains are called *cartridge loaded*. Large grains are cast in segments; the segments are then stacked to form a motor. The motors for the Space Shuttle and Titan IV are made in this way; Fig. 4.49 shows the segmented Space Shuttle motor.

Segmented grains solve the difficult motor transport problem; however, they also require high-temperature case joints, which can be a serious failure source. The aft field joint, shown in Fig. 4.49, was the cause of the Space Shuttle *Challenger* explosion.[27]

Liner is used at the grain to case interface to inhibit burning as the flame front arrives at the case wall. The liner composition is usually binder without added propellants. In areas of the case where there is no grain interface, usually at the ends of the motor, insulation is used to protect the case for the full duration of a firing.

4.7.2.1 Grain shape. Grain shape, primarily cross section, determines the surface burning area as a function of time. Burning area, along with burning rate, determine thrust. For a given propellant, surface area vs time determines

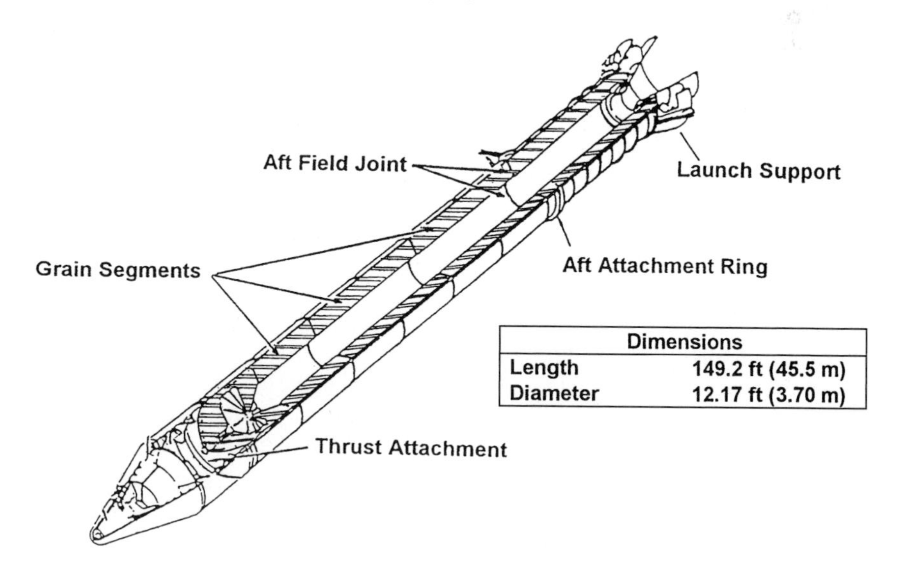

Dimensions	
Length	149.2 ft (45.5 m)
Diameter	12.17 ft (3.70 m)

Fig. 4.49 Segmented Space Shuttle solid motor.[27]

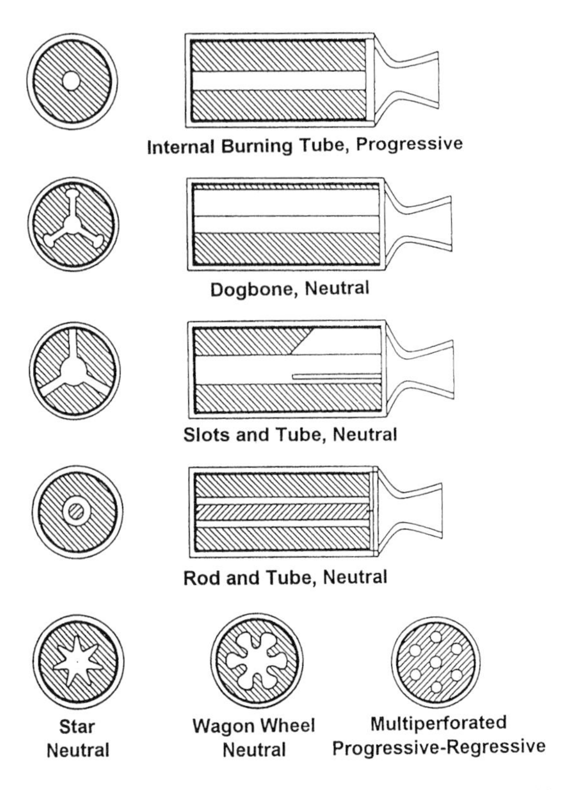

Fig. 4.50 Grain configurations. (*Rocket Propulsion Elements*, Fifth Ed., George P. Sutton, Copyright John Wiley and Sons, Inc., 1986. This material is used by permission of John Wiley and Sons, Inc.)

the shape of the thrust time curve. A cylindrical grain inhibited on the sides, a "cigarette burner" or "end burner," would have a constant burning area and a constant thrust. A grain with an essentially constant burning surface area and thrust time curve is called *neutral burning*. End-burning grains were used by the thousands for JATO rockets in the 1940s.

A grain that has increasing surface area with time is called *progressive*. The primary disadvantage of progressive burning is that as the mass decreases the thrust increases. The resulting loads on a spacecraft increase sharply with time, which is not conducive to an efficient structure. A cylindrical grain inhibited at the ends and burning from the sides would be regressive. Neutral burning is the most common design for spacecraft. Grain cross sections are shown in Fig. 4.50.

A crack in the grain provides an unplanned increase in exposed area with a resultant sudden increase in pressure, which can lead to failure. Numerous precautions must be taken to prevent grain cracks; these include 1) limits on grain temperature extremes, 2) limits on shock loads on the grain and bond line, and 3) a final set of grain x rays at the launch site just before launch.

A star grain is the most commonly used shape and is nearly neutral. The progression of the burning surface in a star grain is as shown in Fig. 4.51. The flame

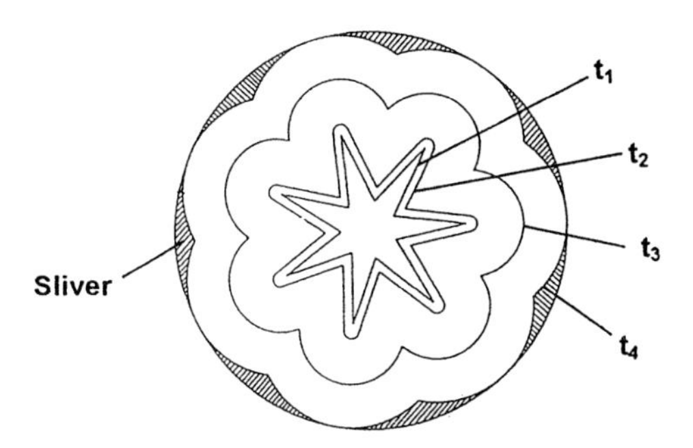

Fig. 4.51 Burning surface progression in a star grain. (Reproduced with permission of McGraw–Hill; Ref. 3, pp. 19–24.)

front does not reach the liner in all locations at the same time. When the flame front first reaches the liner, chamber pressure starts to decay. At a given pressure aggressive burning or deflagration can no longer occur. The residual, useable propellant, which remains when combustion stops, called *sliver*. Sliver will be about 2% of the grain weight.

4.7.2.2 Shelf life. The shelf life of a motor is a conservative estimate of the acceptable length of time from the pour date to the firing date. A typical shelf life is three years; therefore, the pour date must not be more than three years from the on-mission firing. Thus the pour date becomes a major program milestone.

4.7.2.3 Offload capability. Many solid motors have been demonstrated to perform properly over a range of propellant loads. This is called *off-load capability* and is useful in the matching the required total impulse with the available total impulse. Off loading is accomplished by pouring the motor case less than full. The Star 48, for example, is normally loaded with 1997 kg of propellant; propellant can be off loaded down to 1600 kg.

4.7.3 Thrust Control

A spacecraft must make provision to start and terminate thrust as well as align the nominal vector through the center of mass and move the vector through small angles for vehicle stability.

4.7.3.1 Ignition. During most of a spacecraft assembly and flight, the objective is to prevent solid motor ignition. During times when people are around the motor, it is a common requirement for the ignition system to be two-fault tolerant. The system must be designed such that unwanted ignition cannot occur after any two conceivable failures. If the spacecraft is launched on the Space Shuttle, this requirement is in effect until deployment. When ignition is required in the mission, it is common to require single fault tolerance.

A typical solid rocket motor ignition system is shown schematically in Fig. 4.52. It requires careful use of redundancy to meet the failure criteria just mentioned. Redundancy is not shown in Fig. 4.52 in the interest of simplicity.

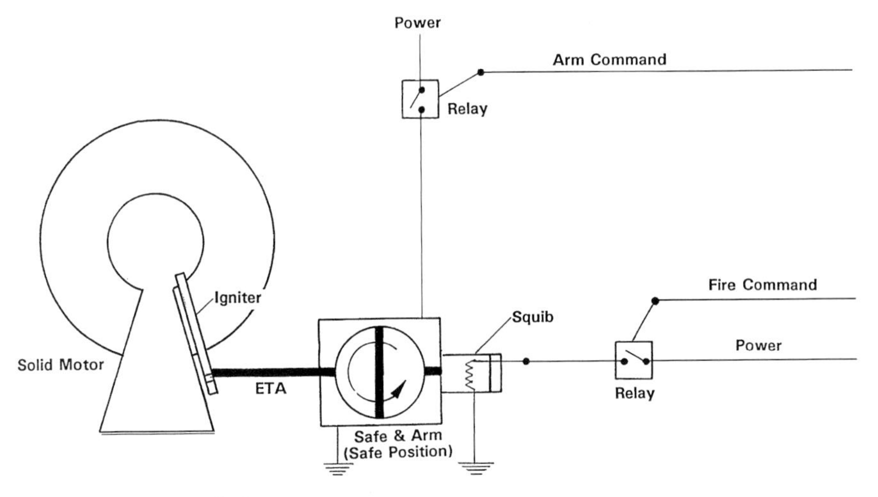

Fig. 4.52 Solid motor ignition system.

A typical ignition system consists of the following:

1) *Squibs* are cylinders that contain a small chemical charge so sensitive that the heat from an electrical filament will cause a substantial energy release. The squibs have an electrical connector, which receives the start signal. Squibs are required to have a certain all-fire current and a certain no-fire current. The electrical filament is called a *bridge wire*. Some squibs have redundant bridge wires to receive redundant signals.

2) The *explosive transfer assembly* (ETA) is an explosive train that runs from the safe and arm to the igniter. An ETA has the same function as a dynamite fuse but is much, much faster.

3) The *safe and arm* (S&A) is a dc motor that houses the squibs on one side and the ETAs on the other with a rotating cylinder between. A section of the explosive train is embedded in the rotating cylinder. The flame front that starts in the squibs must pass through the cylinder to reach the ETAs. When the cylinder is rotated to the safe position, as shown in Fig. 4.52, the train is interrupted. In the armed position the train is continuous; the cylinder is rotated 90 deg from the position shown in Fig. 4.52. There is a physical flag on the exterior of the S&A, which shows position. In addition, there is a position signal sent to the command and data system so that the position can be telemetered to the ground operations team. It is typical to use dual S&As and dual igniters.

4) The *igniter* is similar to a small solid motor with a propellant mixture, which is readily ignited. The igniter raises the pressure in the motor port and raises the grain temperature to the ignition point.

At the appointed time the spacecraft command system sends the "arm" command to a relay, which powers the S&A motor. (In the Magellan mission this point was 15 months after launch.) The S&A rotates and signals that the armed position has been achieved.

Table 4.17 Common methods of thrust vector control

Type	Flight use
1) Jet vanes	Jupiter, Juno
2) Jetavators	Polaris, Bomarc
3) Ball and socket nozzle	Minuteman
4) Flexible bearing nozzle	IUS, Shuttle
5) Liquid injection	Titan III, IV, Minuteman
6) Auxiliary propulsion	Magellan

The "fire" command is sent to a separate relay, which powers the squib bridge wires lighting the explosives in the squib. The flame front progresses through the S&A and the ETA and lights the motor igniters. The igniters bathe the grain surface in flame and the motor ignites. From signal to grain ignition takes about 150 ms; full thrust takes about another 50 ms.

4.7.3.2 Thrust vector control. Moving the net thrust vector through small angles for vehicle stability and maneuvering is called thrust vector control (TVC). The methods in common use are summarized in Table 4.17. All of the thrust vector control schemes in Table 4.17, except jet vanes, require auxiliary equipment to achieve roll control. All types can provide pitch and yaw control with a single nozzle.

Jet vanes and jetavators operate with low actuator power and provide high slew rates; however, there is a thrust and I_{sp} loss of up to 2%. Jet vanes, jetavators, and movable nozzles, types 1 through 4 in Table 4.17, all require an actuator system. Hydraulic and electromechanical actuator systems are used. The IUS uses an electromechanical system consisting of the following.

1) A controller which receives analog pitch and yaw commands from the guidance system. The analog signals are converted into pulse-width modulated voltages for the actuator motors.

2) Actuator control motors (one pitch, one yaw), which drive the ball screw mechanisms.

3) A ball screw mechanism, which positions the nozzle (one pitch, one yaw)

The most common TVC methods are movable nozzles and fluid injection, types 3, 4, and 5 in Table 4.17. Figure 4.53 shows the Space Shuttle solid motor nozzle incorporating gimbaled movement with an flexible bearing nozzle.

A flexible bearing movable nozzle has no hot moving parts and provides repeatable, albeit high, actuator forces.

The fluid injection systems inject a liquid into the nozzle, which causes an oblique shock in the exhaust stream deflecting the thrust vector as shown in Fig. 4.54. The thrust vector is diverted in the direction opposite the injection point.

The advantages of a liquid injection system are high slew rate and increased motor performance. The increased performance can essentially offset the added weight. The disadvantage is the additional equipment required and the liquid propellant required. The Titan III and IV, stage 0, solid rocket motors use liquid

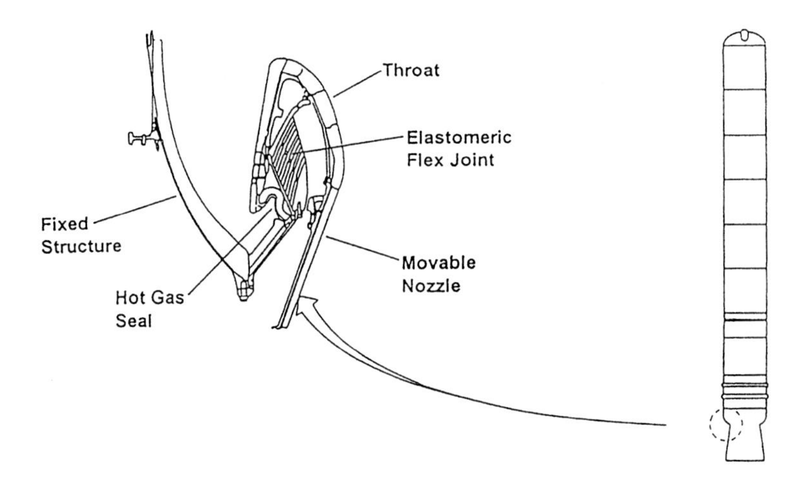

Fig. 4.53 Flexible bearing gimbaling nozzle. (Reproduced courtesy of NASA.)

injection; nitrogen tetroxide is used as the injectant. The fluid system is shown in Fig. 4.55.

The pyroseal valve shown in Fig. 4.55 seals the injectant prior to ignition. At ignition the exhaust melts the exposed end of the valve allowing injectant flow. The TVC injectant tank on either of the two Titan IV, stage 0, rocket motors is bigger than the V-2.

An alternative approach to controlling the motor thrust vector is to provide an auxiliary propulsion system to stabilize the spacecraft. This was the approach used by Magellan. The monopropellant attitude control system was designed to provide three-axis control with or without the solid motor firing. Redundant, aft-facing 445-N thrusters provided pitch and yaw control during the firing. The system also provided vernier impulse adjustment after solid motor shut down.

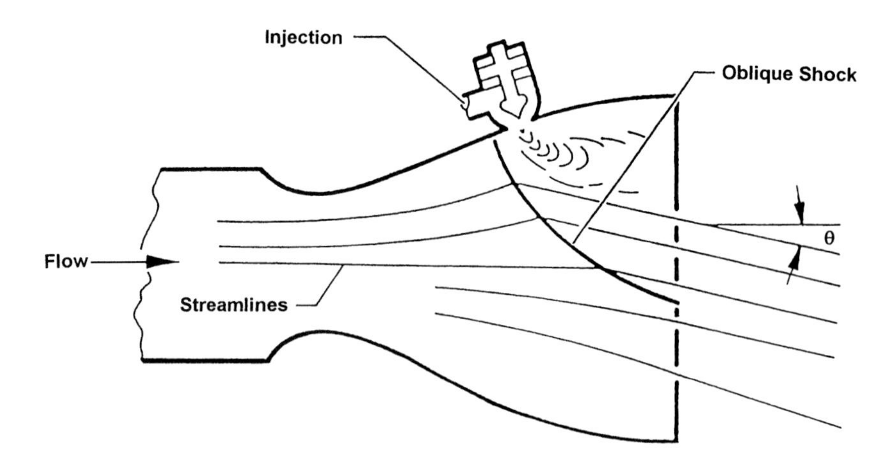

Fig. 4.54 Fluid injection thrust vector control.

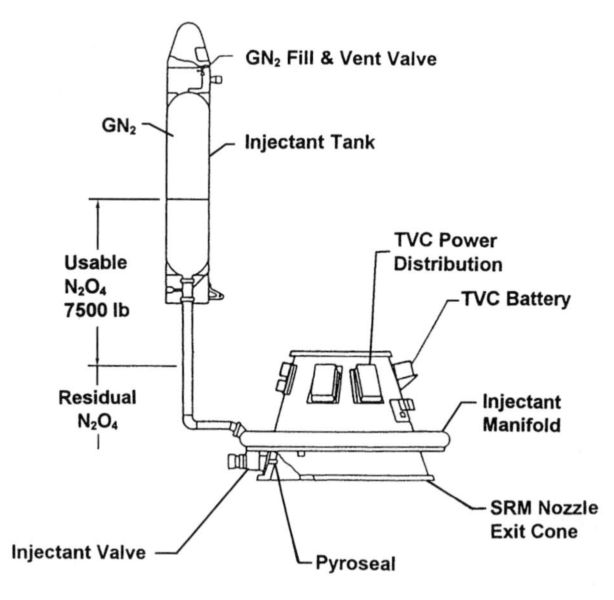

Fig. 4.55 Titan fluid injection thrust vector control system. (Reproduced with permission of Lockheed Martin; Ref. 6.)

4.7.3.3 Thrust vector alignment. For three-axis stabilized spacecraft the solid motor must be installed with the thrust vector accurately aligned with the spacecraft center of mass. The severity of this requirement depends on the magnitude of attitude control authority, the accuracy of c.g. location and the accuracy of the thrust vector location. For a geometrically perfect nozzle the thrust vector lies on the centerline. The thrust vector is very nearly on the centerline in real nozzles as shown in Table 4.18, for the STAR48B™ solid motor.

The shape and size of the nozzle changes during the burn. The resulting variation in thrust vector location is more difficult to determine. The center of mass of a motor moves during the burning process; the tolerance about nominal movement is also difficult to determine.

In the Magellan spacecraft design it was necessary to use 8 lb of ballast to bring the center of mass to within the required 0.02 in. of the predicted thrust vector location. It is difficult to obtain an actual spacecraft center of mass. It must be

Table 4.18 STAR48B™ nozzle alignment data[a]

Alignment	Radial offset, in.	Angular error, deg
Population range	0.003–0.011	0.004–0.059
Average	0.0072	0.0247
One sigma	0.0026	0.0152
Magellan motor	0.0085	0.01761

[a]Data copyright AIAA, reproduced with permission; Ref. 28.

Table 4.19 Thrust termination techniques

Action	Result
Venting forward ports	Balance thrust/reduce pressure
Chamber destruction	Reduced chamber pressure
Liquid quenching	Extinguish flame
Nozzle ejection	Reduced chamber pressure

done late in the process at the launch site, and even then certain equipment will not be installed; for example, liquid propellants will not be loaded, the solid motor will be represented by an inert motor, igniters will not be installed, and remove before flight equipment will still be installed A spin-stabilized spacecraft avoids this set of problems because rotation nullifies thrust vector alignment errors.

4.7.3.4 Thrust termination. Thrust termination is desirable because the delivered impulse cannot be predicted exactly. Delivered impulse depends on the temperature of the grain, the actual propellant weight, the exact composition, and the weight of inert parts consumed. Thrust termination allows a spacecraft to measure velocity gained and shut down when the desired velocity is reached. The impulse uncertainty with thrust termination is reduced to the shutdown impulse uncertainty. Table 4.19 lists demonstrated thrust termination methods.

Thrust termination can be done by suddenly reducing chamber pressure below certain limits, quenching the combustion, or nullifying thrust. Titan, for example, terminates stage 0 thrust by blowing out forward ports. Magellan and the Space Shuttle are designed so that thrust termination is not required. In both of these cases the solid motor burns out before the desired impulse is achieved. The deficit is made up by liquid engines, which can be readily shut down on command.

4.7.4 Mounting

There are myriad mounting provisions for solid rocket motors. Spacecraft loads can be carried through the case (IUS motor), or spacecraft loads can be carried around the motor (Magellan). Mounting pads (Space Shuttle) or mounting rings (IUS) can be used. Mounting pads carry a point load, and an interface truss is desirable. Mounting rings carry a distributed load, and an interfacing shell structure is desirable. For large spacecraft motors the dominant mounting design is a mounting ring. Figure 4.56 shows the Magellan SRM adapter structure designed to mount the STAR48B™.

The SRM adapter structure attaches at the forward mounting ring of the STAR48B™. The adapter is a conical aluminum honeycomb structure, which mates with a truss the Magellan central structures.

After burnout it is desirable to jettison the spent motor case and as much support structure as possible in order to reduce vehicle mass. Reduced mass reduces the energy required by the attitude control system and reduces the energy required for any subsequent maneuvering. There are two basic system types used for jettisoning: 1) linear shaped charge and 2) explosive bolts or nuts. Linear shaped charges are

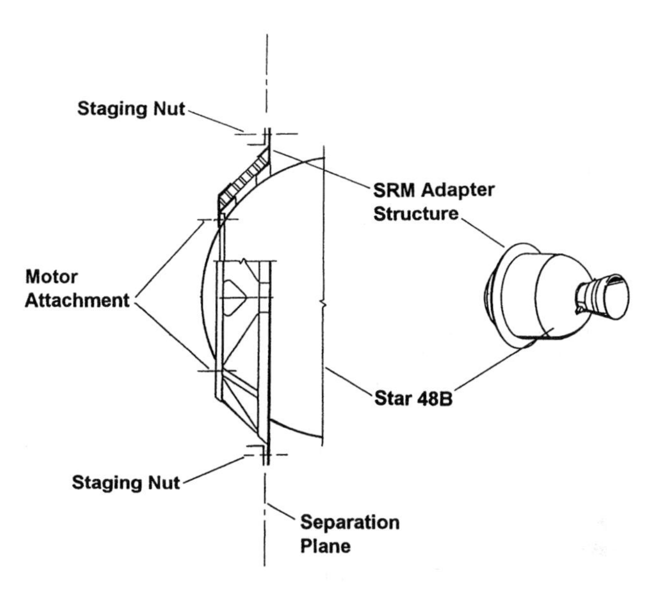

Fig. 4.56 Magellan SRM adapter structure. (Reproduced with permission of Lockheed Martin; Ref. 6.)

used to cut all of the way around the girth of a shell structure. The freed motor case and attached structure are pushed away by springs or by gas from the linear charge.

Explosive nuts (or bolts) release a bolted joint in the structure releasing the spent motor and support structure. The Magellan design used explosive nuts. In Fig. 4.56 the separation plane is between the main structure and the SRM adapter. This joint is held together by explosive bolts. When the bolts are blown, the spent Star 48B and the SRM adapter structure are pushed away with springs. The resultant reduction in spacecraft mass was 163 kg.

4.7.5 Performance

Solid motor performance calculation must be approached somewhat differently than liquid system performance as detailed in this section.

4.7.5.1 Burning time and action time. Burning time and action time for a solid rocket motor can be estimated from internal ballistics using strand burning rates or measured on a test stand. There are two times of interest, as shown in Fig. 4.57: 1) the burning time, which is measured starting from 10% thrust at ignition to 90% thrust during shutdown, and 2) the action time, which is measured starting at 10% thrust during ignition to 10% thrust during shutdown. (The thrust level from which the 10 and 90% points are calculated is the maximum thrust level.) The 90% thrust point on shutdown is the approximate time when the flame front reaches the inhibitor at the case wall. From the 90% point down to the 10% point, the slivers are being consumed (less efficiently than the combustion at full chamber pressure).

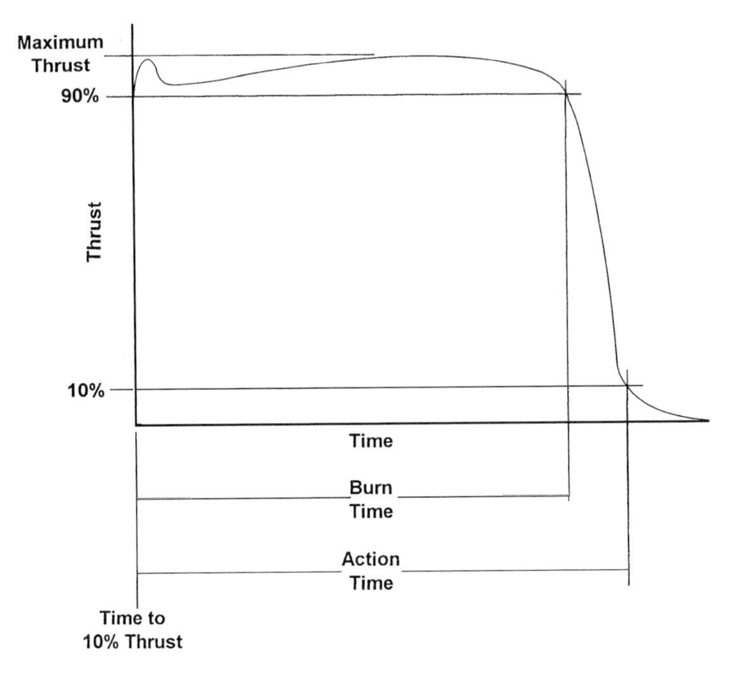

Fig. 4.57 Burning time and action time.

For a motor firing in a vacuum, thrust will continue for a substantial period of time after the 10% thrust point on shutdown. During this tail-off period, a low chamber pressure is maintained by burning of sliver, outgassing of inert materials, and low level smoldering of hot insulation, nozzle materials and liner. The impulse, thrust level, and time-to-zero thrust are nearly impossible to measure in the atmosphere because a hard vacuum cannot be maintained with a motor outgassing. It is very important in the design of the spent case staging sequence to provide an adequate period before release in order to make sure the case will not follow the spacecraft.

4.7.5.2. Effective propellant weight. The *effective propellant weight* is the propellant actually consumed by the motor. Effective propellant is the difference between the initial total weight of the motor and the post burn total weight of the motor:

$$W_{\text{eff}} = W_i - W_f \tag{4.90}$$

Effective propellant weight can be either more or less than the total propellant load. There are two effects at work: 1) sliver, which tends to reduce effective propellant, and 2) consumed inert parts, which tend to increase effective propellant. Consumed inert weights are nozzle materials that ablate, particularly in the throat area, as well as volatiles forced out of the insulation and liner. In a vacuum motor it is very likely that the sliver is consumed during the shutdown or tail off; thus, spacecraft motors usually have an effective weight greater than the total propellant loaded. Table 4.20 shows two examples.

Table 4.20 Effective propellant weight

Weight	Star 48B	Minuteman stage I
Initial weight, lb	4,660	50,550
Burnout weight, lb	235	4,264
Effective propellant, lb	4,425	46,286
Loaded propellant, lb	4,402	45,831
Inerts consumed, lb	23	455

4.7.5.3 Impulse. Two impulse values result from the two thrust times just described. If thrust is integrated from time zero to time $= t_b$, the resulting impulse I_b is called *burning time impulse*:

$$I_b = \int_0^{t_b} F \, \mathrm{d}t \tag{4.91}$$

If action time is used I_a, action time impulse results:

$$I_a = \int_0^{t_a} F \, \mathrm{d}t \tag{4.92}$$

Action impulse is often called *total impulse*, which is true in the sense that the tail-off impulse is essentially unusable. However, the tail-off impulse must be provided for as just noted.

4.7.5.4 Thrust. Two versions of time-average thrust arise from the preceding dual-time standard:

$$\bar{F}_b = \frac{I_b}{t_b} \tag{4.93}$$

where t_a is the action time, s; I_b the burning time total impulse, N-s; and I_a the action time total impulse, N-s.

The burn time average thrust $\bar{F}_b$ is the most often used and is frequently referred to simply (and imprecisely) as average thrust. Maximum thrust is critical to the design of the case. It is calculated from internal ballistics for a hot grain using maximum burning rate.

4.7.5.5 Specific impulse. Theoretical specific impulse for the propellant can be calculated from theoretical thermodynamic properties as described in Section 4.2. Propellant I_{sp} can also be obtained from thrust time test data because

$$I_{\mathrm{sp}} = \frac{I_a}{w_p g_c} \tag{4.94}$$

where I_{sp} is the propellant specific impulse, s; w_p the propellant weight loaded, kg; and g_c the gravitational constant, 9.80665. In calculating propellant I_{sp} from Eq. (4.96), the loaded propellant weight is used.

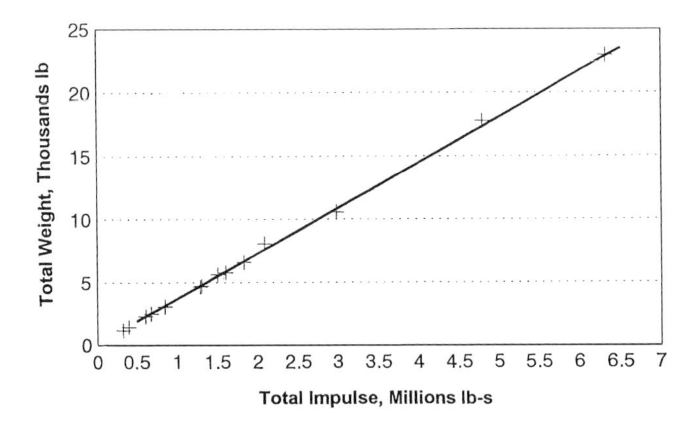

Fig. 4.58 Solid rocket motor weight.

4.7.5.6 Motor weight. The total weight of a solid rocket motor can be expressed as

$$W = \frac{w_p}{\eta} \tag{4.95}$$

where W is the total motor weight, kg and η the propellant mass fraction, W/w_p. The total weight of 23 spacecraft solid rocket motors of current design is shown in Fig. 4.57 as a function of total impulse. For these motors the average propellant I_{sp} is 290.

Figure 4.58 shows that the average propellant mass fraction of these 23 current motors is 0.93 with a correlation coefficient of 0.954. The results of Fig. 4.58 can be expressed in a useful equation form

$$W = \frac{w_P}{0.93} \tag{4.96}$$

where w_p is propellant weight and W is the total motor weight.

Example 4.11 Solid Motor Performance, Size, and Mass

Very frequently it is necessary to derive solid motor characteristics from the top down from the requirements of a mission. For example, given the following mission requirements what would the solid motor size, weight, and total impulse be (the key parameter in motor selection is total impulse): required velocity change is 1.831 km/s, and maximum spacecraft weight at burnout (including the spent solid motor) is 1500 kg, (3307 lb). (These are typical geosynchronous spacecraft requirements.)

It is fair to assume that today's solid motor technology will yield a specific impulse of about 290 s. From the Tsiolkowski equation (4.21), the expected propellant weight would be

$$w_P = (1500)\left\{\exp\left[\frac{(1.831)(3280.84)}{(32.1741)(290)}\right] - 1\right\} = 1356\,\text{kg}\,(2989\,\text{lb})$$

The *total impulse* of the motor is

$$I = w_p I_{sp} g_c = (1356)(290)(9.8066) = 3{,}856{,}347\ \text{N-s}\ (866{,}950\ \text{lb-s})$$

Assuming a mass fraction of 0.93, the motor *total weight* will be about

$$W = \frac{w_p}{0.93} = 1458\ \text{kg}\ (3214\ \text{lb})$$

The weight available for the spacecraft bus is 1500 − (1458 − 1356) = 1398 kg (3082 lb).

Motor volume can be obtained from the *volumetric loading fraction*, which is the ratio of the propellant volume to the motor volume excluding the nozzle. A volumetric loading fraction of 90% is typical for a spacecraft motor. A typical propellant density is 1770 kg/m^3 from Table 4.18:

$$V_m = \frac{W_p}{\rho\, V_f} = \frac{1356}{(1770)(0.90)} = 0.85\ \text{m}^3$$

4.7.6 Selecting a Solid Motor

It would be highly unusual to develop a custom solid motor for a spacecraft. In practice a motor is selected from the wide array of off-the-shelf motors offered by the industry. "Off the shelf" means that the design, manufacturing processes, tooling and test equipment exist, have been qualified, and motors have a flight history. It does not mean that motors are immediately available in a warehouse. The flight motor will be poured specifically for the program.

The key parameter in the selection of a solid motor is the total impulse requirement for the mission. The industry now offers solid motors for almost any total impulse starting with a few hundred newtons-second and going up to the 1.4 billion N-s supplied by each Space Shuttle booster rocket. Many solid motors are designed to be offloaded slightly; thus, motors can be tailored somewhat to fit the desired total impulse. The spacecraft design can also be tailored to fit the motor. For example, Magellan was designed to fit the STAR48 by providing a monopropellant trim burn that completed the required impulse and improved the impulse accuracy needed. The only change to the STAR48 was to add redundancy to the safe and arm system.

4.7.7 Some Flight Motors

In this section the characteristics of some solid motor designs with successful flight history are described.

4.7.7.1 Explorer I. Explorer I, the first U.S. spacecraft, was also the first to carry a solid rocket motor. A small solid motor, integral with the spacecraft structure, inserted the spacecraft into orbit. The motor, shown in Fig. 4.59, was originally used as a scale model in the development of the Sergeant missile. Table 4.21 summarizes the performance of the motor.

Research at the Jet Propulsion Laboratory in the 1940s leads to the composites in use today. The polysulfide-based propellant used on Explorer I is an early example of composites. Explorer I was developed by the Jet Propulsion Laboratory in

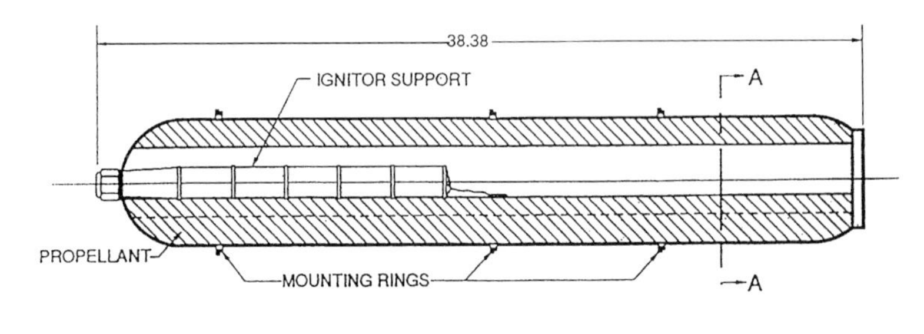

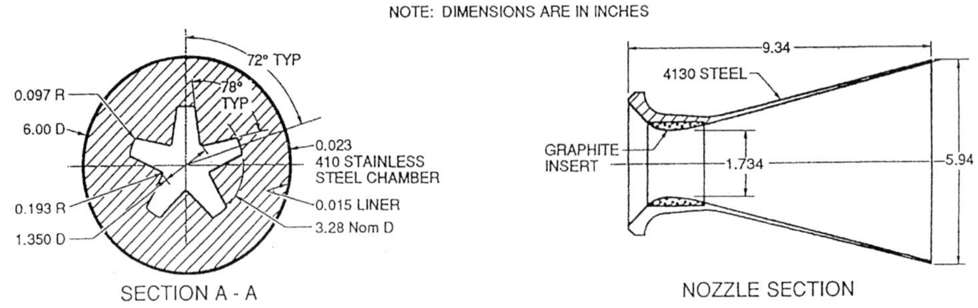

Fig. 4.59 Explorer I solid motor. (Copyright AIAA, reproduced with permission; Ref. 29.)

the explosive political atmosphere following the Sputnik. It was launched on a Jupiter C on 31 January 1958. Its primary scientific result was the discovery of the Van Allen belt.

4.7.7.2 STAR48B™. The STAR48B™, Fig. 4.60 and Table 4.22, was developed by ATK Thiokol for the McDonnell Douglas Payload Assist Module. In addition, the STAR48B™ has flown successfully on numerous Earth-orbiting missions as a kick stage. The STAR48B™ was chosen to put Magellan in orbit around Venus in 1990. It contains 4430 lb of propellant and delivers 5.8 million N-s of impulse. It has off-load capability down to 1740 kg of propellant.

4.7.7.3 Inertial upper stage. The inertial-upper-stage (IUS) motors (Fig. 4.61) are built by UTC for Boeing Aerospace Company. It is used with both

Table 4.21 Performance of the Explorer I solid motor[a]

Parameter	Value
Effective burning time	5.35 s
Effective chamber pressure	3420 kPa (496 psi)
Maximum vacuum thrust	9400 N (2100 lb)
Total impulse	47360 N-s (10566 lb-s)
Vacuum specific impulse	219.8 s

[a]Data copyright AIAA, reproduced with permission; Ref. 29.

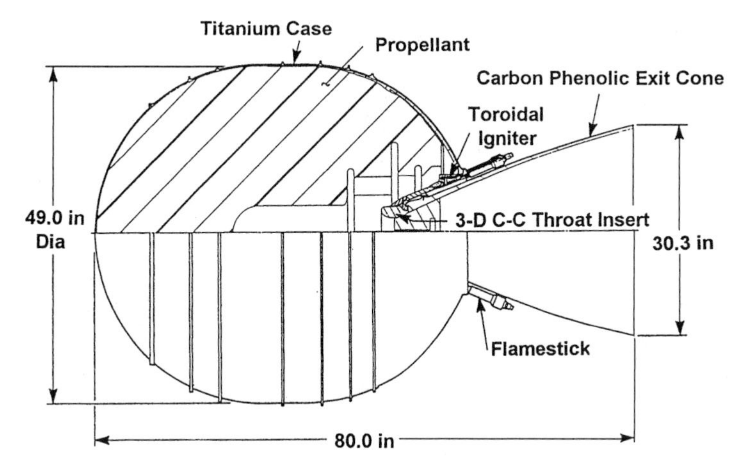

Fig. 4.60 STAR48B™ solid rocket motor. (Reproduced with permission of ATK Thiokol; Ref. 6.)

the Space Shuttle and the Titan family of launch vehicles. As shown in Fig. 4.61 and Table 4.23, the vehicle incorporates two solid rocket motors. Stage I contains 9710 kg of propellant, and stage II contains 2750 kg. Both stages have gimbaled nozzles operated by hydraulic actuators. The upper stage has an extendible nozzle. Both motors have carbon/carbon integral throats and two-dimensional carbon/carbon exit cones and redundant ignition systems.

The grains have a tubular design that allows 50% offload capability. The motors are cast on a mandrel and then machined to nominal or offload requirements. Offload machining follows the grain burning pattern. This approach does not alter the thrust profile. The igniter can accommodate 50% offloading. Roll control for both motors is provided by a monopropellant reaction control system on the second stage. (The monopropellant system also provides three-axis control when the motors are not firing.) Stage II can be flown with or without its extendible exit cone (EEC). The increase in area ratio provides an additional 14.4 s in specific impulse.[30]

4.8 Cold-Gas Systems

Cold-gas systems are the simplest and oldest type used for attitude control thrusters. In the 1960s-era cold gas was the most common system type; the system is still used in cases where the total impulse needed is less than about 1000 lb-s.

4.8.1 Design Considerations

Figure 4.62 shows a typical cold-gas system. A typical system includes the gas storage container, a gas loading valve, filtration, pressure regulation, pressure relief, and a series of thrusters with valves.

Table 4.22 STAR48B™ motor characteristics (Courtesy ATK Thiokol)

Parameter	Value
Motor performance	
Total impulse, lb-s	1,303,705
Maximum thrust, lb	17,490
Burn time ave. thrust, lb	15,430
Action time ave. thrust, lb	15,370
Propellant specific impulse, s	294.2
Effective specific impulse, s	292.1
Burn time/action time, s	84.1/85.2
Ignition delay, s	0.099
Burn time ave. chamber pressure, psia	579
Action time ave. chamber pressure, psia	575
Maximum chamber pressure, psia	618
Case	
Material	6Al-4V Titanium
Minimum ultimate strength, ksi	165
Minimum yield strength, ksi	155
Hydrostatic test pressure, psia	732
Minimum burst pressure, psia	860
Nominal thickness, in.	0.069
Nozzle	
Exit cone material	Carbon Phenolic
Throat insert	3D Carbon/Carbon
Internal throat diameter, in.	3.98
Exit diameter, in.	29.5
Expansion ratio, initial/average	54.8/47.2
Exit cone half-angle, exit/eff., deg	14.3/16.3
Type	Fixed, Consoured
Liner	
Type	TL-H-318
Density, lb/in.^{3}	0.038
Igniter	
Min. firing current, amp	5.0
Circuit resistance, ohm	1.1
No. of detonators and TBIs	2
Weight, lb	
Total	4721
Propellant	4431
Case assembly	129
Nozzle assembly	97
Internal insulation	60
Liner	2
SA, ETA	5
Miscellaneous	3
Total inert	290
Burnout	258
Propellant mass fraction	0.939
Propellant	
Propellant designation	TP-H-3340
Formulation: Al, 18%	
AP, 71%	
HTPB binder, 11%	
Propellant configuration	
Type	internal burning, radial slotted star
Web thickness, in.	20.47
Web fraction, %	84
Sliver fraction	0
Propellant volume, in.^{3}	68,050
Volumetric loading density, %	93.1
Web ave. burning surface area, in.^{2}	3325
Initial surface to throat area ratio, K	226
Propellant characteristics	
Burn rate @ 1000 psi, in./s	0.228
Burn rate exponent	0.30
Density, lb/in.^{3}	0.0651
Temperature coefficient of pressure	0.10
Characteristic velocity, ft/s	5010
Adiabatic flame temperature, °F	6113
Ratio of specific heats: chamber	1.14
Nozzle exit	1.18

Gas is loaded through a ground fitting, V1, and filter, F1, (typically 20 μ) into a titanium tank (see Section 4.6 for tank design information). Relatively high pressures are used—34,500 kPa, for example. The gas might be isolated by an ordnance valve V2 until release from the launch vehicle. Gas then flows through filter F2, which protects the regulator and flows at a reduced pressure through the low-pressure filter F3 to the thruster valves V3. On command, the thruster valves are opened in pairs to produce attitude control torques. The small converging/diverging

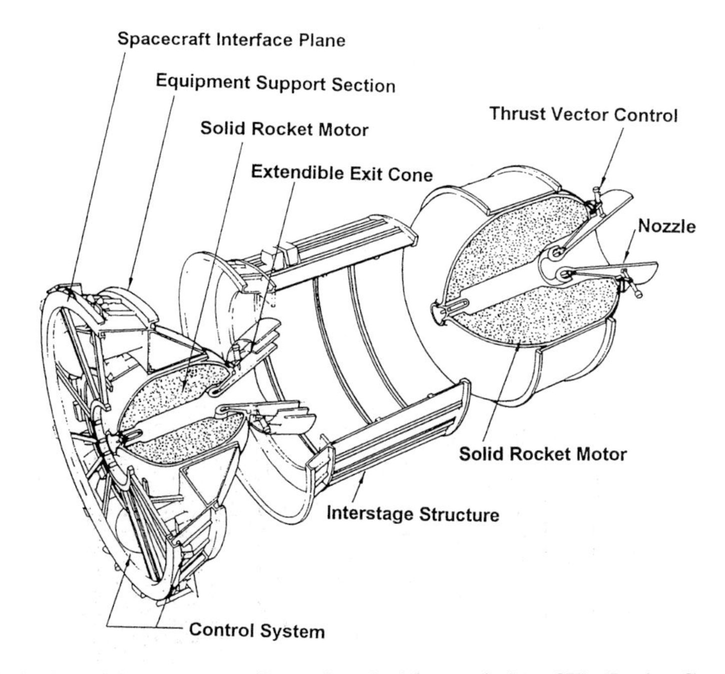

Fig. 4.61 Inertial upper stage. (Reproduced with permission of The Boeing Company; Ref. 18.)

Table 4.23 IUS solid rocket motors[a]

Characteristic	Stage I	Stage II
Motor total weight, kg	10,424	3009
Total impulse, kN-s	28,350	78,540
Total impulse (with EEC)	—	8,245
Nominal propellant load, kg	9,710	2,750
Minimum propellant load, kgb	4,853	1,360
Burn time (no offload), s	152.02	103.35
Average thrust, N	186,510	75,991
Average thrust (with EEC), N	—	79,780
Average chamber pressure, kPa	3,992	4,206
Delivered I_{sp} (without EEC), s	295.5	289.1
Delivered I_{sp}(with EEC), s	—	303.5
Throat diameter, cm	16.46	10.67
Area ratio	63.80	49.33
Area ratio (with EEC)	—	181.60
Exit diameter (without EEC), cm	131.5	74.9
Exit diameter (with EEC), cm	—	143.8
Motor outside diameter, cm	234	161
Overall length (without EEC), cm	356	169
Gimbal movement, deg	4	7

[a]Courtesy of The Boeing Company; Ref. 18.

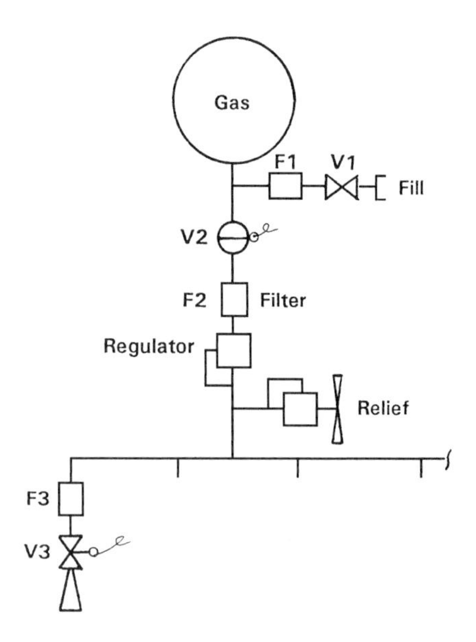

Fig. 4.62 Cold-gas system.

nozzles can be integral with the valve body. A relief valve protects the low-pressure system from regulator leakage or failure possibilities. The relief exhaust can be split to provide zero torque on the vehicle.

The gases that have been used or tested are shown in Table 4.24. Helium, nitrogen, and Freon 14 have flown. Helium has the best performance and lowest gas weight; however, a leakage failure is less likely with nitrogen or freon. From Table 4.24, note that the measured specific impulse is about 90% of theoretical, less than you would expect of larger thrusters. This lower performance is primarily caused by the conical nozzles commonly used, as opposed to the contoured nozzles on larger thrusters.

Table 4.24 Cold gases used

Gas	M	k	R	I_{sp} Theo[a]	I_{sp} Meas.	Reference[b]
Helium	4	1.659	2079 (386)	176	158	26
Nitrogen	28	1.4	296.9 (55.1)	76	68	26
					67	31
Freon-14	88	1.22	94.5 (17.6)	49	46	26, 31
Freon-12	121	1.14	68.7 (12.8)	46	37	26
Ammonia	17	1.31	489 (90.8)	105	96	26
Hydrogen	2	1.40	4157 (767)	290	260	26
Nitrous oxide	44	1.27	188.9 (34.9)	67	61	31

[a]Theoretical specific impulse is for vacuum, frozen equilibrium, area ratio = 100, gas temperature = 311°K (560°R).

[b]Reference applies to the measured I_{sp} values.

Specific impulse can be calculated from ideal rocket thermodynamics: Theoretical C_f is

$$C_f = \sqrt{\frac{2k^2}{k-1}\left(\frac{2}{k+1}\right)^{\frac{k+1}{k-1}}\left[1-\left(\frac{P_e}{P_c}\right)^{\frac{k-1}{k}}\right]} + \left(\frac{P_e - P_a}{P_c}\right)\frac{A_e}{A_t} \tag{4.13}$$

Thrust is

$$F = P_c A_t C_f \tag{4.12}$$

The gas volume needed can be obtained from the total impulse requirement and the equation of state:

$$W_p = \frac{I}{I_{sp} g_c} \tag{4.15}$$

and

$$V = \frac{W_p RT}{P} \tag{4.68}$$

Design of the tankage is discussed in Section 4.4.4.

Example 4.12 Cold-Gas System Design

Design a cold-gas system to meet the following requirements: total impulse is 4482 N-s, minimum impulse bit is 0.004 N-s, area ratio is 100, and thruster temperature is maintained at 37.8°C.

Use helium as a working fluid.

Assuming a minimum valve response time of 20 ms, the thrust level can be found as follows:

$$F = \frac{0.004}{0.02} = 0.2\,\text{N}$$

A thrust level of 0.2 N is within the capability of cold-gas thrusters. If the spacecraft is three-axis stabilized, 12 thrusters will be required to provide pure couples.

Note from Table 4.24 that a specific impulse of about 158 s can be expected for helium.

The helium weight required can be obtained from total impulse [Eq. (4.15)]:

$$W = \frac{4482}{(158)(9.80665)} = 2.893\,\text{kg}$$

The tank volume needed to contain 2.893 kg of helium plus 0.23 kg of unusable helium is obtained from the equation of state; R for helium is 2078.5, temperature 311°K, tank pressure is 20684271 Pa (3000 psi).

$$V = \frac{(3.123)(2078.5)(311)}{(20684271)} = 0.0976\,\text{m}^3$$

Tank design is covered in detail in Section 4.4.6. A spherical titanium tank, stressed to 100,000 psi maximum for 6.6 lb of helium at a maximum pressure of 3000 psi, weighs about 22 kg. Allowing 5 kg for valves, lines, filters, and thrusters, the cold-gas system weighs 27 kg.

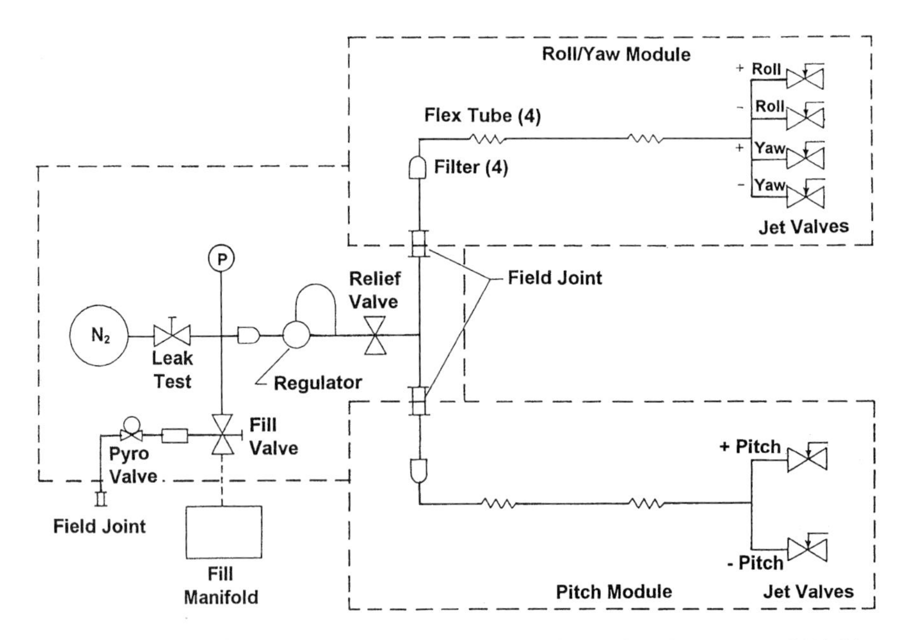

Fig. 4.63 Viking Orbiter reaction control system. (Reproduced courtesy of NASA; Ref. 32, p. 189.)

4.8.2 Flight Systems

4.8.2.1 Viking Orbiter. The Viking Orbiter reaction control system provided three-axis control of the spacecraft when the main engine was not firing. There were two redundant systems; one is shown in Fig. 4.63.

The systems used nitrogen gas as the propellant. Each tank held 7.03 kg of nitrogen at an initial supply pressure of 1.33×10^{11} Pa (4400 psig) and a temperature of 37.8°C (100°F). The specific impulse was 68 s at 20°C (68°F). Nitrogen was filtered enroute to six thrusters. The thrusters were mounted on the solar panel tips to maximize moment arm and canted at 25 deg to avoid impingement on the panels. The relief valve exhaust was branched to produce zero net torque.[32]

The system was designed to operate on helium borrowed from the main propulsion system in the event of failure. The system performance using nitrogen and helium is summarized in Table 4.25 (Ref. 32).

Table 4.25 Viking Orbiter performance

Performance	Nitrogen (primary mode)	Helium (backup mode)
Thrust per jet, N (lb)	0.13 (0.030)	0.12 (0.0278)
Specific impulse at 68°F, s	68	158
Minimum impulse bit at 22 ms, N-s (lb-s)	0.0028 (0.00063)	—

The Viking Orbiter gives a rare opportunity to compare the performance of two gases in the same thrusters. The minimum impulse bit was not measured with helium; however, from Eq. (4.44) I_{min} is estimated to be 0.0027 N (0.00061 lb-s) for helium.

4.8.2.2 LANDSAT 3. The LANDSAT 3 was the third of a series of Earth resource surveyor spacecraft. The 950-kg spacecraft provided complete multi-spectral images of the Earth's surface every 18 days. The attitude control system provided three-axis stabilization and nadir pointing for the spacecraft. Attitude control torques were provided by six Freon-14 thrusters. The 7870-cc propellant tank was initially charged with 5.5 kg of Freon-14 at 13789 kPa (2000 psia) at 20°C. Freon was fed to the thrusters through a pressure regulator. Each thruster provided about 0.45-N thrust. The system provided a total impulse of 2523 N-s at a measured specific impulse of 46.2 s (Ref. 12).

References

[1]Ring, E., *Rocket Propellant and Pressurization Systems*, Prentice–Hall, Upper Saddle River, NJ, 1964.

[2]Sutton, G. P., *Rocket Propulsion Elements*, Wiley, New York, 1986.

[3]Koelle, H. H. (ed.), *Handbook of Astronautical Engineering*, McGraw–Hill, New York, 1961.

[4]Seifert, H. S. (ed.), *Space Technology*, Wiley, New York, 1959.

[5]Brown, C. D., *Spacecraft Mission Design*, 2nd ed. AIAA, Reston, VA, 1998.

[6]Brown, C. D., *Spacecraft Propulsion*, AIAA, Reston, VA, 1996.

[7]*Monopropellant Hydrazine Design Data*, Rocket Research Corp. Seattle, WA.

[8]Morningstar, R. E., Kaloust, A. H., and Macklis, H., "*An Operational Satellite Propulsion System Providing for Vernier Velocity , High and Low Level Attitude Control and Spin Trim*," AIAA, Paper 72-1130, 1972.

[9]Sackheim, R. L., "Survey of Space Applications of Monopropellant Hydrazine Propulsion Systems," AIAA Paper 74-42367, 1974.

[10]Hearn, H. C., "*Design and Development of a Large Bipropellant Blowdown Propulsion System*," AIAA Paper 93-2118, 1993.

[11]Schmitz, B. W., Groudle, T. A., and Kelley, J. H., "*Development of the Post Injection System for Mariner C Spacecraft*," Jet Propulsion Lab. Technical Rept. 32-830, California Inst. of Technology, Pasadena, CA, 1966.

[12]"*LANDSAT 3 Reference Manual*," General Electric Space Div. Philadelphia, PA, 1978.

[13]Holmberg, N. A., Faust, R. P., and Holt, H. M., "*Viking '75 Spacecraft Design and Test Summary*," NASA RP 1027, 1980.

[14]Frazier, R. E., "*HEAO Case Study in Spacecraft Design*," TRW Defense and Space Systems Group, Rept. 26000-100-102, AIAA Case Study, July 1981.

[15]Schatz, W. J., Cannova, R. D., Cowley, R. T., and Evans, D. D., "Development and Flight Experience of the Voyager Propulsion System," AIAA Paper 79-1334, June 1979.

[16]*Pioneer Venus Case Study in Spacecraft Design*, AIAA, New York, 1979.

[17]Rausch, R. J., Johnson, J. T., and Baer, W., *Intelsat V Spacecraft Design Summary*, Spacecraft Design Case Studies, AIAA, New York, 1981.

[18]"*Inertial Upper Stage Users Guide*," Boeing Aerospace Co., D290-11011-1, Seattle, WA, Jan. 1984.
[19]"*Magellan Spacecraft Final Report*," Lockheed Martin, Flight Systems, MGN-MA-011, 1995.
[20]Neer, J. T., Hoeber, C. F., and Marx, S. H., *Intelsat V, a Case Study in Spacecraft Design*, AIAA, New York, 1981.
[21]Gale, Maj. H. W., and Adams, Capt. N., *Post-Boost Propulsion Experience*, 1974.
[22]Vote, F. C., and Schatz, W. J., "*Development of the Propulsion System for the Viking 75 Orbiter*," AIAA Paper 73-1208, 1973.
[23]Barber, T. J., Krug, F. A., and Froidevaux, B. M., "Initial Galileo Propulsion System In-Flight Characterization," AIAA Paper 93-2117, June 1993.
[24]Virdee, L. S., Chang, C. P., and Dest, L., "In-Orbit Performance of Intelsat Liquid Propulsion System," AIAA Short Course, IAA/SAE/ASME/ASEE 29th Joint Propulsion Conference, Monterey, CA, June 1993.
[25]"*Solid Rocket Motor Performance Analysis and Prediction*," NASA SP 8039, June 1971.
[26]Greer, H., and Griep, D. J., "Dynamic Performance of Low Thrust, Cold Gas Reaction Jets in a Vacuum," *Journal of Spacecraft and Rockets*, Vol. 4, 1967, pp. 983–990.
[27]Rogers, W. P., "Report of the Presidential Commission on the Space Shuttle Challenger Accident," GPO, Washington, DC, June 1986.
[28]Hamlin, K., McGrath, D. K., and Lara, M. R., "*Venus Orbit Insertion of the Magellan Spacecraft Using a Thiokol STAR 48B Rocket Motor*," AIAA Paper 91-1853, June 1991.
[29]Sola, F. L., "Development of the Explorer Solid Rocket Motor," AIAA Paper 89-2954, July 1989.
[30]Isakowitz, S. J., Hopkins, J. P., Jr., and Hopkins, J. B., *International Reference Guide to Space Launch Systems*, 3rd ed. AIAA, Washington, DC, 1999.
[31]Hall, S. E., Lewis, Mark J., and Akin, D. L., "Design of a High Density Cold Gas Attitude Control System," AIAA Paper 93-2583, 1993.
[32]Holmberg, N. A., Faust, R. P., and Holt, H. M., "*Viking '75 Spacecraft Design and Test Summary*," NASA RP 1027, Nov. 1980.
[33]Bell, T., "Magellan OTM1 Propulsion Report," (rev. A), Lockheed Martin, Denver, CO, Feb. 1992.

Problems

If you need propellant property data for these problems, it can be found in Appendix B and elsewhere.

4.1 You are considering using N_2O_4/N_2H_4 bipropellant engines in your design. One of the potential engine manufacturers claims that he has an engine which will produce 2224 N thrust and a specific impulse of 327 s in a vacuum. Noticing your skepticism, he provides you with the following test data from a near-vacuum firing of the engine: a) pressure in the test chamber during firing = 2.482 kPa, b) measured thrust = 2157 N, c) measured propellant flow rate = 0.7049 kg/s, d) measured chamber pressure = 1965 kPa, and e) measured gas total temperature = 3283°K. The engine has a throat area of 6.5374 sq cm and an area ratio of 65. From other sources you know that the exhaust gas produced by this propellant combination has a molecular weight of 18.9 and the ratio of specific heats of 1.26. Do you believe this engine will achieve the expectations of the manufacturer? To decide:

(a) Calculate the theoretical vacuum specific impulse of this engine, using frozen equilibrium assumptions.
(b) Adjust the test data, thrust, and specific impulse, to vacuum conditions.
(c) Compare adjusted specific impulse from the test to theoretical values.
(d) Compare the adjusted specific impulse and thrust to claimed values.

4.2 If a propulsion system delivers a thrust of 4448 N with a propellant flow rate of 1.474 kg/s, what is the specific impulse? If a propulsion system delivers 1000-lb thrust with a propellant flow rate of 3.25 lb/s, what is the specific impulse?
Solution: 308 s.

4.3 The measured characteristics of an engine are as follow: $F = 82.78N$, $I_{sp} =$ 289 s, P_c = 668.8 kPa, area ratio = 40, and A_t = 0.761 cm^2. The thrust and I_{sp} were measured in a test chamber with an ambient pressure of 1944 Pa. What vacuum thrust and I_{sp} would you expect of this engine? What is the vacuum thrust coefficient? Is this a monopropellant engine, bipropellant, or solid motor?

4.4 A spacecraft has a monopropellant propulsion system that delivers an I_{sp} of 225 s. How much propellant would be consumed to trim the orbit if a ΔV of 200 m/s were required and the spacecraft weighed 950 kg at the end of the burn?
Solution: 90.1 kg.

4.5 A spacecraft has the following characteristics: a) thruster pair, each located at 1.8-m radius from the center of mass; b) moment of inertia of 2700 kg-m^2 about the Z axis; and c) thrust of each engine is 0.9 N

What is the minimum time for the spacecraft to maneuver through a 65-deg rotation about the z axis? How much propellant is consumed at a specific impulse of 185 s?

4.6 A spacecraft must be maneuvered through an angle of 60-deg in 30 s. The spacecraft moment of inertia is 8000 kg-m^2; the thrusters are located at a radius of 2.2 m from the center of mass. What is the minimum thrust level for the thrusters?
Solution: 1.90 lb_f, 8.46 N.

4.7 A spacecraft with the following characteristics is in a limit-cycle operation about the z axis: angular control requirement = ± 0.5 deg, moment of inertia about z = 12,000 kg-m^2, minimum impulse bit = 0.01 N-s (0.5-N engines with *t-on* = 0.02 s), pulsing specific impulse = 120 s, and two thrusters mounted 2 m from the center of mass. If external torques are negligible, what is the time rate of fuel consumption?

4.8 A 962-kg communications satellite has attained an orbit in the equatorial plane; however, the apogee altitude is 41,756 km, and the eccentricity is 0.0661. What is the minimum ΔV and amount of monopropellant to place the satellite in a geosynchronous orbit? The final orbit altitude must be 35,786 $\pm$ 10 km, and the system specific impulse is 225 s.
Solution: $\Delta V = 0.16$ km/s, and $w_p = 67.3$ kg.

4.9 Prepare a propellant inventory for a monopropellant system to meet the following requirements:

a) Translational maneuver $\Delta V = 200$ m/s at an average I_{sp} of 225 s.
b) Spacecraft wet weight = 1100 kg.
c) Attitude control total impulse = 15,900 N-s in pulse mode at an average I_{sp} of 130 s.
d) Propellant reserve = 35% of usable.

4.10 The Centaur upper stage has the following characteristics[30]: stage dry mass = 2,775 kg; usable propellant mass = 20,950 kg; unusable propellant mass = 16 kg; and vacuum specific impulse = 444 s.

What is the maximum velocity increase that the upper stage will impart to a spacecraft with a mass of 3592 kg?

What is the vehicle burn out mass before spacecraft separation?

Solution: $\Delta V = 6.26$ km/s, and Burnout mass = 6527 kg.

4.11 What is the weight of nitrogen remaining in a 0.288-m^3 propellant tank assuming that all of the propellant is consumed, the tank pressure at burnout is 2413 kPa, and the nitrogen temperature is 15°C?

Solution: 1.17 kg.

4.12 Estimate how much a titanium pressurant sphere will weigh if the allowable stress is 6.895×10^8 Pa (100,000 psi), the internal volume is 0.00385 m^3, and the maximum operating pressure is 2.0684×10^7 Pa (3000 psi)? What weight of helium does it contain at maximum operating pressure and 21°C? (This is the Magellan repressurization sphere. The actual tank weight was 1.51 kg, and it held 0.126 kg of helium. Your estimates will be slightly different.)

4.13 Design a hydrazine tank to meet the following requirements:

a) Propellant load = 350 lb.
b) Maximum system temperature = 110°F.
c) Minimum system temperature = 40°F.
d) Blowdown ratio = 4.5.
e) Maximum operating pressure = 490 psia.
f) Propellant control device = 0.070-in. thick elastomeric diaphragm.
g) Allowable stress in titanium = 110,000 psi.

What is the required volume of the tank and the estimated tank assembly weight? What is the weight of the helium loaded, assuming no absorption?

4.14 Estimate the mass of a titanium barrel tank given the following: a) allowable stress = 8.6184E5 kPa, b) internal volume = 0.71726 m^3, c) maximum tank pressure = 6136 kPa, and d) inside diameter = 88.9 cm. (This is the Viking Orbiter tank; its actual length was 1.4478 m and actual weight was 50.53 kg. Your estimate will be slightly different.)

4.15 Design a monopropellant hydrazine system for the following spacecraft: spacecraft dry weight = 1200 lb; $Ix = 1500$ slug-ft^2, $Iy = 3000$ slug ft^2,

$Iz = 8000$ slug-ft^2; and maximum moment arm for thrusters = 6.5 ft (shroud internal radius).

The system must meet the following requirements: translational velocity change = 900 m/s (steady-state burns); reaction wheel unloading impulse = 9800 lb-s (steady-state burns); attitude control maneuvers = 7000 lb-s (pulsing); steady state I_{sp} = 230 s; pulsing I_{sp} = 120 s; minimum pulse width = 20 ms; maximum time available for pitch, yaw, or roll maneuvers = 90 deg in 5 min; blowdown pressurization; blowdown ratio = 4.5; nitrogen pressurant; maximum tank pressure = 650 psia; diaphragm propellant control; maximum stress in titanium = 100,000 psi; and no single, nonstructural, malfunctions can cause loss of the mission.

Prepare a propellant inventory. Select the number of tanks. Select the thrust levels. Size the tank or tanks. Size the diaphragm/s. Prepare a schematic; include instrumentation and first pulse provisions. Prepare a propulsion weight statement. Prepare a failure modes and effects analysis.

4.16 Determine the burning time, action time, and maximum thrust from the solid motor test data shown in Fig. 4.64. Estimate the action time impulse and the burning time impulse from Fig. 4.64. What is the burning time average thrust and the action time average thrust?

4.17 Compare the propellant mass for a maneuver requiring a ΔV of 1.819 km/s with the following three propulsion candidates: a) a solid motor system with I_{sp} of 290 s (HTPB/AP/AL), b) a monopropellant system with I_{sp} of 235 s (N2H4), and c) a bipropellant system with an I_{sp} of 320 s (N2H4/N2O4).

The vehicle mass at the start of the burn is 3600 kg.

Solution a: 1700 kg.

4.18 Assume that you have chosen monopropellant hydrazine thrusters for attitude control propulsion and four roll thrusters (two roll clockwise and two roll counterclockwise) for pure couples placed at 1.8 m from the center of gravity. Find the minimum thrust level for the attitude control thrusters in order to roll 90 deg in 66 s. The roll moment of inertia is 1500 kg-m^2.

Solution: 0.60 N.

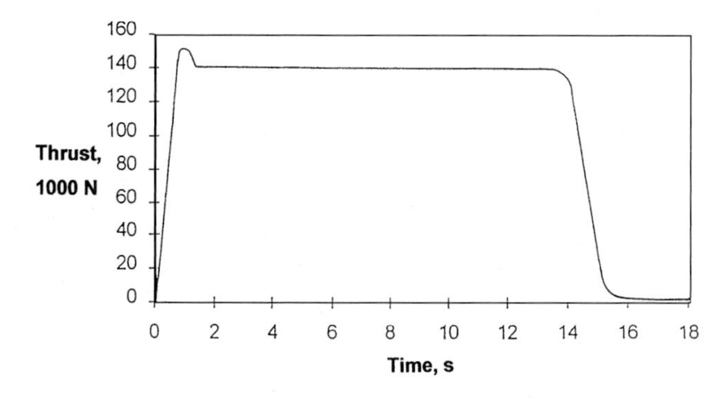

Fig. 4.64 Solid motor thrust time curve.

4.19 It is necessary to unload the three 17 N-m-s wheels once a day. Assume the wheels are 75% loaded. What propellant weight would be required over the five-year life of the mission if the specific impulse of the 112-N thrusters is 185 s for short burns of this type and the thrusters are 1.8 m from the c.g.?

4.20 Calculate the propellant tank volume for the following requirements: 75 kg of hydrazine propellant; blowdown pressurization, blowdown ratio = 4, using helium; elastomer diaphragm propellant control device, thickness 0.19 cm; and propellant temperature limits 5 to 30°C.

4.21 Consider a solid rocket motor as an alternative geosynchronous orbit insertion system. Use a ΔV requirement of 1819 m/s nominal Design a motor using the following requirements; do not use thrust termination in the motor design: HTPB/AP/AL propellant, $I_{sp} = 295$ s; and spacecraft initial weight = 3600 kg (at start of burn).

(a) Calculate the propellant required.

(b) Calculate the spacecraft weight at burnout.

(c) Calculate total impulse required of the motor.

(d) Estimate a spherical solid motor weight.

(e) Assume ±2% variations in the delivered velocity change or ±36 m/s. How much monopropellant mass will be required to correct the error, assuming an I_{sp} of 235 s?

f) What thrust and impulse would be required for thrust vector control in pitch if the solid motor thrust vector was offset from the center of gravity by 2.1 cm in the pitch plane. Assume a solid motor thrust of 80,000 N, a burn time of 60 s, and a thruster moment arm of 2.1 m. Assuming the correction is made with a monopropellant thruster with an I_{sp} of 235 s, how much propellant is consumed.

5
Attitude Control
E. M. Dukes

Attitude control deals with the *orientation* of the spacecraft axes with respect to an inertial reference frame. An instantaneous spacecraft *attitude* is commonly described by a *pitch angle*, a *roll angle*, and a *yaw angle*. The most frequently used inertial reference frame is shown in Fig. 5.1. The **+*X*** vector is parallel to the spacecraft velocity vector. The +*X* and +*Z* axes are contained in the orbit plane, as is the velocity vector. The **+*Z*** is "up," opposed to the gravity vector. The **+*Y*** vector is perpendicular to the orbit plane. Spacecraft attitude is measured as an angular deviation of the spacecraft body axes from these inertial coordinates. The attitude control system (ACS), controls the spacecraft body axes such that the errors in pitch, yaw, and roll angles are within defined limits. Note the distinction between *attitude control*, which deals with rotation about spacecraft axes, and *guidance and control*, which deals with an objects position in geographical coordinates, usually latitude and longitude.

Any uncontrolled body in space, for example, an asteroid, will tumble about all axes in response to natural forces, notably solar pressure, gravity gradients, and magnetic torques. Sputnik I, the first spacecraft, was allowed to tumble in this way. However, natural tumbling is not normally acceptable for a spacecraft because solar panels must be pointed at the sun for power, antennas must be pointed at an Earth station for communication, and science instruments must be pointed at their targets. As a result, spacecraft attitude must be controlled. From this simple requirement comes arguably the most technically sophisticated subsystem in the spacecraft.

As Malcom Shuster[1] so beautifully put it, "The baggage of attitude studies is very extensive and not particularly easy to acquire." In this section we will describe what this mathematical baggage does in general and leave more in-depth understanding to the discretion of the reader.

Explorer I, the first U.S. spacecraft, had the simplest form of attitude control; it was spun about one axis. A spinning spacecraft tends to hold one axis, the spin axis, fixed in inertial coordinates. Having one axis fixed is adequate for many missions. For example, if the spin axis is perpendicular to the sun vector the body can be populated with solar cells to provide power needs (see Fig. 5.2). Omnidirectional antennas can be used, and some payloads need no more than one fixed axis. This attitude control method is still in use today; however, it is not as simple as it sounds.

In today's complex missions, many spacecraft have requirements too diverse for the one-axis control. It is not unusual for a spacecraft to need solar panels pointed to the sun, science payload (a camera for example) pointed at the planet surface, and a high-gain antenna pointed at a ground station. For complex situations three-axis control is required (see Fig. 5.3).

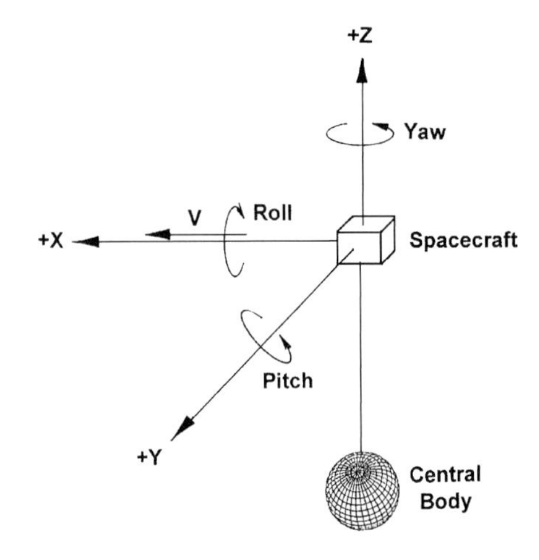

Fig. 5.1 Attitude orientation of a spacecraft.

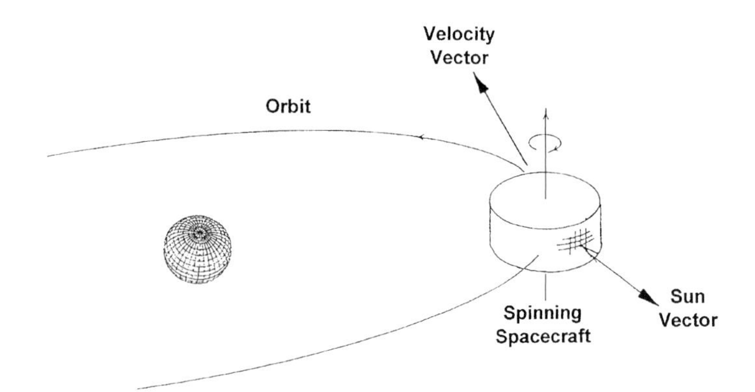

Fig. 5.2 Spinning spacecraft.

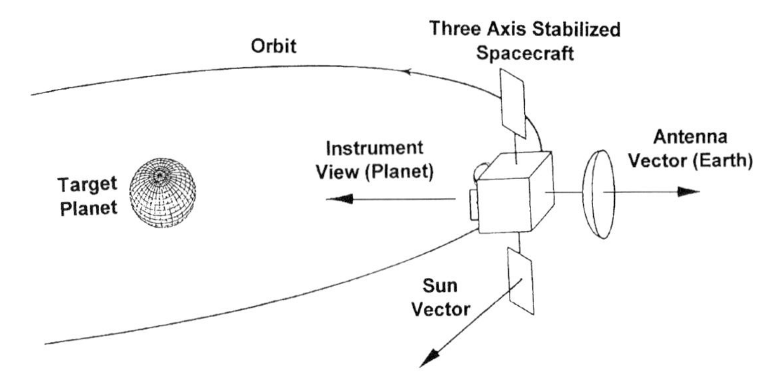

Fig. 5.3 Three-axis-stabilized spacecraft.

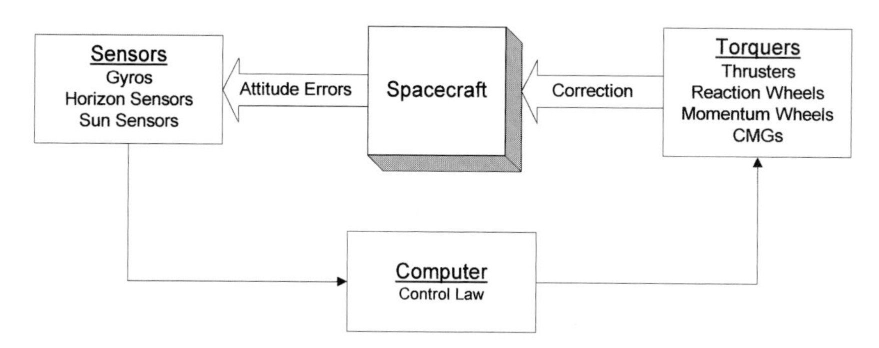

Fig. 5.4 Attitude control operation.

The attitude control task can be divided into three subtasks: 1) measuring attitude, which is done by attitude *sensors* such as gyroscopes; 2) correcting attitude, which is done by *torquers* or *actuators* such as thrusters; and 3) a *control law*, which is software that determines the magnitude and direction of torque in response to a given disturbance. The ACS, conceptually simple, is a classic feedback control system as shown in Fig. 5.4. If the spacecraft drifts off the desired attitude, the sensors detect the error, the control law determines the magnitude of the response and directs a torquer to correct it.

In Section 5.1 the requirements definition for an attitude control system is described. The most important of these is the selection of the basic system type. This choice has significant impact on all of the subsystems. The disturbances that make attitude control necessary are torques from solar pressure, aerodynamics, magnetic fields, gravity gradients, and spacecraft activities (rotating antennas, pointing cameras, etc.). These disturbances are discussed in Section 5.2. The first step in control is the determination of current attitude as discussed in Section 5.3. The actual control process is discussed in Section 5.4. The attitude control system also controls spacecraft maneuvers to new attitudes; these maneuvers are discussed in Section 5.5, and attitude control hardware is discussed in Section 5.6.

5.1 Requirements Definition

The steps in setting ACS requirements are as follows:

1) Summarize mission pointing direction requirements. This is usually defined by the payload as well as the mission design [usually nadir (or Earth) pointing, inertial target, or scanning].

2) Summarize mission and payload pointing accuracy requirements. Pointing accuracy is a function of both knowledge accuracy and control accuracy. Knowledge accuracy is a function of the inherent accuracy of the sensors used and the attitude determination method used. Control accuracy is a function of the capability of the actuators and the accuracy of the control law. Typical pointing accuracy requirements: a) solar array: 4 to 10 deg; b) high-gain antenna: 0.1 to 0.5 deg; c) optics, telescopes, and cameras: 0.001 to 0.1 deg.

3) Define the translational and rotational maneuvering required for the mission including angles, durations, rates, and frequency. (The propulsion requirements come from this activity.)

4) Select the ACS type. The selection of system type is a major driver for the entire spacecraft design. Type selection requires knowledge of a) payload pointing requirements (usually the major driver); b) thrust vector control requirements; c) control authority and allowable attitude error, for translational ΔV maneuvers; d) maneuver slew rates, a driver if greater than about 0.5 deg/s; e) approximate solar panel area. (If panel area is large compared to the spacecraft body area, three-axis stabilization is indicated.)

5) Quantify the internal and external disturbance torques.

6) Select and size the major hardware elements.

7) Define the control law and attitude determination method.

8) Conduct trade studies to improve the requirements and selections.

Figure 5.5 shows the repetitive cycle of these tasks as the system design is refined and improved.

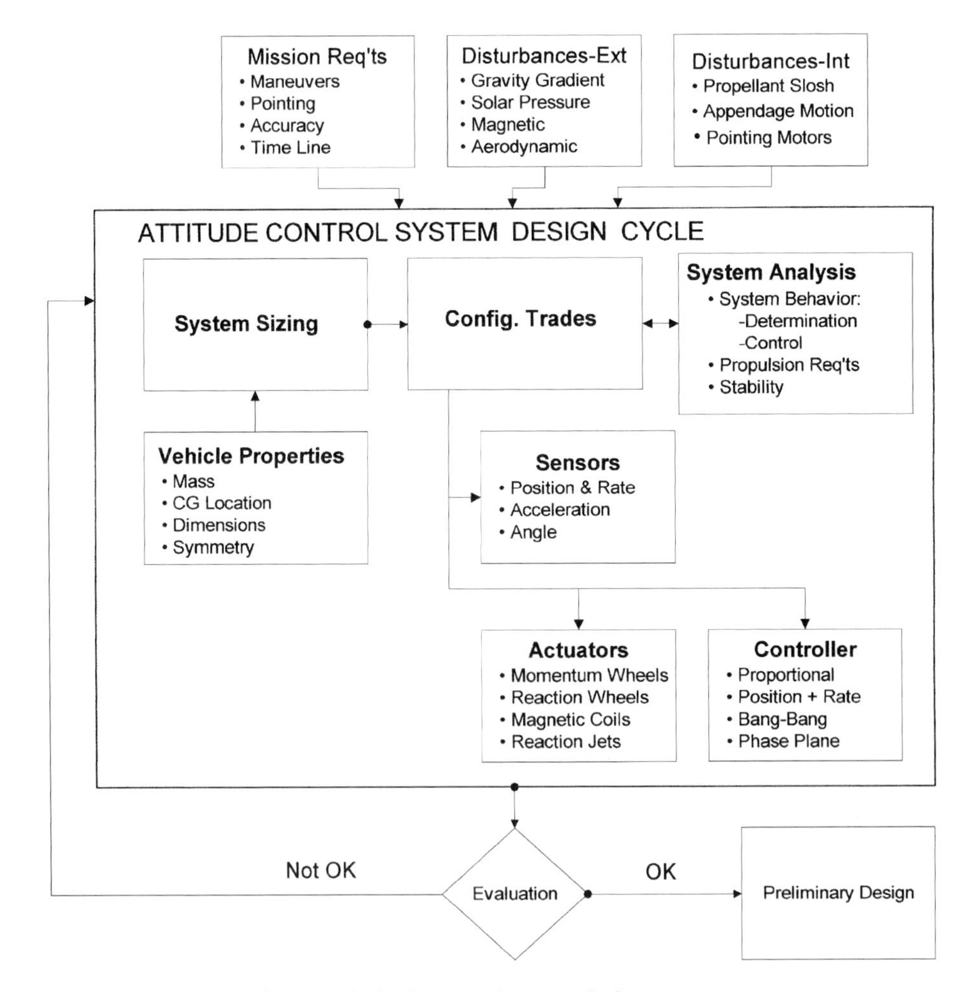

Fig. 5.5 Attitude control system design process.

The most common attitude control systems types are as follows: 1) spin stabilized, 2) dual spin, 3) three-axis stabilized, 4) momentum bias, and 5) gravity gradient. A spin-stabilized spacecraft is one in which the entire spacecraft spins around the axis with the highest moment of inertia. A dual-spin spacecraft has a spinning segment and an inertially fixed section. A three-axis controlled spacecraft is one which actively controls the inertial position of all three axes. Gravity-gradient control is completely passive and takes advantage of the spacecraft tendency to align the long axis with the gravity gradient. A momentum bias system uses a momentum wheel to provide stiffness in two axes and wheel speed to control the third axis. Several other types have been successfully used but are uncommon.

Sometimes different methods are used on the same spacecraft for different mission phases which have different control requirements.

Table 5.1 compares typical hardware implementation of each system type. Table 5.2 summarizes the advantages and disadvantages of the types of stabilization. Table 5.3 compares the suitability of the primary attitude control system types to various missions. It is assumed in Table 5.3 that the three-axis control system has both thrusting capability and reaction wheels.

5.1.1 Spin-Stabilized System

A spin-stabilized spacecraft (Fig. 5.2) takes advantage of the inherent resistance of a spinning body to disturbance torques. If no disturbance torques are experienced, the momentum vector remains constant in magnitude and fixed in inertial space. If a disturbance torque occurs that is parallel to the momentum vector, the spin rate will be affected but not the attitude. Thrusters are used to correct the spin rate. Disturbance torques perpendicular to the momentum vector will cause the spin axis to precess; thruster force can be used to remove precession. Rotational maneuvers are performed by precessing the spin axis. Translational maneuvers are always made parallel to the spin axis. Maneuvering is a slow and energy-consuming process because of the inherent stability of the vehicle.

5.1.1.1 Advantages. Spin stabilization is useful in situations that require simplicity, low cost, modest pointing accuracy, and minimal maneuvering. Stabilization about transverse axes is passive for long periods of time.

Sensor gyros, momentum exchange devices, and onboard computers are unnecessary on a spinner. Substantial cost and mass savings result.

Spin stabilization greatly simplifies the use of solid rocket motors for translational maneuvers by eliminating the need for thrust vector control. INTELSAT VI, for example, is three-axis-stabilized during its mission; however, it was spun during solid motor ΔV burns to control the thrust vector.

The centrifugal force generated by spinning provides a bottoming force on propellants and makes 0-g propellant control devices unnecessary.

Spinning also supplies a scanning motion very desirable for some instruments.

5.1.1.2 Disadvantages. Pointing accuracy is low, 0.3 to 1 deg. Tight control of the moments of inertia is required. The moment of inertia about the spin axis must be substantially greater than that about the transverse axis or the vehicle will reverse axes. The moment of inertia ratio must be well above 1; a ratio of 1.2

Table 5.1 Typical hardware implementation

Hardware	Gravity gradient	Spin	Dual spin	Three axis	Momentum bias
Sensors	Sun or horizon	Star, horizon, or sun	Gyros, star, or horizon scanner	Precision gyros, sun sensor, star tracker, or horizon sensor	Sun sensors, horizon sensor
Control	Control electronics	Control electronics, damper	Control electronics, damper, programmable computers, I/O and software	Control electronics, programmable computers, I/O and software	Control electronics, programmable computer
Torquers	Boom, momentum wheel	Thrusters	Thrusters	Thrusters, reaction wheels, magnetic torquers	Momentum wheel, thrusters
Mechanisms	None	Dampers	Despin drive, dampers, slip rings	Antenna pointing, solar array pointing	Antenna pointing, solar array pointing, slip ring

Table 5.2 **Stabilizing techniques summary (Ref. 2)**

Stabilization type	Advantages	Disadvantages
Gravity gradient	Simple, passive Low cost Long life Provides nadir pointing	Low accuracy Poor maneuverability Poor yaw stability
Spin stabilized	Simple, passive Low cost Long life Provides scan motion Provides propellant control Provides thrust vector control for ΔV burns	Poor maneuverability Does not provide pointing Limited to scan motion Only one axis fixed Low solar panel efficiency, 32% maximum Requires control of moments of inertia
Dual spin	Provides both scanning and pointing capability Provides propellant control Provides thrust vector control for ΔV burns	Sensitive to mass properties Articulated elements require balance compensation Cost and complexity can equal or exceed three axis if accuracy is required Complex nutation and flexibility dynamics Low solar panel efficiency, 32% maximum Requires control of moments of inertia
Three axis	High pointing accuracy, limited only by sensors Unlimited payload pointing Most adaptable to changing mission requirements Can provide rapid maneuvering Can accommodate large power requirements	Most expensive method Complicates thrust vector control Propellant control required High weight and power Fault protection complex Thrust vector control complex

Table 5.3 **Mission suitability matrix (Ref. 2)**

Requirement	Gravity gradient	Spin	Dual spin	Three axis	Momentum bias
Nadir pointing	Yes	No	Poor	OK	OK
Geosynchronous	No	OK	OK	OK	OK
Planetary	No	OK	OK	OK	NO
Thrust vector control	No	Good	Good	OK	NO
Maneuvering	No	Limited	Limited	Good	Poor
Pointing accuracy, deg	5	1	0.1	0.001	0.1 to 3
Relative cost	—	1.00	1.19	2.10	1.45

is a common requirement. The inertia ratio requires constant monitoring in the same manner as mass and power margins.

Virtually the only location for solar arrays is the spinning body exterior. Total power available is limited to that which can be obtained from the body surface. In addition, this area is not in the sun all of the time. A given area on a cylindrical spinning body gets only 32% of the solar intensity that would fall on a pointed planar array. Power is therefore a scarce commodity on a spinner.

The maneuver rate is limited. A maneuver is made by precessing the spin axis, which is a relatively slow process. Maneuver slew rates greater than 0.5 deg/s indicate three-axis stabilization.

Body pointing of payload sensors and antenna is not possible.

Examples of spinning spacecraft are Explorer I, Pioneer Venus, and INTELSAT I, II, and III.

5.1.2 Dual-Spin System

Dual spinning is a compromise design, which has some of the simplicity of a spinner and some of the pointing accuracy of a three-axis vehicle. As shown in Fig. 5.6, the major mass of the spacecraft spins providing gyroscopic stiffness, while an instrument platform is despun to point at instruments or an antenna.

5.1.2.1 Advantages. The vehicle is stable about the transverse axes for long periods of time. Sensing gyros and onboard computers are not required.

The spinning body provides a built-in scan for sensors and provides a centrifugal bottoming force for any liquid propellants. Thrust vector control is not required for ΔV maneuvers.

The despun platform provides pointing for antenna and instruments.

5.1.2.2 Disadvantages. The despin drive assembly (motor, bearings, slip rings) is expensive and failure prone.

Nadir tracking is not practical except at very high altitudes, geosynchronous and above. Solar-array efficiency is limited because a given cell is illuminated 32% of the time.

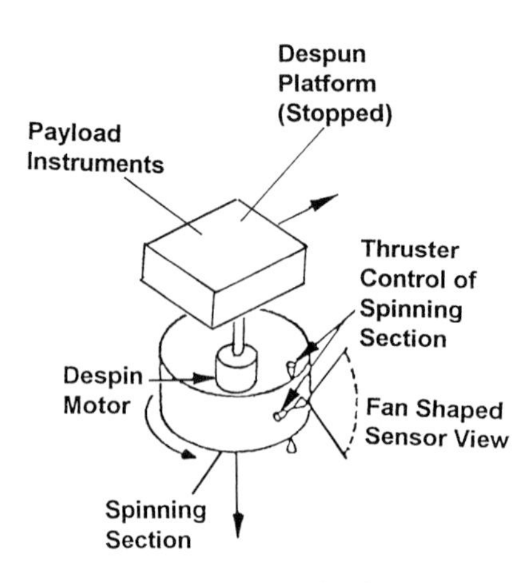

Fig. 5.6 Dual spin.

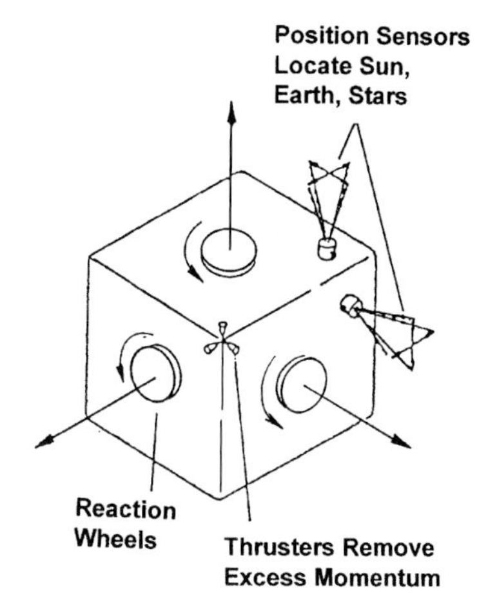

Fig. 5.7 Three-axis stabilized.

Complex nutation dynamics must be dealt with. For stability a dual spinner places constraints on the inertial properties and damping in the spun and despun sections. The energy dissipation of the spun section must be greater than that of the despun section. This turned out to be an expensive issue late in the development of Galileo.

Examples of dual-spin spacecraft are Galileo and INTELSAT VI.

5.1.3 Three-Axis-Stabilized System

A three-axis-stabilized system (Fig. 5.7) actively maintains the vehicle axis system aligned with a reference system, usually inertial reference or nadir reference. A typical system uses gyros as inertial reference and updates the gyros periodically using star scanning or horizon scanning. Attitude errors are removed by torquing reaction wheels, which are periodically unloaded using thrusters. The thruster layout provides pure torque about all three axes and positive or negative translation along each axis. An attitude control computer houses the software, which controls the process.

5.1.3.1 Advantages. The system provides unlimited pointing capability in any direction—nadir, inertial, sun, scanning—and provides the best pointing accuracy, limited only by sensor accuracy. Pointing accuracy of greater than 0.001 deg can be achieved.

Solar panels can make full use of the available solar energy. Solar panel size is not restricted. This type is the most adaptable to changing requirements.

5.1.3.2 Disadvantages. ACS hardware (gyros, reaction wheels, star scanners, computers) are complex, heavy, high-power consumers, failure sources, and expensive. Active thrust vector control is required for ΔV burns. Propellant tanks require 0-g propellant control devices.

Redundancy schemes are more complex; for example, what watches for computer error, another computer? Two out of three? Mechanical gimbals are

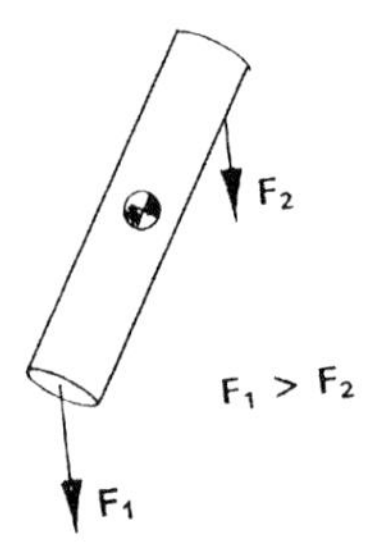

Fig. 5.8 Gravity gradient.

required for scanning instruments. Examples of three-axis-controlled spacecraft are Magellan, INTELSAT VIII, the Hubble Telescope, and GPS.

5.1.4 Gravity-Gradient System

Gravity-gradient stabilization takes advantage of the tendency of a spacecraft to align its long axis with the gravity vector; gravity force F_1 is greater than F_2 in Fig. 5.8. For this stabilizing technique to work, it is necessary that the gravity-gradient torques are greater than any disturbance torque; this criteria can usually be met in orbits lower than 1000 km. It is necessary for the moment of inertia about x and y axis to be much greater than the moment of inertia about the z axis. Deployed booms have been used on the long axis to improve the inertial properties. Gravity gradient stabilizes the pitch and roll axes only and not the yaw axis. It is common practice to use a momentum wheel with its axis perpendicular to the orbit plane to provide stiffness in yaw. Gravity-gradient control torques are small at best, and active damping might be required to prevent slow oscillations of as much as 10 deg.

Gravity-gradient stabilization is useful when long life and high reliability are required and the pointing requirements are modest. Spacecraft that have used gravity-gradient stabilization include ATS-5, GEOSAT, and ORBCOM.

5.1.5 Momentum Bias System

A momentum-bias system (Fig. 5.9) uses a momentum wheel to provide inertial stiffness in two axes and control of wheel speed provides control in the third axis.

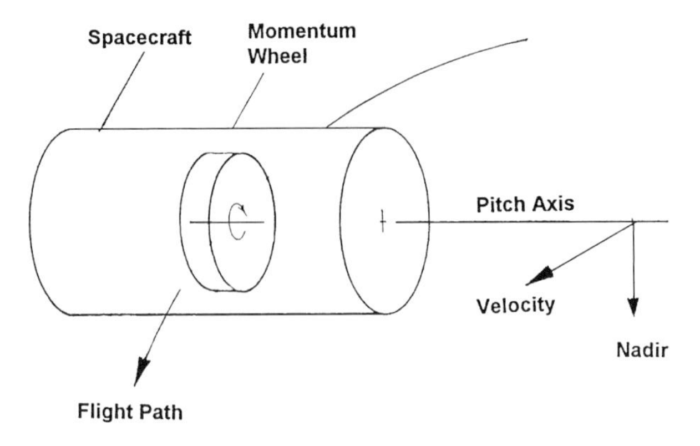

Fig. 5.9 Momentum bias control.

This system is particularly useful for a nadir pointing spacecraft using wheel speed to hold z axis on nadir.

The system is relatively simple and good for long-life missions. It is cheaper than a three-axis system. It offers good pointing in one axis (usually pitch) and poor accuracy in the wheel axes (usually roll/yaw). Momentum bias cannot achieve the pointing accuracy of three-axis control. Maneuvering capability is very restricted, and supplementary schemes are required for orbit insertion. INTELSAT VIII, for example, used spin stabilization for launch and orbit insertion. It does not provide adequate torque authority for thrust vector control. Seasat and more recently INTELSAT VIII used this technique.

5.2 Disturbance Torques

If a spacecraft were released into Earth orbit without attitude control, it would tumble in response to five different kinds of environmental torques: 1) drag torque—orbits below about 500 km; 2) gravity-gradient torque—orbits 500 to 35,000 km; 3) magnetic torque—orbits 500 to 35,000 km; 4) solar torque—dominant geosynchronous and above; and 5) spacecraft-generated torques.

Torque is generated by asymmetry of the spacecraft relative to the disturbance. For example, a spherical-shaped body would experience drag that is important for orbit determination, but there would be no drag torque resulting in an attitude disturbance. The physical properties of the surfaces, which will be discussed in more detail below, in combination with their symmetry relative to the disturbance, determine the impact of the various disturbance torques.

5.2.1 *Solar Torque*

When a solar photon strikes a spacecraft surface, there is a small momentum exchange, and a force is exerted on the surface as shown in Fig. 5.10. The pressure produced is proportional to the projection of the surface area perpendicular to the sun and the solar intensity, which is inversely proportional to the square of the distance from the sun. The pressure produced also depends on whether the photon was absorbed, specularly reflected (as a mirror), or diffusely reflected.

5.2.1.1 Absorption. If the solar radiation impinging on a surface is totally absorbed (Fig. 5.10a), the force on the surface will be aligned with the sun vector and of magnitude

$$F = P_s A \cos\alpha \tag{5.1}$$

where

$$P_s = \frac{I_s}{c} = \frac{1376\ \mathrm{W/m^2}}{2.998 \times 10^8\ \mathrm{m/s}} = 4.59 \times 10^{-6}\ \mathrm{N/m^2}\ \text{(near Earth)}$$

and

I_s = incident solar radiation, W/m^2
c = speed of light, m/s
P_s = solar pressure, N/m^2

Table 5.4 shows solar pressures for each planet at mean solar radius.

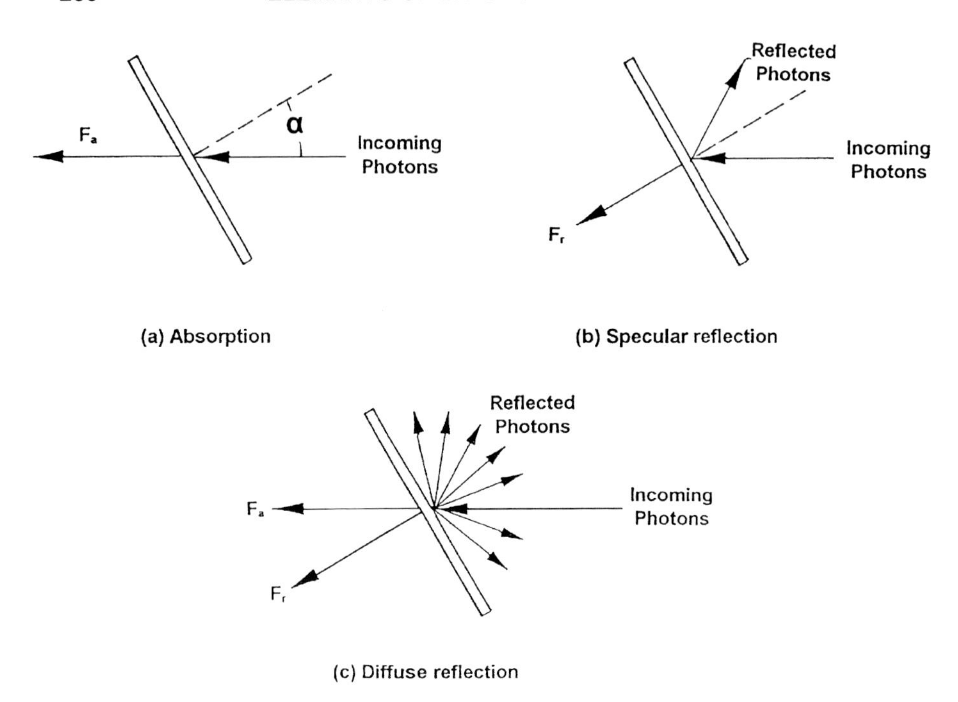

Fig. 5.10 Solar pressure.

5.2.1.2 Specular reflection. The force resulting from impingement on a specularly reflective surface is normal to the surface regardless of sun line (Fig. 5.10b) and is an elastic collision with twice the magnitude of that of an absorbing surface:

$$F = 2P_s A \cos \alpha \tag{5.2}$$

5.2.1.3 Diffuse reflection. A diffusely reflective surface (Fig. 5.10c) can be considered to be an absorption and a reradiation uniformly distributed over a hemisphere. The absorption component is aligned with the sun vector with a

Table 5.4 Solar pressure P_s at each planet

Planet	Solar pressures, N/m^2
Mercury	3.05×10^{-5}
Venus	8.77×10^{-6}
Earth	4.59×10^{-6}
Mars	2.0×10^{-6}
Jupiter	1.70×10^{-7}
Saturn	5.10×10^{-8}
Uranus	1.24×10^{-8}
Neptune	5×10^{-9}
Pluto	3×10^{-9}

magnitude given by Eq. (5.1). The net force vector resulting from the reflected component is normal to the surface; all tangential components cancel. The magnitude of the diffuse component is two-thirds that of the absorption component [two-thirds of Eq. (5.1)].[3]

Note that the extent to which solar energy is absorbed, diffusely reflected, or specularly reflected is a surface property of the material upon which the sunlight falls.

5.2.1.4 Solar torque. The solar torque on the spacecraft is the sum of all of the forces on all elemental surfaces times the radius from the centroid of the surface to the spacecraft center of mass. The reflectivity is represented by q, where $q = 1$ is a specular reflector and $q = 0$ is an absorber. The reflectivity q of the various spacecraft elements is a known (measured) property of the materials. The relative sun angle is accounted for in the area calculation, where A is the projected area normal to the sun vector. The following equation represents the total solar torque on the body caused by a spacecraft surface:

$$T_s = PAL(1 + q) \tag{5.3}$$

where

T_s = solar torque on the spacecraft caused by a surface A, N-m
A = area of surface projected to sun line normal, m^2
L = distance from the centroid of the surface to the center of mass of the spacecraft, m
q = reflectance factor between 0 and 1, unitless. Spacecraft bodies tend to be reflectors; a q of 0.5 is representative[4]; solar panels tend to be absorbers, a q of 0.3 is representative

Solar torques increase in relative importance with orbital altitude. Above 1000-km solar pressure is usually the dominant external torque on the vehicle. In geosynchronous orbit solar torque is the disturbance, which sizes the station-keeping propellant load and hence the orbital life of the spacecraft. Solar pressure is also important to the design of planetary vehicles, especially for the inner planets. Solar pressure was used, on an emergency basis, to control roll on Mariner Venus–Mercury (Mariner 10)[3]. In the 1970s solar sailing was seriously considered as a source of continuous, low-level thrust for interplanetary travel. Feasibility was established; however, the practical problems of deploying and controlling large sails are considerable.

Example 5.1 Solar Pressure

Part I: What is the solar force on a 9 m^2 solar panel inclined at 20 deg to the sun with a reflectance factor q of 0.3 if the vehicle is in Earth orbit? From Eq. (5.3),

$$F_s = (1 + 0.3)(4.59 \times 10^{-6})(9)(\cos 20)$$

$$F_s = 4.0 \times 10^{-5} \text{ N}$$

Part II: What is the resultant solar torque for a spacecraft with a single square solar panel of this size with a 1-m body with the c.g. in the center of the body and a

0.25-m boom to attach the panel?

$$T = (4.0 \times 10^{-5})(2.25)$$

$$T = 9.0 \times 10^{-5} \text{ N-m}$$

To calculate to the total solar torque acting on the spacecraft, the sum of the solar torques acting on each surface must be taken into account with strict attention to the direction (sign) of the individual torques. Also, this calculation would be performed in three dimensions unlike the simple example.

5.2.2 Magnetic Torques

The Earth and several other planets have a magnetic field that produces torque on spacecraft. The torque of any magnetic field on a current-carrying coil is

$$T = NIAB \sin\theta \tag{5.4}$$

where

T = magnetic torque, N-m
N = number of loops in the coil
I = current in the coil, amp
A = coil area, m^2
B = Earth's magnetic field, tesla (still encountered is the cgs unit for B, which is gauss, 1 G is 10^{-4} tesla)
θ = angle between the magnetic field lines and perpendicular to the coil

The residual magnetic field of a spacecraft is the result of current loops and residual magnetism in the metal parts. It can be expressed as the product

$$M = NIA \tag{5.5}$$

and

$$T = MB \sin\theta \tag{5.6}$$

An accurate value of the residual dipole of a spacecraft must be obtained by test. The value can be as little as 0.2 A-m^2 or as much as 20 A-m^2 depending on spacecraft size, with 2 A-m^2 a representative value.[4] Onboard compensation can be used to reduce the residual dipole if necessary.

The Earth's magnetic field is tilted 11 deg with respect to the Earth's rotational axis and is centered about 400 km from the Earth's geometric center; hence, at a given altitude the field is stronger over the Pacific than the Atlantic.[1] The magnetic field strength varies as one over radius from the Earth's center cubed; thus, the higher the orbit the less disturbance. The field strength also varies by a factor of 2 depending on latitude; with the highest value at the poles, from Piscane and Moore[1]

$$B = \frac{B_0 r_0^3}{r^3}(3 \sin^2 L + 1)^{1/2} \tag{5.7}$$

Table 5.5 Magnetic field strength at given radii

Planet	Magnetic field strength, T	At radius, km
Mercury	3.5×10^{-7}	2,440
Venus	$<3 \times 10^{-8}$	6,051
Earth	3.1×10^{-5}	6,378
Mars	5×10^{-8}	3,397
Jupiter	4.3×10^{-4}	71,372
Saturn	2.1×10^{-5}	60,330
Uranus	2.28×10^{-5}	25,600
Neptune	1.33×10^{-5}	24,765
Pluto	??	2,500

where

B = Earth's magnetic field strength at any altitude or latitude
B_0 = Earth sea-level magnetic field strength, approximately 3×10^{-5} tesla
r = spacecraft orbital radius, m
r_0 = Earth surface radius 6,378,000 m
L = latitude in magnetosphere, deg (geographic latitude is adequate for approximate calculations)

Note from Eq. (5.7) that the Earth's magnetic field strength at the poles is twice the equatorial strength.

The magnetic field strength of the planets at a given radius is shown in Table 5.5 (from Lang[5]). Note the intensity of the Jupiter field and that Venus and Mars have negligible fields.

Example 5.2 Magnetic Torque

Consider a spacecraft with a residual dipole of 2 A-m^2 in a circular equatorial orbit at an altitude of 400 km. What is the magnitude of magnetic moment on the spacecraft?
Calculating the orbit radius,

$$r = (400 + 6378)(10^3) = 6{,}778{,}000\,\text{m}$$

The Earth's magnetic field in a 400-km equatorial orbit is approximately, from Eq. (5.7),

$$B = \left(\frac{6{,}378{,}000}{6{,}778{,}000}\right)^3 3 \times 10^{-5}(1)^2$$

$$B = 2.5 \times 10^{-5}\ \text{tesla}$$

The magnitude of magnetic torque on the vehicle is approximately, from Eq. (5.6),

$$T_m = (2.5 \times 10^{-5})(2) = 5 \times 10^{-5}\ \text{N-m}$$

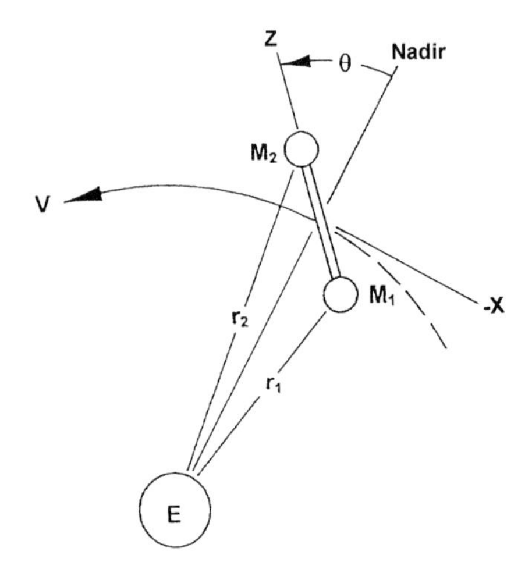

Fig. 5.11 Gravity-gradient torque.

5.2.3 Gravity-Gradient Torque

Gravity-gradient torques arise from the fact that the lower extremities of the spacecraft are subjected to exponentially higher gravity forces than the upper extremities. The effect is particularly pronounced for long slender spacecraft; the effect acts to align the long axis with the Earth radius vector. Figure 5.11 shows a dumbbell-shaped spacecraft with the mass equally divided into two spheres. The net forces on the spacecraft are in equilibrium at the center of gravity. The gravitational acceleration on the lower mass is

$$g = \frac{GM}{r_1^2} \tag{5.8}$$

and on the upper mass

$$g = \frac{GM}{r_2^2} \tag{5.9}$$

Because r_2 is greater than r_1, the gravitational force is greater on the lower mass than the upper mass, and the net force tends to hold the spacecraft upright. It is convenient to assign a gravitational constant μ, which is the product of central body mass M and gravitational constant G. The resulting torque is (assuming that small angles, in radian measure, are equal to their sine)

$$T_g = \frac{3\mu}{r^3}|I_z - I_y|\theta \tag{5.10}$$

where

T_g = gravity-gradient torque, N-m
μ = gravitational parameter, 398,600.4 km^3/s^2 for Earth (see Chapter 3, "Orbital Mechanics") = GM
r = radius from spacecraft center of mass to central body center of mass, km
I_z = moment of inertia about the z axis

I_y = moment of inertia about the y axis (for maximum torque use the least of I_y or I_x)
θ = angle between spacecraft z axis and nadir vector, rad

Note that gravity gradient produces torques in two axes. Note also that it requires an angular error to produce a torque. This characteristic is the basis for gravity-gradient control. In this case it is desired to maintain the z axis to nadir. Disturbances cause a displacement of the z axis, which in turn results in a restoring gravity torque.

Example 5.3 Gravity Gradient

The Skylab was the first space station launched by the United States and the largest U.S. spacecraft launched to that time (1973). Estimate the gravity-gradient torque on Skylab given[4] the following characteristics.

Vehicle properties:

$$\text{Mass} = 90{,}505 \text{ kg}$$
$$\text{Height} = 35 \text{ m}$$
$$\text{Diameter} = 5.4 \text{ m}$$
$$\text{Radius} = 2.7 \text{ m}$$

Orbit:

$$\text{Altitude} = 442 \text{ km circular}$$
$$\text{Radius} = 5820 \text{ km}$$

Attitude error:

$$5 \text{ deg} \quad \text{or} \quad 0.087266 \text{ rad}$$

Calculate moments of inertia assuming a uniform density right circular cylinder (see Chapter 2):

$$I_z = \frac{Wr^2}{2} = \frac{(90{,}505)(2.7)^2}{2} = 329{,}890.7 \text{ kg-m}^2$$

$$I_{x,y} = \frac{W}{12}(3r^2 + h^2) = \frac{90505}{12}[3(2.7)^2 + (35)^2] = 9{,}403{,}997.5 \text{ kg-m}$$

Calculate the resulting gravity torque at 5 deg:

$$T_g = \frac{3(398{,}600)}{(6820)^3}(9{,}403{,}997.5 - 463{,}902)(0.087266) = 2.985 \text{ N-m}$$

5.2.4 Aerodynamic Drag

Aerodynamic drag is a source of torque as well as velocity reduction for spacecraft in low Earth orbits. Drag force can be estimated from the classic relationship

$$D = \tfrac{1}{2}\rho V^2 C_d A \tag{5.11}$$

where

D = drag force (always aligned with the velocity vector and opposite in sign)
ρ = atmospheric density, kg/m^3

V = spacecraft velocity, m/s
C_d = drag coefficient depends on shape but usually about 2.5
A = area normal to velocity vector, m^2

Drag force is the ultimate cause of orbit decay and reentry of low-Earth-orbit spacecraft. Although the atmospheric density is very low, the spacecraft velocity is very high; over time inactive spacecraft below about 1000 km will reenter. The greatest uncertainty in a spacecraft drag analysis is in the atmospheric density. In addition to the variation with altitude, at high altitudes the density varies with temperature and therefore, time of day. It also varies with the intensity of solar radiation activity, which has an approximate 11-year cycle. High density is associated with high solar activity. It also varies with the season, with the latitude, with the rotation of the atmosphere, and with atmospheric tides.

For attitude control, it is important how the drag acts on various surfaces, much like the solar pressure torques. The torque results from a moment arm between the center-of-pressure and the c.g. If the spacecraft is symmetric, i.e., the center-of-pressure is coincident with the c.g., then no aerodynamic torque is produced. However, it is rarely the case that the center-of-pressure and the center of gravity are coincident for all three axes. The drag torque is calculated as

$$T = DL \tag{5.12}$$

where L is the distance between the center of pressure and the c.g.

Example 5.4 Drag Force

Part I: Consider a spacecraft in a 400-km circular Earth orbit. What is the drag force on a solar panel with 9 m^2 of surface area normal to the velocity vector?

The velocity of a spacecraft in a 400-km altitude Earth orbit is 7.669 km/s (see Chapter 3 on orbital mechanics). Assuming the atmospheric density at 400 km is 1.2×10^{-11} kg/m^3, under average conditions, and the drag force is

$$D = 1/2(1.2 \times 10^{-11})(7669)^2(2.5)(9) = 7.9 \times 10^{-3}\ \text{N}$$

Part II: Using the same spacecraft as in the solar torque example, assume that the body and solar arrays are each uniform such that their center of pressures are at the centroid. The torque on the main body would be zero because the c.g. is at the centroid. The drag torque caused by the solar array is

$$T = (7.9 \times 10^{-3})(2.25\ \text{m}) = 1.78 \times 10^{-2}\ \text{N-m}$$

The area of high-altitude drag is one in which large margins should be used as result of the large uncertainty of atmospheric density and uncertainty in drag coefficient. It is also important that the spacecraft is symmetric with the center of pressure close to the c.g. Otherwise, major resources such as propellant or power for large reaction wheels will be consumed in counteracting large drag torques.

5.2.5 Spacecraft-Generated Torques

In addition to the environmental torques, there are a variety of spacecraft-generated torques of concern to the operation of the attitude control system. These torques are generally much smaller than the external torques but need to be accounted for especially for a high-precision pointing system. The common causes of internal torques are as follows: 1) *pointing* rotation of solar arrays, antennas, or cameras; 2) *deployment* of antennas, solar arrays, instruments, booms; 3) *parts jettison*, which means that the spacecraft will react to jettison of parts such as covers, doors, and solid-rocket-motor cases; 4) *propellant slosh*, which can cause motion of the vehicle and c.g. (slosh is greatly attenuated by bladders and diaphragms); 5) *flexible appendages* can cause motion by thermal distortion or by dynamic interaction with the attitude control system; 6) *reaction wheel imbalance*, which is caused by small misalignments in the reaction wheels.

5.2.6 System Sizing

Once all of the disturbance torques have been identified and quantified, the actuator sizing can be determined. The first consideration is the magnitude of the torques. The actuator must have sufficient torque authority to conteract the disturbance torques. The difference in the magnitude of the disturbance torque and the actuator torque capability is a measure of the control authority. Typically, the control authority is expressed as a percentage such that an actuator with twice the capability of the disturbance torques would have a 100% control authority margin.

After the actuator with sufficient control authority is chosen, then the system resources over time must be considered. In the case of a three-axis-stabilized spacecraft with reaction wheels, the storage time of the reaction wheels must be considered. Reaction wheels store momentum until they reach their maximum specified speed, at which point they are saturated and must be *desaturated* by another torque. In low Earth orbit, magnetic torquers are typically used for wheel desaturation. For higher orbits or interplanetary spacecraft the reaction control system or thrusters are used. Considerations for the frequency of desaturation are the disturbances caused by the event and, in the case of thruster desaturation, the amount of propellant consumed. Magnetic torquers require electrical power but avoid the use of consumable propellant. If thruster control is selected, the disturbances caused by the thruster firings and the amount of propellant consumed must be considered.

Example 5.5 Reaction Wheel Sizing

Consider the spacecraft from the solar torque and drag force examples and determine the size, that is, the torque capability and momentum storage capability of a reaction wheel required to maintain position and require desaturation once per 98-min orbit.

In this case the solar torque and the drag torque are acting in the same direction, so

$$T = \text{Solar torque} + \text{Drag torque} = 9.0 \times 10^{-5} + 1.78 \times 10^{-2} = 1.789 \times 10^{-2}\ \text{N-m}$$

So the reaction wheel torque capability, ignoring maneuvers, is $>1.85 \times 10^{-2}$ N-m.

The momentum buildup over one orbit is given by

$$M = T \times t = 1.85 \times 10^{-2} * (98 * 60) = 108.6 \, \text{N-m-s}$$

As will be discussed later in Section 5.6.2, it would require fairly large and heavy reaction wheels to accommodate this momentum buildup. A better alternative might be to have two, smaller solar panels placed symmetrically about the central body. However, there may be reasons, such as instrument field of view, that necessitate the single panel and require the relatively large reaction wheels.

5.3 Attitude Determination

Attitude determination is the process of estimating the orientation of the spacecraft with respect to a reference frame, on a more or less continuous basis, and location of the sun, Earth, and other objects in the same frame. The process is conceptually simple, as shown in Fig. 5.12, but is mathematically complex to the point of obscurity. Only the most important parts of that baggage will be discussed in this section.

Attitude is determined by the following:

1) Gathering data from the onboard sensors. These data are corrected for errors and biases, then analyzed by mathematically sophisticated methods to determine the attitude estimate. (The mechanization of common sensors is discussed in Section 5.6.)

2) Body frame axes or spin axis location is determined from the sensor data.

3) The instantaneous attitude, or state vector, is expressed with respect to a reference frame, usually inertial or geocentric, as a set of Euler angles, a direction cosine matrix, or a quaternion.

4) The attitude estimate is the basis for correcting the attitude (see Section 5.4).

5.3.1 Attitude Determination Methods

The spacecraft axes must be located with respect to a reference frame (and vice versa) because the target of the spacecraft payload and/or sensors are specified relative to a reference frame. The payload can be pointed to specific points on the Earth's surface or at a specific celestial object. These targets are typically specified in an inertial reference frame so the relationship of the spacecraft body,

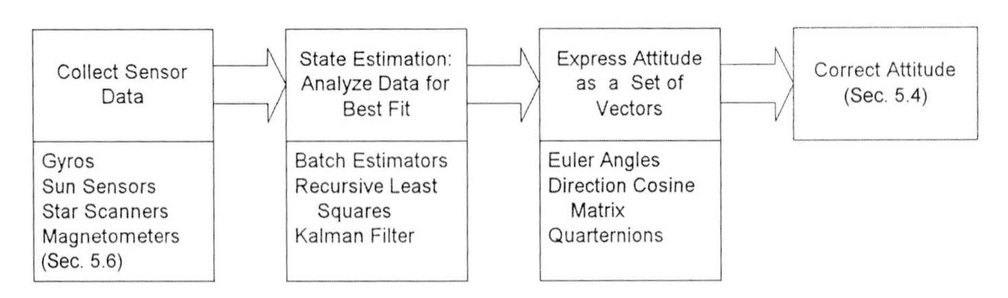

Fig. 5.12 Attitude determination process.

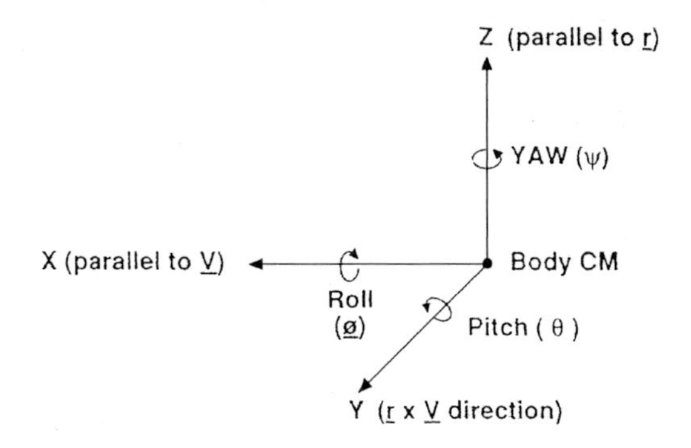

Fig. 5.13 Spacecraft axes and pitch, yaw, and roll angles.

and subsequently the sensor body, is an important relationship. Length dimensions from the origin are not important to attitude determination or control; only angular relationships are important. Consequently, unit vectors are used for computation.

The conventional spacecraft frame of reference points the spacecraft **+*Z*** axis anti-nadir, and the **+*X*** axis in the direction of motion is shown in Fig. 5.13 along with the angle convention for pitch, yaw, and roll.

The relationship between the reference frame and the spacecraft frame are defined by three rotation angles known as *Euler angles*. Euler angles are a set of three angles and a sequence of rotation such that one coordinate system can be rotated into another. Both the magnitude of the angles and the sequence of rotation are important; if another sequence is used, the result will be different. It can be shown that there are 12 different Euler sets that describe the same relative position. Figure 5.14 shows an Euler angle rotation sequence.

A *direction cosine matrix* (DCM) can be defined as the product of the three Euler rotations. Consider the rotation of a spacecraft frame into an inertial reference frame; make the rotations in the sequence yaw, pitch, roll.

Define R_1, R_2, and R_3 as the three rotation matrices that correspond to the three Euler angles ψ, θ, and ϕ. Then the total rotation ψ, θ, and ϕ can be expressed as

$$R = R_3 * R_2 * R_1 \tag{5.13}$$

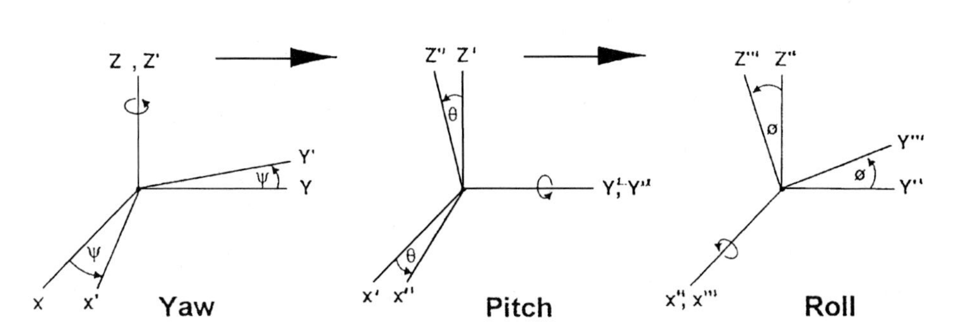

Fig. 5.14 Euler-angle rotation sequence.

or

$$R(\psi\theta\phi) = R_3(\phi)R_2(\theta)R_1(\psi) \tag{5.14}$$

and

$$\begin{bmatrix} X'_b \\ Y'_b \\ Z'_b \end{bmatrix} = R_1 \begin{bmatrix} X_b \\ Y_b \\ Z_b \end{bmatrix} \tag{5.15}$$

where

$$R_1 = \begin{bmatrix} \cos\psi & \sin\psi & 0 \\ -\sin\psi & \cos\psi & 0 \\ 0 & 0 & 1 \end{bmatrix} \tag{5.16}$$

The rotations R_2 and R_3 can be specified similarly to yield

$$DCM = R$$
$$= \begin{bmatrix} \cos\theta\cos\psi & \cos\theta\sin\psi & -\sin\theta \\ -\cos\phi\sin\psi + \sin\phi\sin\theta\cos\psi & \cos\phi\cos\psi + \sin\phi\sin\theta\sin\psi & \sin\phi\cos\theta \\ \sin\phi\sin\psi + \cos\phi\sin\theta\cos\psi & -\sin\phi\cos\psi + \cos\phi\sin\theta\sin\psi & \cos\phi\cos\theta \end{bmatrix} \tag{5.17}$$

Transformational matrices, such as Eq. (5.17), are orthogonal so that the inverse transformation, in this case from inertial coordinates to body coordinates, is easily found by transposing the original

$$R_{I/B} = \frac{1}{R_{B/I}} \tag{5.18}$$

In addition the Euler angles can be extracted from the direction cosine matrix as follows:

$$\psi = \arctan\left(\frac{r_{1,2}}{r_{1,1}}\right) \tag{5.19}$$

$$\theta = \arctan\left(\frac{-r_{1,3}}{\sqrt{1 - r_{1,3}^2}}\right) \tag{5.20}$$

$$\phi = \arctan\left(\frac{r_{2,3}}{r_{3,3}}\right) \tag{5.21}$$

where $r_{i,j}$ are the corresponding elements of the direction cosine matrix and where i is the row and j the column. When attitude determination involves very small angles, the small-angle approximation can be made, yielding

$$R = \begin{bmatrix} 1 & \psi & -\theta \\ -\psi & 1 & \phi \\ \theta & -\phi & 1 \end{bmatrix} \tag{5.22}$$

If the angles are infinitesimal, it can be shown that matrix multiplication is commutative, and the rotations can be made in any order.

Performing a rotation using a direction cosine matrix [Eq. (5.17)] requires 27 multiples, 15 adds, and, if the small-angle simplification cannot be used (as in Eq. (5.22)], 29 trigonometric evaluations. Therefore, direction cosine matrices require a large amount of memory and are computationally intensive.

5.3.1.1 Quaternions. One widely used alternative to the direction cosine matrix is the quaternion, sometimes referred to as *Euler symmetric parameters*, which has no singularities and no trigonometric functions. Quaternions make use of Euler's theorem, which states that *any series of rotations of a rigid body can be expressed as a single rotation about a fixed axis*. The orientation of a body can be defined by a vector giving the direction of a body axis and a scalar angle specifying a rotation angle about that axis. Thus, quaternions express the same information as a direction cosine matrix, that is, a rotation from one frame to another, with four elements. Three elements describe the vector of rotation, and the fourth element is the angle of rotation. Quaternion mathematics was actually formulated by Hamilton[7] in 1843 but was not applied to coordinate rotations until Whittaker's work[8] first published in the 1937. A rigorous derivation of Euler symmetric parameters may be found in Wertz.[9]

Let the four elements of a quaternion be q_1, q_2, q_3, q_4. Then, the quaternion Q is

$$Q = iq_1 + jq_2 + kq_3 + q_4 \tag{5.23}$$

where i, j, and k satisfy the following conditions:

$$i^2 = j^2 = k^2 = -1 \tag{5.24}$$

$$ij = -ji = k \tag{5.25}$$

$$jk = -kj = i \tag{5.26}$$

$$ki = -ik = j \tag{5.27}$$

The quaternion is often written in the general form

$$Q = (q_1, q_2, q_3, q_4) \tag{5.28}$$

If a set of four Euler symmetric parameters corresponding to a rigid-body rotation are the components of Q, then Q is the representation of the rigid-body rotation.

Two quaternions, Q and P, are equal if and only if all of the elements are equal, that is,

$$q_1 = p_1, \quad q_2 = p_2, \quad q_3 = p_3, \quad q_4 = p_4 \tag{5.29}$$

Quaternion multiplication is not commutative, and so like DCMs the order of operation must be taken into account:

$$R = QP = (iq_1 + jq_2 + kq_3 + q_4)(ip_1 + jp_2 + kp_3 + p_4) \tag{5.30}$$

and

$$R = (-q_1p_1 - q_2p_2 - q_3p_3 + q_4p_4) + i(q_1p_4 + q_2p_3 - q_3p_2 + q_4p_1)$$
$$+ j(-q_1p_3 + q_2p_4 + q_3p_1 + q_4p_2) + k(q_1p_2 - q_2p_1 + q_3p_4 + q_4p_3) \tag{5.31}$$

The conjugate of Q is given by

$$Q^* = -iq_1 - jq_2 - kq_3 + q_4 \tag{5.32}$$

The inverse of Q is another quaternion P, which satisfies

$$QP = 1 \tag{5.33}$$

or

$$QQ^{-1} = 1 \tag{5.34}$$

The norm (or magnitude) of a quaternion is a scalar quantity defined as

$$N = |Q| = \sqrt{QQ^*} = \sqrt{q_1^2 + q_2^2 + q_3^2 + q_4^2} \tag{5.35}$$

A quaternion with $N = 1$ is said to be "normalized," and the inverse is equal to the conjugate:

$$QQ^* = 1 = QQ^{-1} \tag{5.36}$$

and

$$Q^* = Q^{-1} \tag{5.37}$$

This property is very useful in practice; thus, attitude determination is usually done with normalized quaternions.

5.3.1.2 Relating DCMs to Euler axis and quaternions. Remembering Euler's theorem, the Euler axis $\boldsymbol{E}$ and the Euler angle Φ can be related to the direction cosine matrix by

$$\boldsymbol{E} = e_1i + e_2j + e_3k \tag{5.38}$$

$$e_1 = \frac{R_{23} - R_{32}}{2\sin\phi} \tag{5.39}$$

$$e_2 = \frac{R_{31} - R_{13}}{2\sin\phi} \tag{5.40}$$

$$e_3 = \frac{R_{12} - R_{21}}{2\sin\phi} \tag{5.41}$$

And, if $\sin\phi \neq 0$, then

$$\cos\phi = \tfrac{1}{2}[T(A) - 1] \tag{5.42}$$

where T is the trace of the matrix

$$T = R_{11} + R_{22} + R_{33} \tag{5.43}$$

The quaternion elements can be similarly related to the Euler axis and angles as defined by

$$q_1 = e_1 \sin(\Phi/2) \tag{5.44}$$

$$q_2 = e_2 \sin(\Phi/2) \tag{5.45}$$

$$q_3 = e_3 \sin(\Phi/2) \tag{5.46}$$

$$q_4 = \cos(\phi/2) \tag{5.47}$$

Note that there is a sign ambiguity in ϕ because of a rotation of ϕ about $\boldsymbol{E}$ is equivalent to a rotation of $-\phi$ about $-\boldsymbol{E}$. During development, a convention must be chosen for consistency. Usually the positive value is chosen, and that will be the choice throughout this text.

Alternatively Q can be written as

$$Q = [\boldsymbol{E} \sin(\Phi/2) + \cos(\Phi/2)] \tag{5.48}$$

To relate direction cosine matrix C to the equivalent quaternion Q, let

$$C = \begin{bmatrix} C_{11} & C_{12} & C_{13} \\ C_{21} & C_{22} & C_{23} \\ C_{31} & C_{32} & C_{33} \end{bmatrix} \tag{5.49}$$

Using Eqs. (5.39–5.47) and performing the appropriate substitutions, it can be shown that C relates to the quaternion elements as follows:

$$C = \begin{bmatrix} (2q_4^2 + 2q_1^2 - 1) & (2q_1q_2 - 2q_4q_3) & (2q_1q_3 + 2q_4q_2) \\ (2q_1q_2 + 2q_4q_3) & (2q_4^2 + 2q_2^2 - 1) & (2q_2q_3 - 2q_4q_1) \\ (2q_1q_3 - 2q_4q_2) & (2q_2q_3 + 2q_4q_1) & (2q_4^2 + 2q_2^3 - 1) \end{bmatrix} \tag{5.50}$$

It is simply a matter of algebra to draw out the equations for the quaternion elements in terms of the direction cosine matrix elements. Isolating the diagonal terms and using Eq. (5.43) results in the following:

$$q_4^2 = (T + 1)/4 \tag{5.51}$$

$$q_i^2 = (2C_{ii} + 1 - T)/4, \qquad i = 1, 2, 3 \tag{5.52}$$

Manipulating the off-diagonal term yields

$$q_4 = \tfrac{1}{2}(1 + T)^{\frac{1}{2}} \tag{5.53}$$

$$q_1 = \tfrac{1}{4_{q_4}}(C_{23} - C_{32}) \tag{5.54}$$

$$q_2 = \tfrac{1}{4_{q_4}}(C_{31} - C_{13}) \tag{5.55}$$

$$q_3 = \tfrac{1}{4_{q_4}}(C_{12} - C_{21}) \tag{5.56}$$

The preceding equations can be used to solve for the quaternion elements. The procedure is to select any q to start the calculation using Eq. (5.53) assigning, for convenience, the positive square root. Then use Eqs. (5.54–5.56) to establish the signs of the remaining elements.

Example 5.6 Conversion of DCM to Quaternion

The following DCM relates a spacecraft body to the J2000 inertial frame:

$$DCM = \begin{bmatrix} -0.0614637 & 0.3985394 & -0.9150894 \\ -0.7304249 & 0.6068640 & 0.3133615 \\ 0.6802215 & 0.6876644 & 0.2538028 \end{bmatrix}$$

The equivalent quaternion is given by the following.

From Eq. (52):

$$q_4 = \tfrac{1}{2}(0.7992031 + 1)^{\frac{1}{2}} = 0.6706719$$

From Eq. (53):

$$q_1 = \frac{0.3133615 - 0.6876644}{4(0.6706719)} = -0.1395253$$

From Eq. (54):

$$q_2 = \frac{0.6802215 + 0.9150894}{4(0.6706719)} = 0.5946689$$

From Eq. (55):

$$q_3 = \frac{0.3985394 + 0.7304249}{4(0.6706719)} = 0.4208333$$

and

$$Q = (-0.1395253, 0.5946689, 0.420833, 0.6706719)$$

5.3.1.3 Relating frames. By convention, the quaternion relating frame A to frame B is written Q_b^a. The transformation of the general vector in frame A, $\boldsymbol{V}_A$ to a vector in frame B is given by

$$\boldsymbol{V}_B = Q_b^a \boldsymbol{V}_A \left(Q_b^a\right)^* \tag{5.57}$$

and the general vector in frame B, V_B to a vector in frame A is given by

$$V_A = \left(Q_b^a\right)^* V_B Q_b^a \quad \text{(premultiply by inverse in this case)} \tag{5.58}$$

Quaternions can be combined such that Q_b^a relates frame A to frame B, and Q_c^b relates frame B to frame C, then the quaternion relating frame A to frame C is given by

$$Q_c^a = Q_c^b Q_b^a \tag{5.59}$$

Table 5.6 Comparison of three-axis-attitude representations

Method	Advantages	Disadvantages	Common use
Euler angles (yaw, pitch, roll)	No redundant parameters Clear physical interpretation	Singularities Trigonometric functions No convenient product rule	Analytical study Onboard control of three-axis spacecraft, e.g., Space Shuttle
Direction cosine matrix	No singularities No trigonometric functions (small angles only) Good physical representation Convenient product rule	Six redundant parameters Many trigonometric functions	To transform vectors from one reference frame to another
Quaternions	No singularities No trigonometric functions Convenient product rule	One redundant parameter No physical meaning	Onboard control of three-axis spacecraft

The computation of Q_c^a requires 15 multiplies and 12 adds. It does not require any trigonometric evaluations. Thus, the use of quaternions offers a significant advantage in both computer memory and computing time.

The common methods of representing three-axis attitude are compared in Table 5.6.

5.3.2 Axis Location from Sensor Output

The most basic methods of axis location are deterministic in that they use the same number of observations as variables. Deterministic methods are most useful when the orientation of only one axis needs to be specified. A single axis requires two parameters. Generally, for a spin-stabilized spacecraft, specifying the orientation of the spin axis with respect to some inertial coordinate frame is sufficient along with knowledge of the spin rate.

5.3.2.1 Spin-axis method. A deterministic, two-component attitude solution for a spin-stabilized spacecraft requires two measurements: 1) the sun angle relative to the spin axis and 2) the Earth radius vector angle relative to the spin axis. The intersection of these two cones is the spin axis, as shown in Fig. 5.15.

There are two solutions for the intersection: the real and the imaginary spin vector. The ambiguity can be resolved by taking the measurements again at a later time. The true solution will remain unchanged, whereas the imaginary solution changes.

Assume that the inertial frame of interest is the celestial system of right ascension α and declination δ, as shown in Fig. 5.16.

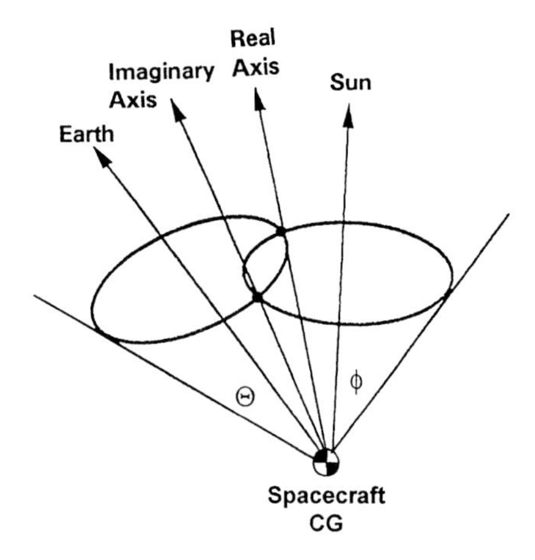

Fig. 5.15 Locating spin axis.

The sun vector and the Earth vector in inertial coordinates are parameters onboard from ephemeris tables or polynomials. The position of celestial bodies is determined using ephemeris that is calculated on the ground based on the reduction of tracking data, which is typically in either table form or polynomial form. Ephemeris tables are usually updated regularly from the ground; they require large amounts of onboard computer memory and are highly accurate if the table is large enough. Polynomials require less memory and more computation time; they also require coefficient updates from the ground. High accuracy can be obtained by using high ($\approx$eighth-order) polynomials. For example, Magellan used eighth-order

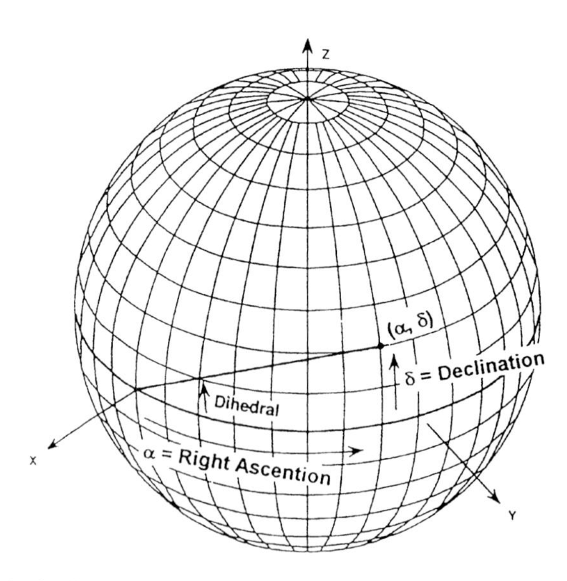

Fig. 5.16 Celestial frame of reference. (Copyright AIAA, reproduced with permission; Ref. 14.)

LaGrange polynomials for the sun and Earth positions. Polynomial coefficients were updated about every 50 days to maintain knowledge within 0.1 deg.

The right ascension and declination of the spin vector can then be obtained by solving the following equations:

$$\cos\phi = \sin\delta_\omega \sin\delta_S + \cos\delta_\omega \cos\delta_S \cos(\alpha_\omega - \alpha_S) \tag{5.60}$$

$$\cos\theta = \sin\delta_\omega \sin\delta_E + \cos\delta_\omega \cos\delta_E \cos(\alpha_\omega - \alpha_E) \tag{5.61}$$

where

α_S = sun right ascension
δ_S = sun declination
α_E = Earth right ascension
δ_E = Earth declination
α_ω = spin-axis right ascension
δ_ω = spin-axis declination

The spin axis is completely described by its right ascension and declination; however, sometimes knowledge of the pointing direction of a fixed body axis is also required. The azimuth angle measures the body rotation about the spin axis and is effectively measured each time the sun or Earth sensors trigger.

The accuracy of the solution obtained depends on many factors such as measurement uncertainties, biases, systematic errors, and sensor misalignments. One way to decrease the uncertainties is to calculate the attitude using several different methods and using a weighted average of the results. For a spin-stabilized spacecraft with one sun sensor and two Earth sensors mounted at different angles, there are six methods of determining attitude.

A method similar to that used for the spin-stabilized spacecraft can be used when there is a particular body axis about which there is preferential attitude data such as the spin axis of a momentum wheel. The spinner method can also be used for attitude acquisition, when a particular body axis is fixed inertially; however, the accuracy is usually insufficient for the more precise pointing normally required for a three-axis-stabilized vehicle.

5.3.2.2 Two-vector method. Three-axis-attitude determination is more complex because it requires the complete specification of the attitude matrix (DCM). Attitude can be determined by solving for the rotation matrix R by using two vectors measured in the body frame. Any two vectors define an orthogonal coordinate system, provided that the vectors are not parallel. Let $\boldsymbol{u}$ and $\boldsymbol{v}$ define the following coordinate system (after the method of Markley, Ref. 9, p. 424):

$$\boldsymbol{x} = \boldsymbol{u} \tag{5.62}$$

$$\boldsymbol{y} = \frac{\boldsymbol{u} \times \boldsymbol{v}}{|\boldsymbol{u} \times \boldsymbol{v}|} \tag{5.63}$$

$$\boldsymbol{z} = \boldsymbol{x} \times \boldsymbol{y} \tag{5.64}$$

Let $\boldsymbol{u}_b$ and $\boldsymbol{v}_b$ be two vectors measured in the spacecraft body frame such that a sun vector is measured by a two-axis sun sensor and a star vector measured by a star tracker. Assume the X body axis is aligned to the sun vector and the Y body axis is perpendicular to the sun-star plane, then the spacecraft body matrix can be determined as

$$\boldsymbol{x}_b = \boldsymbol{u}_b$$

$$\boldsymbol{y}_b = \frac{\boldsymbol{u}_b \times \boldsymbol{v}_b}{|\boldsymbol{u}_b \times \boldsymbol{v}_b|}$$

$$\boldsymbol{z}_b = \boldsymbol{x}_b \times \boldsymbol{y}_b$$

$$M_b = \begin{bmatrix} \boldsymbol{x}_b \\ \boldsymbol{y}_b \\ \boldsymbol{z}_b \end{bmatrix}$$

M_b is the direction cosine matrix for the measured estimate of the pointing of the body in the sun–star reference frame. For this application the J2000 reference frame is the most commonly used. J2000 is an Earth–mean–equator of epoch 2000 reference frame. This estimate of attitude does not take into account the relative accuracy of the two measured vectors except that $\boldsymbol{u}$ was used as a basis, implying that $\boldsymbol{u}$ is the more accurate measurement. However, error sources, such as measurement error, biases, and misalignments, have not been accounted for.

The next step is to determine how well the spacecraft is pointed relative to the desired position. A reference, or desired matrix, can also be determined by using a sun vector from ephemeris and a star vector from a star catalog to yield

$$\boldsymbol{x}_r = \boldsymbol{u}_r$$

$$\boldsymbol{y}_r = \frac{\boldsymbol{u}_r \times \boldsymbol{v}_r}{|\boldsymbol{u}_r \times \boldsymbol{v}_r|}$$

$$\boldsymbol{z}_r = \boldsymbol{x}_r \times \boldsymbol{y}_r$$

$$M_r = \begin{bmatrix} \boldsymbol{x}_r \\ \boldsymbol{y}_r \\ \boldsymbol{z}_r \end{bmatrix}$$

Then, the attitude error between the measured pointing M_b and the desired pointing M_r is given by

$$RM_r = M_b \tag{5.65}$$

$$R = M_b M_r^{-1} \tag{5.66}$$

5.3.3 Quaternion Parameterization of the Attitude Error

Quaternions simplify the computational work done by the spacecraft computer in implementing a control law as described in this section. (That is not to say that they make the computation easier to understand.) The *position error* is computed from sensor output as follows:

$$Q_b^d = Q_b^i (Q_d^i)^{-1} = \begin{bmatrix} -e_1 \sin(\Delta\theta/2) \\ -e_2 \sin(\Delta\theta/2) \\ -e_3 \sin(\Delta\theta/2) \\ \cos(\Delta\theta/2) \end{bmatrix} \tag{5.67}$$

where

Q_b^i = current, measured body attitude expressed as a quaternion
Q_d^i = desired body attitude expressed as a quaternion
Q_b^d = difference between current and desired attitude expressed as a quaternion
$e_{1,2,3}$ = error

In this case both the current and the desired quaternions define a rotation from the spacecraft body to inertial space but for different spacecraft orientations. The delta quaternion (difference between current and desired) expresses the spacecraft rotation required to achieve the desired attitude. In effect the inertial components cancel, and what remains is the current body to desired body rotation. The magnitude of the rotation required is then

$$\theta = 2\cos^{-1} q_4 \tag{5.68}$$

The errors in the current body frame are then extracted by taking the projection of the Euler axis in the body frame as follows:

$$\varepsilon(j) = \frac{-2\theta\, Q_b^d(j)}{\sin(\theta/2)} \qquad j = 1, 2, 3 \tag{5.69}$$

By using the small-angle approximation, $\sin\theta \approx \theta$, small errors, $\varepsilon(j)$, can be expressed as

$$\varepsilon(j) = -2Q_b^d(j), \qquad j = 1, 2, 3 \tag{5.70}$$

The *rate error* can be determined using quaternions as follows:

$$Q_p^b = Q_p^i\left(Q_b^i\right)^{-1} \tag{5.71}$$

where

Q_p^i = already measured body attitude
Q_p^b = delta quaternion between preceding and current attitude
$\Delta\tau$ = control cycle period or time between attitude measurements

For small rotations, again using the small-angle approximation,

$$\varepsilon(j) = \frac{-2Q_p^b(j)}{\Delta\tau} \qquad j = 1, 2, 3 \tag{5.72}$$

5.3.4 State Estimation Methods

State estimation methods successively correct estimates of the attitude by using the partial derivatives of the observables with respect to the solved-for attitude parameters. The collection of attitude parameters are called the *state vector*. The state vector can represent many attitude parameters such as rates, biases, temperature dependencies, and orbit profile effects. Parameters can be weighted based on their known accuracy and a judgment of quality or importance. The process of determining the state vector is called *state estimation* or *filtering*. An observation is a sensor measurement or combination of measurements. The measurement can be direct or indirect. State estimation can provide a more accurate attitude estimate—one that is statistically optimal.

There are many methods of state estimation based on statistical methods and estimation theory, many more than can be covered here. However, a brief overview of the most common methods are presented to aid in the system level design of the attitude determination system. These methods can be sequential or batch estimators. A *sequential estimator*, also called a *recursive estimator*, obtains a new state estimate after each observation. A *batch estimator* processes all observations concurrently to produce a new estimate of state vector.

There are two types of estimators in common use: *least-squares* and *Kalman filter*. The least-squares method determines the state vector, which minimizes the square of the difference between observed data and computed data from a dynamics model. An assumption is made that the error or uncertainty has a Gaussian distribution. This type of estimator can be used either sequentially or by batch. The Kalman filter method makes a sequential minimum variance estimation. No assumptions are made about the type of error distribution. This filter is named for R. E. Kalman who developed this method in the early 1950s.

5.4 Attitude Control Systems

An attitude system is composed of three major parts: 1) attitude sensors, which provide direct measurements of spacecraft attitude and were discussed in a preceding section; 2) a feedback control system, which corrects measured attitude to desired attitude; and 3) actuators, also called effectors or torquers, that provide the desired control torques. In preceding sections we have seen how the current attitude of spacecraft, its angular position with respect to a reference system, is determined. In this section we will describe the process by which known or current attitude is corrected to the desired attitude. We will first review feedback system design in general and specifically for a spacecraft. The key element of a feedback system is the control law, which will be considered in more detail. The actuators, the hardware elements which drive the spacecraft, will be discussed later.

A spacecraft can have several feedback control systems. This section will deal with the control of the spacecraft about each of three axes. In many spacecraft similar three-axis feedback control is also required to point the communications antenna, science platforms and instruments, solar arrays, and other sensors.

5.4.1 Fundamentals of Feedback Loop Systems

An open-loop control system is one where the desired position is commanded with no feedback to indicate effectiveness of the commanded action. This method is used where low precision is required. Examples of open-loop control are as follows:

1) It is common practice to command solar arrays to an angle based on an ephemeris sun vector.

2) Tracking and Data Relay Satellite System (TDRSS) station keeping is performed from the ground where an analyst uses trend data to determine the thrusters and firing duration that are needed for correction. The thruster commands are uplinked to the spacecraft.

A closed-loop control system, shown in Fig. 5.17, includes a feedback loop that provides data on the difference between the desired and actual attitude. This

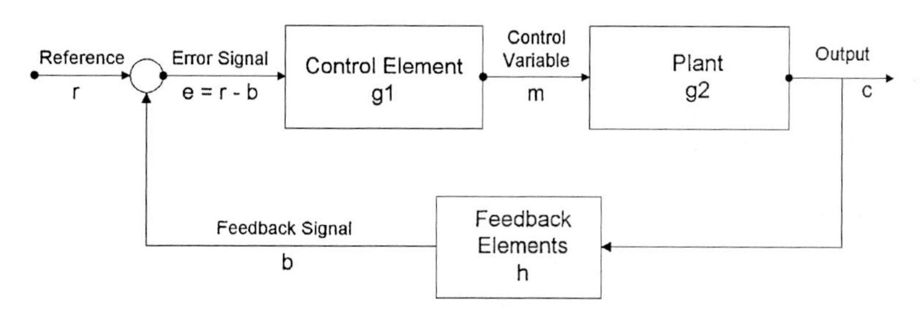

Fig. 5.17 Generic closed-loop control system.

difference is the error signal, which is an input to the control law. The control law operates to drive the error signal to zero within some limit. The system consists of the object to be controlled, called the *plant* (in our case the spacecraft), a reference input r, control elements g, feedback elements h.

The small circle in Fig. 5.17 is an algebraic summing point; signals arriving at this point are either added or subtracted. The boxes, or *transfer elements*, represent functions, not necessarily hardware elements, of the system. The system operates to drive the plant output c toward the reference parameter r. Figure 5.17 can readily be rearranged to describe a spacecraft feedback loop for a single axis as shown in Fig. 5.18.

In Fig. 5.18 the subscript a indicates actual conditions, r indicates reference, and e indicates error. The plant, the object operated on, becomes the spacecraft, and the control element becomes the control law, resident in the computer and the control actuators. The *Feedback Elements* box shown in Fig. 5.17 is not needed for the attitude control diagram (Fig. 5.18) because θ_a can be compared directly with θ_r. (In actuality, the θ_a signal is filtered and smoothed before it is compared to the reference angle, but these operations do not affect the feedback process.)

The angle between the spacecraft axis and the reference coordinate is θ_a; the desired angle is θ_r. An error signal is calculated by subtracting θ_a from θ_r. The error signal is used by the control law to calculate the necessary control torque T_c. The calculated control torque is generated by an actuator, for example, a reaction wheel. In a perfect world the calculated control torque would rotate the vehicle to the reference angle and that would be the end of it. In the real world the control

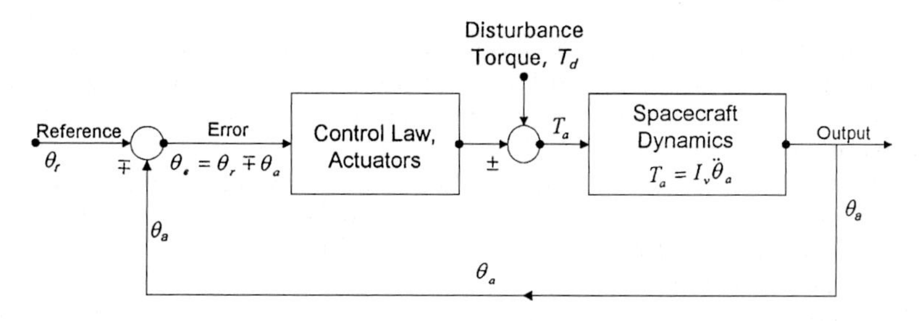

Fig. 5.18 Attitude control feedback system.

torque is summed with any disturbance torque present. The sum of the torques T_a creates vehicle rotation, θ_a in accordance with the relation

$$T_a = I_v \ddot{\theta}_a \tag{5.73}$$

where

T_a = actual torque on the vehicle about a given axis
I_v = moment of inertia of the vehicle about the axis of rotation
$\ddot{\theta}_a$ = actual rotational acceleration (second derivative of angle with respect to time) of the vehicle about the given axis

The actual angle is measured, fed back, and compared to the reference angle. The error caused by the disturbance torque is detected, and the loop repeats. More complex systems compensate for known errors by including detailed models and predictive capability to the spacecraft dynamics. For example, thruster alignment to the control axes is not perfect, causing cross coupling into the other axes. This misalignment can be measured on the ground, and this knowledge can be included in the modeling of the spacecraft dynamics to include a compensating torque rather than sensing the resulting motion and feeding back the error.

The characteristics of the control element and the spacecraft are described by differential equations as a function of time. The transfer function of the spacecraft can be obtained by taking the Laplace transform of the spacecraft dynamic equation. The Laplace transform of a linear differential equation has the form of a polynomial in s, where s is an imaginary number of the form $\sigma + i\omega$ or $\sigma + j\omega$. The function s is sometimes written $R(s)$. The transformed quantities are denoted by capital letters R, G, H. The ratio of Laplace transform of input to the Laplace transform of output of a box is called a *transfer function*.

$$\ddot{\theta} I_v = T_a \tag{5.74}$$

Assuming the moment of inertia I_v is constant, the Laplace transform is

$$I s^2 \mathcal{L}(\theta) = \mathcal{L}(T) \tag{5.75}$$

Recalling that the feedback transfer function is one, the transfer function $G(s)$ of the spacecraft element is

$$G(s) = \frac{\mathcal{L}(\text{output})}{\mathcal{L}(\text{input})} = \frac{\mathcal{L}(\theta)}{\mathcal{L}(T)} = \frac{1}{I s^2} \tag{5.76}$$

A feedback system is stable if its response to a disturbance approaches zero as time approaches infinity. Alternatively, a stable system can be defined as one that responds stably to impulses less than a specified magnitude. An attitude control system is stable if every bounded disturbance produces a bounded response.

The degree of stability is also important. If it is stable, how close to instability is it? From what angular excursion will it recover? The transforms can be used to study the stability as well as performance of the control system. Stability is most frequently studied using Nyquist or Bode analysis. Both are graphical procedures for determining absolute and relative stability of a feedback system.

The subject of feedback control systems is an extensive and deep topic. There are numerous references on the subject; notably Refs. 10 and 11. Additional information on feedback systems as they apply to spacecraft control can be found in Wertz,[9] Agrawal,[12] and Kaplan.[13]

5.4.2 Control Laws

The transfer function of the control element is the control law that relates the control torque to the error signal. Several basic types of control law are discussed in this section.

5.4.2.1 Proportional control. The simplest control lawis proportional control, described by

$$T_c = -K\theta \tag{5.77}$$

where T_c = control torque, K = system gain, and θ = error signal.

Proportional control is seldom used because it allows large angular excursions.

5.4.2.2 Bang-bang. Another type of proportional control is a bang-bang controller, which is sometimes used with thruster control. A thruster pulse is output, and the direction is determined by the sign of the error signal:

$$T_c = T_p \sin\theta \tag{5.78}$$

Addition of a dead band improves the performance of the simple bang-bang controller (see Fig. 5.19). The error signal is compared to a limit, and a pulse is fired only if the error exceeds the limit. The region where no firings occur is called the *dead band.*

$$T_c = T_p \qquad \text{for } \theta > \theta_{\text{lim}} \tag{5.79}$$

$$T_c = 0 \qquad \text{for } -\theta_{\text{lim}} < \theta < \theta_{\text{lim}} \tag{5.80}$$

$$T_c = -T_p \qquad \text{for } \theta < -\theta_{\text{lim}} \tag{5.81}$$

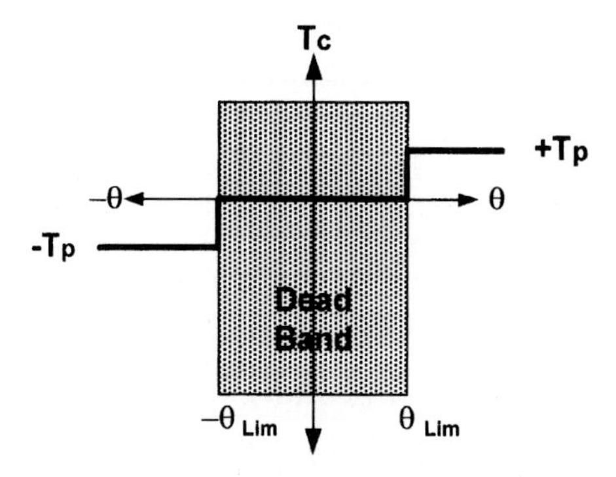

Fig. 5.19 Bang-bang control with a dead band.

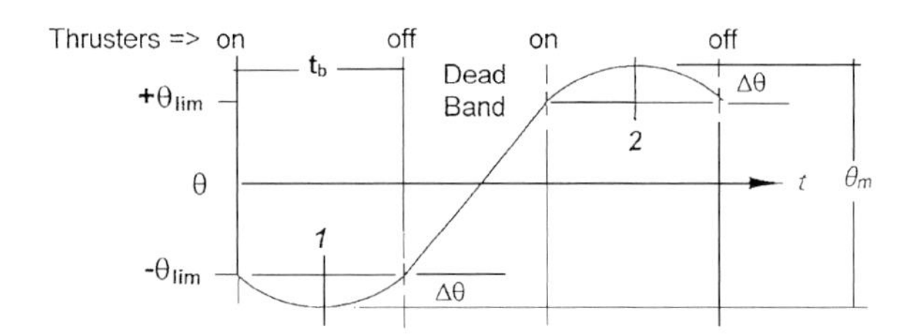

Fig. 5.20 Vehicle motion with thruster control.

5.4.2.3 Vehicle motion. (Also see Section 4.3.4 in Chapter 4.) Figure 5.20 shows the rotation of a spacecraft using bang-bang control with negligible external torque. As the vehicle rotates, it reaches the limit displacement $-\theta_{\text{lim}}$ at which point correcting torque is commanded by firing a selected thruster pair for a selected duration. The thruster firing stops the rotation of the vehicle at point 1, and the second half of the burn reverses the vehicle rotation. At the end of the pulse, the angular acceleration stops, and the vehicle rolls at a constant rate ω until it reaches the opposite limit $+\theta_{\text{lim}}$ and the process repeats.

During the burn, the vehicle rotated beyond the control limit by an amount $\Delta\theta$. The overshoot magnitude can be found as follows:

At point 1 the rotation of the vehicle ω is zero, and

$$\Delta\theta = \tfrac{1}{2}\alpha t^2 \tag{5.82}$$

where

$$\alpha = \frac{T}{I_v} \quad \text{and} \quad t = \frac{t_b}{2} \tag{5.83}$$

Therefore,

$$\Delta\theta = \frac{T t_b^2}{8I} \tag{5.84}$$

Noting that torque on the vehicle $= nFl$, then substituting produces

$$\Delta\theta = \frac{nFlt_b^2}{8I} \tag{5.85}$$

where

θ_{lim} = the angular limits of the dead band, rad
θ_m = the maximum and minimum rotational positions of the vehicle, rad
$\Delta\theta$ = the overshoot in angular rotation, rad $= \theta_m - \theta_{\text{lim}}$
α = angular acceleration, rad/s^2
t_b = the thruster burn time or pulse width, s
I_v = the moment of inertia of the spacecraft about the axis of rotation, N-m^2
T = torque on the spacecraft, N-m
n = number of thrusters firing (usually two)
F = thrust level of the thrusters, N
l = moment arm from spacecraft c.g. to thruster centerline, m

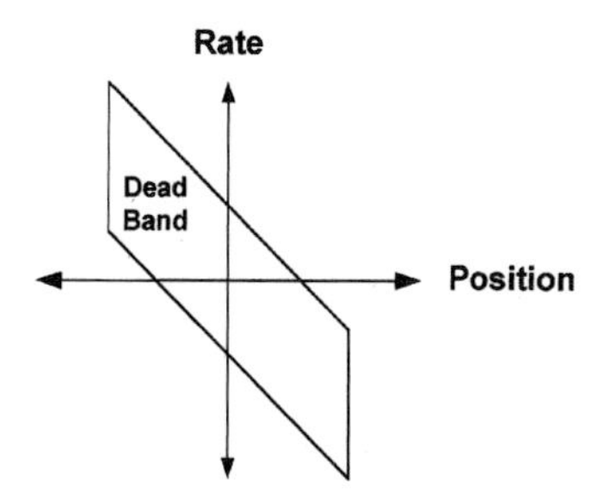

Fig. 5.21 Phase plane controller.

Figure 5.20 illustrates another interesting facet of thruster control. The pointing accuracy attained θ_p is

$$\theta_p = 2\theta_m \tag{5.86}$$

or

$$\theta_p = 2\theta_{\text{lim}} + 2\Delta\theta \tag{5.87}$$

and pointing accuracy cannot be better than the case where the dead band $2\theta_{\text{lim}}$ is zero, and

$$\theta_p = 2\Delta\theta \tag{5.88}$$

The relative inaccuracy of bang-bang control pointing is the reason it is seldom used.

5.4.2.4 Phase plane. A more commonly used, and more complicated, control law for thruster control is the phase plane, shown in Fig. 5.21. The phase plane can be tailored to avoid large excursions and minimize thruster firings, hence minimizing propellant usage.

Figure 5.22 shows an actual phase-plane controller for the Magellan X-axis thrusters. The plane is divided into seven regions. The thruster response is different for an error in each region. In region 0, the "dead band," there is no thruster response. In region -1 thrusters 7 and 8 are fired for 11.1 ms; in region $+1$ the duration would be the same, but the thrusters selected would be different. As errors increase, the firing time is increased to provide more correction.

5.4.2.5 Position-plus-rate controller. A more commonly used control law is a position-plus-rate controller, described by

$$T_c = -K_1\theta - K_2\dot{\theta} \tag{5.89}$$

where $\dot{\theta}$ is the time derivative of the error signal. This control law is also called *position-plus-derivative* or a PD controller. In this case the error signal is a position error, and the time derivative is a rate error. The rate term provides damping and reduces the angular excursions.

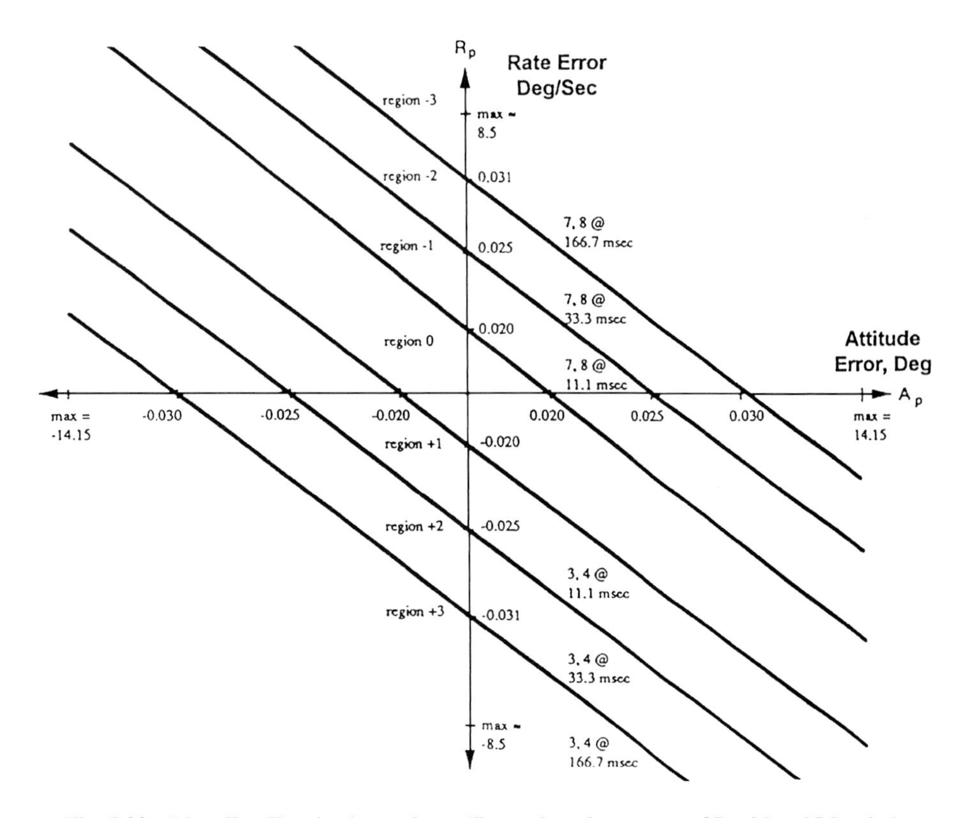

Fig. 5.22 Magellan X-axis phase plane. (Reproduced courtesy of Lockheed Martin.)

5.5 Maneuver Design

The elemental action in any maneuver is applying a torque to the spacecraft about an axis. From kinetics,

$$\theta = \tfrac{1}{2}\alpha t_b^2 \tag{5.90}$$

$$\alpha = \frac{T}{I_v} \tag{5.91}$$

$$\omega = \alpha t_b \tag{5.92}$$

$$H = I_v \omega \tag{5.93}$$

$$H = T t_b \tag{5.94}$$

where

θ = angle of rotation of the spacecraft, rad
ω = angular velocity of the spacecraft, rad/s
α = angular acceleration of the spacecraft during a firing, rad/s^2

I_v = mass moment of inertia of the vehicle, kg-m^2
t_b = duration of acceleration or deceleration, s
H = spacecraft angular momentum, kg-m^2/s
T = torque, N-m.

Angular acceleration and angular velocity are actually vector quantities, but so long as the axis of rotation is fixed their vector properties are not important. Angular momentum is also a vector quantity, and it is frequently necessary to account for vector properties.

Maneuver design is also covered, from the propulsion point of view, in Chapter 4.

5.5.1 Spin-Stabilized Spacecraft

For a spin-stabilized spacecraft a maneuver consists of reorienting the spin axis. Reorientation of the spin axis constitutes a change in the direction of the momentum vector, as shown in Fig. 5.23.

Putting a torque on the spacecraft rotates the momentum vector through angle ϕ

$$\phi \approx \frac{H_a}{H_i} = \frac{nFLt}{I_y\omega} \tag{5.95}$$

where

H_a = momentum vector added, kg-m^2/s
H_i = initial spin momentum, kg-m^2/s

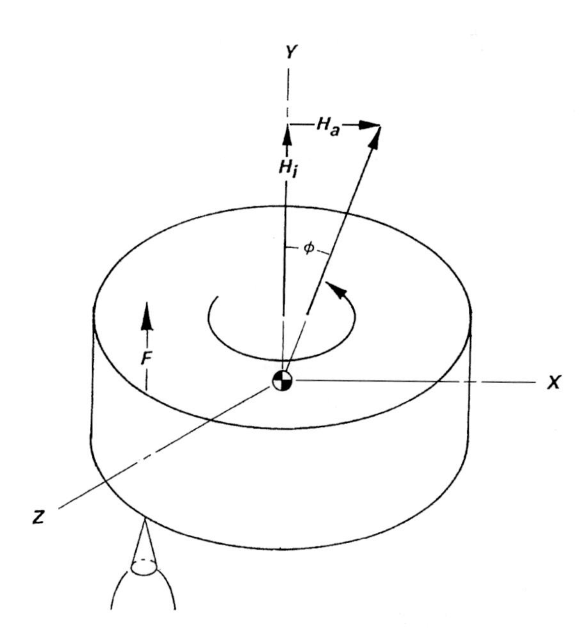

Fig. 5.23 Precessing a spin axis.

I_y = moment of inertia about the spin axis, kg-m^2
ϕ = precession angle, rad
ω = angular velocity of the spacecraft, rad/s
L = lever arm, distance from thrust vector to spacecraft center of gravity, m

The pulse width t must be short compared to the period of spin. A pulse during an entire revolution would result in $H_a = 0$.

The spin axis will continue to precess until a second pulse of equal magnitude and opposite direction is fired. The spin axis can be repositioned by selecting the timing of the second pulse.

5.5.2 Three-Axis-Stabilized Maneuvers

In a three-axis-stabilized system an attitude maneuver consists of rotations about each of the spacecraft axes. The rotation about each axis consists of three parts: 1) angular acceleration, 2) coasting, and 3) braking.

Angular acceleration is produced by a thruster pair firing or by reaction wheels; braking is caused by a firing of the opposite pair or reversing the reaction wheel torque. Figure 5.24 shows a one-axis maneuver controlled by thrusters.

The total angle of rotation is

$$\theta_m = \theta(\text{accelerating}) + \theta(\text{coasting}) + \theta(\text{braking})$$

The time t_b is usually fixed and is determined by the capability of the actuator. If the actuator is a thruster pair, then the torque produced is

$$T = nFL \tag{5.96}$$

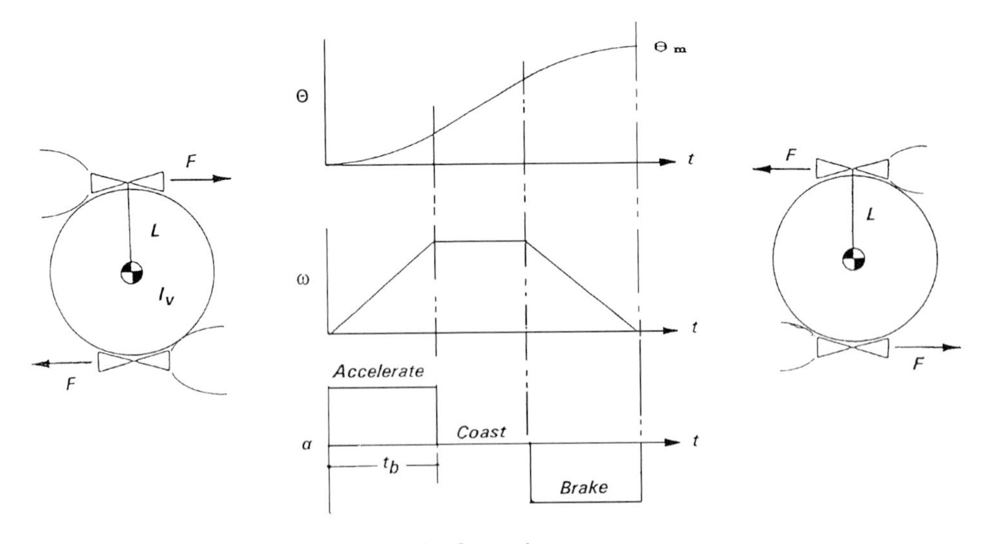

Fig. 5.24 One-axis maneuver.

where

T = torque on the spacecraft
F = thrust of a single motor
n = number of motors firing usually two (must be a multiple of two for pure rotation)
L = radius from the vehicle center of mass to the thrust vector

If the actuator is a reaction wheel, the torque produced is

$$T = I_w \alpha_w \tag{5.97}$$

where I_w is the moment of inertial of the reaction wheel and α_w the angular acceleration of the wheel. The acceleration of the wheel and duration of acceleration will be limited by the wheel inertia and maximum speed. The torque produced by thrusters is limited by the engine thrust level and by the moment arm. The time t_b required for a given acceleration is

$$t_b = \frac{\omega}{\alpha} = \frac{\omega I}{T} \tag{5.98}$$

$$\theta = 2\left(\tfrac{1}{2}\alpha t^2\right) + \omega(t_m - 2t_b) \tag{5.99}$$

where t_m is the maneuver time. Substituting for t_b and solving for t_m yields

$$t_m = \frac{\theta + \omega^2/\alpha}{\omega} \tag{5.100}$$

Alternatively, the largest maneuver that can be performed in a fixed time interval has a zero coast time, and

$$\theta = \alpha t_b^2 \tag{5.101}$$

$$\theta = \frac{nFL}{I} t_b^2 \tag{5.102}$$

Note that the maneuver time is twice the burn time.

Additional information concerning maneuver design can be obtained from the propulsion system requirements section.

Example 5.7 Reaction Wheel Maneuver

A spacecraft has three reaction wheels aligned with the body axes. The inertias are as follows.

Spacecraft:

$$I_x = 1500\,\text{kg-m}^2 \quad I_y = 1500\,\text{kg-m}^2 \quad I_z = 2000\,\text{kg-m}^2$$

Reaction wheels:

$$I_w = 0.06\,\text{kg-m}^2$$

The wheels produce 0.18 N-m of torque each. Of the total storage capacity of 25 N-m-s, 15 N-m-s are allocated to maneuvers. (The remaining 10 N-m-s are allocated to momentum storage.)

The acceleration duration, or ramp-up time, is

$$t = \frac{15}{0.18} = 83.3\text{ s}$$

Acceleration is limited by the greatest inertia I_z:

$$\alpha = \frac{T}{I_z} = \frac{0.18}{2000} = 9 \times 10^{-5}\text{ rad/s}^2$$

The maximum rotation rate is

$$\omega = \alpha t = (9 \times 10^{-5})(83.3) = 7.5 \times 10^{-3}\text{ rad/s}$$

To this point one-axis maneuvers have been discussed. Most spacecraft maneuvers require rotation about three axes, and often the spacecraft activity requires three-axis maneuvers to a new position and three-axis maneuvers to return to normal attitude. Consider for example a Magellan star scan maneuver. The normal attitude for Magellan was nadir pointing for acquiring surface images of Venus. Once per orbit (every 3.15 h) it made a three-axis maneuver to point at a given star. It then made a roll such that the star scanner crossed the star position. The two-axis data from the star scanner were used to update the gyros while the spacecraft made a three-axis maneuver to return to nadir pointing.

There are always constraints on the attitudes the spacecraft is allowed to take. These constraints come from requirements such as 1) keeping bright objects, sun and planet, from the star scanner field of view, or camera field of view; 2) keeping the solar panels pointed at the sun; 3) keeping the high-gain antenna pointed at the ground station; and 4) keeping rotation within gyro constraints. Maneuvers must be designed to rotate all axes without violating these exclusion zones. In addition, there are often time constraints on maneuver; for example, maneuvers for star scanning are timed for the star to be in rough position, and maneuvers to accomplish planetary orbit insertion have tight time limits.

5.6 Attitude Control Hardware

In this section, the specialized equipment used in an attitude control system is described. This information provides the reader with the general performance of current equipment for student designs and the like. The performance of aerospace equipment like this changes rapidly with time; for current design information you should contact the equipment manufacturers. Note also that by no means all of the equipment manufacturers are included here and the names and corporate affiliations of these manufacturers are changing rapidly.

Attitude control requires three types of specialized equipment:

1) Sensors are used to measure attitude of the spacecraft with respect to known quantities such as the sun, stars, or Earth.

2) Actuators are used to provide a torque to the spacecraft to correct measured attitude to desired attitude.

3) Computers are used to perform the sensor processing, attitude determination, control law, attitude, and maneuver calculations.

Table 5.7 Sensors and typical performance

Sensor	Typical performance, deg	Mass, kg	Power, W	Typical application
Horizon (Earth)	0.02–0.1	0.6–5	1–8	Earth orbiters, LEO to GEO, spinner or three-axis, depending on type
Magnetometer	0.5–1.0	0.2–0.7	<1	Earth orbiters, three-axis
Sun	0.2–1	0.04–0.5	<1	All
Star trackers and cameras	0.0002–0.08	1.2–16.5 (typical 3–7)	4–32	Three-axis stabilized, high precision pointing
Star scanner	0.003–0.1	2–4.5	0.5–10	Spinner, 1–20 rpm
GPS receiver	—	3	2	Navigation, may be used for attitude on large space structures

5.6.1 Sensors

Sensors provide "sensed" data about the position of the spacecraft relative to known quantities. A spacecraft uses sensors to 1) detect and measure rotation about three axes, 2) locate the spacecraft–sun vector, 3) measure rotational and linear acceleration, and 4) detect or track stars.

Typical sensors and performance are compared in Table 5.7.

As already discussed, a satellite using a spin-stabilized attitude control system will employ a sun sensor, horizon sensor, or a star scanner. Typically, a combination of at least two of the three sensors is used to provide more information and to provide an alternate data source when data are not available such as eclipse-causing sun-sensor outages. Gyros are rarely used on a spin-stabilized satellite because repetitive measurements from the inertial sensors provide the spin rate to a sufficient accuracy.

A three-axis-stabilized spacecraft uses a combination of inertial sensors and rate sensors. One typical configuration would be gyros, a star tracker, and sun sensors. A combination of gyros and sun sensors would be used for attitude acquisitions and safe-hold mode, which requires lower accuracy. The gyros and star tracker would be used for the primary mode with the gyros propagating attitude based on sensed rate between star measurements. Another typical configuration replaces the star tracker with horizon sensors.

As the sensors, algorithms, and computers become more sophisticated and capable, it becomes possible to rely on a star camera or horizon sensors only for long periods of time. Rate information is reduced from successive readings of the inertial sensor when the spacecraft is relatively quiescent. The higher bandwidths now available and greater computing power, especially image processing, make these advances possible. Usually, this configuration is flown for power savings because gyros require a large amount of power relative to inertial sensors. However, gyros

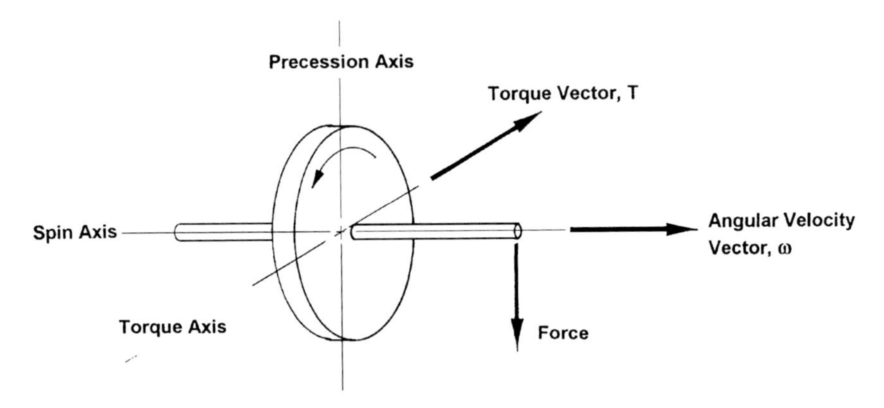

Fig. 5.25 Precession.

are still included to accommodate maneuvers, which still require the higher rate capability.

5.6.1.1 Rate sensors. Gyros are precision instruments that detect small rotations with respect to inertial space. Spacecraft gyros have a small angular range and high accuracy as opposed to aircraft instrument gyros, which have a wide range and low accuracy. Gyros operate on the basic principle that when a torque is applied to a spinning wheel the wheel precesses. Figure 5.25 shows the precession torque T caused by force F on a spinning wheel.

The axes that describe precession are the spin axis, the torque axis, and the precession axis. The angular velocity vector and the momentum vector are aligned with the spin axis; the vector direction is determined by the right-hand rule. Precession is in the direction of torque (see Fig. 5.25), as is the angular acceleration of the wheel. If the wheel in Fig. 5.25 were a single-axis gyro, the torque axis would be called the *input axis* and the precession axis would be called the *output axis*. The angular rate of precession ω_p is directly proportional to the input torque and inversely proportional to the momentum of the gyro,

$$\omega_p = \frac{T_{\text{in}}}{I_w \omega_w} = \frac{T_{\text{in}}}{H_w} \tag{5.103}$$

Inertial systems. Gyros can be used in two major ways to effect attitude control:

1) Gyros and accelerometers are mounted on a servo-driven stable-platform; gyro output is used to keep the platform inertially fixed. Vehicle control information is obtained in inertial coordinates by comparison with the stable platform position.

2) The increasingly more popular method uses *strap-down* gyros, which are fixed to the spacecraft body. The gyros read disturbance in body-fixed coordinates, and the ACS computer is used to relate body-fixed output to inertial reference frame. A strap-down system offers hardware simplicity, less power and weight at a cost of computational complexity.

Gyros are not usually bought individually but are packaged with electronics. Gyro performance is a combination of the sensitivity of the instrument, the range of the instrument, and the magnitude of the error sources and is determined by the type of gyro and its electronics. The sensitivity is usually inversely proportional to the range. Some units compensate for this problem by having selectable modes for high rate, with lower sensitivity, and low rate with higher sensitivity. Recent advances in technology have made many types of high-quality gyros available from a wide selection of vendors.

Gyro errors or drift. Gyro errors or drift result from imperfections in the gyro. Some forms of drift are proportional to acceleration or acceleration squared. These are important during thrusting flight or high rate maneuvers. These errors are specified as scale-factor error and are calculated as the difference between the measured angle of rotation and the known angle. These errors are quantified during unit calibration, but they can change or drift over the lifetime and may require periodic measurement in flight. Gyro bias stability is a measure of noise in the gyro. Conceptually, the gyro should have zero output if the spacecraft is holding perfectly still, but because all gyros have some type of moving parts there is some signal or bias. This bias changes over time. The amount that the bias changes is specified as the bias stability. A third type of error is specified as random walk and is exactly what the name implies, a random error.

Some repeatable forms of drift can be measured and removed by calibration corrections. The random forms of drift require periodic measurement and updates to remove. Stellar reference is the most common method of absolute reference for gyro update. For example, the Magellan spacecraft used a star calibration, which consisted of a three-axis maneuver to place the roll axis perpendicular to the expected position of the chosen guide star. A roll maneuver then swept the star scanner across predetermined stars providing a star update, which was compared to the predicted star position to provide an estimate of gyro bias drift. Special calibrations are required to separate bias error from scale-factor error.

Gyro drift is an important consideration when choosing the ACS hardware. How much drift can be tolerated depends on how the gyro data will be used in conjunction with the other sensors onboard. If the gyros will be relied upon to propagate attitude during other sensor outages, then high performance with a low drift rate would be indicated. Once the measurement performance requirement is established, other considerations such as power, mass, and lifetime need to be considered.

One-degree-of-freedom gyros. Figure 5.26 is a schematic of a one-degree-of-freedom gyro; it has a single gimbal, which is free to rotate about the output axis and is otherwise fixed. The reference axis is body fixed and coincident with the gyro spin axis at null. The gyro input axis is also body fixed and is perpendicular to the reference axis and the output axis. Gyro input is a rotation of the case about the input axis, shown as the spacecraft x axis in Fig. 5.26. This rotation is transmitted to the wheel and axle through the gimbal bearings and constitutes a forced precession of the gyro, which causes a rotation θ about the output axis. The instrument measures the error angle, which can then be used to correct the error. Depending on the way disturbing torque error is handled, the gyro is termed an

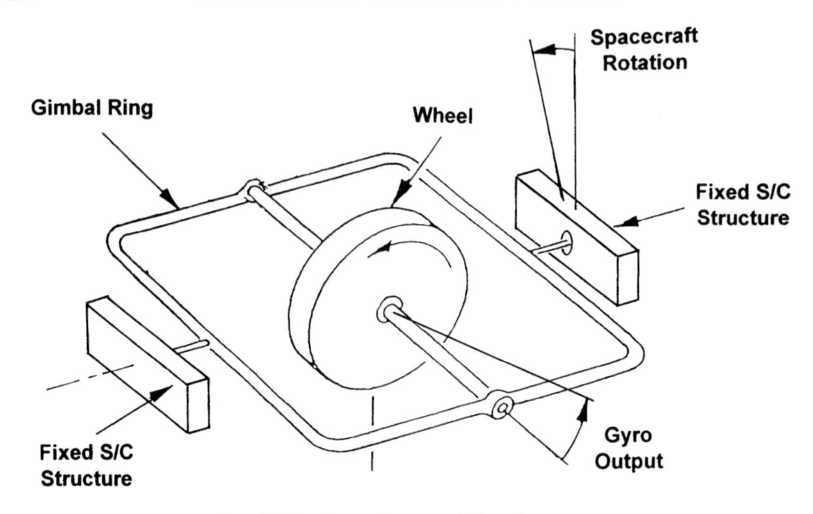

Fig. 5.26 One-degree-of-freedom gyro.

integrating gyro, an undamped gyro, or a rate gyro.

$$\theta_{\text{out}} = \frac{H}{C} \int_0^t \dot{\theta}_{\text{in}} \, \text{dt} \tag{5.104}$$

$$\theta_{\text{out}} = \frac{H}{C} \theta_{\text{in}} \tag{5.105}$$

where

θ_{in} = angular input error caused by spacecraft motion
θ_{out} = sensor output angle
H = momentum of gyro rotor
C = viscous damping around the output axis
H/C = gyro gain

Floated gyros can sense rates as small as 0.015 deg/h and were used on the Apollo, Polaris, Titan, IUS, and Peacekeeper. The Titan III inertial guidance system used a rate-integrating, strap-down, one-degree-of-freedom, floated gyro; the rotor was 2 in. in diameter, weighed about 2 lb, and rotated at 24,000 rpm.

Two-degree-of-freedom gyro. Figure 5.27 is a schematic of a two-degree-of-freedom gyro. The gyro wheel is set on bearings in the inner gimbal. The inner gimbal is set in bearings in the outer gimbal, which in turn is set in bearings in the gyro case. The gimbal arrangement allows the case rotational freedom with respect to the inner gimbal. The motion of the inner gimbal is measured in two sensitive axes with respect to the case. Any small motion about the sensitive axes (x and y in Fig. 5.27) is measured and used to correct the disturbing torques.

Two-degree-of-freedom gyros use a variety of designs to reduce friction and, hence, improve accuracy. The *electrostatic gyro* uses a spherical beryllium rotor that rotates at 100,000 rpm and is supported electromagnetically in a cavity to reduce friction. Aircraft and submarine systems use this type of gyro.

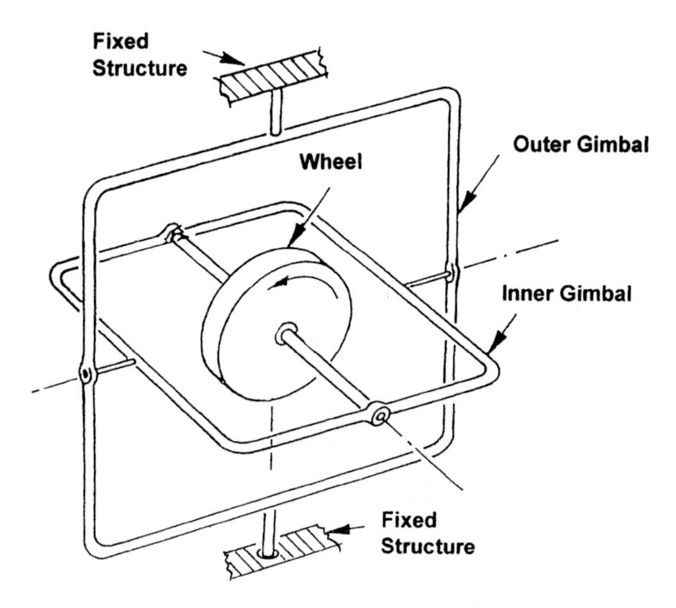

Fig. 5.27 Two-degree-of-freedom gyro.

A *gas-bearing gyro* uses a spherical rotor supported in a cavity by a thin layer of gas under pressure. Gas-bearing gyros were used on the Minuteman IMU. A *dry-tuned gyro* spins at the precise frequency at which the forces and torques offset each other, leaving a nearly frictionless system. The Trident, Space Shuttle, and Magellan use dry-tuned gyros.

Ring-laser gyro. The ring-laser gyro is more recent technology that promises greater reliability; there are few moving parts and greater stability because they use spinning light instead of spinning mass. The gyro detects and measures angular rates by measuring the frequency difference between two contrarotating laser beams (see Fig. 5.28). The two laser beams circulate in a triangular cavity simultaneously. Mirrors are used to reflect the beam around an enclosed area. The resonant frequency of a contained laser beam is a function of its optical path length.

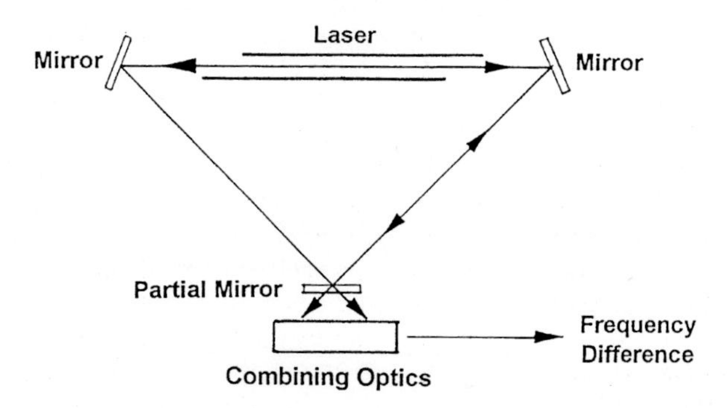

Fig. 5.28 Ring-laser gyro schematic.

Consequently, the two laser beams have identical frequencies when the gyros are at rest. When the gyro is subjected to an angular rotation around an axis perpendicular to the plane of the two beams, one beam then has a longer optical path, and the other beam has a shorter optical path. Therefore, the two resonant frequencies change, and the frequency differential is directly proportional to the angular turning rate. The frequency difference is measured by optical means, resulting in a digital output. Pulse rate is proportional to the input angular rate, and the pulse count is a measure of the angle turned.

There are two main design variants on Fig. 5.28. Fiber-optic laser paths can be used; in this design the optic path can be made very long, without awkward unit geometry, by coiling the fiber. Many turns can be used. The other major variant produces a phase difference between the two laser beams rather than a frequency difference.

Honeywell International, Inc. uses their ring laser gyros as the basis for their miniature inertial measurement unit (MIMU), which provides temperature compensated three axis angular rate output. The radiation-hardened inertial reference unit (IRU) contains three orthogonal gyros with an available inertial measurement unit (IMU) option that contains three orthogonal accelerometers. The primary power inputs range from 28 vdc with telemetry on/off commands. The IMU is currently flying on the Odyssey and the Stardust spacecraft. The radiation hardened IRU and its characteristics are shown in Fig. 5.29. The IRU weighs less than 4.7 kg

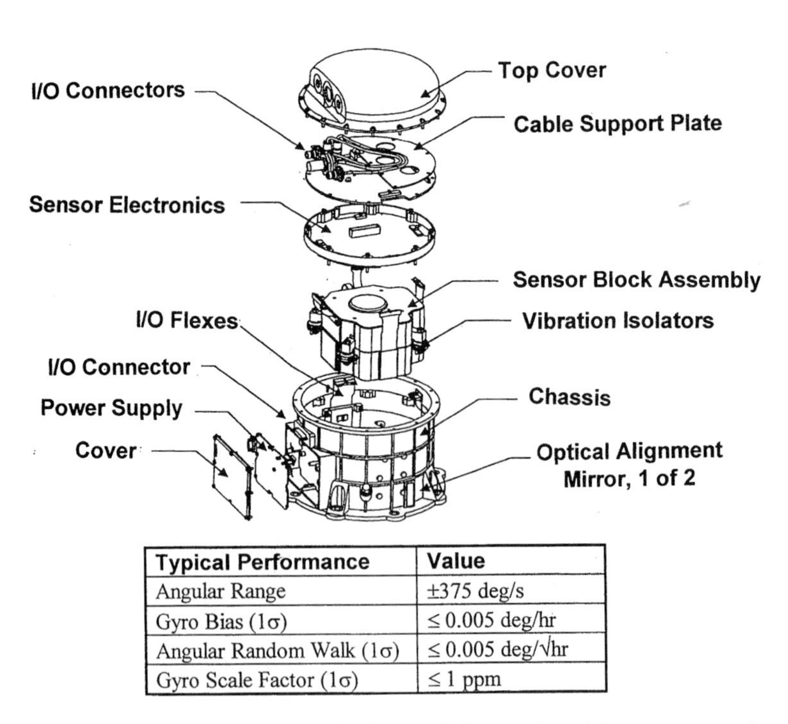

Typical Performance	Value
Angular Range	±375 deg/s
Gyro Bias (1σ)	≤ 0.005 deg/hr
Angular Random Walk (1σ)	≤ 0.005 deg/√hr
Gyro Scale Factor (1σ)	≤ 1 ppm

Fig. 5.29 Honeywell International, Inc., miniature inertial measurement unit. (Reproduced with permission of Honeywell International, Inc.)

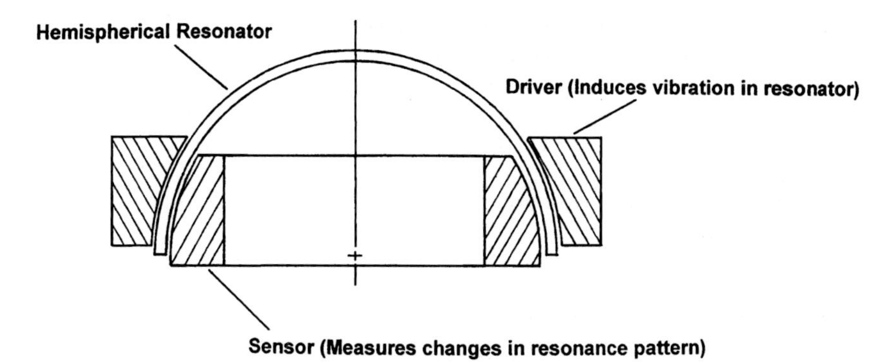

Fig. 5.30 Hemispheric resonator gyro.

and requires less than 32 W. The overall envelope is 233 mm in diameter by 169 mm in length.

Hemispherical resonator gyro. The hemispherical resonator gyro (HRG) is another advanced gyro concept that offers small volume and mass along with simple operation and no wearout components. The operating principle is shown in Fig. 5.30. The unit consists of a hemispherical resonator that is driven at its resonant frequency. Disturbance torques produce measurable changes in the resonance pattern of the hemisphere, which are detected by a collar surrounding the resonator.

The HRG can be configured to operate as a rate gyro or an integrating gyro. It flew on the NEAR spacecraft and is currently flying on the Cassini spacecraft.

Accelerometers. All spacecraft linear thrusting maneuvers require accurate velocity change measurement for the shutdown signal. There are three ways of controlling the burn: 1) use a calibrated propulsion system and time the burn, 2) measure acceleration on the ground from range rate, 3) measure acceleration onboard. Onboard accelerometers are the most common approach. An accelerometer (Fig. 5.31) is a relatively simple instrument that measures force on a known mass.

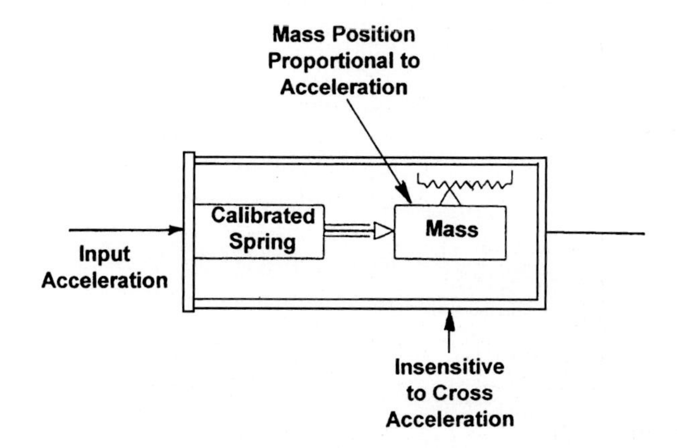

Fig. 5.31 Accelerometer.

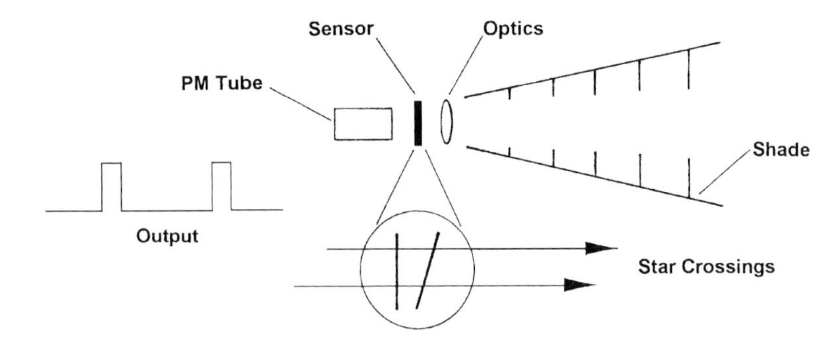

Fig. 5.32 Star scanner.

5.6.1.2 Inertial sensors. *Star scanners.* Star scanners are rotated past the calculated position of a guide star. The difference between the calculated position and the measured position of the guide star is used to calculate an attitude update. Figure 5.32 shows a schematic of a star scanner.

As the scanner is rotated across the spacecraft–star vector, light enters the optical shade and passes through the slit pattern to the sensor photomultiplier tube, which produces a pulse as the star passes each slit. The radial position of the star in the roll plane is given by the leading edge of the first pulse. The star position perpendicular to the roll plane is given by the time between pulses.

Star trackers. A star tracker is more like a camera using a charge-coupled device (CCD) array. It provides a horizontal and vertical position (in tracker coordinates), which is then converted to a position error. Trackers can provide information about several stars simultaneously. The term "tracker" refers to the ability to provide continuous position updates as the star moves through the instruments field of view (see Fig. 5.33). Star trackers require an a priori estimate of attitude in order to identify the stars in the field of view.

Star cameras. The Clementine spacecraft was the first to fly a star camera, a new inertial sensor, which uses a CCD combined with image processing performed in the onboard computer. The star camera observes a segment of sky and deduces position by pattern matching. It is not necessary to scan or track given stars and

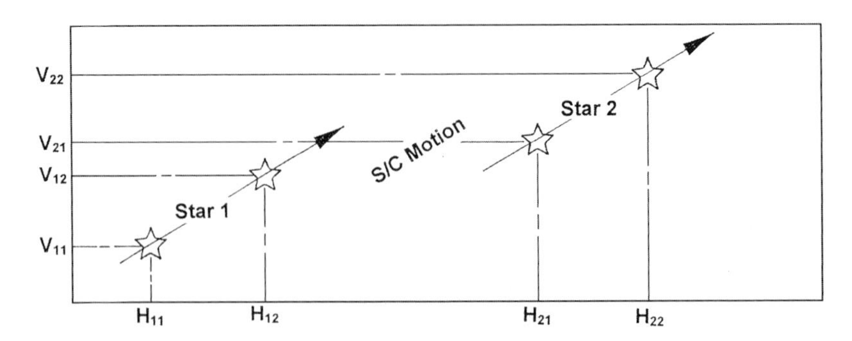

Fig. 5.33 Star tracker field of view.

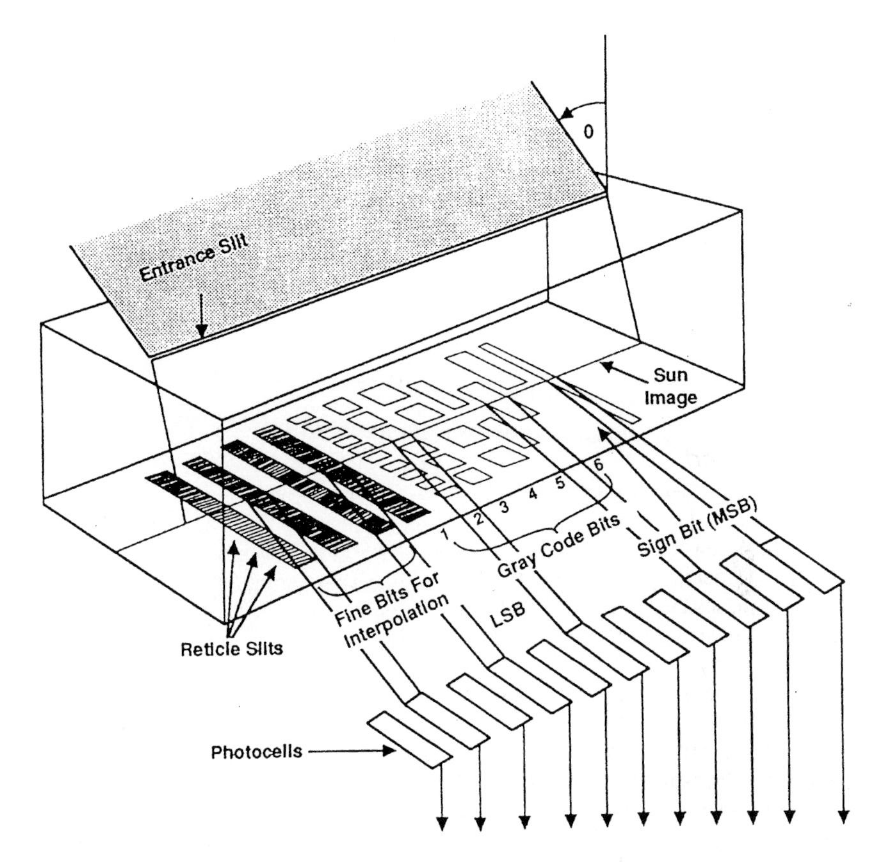

Fig. 5.34 Sun-sensor measurement operation. (Reproduced with permission of Adcole, Inc.)

therefore requires no a priori knowledge. The characteristics of the Clementine OCA star camera are field of view, 28.9 × 43.4 deg; accuracy, 96 arc-sec; weight, 1.2 kg.

Sun sensors. Sun sensors are a wide-angle measurement used primarily to point solar panels or in attitude initialization. They are a much smaller, more rugged instrument than star scanners. They weigh about 0.3 kg and consume about 0.5 W continuously. Magellan used one on each panel. The operation of a typical sun sensor is shown in Fig. 5.34. The reticule pattern is designed to convert the sunlight line entering the instrument into a digital code representing the sun angle within 0.25 deg.

Horizon sensors. Horizon sensors are particularly useful for finding the nadir vector (see Fig. 5.35). Horizon scanners are an infrared (IR) sensing instrument; the location of the horizon is sensed by the dramatic difference in IR emission of the Earth disk and cold deep space. Because they are a scanning instrument, they are particularly suitable for spinning spacecraft; if they are used on a three-axis spacecraft, the instrument must have a scanning head. Used in pairs they can provide nadir within about 0.1 deg. They weigh about 2–7 kg and require 5–10 W.

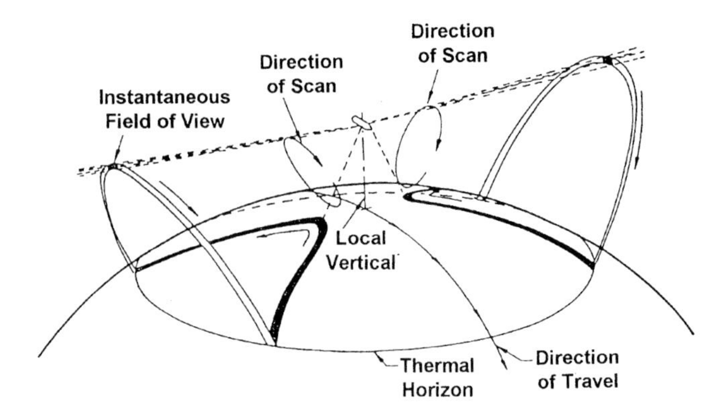

Fig. 5.35 Conical scanning Earth sensors. (Reproduced courtesy of NASA; Ref. 6.)

Global positioning system receiver. The advent of the global positioning system (GPS) provides spacecraft with a small, light, very accurate instrument for determining position. (Technically, position determination is *navigation* rather than *attitude control*; however, the instrument is usually assigned to the ACS system.) The GPS receiver used on the SNOE spacecraft is typical; its characteristics are as follows: mass, 3.2 kg; power, 2.1 W (SNOE reduced average power by using the instrument intermittently); size, $2.5 \times 4.5 \times 2.0$ in.

5.6.2 Actuators

5.6.2.1 Wheels. Several types of rotating flywheels are used in attitude control systems. These are all similar in construction but are used in different ways to provide stiffness to resist disturbing torques, to exchange momentum to correct error, to allow operation at one revolution per orbit for Earth-oriented missions, to absorb cyclic torques, and to transfer momentum to spacecraft for slewing maneuvers. Each type is described in this section.

Control moment gyros. Control moment gyros (CMG) are gimbaled wheels spinning at a constant rate. A commanded force on the input axis of the gyro causes a control torque to the spacecraft on the output axis. They are larger and heavier than reaction wheels and consume more power. Skylab used control moment gyros; they are seldom used on smaller spacecraft because of weight and power requirements.

Momentum wheels. Momentum wheels are flywheels designed to operate at a biased, nonzero speed. Momentum is exchanged with the wheel by changing wheel speed. They are usually body fixed. Momentum wheels and reaction wheels differ only in speed bias. INTELSAT VIII used momentum wheels.

Reaction wheels. Reaction wheels (see Fig. 5.36) are simply small flywheels powered by a dc motor, which exchanges momentum with the spacecraft by changing wheel speed. For example, when a clockwise disturbance torque is imposed on the spacecraft the attitude control system holds spacecraft attitude constant by rotating a reaction wheel counterclockwise. In the reverse case the spacecraft can be rotated by using the dc motor to slow wheel speed and cause the spacecraft to

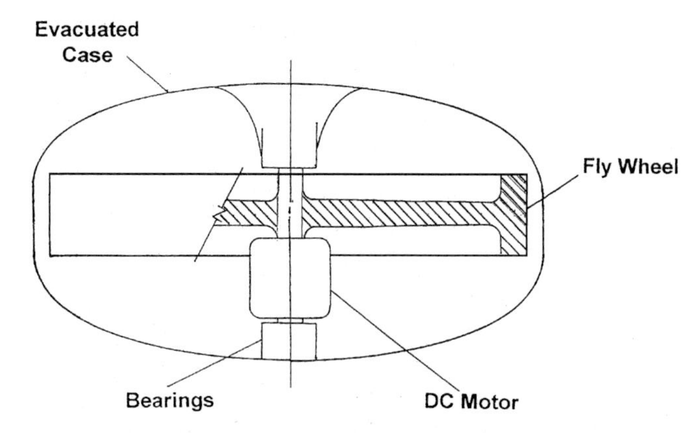

Fig. 5.36 Reaction wheel.

rotate. Reaction wheels are particularly suited to situations that require positive and negative sets of momentum exchanges, for example, limit-cycle maneuvers or a maneuver from cruise attitude to target and return to cruise attitude. For maneuvers of this type, momentum is borrowed from the wheels and then returned to the wheels. Except for wheel friction there is no net loss of momentum.

Momentum exchanges are very efficient; however, eventually the wheel momentum must be adjusted by changing wheel speed. Also situations with a constant disturbance torque, solar pressure for example, result in constant wheel speed increase to hold the spacecraft. Typically wheel speed is reduced (momentum is "dumped") by holding the spacecraft stationary with thrusters, or magnetic torquers, while the dc motor reduces wheel speed.

If reaction wheels are placed on each of the three principle spacecraft axes, the control law for each axis will be linear; however, it is common practice to provide wheel redundancy by placing a fourth wheel in a position oblique to all axes. With this arrangement any single wheel failure can be accommodated by the oblique wheel, which is far more efficient than providing three redundant wheels. The control law for oblique wheel will be slightly more complex and nonlinear. The wheel control laws must be designed to avoid wheel operation near 0 rpm. Wheel friction is nonlinear in this region, and wheel motion becomes jerky and imparts an attitude error. This effect is sometimes called "stiction."

Reaction wheels have a deep application heritage and come in many sizes, ranging from the HR01 from Honeywell International, Inc. used on DSCS (1.9 N-m-s momentum and 0.05 N-m output torque) to the large HR195 (264 N-m-s momentum and 0.7 N-m torque) used on the Hubble Space Telescope. A sample of a wide selection of current production wheels from Honeywell International, Inc. is shown in Table 5.8.

Torque rods. Torque rods take advantage of the Earth's magnetic field to generate a correcting force on a spacecraft. A torque rod is simply a wire coil wrapped around a rod, usually a few centimeters in diameter and about a meter in length. When a current is sent through the coil a torque is generated by the interaction of

Table 5.8 Reaction wheel characteristics[a]

Characteristic	Mini-wheel	HR 0610	HR 12	HR 14	HR 16	HR 4820	HR 2010	HR 2020	HR 2030	HR 4520
Angular momentum, N-m-s	0.2 to 1.0	4 to 12	12 to 50	20 to 75	75 to 150	65	33.2 to 68.4	27	19.5 to 45.6	60.75
Output torque, N-m	>0.028	0.07 to 5	0.1 to 0.2	0.1 to 0.2	0.1 to 0.2	0.14	0.1	0.13	0.21	0.135
Wheel rpm, ±	9000	6000	6000	6000	6000	6000	6000	6500	6000	5400
Power, W[b]	>6	<15	22	22	22	20	17	35	20	35
Bus voltage, dc	12 to 34	14 to 35	23 to 57	23 to 57	23 to 57	22 to 36	27 to 44	70	27.7 to 31.3	51
Mass, kg	1.3	3.6 to 5.0	7.0	8.5	12	10.2	9.2 to 10.9	7.9	8.9 to 11.2	11.1
Integral electronics	Y	Y	Y	Y	Y	Y	N	Y	Y	Y
Diameter, mm[c]	108	267	316	368	418	405	406	300	305	406
Height, mm	54	12.0	159	159	178	214	235	172	191	215
Op temperature,[d]										
Low	−25	−15	−30	−30	−30	−15	−15	−13	−15	−24
High	+60	+60	70	+70	+70	+71	+70	+75	+80	+61

[a]Reproduced with permission of Honeywell International, Inc.
[b]Power values are steady-state power at maximum wheel speed, W.
[c]Dimensions are overall envelope, mm.
[d]Temperature ranges are qualification limits, operating, °C.

the current and the Earth's magnetic field,

$$T = NBAI \sin \theta \tag{5.106}$$

where

T = torque, N-m
N = number of loops in coil
B = magnetic field of central body, T (see Table 5.5)
A = area of the coil, m^2
I = current A
θ = angle between the Earth's magnetic field lines and coil centerline

Magnetic torquers are very useful for momentum management, often in conjunction with reaction wheels. Over 100 low-Earth orbiters have used magnetic torquers; these include SAGE, NIMBUS, LANDSAT, SATCOM, Iridium, and Globalstar. Their usefulness decreases with orbit altitude, as the Earth's magnetic field decreases. They are not useable for deep space missions.

For a small spacecraft directly inserted into LEO, propulsion can be avoided by using the magnetic torquers to provide momentum desaturation to the reaction wheels, eliminating an entire subsystem. The SOURCE spacecraft design takes advantage of this configuration.

5.6.2.2 Thrusters. Most spacecraft use thrusters as actuators; they provide momentum to the spacecraft by ejecting mass overboard in the form of high velocity exhaust gas. Three types of thrusters are in use: 1) cold gas, 2) monopropellant hydrazine, 3) bipropellant. Cold-gas systems are the simplest and are used for small spacecraft with impulse requirements of a few hundreds newtons/second. Bipropellant systems are the most expensive and complex; they are used for trajectory control and infrequently for attitude control. The dominant choice is monopropellant hydrazine, which gives midrange specific impulse with a simple system. It requires 12 thrusters to provide pure moments about three axes. More detailed information is given in Chapter 4.

Thrusters can be used directly to control the spacecraft attitude or used as momentum desaturation actuators for the reaction wheels. Primary thruster control can be very costly in terms of propellant mass if there are noticeable disturbance torques.

5.6.3 Computers

Computers are necessary for the attitude control system and for the data handling system. They are discussed in Chapter 8.

The decision of a central computer for both C&DH and attitude control vs individual computers for each subsystem is a significant early trade in the spacecraft design process. The considerations are as follows:

1) A central computer is lighter and cheaper.

2) A central computer is a simpler system from the hardware standpoint; however, the software system may be more complex.

3) The dominant requirements of the two systems are different. The attitude control system needs a very fast computer; the data system needs large memory and

file handling features. Specialized computers offer better performance; a central computer is a performance compromise.

4) Two different computers make a more comprehensive fault protection system possible. The ACS computer can monitor the C&DH computer and vice versa. To take full advantage of cross monitoring requires redundant computers of each type so that a failed computer can be taken off line. A fault protection system that incorporates cross monitoring requires substantially more complicated software.

As computers become more and more capable and able to perform more calculations faster, the attitude control software has become more capable and more complex. The amount of data processed from the various sensors and the attitude determination methods used to process that data have increased in size and complexity. These advances have led to the increased importance of the attitude control software development and testing.

5.6.4 Example Systems

The hardware selections made by four flight systems are described in this section.

5.6.4.1 Space Shuttle Orbiter. 1) Attitude sensing consists of three independent inertial platforms, each with four fixed-rate gyros and four fixed accelerometers, and two star trackers.

2) Computers needed are three dedicated computers and two backup computers.

3) Control involves three gimbaled main engines (10.5-deg pitch, 8.5-deg yaw), throttleable (65 to 109%); two orbital maneuvering systems, each with 26,700 N engines, gimbaled (6-deg pitch, 7-deg yaw) and 44 control thrusters, 38 at 3870 N each and 6 at 100 N each.

5.6.4.2 Inertial upper stage. 1) Attitude sensing consists of redundant inertial measuring units, each with five rate-integrating one-degree-of-freedom, gas-bearing gyros; five linear accelerometers; and one star scanner to update the gyros.

2) Computers needed are two redundant computers.

3) Control involves gimbaled first- and second-stage solid rocket motors and 12 monopropellant thrusters.

5.6.4.3 Stardust. 1) Attitude sensing consists of three ring-laser gyros used only for trajectory maneuvers and comet encounter; one star camera (+ one redundant) for inertial attitude determination "all stellar"; and two analog sun sensors for backup modes.

2) Computers needed are redundant all-purpose onboard computers RAD6000.

3) Control involves eight 0.9 N thrusters for attitude control and eight 4.5 N thrusters for trajectory maneuvers and backup control for comet encounter.

5.6.4.4 INTELSAT VIII. This system is spin stabilized with active nutation control, with deployment through geosynchronous orbit insertion and three-axis-stabilized nadir pointing momentum bias during on station operation to 0.017 deg.

1) Attitude sensing consists of earth sensing and roll yaw gyros.

2) Computer needed is a dedicated computer with firmware.

3) Control involves 110 N-m-s, 6000-rpm momentum wheel; magnetic torquing for roll, yaw; and dual-mode thruster backup for all modes.

References

[1]Piscane, V. L., and Moore, R. C. (eds.), *Fundamentals of Space Systems*, Oxford Univ. Press, 1994.

[2]McGlinchey, L. F., and Rose, R. E., "Pointing and Control of Planetary Spacecraft—The Next 20 Years," *Astronautics and Aeronautics*, Vol. 17, No. 10, Oct. 1979.

[3]Accord, J. D., and Nicklas, J. C., "Theoretical and Practical Aspects of Solar Pressure Attitude Control for Interplanetary Spacecraft," *Guidance and Control II, Progress in Astronautics and Aeronautics*, Vol. 13, edited by R. C. Langford and C. J. Mundo, Academic Press, New York, 1964.

[4]Wertz, J., and Larson, W., *Space Mission Analysis and Design*, 3rd ed., Kluwer Academic, Dordrecht, The Netherlands, 1998.

[5]Lang, K. R., *Astrophysical Data: Planets and Stars*, Springer-Verlag, New York, 1991.

[6]Corless, W. R., "*Scientific Satellites*," NASA SP133, 1967.

[7]Hamilton, W. R., Sr., *Elements of Quaternions*, Longmans, Green & Company, 1866.

[8]Whitaker, E. T., *A Treatise on the Analytical Dynamics of Particles and Rigid Bodies*, Cambridge Univ. Press, 1961.

[9]Wertz, J. R. (ed.), *Spacecraft Attitude Determination and Control*, Kluwer Academic, Dordrecht, The Netherlands, 1978.

[10]Di Stefano, J. J., Stubberud, A. R., and Williams, I. J., *Feedback and Control Systems*, Schaum's Outline Series in Engineering, McGraw–Hill, New York, 1967.

[11]Thaler, G. J., and Brown, R. G., *Servomechanism Analysis*, McGraw–Hill, New York, 1953.

[12]Agrawal, B. N., *Design of Geosynchronous Spacecraft*, Prentice–Hall, Upper Saddle River, NJ, 1986.

[13]Kaplan, M., *Modern Spacecraft Dynamics and Control*, Wiley, New York, 1976.

[14]Griffin, M. D., and French, J. R., *Space Vehicle Design*, AIAA Education Series, AIAA, Washington, DC, 1991.

[15]Seifert, H. S. (ed), *Space Technology*, Wiley, New York, 1959.

[16]Draper, C. S., Wrigley, W., and Grohe, L. R., "The Floating Integrating Gyro and Its Applications to Geometrical Stabilization Problems on Moving Bases," Inst. of the Aeronautical Sciences, Preprint 503, 1955.

[17]Kalman, R. E., "A New Approach to Linear Filtering and Prediction Problems," *Journal of Basic Engineering*, Vol. 82, 1960, pp. 35–45.

Problems

Some of these problems require data and equations from Chapter 4 as well as Chapter 5.

5.1 Which of the attitude sensors requires a nonzero body rate, in inertial space, to provide a useful output?

5.2 For the axis of interest, the spacecraft inertia is 6000 kg-m^2. The effectors are reaction wheels and a thruster pair located at a 1.52-m radius. Each thruster is rated at 0.89 N and is pulsed on for 0.025 s.

(a) Determine the reaction wheel momentum necessary to cause a spacecraft maneuver rate of 0.004 rad/s.

(b) For some mission segments, control will be with thrusters alone. What pointing accuracy can be achieved if the dead band is 0.001 radians and there are no external torques?

5.3 What minimum torque is required to rotate a spacecraft 0.50 deg in 5 s and hold the new position? The spacecraft moment of inertia is 12,000 kg-m^2 about the axis of rotation. There is no significant external torque. What is the total momentum exchange during the maneuver?

5.4 A spacecraft, with a moment of inertia of 1300 kg-m^2 around the roll axis, is maneuvered using a pair of 2.2 N thrusters on a 2 m radius. What is the minimum time required to roll this spacecraft 90 deg if the maximum roll rate is 4 deg/s?

5.5 How frequently will a 20 N-m wheel require unloading if the unbalanced torque on the spacecraft, about the axis of interest, is 5.43×10^{-6} N-m? Assume that the wheel is allowed to saturate.

5.6 The spacecraft is at an attitude, specified by the inertial to body quaternion: $Qi = 0.75919795, -0.06182982, -0.64003044, 0.1007807$. A maneuver to a new attitude is commanded: $Qf = 0.14432141, 0.50015652, 0.59227353, 0.61500162$.

(a) What is the magnitude of the maneuver?

(b) What is the sun-in-body vector at the initial attitude? The inertial sun vector is $Si = 0.91569, 0.38498, 0.11524$.

(c) What is the sun-in-body vector at the final attitude?

(d) Describe the maneuver from Qi to Qf in physical terms. What do (b) and (c) indicate about the maneuver?

5.7 You are selecting the reaction wheels for a new spacecraft design. The spacecraft has the following parameters: S/C mass = 4000 kg; orbit altitude = 500 km; center of mass = 0.1, 0.3, 0 cm; I_{xx} = 3000 kg-m^2, I_{yy} = 3500 kg-m^2, I_{zz} = 4200 kg-m^2; fixed attitude: z-axis nadir pointing ± 5 deg.

(a) Assuming that gravity is the only environment torque, what is the momentum buildup per orbit as a result of external torque?

(b) The mission design has a reaction wheel desaturation every fourth orbit. How much reaction wheel momentum must be allocated to external torques?

(c) The mission also requires that the spacecraft be able to rotate 90 deg in 10 min. Given the following reaction wheel specifications, which one would you choose for the spacecraft and why?

Vendor	Torque, N-m	Momentum storage
A	0.08	5.69 N-m-s @ 2000 rpm
B	0.48	40.0 N-m-s @ 2000 rpm
C	0.18	27 N-m-s @ 4500 rpm

5.8 You have a spacecraft with thrusters that are rotated about the body z axis by +30 deg. The body axes are given by DXc, Yc, Zc, and $zb \equiv zc$.

(a) What is the direction cosine matrix that describes the rotation from the spacecraft body frame to the thruster frame?

(b) What is the quaternion that describes the same rotation?

(c) A sensor on the spacecraft measures the attitude error in the body frame. If the measured error is $eb = (0.01, 0.02, 0.01)$, what is the attitude error in the thruster frame?

(d) Show that the answer to (c) is the same whether the direction cosine matrix or the quaternion method is used.

(e) What is the advantage of using the quaternion method?

6
Power System

The power system supplies the life blood of the spacecraft. As long as the spacecraft has power, it can perform its mission. Almost all other failures can be worked around by ground operations, but a loss of power is a fatal heart attack for the spacecraft. In the early years of spaceflight, the power system was also the limiting factor in mission duration. Sputnik I consisted of a structure, a battery, a transmitter, and an antenna. The highly successful mission lasted exactly as long as the battery: 21 days. Explorer I, the first U.S. spacecraft, lasted longer but also ended abruptly with battery depletion. One-shot battery systems of the type Sputnik and Explorer I used are called *primary batteries* for reasons lost in space history. The short duration of time provided by primary batteries was clearly unsatisfactory for most missions. Solar panels were developed to convert the sun's energy into power, and solar panel–battery power systems became the backbone of unmanned spacecraft design from 1959 to the present.

As planetary missions outbound beyond Mars were considered, it became necessary to develop power sources independent of solar energy because of the great solar distances involved. (Recall that incident solar energy decreases as the square of the distance from the sun.) The development of radioisotope thermoelectric generators (RTGs) resulted. RTGs convert the energy released by the decay of a radioisotope into electrical power. The power levels of an RTG are moderate, but the life of an RTG is very long, depending only on the half-life of the radioisotope used.

Nuclear reactors were briefly considered as a power source, providing both high power and long life. The former Soviet Union flew nuclear reactors on a routine basis; the United States flew only one experimental system, the SNAP 10A, in 1965. The system was a mercury-Rankine-cycle and operated successfully at full power generating 500,000 W-h electrical. The limiting factor in the use of reactors and RTGs is the human and political consequences of nuclear safety.

Manned systems require large amounts of power for short durations. Fuel cells are the system of choice for Gemini, Apollo, and the Space Shuttle for this requirement. Fuel cells derive electrical power from the energy released in the reaction of oxygen and hydrogen to form water. One method is to release the chemicals through porous rods into an aqueous solution. The reaction takes place at the electrodes, ions migrate through the solution, and electrical power flows through external circuits attached to the porous rods. The operating regimes of these basic power sources are depicted in Fig. 6.1.

A typical unmanned solar panel–battery power system is shown in Fig. 6.2. When the spacecraft is in the sun, power is generated by photoelectric conversion of sunlight in the cells of the solar panel. The power is distributed directly to all of the loads on the spacecraft, and a portion is diverted to the battery for charging. The battery charger controls the rate at which the batteries are recharged. This process is more complicated than you might think and is discussed in more detail later.

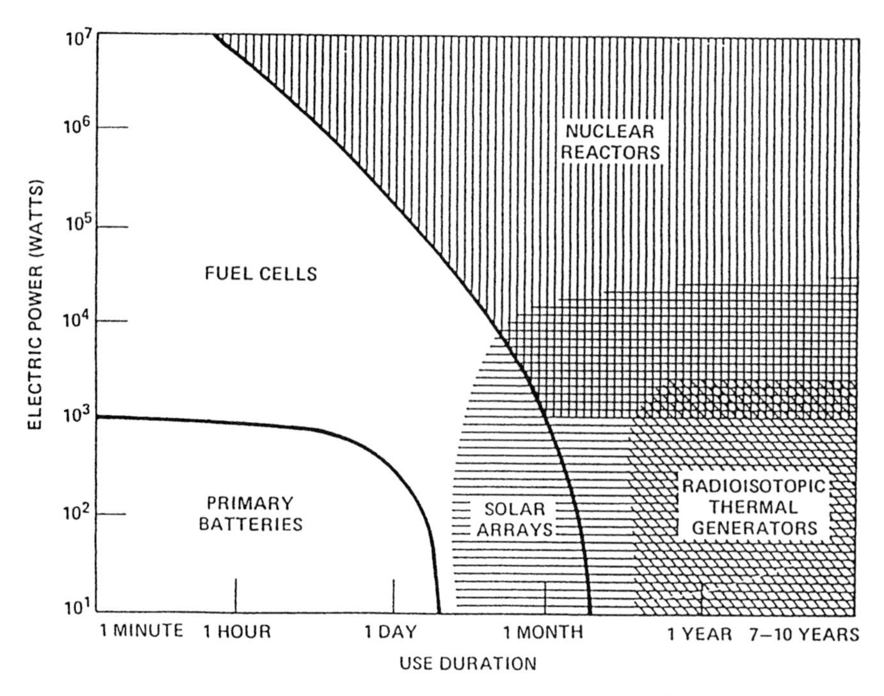

Fig. 6.1 Operating regimes of spacecraft power sources. (Copyright AIAA, reproduced with permission; Ref. 1.)

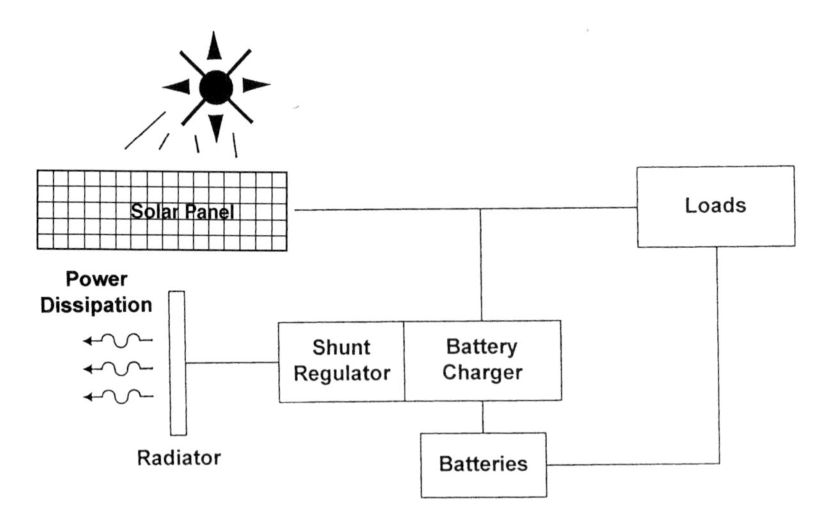

Fig. 6.2 Typical solar panel–battery power system (direct energy transfer architecture).

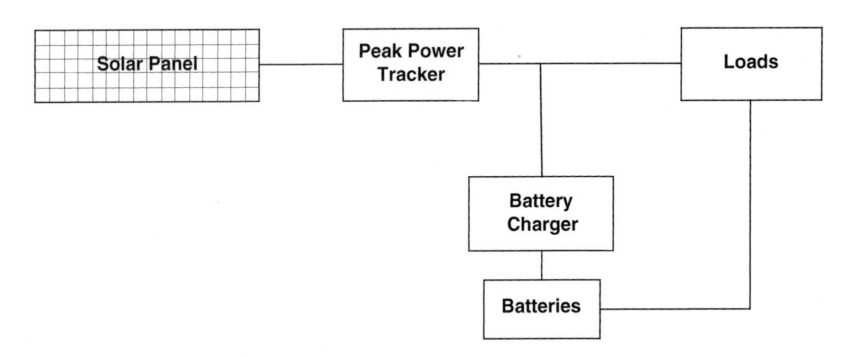

Fig. 6.3 Peak power tracker architecture.

The shunt regulator diverts any excess power from the bus to the shunt radiator to be radiated into space. When the spacecraft is in the shade, power from the solar panels drops to zero, and power is taken from the batteries to spacecraft loads. The shunt regulator and battery charging functions shown in Fig. 6.2 are normally packaged with other control functions in a power control unit, which also controls the voltage levels on the bus or buses, and turns power on and off to specific items of equipment [at the command of the command and data system (CDS)]. This general architecture is called a *direct energy transfer system* (DET) because power is sent directly from the solar panels to the loads without intervening control equipment. Direct energy transfer is the most common power system type.

Geosynchronous communication spacecraft sometimes use a variant of the DET system, segregated a portion of the solar array for battery charging and the remainder of the array for spacecraft loads. The characteristics of the geosynchronous day/night cycle make it an attractive alternative, as will be shown later.

Some spacecraft use a different architecture called a *peak power tracker* (PPT). The system schematic is similar to a direct energy transfer system with the addition of a peak power tracker box between the solar panels and the loads (see Fig. 6.3). The peak power tracker holds the solar panel output voltage on peak power point of the cells. The PPT system delivers the maximum power that the panels can provide at the expense of additional complexity, additional power loss from the panels to the loads, and potential failure modes in series with the main power bus. Peak power tracker architecture is used infrequently and usually on smaller spacecraft.

Subsequent sections will describe the steps in the preliminary design of a power system, which are as follows:

1) Select the power source, as just described.

2) Establish the power requirements, which are described in Section 6.1: a) determine power consumption for each mission mode; b) prepare an energy balance for each mode; c) set the requirements for the power source, usually solar arrays; and d) set the requirements for the battery system.

3) Size the solar arrays or RTGs; see Section 6.2.

4) Size the battery system; see Section 6.3.

5) Establish requirements for power distribution and control; see Section 6.4.

6) Prepare mass and power estimates for the power system; see Section 6.5.

7) Conduct trade studies in search of design improvements, and refine the power requirements.

6.1 Power System Requirements

In this section the functional requirements, power budgets, power margins, energy balance, and derived requirements for battery and solar panel are discussed.

6.1.1 Functional Requirements

Beyond the obvious requirement to generate and store power, the functions of a power system are as follows:

1) *Regulate power.* It would be ideal, from a power system standpoint, if none of the payload instruments or spacecraft equipment required voltage regulation more exact than the unregulated dc bus provides, normally about 24 to 33 V. This range is clearly adequate for most equipment and for all heaters, but probably not for the spacecraft computers and possibly not for the payload instruments. Usually a spacecraft requires a regulated bus as well as an unregulated bus. Some equipment may require ac, which would entail an inverter and an ac bus be added to the power system. (The Magellan power system supplied 50 ± 1.5 V ac at 2.4 kHz to the CDS.)

2) *Distribute power.* The task of distributing power entails running wiring from each power relay in the power control system to each piece of equipment to be powered. The result is a set of cable bundles routed to every corner of the spacecraft. To appreciate how difficult it is to do this job well, look at the cable mess behind your computer.

3) *Provide power margin.* It is essential to provide power margin over and beyond usage at worst anticipated conditions. This important topic is discussed in detail in Chapter 2, "System Engineering." Power margin is also necessary during the mission to accommodate unplanned situations or maneuvers. It is not acceptable to power down essential equipment to achieve in-flight margin. Most electronic failures occur at power turn on. The best practice is to turn it on and leave it on.

4) *Establish uninterruptable power.* There can be no physical means to remove power from the bus. In addition, it must be physically impossible to remove all batteries from the bus. A Mars-bound spacecraft was lost by inadvertent opening of a main power switch in flight. All power was lost, including power to the command receiver. There is no way to recover from this condition; an otherwise healthy spacecraft was lost. Inability to turn power off is an inconvenience during ground test, but it is an essential requirement.

5) *Run ground cable.* It is highly desirable to run ground cable rather than grounding through the structure because it is difficult to maintain low resistance throughout the structural joints. In either case it is necessary to maintain electrical continuity between structural elements and between thermal blankets.

6) *Keep memory alive.* Under all circumstances power must be supplied to the computer memories and data memories. A power loss to the memories would be difficult and dangerous, if not impossible, for the Mission Operations team to repair.

7) *Perform equipment switching.* All equipment on the spacecraft that can be switched on and off is switched by the power system using banks of relays. Switching should be done on the positive side with the negative side hard wired. When the CDS reads a command to shut down a given piece of equipment, a camera, for example, the command is sent to the power system, which opens the proper

relay to remove power from the camera. Note that there is a class of equipment which can never be turned off, for example, the power system and the memory just mentioned. This critical equipment is hard wired to power.

8) *Establish fuse protection.* The fuses for all spacecraft equipment are contained in the power system.

9) *Establish an emergency load shed.* If the bus voltage drops below a given number (usually 24 V), it is assumed that there is a short somewhere in the system. In this case all noncritical loads are shed (powered off) automatically, and an emergency message is sent to the Mission Operations team. The loads to be shed are preselected during the design phase.

10) *Establish power on reset (POR).* If power has been removed, as already described, and subsequently reset, an unambiguous message must be appended to the engineering data stream for the next download.

11) *Regulate pyrotechnic firing.* All of the pyrotechnic devices on the spacecraft—explosive nuts, squib valves, separation ordinance, and the like—are fired from the power system. The usual sequence of events is as follows: a) command to arm ordnance is issued by the CDS, and the power system closes relays making firing of the ordnance device possible; and b) command to fire ordnance is issued by the CDS, usually based on a time tick from the spacecraft clock, then the relay that sends power to the ordinance device is closed; the power firing circuits usually are redundant such that each pyro device receives firing current from two independent sources.

6.1.2 System Level Considerations

The characteristics of some representative flight power systems are shown in Table 6.1.

Table 6.1 is arranged in order of flight date with older systems on the left progressing to more modern systems on the right. There are three key observations to be made from Table 6.1:

1) Silicon solar cells were the work horses of the first three decades of the space age. They are now being replaced in flight systems by gallium arsenide (GaAs) cells, which offer greater energy conversion efficiency and higher temperature limits.

2) Similarly, nickel–cadmium (NiCd) batteries have been the batteries of choice and are now being replaced by nickel–hydrogen (NiH_2) cells, which offer improved cycle life and reduced mass. NiH_2 batteries were first used on communications spacecraft in geosynchronous orbit where battery cycling is at a minimum; they are now being used in low Earth orbit, which requires higher cycle life. There are two types on NiH_2 cells: individual pressure vessels (IPV) and common pressure vessels (CPV) with CPV the lightest and latest design.

3) Most spacecraft use direct current only. A few require ac and, as result, require an inverter.

All of the spacecraft in Table 6.1 are three-axis stabilized except Pioneer Venus, which was spin stabilized. Three-axis systems can point the panels directly at the sun continuously. Spin-stabilized spacecraft can not readily provide sun-pointed panels. The common design solution is to place the spin axis perpendicular to the sun and put the solar cells on a drum-shaped exterior (see Fig. 6.4). With a drum spinner the solar cells are in and out of the sun each revolution. With this design the

Table 6.1 Flight qualified power systems

Characteristic	Viking Orbiter	HEAO-B	Pioneer Venus	INTELSAT V	Magellan	Mars Observer	Clementine	MGS[a]
Launch year	1976	1977	1978	1980	1989	1992	1994	1996
System total load, W	674	565	—	1011	760	1094	278	300
Power system type	DET	PPT	DET	DET	DET	DET	—	DET
Solar-cell type	Si	Si	Si	Si	Si	Si	GaAs	GaAs/Si
Cell platform	Panel	Panel	Drum	Panel	Panel	Panel	Panel	Panel
Battery type	NiCd	NiCd	NiCd	NiCd	NiCd	NiCd	NiH_2 IPV[b]	NiH_2 CPV[c]
Inverter	Y	N	N	N	Y	N	N	N

[a]MGS = Mars Global Surveyor.
[b]IPV = independent pressure vessel (for each cell).
[c]CPV = common pressure vessel (for all cells).

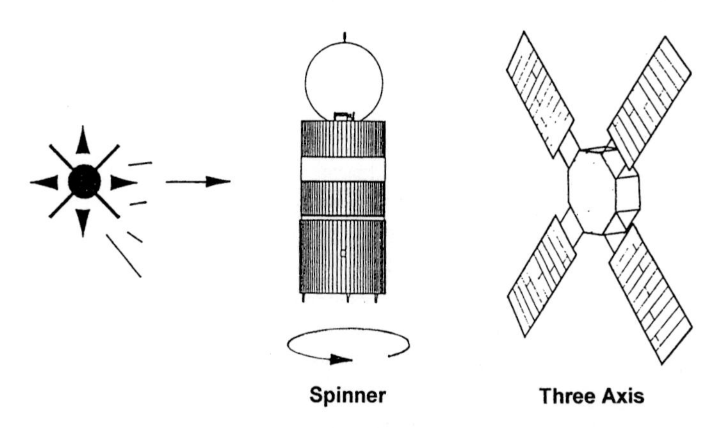

Fig. 6.4 Three-axis and spin-stabilized solar panels.

solar cells produce $1/\pi$ times, or about one-third, as much power as a sun-pointed panel of the same. High-power missions are difficult for a spinning spacecraft.

6.1.3 Spacecraft Total Power Consumption

The primary power system requirement is the power consumption of the spacecraft during each mission mode. In the early stages of a design, the total spacecraft power requirement is not available; however, it can be estimated from payload power, which is usually known. In this section we will consider estimating relationships from which early estimates of total spacecraft power can be made. These data are based on prior spacecraft designs and by their very nature do not take in account improvements in technology; however, they are useful in the early phases where there is little else to go on. Figure 6.5 shows a power-estimating relationship between payload power and total spacecraft power for 40 spacecraft of various missions.

The best statistical fit of the data in Fig. 6.5 is

$$P_t = 1.13P_{\text{pl}} + 122 \tag{6.1}$$

where

P_t = total spacecraft power to the load, including payload power, W
P_{pl} = payload power, W

6.1.3.1 Comsats. Figure 6.5 considers all spacecraft without regard to the mission. An improvement in prediction accuracy can be obtained by separating spacecraft into types by mission, specifically, communication, meteorological, planetary spacecraft. Figure 6.6 shows payload power vs total power for 10 communication spacecraft. A linear regression fit of the data from Fig. 6.6 yields the following relation for communication spacecraft:

$$P_t = 1.17P_{\text{pl}} + 56 \tag{6.2}$$

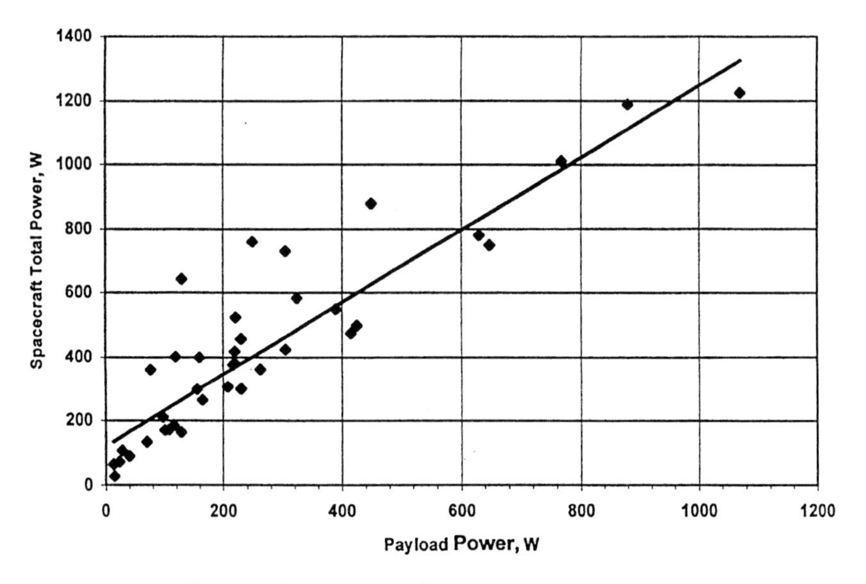

Fig. 6.5 Total spacecraft power vs payload power.

6.1.3.2 Metsats. Similarly, Fig. 6.7 shows the relation of payload to total power for five meteorology spacecraft. A regression fit of the data from Fig. 6.7 yields the following relation for total power and payload power for meteorology spacecraft:

$$P_t = 1.96 P_{\text{pl}} \tag{6.3}$$

Planetary spacecraft are more diverse in power system design because of the large variation in solar intensity encountered and because of the use of RTGs for outer

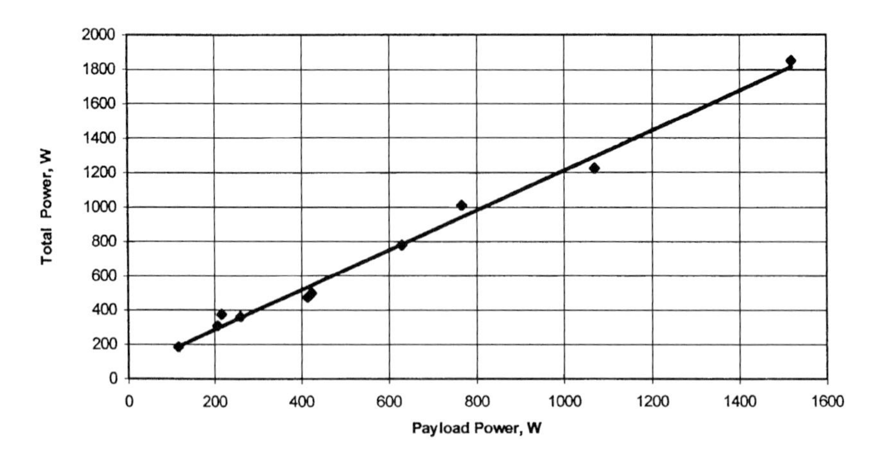

Fig. 6.6 Communication spacecraft total power vs payload power.

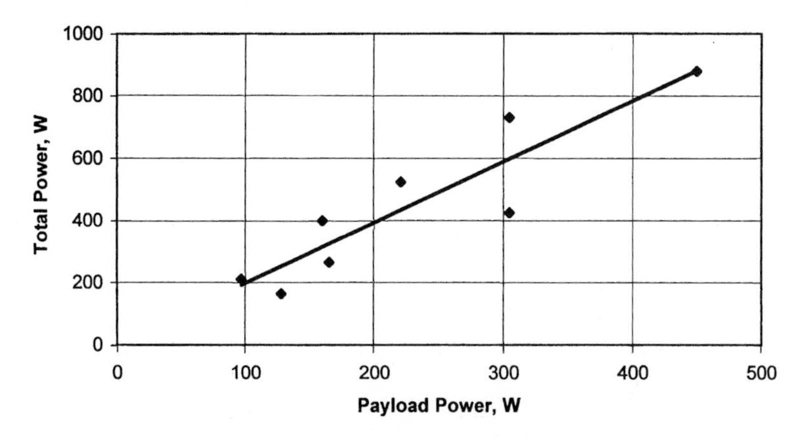

Fig. 6.7 Meteorology spacecraft total power vs payload power.

planet missions. Ignoring these difficulties, Fig. 6.8 relates spacecraft power to total power for three inner planetary spacecraft.

A regression fit of the data from Fig. 6.8 yields the following relation for total power and payload power for planetary spacecraft:

$$P_t = 332.93\ell n(P_{\text{pl}}) - 1047 \tag{6.4}$$

In summary, Table 6.2 shows the spacecraft power estimating relationships.

As a spacecraft design progresses, the early power estimates made with the relations in Table 6.2 will be replaced with data from the design process. The purpose of Table 6.2 is to provide a rational starting point for the design. For metsats with payload power less than 150 W, use the "All missions" estimating relationship from Table 6.2.

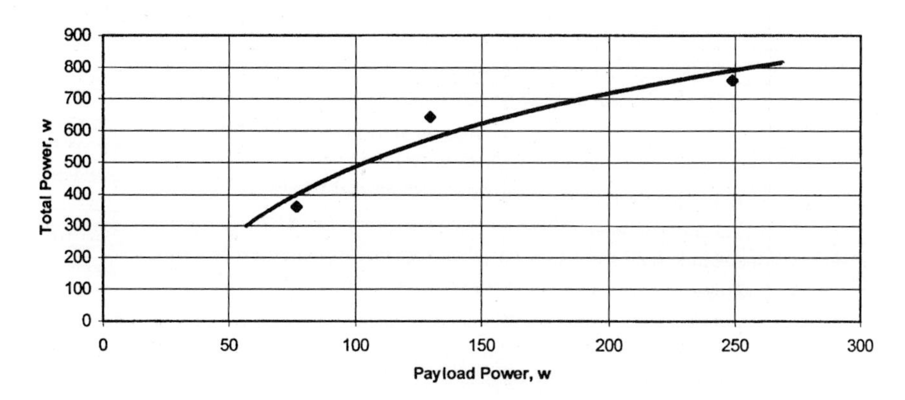

Fig. 6.8 Planetary spacecraft total power vs payload power.

Table 6.2 Total spacecraft power-estimating relationships

Spacecraft mission	Power-estimating relationship
Communications	$P_t = 1.17P_{pl} + 56$
Meteorology	$P_t = 1.96P_{pl}$
Planetary	$P_t = 332.93\ln(P_{pl}) - 1047$
All missions	$P_t = 1.13P_{pl} + 122$

6.1.4 Subsystem Power Consumption

Once an estimate has been made for the total spacecraft power requirement, the total power available to the subsystems can be obtained:

$$Psubsystems = P_t - P_{pl} \tag{6.5}$$

Subsystem total power can be allocated to each subsystem on a historical percentage basis as shown in Table 6.3.

In Table 6.3, cable losses are included in the power system allotment. For communication satellites the engineering and command communication equipment is included in the command and data (CDS) system. (In the communication satellite industry the command and data system is generally called the TT&H subsystem.) Power allocations from Table 6.3 provide a starting point for the design process. These allocations will be refined as the design progresses.

6.1.5 Mission Modes

Subsystem power allocations and reserves must be made for each mission mode. Every spacecraft has at least four mission modes: 1) launch, 2) postseparation, 3) daytime, and 4) nighttime. Daytime and nighttime are also called full sun and eclipse phases. The allocations from Table 6.3 are full sun power allocations. During nighttime, the thermal control power must be increased substantially to account for additional heater power; a factor of two is a good initial estimate. The remaining subsystems probably remain at the daytime levels. The daytime allocations must account for battery charging power, not shown in Table 6.3. Battery charging power

Table 6.3 Subsystem power allocation guide

	% of subsystem total			
Subsystem	Comsats	Metsats	Planetary	Other
Thermal control	30	48	28	33
Attitude control	28	19	20	11
Power	16	5	10	2
CDS	19	13	17	15
Communications	0	15	23	30
Propulsion	7	0	1	4
Mechanisms	0	0	1	5

Table 6.4 Magellan power summary table[a]

Characteristic	Cruise, near Earth	Venus mapping, full sun	Venus playback, eclipse
Time in mode, min	—	37.20	19.50
Attitude control, W	138.80	159.80	142.80
Command and data, W	47.96	63.41	65.23
Power system, W	42.70	50.18	53.68
Propulsion, W	0.76	7.96	5.74
Thermal control	313.00	126.58	239.75
Communication, W	43.80	61.80	110.90
Payload, W	0.00	248.78	18.54
Total power to loads, W	587.02	718.51	636.64
Total wire loss, W	71.25	40.99	34.23
Total power required, W	658.27	759.50	670.87

[a]Courtesy of Lockheed Martin; Ref. 2.

can be added as a separate line item or added to the power subsystem allocation; the former is better practice. All spacecraft have a power mode associated with the launch vehicle ride. Typically the in-shroud mode is a low power level, which is provided, all or in part, by the launch vehicle. After separation from the launch vehicle, the spacecraft has a number of chores to perform in rapid order before the solar panels can be deployed and pointed to the sun. The battery energy must provide all of the power required for this postseparation period, which is not necessarily short.

As the design progresses, numerous other modes will be added, downlink-on and downlink-off, for example. In the late phases of a program, the power modes become so numerous that an analytical model is usually built, which will determine the power consumption in complex spacecraft maneuvers with individual equipment items switched on or off.

Table 6.4 is the power summary table for Magellan showing three of the mission modes.

6.1.6 Power Margin

No power budget is complete without margins. In the earliest phases of the spacecraft design, the loads will be not much more than a power allocation, and the power margin must be high. As the design matures, the power requirements become more refined, and the margins can be reduced. AIAA[3] conducted a review of historical data from prior NASA and U.S. Department of Defense programs to establish the industrial guidelines for power margin; these are discussed in detail in Chapter 2. These history-based data indicate that margins as high as 90% of estimated power might be necessary for a new design spacecraft in the conceptual stages; lower percentages are possible for mature designs with better definition.

Adequate power margin in the early design stages is particularly important to the program because an increase in power requirement means an increase in

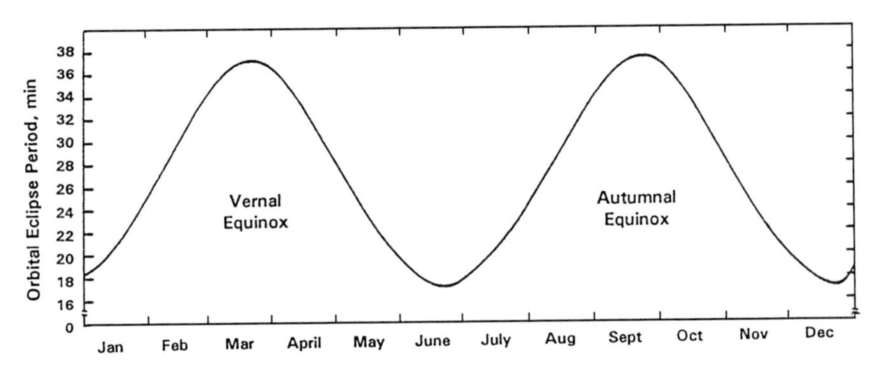

Fig. 6.9 Eclipse period for a low Earth orbit. (Reproduced courtesy of NASA/JPL/Caltech; Ref. 4.)

solar-panel size, and the solar panels are normally long lead items. Their size and power output are set early in the program. If it later becomes necessary to increase panel size, the schedule impact can be severe. A little to much panel is much better for the program than a little too little.

At liftoff there should be a power margin of at least 10% in mission critical modes to accommodate unforeseen changes in planned usage and resultant changes in planned power profile.

6.1.7 Eclipse Period

The maximum eclipse period T_n is the design point for both the solar array and the battery system. Figure 6.9 shows the eclipse period for a 425-km, circular, polar, Earth orbit with a period of 93 min. The eclipse period over a semiannual cycle varies from a maximum of 37 min to a minimum of 18 min.

In geosynchronous orbit there are long periods without an eclipse, as shown in Fig. 6.10; however, during the eclipse season the eclipse period reaches a maximum of 1.2 h.

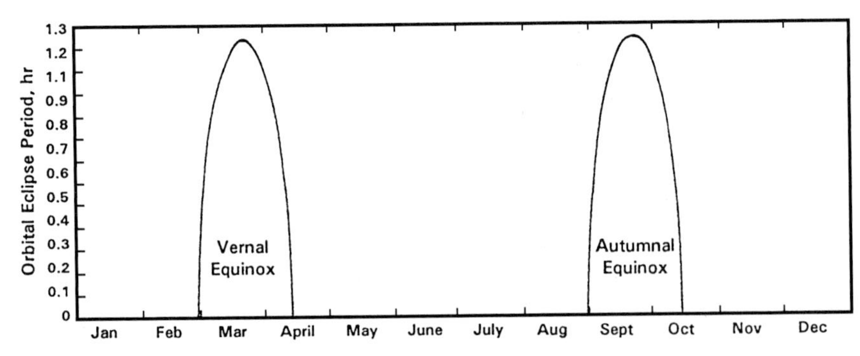

Fig. 6.10 Eclipse period in geosynchronous orbit. (Reproduced courtesy of NASA/JPL/Caltech; Ref. 4.)

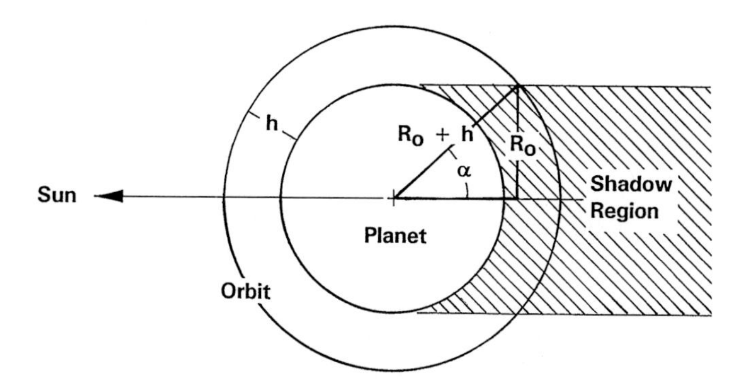

Fig. 6.11 Maximum eclipse period, circular orbit.

Figure 6.11 shows the maximum eclipse period condition, which occurs when the sun is in the orbit plane. From Fig. 6.11

$$\alpha = \sin^{-1}\left(\frac{R_0}{R_0 + h}\right) = \sin^{-1}\left(\frac{1}{1 + h/R_0}\right) \tag{6.6}$$

where

h = orbit altitude in the same units as orbit radius
R_0 = radius of the central body, usually Earth

For a circular orbit

$$\frac{T_n}{P} = \frac{2\alpha}{360} \tag{6.7}$$

where

T_n = maximum eclipse period for a circular orbit, in the same units as orbit period
P = orbit period
α = shadow region half-angle, deg (see Fig. 6.11)

For a circular orbit the maximum eclipse period is

$$T_n = \frac{P}{180}\sin^{-1}\left(\frac{1}{1 + h/R_0}\right) \tag{6.8}$$

If α is in radians, modify Eq. (6.7) by replacing 360 with 2π and Eq. (6.8) by replacing 180 with π.

Figure 6.11 shows a special case where the orbit plane is parallel to the sun vector. Not all orbit inclinations will produce an eclipse. Figure 6.12 is a more general case with a spacecraft in an orbit inclined to the sun vector. The angle β is the angle between the sun vector and the orbit plane. The maximum β angle beyond which the orbit will have eclipses is shown.

By the same logic leading to Eq. (6.8), the maximum β angle for an eclipse-free orbit is

$$\beta_{\max} = \sin^{-1}\left(\frac{1}{1 + h/R_0}\right) \tag{6.9}$$

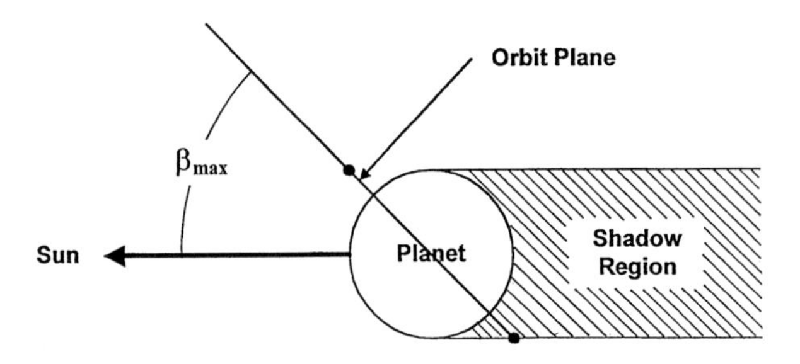

Fig. 6.12 Maximum beta angle for an eclipse-free orbit.

At sun angles less than $\pm\beta_{\max}$, there will be an eclipse period. For the general case of any β angle between $\pm\beta_{\max}$, the eclipse time for a circular orbit is

$$T_n = \frac{P}{180}\left[\cos^{-1}\left(\frac{\cos\beta_{\max}}{\cos\beta}\right)\right] \tag{6.10}$$

Note that β is measured with respect to the ecliptic plane not the equatorial plane. Equation (6.10) is for circular orbits, but it can be used to approximate eclipse time for elliptical orbits of low eccentricity. If $\beta_{\max}$ is in radians, modify Eq. (6.10) by replacing 180 with π.

6.1.8 Solar-Panel Power Requirement

The solar-panel power requirement and the battery capacity requirement can be determined from the energy balance for the spacecraft. As shown in Fig. 6.13, all electrical energy consumed by the spacecraft, day or night, must be collected by the solar panels during the daylight period of the orbit.

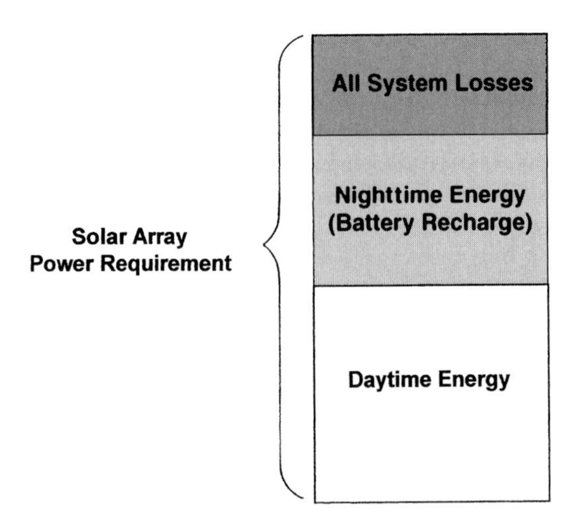

Fig. 6.13 Solar-array power requirement.

The energy required from the solar panels consists of four parts: 1) energy required for the daytime loads; 2) energy required to charge the battery for the nighttime loads; 3) all of the energy losses involved in the system including a) power losses of solar panel to the daytime loads, b) power losses of solar panel to the battery, c) battery charging losses, d) power losses of battery to the nighttime loads; and 4) all energy reserves, day and night.

6.1.8.1 Without losses. The solar-panel power requirement can be derived from an energy balance for one orbit day–night cycle. The power consumed can temporarily exceed the solar-panel output, but energy must balance. Without losses the total system energy balance is the sum of the energy to charge the battery and the energy to daytime loads:

$$E_{\mathrm{sa}} = E_n + E_d \tag{6.11}$$

where

E_{sa} = required solar-array energy
E_n = energy consumed at night including reserves
E_d = energy consumed in daylight including reserves

The individual terms of the energy balance equation can be expressed as the product of an average power and a time increment, or

$$P_{\mathrm{sa}} T_d = P_n T_n + P_d T_d \tag{6.12}$$

where

P_{sa} = average output power from the solar array over daylight period, W
T_d = spacecraft daylight, the period in sunlight, h
T_n = spacecraft night, the period without solar energy, h
P_n = average power consumed by spacecraft loads during the spacecraft night, W
P_d = average power consumed during the spacecraft loads during spacecraft daylight, W

6.1.8.2 With losses. The solar array must supply the energy for all losses as well as the loads. It is convenient and relatively general to group the losses into three fractions: X_{a-l} = power transfer efficiency, solar array to daytime loads; X_{a-b} = power transfer efficiency, solar array to battery, including battery charging efficiency; X_{b-l} = power transfer efficiency, battery to nighttime loads.

(Note that if the power loss = 10% the power transfer efficiency = 0.9.)

With losses Eq. (6.12) becomes

$$P_{\mathrm{sa}} = \frac{P_n T_n}{X_{a-b} X_{b-l} T_d} + \frac{P_d}{X_{a-l}} \tag{6.13}$$

P_{sa} is the minimum required power from the solar array at the end of life, and P_d and P_n are the maximum, time-averaged electrical loads from the power requirements table (including margin). The sunlight and eclipse time durations (T_d and T_n)

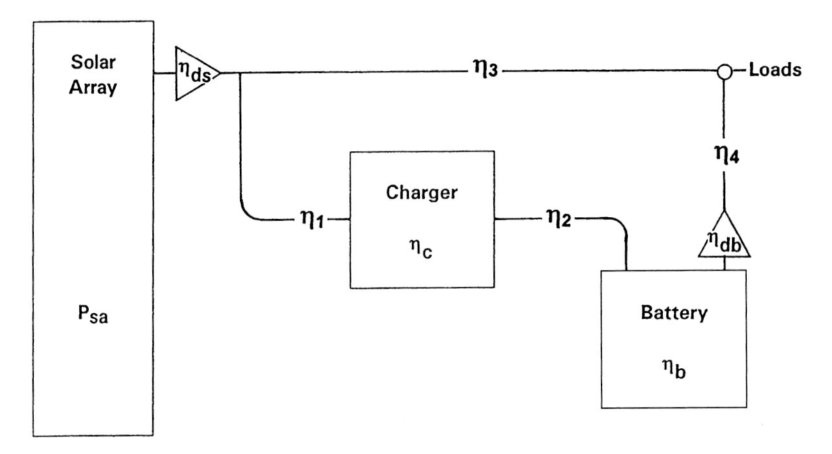

Fig. 6.14 Power losses in a typical direct energy transfer system.

vary from mission to mission and from day to day during a given mission (see Section 2.5). Maximum nighttime period is used for computation of the solar-array power requirement.

The power losses in a typical direct energy transfer system are shown in Fig. 6.14. For the system shown in Fig. 6.14, the power transfer efficiencies would be

$$X_{a-l} = \eta_{ds}\eta_3 \tag{6.14}$$

$$X_{a-b} = \eta_{ds}\eta_1\eta_c\eta_2\eta_b \tag{6.15}$$

$$X_{b-l} = \eta_{db}\eta_4 \tag{6.16}$$

where

$\eta_1, \eta_2, \eta_3, \eta_4$ = power transfer efficiencies in the power cables as shown
η_c = charger efficiency
η_b = battery charging efficiency.

Equations (6.14–6.16) are not general; the constituents of X_{a-l}, X_{a-b}, and X_{b-l} depend entirely on the detailed schematic of the system under study and will vary widely.

6.1.9 Battery Requirements

The traditional definition of battery capacity is the current that can be supplied by the battery multiplied by the time from fully charged to depletion in ampere-hours. Battery ampere-hour capacity and energy capacity are proportional. It is convenient to define battery energy capacity from the energy balance and convert energy capacity to ampere-hours. The battery energy capacity is the average night-time power multiplied by the maximum eclipse time divided by the transmission efficiency battery to loads

$$E_b = \frac{P_n T_n}{X_{b-l}} \tag{6.17}$$

where

X_{b-l} = power transmission efficiency from battery to nighttime loads
E_b = energy supplied by the battery in a single nighttime cycle, measured at the battery terminals

For the direct energy transfer system shown in Fig. 6.14, X_{b-l} is the product of battery diode loss, η_{db} times the line loss battery to loads η_4, and

$$E_b = \frac{P_n T_n}{\eta_{\text{db}} \eta_4} \tag{6.18}$$

6.1.9.1 Depth of discharge. Note that E_b, as just defined, is the energy supplied by the battery in a normal day–night cycle. It is not the total energy capacity of the battery because the battery is never allowed to discharge completely. The percentage of energy removed from the battery in a discharge cycle is called the *depth of discharge* (DOD). By definition DOD is

$$DOD = \frac{E_b}{E_{\text{Bcap}}} \tag{6.19}$$

where E_{Bcap} is the total energy capacity of the battery or battery system, W-h.

Note that depth of discharge is normally discussed in percentage but must be used in equations as a fraction.

6.1.9.2 Cycle life. Depth of discharge and temperature determine the cycle life of a battery. Cycle life characteristics are peculiar to the cell type and are discussed in Section 6.4 for NiH_2 and NiCd batteries. Conversely, the required cycle life of a battery determines the maximum allowable DOD.

6.1.9.3 Battery capacity requirement. From Eqs. (6.17) and (6.19) the minimum energy capacity requirement for a battery system is

$$E_{\text{BCap}} = \frac{P_n T_n}{X_{b-l} DOD} \tag{6.20}$$

Because $P = I^* V$ for direct current systems, the ampere-hour capacity of a battery system is

$$C = \frac{E_{\text{BCap}}}{V_d} = \frac{P_n T_n}{X_{b-l} DOD \; V_d} \tag{6.21}$$

where

C = battery capacity, A-h
V_d = average battery discharge voltage, usually about 28 V
P_n = nighttime power required (including margin), W
T_n = eclipse period, h

And for a direct energy transfer system the battery capacity requirement is

$$C = \frac{P_n T_n}{V_d DOD \eta_{\text{db}} \eta_4} \tag{6.22}$$

Battery charging is specified in terms of charging rate

$$C_R = \frac{C}{T_c} \tag{6.23}$$

where C_R is the charging rate and T_c the time of charging, h. A charging rate of $C/15$ means a charge of one-fifteenth of full capacity in 1 h. Charging efficiency varies as a function of charge rate and temperature; low charging rates produce low losses.

Battery requirements and design are described in more detail in Section 6.4.

6.2 Solar Arrays

Solar arrays are the power source of choice for the vast majority of spacecraft. A solar array consists of solar cells interconnected in strings, a diode set, and a substrate. A spin-stabilized spacecraft usually has body-mounted solar arrays. A three-axis-stabilized spacecraft normally has deployed and articulated panels. Table 6.5 shows the performance of some flight solar arrays.

The weights in Table 6.5 include the substrate; articulating mechanism weights are not included. The present level of solar-cell efficiency is about 15% at air mass zero, and for an axis-stabilized vehicle a specific mass of about 100 W/m^2 can be achieved. After accounting for all losses and degradation, an articulated

Table 6.5 Solar-array performance[a]

Spacecraft	Launch year	Power EOL, W	Mass, kg/m^2	Watts/ kg	Watts/ m^2	Body-mounted: no. cells/ panel-mounted: area, m^2	Cell type	Mount type
Vela	1963	90	4.19	5.8	24.3	13,236	Si	Body
INTELSAT III	1968	167	3.36	11.4	38.3	10,720	Si	Body
INTELSAT IV-A	1971	491	3.65	6.6	24.1	45,012	Si	Body
INTELSAT VI	1989	2204	—	—	—	39,658	Si	Body
OGO A	1964	710	6.46	14.9	96.2	7.4	Si	Panel
Mariner 3, 4	1964	680	4.9	21.2	104.4	6.5	Si	Panel, 4
OGO D	1967	745	6.43	15.7	100.9	7.4	Si	Panel
Nimbus	1969	470	7.98	13.2	105.4	4.5	Si	Panel, 2
NOAA-1	1970	393	6.64	14.1	93.6	4.2	Si	Panel, 2
INTELSAT V	1980	1354	3.59	20.8	74.7	18.1	Si	Panel
Magellan	1989	1348[b]	4.39	22.1	108.1	12.6	Si	Panel, 2
Hubble ST	1989	4000	2.45	53.6	131.1	30.5	Si	Flex, 2
TDRSS 6	1993	3094[c]	4.55	23.0	104.6	29.6	Si	Panel, 2
Clementine	1994	360	—	—	156.5	—	GaAs	Panel, 2
INTELSAT VIII	1995	4800	—	—	98.8	48.6	Si	Panel, 2
Inmarsat-3	1995	2800	—	—	91.9	30.47	Si	Panel, 2
Deep Space 1[d]	1998	2600	8.10	44.8	363	7.15	2 types	Panel, 4

[a]Data, in part, courtesy of U.S. Air Force; Ref. 12, pp. 4–7.
[b]EOL at Venus. [c]BOL. [d]Fresnel concentrator arrays.

panel can deliver a specific power of about 100 W/m^2. Body-mounted solar-cell systems have specific mass and power considerably less than articulated panels, as would be expected. Three-axis spacecraft deliver higher performance because the panels are articulated and sun pointed. However, articulated panels are not free; they require 1) an additional attitude control function, 2) sun sensors to position the panels, 3) deployment and rotation mechanisms, and 4) a means of transferring power through the mechanism, either slip rings or cable wrap with attendant failure modes. The high specific power of Deep Space 1 panels is because of an advanced Fresnel lens concentrator design.

A solar array has three primary constituents: 1) solar cells connected electrically in a series parallel arrangement, 2) a substrate, which is the structural foundation for the array, and can be rigid or flexible, rigid substrate being the most common; and 3) a deployment mechanism. These parts will be discussed in subsequent sections.

6.2.1 Solar Cells

Solar cells convert the incident solar energy into electrical energy. There are two types of cells with a flight history: 1) silicon cells, which have by far the most flight history, and 2) gallium arsenide cells, which are a more recent development. A solar cell is a sandwich of n-type (electron rich) and p-type (electron poor) semiconductors. For silicon cells phosphorous-doped silicon forms the n layer and boron-doped silicon forms the p layer. When photons are absorbed in the vicinity of the p-n junction, electron-hole pairs are formed, and an electromotive force is established across the junction (see Fig. 6.15). Current will flow through an external circuit connected across the junction. Bell Telephone Laboratories was the first to demonstrate this conversion in 1954. There are two ways to make a solar cell: one is with a thin layer of n-type silicon on a layer of p-type silicon; the other method is just the reverse with p-on-n. Both types were made in the early

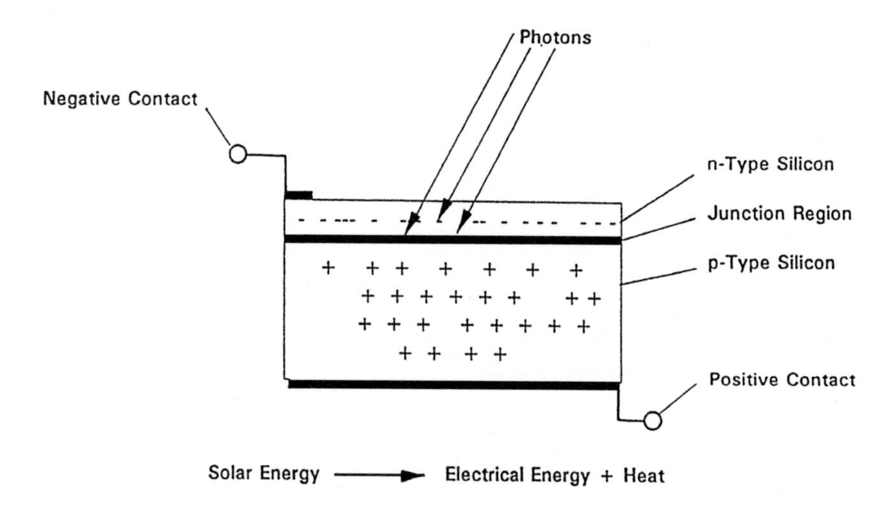

Fig. 6.15 An n-on-p solar cell.

experiments by Bell Laboratories. Because p-on-n gave slightly higher output, all of our early spacecraft used p-on-n. In 1960 it was found that n-on-p gave substantially better radiation resistance. All U.S. spacecraft since 1960 have used n-on-p; Russian spacecraft have used n-on-p since the beginning.

Silicon solar cells are sliced from a crucible-grown, p-type, silicon crystal. The crystals are cut into slices about 0.5 cm thick. (From a cell performance view-point they need be only $\sim$0.002 cm thick.[5]) They are then cut into cells: common sizes are 2×2 cm, 2×4 cm, and 2×6 cm; earlier cells were 1×2 cm. The n-type impurity is infused into the cell at high temperature, forming a junction less than 1 μm thick. The impurity is removed from all surfaces except the top or light sensitive surface. Contacts are applied to the n and p surfaces. Normally the front surface has grid lines to reduce shadowing, and the back surface has full contact. An antireflective coating is placed on the light sensitive area. The back surface may be coated to enhance power and reduce temperature. A cover slide, usually fused silica, is placed over the solar cell; the cover slide serves several purposes:

1) The upper surface is systematically roughened and coated to reduce energy loss caused by reflection. (The antireflective coating gives solar arrays their characteristic blue color.)

2) The lower surface is coated to reflect UV, which is not converted to electrical energy by the solar cell. In addition, UV degrades the adhesives and coatings in the cell.

3) It protects the cell from radiation damage.

4) It protects the cell from physical damage and dust.

6.2.1.1 I-V curve. The voltage-current characteristics of an illuminated solar cell are shown in Fig. 6.16. The *I-V* curve expresses the fact that, at a given voltage, a solar cell can deliver only one current. There are three significant points

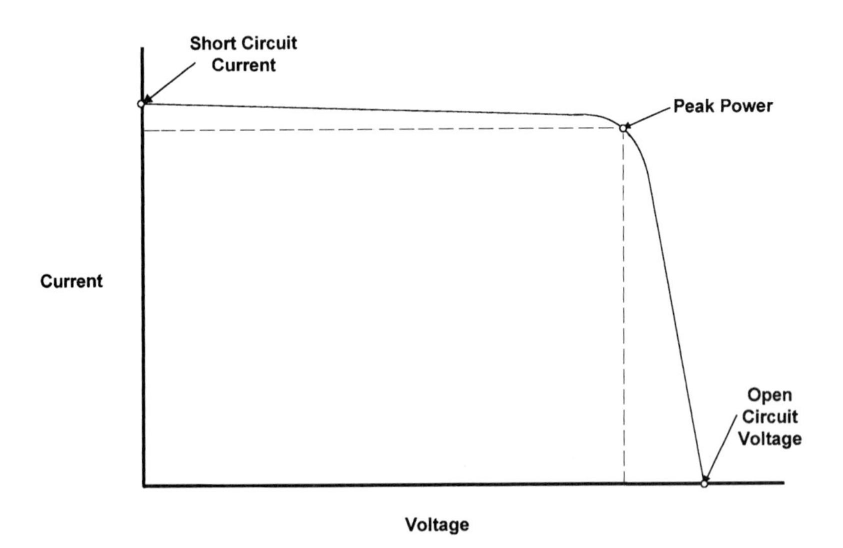

Fig. 6.16 Solar cell *I-V* curve.

on the curve: 1) the maximum power point, 2) the open-circuit voltage, and 3) the short-circuit current. Starting at zero voltage, a cell delivers a current, characteristic of the cell, called the short-circuit current. As voltage is increased, current drops slowly until the peak power point is reached. As voltage is further increased, current drops dramatically to zero at the open-circuit voltage. The power delivered by the cell is the product I^*V. The maximum power output of a cell is the maximum I^*V product. It is also the largest rectangle that can be placed under the I-V curve as shown in Fig. 6.16.

For a typical GaAs solar cell the peak power point is about 25 mW/cm^2 at 0.9 V. The open-circuit voltage is about 1 V, and the short-circuit current is about 0.30 mA/cm^2. For a typical silicon cell the peak power point is about 20 mW/cm^2 at 0.5 V.

6.2.1.2 Shaded cells. Cells that are not illuminated become open circuits; therefore, if a shadow falls across a string the entire string stops generating power. Blocking diodes are placed between parallel strings to prevent current from circulating uselessly through shaded cells. Preventing shadowing of the solar array is a constraint on the general arrangement of the spacecraft. For spinning spacecraft, strings must be arranged to pass into shadow as a unit.

Cell efficiency is defined as the peak power output of the cell divided by the incident solar power, at a given set of conditions

$$\eta_c = \frac{P_{\text{out}}}{P_{\text{inc}}} \tag{6.24}$$

The direct solar flux incident on a spacecraft in low Earth orbit is 1371 ± 5 W/m^2 on a surface normal to the sun. (This value was recommended for spacecraft design by NASA in Ref. 6.) The incident power is inversely proportional to the square of the distance from the sun; therefore, the solar flux near Earth varies from about 1310 to 1399 W/m^2 as the Earth moves in an elliptical orbit around the sun. Table 6.6 shows the incident solar power available at the planets (mean planetary radius).

Table 6.6 shows why solar panels are not practical for use near the outer planets beyond Mars. A 100% efficient 2 × 2 cell would generate 548 mW normal to the

Table 6.6 Incident solar power at the planets

Planet	Solar power, W/m^2
Mercury	9150
Venus	2620
Earth	1371
Mars	590
Jupiter	50.6
Saturn	15.1
Uranus	3.7
Neptune	1.5
Pluto	0.9

sun, near Earth. The early Bell Telephone Laboratories cells were 5% efficient. By 1972 multiple improvements in cell construction had raised cell efficiency to 10%, and modern black silicon cells had reached 15.5%[2] or 82 mW per 2 × 2 cell. Gallium arsenide cells have reached 18% efficiency. Efficiencies are normally measured under air mass zero (AM0) sunlight at 25°C.

The power output of solar cells is at its peak when the solar radiation is normal to the cell (incidence angle = 0). At incidence angles other than 0 deg, the power is reduced proportionally to the cosine except at low angles where the increase in reflection causes power loss greater than the cosine would indicate. Solar-panel pointing need not be better than ±5 deg; at 5 deg the loss is less than 0.5%.

6.2.2 Temperature Effect

The approximately 85% of the solar energy that is not converted to electrical power is either reflected or converted into heat. The heat in turn is reradiated into space; the net effect is to raise the solar-panel temperature. The power output and voltage of solar cells drop with increasing temperature, as shown in Fig. 6.17.

Peak power and efficiency of solar cells are measured at a standard temperature, usually 25°C. The power loss is a linear function of temperature for silicon cells and is approximately 0.5%/°C; thus,

$$P_2 = P_1[1 - 0.005(T_2 - T_1)] \tag{6.25}$$

where P_1 and P_2 are the power levels for a cell (or a solar array) at T_1 and T_2, respectively.

Solar cells are designed to reflect solar energy in the frequencies that do not convert to power, and the solar array must be designed to reject as much heat as possible. Body-mounted cells on a spinning spacecraft operate at much lower temperatures than articulated solar panels on three-axis spacecraft. The Magellan solar panels, which were designed to withstand the solar intensity at Venus, used optical solar reflectors (OSRs are small thin mirrors) to reject heat. The entire back surface of the panel was covered with OSRs as was 35% of the front surface.

Gallium arsenide cells perform better than silicon cells at high temperature, as shown in Fig. 6.18.

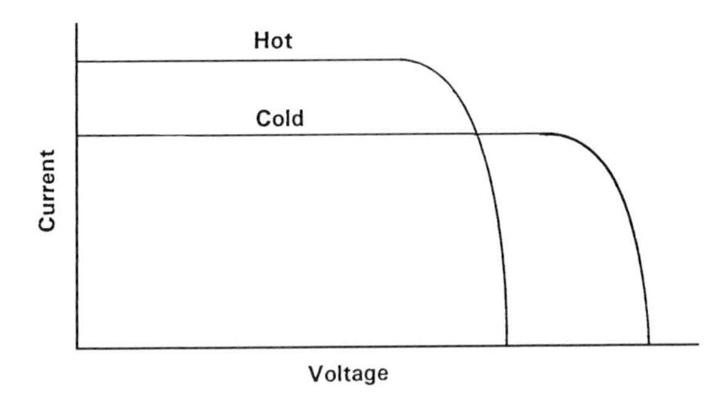

Fig. 6.17 Effect of cell temperature on power.

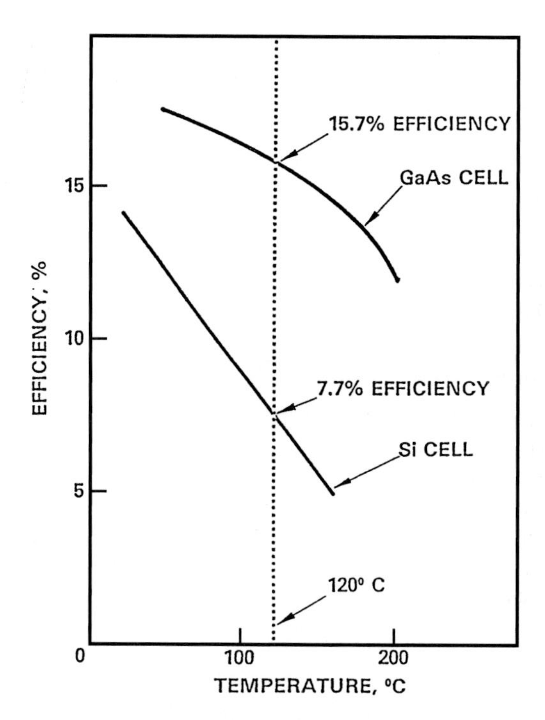

Fig. 6.18 Comparison of temperature performance of silicon and gallium arsenide cells. (Courtesy of The Boeing Company and AIAA; Ref. 7.)

The most severe temperature environment for a solar array is the temperature cycle as a spacecraft flies through a planet's shadow. Figure 6.19 shows the temperature cycling of the Magellan solar panel as it passes through the shadow of Venus; the orbit period is 3.2 h.

The design temperature limits for the Magellan panels were −120 to 115°C. The electrical interconnects between the solar cells require particular design attention to withstand this kind of cycling.

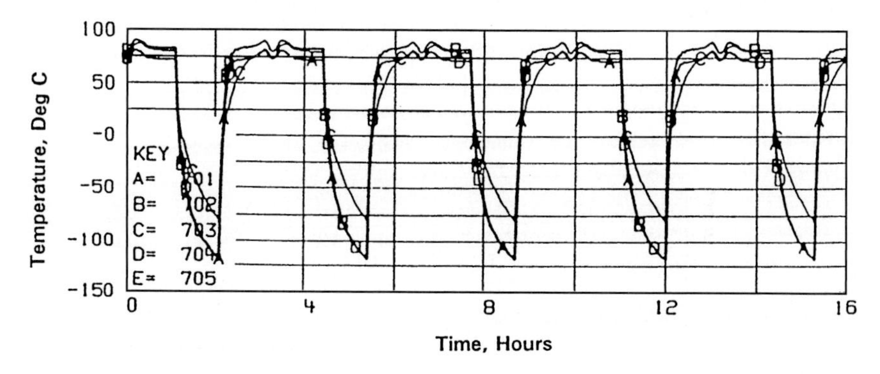

Fig. 6.19 Magellan solar-array eclipse temperature cycle. (Reproduced with permission of Lockheed Martin; Ref. 2.)

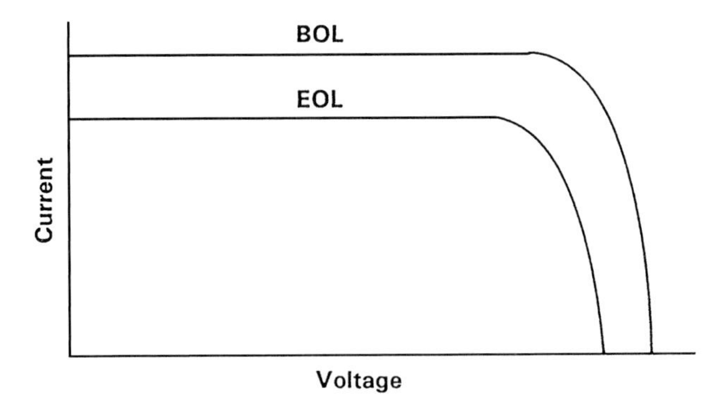

Fig. 6.20 Effect of radiation on solar cells.

6.2.3 Radiation Degradation

Solar-cell power, short-circuit current, and open-circuit voltage are degraded by radiation, as shown in Fig. 6.20.

The power system is designed such that the end-of-life (EOL) power is adequate for the mission. Beginning-of-life (BOL) power is set by the estimate of radiation damage over the life of the spacecraft.

Radiation damage to solar cells is caused by high-energy protons from solar flares and from trapped electrons in the Van Allen belt. Figure 6.21 shows the fluences of these two particles; the radiation has been normalized to equivalent 1 Mev fluence, electrons/cm^2/year. The fluence is higher in orbits with 0-deg inclinations and altitudes above 1000 km; electron radiation is dominant at altitudes above 15,000 km. Proton radiation is negligible at geosynchronous altitudes.

Using the fluences from Fig. 6.21, Nagler[4] calculated the power losses shown in Table 6.7.

Agrawal[5] lists power loss of approximately 25% for spacecraft in geosynchronous orbit for seven years and 30% for 10 years. Figure 6.22 shows the effect of radiation on the peak power for three types of solar cells.[5]

Radiation damage effects for planetary spacecraft in deep space are caused primarily by the proton flux from solar flares. Table 6.8 shows the estimated degradation of the Viking Orbiter solar panels over an 18-month cruise to Mars.[8]

The Magellan design power loss factors for a three-year Venus mission are shown in Table 6.9.

Both spacecraft conducted successful multiyear missions at the respective planets. Gallium arsenide solar cells have slightly different degradation results as shown in Table 6.10.

The loss of power under radiation is caused in part by darkening of the cover glass and adhesive; these losses amount to 4–10% during the first year and very little thereafter.[5] The cell damage rate decreases with time also, with damage in the first two years about the same as that in the next five years. The degradation of body-mounted arrays is slower than deployed arrays because body-mounted cells are better shielded on the back side.

Table 6.7 Radiation degradation[a]

Orbit altitude, km	Fluence, 1 Mev e/cm^{2-yr}	Orbit inclination, deg	Power loss, 1 yr, %	Power loss, 3 yr, %
556	1.89×10^{10}	0	Negligible	Negligible
883	9.78×10^{12}	0	2	4
1,480	7.38×10^{14}	0	26	33
33,300	2.03×10^{13}	0	4	9
556	6.11×10^{12}	90	1	3
833	2.55×10^{13}	90	4	10
1,480	3.08×10^{14}	90	19	27
33,300	4.01×10^{12}	90	1	2

[a]Courtesy of NASA/JPL/Caltech; Ref. 4, p. 38.

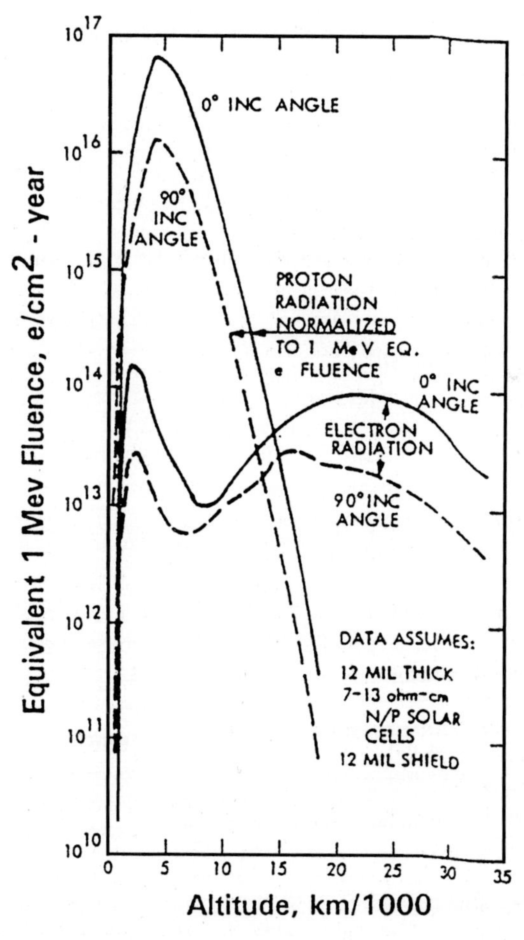

Fig. 6.21 Trapped radiation fluence. (Reproduced courtesy of NASA/JPL/Caltech; Ref. 4, p. 37.)

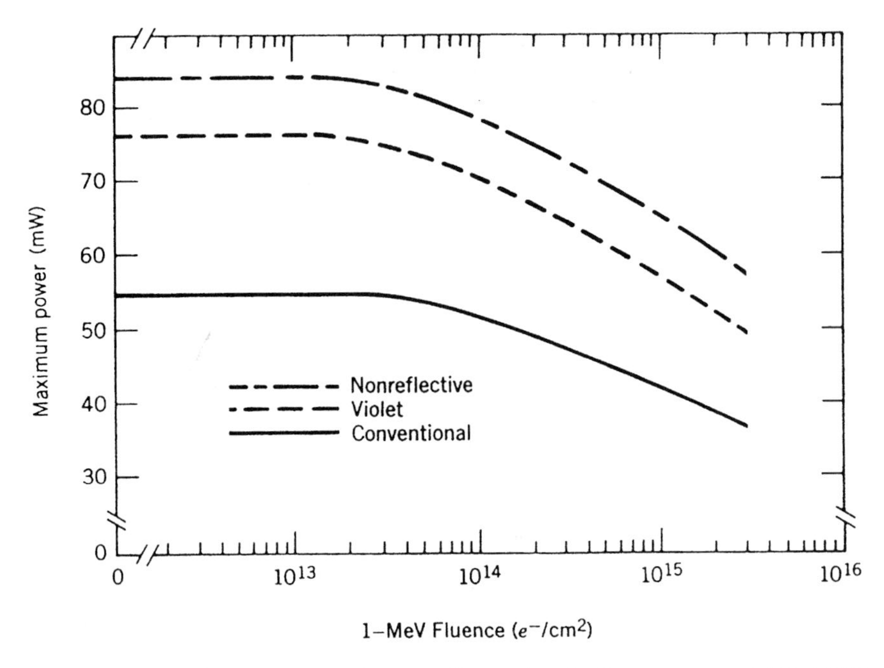

Fig. 6.22 Radiation effect on peak power. (Reproduced with permission of COMSAT Technical Review; Refs. 21, 22.)

6.2.4 Solar-Array Design

Solar cells with a cover glass are interconnected electrically and bonded to a structural substrate to form a solar array. Figure 6.23 shows a cross section through a typical rigid solar array. An aluminum honeycomb substrate is shown in Fig. 6.23. The cells are connected electrically, bottom to top, with an interconnect. Even the lowly interconnect must be designed with care to take the expansion and contraction of the wide temperature excursions a solar array is subjected to.

6.2.4.1 String design. Individual solar cells generate small power, voltage, and current levels. A solar array uses cells connected in series to boost voltages to desired levels, which are called *strings*. Strings are connected in parallel to produce

Table 6.8 Cell degradation, Mars mission[a]

Degradation source	Degradation, %		
	Current	Voltage	Power
Proton and UV	10.6	5.7	14.0
Neutron	1.0	0	1.0

[a]Courtesy of NASA; Ref. 8.

Table 6.9 Cell degradation factors, Venus mission[a]

Degradation source	Power loss factor
Charged particle	
Solar wind	1.000
Van Allen belt	0.997
Solar flares	0.971
Additional particle	0.956
Ultraviolet	0.980
Micrometeoroid	0.990
Thermal cycling	0.990
Contamination	0.990
Total	0.880

[a]Courtesy of Lockheed Martin.

desired current levels (see Fig. 6.24). The series-parallel arrangement of cells and strings is also designed to provide redundancy or string failure protection. The strings for battery charging are sometimes segregated from the remainder of the array and designed with a different series-parallel arrangement specifically tailored for battery charging.

6.2.4.2 Minimum string voltage. The minimum voltage produced by a string must be greater than the maximum voltage required on the power bus, which is normally the battery charging voltage. The number of cells in series required to produce a given voltage is

Number of cells = (Required string voltage)/(Minimum operational cell voltage)

A silicon cell produces about 0.5 V; therefore, it takes a 60-cell string to reach 30 V.

6.2.4.3 Minimum current. The desired minimum current levels are reached by connecting strings in parallel:

Minimum current
= (Minimum operational cell current) × (Number of strings in parallel)

Table 6.10 Effect of radiation on GaAs solar cells[a]

Fluence, e/cm^2	Peak power loss, %
10^{14}	8
10^{15}	12
10^{16}	45

[a]Courtesy of The Boeing Company; Ref. 7.

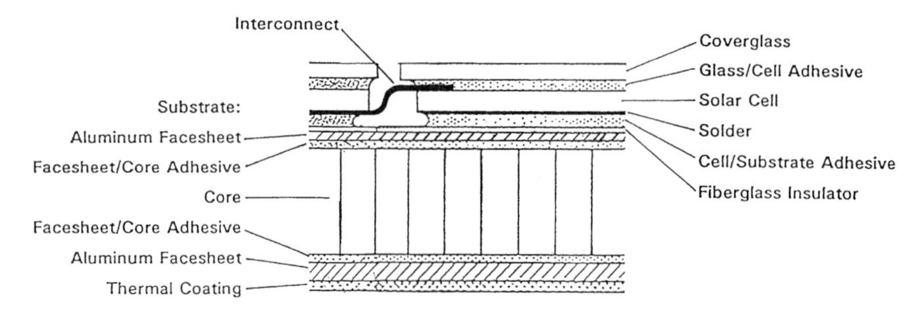

Fig. 6.23 Cross section of a typical solar array.

6.2.4.4 Nominal bus voltage. Since Explorer I, spacecraft have used an average bus voltage of 28 V dc. This selection was a direct descendant of aircraft design practice that provided a wealth of equipment designed and qualified to operate at this voltage. A few modern spacecraft are using higher voltages. The considerations are as follows:

1) The higher the spacecraft power consumption, the higher the transmission losses and the more attractive higher bus voltages are. For this reason, several modern communication spacecraft have gone to higher bus voltages; INTELSAT VIII, for example, uses a 100-V bus with a 4.5-kW system. Communication spacecraft also have the potential of multiple sales to help justify the expense of developing and qualifying high-voltage equipment.

2) The larger the spacecraft the greater the cable losses and more attractive higher voltages become. The Space Station uses a 120-V bus at power levels of 100 kW. (In manned spacecraft astronaut safety considerations place an upper limit of about 120 V.)

These spacecraft are the exception; the vast majority of spacecraft still use a 28-V bus.

6.2.4.5 Maximum and minimum operational power points. The maximum power for a cell under operational power points at BOL during design minimum temperature conditions is shown in Fig. 6.25.

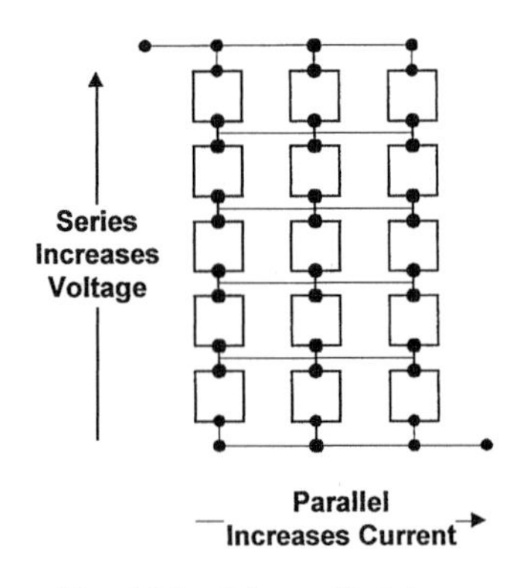

Fig. 6.24 Solar-cell strings.

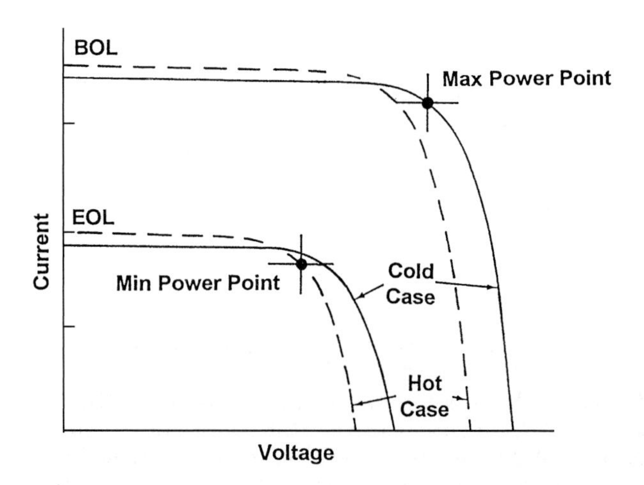

Fig. 6.25 Cell minimum and maximum operational power points.

The minimum power condition, which is a design point for a solar array, occurs at EOL during maximum temperature conditions. The minimum cell current and voltage also occur at the minimum power point. For a planetary spacecraft the change in solar intensity with distance must also be considered in establishing maximum and minimum power points. For example, the minimum power point for a Mars orbiter would occur at EOL, hot, at Mars. For a Venus orbiter like Magellan, the maximum power point occurred at Venus under cold conditions after the cruise degradation.

6.2.4.6 Area. Solar-array area is determined by the number of cells required to meet the spacecraft power requirement at the minimum power point. Laboratory data on solar-cell performance must be adjusted to EOL operational conditions, and the solar flux expected. In addition, the cell density on the panel and power losses in the array must be considered as follows:

$$P_c = P_L(\eta_{\text{rad}}\eta_{\text{uv}}\eta_s\eta_{\text{cy}}H_I\eta_m\eta_l\eta_{\text{con}}\eta_t L_p) \tag{6.26}$$

where

P_c = power from one cell under operational conditions installed in the array, W
P_L = power delivered by one cell under laboratory conditions, W
η_{uv} = power loss caused by UV discoloration of cell materials (~0.98)
η_{cy} = power loss caused by thermal cycling (~0.99)
η_m = power loss caused by cell mismatch (~0.975)
η_l = power loss caused by resistance in cell interconnects (~0.98)
η_{con} = power loss caused by contamination from all sources (~0.99)
η_s = power loss caused by shadowing
η_{rad} = power loss caused by radiation damage
η_t = power adjustment for operation temperature
H_I = adjustment for solar intensity at orbit position, see Eq. (6.27)
L_p = array pointing loss factor

Table 6.11 Mean solar distance from the planets

Planet	Mean distance, km × 10^6	Planet	Mean distance, km × 10^6
Mercury	57.9	Saturn	1433.3
Venus	108.2	Uranus	2882.8
Earth	149.6	Neptune	4516.8
Mars	228.0	Pluto	5890
Jupiter	778.4		

For cylindrical spin-stabilized spacecraft with body-mounted cells $L_p = 1/\pi$. For sun-oriented flat arrays $L_p = \cos\alpha$, where α is the solar incidence angle. (Recall that incidence is measured from the normal to the array.) Solar-cell power output decays as the cosine of the incidence angle from $\alpha = 0$ to about $\alpha = 45$ deg. The decay is worse than the cosine would indicate at angles greater than 45 deg, and output goes to zero about 85 deg.

The product of mismatch loss η_m and interconnect loss η_l is called the *assembly factor*.

The effect of distance on solar intensity is an inverse square law, where R is distance from the sun in millions of kilometers:

$$H_I = \left(\frac{149.6}{R}\right)^2 \tag{6.27}$$

The mean solar distance from the planets is shown in Table 6.11.

The effect of temperature is

$$\eta_t = 1 - 0.005\,(T - t) \tag{6.28}$$

where T is the temperature of the cells in orbit, °C; and t is the temperature at which cells were tested, usually 25–28°C.

The number of cells needed is

$$N_c = \frac{P_{\text{sa}}}{P_c} \tag{6.29}$$

where N_c is the number of cells in the solar array and P_{sa} the power produced by the solar array, W.

The density of the cells installed in the array D_c can be estimated from the packing factor, which is usually about 88%, and the area of the array is then

$$A_{\text{sa}} = \frac{N_c}{D_c} \tag{6.30}$$

where A_{sa} is the area of the array, m^2; and D_c is the installed density of the cells, cells/m^2.

6.2.4.7 Deployment. The Magellan solar-panel deployment method (Fig. 6.26) is typical for three-axis-stabilized spacecraft. For launch the solar panels were stowed folded downward and latched in position with explosive pins at the midpoint. There were spring-loaded hinges located at the bus attachment point.

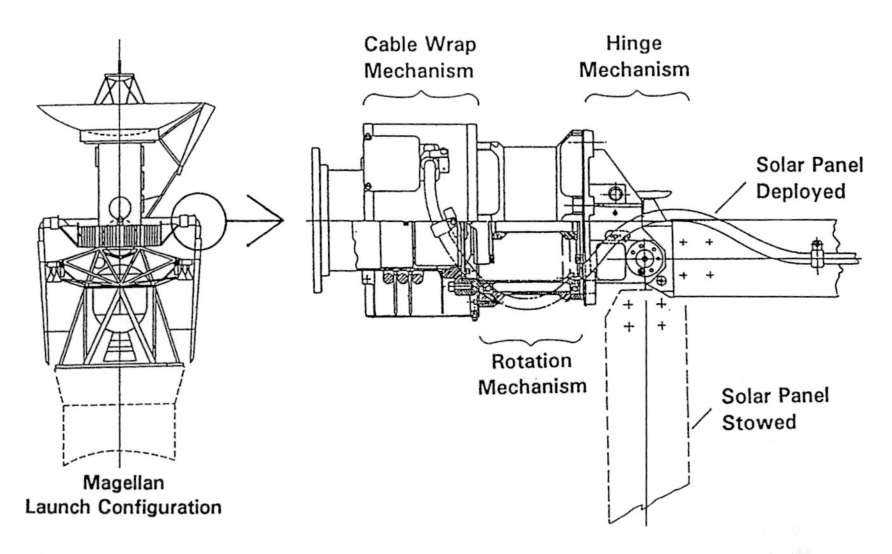

Fig. 6.26 Magellan solar-panel deployment/rotation. (Reproduced with permission of Lockheed Martin.)

When the panel was released by the pin pullers, the springs raised the panels into the flight position where they were locked in place.

6.2.4.8 Articulation. Normally the solar panels of three-axis-stabilized spacecraft must be articulated in one or two axes to keep the panels normal to the sun. The Magellan mission was designed such that one axis or rotation would be adequate. The rotation mechanism (Fig. 6.26) had two components: 1) a rotation mechanism, which was a motor bearing set, and 2) a cable wrap system for power transfer.

6.2.4.9 Power transfer. As the solar panels rotate, in one axis or two, it is necessary to transfer power from the rotating panels to the stationary bus. There are two transfer methods in common use: 1) The first method is a cable wrap mechanism, which wraps and unwraps a continuous cable with one end fixed to the bus and one fixed to the panel. This method requires programmed panel unwinds. (Magellan used cable wrap.) 2) The second method involves slip rings, which transfer power from a rotating ring to a fixed brush. Slip rings were the cause of the high-profile SEASAT failure. As a direct result of the failure, NASA Ames Research Center developed an innovative third method of power transfer using inductive components. The inductive components form a magnetic slip ring that functions like a power transformer, but makes use of a rotating coil to accommodate rotation without contacts. A constant level of transfer power can be maintained regardless of panel position.

6.2.4.10 Mass. The mass of a three-axis-stabilized solar array is about 4.0 kg/m^2 including the substrate; 3.6 kg/m^2 has been achieved (see Table 6.5). For body mounted solar arrays on a spinning spacecraft 3.4 kg/m^2 has been achieved frequently. The Magellan substrate was a vented aluminum honeycomb core with aluminum face sheets and a single central support beam running the length of the panel. The mass of the support structure, which mounted the panel to the bus, was 15% of the supported mass. The mass of deployment hinge and the single-axis rotation mechanism together was 13.8 kg/panel. There are other ways of doing deployment and rotation; however, any three-axis vehicle must have some solution.

Example 6.1 Solar-Array Design

Design a solar array for a three-axis-stabilized, geosynchronous spacecraft, which will deliver 1200 W at the end of a 10-year life. The solar-panel pointing system controls to within ±5 deg. The design temperature range of the panels is −120 to +115°C. The chosen 2 × 4 silicon cell will deliver a minimum of 14.9% efficiency at 28°C and a solar flux of 1370 W/m^2 normal to the cell.

In the laboratory the power output of a minimum cell is

$$P_L = (0.137)(0.149)(8) = 0.163 \text{ W/cell}$$

After Agrawal[7] the power loss caused by radiation after 10 years in a geosynchronous orbit is about 30%, and the loss caused by UV discoloration is about 2%; therefore, $\eta_{\text{rad}} = 0.70$, and $\eta_{\text{uv}} = 0.98$. Assume that the general arrangement is designed such that shadowing is zero and $\eta_s = 1$. No adjustment of the solar flux is required because the orbit is near Earth. The adjustment for panel temperature is

$$\eta_t = 1 - 0.005(115 - 28) = 0.565$$

Assuming maximum pointing error and using the generic values for the remainder of the factors in Eq. (6.26) produces

$$P_c = (0.163)[(0.70)(0.98)(1.0)(0.99)(1.0)(0.975)(0.98)(0.99)(0.565)(0.996)]$$
$$= 0.0589 \text{ W/cell}$$

The number of cells required to provide 1200 W is

$$N_c = \frac{1200}{0.0589} = 20374 \text{ cells}$$

The cell density with a packing factor of 88% is

$$D_c = \left(\frac{10000}{8}\right)(0.88) = 1100 \text{ cells/m}^2$$

And the total area of the array is

$$A_{\text{sa}} = \frac{20374}{1100} = 18.5 \text{ m}^2$$

6.2.5 Array Configurations

There are three primary array configurations: 1) body-mounted array, 2) rigid planar array, and 3) flexible array. *Body-mounted* arrays are used primarily with spinning spacecraft and were the first type of arrays used. Typically the spacecraft

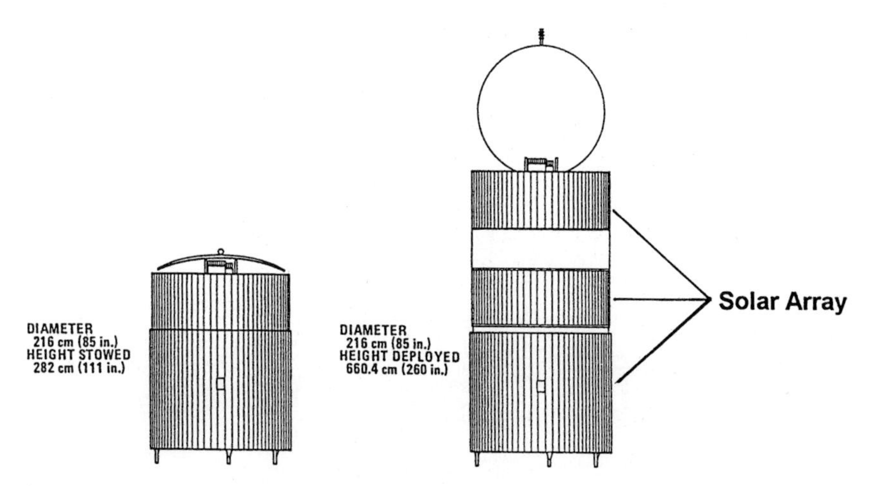

Fig. 6.27 WESTAR IV showing telescoping solar array. (Courtesy of The Boeing Company.)

is drum shaped with the cells mounted on the cylindrical surface parallel to the spin axis (see Fig. 6.27).

The advantages of this configuration are 1) lightweight—the spacecraft structure serves as the substrate; 2) the cells are shielded from radiation on the back side; 3) simplicity—no deployment or articulation mechanism is required. The disadvantages are as follows: 1) Cells are rotated through sun and shadow; hence, cell output is reduced by a factor of π. 2) The surface area of the array and power are limited by the body size of the spacecraft. Hughes Aircraft Company has used this configuration or several generations of spacecraft; in later generations they gained area by the use of telescoping concentric cylinders. The specific power of the body-mounted, spinning solar array of INTELSAT IV was 6.2 W/kg, EOL. This compares to a specific power of 21.2 W/kg, EOL, for three-axis-stabilized INTELSAT V.[9] Both INTELSAT IV and V used the same solar cells; the specific power gain is from pointing the solar panels.

Rigid planar arrays are the most common configuration for three-axis-stabilized spacecraft. The advantages of this configuration are that 1) the array can be actively pointed at the sun so that the area of the panel is used efficiently and 2) area is not constrained, thus large power demands can be met. The disadvantages are that 1) the panels must be stowed at launch and deployed after separation, 2) the panels must be articulated, 3) power must be brought out of the panel through the articulation mechanism, which requires a slip joint or a cable wrap. The Viking Orbiter (Fig. 6.28) used four planar arrays, which were double folded for launch.

Once the panels were deployed on Viking Orbiter, the spacecraft was sun oriented, and no further articulation of the panels was required.

Flexible arrays are a relative newcomer to spacecraft design, although the desirability of a flexible, roll-up array was recognized very early. The major advantages this design offers are 1) reduction of substrate weight and 2) flexible stowage. The resulting array opens the door for very high-power arrays. Numerous configurations are in the advanced stages of development; these are summarized in

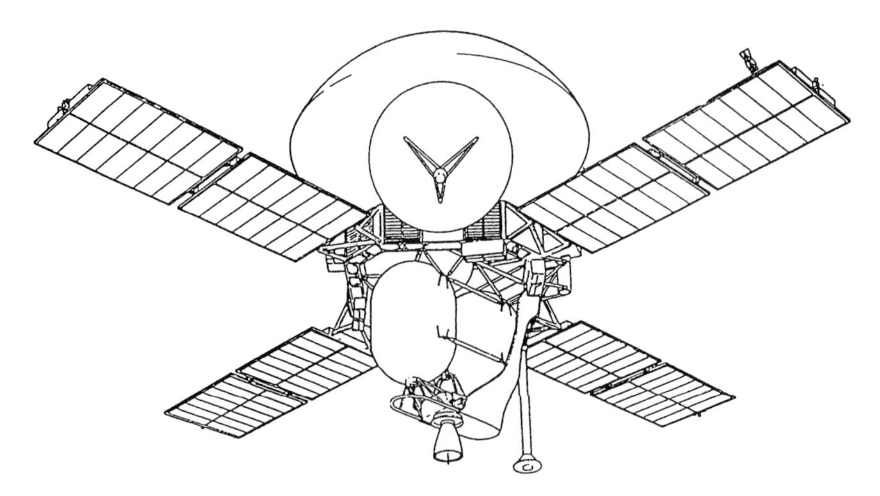

Fig. 6.28 Viking Orbiter showing fold-out arrays. (Reproduced courtesy of NASA/JPL/CalTech; Ref. 8.)

Fig. 6.29. Table 6.12 shows the performance demonstrated with each of the solar-array types.

The roll-out array used on the Hubble Telescope is shown in Fig. 6.30. The solar arrays are two rectangular wings of retractable solar-cell blankets fixed between a two stem frame. The blanket unfurls from a cassette in the middle of the wing. A spreader bar at each end of the wing stretches out the blanket and maintains tension.

The wings are on arms that connect to a drive assembly on the telescope shell at one end and to the cassette on the other end. The total length of the cassette, arm, and drive is 4.8 m. Each wing has 10 panels, five on each half of the wing, that roll out from the cassette. The small panels are made up of 2,438 solar cells attached to a glass fiber/Kapton surface, with silver mesh wiring underneath that is covered by another layer of Kapton. The blankets are less than 500 μm thick so that they can roll tightly when the wings are stowed. Each wing weighs 7.7 kg (17 lb) and at full extension is 12.2 m long and 2.5 m wide.

Table 6.12 Flexible and rigid solar-array performance[a]

Characteristic	Rigid fold-out INTELSAT V	Flex fold-out Hermes	Flex roll-out FRUSA
BOL power, W	1564	1332	1500
EOL power, W	1288	—	—
Array area, m^2	18.12	16.9	16.66
Array mass	64.1	60.0	31.7
Electrical, kg	22.8	46.06	13.1
Structural, kg	41.3	13.94	18.6
Mass/area, kg/m^2	3.537	3.55	1.90
Sp. power, W/kg	20.0 EOL	22.2 BOL	47.4 BOL
Primary reference	17	2	2

[a]Data in part courtesy of B. N. Agrawal; Ref. 5.

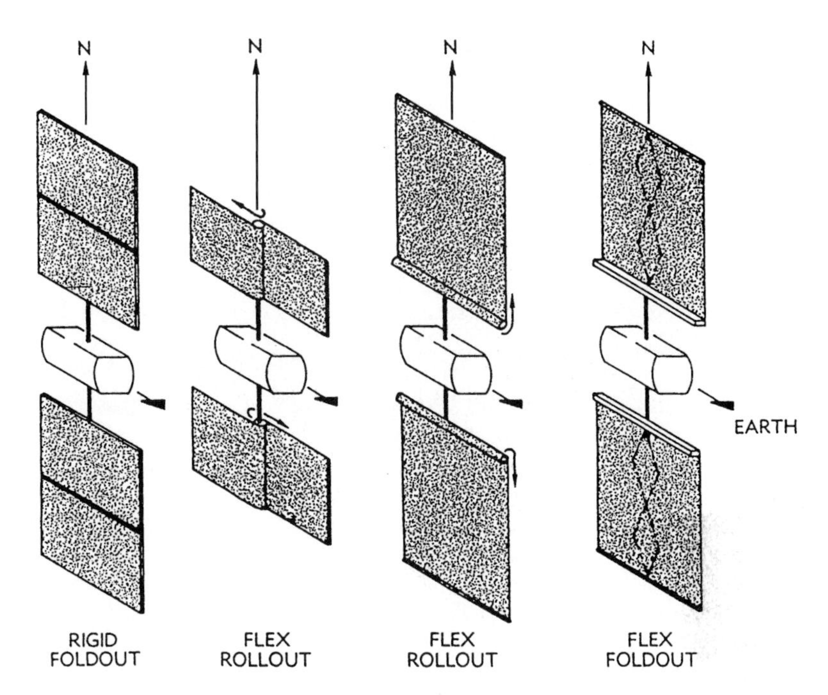

Fig. 6.29 Flexible array concepts. (Reproduced with permission, copyright IEEE, 1973; Ref. 10.)

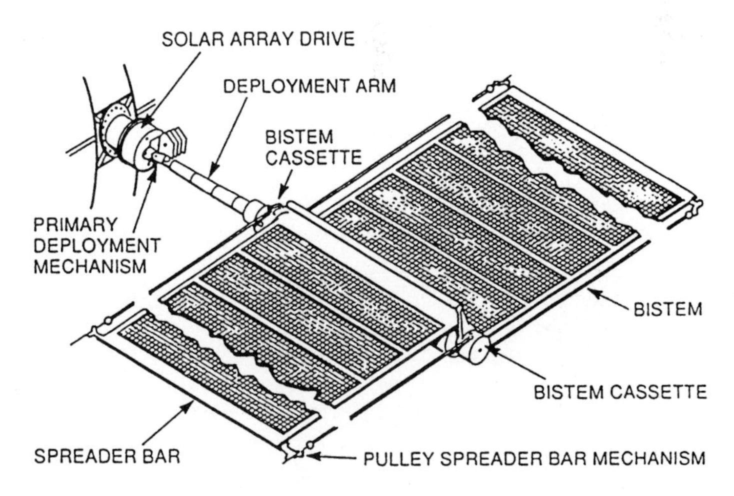

Fig. 6.30 Flexible roll-out array—Hubble Telescope. (Reproduced with permission of Lockheed Martin; Ref. 11.)

A primary deployment mechanism raises the solar-array mast to a position perpendicular to the telescope. There are two mechanisms, one for each wing; each mechanism has motors to raise the mast and supports to hold the mast in place. Once the mast is raised, the secondary deployment mechanism unfurls the wing blankets. The assembly rolls out the blanket, applies tension evenly so that the blankets stretch, and transfers data and power along the wing assembly. The solar-array drive rotates the panels to track the sun as the telescope moves in orbit. Each blanket can roll out completely or part way. The mechanisms can be operated manually by an astronaut if necessary.

The arrays were designed by the European Space Agency and built by British Aerospace.

6.3 Radioisotope Thermoelectric Generators

Radioisotope thermoelectric generators (RTGs) offer a spacecraft power source, which is independent of the sun, and there are missions where this independence is mandatory, for example, the Viking Lander, which operated on the surface of Mars, or the Voyager, which visited the outer planets.

The development of RTGs was assigned to The Atomic Energy Commission (AEC), now ERDA, in 1955. They were developed under the Systems for Nuclear Auxiliary Power (SNAP) program. The history of the SNAP RTGs is summarized in Table 6.13. The odd-numbered devices, such as SNAP-3, used a radioisotope as an energy source, and the even-numbered devices, such as SNAP-8, used a nuclear reactor as an energy source. (Only one U.S. reactor device, the SNAP 10A, ever flew,[12] although the former USSR flew reactor power supplies on a routine basis.[1]) The first RTG to fly was SNAP-3 in 1961. The early RTGs had a power density of slightly better than 1 W/kg. SNAP-9A reached 2.0 W/kg, and Galileo reached 5.4 W/kg. All specific powers are for new isotopes.

Table 6.13 History of RTG power[a]

RTG	Spacecraft	Isotope	Power, W	Mass, kg	Power density, W/kg	Life	Status
SNAP-1	—	Ce-144	500	272.0	1.84	60 d	Cancelled '58
SNAP-1A	—	Ce-144	125	79.0	1.58	1 y	Cancelled '59
SNAP 3	Nav Sat	Pu-238	2.7	2.1	1.29	5 y	2 flew '61
SNAP-9A	Nav Sat	Pu-238	25	12.2	2.04	6 y	3 flew '63
SNAP-19	Nimbus	Pu-238	30	13.6	2.21	5 y	2 flew '69
SNAP-19	Pioneer	Pu-238	30	13.6	2.21	5 y	1 flew '72
SNAP-19	Viking	Pu-238	40	15.9	2.52	5 y	4 flew '76
SNAP-27	Apollo LSEP	Pu-238	60	20.9	2.87	3 y	5 flew
SNAP-29	—	Po-210	400	181.0	2.21	90 d	Cancelled '69
MHW	Lincoln Labs	Pu-238	150	36.2	4.14	5 y	1 flew '76
MHW	Voyager	Pu-238	160	37.6	4.25	12 y	6 flew '78
MHW	Galileo	Pu-238	298	55.7	5.35	12 y	2 flew '89
GPHS	Cassini	Pu-238	296	60.0	4.93	—	2 flew '97

[a]Data, in part, courtesy of C. D. Cochran, U.S. Air Force, Ref. 12, pp. 4–11.

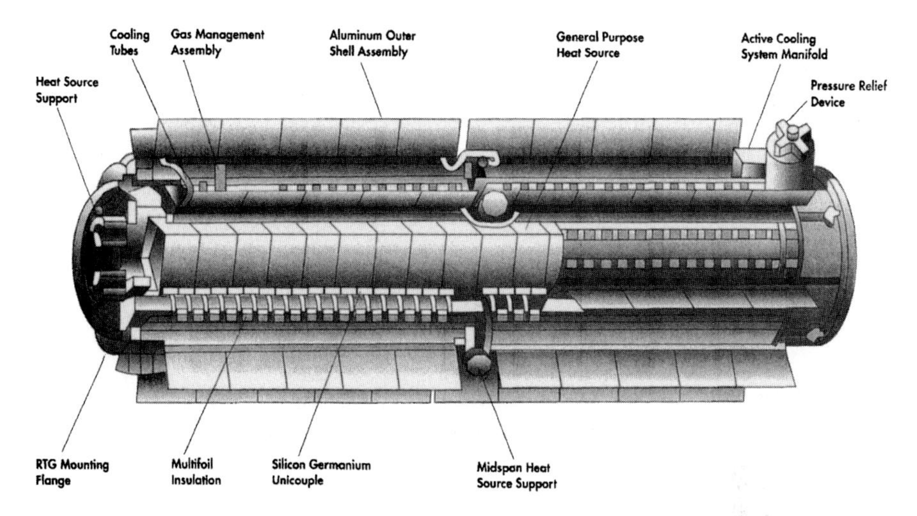

Fig. 6.31 Cassini RTG. (Reproduced courtesy of NASA/JPL/Caltech.)[24]

RTGs convert the energy released during the decay of an isotope, usually Pu 238, into electricity using the thermoelectric effect. In a typical RTG, shown in Fig. 6.31, heat source blocks in the central core contain the isotope capsules, each encased in an individual reentry heat shield. Each of the heat source blocks are in thermal contact with the hot junction of thermoelectric units. Power is collected from the thermoelectric units in a series-parallel arrangement and brought out of the unit through a connector. The waste heat generated, about 90% of the energy, is radiated to space by a set of fins.

The nominal, steady-state power, voltage, and current produced by one of the Viking SNAP 19 units is shown in Fig. 6.32. The two operating modes of the RTG were shorted and on load. In the shorted mode the RTG output voltage was less than 1 V, and the output current was about 22 A. While shorted, the degradation of the thermoelectric units was minimized as a result of reduced hot junction temperature. When the RTG was on load, the power control equipment maintained the voltage at 4.4 ± 0.1 V dc. In this configuration the RTG could produce from 35 to 47 W depending on fin root temperature and RTG age.

The main advantage of RTGs is independence from the sun. The disadvantages of RTGs are as follows: 1) cost of the units, particularly the cost of the isotope; 2) cost of the labor involved in demonstrating compliance with safety requirements; 3) cost of safety provisions for ground crew; 4) power decreases with time because of radiation damage to the units as well as nuclear decay; 5) neither the power output nor the heat output can be turned off; and 6) emitted radiation is detrimental to electronics and instruments; an extended boom is often required. In sum, RTGs are not used when solar arrays will suffice.

6.4 Batteries

All spacecraft require batteries during eclipse periods, peak load periods, and during launch operations. Table 6.14 compares battery types that have a successful flight history.

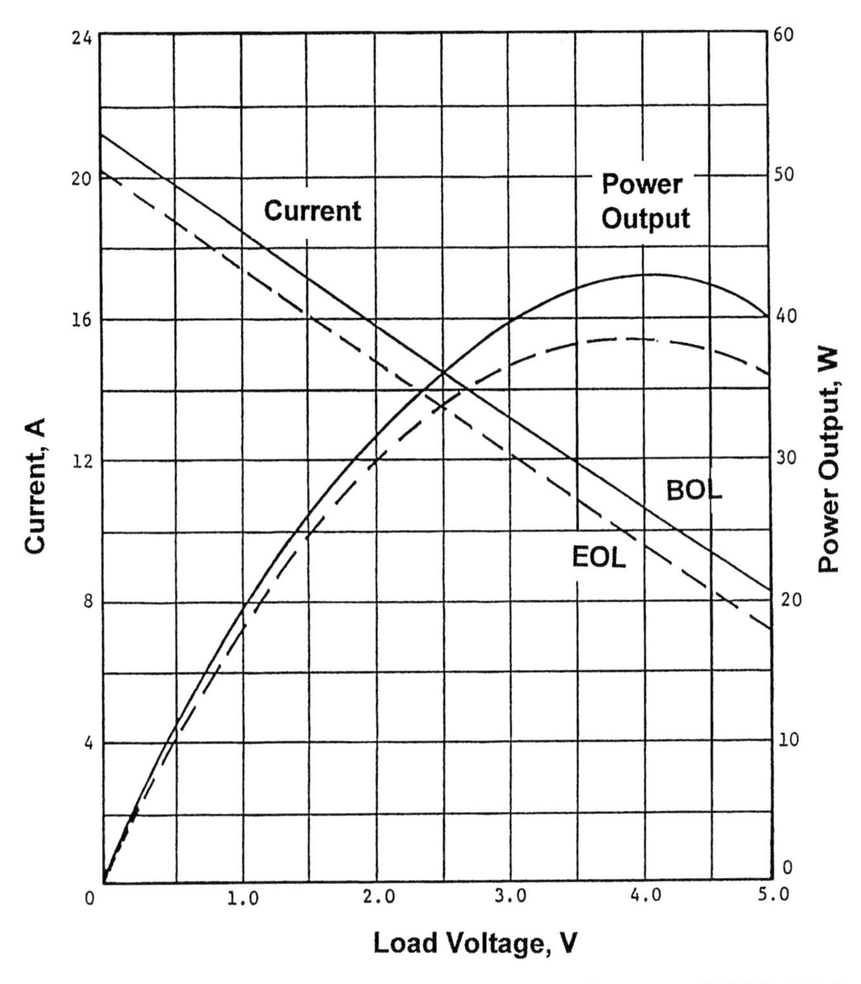

Fig. 6.32 Viking Lander RTG performance. (Reproduced courtesy of NASA; Ref. 8, Vol. 1, p. 109.)

Batteries are categorized as *primary* if they are used as the primary power supply and are not rechargeable. Primary batteries are used by launch vehicles and were used in the early short-lived spacecraft like Explorer, Pioneer, and Vanguard.

A *secondary* battery is used to store energy rather than as an energy source; secondary batteries are rechargeable for many cycles. Nickel–cadmium, or NiCd batteries, were the dominant spacecraft battery for decades; however, nickel–hydrogen batteries, which first flew in 1976, are rapidly becoming the battery of choice. Lithium batteries, with high power densities, are an emerging technology with several chemical combinations being developed. A Lithium–sulfur dioxide battery flew on Galileo as the power supply of an experiment.

Fuel cells are a primary source of power for spacecraft requiring power in the kilowatt range for periods up to a month. They are the power source of manned

Table 6.14 Battery types with a flight history

Type	Electrode materials	Cell voltage	Specific energy, W-h/kg
Nickel–cadmium	Nickel oxide–cadmium	1.25	24
Nickel–hydrogen	Nickel–hydrogen	1.25	55
Lithium	Lithium–sulfur dioxide	2.7	220
Silver–zinc	Silver peroxide–zinc	1.55	175
Mercuric oxide	Mercuric oxide–zinc	1.20	97
Fuel cells	Oxygen–hydrogen	0.8	—

spacecraft; they have flown on Gemini, Apollo, and the Space Shuttle. Energy is supplied by the chemical reaction of hydrogen and oxygen producing water as a waste product. The hydrogen and oxygen are stored separately and fed to the fuel cells at a controlled rate. The water is removed from the cells by capillary devices. The Gemini system generated 320 W/kg and produced a kilowatt of power from a pound of reactants.[13]

6.4.1 Nickel–Cadmium Batteries

The NiCd cell (Fig. 6.33) has four main components: 1) the negative plate made of cadmium, which supplies electrons for the external loads when it is oxidized during discharge; 2) the positive plates made of nickel, which accepts the electrons returning from external loads; 3) the electrolyte, an aqueous potassium hydroxide solution, which completes the circuit internally; and 4) a fibrous plastic fabric separator, which holds the electrolyte in place and isolates the positive and negative plates. The chemical reactions are reversible; hence, the cells can be cycled. The units are hermetically sealed and must not be overcharged; overcharging releases hydrogen gas.

NiCd cell voltage is about 1.1 V fully discharged and about 1.4 V fully charged. Cells are connected in series to get the desired voltage. The average discharge voltage is about 1.25 V times the number of cells in the battery; for a typical 22 cell battery the average discharge voltage V_{ave} is 27.5 V. Figure 6.34 shows the electrical characteristics of a NiCd battery.

6.4.1.1 Cycle life. Depth of discharge and operating temperature are the primary life-limiting factors for a cell. Batteries in a low Earth orbit are cycled approximately 6000 cycles per year; as a result, the depth of discharge must be low—20% is common. Geosynchronous spacecraft cycle the battery only about 100 cycles per year, and a depth of discharge of 50% is acceptable. Figure 6.35 shows cell cycle life as a function of depth of discharge for post 1970 NiCd cells operating between 0 and 25°C.

6.4.1.2 Temperature. Batteries are the most temperature-sensitive components of a spacecraft. Low temperatures reduce the recombination rate of oxygen in the cell, which results in incomplete recharging and high cell pressure. At high

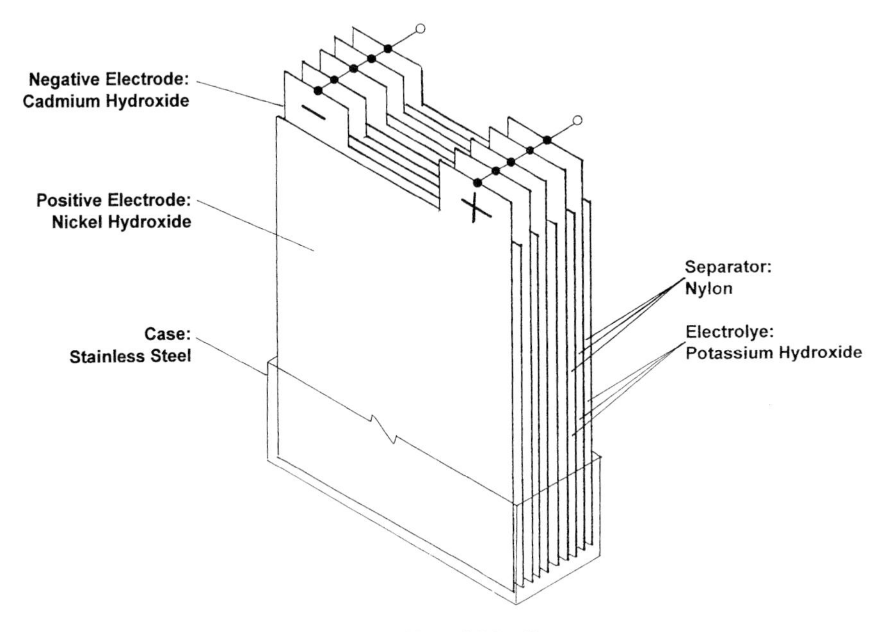

Fig. 6.33 NiCd cell.

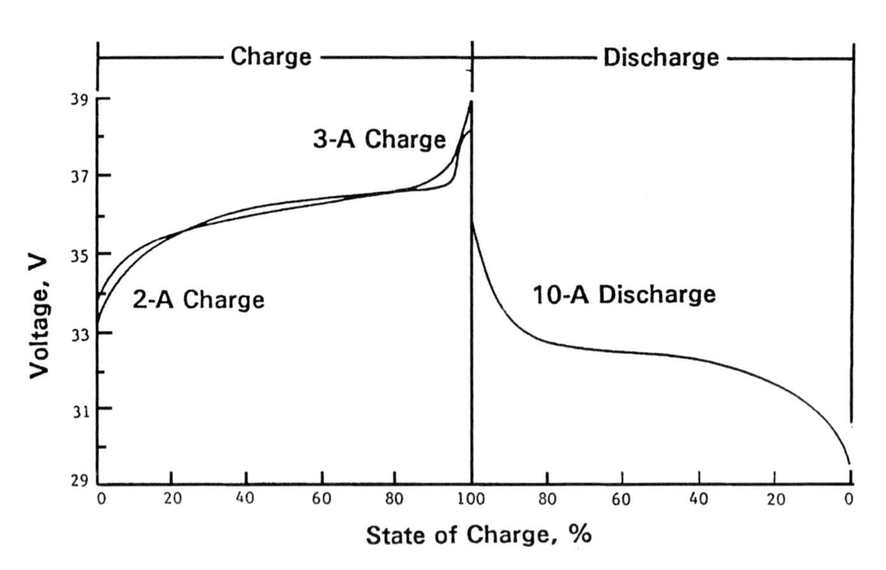

Fig. 6.34 NiCd battery electrical characteristics. (Courtesy of NASA; Ref. 8.)

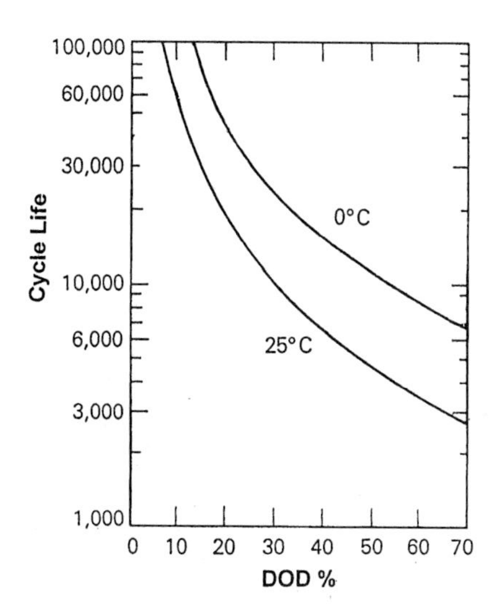

Fig. 6.35 Estimated cycle life of post-1970 NiCd cells. (Data courtesy of Piscane and Moore; Ref. 13, p. 395.)

temperatures the life of the separator plastic is reduced. Typical NiCd battery temperature limits are 5 to 20°C.

6.4.1.3 Battery charging. Charging efficiency varies as a function of charge rate and temperature; low charging rates produce low losses. At normal temperatures a charging rate of C/100 is about 50% efficient; for charging rates in the range of C/10 to C/15, the charging process is about 80% efficient (see Fig. 6.36).

Very high charging rates result in high battery temperatures. The normal range of charging rates is C/15 to C/5, and several rates are usually provided. The three charging rates for the 30 A-H Viking Orbiter battery are typical: C/40 (trickle charge), C/15 (normal charge), and C/10 (maximum charge rate).

6.4.1.4 Charging efficiency. The energy used in charging a battery goes to stored electrical energy and to heat. The percentage of total charging energy that becomes electrical energy in the battery is the charging efficiency. Charging efficiency is a strong function of battery temperature and charging rate (Fig. 6.36).

6.4.1.5 Dual-mode charging. For low Earth orbiters the time available for recharge is too short for simple charging schemes. At maximum eclipse the sunlight time is about 60 min, and eclipse is about 40 min. If a 33% DOD were allowed, a constant charging rate of C/3 would be required to recharge the battery. A dual charging mode is used to charge the battery in the time available. Just emerging from eclipse the solar panels are cold, and the power output is at the orbital maximum; the battery voltage is at an orbital minimum. During this period,

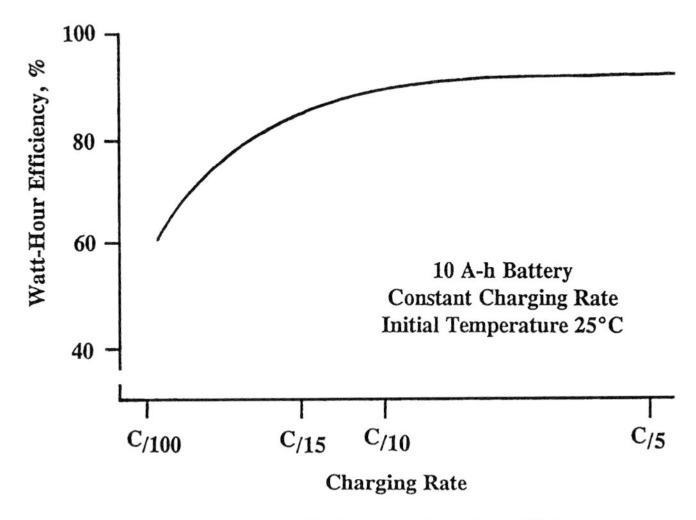

Fig. 6.36 Typical NiCd battery charging efficiency.

power limited charging is used; all of the excess power from the panels is used to charge the batteries. The time spent in the high charge mode is limited by a battery temperature–battery voltage relationship. When a preset battery temperature and voltage are reached, the mode is changed to battery voltage limited charging, which is done at a much lower rate (trickle charging). A number of selectable battery temperature–voltage curves are provided, in the control electronics, for battery charging. Typical control curves are shown in Fig. 6.37.

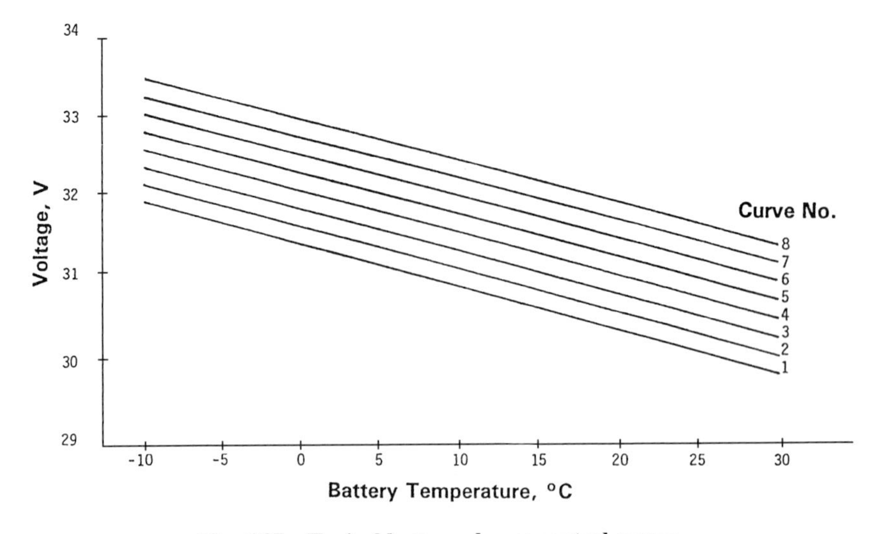

Fig. 6.37 Typical battery charge control curves.

Table 6.15 Characteristics of flight NiCd battery systems

Spacecraft	No. cells	Capacity, A-h	Capacity, W-h	Mass, kg	Sp. energy W-h/kg
INTELSAT-IV-A	—	18	428	19.5	22
Viking Lander	—	10	265	11.5	23
HEAO-2	22	20	550	29.9	18.4
Magellan	22	26.5	729	30.7	24

6.4.1.6 Reconditioning. Battery voltage gradually drops with cycling. Performance can be essentially restored by reconditioning the battery. Reconditioning consists of a very deep discharge followed by recharge at a high rate.

6.4.1.7 Battery mass. The specific energy of some NiCd battery systems is shown in Table 6.15. The battery mass in Table 6.15 is the installed mass. The average specific energy of these four systems is 24 W-h/kg.

Example 6.2 Battery System Design

Design a NiCd battery system to provide an average eclipse load of 567 W (including margin) at 28 V for a duration of 38 min and a battery-to-load power loss of 3%. The required cycle life is 10,000 cycles at 25°C. Use 20 A-h cells, and provide battery-out capability.

From Fig. 6.35 the maximum DOD is 30%; from Eq. (6.21) the battery capacity requirement is

$$C = \frac{567(38/60)}{(0.97)(0.30)(28)} = 44.07 \text{ A-h}$$

Using 20 A-h cells, three batteries would be required for normal operation, and four would be required for battery-out capability. The energy capacity of the four battery system is

$$E_b = (80)(28) = 2240 \text{ W-h}$$

From Table 6.15, the specific energy of an installed NiCd battery is about 24 W-h/kg; therefore, the installed mass of the four battery system would be about $2240/24 = 93.3$ kg. Assuming 1.25 V per cell, 23 cells per battery would be required for a 28-V bus.

The time in the sun, at maximum eclipse, for this orbit is 55 min; dual-mode charging electronics would be required to charge the batteries in the 55 min available.

Note that designing the battery system for maximum eclipse duration each orbit is conservative because most eclipse periods are less than maximum (see Fig. 6.9). In the preceding example it would be reasonable to argue that three batteries are adequate because one battery out would result in a maximum DOD of 33% and only a slight cycle life reduction in the case of a battery failure.

6.4.2 Nickel–Hydrogen Batteries

Nickel–cadmium batteries have been the spacecraft battery of choice for three decades. They are in the process of being replaced by nickel–hydrogen batteries, which offer improved cycle life and reduced mass. Table 6.16 shows the early flight history of NiH_2 batteries.

There are currently more than 50 spacecraft in orbit using NiH_2 batteries.[14] The coming dominance of Ni-H batteries is even clearer looking downstream at soon-to-be-launched designs. Nickel hydrogen batteries are replacing NiCd batteries for all geosynchronous spacecraft[15] and planetary spacecraft requiring power over 1000 W.

The nickel–hydrogen batteries use the most reliable electrodes from a fuel cell and a NiCd battery. The nickel electrode in the battery is a conventional reversible electrode as in the NiCd battery; however, the opposing hydrogen electrode is similar to a fuel cell electrocatalytic diffusion electrode. The anodic active material is hydrogen gas, and the gaseous reactions are catalyzed by the electrode. The

Table 6.16 Partial flight history of NiH_2 batteries[a]

Spacecraft	Number of batteries	Orbit type	Number of cells	Capacity A-H	Pressure vessel	Launch
NTS-2 (Navy)	—	HEO[b]	14	35	IPV[c]	1976
AF Experiment (LMSC)	—	LEO[d]	21	50	IPV	1976
INTELSAT V (Ford)	2	GEO[e]	27	30	IPV	1983
INTELSAT V (Ford)	1	GEO	27	30	IPV	1984
SPACENET (RCA)	2	GEO	22	40	IPV	1984
INTELSAT V (Ford)	3	GEO	27	30	IPV	1985
SATCOM K (RCA)	1	GEO	22	50	IPV	1985
American SAT (RCA)	1	GEO	22	35	IPV	1985
G-STAR (RCA)	1	GEO	22	30	IPV	1985
G-STAR (RCA)	1	GEO	22	30	IPV	1986
INTELSAT V (Ford)	1	GEO	27	30	IPV	1986
Hubble Space Telescope	6	LEO	23	88	IPV	1990
EUTELSAT II (ETSO)	—	GEO	—	58	IPV	1992
INTELSAT K (GE)	—	GEO	—	50	IPV	1992
PANAMSAT (LM)	—	GEO	—	35	IPV	1994
MILSTAR (LM)	—	GEO	—	35	IPV	1994
Clementine (NRL)	1	Lunar	22	15	CPV[e]	1994
Mars Global Surveyor (LM)	—	Planetary	—	20	CPV	1996
Mars Climate Orbiter	1	Planetary	12	16	CPV	1998
Intl. Space Station	24	LEO	—	81	IPV	1998
Odyssey	1	Planetary	—	16	CPV	2001

[a]Data in part courtesy of NASA, Ref. 17; Piscane and Moore, Ref. 13, p. 398; and Eagle Pitcher Industries, Ref. 14.
[b]HEO = high Earth orbit
[c]IPV = independent pressure vessel
[d]LEO = low Earth orbit
[e]GEO = geosynchronous orbit
[f]CPV = common pressure vessel or single pressure vessel

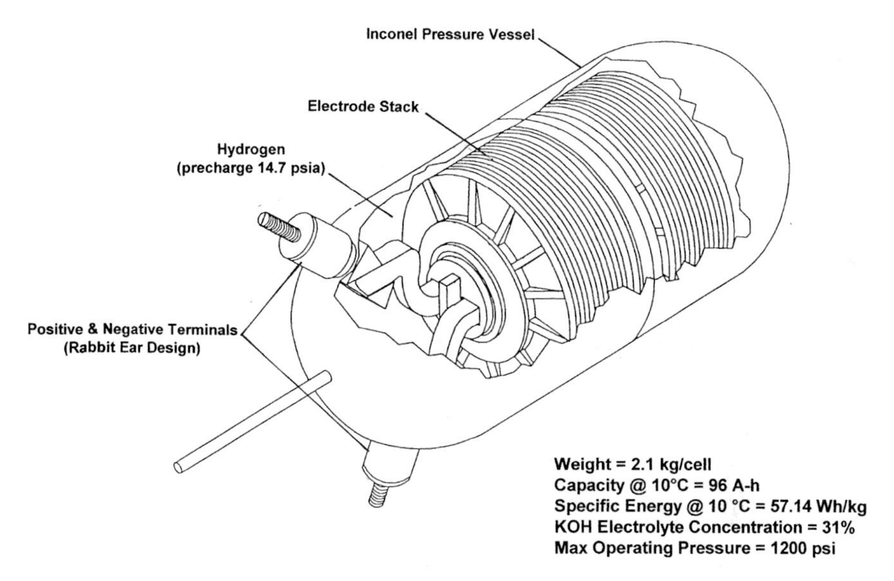

Fig. 6.38 Hubble NiH_2 cell. (Courtesy of NASA; Ref. 17.)

battery operates essentially as a sealed fuel cell. When the battery is charged, nickel is oxidized in the conventional manner, and molecular hydrogen gas is evolved at the hydrogen electrode. This hydrogen gas is contained in a cell pressure vessel. When the cell is discharged, nickel is reduced at the nickel electrode, and hydrogen gas is consumed at the hydrogen electrode. The cell construction, shown in Fig. 6.38, consists of positive nickel electrodes and negative platinum electrodes separated by asbestos mat. The electrode stack is surrounded by hydrogen gas under pressure. The platinum electrode consumes hydrogen gas on discharge and releases hydrogen on recharge. The hydrogen pressure in a cell is a direct linear function of the state of charge and is used as a primary or backup measurement of this parameter.

A typical multiple-cell NiH_2 battery consists of a number of cells, connected in series, to deliver the required voltage. Each cell is sized to deliver the ampere-hour capacity required for the spacecraft loads. This process is identical to that used for NiCd batteries.

Nickel–hydrogen batteries offer improved mass, cycle life, and reduced failure modes compared to NiCd batteries. The performance of flight NiH_2 battery systems are shown in Tables 6.16 and 6.17.

Batteries designed for low Earth orbit (Table 6.18) are substantially different from their GEO counterparts, particularly in the DOD used and specific energy obtained.

Specific energy increases as capacity increases but tapers off above 50 Ah and approaches an upper limit of about 63 Wh/kg for 8.89-cm-diam cells. Energy density is primarily a function of cell pressure range. The cells in the 1980s, the COMSAT design, and U.S. Air Force design cells had a pressure range of 500 to 580 psi. The cells in the 1990s have a pressure range of 800 to 850 psi with attendant higher energy densities. Note that Tables 6.17 and 6.18 give specific energy for complete IPV batteries. Specific energy is often given for the *cells* rather than the

Table 6.17 GEO performance of NiH_2 batteries[a]

Battery design	INTELSAT V	Spacenet	GSTAR	Super Bird	INTELSAT VI	EUTELSAT II	INTELSAT VII
First launch	1983	1984	1985	1988	1989	1990	1993
Nameplate capacity, A-h	30	40	30	83	44	58	85.5
Measured capacity, A-h	36	48.3	35.4	91.5	59	70	97.0
Length, cm	51.8	58.5	58.5	51.0		60	51.0
Width, cm	52.1	26.7	22.86	51.0		45	51.0
Height, cm	22.2	19.7	19.7	30.4	26.9	27	30.4
No. cells	27	22	22	15	32	27	15
Battery mass, kg	30.12	32.6	25	63.9	66.3	48.60	66.7
Mass-one cell, kg		1.13	0.89	1.867		1.385	1.867
Mass cells/battery, %	80	76	78	79	77.9	76.9	75.6
Battery energy, Wh	1215	1328	974	3088	2303	2363	3273
Specific energy, Wh/kg	40	41.16	38.96	48.3	34.7	48.6	49.1
DOD, %	56	60	60	75	70	74	70
Energy density, Wh/cc	0.020	0.0216	0.0185	0.022	0.00426	0.0324	0.024

[a]Courtesy of NASA Goddard Space Flight Center; Ref. 17, pp. 3-20 to 3-40.

Table 6.18 LEO performance of NiH_2 Batteries[a]

Battery design	Air Force Experimental	Hubble Space Telescope	Space Station
First launch	1977	1990	2000
Nameplate capacity, A-h	50	88	81
Measured capacity, A-h	55	96	89
Length, cm	64.64	51.8	91.44
Width, cm	43.89	52.1	101.6
Height, cm	26.82	22.2	44.45
No. cells	21	22	38
Battery mass, kg	50	30.12	172.9
Mass-one cell, kg	1.89		2.16
Battery energy, W-h	1444	1215	4227
Specific energy, W-h/kg	28.87	37.89	24.4
DOD, %	7	8	35
Energy density, W-h/cc	0.0182	0.020	0.010

[a]Courtesy of NASA Goddard Space Flight Center; Ref. 17, pp. 3-36 to 3-57.

battery. IPV batteries weigh about 1.25 times the sum of the cell weights. CPV batteries weigh about 1.13 times the cell weights.

6.4.2.1 Charging efficiency. The charging efficiency of NiH_2 batteries is shown as a function of temperature in Fig. 6.39. These values were calculated from data in Dunlop et al. (Ref. 17, pp. 5–15).

6.4.2.2 Pressure vessel. Rapid improvements are still being made in the NiH_2 battery designs. The major design variants are IPV and the CPV. An IPV battery has a number of cells like those shown in Fig. 6.38; each cell has its own pressure vessel. A CPV battery consists of a number of individual cells connected electrically and contained in a single pressure vessel. The CPV design has several advantages over conventional IPV designs, including a reduction in mass (~10%),

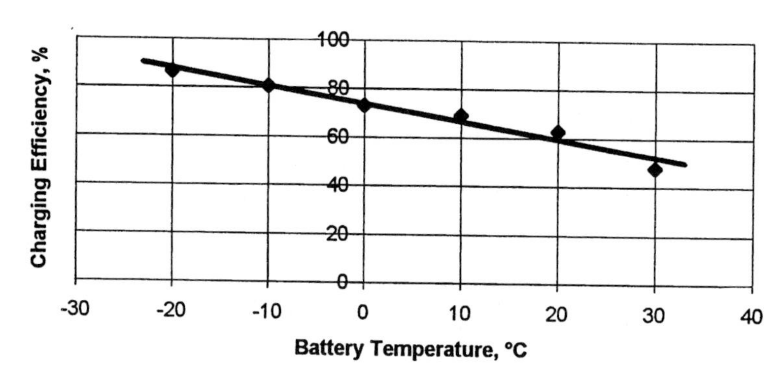

Fig. 6.39 NiH_2 battery charging efficiency. (Calculated from Ref. 17, pp. 5–15.)

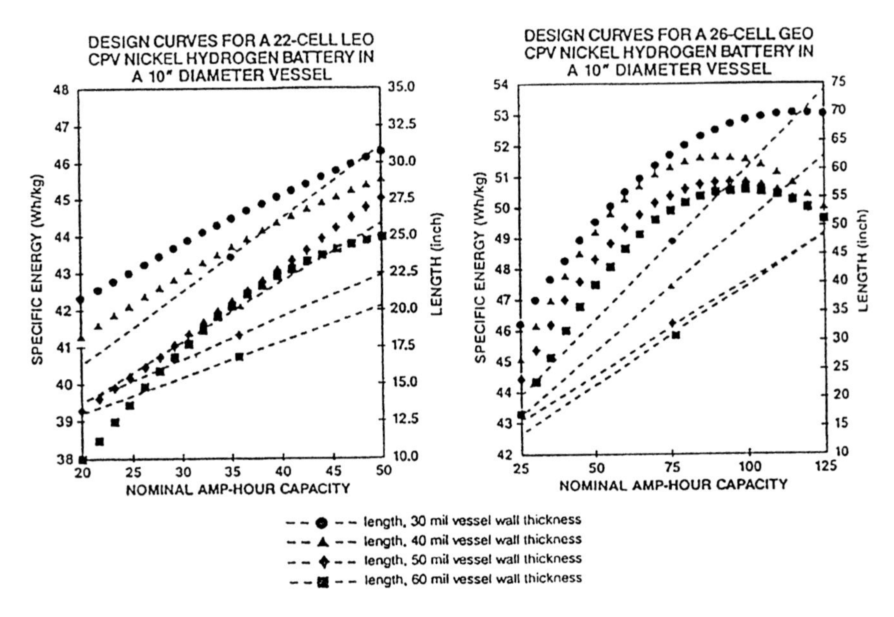

Fig. 6.40 NiH_2 CPV battery specific energy and size. (Reproduced with permission of Eagle Pitcher; Ref. 23.)

a reduction in footprint (~50%), and a reduction in volume (~30%). The size and specific energy of CPV battery designs are given in Fig. 6.40

All of the early NiH_2 batteries were IPV designs; Clementine was the first spacecraft to fly a CPV battery in 1994, as shown in Table 6.16. All of the batteries listed in Tables 6.17 and 6.18 use individual pressure vessels. The Mars Global Surveyor, launched in 1996, and several soon-to-be-launched spacecraft are using CPV technology.

6.4.2.3 Charge control. The most frequently used method of charging NiH_2 batteries in LEO applications is to charge at a high rate to a preselected voltage–temperature limit and to trickle charge thereafter. The rate of heat dissipation during charging increases markedly after 85% of capacity is reached. To minimize dissipation losses in LEO applications, batteries are not normally charged above 85% of capacity.

In GEO applications NiH_2 batteries are normally coulometrically charged at a fixed c/d ratio (c = ampere-hours returned on charge and d = ampere-hours removed during discharge), returning 105–115% of the capacity removed on discharge at a high rate of C/10 to C/25. The batteries are then switched to a low trickle charge rate of C/60 to C/25 to bring the batteries to full charge.[17]

6.4.2.4 Efficiency. The best watt-hour efficiency achieved by current NiH_2 batteries is 85% (similar to NiCd batteries).

6.4.2.5 Depth of discharge and cycle life. Batteries for spacecraft in geosynchronous orbit require about 100 discharge cycles per year. Low Earth and planetary orbits are much more demanding, requiring about 6000 cycles/year.

Long cycle life in any battery can be achieved at the expense of depth of discharge. (Recall that battery system mass increases in inverse proportion to DOD.) The goal of NiH_2 battery development is to reach 40,000 cycle life at a DOD of 35%,[17] which is slightly better than current batteries.

The Hubble Space Telescope, the first major program to use NiH_2 batteries in LEO, is using them at the very conservative DOD of 10%. Significant test programs are being conducted at Eagle Pitcher, COMSAT, Hughes, Lockheed Martin, NASA Lewis Research Center, and elsewhere to determine the cycle life of NiH_2 cells as a function of DOD and temperature.

6.4.2.6 Temperature control. The recommended temperature range for NiH_2 batteries is -5 to 25°C during operation and 0 to 10°C during charging.[17]

6.4.2.7 Reconditioning. The INTELSAT V spacecraft has two 27-cell, NiH_2 batteries, which are reconditioned two weeks before each eclipse season. The batteries are conditioned one at a time to keep one battery fully charged for power in case of emergency. It takes about a week to recondition each battery. The procedure for reconditioning is as follows[17]:

1) The battery is discharged through a 50-ohm resistor.

2) Individual cell voltages are scanned once per minute. Battery discharge is terminated when the first cell in the battery reaches 0.5 V.

3) The battery is recharged at the high rate (C/21 to C/25) until 115% of the amp-hour capacity removed has been returned ($c/d = 1.15$).

4) The battery is placed back on trickle charge until the beginning of the eclipse season at the C/96 rate.

There is some evidence that NiH_2 batteries do not have a memory and hence do not need reconditioning.

6.5 Power Distribution and Control

Power control electronics performs three functions: 1) *source control*, which is controlling the solar array; 2) *storage control*, which is controlling the battery; and 3) *output control*, which is controlling power served to the loads. Figure 6.41 shows these power control functions superimposed on a direct energy transfer system.

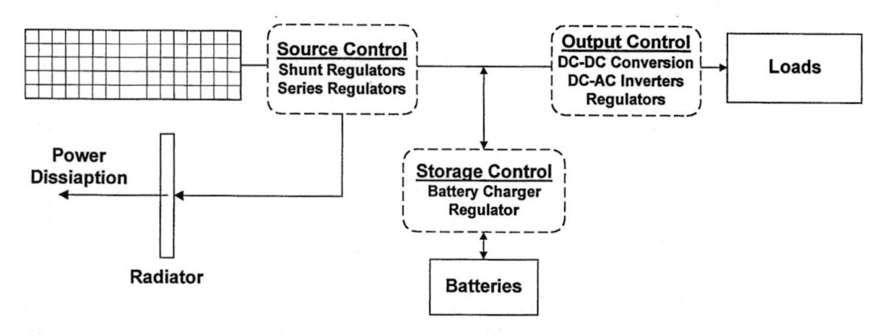

Fig. 6.41 Power control functions.

6.5.1 Source Control: Controlling the Array Output

There are two common approaches to source control: 1) A peak power tracker operates as a dc-dc converter in series with the solar array. It extracts the exact power that the loads require, up to the peak power of the array, and transforms the output voltage to the bus operating level. The PPT system makes efficient use of the array, but it consumes about 5% of the array power.

2) In a direct energy transfer system the panel power output in not controlled. Excess power generated by the panels is diverted to a shunt radiator, which converts the power to heat in a resistor bank and radiartes the generated heat to space. The shunt radiator and associated equipment is equipment is not needed in a PPT system.

6.5.2 Output Control: Controlling Bus Voltage

Selecting the nominal bus voltage is basically a tradeoff between the availability and flight history of 28-V equipment vs distribution power losses, as already discussed. Controlling the *variation* of bus voltage is a job of the power control electronics. A typical unregulated bus varies from 24 to 48 V. The lower voltage limit occurrs at the end of battery charging. The high voltage occurs when a cold solar array emerges from shadow into full sun (see Fig. 6.17). Much spacecraft equipment can tolerate unregulated bus voltage swings; the thermal control heaters for example. Some equipment cannot, notably the computers and science instruments.

There are three major options for regulating bus voltage: First is the *unregulated bus*, where sensitive equipment provides its own regulation with an unregulated input voltage. The attractiveness of this option depends on how much sensitive equipment there is. Second is *partial regulation*, where bus voltage is regulated during solar-panel operation in sunlit periods; battery discharge voltage is not regulated. Third is the *fully regulated bus*, where voltage from the solar panel and the battery is controlled. In large systems a fully regulated bus may be a lighter option than individual equipment regulation.

The power control system might also be required to deliver special power types to some equipment, for example, dc to dc voltage conversion or dc to ac.

6.6 Estimating Subsystem Mass

The mass-estimating relationships for the power subsystem are summarized in this section. The big mass items in the power subsystem are the batteries, solar array, control electronics, and cabling, as shown by the Magellan power system mass statement (Table 6.19). It is common practice to account for power and cabling as separate items at the subsystem level.

The solar-panel substrate, mechanisms, and bus attachment, amounting to 95.1 kg, are normally budgeted as part of the spacecraft structure; however, these masses are a direct function of solar-array size and must be considered in the power system design choices. The solar-cell assembly includes the cells, interconnects, adhesives, wiring, and diodes; in the Magellan case it also includes optical solar reflectors, which are small thin mirrors used to reflect the sun and keep panel temperature down. Thirty-five percent of the Magellan panel area was populated with OSRs instead of solar cells. This unusual protection was made necessary by the high solar intensity at Venus.

Table 6.19 Magellan power system mass statement[a]

Unit	Quantity	Mass, kg
Batteries, 26.5 A-h	2	31.25
Total		**62.5**
Solar array, 6.31 m^2 each	2	
Solar-cell assembly		23.6
Substrate	2	53.6
Mechanism	2	29.6
Bus attach structure	2	11.9
Total		**118.7**
Control equipment		
Preregulator	1	10.8
Power control unit	1	12.9
Power distribution unit	1	6.7
Pyro switching unit	1	4.1
Relay board	1	6.7
2.4 KHz inverter	2	4.6
Shunt regulator	2	8.6
Shunt radiator, 2.34 m^2 total	2	3.1
Signal processing unit	1	2.7
Fuse unit	1	3.5
Total		**63.7**
Cabling	—	55.6
Grand total		**300.5**

[a]Data courtesy of Lockheed Martin.

6.6.1 Solar Array

In comsat design it is traditional to use specific power to estimate solar-array mass. This practice works well because the geosynchronous environment is well defined. In considering a broader range of missions, it is better practice to calculate the required solar-array area under the expected conditions and then compute solar-array mass using a mass per unit area which describes your planned construction. This latter technique works for LEO, GEO, comsat, metsat, Earth observer, planetary, any cell type, spinner, or three-axis spacecraft. Table 6.20 shows solar-panel mass per unit area for various flight spacecraft.

The array mass in Table 6.20 includes cells, interconnects, panel wiring, and substrate. For three-axis-stabilized spacecraft Table 6.20 shows that flight solar panels weigh about 4.5 kg/m^2. One spacecraft, INTELSAT V, achieved 3.6 kg/m^2 with aluminum honeycomb substrate and graphite–epoxy face sheets. Laboratory panels[18] are achieving 3.5 kg/m^2. For body-mounted solar-arrays on spinning spacecraft, 3.4 kg/m^2 has been achieved several times.

All of the three-axis spacecraft shown in Table 6.20 used rigid, folding panels. Some flight experience is available with flexible roll-out arrays and flexible fold-out arrays (FRUSA). Table 6.21 shows solar-panel specific mass, which can be expected for future designs of each array type.

Table 6.20 Solar-panel mass[a]

Spacecraft	Launch year	Mass kg/m^2	Type	Mount type
Vela	1963	4.19	Si	Body
INTELSAT III	1968	3.36	Si	Body
INTELSAT IV-A	1971	3.65	Si	Body
OGO A	1964	6.46	Si	Panel
Mariner 3, 4	1964	4.9	Si	Panel, 4
OGO D	1967	6.43	Si	Panel
Nimbus	1969	7.98	Si	Panel, 2
NOAA-1	1970	6.64	Si	Panel, 2
INTELSAT V	1980	3.59	Si	Panel
Magellan	1989	4.39	Si	Panel, 2
TDRSS 6	1993	4.55	Si	Panel, 2

[a]Data, in part, courtesy of U.S. Air Force; Ref. 4, pp. 4–7.

Example 6.3 Solar-Array Mass

Calculate the mass of a two-panel solar array with a total area of 18.5 m^2. The panel construction is rigid foldout using honeycomb substrate; the spacecraft is three-axis stabilized. The estimated mass of the solar array system is as follows:

Solar cells, interconnects, cabling, and substrate, kg	(18.5)(4.0) =	74.0
Deployment and one-degree-of-freedom pointing mechanism, kg	(15)(2) =	30.0
Bus attachment structure, kg	(104)(0.15) =	15.6
Total, kg		**119.6**

6.6.2 *Batteries*

Specific energy of battery systems are shown in Table 6.22. The specific energies shown in Table 6.22 are for batteries. Specific energy for *cells*, especially NiH_2

Table 6.21 Solar-panel specific mass

Mass type	Body fixed (spinner)	Rigid fold-out	Flex roll-out	Flex fold-out (FRUSA)
Array mass, kg/m^2	3.4	4.0	3.6	1.90
Deploy/1 degree-of-freedom mechanism, mass	—	15 kg/panel	15 kg/panel	15 kg/panel
Bus attach mass	—	15% of array	15% of array	15% of array

Table 6.22 Battery specific energy

Orbit type	NiCd	NiH_2-IPV	NiH_2-CPV
LEO, W-h/kg	24	38[a]	42[b]
GEO, W-h/kg	24	49	55[b]

[a]NiH_2 batteries are still uncommon in LEO applications; 38 W-h/kg is based on HST.
[b]CPV batteries are still uncommon in flight use; this specific energy is an estimate from Ref. 19.

cells, is often quoted. The mass of an IPV battery is about 1.25 times the sum of the cell masses and about 1.13 times CPV cell mass.

6.6.3 Control Equipment

The mass of the power control and distribution equipment as a function of EOL solar-panel power output is shown in Fig. 6.42. These three programs indicate that the mass of power control and distribution equipment can be estimated as 0.07 kg/W of EOL power from the solar panels.

6.6.4 Cabling

Cable weights are difficult to estimate in the early phases of a program because of sensitivity to detailed spacecraft geometry and equipment arrangement. Cabling mass is not accurately known until the first cable mock up is made in the fabrication phase; large mass fluctuations are painful at that time. Work is going on investigating computer models for accurate early cable mass predictions, but these are not currently available. In early phases cable weight can best be estimated as a percentage of on-orbit dry weight. Twenty flight spacecraft programs were analyzed, and cable mass as a percentage of on-orbit dry mass showed the best fit and least scatter. Table 6.23 shows these results.

The percentages shown in Table 6.23 are the ratio of cable mass to on-orbit dry mass expressed as a percentage. Comsats are shown separately because all six spacecraft studied showed a lower cable mass percentage. The highest cable mass percentage and lowest percentage in each group are also shown. The comsats studied were INTELSATS IV, V, and VI. The other spacecraft category included

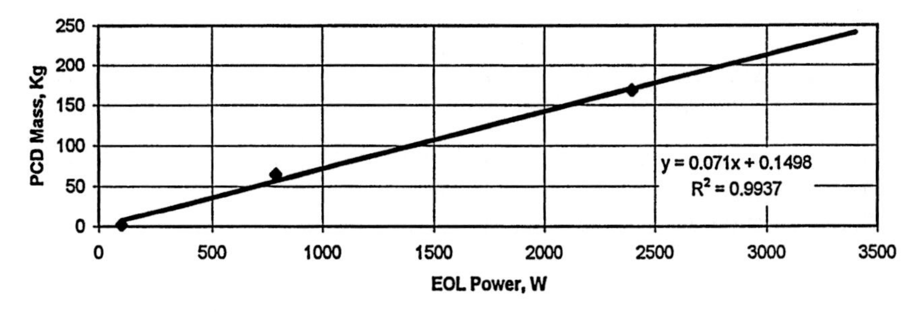

Fig. 6.42 Power control and distribution mass.

Table 6.23 Cabling mass as percentage of on-orbit dry mass

Category	Comsats	All other
Number of spacecraft	6	14
Average % on-orbit dry mass	**4.5%**	**6.8%**
Highest % on-orbit dry	5.8%	9.7%
Lowest % on-orbit dry	3.1%	4.9%

Mariner 2, 4, 5, 7, and 9, Magellan, Mars Orbiter, Mars Global Surveyor, Voyager, Viking Orbiter, Viking Lander, Galileo, DMSP, and NOAA-KL. It is recommended that a cable mass of 4.5% of on-orbit dry mass be used for comsats and a cable mass percentage of 6.8% be used for all other spacecraft types.

Cabling mass is usually carried as a subsystem level entry in the spacecraft mass summary and includes the mass of data cables as well as power cables.

References

[1]Griffin, M. D., and French, J. R., *Space Vehicle Design*, AIAA, Washington, DC, 1991.

[2]"Magellan Spacecraft Final Report," Martin Marietta, MGN-MA-011, Jet Propulsion Lab. Contract 956700, Denver, CO, Jan. 1995.

[3]*Guide for Estimating and Budgeting Weight and Power Contingencies*, AIAA and American National Standards Inst., AIAA-G-020-1992.

[4]Nagler, R. G., *Satellite Capability Handbook*, 624-3, Jet Propulsion Lab., Pasadena, CA, 1976.

[5]Agrawal, B. N., *Design of Geosynchronous Spacecraft*, Prentice–Hall, Upper Saddle River, NJ, 1986.

[6]"Space and Planetary Environment Criteria Guidelines for Use in Space Vehicle Development," NASA TM 82478, 1982.

[7]Wolf, G., and Marcoux, L., "Power System Advances," Hughes Aircraft Co., AIAA Professional Study Series, Culver City, CA, 1981.

[8]Holmberg, N. A., Faust, P. F., and Holt, H. M., "*Viking '75 Spacecraft Design and Test Summary*," NASA RP 1027, Washington, DC, Nov. 1980.

[9]Quaglione, G., "Evolution of the Intelsat System from Intelsat IV to Intelsat V," *Journal of Spacecraft and Rockets*, Vol. 17, No. 2, 1980, p. 67.

[10]Billerbeck, W. S., "Electric Power for State of the Art Communication Satellites," *Proceedings of the National Telemetering Conference '73*, IEEE, New York, Nov. 1973, pp. 19D1–19D12.

[11]"Hubble Space Telescope Media Reference Guide," Lockheed Missile and Space Co., Sunnyvale, CA, 1989.

[12]Cochran, C. D., Gorman, D. M., and Dumoulin, J. D. (eds.), *Space Handbook*, Air Univ. Press, Maxwell AFB, AL, 1985.

[13]Piscane, V. C., and Moore, R. C. (eds.), *Fundamentals of Space Systems*, Oxford Univ. Press, New York, 1994.

[14]Coates, D. K., Fox, C. L., and Miller, L. E., *Hydrogen Based Rechargeable Battery Systems*, Eagle Pitcher Industries, NASA Battery Workshop, Marshall Space Flight Center, Huntsville, AL, Nov. 1993.

[15]Smithrick, J. J., and O'Donnell, P. M., "Nickel Hydrogen Batteries—An Overview," NASA TM 106795, 1995.

[16]Van Ommering, G., Koehler, C. W., and Briggs, D. C., *Nickel-Hydrogen Batteries for Intelsat V*, Ford Aerospace and Communications Corp., AIAA, Washington, DC, 1980.

[17]Dunlop, J. D., Rao, G. M., and Yi, T. Y., "*NASA Handbook for Nickel-Hydrogen Batteries*," NASA RP 1314, Sept. 1993.

[18]Bekey, I., and Marzwell, N., "Rainbow's Array of Promises," *Aerospace America*, Jan. 1999.

[19]Wertz, J. R., and Larson, W. J., *Space Mission Analysis and Design*, Kluwer Academic, Norwell, MA, 1991.

[20]Coates, D. K., Fox, C. L., Standlee, D. J., and Grindstaff, B. K., "Advanced Nickel-Hydrogen Battery Development," Eagle Pitcher Industries, NASA Battery Workshop, Marshall Space Flight Center, Huntsville, AL, Nov. 1993.

[21]Allison, J. F., Arndt, R., and Meulenberg, A., "A Comparison of the COMSAT Violet and Nonreflective Solar Cells," *COMSAT Technical Review*, Vol. 5, No. 2, 1975, pp. 211–223.

[22]Gordon, G. D., and Morgan, W. L., *Principles of Communications Satellites*, Wiley, New York, 1993.

[23]"Single Pressure Vessel Nickel Hydrogen Batteries—The Next Evolutional Step in Hi-Tech Energy Storage," Eagle Pitcher Industries, Joplin, MO.

[24]Spilker, L. J., "Passage to a Ringed World—The Cassini–Huygens Mission to Saturn and Titan," NASA/JPL/Caltech, SP-533, Oct. 1997.

Problems

6.1 A surveillance spacecraft has the unique requirement to stow its solar array for a 20-min period while the spacecraft is in the sun. During this period, there is no power generated by the solar array. The spacecraft is in the sun a total of 60 min per orbit, including the 20-min period with the panel in the stowed condition, and is in the shade 30 min per orbit. The power used by the spacecraft while the solar arrays are operating is 300 W, and the power used while solar the solar panels are not operating is 150 W. The daytime transmission efficiency is 65%, the nighttime efficiency is 54%, and the efficiency with the solar panel stowed is 60%. All of these efficiencies are lumped products of the component efficiencies. Compute the minimum solar array power in watts.

6.2 Determine the depth of discharge following a 28-V, NiH_2, two-battery system. The time in darkness is 35 min/orbit. The power consumed in darkness is 625 W. The rated battery capacity is 8 A-h, and the power transmission efficiency, battery to bus, is 0.95. What depth of discharge is reached?

6.3 A malfunction in a three-axis-stabilized Earth orbiter causes the solar array to run hot at 91°C and to be misaligned. The orientation of the 1×2 m array is such that the solar vector is at 60 deg to the panel normal. What is the total output power of this array if it has the following properties: cell output power at 25 deg = 0.16 W/cell; packing factor; assembly factor = 0.955; no shadowing; loss caused by contamination = 0.99; and loss caused by radiation damage and all other aging effects = 0.88?

6.4 What electrical load can be accommodated by three 25 A-h, 28-V, NiH_2 batteries if the eclipse time is 27 min, the DOD is 30%, and the transmission efficiency from battery to load is 96%?

6.5 The solar-array output power of a low Earth orbiter is 500 W. The time in daylight is 60 min, and the eclipse time is 30 min. The lumped daytime transmission efficiency is 80%, and the lumped nighttime efficiency is 79%. If the average power consumed by the loads during the daytime is 260 W, what is the average power available for nighttime consumption.

6.6 How many 22 A-h batteries are required to power an eclipse load of 462 W at 28 V, for a duration of 46 min, with a maximum depth of discharge of 38%, and a loss, battery to load of 4.2%?

6.7 What solar-array area is required to provide 1500 W, measured after the diode, for the following conditions: $Po = 276$ mW/cell at 115°C; packing factor = 732 cells/m^2; utilization factor = 0.97; assembly factor = 0.96; diode loss = 3%. Assume all other factors are unity.

6.8 A Mars Orbiter solar array generated 1397 W, after the diode, at the end of mission in Mars orbit. There were two panels, each with an area of 0.56 m^2. The following factors apply: diode loss = 3%; assembly factor = 0.94; loss caused by degradation from all sources = 0.86; loss caused by contamination = 0.98; transmission efficiency, bus to shunt radiator = 0.93.

How much excess power was sent to the shunt radiator at beginning of life near Earth if the loads at BOL were 672 W? Assume panel temperatures are the same at BOL and EOL and shadowing is zero in both cases.

6.9 A new design communication spacecraft has a payload power requirement of 647 W. What is the predicted total power requirement for the spacecraft? What power margin would you recommend?

6.10 Calculate the following for a power system with the following parameters: payload power is 72.5 W; total power to subsystems is 307 W; array-to-loads transmission efficiency is 0.95; array-to-battery transmission efficiency is 0.70; battery-to-loads transmission efficiency is 0.96.

(Nighttime payload and subsystem power are the same as those just stated.)

(a) Determine the power margin using the AIAA recommendation in Chapter 2. Nighttime subsystem power and subsystem reserve are the same as daytime values.

(b) Determine the power requirement for full sun and eclipse.

(c) Calculate the maximum eclipse period and the minimum sunlight period for a circular Earth orbit at an altitude of 650 km.

(d) Calculate the required solar-panel output power.

(e) Calculate the required cycle life for the battery system. Use NiH_2 batteries.

(f) Calculate the required energy storage capability of the battery system in watt-hours.

(g) Calculate the required battery A-h capacity, number of batteries, and battery system mass to supply the required energy storage capability. Use 25 A-h, NiH_2

cells, and an IPV battery specific energy of 30 W-h/kg and a bus voltage of 27.5. Provide one battery-out capability.

(h) Prepare a mass estimate for the power system.

6.11 Calculate size and mass of solar panels to supply 1121 W EOL at Mars. Assume a radiation degradation of 18% over the design life; an array operating temperature of 120°C; a 5-deg maximum array pointing error; a laboratory cell output of 0.172 W/cell at 120°C; AM0, new; η_{uv} = UV discoloration degradation = 0.98; η_{cy} = thermal cycling degradation = 0.99; η_m = power loss caused by cell mismatch = 0.975; η_l = power loss caused by resistance in cell interconnects = 0.98; η_{con} = power loss caused by contamination from all sources = 0.99; and η_s = power loss caused by shadowing = 1.0 (none).

Use a mean distance from Mars to the Sun of 228 million kilometers. Use rigid fold-out panels.

6.12 What current is required to trickle charge a 22 A-h battery at a rate of C/38?

6.13 Determine the maximum beta angle for an eclipse free circular Earth orbit given an orbit altitude of 700 km.

6.14 Consider the following 28-V system: transfer efficiency, battery to load = 0.86; maximum DOD = 60%; nighttime power = 842 W; duration of night = 43 min. What is the ampere-hour battery capacity required?

Using a 60-mil vessel wall, what is the battery length and specific energy (see Fig. 6.40)?

7
Thermal Control
Robert K. McMordie*

The purpose of a spacecraft thermal control subsystem is to maintain all of the elements of a spacecraft system within their temperature limits for all mission phases. To provide a design that meets the temperature requirements of the spacecraft, the designer must account for heat inputs from many different sources. These heat sources include the sun, the Earth, and heat dissipation from electrical and electronic components onboard the spacecraft. In most instances the heat inputs are highly variable with time. For example, the magnitude of direct solar heating on a spacecraft surface can vary from 1371 W/m^2 (when the surface normal is parallel to the solar vector) to zero (when the spacecraft is in the shadow of the Earth). Also, the geometry of a spacecraft is generally very complex. These characteristics cause the thermal analyses that define the thermal behavior of spacecraft to be extremely complicated, involving transient solutions for problems with complex geometry. This leads to numerical solutions using specialized computer programs. Therefore, on a typical spacecraft project, the computer usage by the thermal control group will be relatively high. However, the thermal control subsystem accounts for only about 2–5% of the total spacecraft cost and about the same percentage of the weight.

Engineers assigned to spacecraft thermal control groups typically will have a degree in aerospace, mechanical, or chemical engineering. The major engineering discipline used in spacecraft thermal control is heat transfer; however, thermodynamics, applied mathematics, computer science, testing, material science, and instrumentation are also important.

Requirements for a spacecraft thermal control system are generally provided at several levels. Top-level system requirements define temperature margins, overall testing requirements, and environmental definitions (shown in Fig. 7.1), such as the direct solar, Earth-reflected solar (albedo), and Earth-emitted energy flux levels. The spacecraft design personnel develop derived requirements, for example, subsystem weight allocations and cost goals. Finally, temperature limits are defined by the various subsystem groups for their components based on supplier data.

Typical temperature ranges for spacecraft components are given in Table 7.1. Silicon solar cells can generally operate over a wide range; however, it is desirable to operate as low as possible in the range because the lower the operating temperature is the higher the solar cell efficiency. Normally, structures exhibit wide temperature limits. The exception is for a situation involving structures that support equipment, such as cameras, which have extremely accurate pointing requirements.

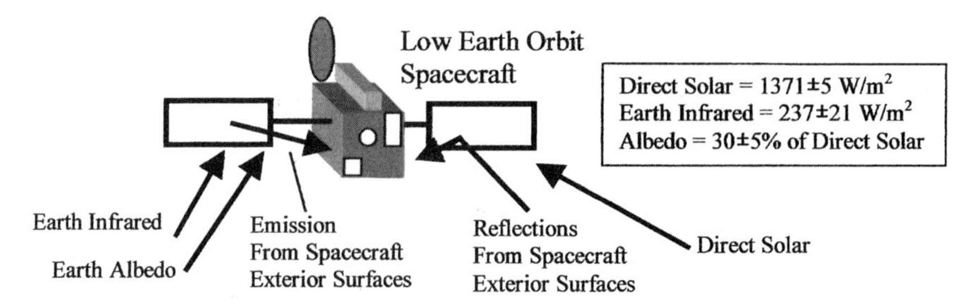

Fig. 7.1 Thermal radiation environment. (Values shown are from Ref. 1.)

In this situation the structural elements might have a temperature variation requirement of $\pm 0.5^{\circ}$C or less in order to minimize differential thermal expansion, which distorts the spacecraft shape and adversely affects pointing. The plastics shown are used for making multilayer insulation blankets. Note that the values given for nylon are an estimate. The temperature limits for hydrazine, a common monopropellant, are dictated by freezing on the low end and decomposition on the high end. Because of this, the typical thermal design for hydrazine lines is to cold bias the lines and use thermostatically controlled heaters to maintain the lines above the lower limit. Cold bias means to design a given region of the spacecraft so that it operates below the upper temperature limit for all equipment in the region. When this is done, usually some of the equipment in the region will operate below their lower temperature limit if nothing is done to prevent this. Therefore, thermostatically controlled heaters, which keep the equipment temperatures above their lower limits, are provided.

7.1 Relationship to Other Subsystems

A spacecraft thermal control subsystem affects and is affected by almost all other spacecraft subsystems. The power subsystem typically has the greatest interaction with the thermal control subsystem. This results from the necessity of accounting for all dissipated electrical energy and transferring this energy to a radiator for rejection to space. Also, batteries generally have a narrow temperature operating range and often require special attention from the thermal control engineers. If infrared (IR) sensors are part of the spacecraft payload, they can be a major problem

Table 7.1 Typical temperature ranges for selected spacecraft components

Component	Typical temperature range, °C	Component	Typical temperature range, °C
Batteries, NiCd	5 to 20	Mylar	−73 to 149
Batteries, NiH	−10 to 20	Nylon	−73 to 150
Electronics	0 to 40	Solar arrays	−100 to 100
Hydrazine	7 to 35	Structures	−46 to 65
Infrared detectors	−200 to −80	Teflon®	−240 to 204
Kapton	−269 to 400	—	—

for the thermal control subsystem. Sensors of this type are often required to operate at temperatures in the cryogenic range, 0–120°K, which demands, generally, a low-temperature mechanical refrigeration system and special low-temperature radiators. If the mission is of short duration, it is sometimes practical to use a stored, expendable cryogen (liquid helium, for example) for a low-temperature heat sink. Finally, spacecraft attitude control can have a significant influence on the thermal control subsystem design. The attitude of the spacecraft determines the thermal radiation inputs from the sun and Earth. The radiation inputs affect radiator performance and spacecraft temperatures.

7.2 General Approaches and Options

7.2.1 Subsystem Classification

Spacecraft thermal control subsystems are classified as active, passive, or semipassive. There are no universally accepted definitions for these three classifications; however, most spacecraft engineers consider pumped-loop systems, heaters controlled by thermostats, and mechanical refrigerators to be active systems. Space radiators thermally coupled to heat sources by conductive paths are generally considered passive subsystems. Systems that use louvers, variable conductance heat pipes, or thermal switches (discussed later) are often referred to as semipassive; however, these will be defined as passive systems for this discussion.

7.2.2 Active and Passive Subsystems

The first decision facing the thermal control engineer with respect to a spacecraft design is whether the subsystem should be active or passive. This is generally an easy decision as about 95% of the time unmanned spacecraft can by controlled passively and a passive design is generally lighter, requires less electrical power, and is less costly than an active design. Active thermal control subsystems are used for manned spacecraft, for situations requiring very close tolerance temperature control (a few degrees), or for components that dissipate large amounts of waste energy, on the order of several kilowatts. Examples of active and passive thermal control systems are shown in Fig. 7.2.

The cold plates shown in Fig. 7.2 are devices used for mounting heat dissipating equipment. The cold plate for the active subsystem is configured with fluid passages integral with the plate. For the active subsystem, thermal energy is dissipated as waste heat in the electrical equipment. This heat is transferred across the bolted interface to the cold plate. The fluid, which is circulated through the cold plate, then transports the waste heat to a radiator where it is rejected to space. The passive subsystem shown in Fig. 7.2 combines the cold plate with the radiator. Also, for the arrangement shown a phase change device has been added. This thermal control device absorbs thermal energy through a solid/liquid phase change and is used when electrical equipment has high, short bursts of power. By incorporating a phase change device for this situation, the phase change material, usually a paraffin, will reduce temperature spikes in proportion to the amount of paraffin used. During the period of heat dissipation, the paraffin melts and absorbs the waste heat. While the equipment is inactive, the phase-change material cools and solidifies. Detailed information concerning phase-change devices is given in Ref. 2.

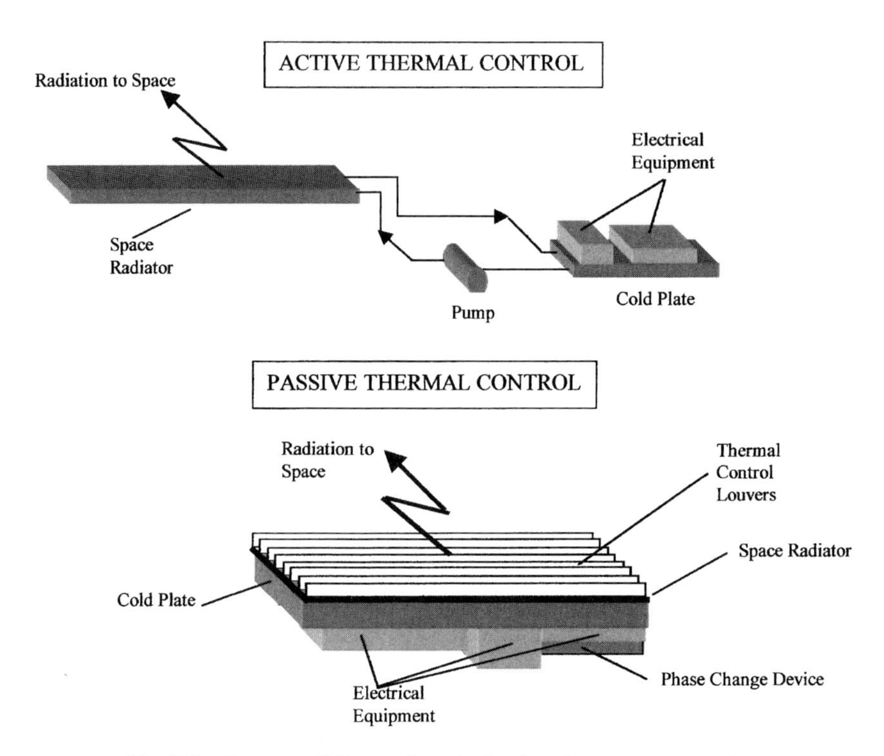

Fig. 7.2 Spacecraft thermal control subsystem arrangements.

7.2.3 Thermal Control Devices

7.2.3.1 Heat pipes. Heat pipes, illustrated in Fig. 7.3, have been flown on numerous spacecraft. These lightweight devices are used to transfer heat from one location to another. For example, a heat pipe might be used to transfer the heat dissipated in an electrical component to a space radiator. The construction of a heat pipe is generally a hermetically sealed tube with a wicking device on the inside surface of the tube. The wicking device might be grooves, as shown in the figure, or a mesh made of fabric or metal. A fluid is enclosed in the tube and fills the open center of the heat pipe with its gas phase and the wick with its liquid

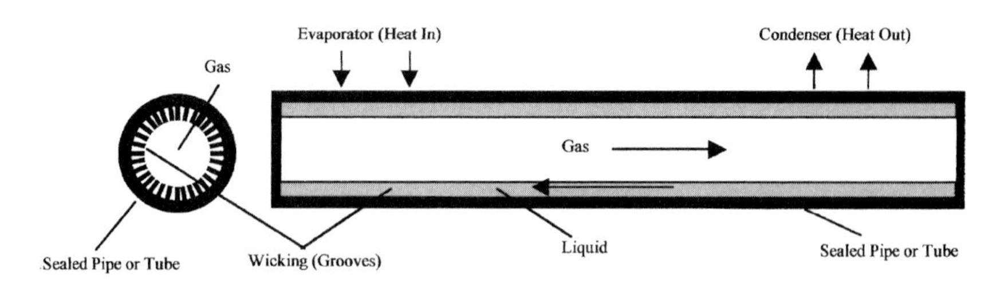

Fig. 7.3 Heat pipe.

phase. Heat applied at one end of the pipe will cause evaporation of the liquid in the wick. The gases formed by the evaporation flow down the center of the heat pipe to the opposite end. At this location heat is transferred from the pipe causing condensation to take place in the wicking material. Capillary forces draw the fluid from the condenser end to the evaporator end of the heat pipe, thus completing a heat-transfer loop. This loop occurs naturally when one end of the heat pipe is maintained at a higher temperature than the other. Because of the relatively high latent heat of evaporation of the heat pipe's working fluid, these devices exhibit high heat-transfer rates with small temperature differences from the evaporator to the condenser end of the heat pipe. Heat pipes can transfer 200–300 times as much energy as a solid copper bar of the same size!

The heat pipe shown in Fig. 7.3 is a fixed conductance pipe. Heat pipes can also be designed to exhibit variable conductance characteristics and, therefore, the ability to automatically modulate heat transfer. Heat pipes that are designed to operate at room temperature generally are made of aluminum with ammonia as the working fluid. The material selection is critical because some combinations of working fluids and pipe materials will cause a slow chemical reaction, which results in the generation of noncondensable gases. These gases collect in the condenser end of the pipe and prevent the normal condensation from taking place. This can severely affect the performance of the pipe. The typical performance of a 1.27-cm-diam heat pipe is about 5080 W-cm. This means that the heat pipe will transfer 508 W over a 10-cm distance. One disadvantage of heat pipes is that their performance in a 1-g environment is affected by the orientation of the axis of the pipe relative to the gravity vector. This means that for testing on Earth the axis of the pipe must be accurately positioned so that it is essentially horizontal. If this is done, the heat pipe's thermal performance in space and on the Earth will be about the same. However, for some heat pipe designs being off horizontal by a degree or less will change their performance significantly, relative to their performance in space.

7.2.3.2 Louvers. Louvers, shown in Fig. 7.2, act similar to venetian blinds and are positioned between a radiator surface and deep space. The rate of heat flow to space is modulated by the opening or closing of the louver blades. Bimetallic springs are typically used to actuate the louver blades. The springs are thermally coupled to the space radiator and react to the radiator temperature. At a temperature of 304°K with the blades wide open, the heat rejection rate will be about 430 W/m^2. Closed and at a temperature of 283°K, the rejection rate is on the order of 54 W/m^2. One problem encountered with louvers is the high temperature that results if the louvers are inadvertently pointed at the sun. Some spacecraft designs have included covers over the louvers to avoid this potential problem; however, a cover significantly decreases the thermal performance of the louvers.

7.2.3.3 Thermal switches. These devices are placed in thermal conduction paths with contacts that open and close in response to a temperature sensor. Opening the contacts breaks the conduction path; closing completes the path. A thermal switch was used on the Viking spacecraft. The switch contacts, which when closed, provided a direct conduction path between the equipment mounting plate (cold plate) and the heat source, a radioactive thermal generator. The actuator that controlled the movement of the contacts was driven by a gas charge that expanded and contracted in response to the temperature of the equipment mounting plate. A special feature of the device was the contacts design. Before Viking, thermal switches had been employed on spacecraft and had failed. The failure mode resulted from foreign particles in the contacts, which prevented the contacts from seating securely and, therefore, restricted the flow of heat at the contacts. The Viking design avoided this potential problem by being fitted with "soft" contacts. The design of the contacts included aluminum blocks coated with silicone grease and covered with a tinfoil.

This design allowed particles on the order of 0.025-cm diam and 0.6 cm long to be present on the contacts interface with no appreciable effect on the thermal switch's performance.

7.2.3.4 Thermal control coatings. These are surfaces that have special radiation properties that provide the desired thermal performance for the surface. Examples are paints, high-quality mirrors, and silverized plastics.

7.2.3.5 Thermal insulation. The primary insulation used on spacecraft is multilayer insulation. A typical arrangement is alternate layers of aluminized Mylar and a course net material. The net acts as a separator for the Mylar layers. Kapton, a plastic which is much stronger and has a higher service temperature that Mylar, is often used for the outer and inner layers of a multiplayer blanket. In some instances all of the layers of the blanket are made of Kapton.

7.2.3.6 Electrical heaters and thermostats. These active devices are generally used to thermally control regions that are cold biased (designed to run cold). The heaters, turned on and off by thermostats, heat a normally cold region and control the temperature within the required limits. This approach is typically used when fine temperature control is required.

7.2.3.7 Pumped loop systems. Pumped loop systems use a fluid that collects waste heat from equipment at a cold plate. The cold plate contains fluid passages, and the heat dissipating equipment is bolted to the cold plate. The fluid transports the waste heat to a space radiator, where it is rejected to space.

7.3 General Design Methodology

The fundamental responsibility of the thermal control engineer on a spacecraft project is to ensure that all elements of the spacecraft remain within their temperature limits during all mission phases. To accomplish this, first the temperature limits of all spacecraft components must be cataloged by the thermal control group with inputs from almost all other spacecraft engineering groups. Next, the thermal control engineer must establish the thermal boundary conditions for the mission. This is accomplished by determining from the mission analysis group the spacecraft attitude and orientation (relative to the Earth and the sun) for all

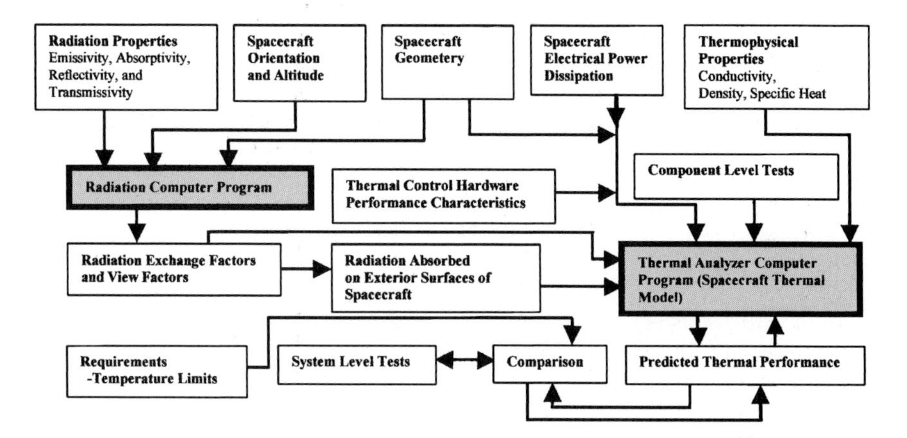

Fig. 7.4 Spacecraft thermal control design methodology.

mission phases. Also, from the power group the electrical power dissipation in all electrical components must be defined.

Once thermal boundary conditions are established, temperature limits defined, and other requirements understood, actual thermal design can be accomplished. The overall design process is illustrated in Fig. 7.4. Analytical tools used in the thermal design are typically two generalized computer programs. The first is a radiation program that defines the absorbed energy on the spacecraft external surfaces. This absorbed energy can come directly from the sun, solar energy reflected from the Earth (albedo), or IR energy emitted from the Earth. This same radiation program is also used to define all of the spacecraft surface radiation exchange factors and view factors. These factors define the radiation interaction between the spacecraft surfaces. After the radiation aspects of the overall problem are established, a heat-transfer model of the spacecraft is assembled. A generalized thermal analyzer is used for this model. This program can be run steady state or transient and accounts for conduction, radiation, convection, internal heat generation, and fluid flow. With this spacecraft thermal model one can inspect component temperatures during the mission and adjust the thermal control hardware elements of the spacecraft until all temperature limits are satisfied. Spacecraft thermal design must keep the temperature limits for a component "inside" the flight acceptance temperature limits of the component. For example, if a component must pass flight acceptance tests at low- and high-temperature levels of 5 and 50°C, with a thermal margin of 5°C, the spacecraft thermal control engineer must maintain the given component at 10 to 45°C, 5°C inside the flight acceptance limits. Thermal margins are analogous to safety factors used in structural design and are generally in the range of 5 to 10°C.

System-level thermal tests are generally required as a necessary part of a spacecraft program, and these tests fall into two major categories. The first of these is a solar balance test, and the other is a thermal vacuum test. Both of these type tests are conducted in a vacuum chamber equipped with nitrogen-cooled walls to thermally simulate deep space. The balance tests are performed usually with a solar simulator and often electric heaters or IR lamps to simulate Earth IR radiation. Solar simulation is accomplished using a light source (often filtered), which simulates the solar spectrum and the solar flux level expected during the mission. The purposes of balance tests are to establish experimentally the worst-case hot and worst-case cold thermal performance of the spacecraft and to validate the analytical thermal model of the spacecraft. If there are significant differences between the test results and the mathematical predictions, studies are performed to identify the reason(s) for the differences. Experience has shown that the differences can be the fault of either inaccuracies in the model or problems with the test. After close correspondence is achieved between analysis and test, the thermal model can be used with confidence in predicting the temperatures of the spacecraft for any flight situation.

Thermal vacuum tests are typically performed using electric heaters or IR lamps, with these heating sources controlled to force spacecraft temperatures to prescribed levels. Thermal vacuum tests are used to verify the workmanship of the spacecraft and provide a means of checking the operation of the spacecraft components in a simulated space environment.

7.4 Basic Analytical Equations and Relations

7.4.1 Introduction

Heat transfer is divided into three major areas: conduction, convection, and radiation. Conduction is associated with thermal energy transfer through matter in the absence of fluid motion. Convection is concerned with thermal energy transfer between a flowing fluid and a solid interface. Radiation is energy transfer via electromagnetic waves. Convection generally plays a minor role in low-Earth orbit or unmanned spacecraft; therefore, this heat-transfer mode will not be discussed. Heat-transfer references are given at the end of this chapter; see Refs. 3 and 9.

7.4.2 Conduction

The fundamental equation for steady-state heat conduction in rectangular coordinates and one dimension is

$$Q = \left(\frac{kA}{\Delta x}\right)(T_1 - T_2) \tag{7.1}$$

In Eq. (7.1) Q represents the energy transfer rate in watts, T the temperature in degrees Kelvin, k the thermal conductivity in W/(m-K), Δx the length of the heat-transfer path in meters, and A the area normal to the heat-transfer direction in square meters.

Figure 7.5 illustrates steady-state conduction for rectangular, cylindrical, and spherical coordinate systems. The relations shown in Fig. 7.5 are used in the development of thermal networks, which will be discussed later in this chapter.

7.4.3 Radiation

7.4.3.1 Radiation fundamentals. All matter radiates electromagnetic energy in the form of "small bundles of energy" called *photons*. Photons travel at the speed of light and have zero mass. The photons are emitted from the surfaces of

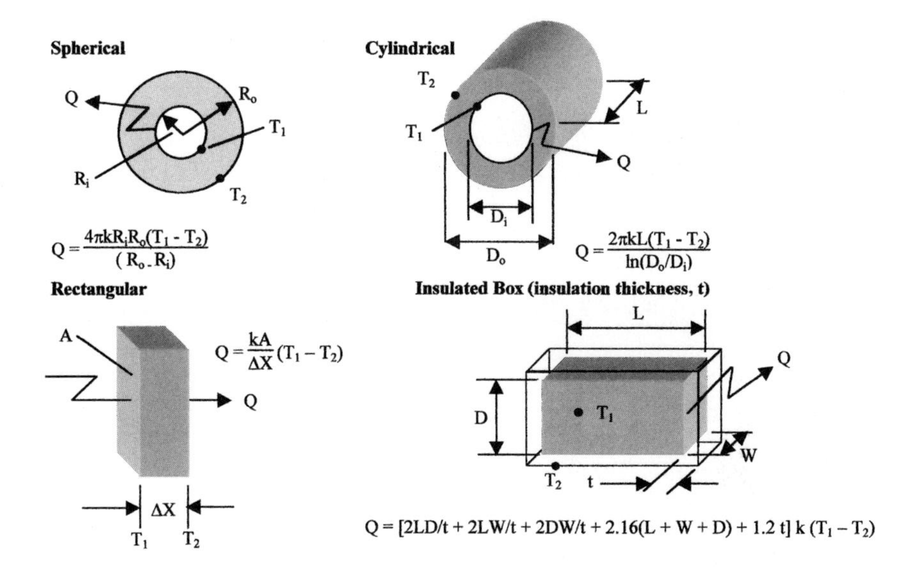

Fig. 7.5 Steady-state conduction equations.

all matter, and the general equation representing the total energy per unit time per unit surface area, $q(\mathrm{W/m^2})$ is given by

$$q = \varepsilon \sigma T^4 \tag{7.2}$$

Equation (7.2) was first enunciated by Stefan who established the relationship experimentally. Later Boltzmann developed the equation theoretically. The numerical value of the constant σ (known as the Stefan–Boltzmann constant) defined by Stefan in 1879 is very close to the currently accepted value of 5.67×10^{-8} W/(m^{2}K^4). The remaining terms in Eq. (7.2) are the emissivity ε, a dimensionless number, and T, the absolute temperature in degrees Kelvin.

Emissivity (or emittance) is a dimensionless thermophysical property that varies between zero and one. A perfect emitter has an emissivity of one and is called a blackbody. Certain black paints have emissivities very close to one. Polished gold or silver surfaces have emissivities in the neighborhood of 0.05.

Surfaces not only emit photons, but they also absorb photons. In an enclosure that is maintained at a uniform temperature, there is emission of radiant energy from all surfaces according to Eq. (7.2). For the objects within the enclosure to remain at a uniform temperature, which we know will indeed happen, there must be an absorption of radiant energy (photons) to balance the emission. This is the case, and it can be proven that at a given wavelength (radiation has a wave characteristic as well as a particle characteristic) the emissivity is identically equal to the absorptivity (Kirchoff's law). Absorptivity is defined as the percentage of incident radiant energy that is absorbed by a surface.

Surface radiation properties, including absorptivity, are described in Fig. 7.6. In this figure a transparent plate is irradiated by an incoming radiation source G. The incoming radiation will be either reflected, absorbed, or transmitted. The

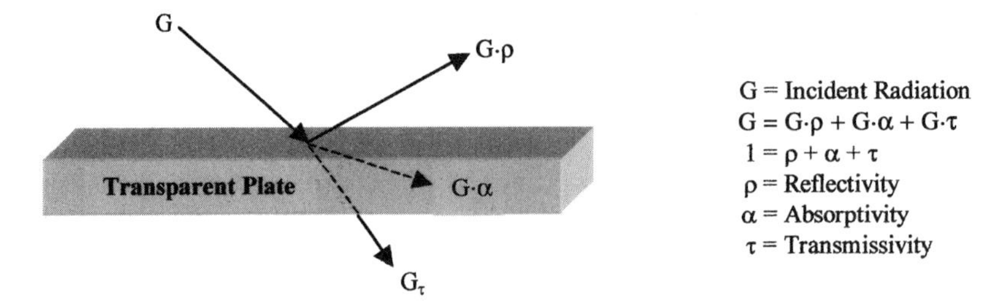

Fig. 7.6 Radiation surface properties.

percentage of incoming radiation that is reflected is defined as the reflectivity, the percentage absorbed is defined as absorptivity, and the percentage transmitted is defined as transmissivity.

As already mentioned, radiation exhibits the characteristics of waves. It is necessary to inspect thermal radiation from the standpoint of electromagnetic waves to be able to understand the principals of spacecraft thermal control.

Let us focus again on the radiant energy emitted from a surface. The equation that defines the monochromatic radiation energy (radiation energy at a given wavelength) $E_{b\lambda}$, emitted from the surface of a blackbody (emissivity $= 1$), is given by Planck's equation:

$$E_{b\lambda} = \frac{C_1 \lambda^{-5}}{\left(e^{\frac{C_2}{\lambda T}} - 1\right)} \tag{7.3}$$

In Eq. (7.3) $E_{b\lambda}$ is the monochromatic emissive power (emissive power at a given wavelength or energy per unit time per unit area per unit wave length), C_1 and C_2 are constants, λ is the wavelength, and T is the absolute temperature. For a nonblackbody the monochromatic emissivity (emissivity at a given wavelength) ε_λ must be included in Planck's equation. To determine the total emission from a surface, Eq. (7.3), along with ε_λ, must be integrated between 0 and ∞, as shown in Eq. (7.4):

$$q = \int_0^{\infty} \frac{\varepsilon_\lambda C_1 \lambda^{-5}}{e^{\frac{C_2}{\lambda T}} - 1} \, \mathrm{d}\lambda \tag{7.4}$$

If ε_λ is a constant, independent of wavelength, Eq. (7.4) reduces to Eq. (7.2), the Stefan–Boltzmann equation.

Equation (7.3) was used to generate the curves shown in the upper part of Fig. 7.7. These curves are plots of normalized emissive power (for a given temperature) vs wavelength. The normalized emissive power is the ratio of the monochromatic emissive power to the maximum monochromatic emissive power for the given temperature. The use of normalized emissive power allows the curves that correspond to the solar temperature level and room temperature level (294°K) to be shown on the same graph. The maximum value for the solar emissive power is several orders of magnitude larger than the room-temperature maximum emissive power, and therefore the room temperature curve would be hard to see on a normal graph. The curve labeled "solar" is a normalized plot of Planck's equation using a temperature consistent with the effective temperature of the Sun, about 5800°K.

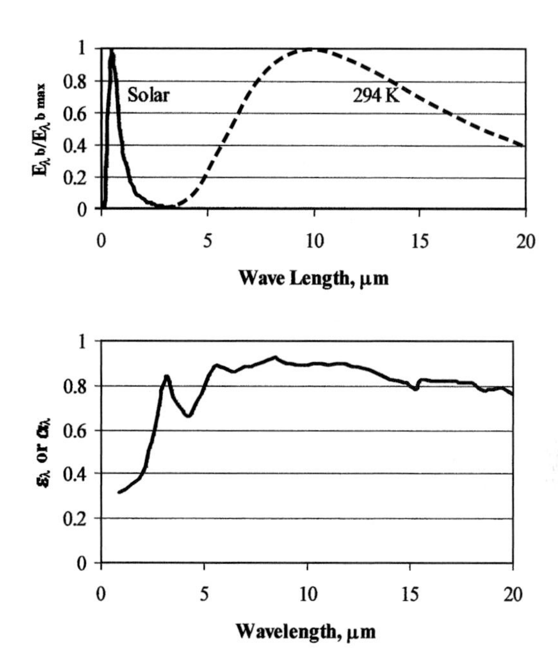

Fig. 7.7 Radiation quantities used to illustrate thermal performance of white painted surface with solar input. (White paint properties courtesy of NASA; Ref. 3, Vol. 1, p. 149.)

The curve labeled "294 K" is the same type of plot representing the normalized emissive power of a surface at a room temperature level. The important feature of thermal radiation illustrated by this plot is that solar radiation is concentrated in the short wavelengths, whereas room temperature emission is concentrated in the long wavelengths.

The emissivity of white paint vs wavelength is shown in the plot at the bottom of Fig. 7.7. The curve is also the absorptivity vs wavelength because monochromatic emissivity is identically equal to monochromatic absorptivity. The white paint properties ε_λ and α_λ will now be used along with the characteristics of the emissive powers curves to show why a white painted surface exposed to solar radiation in space will exhibit a relatively low temperature. To show this, it is necessary to determine the absorbed solar radiation on a white surface and balance this heat input with the radiation emitted from the surface. It is assumed that the back side of the white surface (the side of the plate away from the solar input) is perfectly insulated so that the absorbed solar energy and the emitted (infrared) energy are the only quantities involved in the energy balance on the white surface. To estimate the absorbed radiation by the white paint resulting from incoming solar radiation, one replaces the emissivity in Eq. (7.4) with the absorptivity of the white paint and uses an effective solar temperature:

$$q_{\text{absorbed}} = \int_0^\infty \frac{F\alpha_\lambda C_1 \lambda^{-5}}{e^{\frac{C_2}{\lambda T}} - 1}\, \mathrm{d}\lambda \tag{7.5}$$

The constant factor F in Eq. (7.5) reduces the flux level from the value at the surface of the sun to the value at the Earth and is numerically equal to the ratio of the square of the sun's radius to the square of the distance from the sun to Earth. The results of the integration will yield a relatively low value because the absorptivity in the solar wave band (0 to 3 μm) is low. The absorbed solar energy must be balanced by the emission from the white painted surface.

The radiation emission is determined using Eq. (7.4). The expression for the monochromatic emissivity is taken from the curve for white paint on Fig. 7.7, and the temperature is varied until the results from the integration of Eq. (7.4) equals the absorbed solar radiation on the surface from Eq. (7.5). Results show that the temperature, which satisfies the energy balance, will be relatively low because the monochromatic emissivity for white paint is relatively high over the long (infrared) wavelengths. To summarize, the white paint absorbs weakly in the short solar wave band, but emits strongly in the long infrared wave band. The result is that white surfaces run relatively cool in sunlight.

7.4.3.2 Selective surfaces. Surfaces with radically different emissivity (or absorptivity) values in the solar wave band as compared to the IR wave band are called *selective surfaces*. The mean average value of the absorptivity, over the solar wave band, is called *solar absorptivity* α_s, whereas the mean average value of emissivity over the IR wave band is called *IR emissivity* $\varepsilon_{\mathrm{IR}}$. IR emissivity is a weak function of surface temperature, and generally the temperature must be a few hundred degrees different from room temperature to show a significant change from the room temperature value. Most spacecraft thermal analyses are performed using the mean average values α_s and $\varepsilon_{\mathrm{IR}}$ and do not require use of the integral equations just shown [Eqs. (7.4) and (7.5)].

The type of selective surface that will produce the lowest temperature when irradiated by solar energy in space is an optical solar reflector. This type of surface is shown in Fig. 7.8. An optical solar reflector is composed of a substrate with a highly reflective surface overlaid with a transparent cover. Incoming radiation is partially reflected off the outer surface of the transparent cover, partially absorbed in the cover material, but primarily transmitted to the reflective surface. At the reflective surface the majority of the transmitted radiation is reflected back to space. Highly reflective surfaces are inherently metallic surfaces; however, metallic surfaces inherently exhibit relatively low emissivities. Therefore, the reflective surface will not contribute much to radiation emission. However, the outer surface

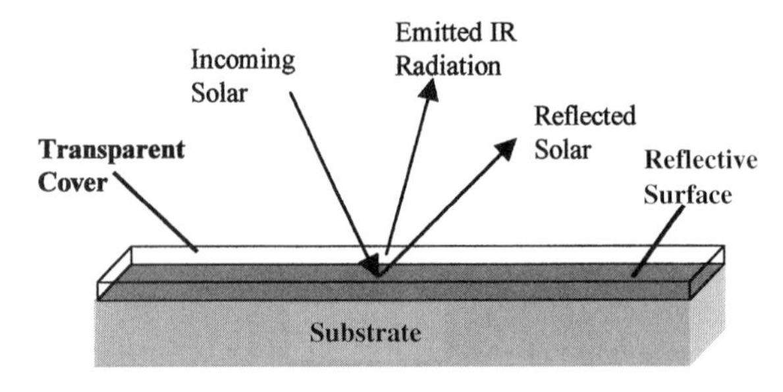

Fig. 7.8 Optical solar reflector.

of the transparent material being a nonmetallic will generally have a high emissivity and, therefore, is a major source of emission. Because the cover is extremely thin, the temperatures of the cover and the substrate are essentially the same. The overall effect is that the optical solar reflector will have a relatively high effective emissivity, primarily a result of the transparent material, but a lowsolar absorptivity resulting from the high reflectivity of the metallic surface. Representative values for optical solar reflectors are an IR emissivity of 0.8 and a solar absorptivity of 0.15.

7.4.3.3 Surface temperature predictions. If a surface with an IR emissivity of 0.8 and a solar absorptivity of 0.15 is perfectly insulated on the back side, placed in space, and irradiated with solar energy normal to the surface, the following energy balance would occur:

$$q_{\text{absorbed}} = q_{\text{emitted}} \tag{7.6}$$

$$G_s\alpha_s = \varepsilon_{\text{IR}}\sigma T^4 \tag{7.7}$$

Solving for the temperature T,

$$T = \left(\frac{G_s\alpha_s}{\varepsilon_{\text{IR}}\sigma}\right)^{\frac{1}{4}} \tag{7.8}$$

$$T = \left[\frac{(1371)(0.15)}{(0.8)(5.67\times10^{-8})}\right]^{\frac{1}{4}} = 259.5^\circ\text{K} \tag{7.9}$$

In this example G_s is the solar flux with a value of 1371 W/m^2, α_s is the solar absorptivity of the optical solar reflector, ε_{IR} is the infrared emissivity of the optical solar reflector, and σ is the Stefan–Boltzmann constant, 5.67×10^{-8} W/(m^2-K^4). The radiation surface properties α_s and ε_{IR} are mean average values. These quantities are used routinely in radiation analysis; values of these properties are given in Table 7.2.

7.4.3.4 Radiator temperatures. Equation (7.7) can be used to estimate the temperature of insulated surfaces with solar heating. If the solar vector is not normal to the surface, multiply the left side of Eq. (7.7) by $\cos\theta$. The angle θ is the angle between the surface normal and the solar vector. Equation (7.7) can also be expanded to provide a means of estimating the performance of space radiators. This is accomplished by adding the waste heat term (Q_W/A_R) to the left side of Eq. (7.7). Q_w is the waste heat rejected by the radiator in watts, and A_R is the radiator area in square meters. The resulting equation is

$$G_s\alpha_s\cos\theta + \frac{Q_W}{A_R} = \sigma\varepsilon_{\text{IR}}T^4 \tag{7.10}$$

This equation can be used to estimate the heat rejection Q_w if the radiator operates at a given temperature T or, conversely, the required temperature to reject a given amount of heat.

Table 7.2 Radiation properties

No.[a]	Material	Measurement temperature, K	Surface condition	Solar absorptivity	Infrared emissivity	Absorptivity/emissivity ratio	Equilibrium[b] temperature, K
1	Aluminum (6061-T4)	294	As received	0.379	0.0346	10.95	717
2	Aluminum (6061-T4)	422	As received	0.379	0.0383	9.90	699
3	Aluminum (6061-T4)	294	Polished	0.2	0.031	6.45	628
4	Aluminum (6061-T4)	422	Polished	0.2	0.034	5.88	614
5	Gold	294	As rolled	0.299	0.023	13.0	749
6	Steel (AM 350)	294	As received	0.567	0.267	2.12	476
7	Steel (AM 350)	422	As received	0.567	0.317	1.79	456
8	Steel (AM 350)	611	As received	0.567	0.353	1.61	444
9	Steel (AM 350)	811	As received	0.567	0.375	1.51	437
10	Steel (AM 350)	294	Polished	0.357	0.095	3.76	549
11	Steel (AM 350)	422	Polished	0.357	0.111	3.22	528
12	Steel (AM 350)	611	Polished	0.357	0.135	2.64	503
13	Steel (AM 350)	811	Polished	0.357	0.155	2.30	486
14	Titanium (6AL-4V)	294	As received	0.57	0.164	3.48	538
15	Titanium (6AL-4V)	422	As received	0.57	0.197	2.89	514
16	Titanium (6AL-4V)	294	Polished	0.448	0.129	3.47	538
17	Titanium (6AL-4V)	422	Polished	0.448	0.148	3.03	520
18	White enamel	294	Al. substrate	0.252	0.853	0.30	291
19	White epoxy	294	Al. substrate	0.248	0.924	0.27	284
20	White epoxy	422	Al. substrate	0.248	0.888	0.28	287
21	Black paint	294	Al. substrate	0.975	0.874	1.12	406
22	Silverized FEP Teflon	295	—	0.08	0.66	0.12	232
23	Aluminized FEP Teflon	295	—	0.163	0.8	0.20	264
24	OSR (quartz over silver)	295	—	0.077	0.79	0.10	220
25	Solar cell-fused silica cover	—	—	0.805	0.825	0.98	392

[a]Items 1–21, see Ref. 4; items 22–24, see Refs. 5 and 6; item 25, see Ref. 7.
[b]Temperature of a perfectly insulated flat plate located in space with a solar heat input of 1371 W/m^2.

7.4.3.5 View factors. The expression for the net radiation exchange Q_{net} between two surfaces that are perfect emitters ($\varepsilon = 1$) and perfect absorbers ($\alpha = 1$), that is, black surfaces, is given by

$$Q_{\text{Net}1-2} = A_1 F_{1-2} \sigma \left(T_1^4 - T_2^4\right) \tag{7.11}$$

$$Q_{\text{Net}2-1} = A_2 F_{2-1} \sigma \left(T_2^4 - T_1^4\right) \tag{7.12}$$

In Eq. (7.12) the absolute temperatures of the surfaces are T_1 and T_2, σ is the Stefan–Boltzmann constant, A is area, and the F terms are view factors. A radiation view factor is defined as that fraction of the radiant energy directly incident on a receiving surface relative to the total radiant energy leaving the sending surface. For example, if the view factor $F_{1-2} = 0.5$, this means that one-half of the radiant energy leaving surface 1 will strike surface 2 via a direct path from surface 1 to surface 2. The term "direct path" means that radiation from surface 1 that arrives at surface 2 because of reflections off of other surfaces is not included in the view factor computation. Two useful characteristics of view factors are that the sum of all of the view factors from a surface must equal 1 and the reciprocity rule $A_x F_{x-y} = A_y F_{y-x}$. View factors, in general, are extremely difficult to evaluate except for the simplest geometry. Fortunately, there are existing solutions for many situations, as illustrated in Fig. 7.9. Also, there are powerful computer programs, such as TRASYS (Thermal Radiation Analysis System), that can be used to evaluate view factors for essentially any geometry. TRASYS was developed by Martin Marietta under contract to NASA Johnson Space Center.

The utility of the data given in Fig. 7.9 can be greatly expanded using view factor algebra. This concept is beyond the scope of this text, however. The development of this technique is discussed in Ref. 9.

7.4.3.6 Oppenheim radiation networks. Equation (7.12) provides a means of calculating radiant energy exchange between "black" surfaces. In Fig. 7.10 an approach is presented that can be used to determine the radiation exchange between nonblack surfaces. The restrictions for this approach, developed by Oppenheim,[10] are as follows:

1) All surfaces are "gray," that is, all surface emissivities are constant over the wavelength bands applicable to the operating temperatures.

2) All surfaces emit and reflect diffusely. (Most surfaces can be considered diffuse except for shiny metallic surfaces.)

3) All surfaces are isothermal.

4) The radiosities are uniform across each surface. Radiosity is defined as the total radiant energy striking a surface including emitted, reflected, re-reflected, etc. (For most radiation problems it is reasonable to assume uniform radiosity.)

The characteristics of the Oppenheim network are such that at each given surface, say surface x, the adjacent conductor is of the form $A_x \varepsilon_x / (1 - \varepsilon_x)$, where A is area and ε is emissivity of the surface. This conductor is connected between the surface node and the dummy J_x node. The potential at the surface node is $\sigma(T_x)^4$ and $\sigma(T_J)^4$ at the J node. The network is completed by connecting each J node to every other J node. The conductors connecting the J nodes are of the form $A_x F_{x-y}$, where A is area and F the view factor. If a given surface is perfectly insulated or if the emissivity of the node is one (a black surface), the conductor

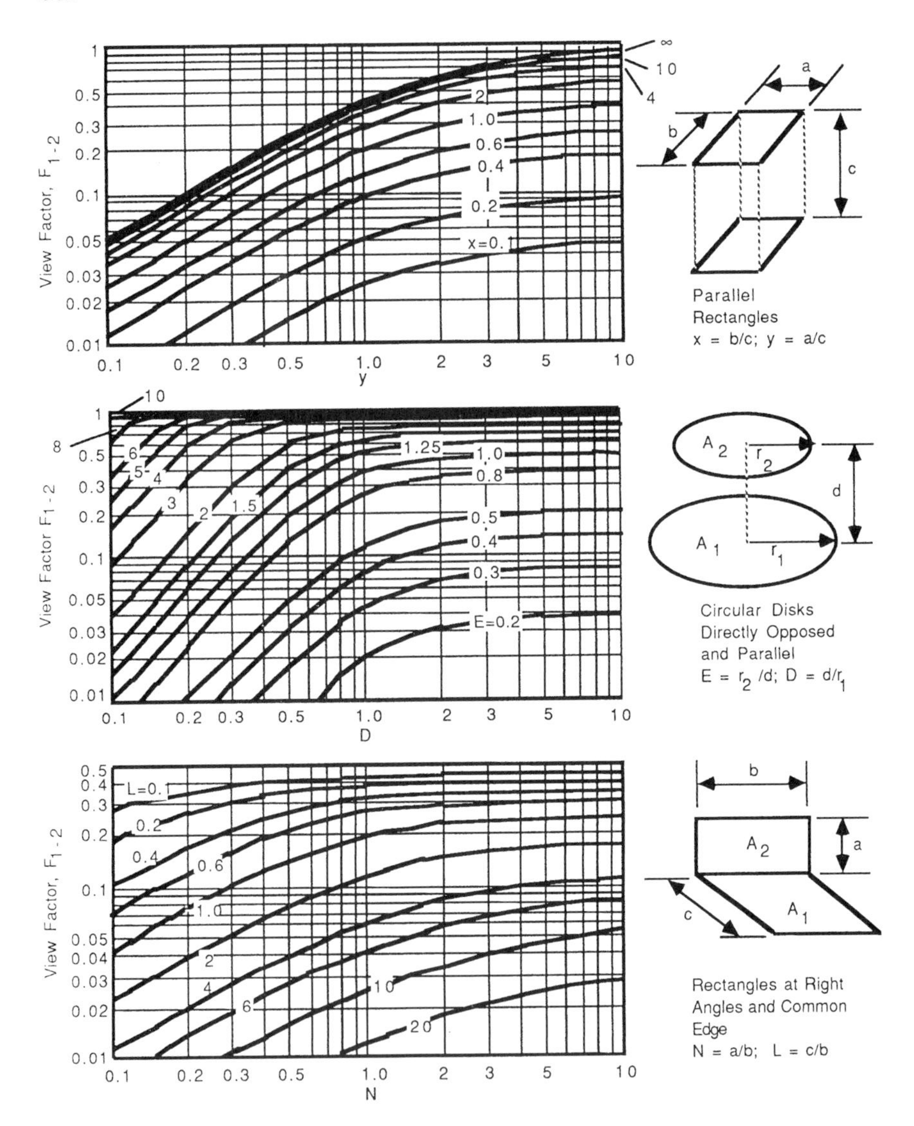

Fig. 7.9 Radiation view factors; see Ref. 8.

$A_x\varepsilon_x/(1 - \varepsilon_x)$ is eliminated. The effect is, therefore, that the surface potential $\sigma(T_x)^4$ moves to the J node location.

7.4.3.7 Sample problem using an Oppenheim network. The Oppenheim approach for analyzing radiation problems is an extremely useful tool. An example problem using this technique is illustrated in Fig. 7.11. The first step in the solution is to determine the necessary view factors. Using Fig. 7.9, the view factor from the radiator to the wall is found to be 0.29. Because the sum of all view factors from the radiator must equal unity, the view factor from the radiator

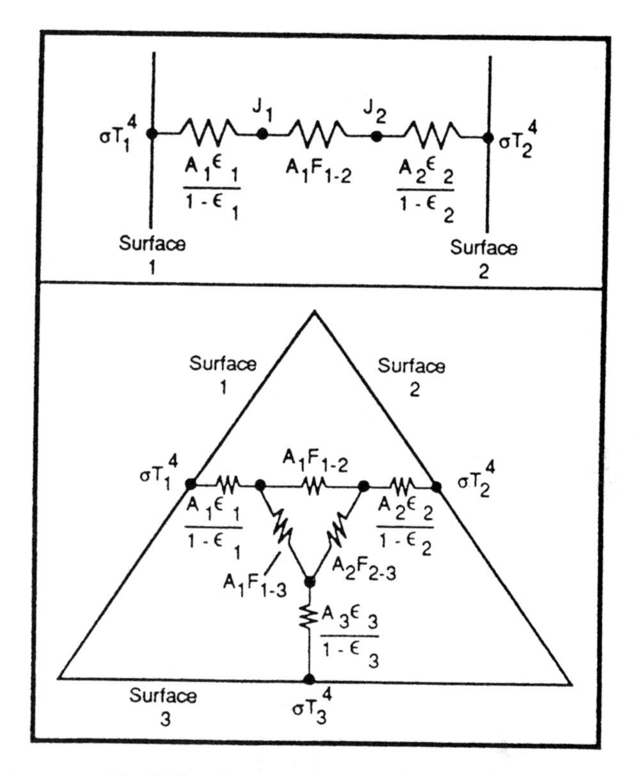

Fig. 7.10 Oppenheim radiation networks.

to space equals $(1 - F_{R-W}) = 1 - 0.29 = 0.71$. The view factor from the wall to space is found by first using reciprocity to find the view factor from the wall to the radiator $A_R F_{R-W} = A_W F_{W-R}$. This results in a value of 0.145 for F_{W-R}. All of the view factors from the wall must sum to unity; therefore, the view factor from the wall to space F_{W-S} equals $(1 - F_{W-R}) = 1 - 0.145 = 0.855$.

The next step in the solution is the creation of the Oppenheim network. The network on the lower left-hand side of Fig. 7.11 is basic representation for the problem with $A\varepsilon/(1 - \varepsilon)$ type conductors connecting the surface nodes (with σT^4 potentials) to the J nodes. The J nodes are connected to each other with conductors made up of the area times the view factor. The network on the left of Fig. 7.11 can be immediately reduced to the network shown on the right side of the figure by noting the following characteristics of Oppenheim networks. First, the net heat flow to a surface is given by the conductor value at the surface node $[A\varepsilon/(1 - \varepsilon)]$ times the adjoining potential difference $\sigma T^4 - J$. If a surface is perfectly insulated, there can be no net heat transfer to the surface; therefore, for this situation $\sigma T^4 = J$ because the conductor $[A\varepsilon/(1 - \varepsilon)]$ is a finite number. This causes the elimination of the conductor $[A_W \varepsilon_W/(1 - \varepsilon_W)]$. If a surface is "black" with an emissivity of one, the conductor adjacent to the surface goes to infinity. This has the effect of shorting the conductor and causes the node surface potential (σT^4) to be one-and-the-same with the J node. This allows the elimination of the $A_s \varepsilon_s/(1 - \varepsilon_s)$ conductor at the space node.

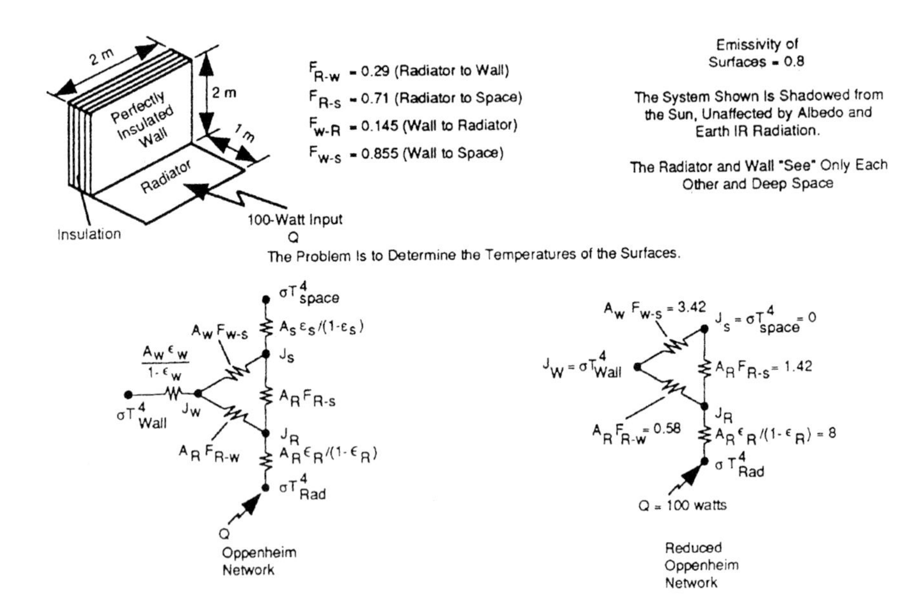

Fig. 7.11 Sample problem using Oppenheim networks.

The reduced Oppenheim network shown in Fig. 7.11 represents a network with all conductors defined and given numerical values. The unknowns are the potentials σT_{Rad}^4, J_R, and J_W. Energy balances are written at each of these points, resulting in the following set of linear equations:

$$100 + 8\left(J_R - \sigma T_{\text{Rad}}^4\right) = 0 \tag{7.13}$$

$$8\left(\sigma T_{\text{Rad}}^4 - J_R\right) + 0.58(J_W - J_R) + 1.42(0 - J_R) = 0 \tag{7.14}$$

$$0.58(J_R - J_W) + 3.42(0 - J_W) = 0 \tag{7.15}$$

The solution of this set of equations results in

$$\sigma T_{\text{Rad}}^4 = 64.69$$

$$J_R = 52.19$$

$$J_W = \sigma T_{\text{Wall}}^4 = 7.568$$

Therefore,

$$T_{\text{Rad}} = \left(\frac{64.69}{5.67 \times 10^{-8}}\right)^{\frac{1}{4}} = 183.8^\circ\text{K}$$

$$T_{\text{wall}} = \left(\frac{7.568}{5.67 \times 10^{-8}}\right)^{\frac{1}{4}} = 107.5^\circ\text{K}$$

7.4.4 Combined Radiation and Conduction Thermal Networks

The network approach of modeling is convenient for organizing and visualizing radiation problems. Conduction heat-transfer problems are also commonly modeled using thermal networks. For conduction problems the potential at node points is the temperature to the first power, not temperature to the fourth power as is the case in radiation. The conductor value is the quantity that is multiplied by the temperature difference in the steady-state conduction equations, shown later in this section. For example, the conductor for rectangular geometry is $kA/\Delta x$ (conductivity times cross-sectional area divided by distance).

For most spacecraft thermal control problems, conduction and radiation heat-transfer modes are interrelated. This forces the coupling of conduction and radiation and makes it highly desirable to combine the conduction and radiation networks. The problem with combining the two networks is that the potentials are vastly different: conduction to the first power and radiation to the fourth power of the temperature. This difference is overcome by manipulating the radiation relationship. In general the radiation equation is given by $q =$ (a coefficient) times $(\sigma T_x^4 - \sigma T_y^4)$. The coefficient is of the form $\varepsilon_x A_x/(1 - \varepsilon_x)$ or $A_x F_{x-y}$. Working with the temperature to the fourth-power term, first the Stefan–Boltzmann constant is factored, and then the remaining temperature term is also factored to yield $[(T_x^2 + T_y^2)(T_x + T_y)](T_x - T_y)]$. This operation transforms the radiation equation to a form that includes the temperature difference to the first power with the remaining terms forming a temperature-dependent conductor. The radiation conductor values are given by

$$\left(\frac{\sigma \varepsilon_x A_x}{1 - \varepsilon_x}\right)\left[(T_x^2 + T_y^2)(T_x + T_y)\right] \tag{7.16}$$

or

$$(\sigma A_x F_{x-y})\left[\left(T_x^2 + T_y^2\right)(T_x + T_y)\right] \tag{7.17}$$

7.4.5 Generalized Thermal Networks

Generalized equations for solving thermal networks will now be developed. For this development the steady-state networks discussed to this point have been extended to include transient conditions. The only difference in the steady-state and transient cases is that a thermal capacitance has been added at the node points. The nodes represent finite volumes or pieces of the physical thermal system, and the thermal capacitance of a node is the node mass times the specific heat of the node material. There are two formulations of the generalized equation in the following paragraphs: an implicit and an explicit solution. The explicit solution allows a step-by-step application of the equation at each node until all nodes are updated and the time step is complete. The implicit solution requires the inversion of a matrix involving all of the nodes at each time step. When using the explicit approach, the size of the time step is limited to a critical time step, and if this limit is exceeded, numerical instabilities might occur. This restriction does not apply to the implicit approach; however, one must be careful and generally not exceed the explicit critical time step by more than a factor of two or three in order to maintain accuracy. Steady-state solutions are readily analyzed using the given transient

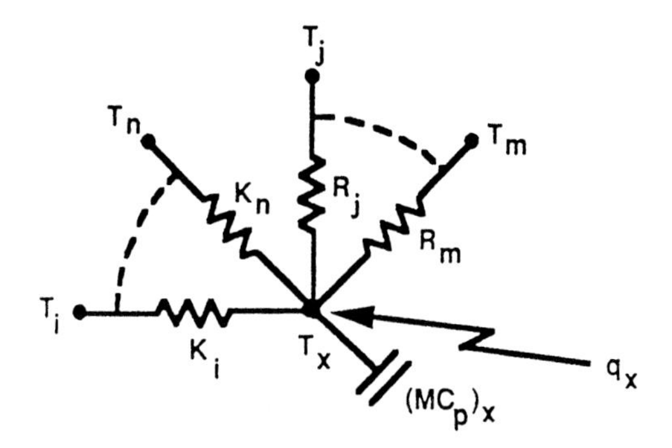

Fig. 7.12 Generalized thermal network.

equations by merely continuing the computations until the temperatures no longer change.

Equations (7.20) and (7.21) correspond to Fig. 7.12 and are used in the generalized thermal analyzer computer programs that thermally model spacecraft. These equations, along with thermal networks, can also be readily input into personal computer spreadsheet programs to support preliminary thermal analyses.

An energy balance on node x in Fig. 7.12 yields

$$\sum_{i}^{n} K_i(T_i - T_x) + \sum_{j}^{m} R_j(T_j - T_x) + q_x = (MC_p)_x \frac{\mathrm{d}T_x}{\mathrm{d}t} \cong (MC_p)_x \frac{T'_x - T_x}{\Delta t} \tag{7.18}$$

$$R = \sigma B\left[\left(T_j^2 - T_x^2\right)(T_j - T_x)\right] \tag{7.19}$$

where

B = $A\varepsilon/(1-\varepsilon)$ or $A_x F_{x-j}$
K = $kA/\Delta x$
k = conductivity
A = area
Δx = distance
M = mass of node
C_p = specific heat
Δt = time step
q_x = heat input, for example, solar heating
T_x is the temperature at the beginning of the time step or at the old time

T'_x is the temperature at the new time after a time step. Solving Eq. (7.18) for T'_x and choosing all of the temperatures on the left side of the equation at the new time, the implicit solution is derived:

$$T'_x = T_x + \frac{\Delta t}{(MC_p)_x}\left[\sum_{i}^{n} K_i(T'_i - T'_x) + \sum_{j}^{m} R_j(T'_j - T'_x) + q_x\right] \tag{7.20}$$

If the temperatures on the left side of Eq. (7.18) are taken at the old time, the explicit solution is derived:

$$T_x' = T_x + \frac{\Delta t}{(MC_p)_x}\left[\sum_i^n K_i(T_i - T_x) + \sum_j^m R_j(T_j - T_x) + q_x\right] \quad (7.21)$$

$$\Delta t_c = \frac{1}{2}\frac{(MC_p)_x}{\sum_i^n K_i + \sum_j^m R_j} = \text{critical time step} \quad (7.22)$$

7.4.6 Example Problem

An example of a thermal control problem that was solved using a spreadsheet program is illustrated shown in Fig. 7.13. The property values are as follows:

$$k = \text{thermal conductivity} = 150\ \text{W/(m-K)}$$

$$c_p = \text{specific heat} = 0.879\ \text{J/(gm-K)}$$

$$\rho = \text{density} = 2.643\ \text{gm/cm}^3$$

$$C = \text{thermal capacitance} = \text{mass times specific heat}$$

$$K = \text{thermal conductance (conduction)} = \text{conductivity} \times \text{area A/distance } \Delta X$$

The elements of the thermal network shown in Fig. 7.13 are determined as follows:

$$R = \text{radiation conductance} = \text{Stefan–Boltzmann constant} \times \text{emissivity} \times \text{surface area} \times \left(T_1^2 + T_1^2\right)(T_1 + T_2)$$

$$q = \text{heat input}$$

$$\begin{aligned} C1 = C2 = C3 &= \text{volume} \times \text{density} \times \text{specific heat} \\ &= 0.2 \times 10 \times 100\,(\text{cm}^3) \times 2.643\,(\text{gm/cm}^3) \times 0.879\,[\text{J/(gm-K)}] \\ &= 464.7\ \text{J/K} \end{aligned}$$

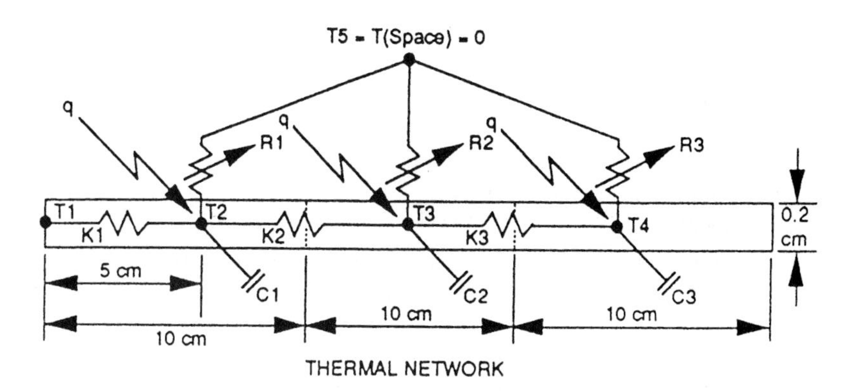

Fig. 7.13 Radiator fin problem—thermal network.

$$K1 = kA/\Delta x = 150\,\text{W/(m-K)} \times 1\,\text{m} \times 0.2\,\text{cm}/5\,\text{cm}$$
$$= 6\,\text{W/K}$$

$$K2 = K3 = kA/\Delta x = 150\,\text{W/(m-K)} \times 1\,\text{m} \times 0.2\,\text{cm}/10\,\text{cm}$$
$$= 3\,\text{W/K}$$

$$R1 = R2 = R3 = \sigma\varepsilon_{\text{IR}}AT^3 = \sigma \times 0.8 \times [10/100 \times 1\,(\text{m}^2)] \times T^3$$
$$= 4.53610^{-9}T^3$$

$$q = \text{solar constant} \times A \times \alpha_s$$
$$= 1371\,\text{W/m}^2 \times [10/100 \times 1\,(\text{m}^2)] \times 0.2$$
$$= 27.42\,\text{W}$$

The thermal network shown in Fig. 7.13 is used to model a space radiator fin. The aluminum fin is 30-cm wide, 0.2-cm thick, and 1-m long (into the plane of the paper). The left edge, or base, of the fin is maintained at a constant temperature of 340 K by the radiator's working fluid. The node representing the base is labeled T1 while the 10-cm wide fin nodes are labeled T2, T3, and T4. The bottom side of the fin is perfectly insulated; the top is exposed to deep space and solar heating. The solar vector is assumed to be normal to the radiator surface, and the surface radiation properties are assumed to be 0.2 for the solar absorptivity and 0.8 for the IR emissivity. The definitions and values for the elements of the thermal network are given above. The values of the radiation conductors R1, R2, and R3 are derived by constructing an Oppenheim radiation network between the surface and deep space. Because the fin node surfaces only "see" deep space, the Oppenheim conductors can be combined to the single conductor shown between the fin surface and space.

The solution of the fin problem is shown in Fig. 7.14. The curve given in the figure represents the steady-state values of the fin temperatures. It was assumed that initially the fin was at a uniform temperature of 340 K. It took approximately 25 min (problem time) for the fin temperatures to gradually change from their

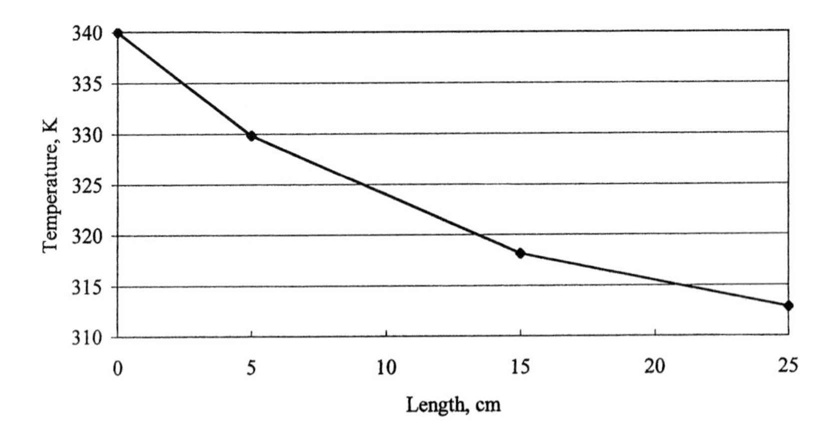

Fig. 7.14 Radiator fin problem—steady-state solution.

initial values to the steady-state values shown in Fig. 7.14. Using a PC and the Microsoft Excel spreadsheet program, it took less than a second of computer time to solve the problem.

7.5 Preliminary Design

7.5.1 Approach

The purpose of Section 7.4 was to provide an introduction to the mathematical techniques used in spacecraft design. The purpose of this section is to develop a technique that will provide a preliminary, first-order estimate of the thermal performance of a spacecraft. This estimate will bound the spacecraft thermal design and provide a basis for estimating the engineering and computer time required to develop the detailed spacecraft thermal design. A description of the preliminary design process is given in Table 7.3, and a numerical example of the preliminary design process is presented later in this chapter.

The specific conditions that are used in the preliminary design process are as follows: 1) the spacecraft directly overhead at high noon and at the equator and 2) the spacecraft in the Earth's shadow with no view of sunlit portions of the Earth's surface. These two situations correspond to worst-case hot and worst-case cold conditions. The basic spacecraft geometry that will be considered in the preliminary design process is spherical. Also, worst-case hot and worst-case cold equations will be developed for a flat plate in space. The flat plate is considered to be a solar array whose surface normal is parallel to the solar vector and passes through the center of the Earth.

7.5.2 Development of Equations

We will first consider a solar array and assume that the array is thin and isothermal; the "top" surface of the array receives direct solar energy while the "bottom" surface receives albedo and Earth IR radiation. The direct solar energy absorbed by the array in space is the product of the solar constant, times the area, times the solar absorptivity of the array surface:

$$q_{a-\mathrm{ds}} = G_s A \alpha_{s-t} \tag{7.23}$$

where $q_{a-\mathrm{ds}}$ is the absorbed direct solar energy, G_s is the solar constant (1371 W/m^2), A is the area of the array (the area of the top of the array is A and area of the bottom of the array is A), and α_{s-t} is the solar absorptivity of the solar cells on the top surface of the array.

The IR energy flux at the Earth's surface is 237 W/m^2. Therefore, the total IR energy leaving the Earth's surface is 237 times the surface area of the Earth. The IR energy flux at the spacecraft altitude is the total IR energy divided by the area of a sphere with a radius equal to the Earth's radius plus the altitude of the spacecraft. Thus,

$$G_{\mathrm{IR}} = q_{\mathrm{IR}}\left[\frac{4\pi R_E^2}{4\pi(H + R_E)^2}\right] = q_{\mathrm{IR}}\left[\frac{R_E^2}{(H + R_E)^2}\right] = q_{\mathrm{IR}} F_{a-e} \tag{7.24}$$

Table 7.3 Preliminary design process for the thermal subsystem

Step	Notes	Where discussed
1) Establish temperature limits.	Payload requirements usually dominate. Batteries often have narrow temperature limits.	Table 7.1
2) Establish electrical power dissipation.	Use powers consistent with worst-case hot and worst-case cold conditions, if available; otherwise use orbital average values.	N/A
3) Determine the diameter of a sphere whose surface area is equal to total outer surface area of actual spacecraft.	First-order estimates of the spacecraft thermal performance can be made by assuming an isothermal and spherical spacecraft.	N/A
4) Select radiation surface property values.	Assume readily available aluminized Teflon® property values unless there is a special reason to use other values. For aluminized Teflon® assume $\alpha_s = 0.316$ (degraded) and $\varepsilon_{\text{IR}} = 0.8$.	Table 7.2
5) Compute spacecraft worst-case hot temperature.	Use high side values for environmental input parameters; direct solar, albedo, and Earth IR emission.	Eq. (7.36), Fig. 7.1
6) Compute spacecraft worst-case cold temperature.	Use low side value for environmental input parameters; Earth IR emission.	Eq. (7.37), Fig. 7.1
7) Compare worst-case hot and cold temperatures with temperature limits (step 1).	If worst-case hot temperature is hotter than upper limit, deployed radiator with pumped loop system will likely be required. If opposite is true, body-mounted radiators can be used.	N/A
8) Required area for body mounted radiator	Use upper temperature limit for radiator temperature, assume no environmental heat inputs, use maximum heat dissipation.	Eq. (7.10)
9) Radiator temperature for worst-case cold conditions	Use the area from step 8 and the minimum heat dissipation.	Eq. (7.10)
10) Heater power required to maintain radiator at lower limit	If temperature found in step 9 is less than the lower limit, assume the radiator temperature is at the lower limit and area is value defined in step 8.	Eq. (7.10)
11) Determine if there are special thermal control problems.	Components with narrow temperature range, high power dissipation, or low temperature requirements	N/A
12) Estimate subsystem weight, cost, and power.	Weight: see Table 7.6; cost: 4% of total spacecraft; power: see step 10.	N/A
13) Document reasons for selections.	Particularly important to document assumptions.	N/A

The radiation view factor, flat plate to Earth, is

$$F_{a-e} = \frac{R_E^2}{(H + R_E)^2} \tag{7.25}$$

and G_{IR} is the Earth IR radiation flux at the altitude H, q_{IR} is the Earth IR emitted energy flux at the surface of the Earth, and R_E is the radius of the Earth. The absorbed Earth IR radiation on the bottom of the array is given by

$$q_{a-\text{IR}} = q_{\text{IR}} F_{a-e} A \varepsilon_{\text{IR}-b} \tag{7.26}$$

where $q_{a-\text{IR}}$ is the absorbed Earth IR radiation on the bottom of the array, A is the area of the bottom of the array, and $\varepsilon_{\text{IR}-b}$ is the IR emissivity of the bottom surface of the array. The quantities q_{IR} and F_{a-e} have already been defined.

The equation for the solar energy reflected off the Earth (albedo) and absorbed by the spacecraft is similar to the preceding equation for Earth IR radiation:

$$q_{a-a} = G_s a F_{a-e} A \alpha_{s-b} K_a \tag{7.27}$$

$$K_a = 0.657 + 0.54\left(\frac{R_E}{R_E + H}\right) - 0.196\left(\frac{R_E}{R_E + H}\right)^2 \tag{7.28}$$

where q_{a-a} is the energy absorbed on the bottom surface of the solar array caused by solar energy reflected from the Earth, G_s is the solar constant (1371 W/m^2), a is the albedo (the percentage of direct solar energy reflected off the Earth, approximately 30%), F_{a-e} is the view factor from the array to the Earth [Eq. (7.25)], and R_E is the radius of the Earth. H is the altitude of the spacecraft, A is the area of the bottom of the array, α_{s-b} is the solar absorptivity of the bottom surface of the array, and K_a is a factor, which accounts for the reflection of collimated incoming solar energy off a spherical Earth.

The total absorbed energy on the solar array is

$$q_{a-t} = G_s A \alpha_{s-t} + q_{\text{IR}} F_{a-e} A \varepsilon_{\text{IR}-b} + G_s a F_{a-e} A \alpha_{s-b} K_a \tag{7.29}$$

$$K_a = 0.657 + 0.54\frac{R_E}{R_E + H} - 0.196\left(\frac{R_E}{R_E + H}\right)^2 \tag{7.30}$$

The emitted radiation energy from the array is given by

$$q_E = \sigma \varepsilon_{\text{IR}-b} A T^4 + \sigma \varepsilon_{\text{IR}-t} A T^4 \tag{7.31}$$

where σ is the Stefan–Boltzmann constant, A equals the area of one side of the array, $\varepsilon_{\text{IR}-b}$ is the IR emissivity of the bottom surface of the array, $\varepsilon_{\text{IR}-t}$ is the IR emissivity of the top surface of the array, and T is the temperature of the array. We can now perform an energy balance on the solar array:

$$q_{\text{absorbed}} - q_{\text{emitted}} - q_{\text{electron generation}} = 0 \tag{7.32}$$

For a solar array the energy balanced equation includes an electrical generation term because the solar cells convert solar energy directly to electrical energy. There is some dissipation associated with solar arrays (diodes, interconnect cabling, and transmission losses); however, these losses are accounted for in the array efficiency

term. This term η is defined as the ratio of the array electrical output to the incident direct solar energy. Substituting into the energy balance equation,

$$G_s A\alpha_{s-t} + q_{\text{IR}} F_{a-e} A\varepsilon_{\text{IR}-b} + G_s a F_{a-e} A\alpha_{s-b} K_a - \sigma\varepsilon_{\text{IR}-b} AT^4$$
$$- \sigma\varepsilon_{\text{IR}-t} AT^4 - \eta G_s A = 0 \tag{7.33}$$

Solving for the worst-case hot temperature,

$$T_{\text{MAX}-A} = \left[\frac{G_s\alpha_{s-t} + q_{\text{IR}} F_{a-e}\varepsilon_{\text{IR-b}} + G_s a F_{a-e}\alpha_{s-b} K_a - \eta G_s}{\sigma(\varepsilon_{\text{IR}-b} + \varepsilon_{\text{IR}-t})}\right]^{\frac{1}{4}} \tag{7.34}$$

The terms used in Eq. (7.34) are defined after Eq. (7.39). If the array is in the shadow of the Earth and is not in view of any portion of the sunlit parts of the Earth, then we have a worst-case cold condition. For this condition there is no direct solar, albedo, or electric power generation. The equation that defines the temperature for this condition is

$$T_{\text{MIN}-A} = \left[\frac{q_{\text{IR}} F_{a-e}\varepsilon_{\text{IR}-b}}{\sigma(\varepsilon_{\text{IR}-b} + \varepsilon_{\text{IR}-t})}\right]^{\frac{1}{4}} \tag{7.35}$$

The terms used in Eq. (7.35) are defined after Eq. (7.39). Worst-case hot-temperature and worst-case cold-temperature expressions can be derived for an isothermal sphere in a similar manner as the preceding equations for a solar array. For the sphere it is assumed that there is uniform energy dissipation over the entire surface of the sphere. Also, it is assumed there is no electrical generation on the spherical surface. The temperature expressions corresponding to worst-case hot ($T_{\text{MAX}-S}$) and worst-case cold ($T_{\text{MIN}-S}$) conditions for a spherical spacecraft are as follows:

$$T_{\text{MAX}-S} = \left[\frac{G_s\alpha_s/4 + q_{\text{IR}}\varepsilon_{\text{IR}} F_{s-e} + G_s a\alpha_s K_a F_{s-e} + Q_W/\pi D^2}{\sigma\varepsilon_{\text{IR}}}\right]^{\frac{1}{4}} \tag{7.36}$$

$$T_{\text{MIN}-S} = \left[\frac{q_{\text{IR}}\varepsilon_{\text{IR}} F_{s-e} + Q_W/\pi D^2}{\sigma\varepsilon_{\text{IR}}}\right]^{\frac{1}{4}} \tag{7.37}$$

$$F_{S-e} = 0.5\left[1 - \frac{(H^2 + 2HR_E)^{0.5}}{H + R_E}\right] \tag{7.38}$$

$$F_{a-e} = \frac{R_E^2}{(H + R_E)^2} \tag{7.39}$$

where

F_{s-e} = view factor, sphere to Earth
F_{a-e} = view factor, flat plate array to the Earth
K_a = factor that accounts for the reflection of collimated incoming solar energy off a spherical Earth
G_s = solar constant = 1371 W/m^2 near Earth
q_{IR} = Earth IR emission = 237 W/m^2

a = albedo = 30%
Q_W = electrical power dissipation, W
σ = Stefan–Boltzmann constant = 5.67×10^{-8} $W/(m^2\text{-}^\circ K^4)$
α_{s-t} = solar absorptivity on top surface of solar array
α_{s-b} = solar absorptivity on bottom surface of solar array
α_s = solar absorptivity of sphere
ε_{IR-t} = IR emissivity on top surface of solar array
ε_{IR-b} = IR emissivity on bottom surface of solar array
ε_{IR} = IR emissivity of sphere
R_E = radius of the Earth = 6378 km
H = altitude of the spacecraft, km
η = solar-array efficiency = ratio of the solar array electrical output to the incident direct solar energy
D = diameter of spherical spacecraft, m

7.6 Preliminary Design Process

Equations (7.34–7.37) define steady-state temperatures. An actual spacecraft will likely exhibit lower maximum temperatures and higher minimum temperatures than those predicted by the equations as a result of transient effects. Nevertheless, these equations can be used to provide reasonable first-order estimates of spacecraft thermal performance. An example of the use of Eqs. (7.34) and (7.35) to predict solar-array temperatures is given in Fig. 7.15. In this figure the maximum altitude is 35786 km, the Earth geosynchronous orbit altitude. The parameters used in the construction of the figure are as follows: incident radiation = 1371 W/m^2, albedo = 30%, Earth IR = 237 W/m^2, array efficiency = 8%, absorptivity (top) = 0.805, absorptivity (bottom) = 0.3, emissivity (top) = 0.825, and emissivity (bottom) = 0.8.

Equations (7.36) and (7.37) are used as the major elements of the preliminary design process shown in Table 7.3. A numerical example of this process follows, and for this example we will assume a spacecraft that is a 1.5-m cube, has a

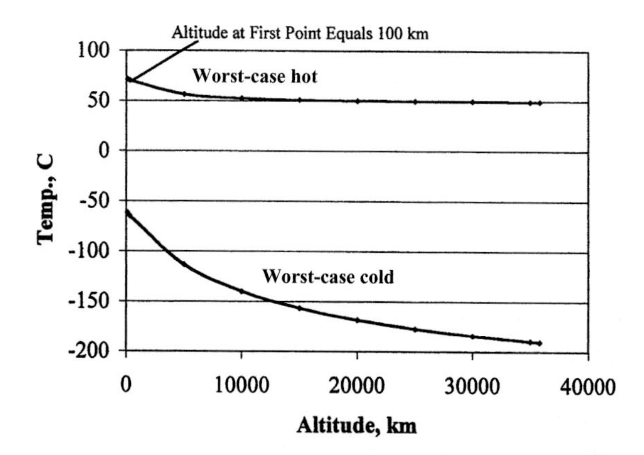

Fig. 7.15 Solar-array temperatures vs altitude.

maximum power dissipation of 170 W, a minimum power dissipation of 80 W, and an altitude of 700 km.

The first step in the process is to establish the equipment temperature limits. Referring to Table 7.1, we find that the typical temperature range for electronics is 0 to 40°C, and for batteries the range is from 5 to 20°C. Next, estimate the electrical power dissipation of the spacecraft. If possible, determine the power consistent with the worst-case hot condition and the power consistent with the worst-case cold condition. If this is not possible to determine, use the orbital average value for both conditions. Compute the diameter of a spherical spacecraft, which has a surface area equal to the actual spacecraft. This is done so that we can use the equations that have been developed for an isothermal sphere and obtain first-order estimates of the spacecraft's thermal performance. For this calculation assume a solar absorptivity of 0.316 and an emissivity of 0.8. These values are consistent with aluminized Teflon® properties and represent radiation property values that are readily obtainable. We have used a value for the solar absorptivity that is higher than the values shown in Table 7.2. This is to allow for UV degradation of the radiation surface, which is discussed later in this chapter.

The results of the performance calculations for the spacecraft are given in Table 7.4. The high side values of the environmental radiation fluxes are used for the worst-case hot conditions, and the low side values are used for the worst-case cold conditions. The results of the analyses are a worst-case hot temperature of −8.5°C and a worst-case cold temperature of −86.5°C. These values tell us that from an overall standpoint the spacecraft will be relatively easy to thermally control because the worst-case hot temperature is well below the upper limit of 20°C for the batteries and 40°C for the electronics. If the worst-case hot temperature had been 50 or 60°C, we would know that the spacecraft could not be controlled using body-mounted radiators and deployed radiators with a pumped loop system would probably be required.

Equation (7.10) is used to assess the performance of a body-mounted radiator. For line 19 of Table 7.4, it is assumed that the radiator temperature is 35°C (40°C upper equipment limit −5°C thermal margin), there are no environmental heat inputs, and the heat dissipation is 170 W. The radiator area required for these conditions is 0.42-m^2. The assumed spacecraft dimensions are a 1.5-m cube; therefore, a 0.42-m^2 body-mounted radiator can easily be accommodated. Line 20 on Table 7.4 illustrates a check on the performance of the 0.42-m^2 radiation consistent with worst-case cold conditions. For this situation it is assumed that the heat dissipation is 80 W, which produces a radiator temperature of −18.4°C. This temperature is considerably below the lower limit of 5°C (0°C lower equipment limit +5°C thermal margin). Therefore, the radiator will require a thermal louver to modulate the heat flow or a thermostat and heater to maintain the temperature above the lower limit. The heater power required to maintain the radiator at 5°C is 33.8 W (computation on line 21 of Table 7.4). This power at the radiator is in addition to the 80 W that is dissipated from the electrical and electronic equipment. We have not attempted to maintain the spacecraft equipment within the battery temperature limits. We will isolate the battery from the other spacecraft equipment and provide thermal control separately for this device. The probable approach will be to cold bias the battery and thermally control the battery temperature with heaters and thermostats.

Table 7.4 Sample calculations illustrating preliminary design process

No.	Item	Value	Source	Comment
1	Spacecraft surface area	13.5 m^2	Assumed	It is assumed that the spacecraft is a 1.5-m cube.
2	Diameter of a sphere that equals spacecraft surface area	2.07 m	$\pi D^2 = 13.5$	—
3	Maximum power dissipation	170 W	Assumed	—
4	Minimum power dissipation	80 W	Assumed	—
5	Altitude, H	700 km	Assumed	—
6	Radius of Earth, R_E	6378 km		—
7	Radiation view factor, F_{s-e}	0.283	Eq. (7.38)	—
8	Albedo correction, K_a	0.993	Eq. (7.28)	—
9	Maximum Earth IR emission, $Q_{\mathrm{IR-max}}$	258 W/m^2	Fig. 7.1	Use maximum value for worst-case hot condition.
10	Minimum Earth IR emission, $q_{\mathrm{IR-min}}$	216 W/m^2	Fig. 7.1	Use minimum value for worst-case cold condition.
11	Direct solar flux	1376 W/m^2	Fig. 7.1	Use maximum value for worst-case hot condition.
12	Albedo	35%	Fig. 7.1	Use maximum value for worst-case hot condition.
13	IR emissivity	0.8	Table 7.2	Assume radiation surface.
14	Solar absorptivity	0.316	Table 7.3	Radiation surface degrades as a result of UV exposure.
15	Worst-case hot temperature	−8.5°C	Eq. (7.36)	—
16	Worst-case cold temperature	−86.5°C	Eq. (7.37)	—
17	Upper temperature limit	35°C	Table 7.1	Assume 5°C thermal margin.
18	Lower temperature limit	5°C	Table 7.1	Assume 5°C thermal margin.
19	Radiator area based on worst-case hot conditions	0.42 m^2	Eq. (7.10)	Assume no solar heat input, dissipation of 170 W, and radiator temperature of 35°C.
20	Radiator temperature based on worst-case cold conditions	−18.4°C	Eq. (7.10)	Assume no solar heat input, an area of 0.42 m^2, and dissipation of 80 W.
21	Heater power required to maintain radiator at lower limit	33.8 W	Eq. (7.10)	Assume no solar heat input, an area of 0.42 m^2, and a radiator temperature of 5°C.

Table 7.5 Thermophysical properties

No.[a]	Material	Conductivity, W/(m-K)	Density, kg/m^3	Specific heat, J/(kg-K)	Measurement temperature, K
1	Silver, pure	418.8	1,0524	234	293
2	Copper, pure	406.7	8,954	383	173
3	Copper, pure	386.0	8,954	383	293
4	Copper, pure	379.0	8,954	383	373
5	Aluminum, pure	228.5	2,707	896	293
6	Aluminum, 87% Al; 13% Si	148.8	2,659	871	173
7	Aluminum, 87% Al; 13% Si	164.4	2,659	871	293
8	Aluminum, 87% Al; 13% Si	174.8	2,659	871	373
9	Brass, 70% Cu; 30% Zn	110.8	8,522	385	293
10	Carbon steel, 1.0% C	43.27	7,801	473	293
11	Stainless steel, 18% Cr; 8% Ni	16.27	7,817	461	293
12	Mercury	8.404	13,543	138	293
13	Glass	0.7615	2,707	837	293
14	Water	0.6040	997.5	4,178	294
15	Air	0.0262	1.177	1,007	300
16	Multilayer insulation, 10 shields	0.0004	—	—	273 cold face, 366 hot face
17	Multilayer insulation, 30 shields	0.00031	—	—	273 cold face, 366 hot face

[a]Thermophysical properties for items 1–15 are from Ref. 11. Multilayer insulation values are from Ref. 12.

7.7 Thermophysical Properties

7.7.1 Multilayer Insulation

Thermophysical properties of selected materials are given in Table 7.5. The values for the multilayer insulation are extremely low. These values are for laboratory conditions and are not representative of values obtainable on an actual spacecraft. Actual performance of multilayer insulation is given in Fig. 7.16. The performance value used in Fig. 7.16 is effective emittance (emissivity), rather than conductivity. Effective emittance is defined as

$$q = \sigma \varepsilon_{\text{eff}} \left(T_x^4 - T_y^4 \right) \tag{7.40}$$

In Eq. (7.40) σ is the Stefan–Boltzmann constant, ε_{eff} is the effective blanket emissivity, and T_x and T_y are absolute temperatures.

The effective emittance is related to conductivity by the following expression:

$$\varepsilon_{\text{eff}} = k_{\text{eff}} \left[\sigma \Delta x \left(T_x^2 + T_y^2 \right) (T_x + T_y) \right] \tag{7.41}$$

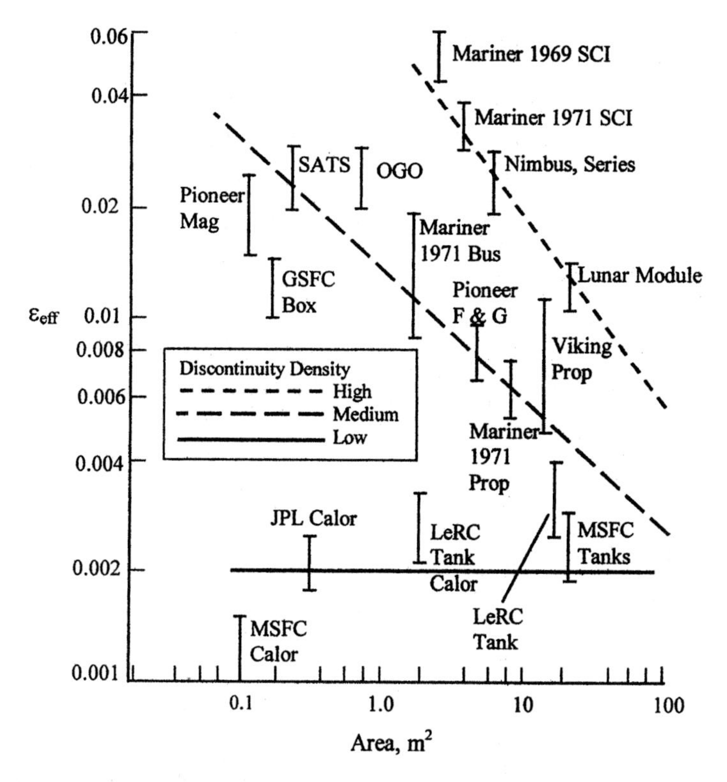

Fig. 7.16 Multilayer insulation performance. (Copyright AIAA, reproduced with permission; Ref. 13.)

In Eq. (7.41) k_{eff} represents the effective blanket conductivity, and Δx is the blanket thickness.

The thermal performance of multilayer insulation is given in Fig. 7.16. In this figure SCI stands for science scan platform, MAG for magnetometer, PROP for propellant, CALOR for calorimeter, and F & G for the Pioneer spacecraft series designations. An important conclusion drawn from Fig. 7.16 is that the effectiveness of multilayer insulation is generally dependent on the size of the object being insulated. Large multilayer insulation systems generally will exhibit a lower effective emissivity than small systems. This occurs because edge effects, joints, seams, and penetrations (electrical wires, structural supports) in multilayer insulation represent relatively high extraneous heat leak paths through the insulation. Generally, large multilayer insulation systems will have a smaller percentage of extraneous heat leak paths than small systems. In Fig. 7.16 calorimeters and cryogenic tanks represent multilayer systems with low discontinuity densities. Most unmanned spacecraft and propulsion systems correspond to medium discontinuity densities, whereas highly complex spacecraft and science platforms generally have high discontinuity densities.

7.7.2 Radiation Properties

Radiation properties are given in Table 7.2. The surface providing the lowest temperature when irradiated by solar energy is an optical solar reflector (OSR). These devices have been used on spacecraft, but are both fragile and costly. One type of optical solar reflector is silver-coated Teflon®. This material does not provide as low a temperature as high-quality quartz optical solar reflectors, but is much less expensive and much more durable. In recent years silver-coated and/or aluminum-coated Teflon® have been used extensively on the outer surfaces of spacecraft.

7.8 Design Considerations

All aspects of spacecraft thermal control, obviously, cannot be covered in this chapter. However, some of the more important items, with regard to spacecraft thermal control design, will be discussed in the remainder of this section.

Table 7.6 provides information for preliminary weight estimates of thermal control hardware items. The values from this table should be used with caution because they are only first-order estimates.

One of the elements of a thermal control subsystem that is difficult to predict is the resistance to heat flow across solid-to-solid interfaces. Thermal control subsystems often have important heat-transfer paths across interfaces, for example, heat dissipating components bolted to cold plates. The thermal performance of an interface is characterized as joint conductance and is defined by the following equation:

$$Q = h_c A \Delta T \tag{7.42}$$

In Eq. (7.42) Q is the heat-transfer rate across the interface in watts, A is the interface area in square meters, ΔT is the temperature difference across the interface in degrees Kelvin, and h_c is the joint conductance, W/(m^2-K).

Table 7.6 Approximate thermal control hardware weights

Item	Weight	Comment
Multilayer insulation	0.03 kg/m^2	12 layers with separators
Heat pipes	0.33 kg/m	Aluminum pipe, 1.27 cm in diameter
Louvers	7.3 kg/m^2	—
Thermostats	0.03 kg	250-W rating
Foam insulation	64 kg/m^3	—
Heaters	2 kg/m^2	Approximate installed weight
OSRs	1 kg/m^2	Approximate installed weight
Radiators	2.707 kg/m^3	Assume aluminum, actual weight depends on strength and conduction requirements
Paint	0.24 kg/m^2	—
Phase-change devices	$0.076 + 0.041Q$ kg	Q is the energy storage in watt-hours; equation is based on use of phase-change material with heat of fusion = 232,800 J/kg

The difficulty in predicting joint conductance is that this property is dependent on the surface finish, hardness of the interface surfaces, the waviness of the interface surfaces, and the interface pressure. A typical value for aluminum surfaces is 4500 W/(m^2-°K); however, this value can vary greatly depending on circumstances. Because of the difficulty in predicting contact conductance, it is necessary to measure this quantity at the component level and then use the measured values in the spacecraft thermal models.

In Table 7.2 the values of solar absorptivity are all beginning-of-life values. This means the materials have not been subjected to the space environment, which will cause degradation of the solar absorptivity with time as a result primarily of UV radiation. The degradation is exponential in nature with an upper limit. For example, the aluminized Teflon® listed in Table 7.2 has an upper limit solar absorptivity of 0.316. The amount of time of solar exposure required for the materials to approach the upper limit, or asymptotic value, is from 3,000 to 12,000 h. In analyzing the thermal performance of a spacecraft, the beginning-of-life and end-of-life values of thermal coatings must be considered to define the temperature extremes during the entire mission.

Multilayer insulation is a primary hardware component for spacecraft thermal control subsystems. This material, however, requires care with regard to installing, modeling, and testing. For example, overlapping multilayer blankets at joints will greatly degrade the thermal performance of the insulation. If penetrations are not accounted for correctly in thermal models, the analytical predictions will not match the test results. This happened on a thermal vacuum test of the Hubble Space Telescope. Also, test times must be long enough for the air within multilayer blankets to vent. Even though multilayer blankets are routinely perforated, it may take several hours in a space chamber for the entrapped air to vent so the blanket performs as it would in space.

In this section many of the hardware elements of a thermal control system have been described. The question remains, which hardware elements should be used, or even considered, for a given spacecraft? This question is often answered in practice by performing trade studies to determine the most effective components to use for a given situation. The criteria used in trade studies usually involves thermal performance, cost, weight, reliability, availability, safety, and durability.

References

[1]*Space and Planetary Environment Criteria and Guidelines*, NASA TM 82478, Vol. 1, 1982.

[2]*Phase Change Materials Handbook*, NASA Technical Brief B72-10464, Marshall Space Flight Center, Huntsville, AL, Aug. 1972.

[3]Siegel, R., and Howell, J. R., *Thermal Radiation Heat Transfer*, Vol. 1, NASA Lewis Research Center, Cleveland, OH, 1968.

[4]*Martin Marietta Thermal Properties Handbook*, ER 13997, Dec. 1965.

[5]Ahern, J. E., and Karperos, K., "Calorimetric Measurements of Thermal Control Surfaces of Operational Satellites," AIAA Paper 83-0075, AIAA 21st Aerospace Sciences Meeting, Reno, NV, Jan. 1983.

[6]Ahern, J. E., Belcher, R. L., and Ruff, R. D., "Analysis of Contamination Degradation of Thermal Control Surfaces on Operational Satellites," AIAA Paper 83-1449, AIAA 18th Thermophysics Conference, Montreal, Canada, June 1983.

[7]*Solar Cell Array Design Handbook*, Vol. II, Jet Propulsion Lab., California Inst. of Technology, Pasadena, California, 1976.
[8]NACA TN 2836.
[9]Chapman, A. J., *Heat Transfer*, Second Edition, Macmillian, New York, 1960, p. 602.
[10]Oppenheim, A. K., "Radiation Analysis by the Network Method," *Transactions of the ASME*, Vol. 78, 1956, p. 725.
[11]Chapman, A. J., *Heat Transfer*, Second Edition, Macmillian, New York, 1960, p. 548.
[12]Streed, E. R, Cunnington, G. R., and Zierman, C. A., "Performance of Multilayer Insulation Systems for the 300 to 800 K Temperature Range," AIAA Paper 65-663, AIAA Thermophysics Specialist Conference, Monterey, CA, Sept. 1965.
[13]Stimpson, L. D., and Jaworski, W., "Effects of Overlaps, Stitches, and Patches on Multilayer Insulation," AIAA Paper 72-285, AIAA Seventh Thermophysics Conference, San Antonio, TX, April 1972.

Problems

7.1 A spherical spacecraft is used for a near solar mission. The outer skin of the spacecraft is made of a material that has a high thermal conductivity, and the spacecraft is spin stabilized. These two conditions cause the temperature of the spacecraft's outer skin to be uniform. The emissivity ε_{IR} and the absorptivity α_s of the outer skin are 0.85 and 0.1, respectively. If steady-state conditions are assumed and there is no heat generation within the spacecraft, what is the skin temperature, in degrees Celsius, if the solar constant is 8? This means that the solar flux is $8 \times 1371 = 10968$ W/m^2.

7.2 A spherical fuel tank is insulated with 8 cm of fiberglass. The tank outside diameter is 0.7 m, and the tank wall is at -195.2°C. If the outer surface of the insulation is a 27°C, what is the heat loss in watts? The thermal conductivity of the insulation is 0.00865 W/(m-K).

7.3 Multilayer insulation is used to thermally isolate a high-temperature component with a spacecraft science instrument. This insulation is made of five layers of aluminum foil, which are separated by a transparent material. The inner layer of the aluminum is at 282°C while the outer layer is at 4.8°C. Neglect all conduction and convection effects and calculate the radiant loss on a square-meter basis. The infrared emissivity ε_{IR} of the foil is 0.1, and it is assumed that the foils are infinite parallel plates. Express answer in watts/square meter.

7.4 A rectangular enclosure has a base of 2×4 m and a height of 6 m. What is the view factor from a) the base to the 2×4 m top, b) from the base to the 2×6 m side, and c) from the base to the 4×6 m side?

7.5 A flat space radiator is mounted on the surface of a spacecraft and perfectly insulated on the inside. The area of the radiator is 0.6×0.6 m with a solar absorptivity of 0.15 and an IR emissivity of 0.8. If the solar vector is at an angle of 45 deg relative to the normal to the radiator, the incident solar flux is 1371 W/m^2, and the radiator temperature is 93.7°C, what is the power rejected by the radiator. Express answer in watts. Neglect albedo and Earth IR radiation.

7.6 Two directly opposed, parallel circular disks are floating in space and 0.6 m apart. The diameter of the disks are both 1.2 m, and they are both shaded from the sun, albedo, and Earth IR radiation. The IR emissivity of both disks is 0.4, and one of the disks (disk a) is maintained at a temperature of 150°C. The other disk (disk b) is thin and highly conductive so that the temperature on both sides is the same. What is the temperature of disk b in degrees Celsius?

7.7 A proposed spacecraft is cylindrical in shape with a diameter of 1.2 m and a length of 2 m. The orbit altitude is 150 km, the maximum power dissipation is 1500 W, and the minimum power dissipation is 500 W. Assume the maximum power occurs when the spacecraft is in the sunlight and the minimum occurs when the spacecraft is in the Earth's shadow. Using the preliminary design process outlined on Table 7.4, estimate the worst-case hot and worst-case cold temperatures for this proposed spacecraft.

8
Command and Data System

8.1 Introduction

The command and data system (CDS), sometimes called the command and data handling (C&DH) system, is the central nervous system for the spacecraft. The system manages three digital data streams, each critical to the spacecraft and each with distinctive characteristics. These data streams are as folllows:

1) Science or payload data. This is the information the spacecraft was sent to get. It is gathered by the spacecraft, processed onboard, and downlinked to the ground. It is usually the highest data rate on the spacecraft.

2) Engineering data. This is the information with which the ground operations team can determine the status and health of each subsystem on the spacecraft. Engineering data are usually downlinked at a medium data rate, substantially lower than the science rate.

3) Commands. The command link is an uplink from the ground at a low data rate. Commands are instructions to the spacecraft dictating configuration, attitude, and actions at particular times. Accuracy is the primary requirement in handling this data stream. A single bit error in the science data stream or the engineering data stream is not serious. It may cause some temporary consternation among the ground team but it is not dangerous. An error in the command stream is dangerous and can be life threatening. Extreme care is taken in the preparation, testing, and transmission of the command stream.

The CDS and the communications systems are closely related, as shown in Fig. 8.1, the block diagram for a typical CDS. The computer is central to the CDS system. The computer may serve all of the spacecraft computational needs, sharing computer time with the attitude control system. Alternatively two computers may be provided: one specialized for data handling and one specialized for the attitude control computations. Magellan is a recent spacecraft that used dual specialized computers, and Mars Observer is a recent spacecraft that used a multipurpose computer for data and attitude-control computation. The CDS has significant interaction with the telecommunication subsystem, the attitude control system, and the power system. In addition, the CDS has command and data-gathering interfaces with all of the subsystems on the spacecraft.

In *commanding*, the CDS system receives command signals from the radio frequency (rf) link through a command detector unit, authenticates the commands, performs onboard processing, and executes both internally generated and externally received commands. The CDS also provides for the generation of precise spacecraft clock data for the spacecraft. Spacecraft time is used to annotate data and to execute time-tagged commands. The *data handling* functions are to collect, digitize, format, encode, route, store, and play back the three major data types. The system provides *digital storage* for all three data types. Storage media is usually

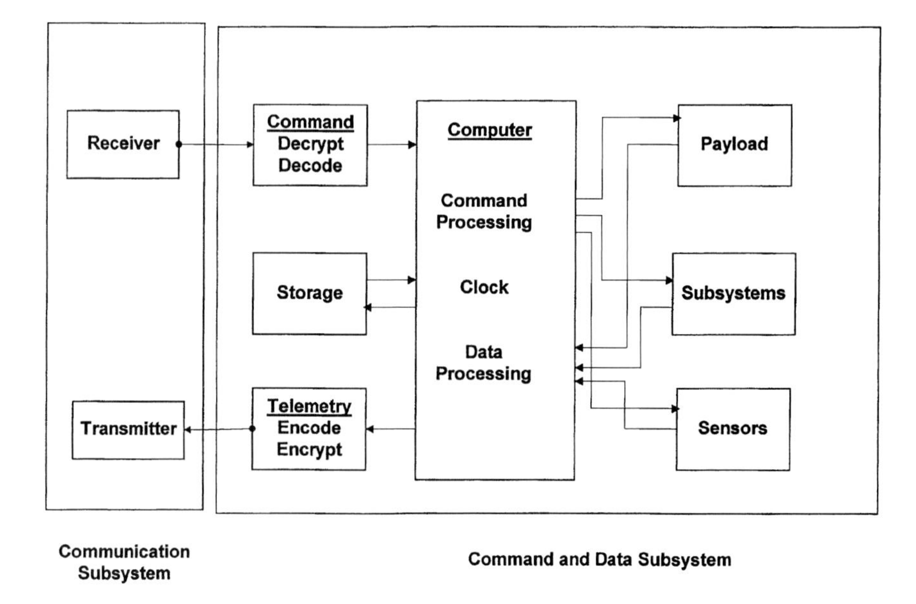

Fig. 8.1 Typical CDS system.

solid-state memory, although tape recorders were used extensively in the recent past.

The International Consultative Committee for Space Data Systems[1] (CCSDS) is an influential force in the design of CDS systems and will be referenced often in this chapter. This international committee has members from the United Kingdom, Canada, France, Germany, European Space Agency, Brazil, Japan, Russia, and the United States. Its purpose is to recommend standard ways of handling data and commands both for spacecraft and ground so that equipment, especially ground equipment, can be standardized. Two benefits are expected: 1) multinational ground stations can be used to receive and handle data from spacecraft of any nation and 2) the considerable cost of ground stations may be reduced. As a result of this committee, the CDS system is the most nearly standardized system on a spacecraft. (The voluminous committee recommendations are posted on the Internet at www.ccsds.org.)

In the following sections the data handling function will be discussed first and then the command handling function, which is simpler in concept but absolutely critical to the spacecraft. The computer and software will be discussed last. (Although computers and software are used elsewhere in the spacecraft, notably in ACS, they are discussed in this chapter only.)

8.2 Requirements

The requirements for the CDS system derive directly from the data rates of the science instruments, the engineering data rates, and the command rates. These, coupled with the downlink schedule and the instrument data schedules, determine the data flow timeline through the system. The timeline determines the data storage requirement and the downlink data rates for telecommunication.

8.2.1 Engineering Data

It is essential for the ground operations personnel to be able to determine the status, health, and performance of each subsystem in near real time. Collectively, the measurements required for this evaluation are called *engineering data*. In the design phase each subsystem is analyzed to determine potential failure modes and the effects of these failures. There are numerous purposes for these analyses. One is to determine the measurements required to evaluate normal and failure situations. These measurements become the engineering measurement list. Typical measurements are temperatures, pressures, voltages, currents, wheel speeds, rotational positions, ACS sensor outputs, and status measurements. Status measurements are inherently digital valve positions, switch on/off positions, relay positions, and safe/arm positions. The engineering data stream is sometimes called telemetry.

8.2.2 Science or Payload Data

The spacecraft is designed around the payload data. For a scientific spacecraft the instruments are as varied as scientific imagination. The Cassini spacecraft, for example, carries six different types of spectrometer, an imaging system that operates in the near-ultraviolet and near-infrared, as well as an imaging radar. It also carries a cosmic dust analyzer, a magnetospheric analyzer, a radio science subsystem, and a dual technique magnetometer. In addition it carries a probe with an additional set of instruments.

In the communication satellite world the payload is a set of communication relay equipment, the selection of which is determined by technology and market considerations. Earth resource satellites carry multifrequency imaging instruments of ever improving efficiency, sensitivity, and size. LANDSAT VII, for example, scans eight frequency bands at a minimum resolution of 15 m with whole Earth coverage. The instrument data rate is as high as 300 Mbps. The large telescopes, like Hubble and Chandra, return images with attendant high data rates. The Chandra downlink rate is from 32 to 1024 kbps.

8.2.3 Data Rates

In general the payload design is reasonably well defined at the start of the spacecraft design process, and the instrument data rates are defined. Command data rates and engineering data rates are difficult to estimate in the early phases of a program, and estimating by analogy to a similar spacecraft may be the best approach. Table 8.1 shows command and engineering data rates for some typical planetary spacecraft.

Table 8.1 shows several spacecraft with commandable rates for the command stream and for the engineering data stream. It is highly desirable to increase both of these rates during critical events in the spacecraft mission. For example, during interplanetary cruise a spacecraft does not make many changes of state, and so high rate commanding is not necessary; a low engineering data rate is acceptable as well.

Table 8.1 Command and engineering data rates—planetary

Spacecraft	Launch	Uplink command rate, bps	Downlink engineering rate, bps
Mariner 9	1971	1	33 1/3
Mariner Venus/Mercury	1973	1	33 1/3
Viking Orbiter	1976	4	1000/331/3
Viking Lander	1976	4	2000/81/3
Voyager	1977	16	40/1200
Pioneer Venus	1978	4	204
Magellan	1989	7.8/62.5	40/1200
Mars Observer	1992	4	250
Mars Global Surveyer	1996	7.8/500	250/2000

Conversely, for near planetary encounter, high data rates and high command rates are essential. The Mars Global Surveyor command and engineering data rates are a good starting place for planetary spacecraft.

It is also desirable to vary the sample rate of many measurements as a function of mission mode. For example, the orbit insertion propulsion system is inert during cruise, and widely spaced measurements are acceptable; once per hour would be adequate. However, near the time of orbit insertion and during the insertion, burn events will be happening in milliseconds and the data system maximum sample rate will be required.

Table 8.2 shows similar data for Earth orbiting spacecraft. Earth orbiters can enjoy much higher command and engineering data rates, as shown in the table Chandra rates would make a reasonable starting point for a modern Earth orbiter.

Table 8.2 Command and engineering data rates—Earth orbit

			Uplink	Downlink	
Spacecraft	Launch	Relay via TDRSS	Command rate, bps	Engineering rate, bps	Science data rate
LANDSAT I	1972	—	128	1040	
FLTSATCOM	1979	—	1000	250/1000	N/A
HEAO	1979	—	200	1000	
GPS	1989	—	1000	500/4000	N/A
HST	1990	Yes	1000	4K/32K	
XTE	1995	Yes		32K	256 kbps
LANDSAT VII	1999	Yes	1000	1K/4.5K/256K	150/300/450 Mbps
Chandra	1999	Yes	2000		32 to 1024K
EO-1	2000	Yes	4000		105 Mbps

8.2.4 Computer Requirements

Computer requirements are derived from timing and sizing analysis; see Section 8.5.

8.2.5 Functional Requirements

Functional requirements come primarily from project and customer top documents. Typical functional requirements are as follows:

1) Direct acting, time critical commands shall not be required for normal operation.

2) It shall not be physically possible to turn off both command receivers at the same time.

3) Critical/irreversible events require two commands for one critical event.

4) To avoid ambiguous spacecraft states, toggle commands shall not be used.

5) The spacecraft shall be commandable to known states without knowledge of current state.

6) The spacecraft shall be capable of telemetering the contents of any computer memory.

7) Onboard computer memory shall be reprogrammable in flight.

8) All enable/disable states must be telemetered.

9) Commandable variation in engineering data sample rate is highly desirable for certain measurements.

10) The fault protection system shall allow ground commanding of enable/disable states.

11) The engineering data stream shall be designed to allow rapid and unambiguous determination of spacecraft state at the start of a data downlink. In particular, it must be possible to quickly determine if the spacecraft has executed a fault protection response.

8.3 Data Handling

Two kinds of data are handled by the CDS: 1) science data or payload data, the information the spacecraft was sent to get, and 2) engineering data, the measurements that portray spacecraft system health and performance. The sequence of data processing, for both engineering and payload data, is shown in Fig. 8.2. Each data measurement is collected at the instrument; analog measurements are converted to an equivalent digital data stream.

Extra bits are added to the data to record the instrument, time, date, and possibly other information. These additional bits are called *overhead.* There are a number of ways the data can be identified. Any method requires additional bits. The overhead requirement is usually expressed as a percentage of the total data rate. The data stream after overhead is calculated as follows.

Consider a data stream of 100,000 bps and an overhead rate of 10%. The overhead bits are added by dividing by 0.9:

$$\text{bit rate with overhead} = \frac{100{,}000}{0.9} = 111{,}111 \text{ bps}$$

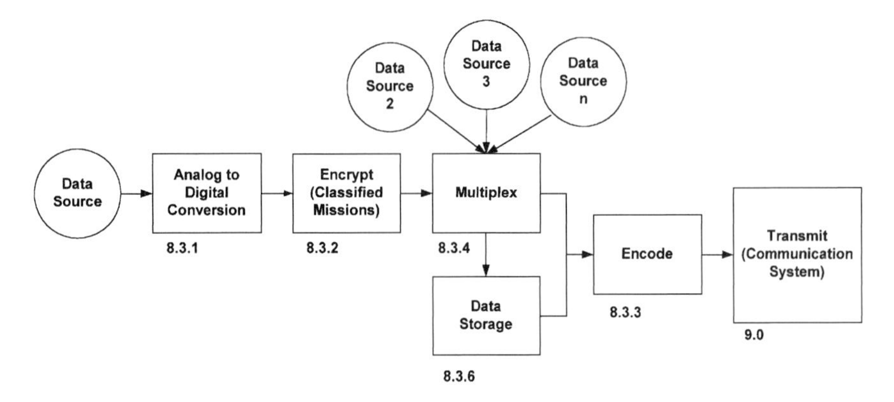

Fig. 8.2 Data handling processes.

It is a common mistake to add 10%, yielding 110,000 bps, which is incorrect and leads to a significant error (11% in this case). The larger the bit rate or the overhead rate, the larger magnitude of the error.

The measurement may then be *encrypted*, if the mission is classified. Next, the data from a single instrument is multiplexed with all other measurements on the spacecraft to form a data block.

The data can be stored for transmission later or transmitted immediately. In either case the data will be *encoded* before transmission. Encoding is a process that adds extra bits for error detection and correction. There are a number of ingenious methods of encoding to be discussed later. Each of these steps is described in more detail in subsequent sections. The section numbers for each step are shown in Fig. 8.2.

8.3.1 Analog to Digital Conversion

Most engineering instruments deliver an analog measurement. Conversely the entire spacecraft is a digital machine. Analog information collected from instruments is converted to digital form before transmission. The steps in conversion are 1) analog data are *sampled* as a function of time, 2) each sample is *quantized* (converted to the nearest digital data number), 3) each data sample is expressed as a digital word, 4) the digital words are inserted into a serial digital bit stream, and 5) the digital stream is transmitted to Earth; see Chapter 9.0.

Nyquist showed in 1928 that, to reconstruct an analog signal from a digital stream, it must be sampled at a frequency at least twice as high as the maximum frequency component in the analog signal. If Nyquist's criteria are not met, the signal cannot be reconstructed. Practical considerations dictate that Nyquist's limit be exceeded with margin. Wertz[2] suggests the sampling rate should be at least 2.2 times the maximum frequency component of the signal. For example, the maximum frequency component of music is about 20 kHz; commercial audio compact disks record music at a digital sampling rate of 44.1 ksps.

Note that when the data are quantized, a *quantization error* is introduced, which is similar to a round off error in computation. Figure 8.3 shows an analog measurement being quantized into a three-bit word. An *n* bit word (or byte) can discriminate

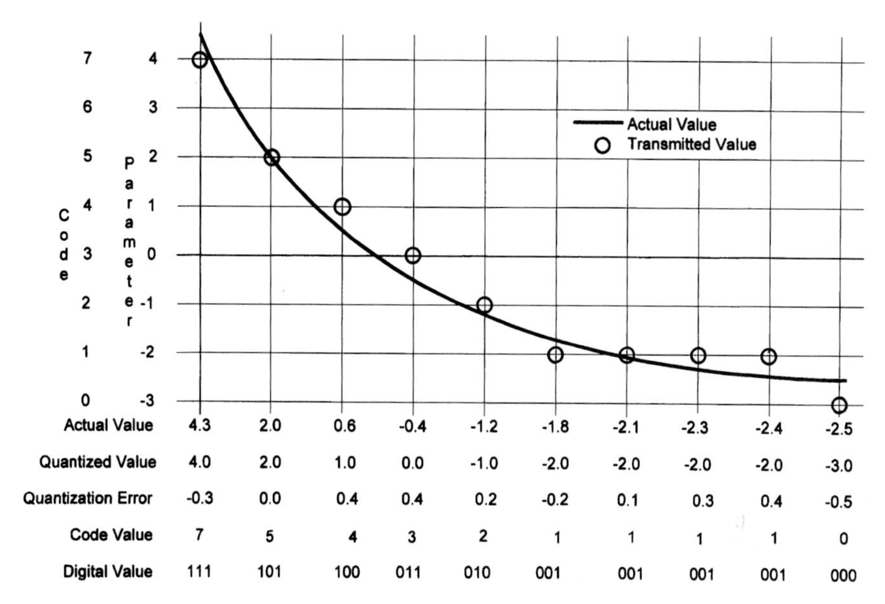

Actual Value	4.3	2.0	0.6	-0.4	-1.2	-1.8	-2.1	-2.3	-2.4	-2.5
Quantized Value	4.0	2.0	1.0	0.0	-1.0	-2.0	-2.0	-2.0	-2.0	-3.0
Quantization Error	-0.3	0.0	0.4	0.4	0.2	-0.2	0.1	0.3	0.4	-0.5
Code Value	7	5	4	3	2	1	1	1	1	0
Digital Value	111	101	100	011	010	001	001	001	001	000

Fig. 8.3 Quantization error.

2^n levels; a three-bit word can discriminate 8 levels. The difference between the natural sample value and the quantized sample value is the quantization error. The quantization error can be reduced by increasing the number of bits used to describe the sample. Table 8.3 shows the quantization error, as a percentage of full scale, as a function of bits per sample.

Three bits per sample is commonly used for engineering data. The Jet Propulsion Laboratory (JPL) Mariner series used 256 shades of gray ($n = 8$) to quantize imaging data. The number of bits per sample times the number of samples per second produces the number of bits per measurement. The sum of the bit rates from all of the measurements on the spacecraft is the spacecraft data rate. For

Table 8.3 Quantization error vs bits/sample

Bits/sample	Quantization error, %
3 (Fig. 8.3)	6.26
4	3.13
5	1.56
6	0.79
7	0.39
8	0.20
9	0.10
10	0.05
12	0.03

any spacecraft there will be a number of mission modes; Magellan examples are cruise mode, orbit insertion, and mapping modes. Each spacecraft mode will have its own data rate. The rate at which data are being collected is not necessarily the downlink rate of the communication system; the data may be either downlinked or stored. In normal cruise mode Magellan collected data and downlinked it at 40 bps (real-time transmission). In mapping mode, data were collected at 806.4 kbps and stored in the tape recorder for later downlink at 268.8 kbps.

8.3.2 Encrypt/Decrypt

Certain missions, notably military missions, require encryption of the downlink and decryption of the uplink. The equipment to perform these functions will normally be supplied by the government and will be long-lead items. They are highly classified and will require special handling. Functionally they will represent the slowest part of the link; commands will come through at a few per second. They represent a single-point failure and the system design should consider both redundancy and bypass capability.

8.3.3 Encoding

The purpose of coding is to locate and correct data bits that are corrupted in the communication process; that is, to reduce the *bit error rate* (BER). The error probability can be reduced by increasing the information bandwidth (data rate). Coding takes advantage of this principle by mapping k data *bits* into n *symbols*, where n is always larger than k. The term "symbols" is used to describe a bit stream that contains coding. For a crude example, data could be transmitted twice and compared on the ground to find corrupted bits. The data rate would be doubled ($n = 2k$); the symbol rate would be twice the data rate. This crude coding technique would find errors; it would not, however, provide enough information to correct errors. It would take a triple transmission to find and correct errors using two out of three voting.

There are a very large number of more elegant and efficient coding schemes available. This chapter will review the most common of the coding schemes that have been studied and used. The error reduction performance of these schemes will be compared along with the attendant data rate increase.[4]

CCSDS recommends the use of Reed–Solomon (255, 223) for the downlink data, both engineering and science. If Reed–Solomon encoding does not adequately reduce the error rate, then convolutional coding may be used in addition. Used this way, the Reed–Solomon code is the outer code and convolutional code is the inner code. The CCSDS recommendation for uplink (commands) is BCH(63, 56) coding. There are two different fundamental approaches to coding: *block codes* and *convolutional codes*. Both are described in subsequent sections.

8.3.3.1 Block coding.

In block coding,[4] the uncoded data are divided into blocks of k bits; the block may or may not be the same length as a word. *Parity* bits are added to each block of k bits, producing a larger block of n symbols. If the number of ones in the block is even, the parity is even. Conversely, if the number of ones in a block is odd the parity of the block is odd. In *even parity* coding, if the number of ones in the block is odd, a parity bit of 1 is added to the block to produce even parity. If the number of ones in a block of data is even, the parity

Table 8.4 Even parity block code, parity bit left[a]

Data block	Parity bit	Symbol block
0 0 0	**0**	**0** 0 0 0
1 0 0	**1**	**1** 1 0 0
0 1 0	**1**	**1** 0 1 0
0 0 1	**1**	**1** 0 0 1
1 1 0	**0**	**0** 1 1 0
0 1 1	**0**	**0** 0 1 1
1 1 1	**1**	**1** 1 1 1

[a]Parity bits in bold.

bit added will be 0, which maintains even parity. A parity bit is added to a data word to make the sum of the bits even for *even parity* code (or odd for *odd parity* code). For example, the sum of the digits in the word 0101110 is even; therefore, the parity bit would be 0 in *even parity* code. Conversely, the sum of the digits in the word 0111000 is odd and the parity bit would be 1.

Table 8.4 shows an even parity (4, 3) block code with the parity bit left of data.

When the bit stream is received, the parity of each symbol block is checked; if parity is odd, the block contains an error. The code in Table 8.4 can detect any single or triple error in a block, and the error is located to the block level in the symbol stream.

Block codes are characterized by the number of symbols bits n and the number of data bits k, and referred to as (n, k) block code. The symbol block in Table 8.4 is (4, 3). The ratio of redundant bits to data bits, $(n - k)/k$, is called the *redundancy* of the code; the redundancy of the code in Table 8.4 is $1/3$. The ratio of data bits to symbol bits k/n is called the *rate* of the code; the rate in Table 8.4 is $3/4$. A block can transmit 2^k distinct words (eight words in this example).

Clever arrangements of parity bits will allow error location and correction at the bit level. If data are arranged in a two-dimensional block, and parity bits added for each row and column, an erroneous bit will produce a parity error in both row and column. Table 8.5 shows an even parity (36, 25) rectangular block code.

Table 8.5 Even parity rectangular block code[a]

Horizontal parity	Data block	Symbol block
1	1 0 1 0 1	**1** 1 0 1 0 1
1	0 0 0 0 1	**1** 0 0 0 0 1
0	1 1 0 0 0	**0** 1 1 0 0 0
1	1 0 1 1 0	**1** 1 0 1 1 0
0	0 1 0 0 1	**0** 0 1 0 0 1
Vertical => parity	**1 0 0 1 1**	**1 1 0 0 1 1**

[a]Parity bits in bold.

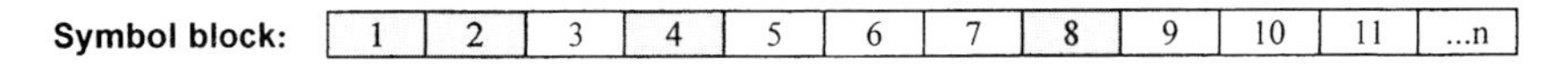

Fig. 8.4 Hamming Code block—parity bits shaded.

Parity errors give row and column of the erroneous bit; therefore, a bit error can be detected, located, and corrected by the decoder.

8.3.3.2 Well-known block codes. Hamming[5] codes are a family of block codes devised by Richard Hamming of Bell Telephone Labs in 1950. These codes place the parity bits in special positions in the symbol block. If the bit positions are numbered starting from 1, then the parity bits are placed in those positions that are powers of two; i.e., positions 1, 2, 4, 8, 16, 32, 64, and so on. Conversely, the data bits are in positions 3, 5, 6, 7, 9, 10, and so on. The symbol block then looks like Fig. 8.4 with the parity bits shaded.

The parity bits make the sum of digits even (or odd) for a given subset of the symbol block. Each position in the symbol block is part of the parity check for those bits whose position numbers add up to the bit position. The symbol position 5 is checked by the parity bit in position 1 and 4 because $5 = 1 + 4$. This causes each position to be checked by a unique set of parity bits. The first parity bit is summed over bit positions 3, 5, 7, 11, 13, etc. The second parity bit is summed over 3, 6, 7, 10, 11, 14, and so on. The pattern of checks is said to be *orthogonal* over the data block.

Devised by Bose–Chadhuri–Hocquehem, BCH codes[4] are a powerful class of Hamming code that allow multiple error correction. A BCH(63, 56) code is recommended by the CCSDS for the all-important job of command transmission. BCH codes outperform other block codes with the same block length and rate. For more data on BCH codes see Sklar,[4] p. 301.

Reed–Solomon (R–S) codes are a special subset of BCH codes that were discovered before the BCH codes.[4] The Reed–Solomon (255, 223) code is the block code selected for standardization by the CCSDS telemetry guidelines. The bit rate increase created by this code is a factor of 255/223, or 1.1434978. The selection of R–S code may well have been because Reed–Solomon codes are particularly suited to correcting burst errors and are effective where the set of input signals is large.[4] The code is capable of correcting any series of t or fewer symbol errors where

$$t = \frac{n - k}{2} \tag{8.1}$$

A symbol stream with an error rate of 10^{-2} can be improved by a factor of 10^4 with R–S coding at $t = 4$.

Two-dimensional Reed–Solomon codes have been devised that will detect and correct very long burst errors. The inner code can correct burst errors of a few bytes in length whereas the outer code can correct very long bursts.

The R–S code has another useful trait. A symbol stream emerging from the encoder presents k bits of unaltered data followed by $k - n$ code correction symbols. The unaltered bits are followed by a trailer of code. Thus the MOS team can read and use the unaltered transmission while the bit stream is being decoded and error

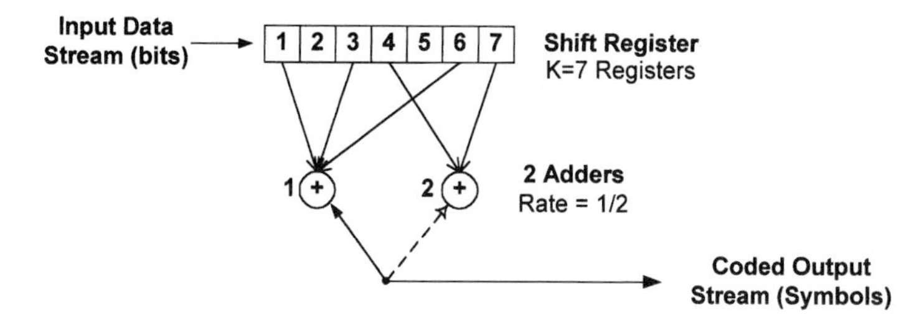

Fig. 8.5 Convolutional encoder, rate $1/2$, $K = 7$.

corrected. This trait is very useful in times of high activity such as the approach to planetary encounter.

8.3.3.3 Convolutional coding. Convolutional coding is a newer type of coding that gives better bit error rate reduction than block coding for a given symbol-to-bit ratio. In addition, the encoder is simpler to implement in hardware than a block encoder. A linear block code is described by two integers, n and k. A convolutional code is described with three integers, n, k, and K, where the ratio k/n has the same significance as with block codes; however, n does not define a block length in convolutional coding. The integer K is known as the constraint length; it is the number of stages in the coding shift register. In practice, n and k are small integers and K is used as the variable to control redundancy. Magellan used rate $1/2$, $K = 7$ convolutional code and rate $1/2$, $K = 7$, is the convolutional code selected for standardization by the CCSDS.

A rate $1/2$, $K = 7$ convolutional encoder is shown schematically in Fig. 8.5, after Edelson,[6] p. 157. The encoder has a seven bit register ($K = 7$); input data bits enter the register from the left. There are two modulo-2 adders (rate $1/2$) that sum selected input bits in the register.

The selection of the bits associated with each adder is arbitrary. There is no current theory to determine the best way to wire the adders. However, any change in the adder wiring to the registers produces a different convolution code.

In each encoding step a single data bit is pushed into the register; existing bits are shifted to the right 1 position; one bit is shifted out of the register. Adder 1 takes the modulo-2 sum of the data bits in registers 1, 3, and 6; then adder 2 takes the modulo-2 sum of the data bits 4 and 7. The rules of modulo 2 addition are the same as the rules of parity:

$$
\begin{aligned}
0 \oplus 0 &= 0 \\
0 \oplus 1 &= 1 \\
1 \oplus 0 &= 1 \\
1 \oplus 1 &= 0
\end{aligned}
$$

where $\oplus$ is the symbol for modulo-2 addition.

Consider a case where the registers have just been flushed and contain all zeros. Now enter the data sequence 1 0 1 1 0 1. Table 8.6 shows the conversion of the data sequence 1 0 1 1 0 1 into a symbol sequence in rate $1/2$, $K = 7$ convolutional code, the CCSDS standard.

Table 8.6 Rate 1/2, K = 7, convolutional encoding example

Step	Register contents 1 2 3 4 5 6 7	Adder 2 4 + 7	Adder 1 1 + 3 + 6	Symbol stream
1	1 0 0 0 0 0 0	0	1	01
2	0 1 0 0 0 0 0	0	0	00 01
3	1 0 1 0 0 0 0	0	0	00 00 01
4	1 1 0 1 0 0 0	1	1	11 00 00 01
5	0 1 1 0 1 0 0	0	1	01 11 00 00 01
6	1 0 1 1 0 1 0	1	1	11 01 11 00 00 01
7	0 1 0 1 1 0 1	1	0	10 11 01 11 00 00 01
8	0 0 1 0 1 1 0	0	0	00 10 11 01 11 00 00 01
9	0 0 0 1 0 1 1	0	1	01 00 10 11 01 11 00 00 01
10	0 0 0 0 1 0 1	1	0	10 01 00 10 11 01 11 00 00 01
11	0 0 0 0 0 1 0	0	1	01 10 01 00 10 11 01 11 00 00 01
12	0 0 0 0 0 0 1	0	0	00 01 10 01 00 10 11 01 11 00 00 01

From Table 8.6, the output symbol sequence is 24 symbols long, line 12. The next register shift would flush the register to all zeros. In this short data stream, 24 symbols were generated for six data bits, or rate $= 1/4$, far short of the expected rate 1/2. This discrepancy is caused by the short data stream. The last seven steps were used shifting the end of the data stream out of the register. If a 10,000 bit stream were shifted through the register, a 20,007 symbol stream would have resulted (much closer to the expected rate 1/2). Convolutional coding has no block length; long uninterrupted data streams are desirable. It is often necessary to force a convolutional code stream into a block structure. When this occurs, it is necessary to append a string of zeros to the data stream to flush the registers (as was necessary in the above example). The trailing zeros reduce the effective code rate.

In addition to the preceding representation, it is possible to represent convolutional code with polynomials, state diagrams, vectors, tree diagrams, and trellis diagrams; see, Ref. 4, pp. 315–377.

8.3.3.4 Shannon's law. Shannon's law places a theoretical limit on the improvement in BER that can be achieved by any coding scheme. In 1949, C. E. Shannon[7] showed that there is a theoretical limit to the capacity of an error-free channel related to the SNR and channel bandwidth:

$$C = B \log_2(1 + \text{SNR}) \tag{8.2}$$

where $C =$ maximum information capacity of a channel, bits/s, and $B =$ bandwidth of the channel, Hz.

It can be shown that for minimum energy/bit, the channel band width B should be about equal to the transmitted bit rate R_b, and this is a common design practice. By setting $R_b = B$, Shannon's law can be rearranged to a more useful form:

$$\frac{E_b}{N_0} = \ln 2 = -1.6 \text{ dB} \tag{8.3}$$

E_b/N_0 is the energy contained in each binary digit in the communication link in watt-seconds or joule. E_b/N_0 can be reduced to a signal-to-noise ratio if bandwidth is equal to bit rate and is a constant for a given link under fixed external conditions. It is a fundamental measure of link performance that will be further discussed in Chapter 9.0. In effect, Eq. (8.3) shows the theoretical limit in error reduction that can be made by coding.

8.3.3.5 Comparison of coding techniques. The fundamental function of coding is to decrease the bit error rate of a communication channel by increasing the symbol rate or bandwidth. In this section the error rate reduction performance and the bandwidth of common coding schemes will be compared. Figure 8.6 compares the BER improvement offered by coding as a function of E_b/N_0. Consider a communication link with an E_b/N_0 of 4 dB. If the information were sent uncoded, the bit error rate, from Fig. 8.6, would be about 10^{-2}. If biorthogonal block coding or Reed–Solomon coding were used, the BER would be reduced by a factor of 10;

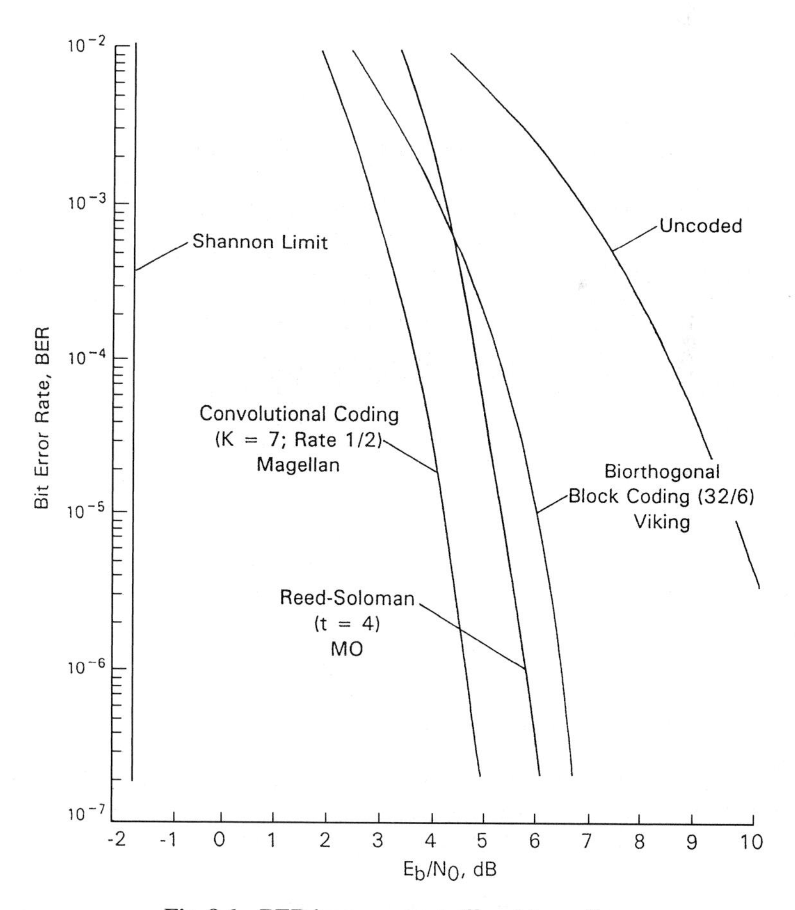

Fig. 8.6 BER improvement offered by coding.

Table 8.7 Increase in information rate caused by coding

Spacecraft	Coding Scheme	Input data bits	Output symbols
Viking	Biorthoganol Block(36/6)	6	32
Magellan	Convolutional K = 7; Rate 1/2	1	2
CCSDS Standard	Reed Solomon (255, 223)	1	1.1434978

with convolutional coding, BER would be reduced by a factor of 1000. The price of encoding is an increase in symbol rate to be transmitted. Table 8.7 shows the increase in information rate produced by three coding schemes.

8.3.3.6 Concatenation. When even greater error correction is needed it is possible to combine coding schemes, called concatenation. The CCSDS recommended concatenation scheme is made by combining convolutional code with Reed–Solomon code, the Reed–Solomon code being the outer layer and convolutional code being the inner layer. The data bits are first convolutionally coded. The output symbol stream from the convolutional coder is Reed–Solomon coded. This scheme provides significant increase in error protection and a significant increase in symbol rate.

8.3.3.7 Typical spacecraft coding plan. *Data storage.* The data are sent to the recorder uncoded. The onboard recorder applies coding, usually a short Reed–Solomon, shorter than the standard (255, 223). The purpose of this code is to protect the data from single-event upsets while in storage. The coding is removed from the data as it is read from the recorder and the detected errors are corrected. This process is internal to the data storage system and invisible to the operations team. The rated capacity of a recorder is in data bits; the extra space for the code symbols is provided beyond the rating.

Computer memory. Onboard processor memory is usually protected by a Hamming code. Data are coded and stored, and code is removed and errors corrected as data are read, similar to data storage.

Downlink. On the downlink the CCSDS coding recommendation is a long Reed–Solomon (255, 223) for engineering and science or payload data. If the standard Reed–Solomon code does not provide adequate error correction, it is concatenated with convolutional rate 1/2 code, the Reed–Solomon code being the outer layer and convolutional code being the inner layer. (Concatenation is seldom needed for the engineering data.)

Uplink. For commands, the CCSDS coding recommendation is a BCH (63, 56) code. Decoded command words are compared to a dictionary of acceptable words; mismatches are downlinked to MOS. In critical cases, the command sequence may be downlinked for a bit check by the MOS team.

In summary, the CCSDS recommended coding strategy is shown in Table 8.8.

Table 8.8 CCSDS coding recommendations (Ref. 1)

Function	Recommended coding	Symbols/bit
Commands	BCH(63, 56)	1.125
Data	Reed–Solomon (255, 223)	1.1434978
Data, very low error rate = concatenation	Reed–Solomon (255, 223) + Convolutional, Rate 1/2, K = 7	2.2869956

8.3.4 Multiplexing

Multiplexing (MUX) is a method by which the measurements from an entire spacecraft can share a single communication channel. MUX is accomplished by sharing the available time division multiplexing (TDM) or frequency division multiplexing (FDM). The most common method, time division multiplexing is shown in Fig. 8.7. Each measurement is sampled on a time schedule controlled by the poll table stored in the spacecraft computer.

Each measurement is placed in sequence in the MUX memory until a data frame is complete. The equipment that samples data measurements in sequence is called a multiplexer or MUX; the inverse equipment is called a demultiplexer or DEMUX.

The data stream is assembled into data frames by the MUX. A major frame usually contains the data downlinked in 1 s. Frame size is a design variable; the frame shown in Fig. 8.8 is 64×16 words and contains 1024 data words (minus overhead). Word size varies typically from 8 to 64 bits.

Each position in the major frame is called a minor frame. The minor frames are used to *subcommutate* measurements. Subcommutate means to sample a measurement at less than the major frame rate. A minor frame may be divided into any number of slots; 16 are shown in Fig. 8.8. Measurements that vary slowly are subcommutated. Measurements that vary rapidly can be *supercommutated* by placing them in multiple frames. A measurement in two minor frames would be sampled at twice the frame rate. The major frame structure can be changed for each mission mode. For example, engine parameters would be sampled during maneuvers but not during cruise.

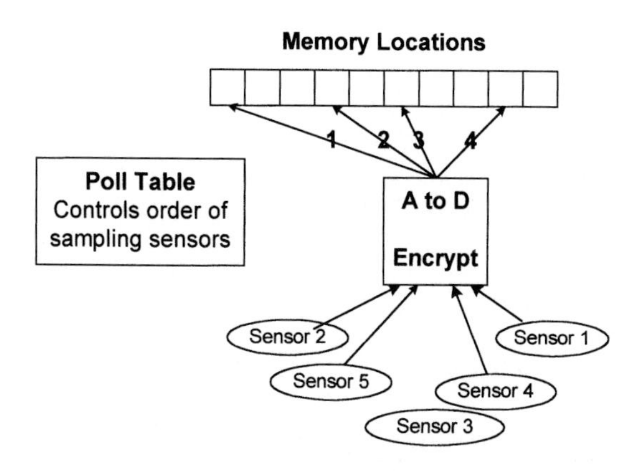

Fig. 8.7 Multiplexing.

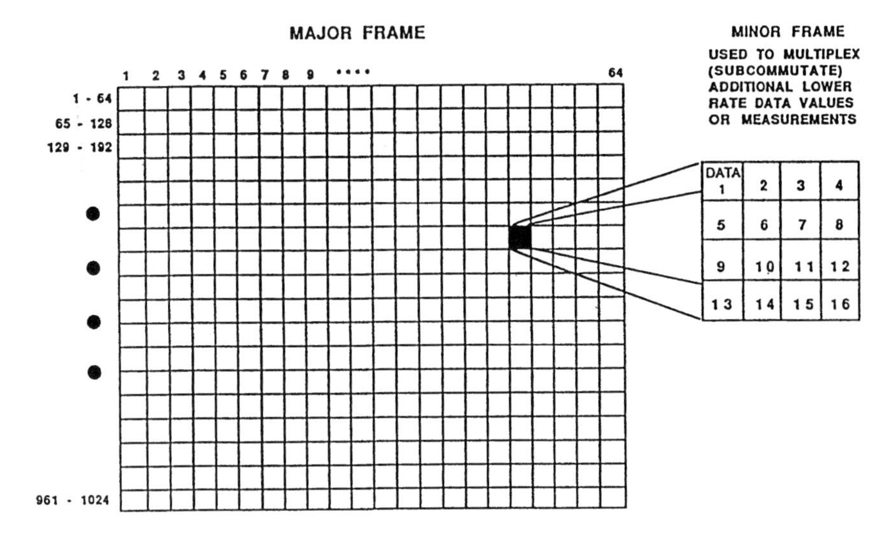

Fig. 8.8 Typical major frame.

Example 8.1 Telemetry Format

Given a data transmission rate of 100,000 symbols/s and the following conditions: major frame = 100 minor frames, minor frame = 100 8-bit data words, four 8-bit words in each minor frame are overhead (synchronization and spacecraft time), and 10 words in each minor frame may be subcommutated. How many separate parameters can be measured in this format?

There are $10 \times 100 = 1000$ subcommutated measurements and there are $100 - 4 - 10 = 86$ measurements that are not subcommutated. Therefore, the maximum number of measurements in this format is 1086 measurements per frame.

What is the sample rate for the measurements that are not subcommutated or supercommutated? There are 800 bits per minor frame; minor frames are transmitted at a rate of $100{,}000/800 = 125$ minor frames/s, major frames are transmitted at a rate of 1.25 frames/s. Therefore, measurements that are not subcommutated or supercommutated occur once per major frame, or 1.25 times/s.

How would you obtain a sample rate of 250 samples/s? Sample the parameter twice in each minor frame = $2 \times$ minor frame rate = 250 samples/s.

Multiple access is similar in principle to multiplexing, already discussed. MUX is suitable for a preplanned, relatively fixed sharing of centralized resource; whereas multiple access techniques are highly flexible in real time and are used with distributed resources. Multiple access is used with local area networks, with distributed data systems, and with, and between, communication satellites.

8.3.5 Packet Telemetry

The CCSDS[1] is attempting to standardize data handling methods on an international basis. If successful, these standards have the promise of making spacecraft

Packet Primary Header	Spacecraft Data Packet
Content and format precisely specified by CCSDS: o Packet identification and version o Packet control data o Packet data length	Contents controlled by spacecraft team o Secondary (spacecraft) Header o Spacecraft Data o Error correction Coding
Primary header length = 6 octets(48 bits) Fixed Length	Data packet length = variable 1 to 65538 Octets

Fig. 8.9 Source packet format.

data systems substantially cheaper, particularly the ground-based system elements. Standardization is also essential to multiagency projects like the Space Station. The CCSDS standard system is called packet telemetry. Standard methods of packet transfer between spacecraft equipment and ground equipment are being recommended. The contents of the packet are entirely up to the spacecraft project. The concept is analogous to the postal system in which the size of letters is standardized as well as the placement of the address, the return address, the stamp and the bar code, but the contents of the letter is entirely up to the sender.

The CCSDS standard source packet structure is shown in Fig. 8.9. The *packet primary header* format is rigidly controlled so that packets can be handled by standardized equipment. The contents of the *packet data field* are entirely up to the spacecraft project. The field can be from 6 to 65542 words in length. If a secondary header is used in the packet data fields, its presence must be noted in the packet primary header. Spacecraft data are collected, digitized, and coded and the standard headers prepared and attached at the front of the data stream.

The headers contain the packet number and the total number of packets in this download. For downlink, these fixed-length packets are assigned to a *virtual channel*. These blocks of data are then called virtual channel data units (VCDU). At the ground station the packets are received, sorted into the correct order, first to last, then routed to the correct spacecraft team using address data from the header.

8.3.6 Data Storage

Data storage can be accomplished by two basic methods: 1) tape recorders or 2) solid-state recorders. Digital tape recorders were the industry standard until the late 1980s when solid-state storage became available. Solid-state recorders offer the advantage of lower mass, power, bit error rate, better temperature tolerance, and are now the industry standard. Table 8.9 compares the characteristics of the two forms of storage.

Solid-state recorders have power and mass advantages compared with tape recorders, as shown in Table 8.10. The memory can be addressed in blocks of selectable size; therefore, data can be read without tape management delays. Bad memory locations can be programmed out of use. Large blocks of redundant memory can be provided. Software can be used to detect and correct errors. Typically the coding is added to the data as it is stored. When data are read out, the code is used to correct any errors that have occurred and the corrected data are forwarded to the communication system for downlink. This process is invisible to

Table 8.9 Data storage methods

Characteristic	Typical tape recorder	Typical solid-state recorder
Total storage, bits	2×10^9	2×10^9
EOL error rate	10^{-6}	10^{-9}
Power		
record, W	30	4
playback, W	54	4
standby, W	2	4
Size, cm^3	16,000	2000
Mass, kg	11.7	4
Temperature range	0 to 35°C	−55 to 100°C

the user and is independent of coding used to improve the downlink. The memory locations where errors occurred are recorded for downlink as engineering data. These records are used to isolate bad memory locations. Table 8.10 summarizes the capabilities of some currently available solid-state data recorders (SSDR). From the table it can be deduced that a circa 2000 solid-state recorder weighs about 0.25 kg/Gbit and requires about 1 W/Gbit.

Data stored in a SSDR is protected by encoding (see Section 8.3.3). As data are read into the recorder, parity bits are added. The encoding provides error detection and correction for the storage period; as the data are read out of storage they are corrected if bit flips have occurred. This whole process is invisible to the user. A recorder rated for 10 Mbits of storage will actually store 10 Mbits plus the required number of parity bits, and 10 Mbits, cleaned and corrected, will be returned to the spacecraft on command.

Table 8.10 Solid state data recorders[a]

Mission	Throughput	Memory	Power	Volume	Weight	Launch
SAMPEX	900 kbps	48 Mbit	12 W	0.6 ft^3	8.2 kg	1993
XTE	1 Mbps	1.1 Gbit	38 W	1.5 ft^3	24.9 kg	1995
Clementine	20 Mbps	1.9 Gbit	4 W	2048 cm^3	4.1 kg	1995
MGS	20 Mbps	8 Gbit	8 W	1920 cm^3	6.9 kg	1996
TRMM	2 Mbps	2.2 Gbit	42 W	2.0 ft^3	37.2 kg	1997
Hubble Space Telescope	1 Mbps	12 Gbit	12 W	0.5 ft^3	11.8 kg	1997[b]
Cassini		1.8 Gbit	8.2 W		15.8 kg	1997
LANDSAT-VII	300 n Mbps	378 Gbit	70 W	3.9 ft^3	100.6 kg	1999
EO-1	840 Mbps	48 Gbit	38 W	36 kcm^3	22 kg	1999

[a]Data in part from Webb, Ref. 8. Performance of these devices, like computers, improves rapidly. Today's performance will be better than that shown here.
[b]Added to HST on the second servicing mission.

8.4 Command Processing

Commands are used to change subsystem configurations, control spacecraft activities, and to load data and instructions into computers. Commands can be discrete or serial digital. Discrete commands can be high level power switching (28 V) or low level logic switching (5 V). Discrete commands may be used to start a series of events such as spacecraft release from the upper stage. A block (or sequence) of commands are used to control a series of related events; they can be as comprehensive as an orbit insertion burn or a complete software upload. Commands can be real-time or stored to be executed at a given time or after a given event.

8.4.1 Command Standardization

Command handling, like data handling, has been standardized by the CCSDS. The standard requirements are summarized in this section. Commands are transmitted in a group or *command link transmission unit* (CLTU) which contains all of the individual commands or *codeblocks* for a major spacecraft event. For example, a maneuver would require commands to 1) rotate the vehicle around each axis by a certain number of degrees, 2) fire a given group of engines until a given velocity change is achieved, 3) shut down the engines, and 4) rotate the vehicle around each axis to return to the initial attitude. Each of these individual commands is formatted into a codeblock. All of the codeblocks associated with the maneuver would be compiled into a CLTU.

The CCSDS-recommended structure of a fixed-length codeblock is shown in Fig. 8.10. The data in a codeblock are encoded with BCH (63, 56) coding. The codeblock length is fixed for a given spacecraft; the recommended length L is 64 symbols. The first 56 bits are the command data bits, the seven parity bits are next. One 0 filler bit is used to make the codeblock an even eight words, 64 bits, long. The command symbol rate is 1.125 times the bit rate.

If the command is to be stored (most are) a memory address and time will be specified. Data needed in conjunction with the command may also be specified, and finally the execution instructions will be given. An incorrect parity bit indicates an error in the command data; in which case a command error message is sent to the telecommunications system. If the error cannot be corrected with certainty, the command is dumped and retransmitted.

A CLTU is the data structure that carries any number of codeblocks in the uplink data stream. The components of a CLTU are shown in Fig. 8.11. The start sequence is a standard 16-bit binary synchronization pattern that announces the start point of

Command Codeblock	
n Command Bits n = 32, 40, 48 or 56	8 Parity Bits

Fig. 8.10 CCSDS command codeblock structure.

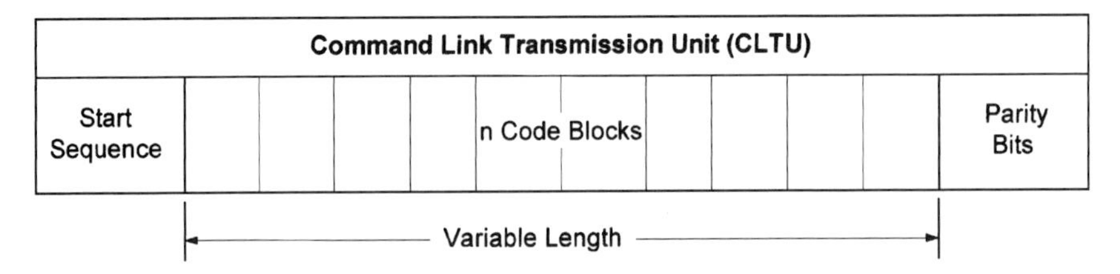

Fig. 8.11 Components of the command link transfer unit.

the CLTU. The "encoded TC Data," shown in Fig. 8.11, is any number of complete codeblocks. The tail sequence is the same length as the codeblocks and consists of a standard binary word repeated eight times for eight-word codeblocks.

Example 8.2 Command Message Timing

1) Given the following, how long will it take to get the message into the spacecraft computer memory assuming no errors are made and that distance is negligible?

The size of the command CLTU is 20,000 16 bit words. The CLTU is to be stored in the spacecraft computer.

The uplink is convolutional coded ($k = 7$, Rate 1/2).

There are 200 bits in the header and trailer.

The link bit error rate is 10^{-5}.

The uplink transmission rate is 2000 symbols/s.

The number of bits in the command CLTU is $20{,}000 \times 16 = 320{,}000$ bits, plus 200 bits of overhead $= 320{,}200$ bits. Rate 1/2 coding increases the transmitted information by a factor of two, thus 320,200 bits become 640,400 symbols. Transmission requires $640{,}400/2000 = 320.2$ s.

1) How much transmission time will it take to have a validated command CLTU in the computer memory assuming the received command CLTU is transmitted to the ground for checking and the erroneous words are retransmitted once?

$640{,}400 \times 10^{-5} = 6.404$ erroneous words can be expected to be received by the spacecraft, round up to seven words. The downlink of the received command will add another seven erroneous words. Therefore, 14 words must be retransmitted.

There are $(14 \times 16) + 200 = 424$ bits or 848 symbols in seven words.

It takes $848/2000 = 0.424$ s to retransmit the erroneous words. The initial transmission takes 320.2 s; the downlink takes 320.2 s. Therefore, it takes $320.3 + 320.2 + 0.424 = 640.824$ s to establish a validated command CLTU in the spacecraft computer, ignoring the time spent by the MOS team between transmissions.

3) How long would it take to establish a validated command CLTU in the computer memory if the spacecraft were 250 million kilometers away in Venus orbit?

The time required for the uplink or downlink signal to travel 250 million kilometers is distance divided by the speed of light in a vacuum: $250E6/3E5 = 833.333$ s. The total time is

$$833.333 + 320.2 + 833.333 + 320.2 + 833.333 + 0.424 = 3140.823 \text{ s}$$

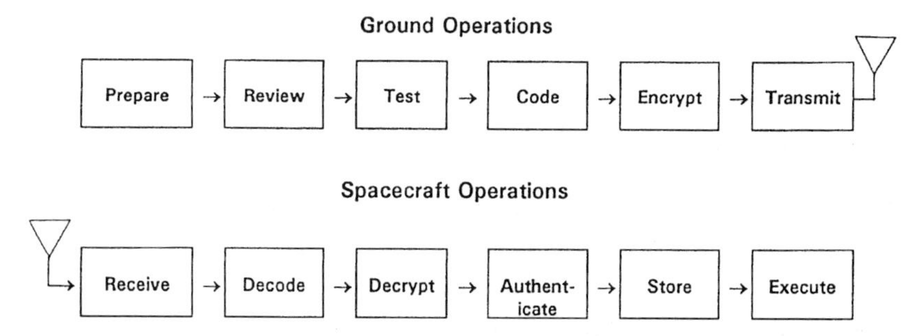

Fig. 8.12 Life cycle of a command.

8.4.2 Command Life Cycle

The life cycle of a command sequence, shown in Fig. 8.12, starts with preparation, review, and testing by the MOS team. Extreme care is required in preparation and testing; command errors are difficult to find and are a leading cause of spacecraft failure. A case in point is the USSR Phobos 1, a Mars spacecraft launched a few years ago. During cruise, a command CLTU was sent to the spacecraft and because of a single bit error, a 0 instead of a 1, the spacecraft went into a "terminate operations" mode, shut itself down and turned off the radio—a mode from which there is no recovery.

Command testing is normally done on a spacecraft simulator; the degree of simulator fidelity varies widely. A low fidelity simulator can miss errors or create bogus alarms. A high fidelity simulator is very expensive. The best simulator is a full-up spacecraft; even then flight simulation is not complete.

When the project is satisfied with a command sequence, it is coded (and may be encrypted) and uplinked to the spacecraft at a low data rate. The low data rate and coding are used to achieve a very low bit error rate; the error rate at threshold is normally less than 10^{-5}. A command is received by the communication system and forwarded to CDS. The CDS system decodes it, decrypts it, and checks to make sure it is for this spacecraft. The command is then authenticated (checked to make sure it is really a command). Authenticity tests are very important. During the first Magellan launch attempt, STS turned on a 10-kW rf beam carrying digitized voice messages at a frequency near the command uplink frequency. The Magellan command receiver locked up on the beam (digitized voice looks much like a random number stream) and over 300 of the random word sequences were accepted by the receiver as commands; the onboard authenticity checks rejected them all. There are many less dramatic ways to get random command-like messages into the spacecraft, particularly near Earth.

The common methods for authenticating a received command are 1) decoding parity checks, 2) verifying each command word against a dictionary of acceptable words, and 3) telemetering the command back down to the MOS team for verification before execution, which is the ultimate check.

8.4.3 Stored Commands

Most command blocks are prepared and tested in advance and stored onboard. For example, Magellan had a pretested emergency periapsis raise maneuver command block stored onboard during the entire aerobraking process. At any time the MOS team could have brought the spacecraft safely up out of the atmosphere simply by activating that command sequence. For planetary spacecraft, with large transmission delay times, real-time commands are essentially not possible. Authentic and verified commands are stored onboard and executed based on spacecraft time.

A portion of the CDS memory is reserved for verified, authenticated, command storage. Commands remain in memory until use. They are protected from bit flips by onboard coding that is removed and errors corrected when it is time to execute the command. After execution the command memory location is reused for new commands.

For stored commands the spacecraft clock is the critical piece of equipment that determines when a command block is to be executed. The clock is typically maintained as part of the CDS system. The clock meters elapsed time since a start point just before launch, by incrementing a time number each second or fraction of a second. It may maintain subordinate fields with smaller granularity. The Magellan clock had four fields and four granularity levels. When the clock number is equal to a value specified in the command uplink, the command sequence is initiated.

In addition, the spacecraft clock is used to time-stamp both science and engineering data as they are taken from the instruments. These time marks are essential to the interpretation of both types of data.

8.5 Spacecraft Computers

8.5.1 Command Sequencers to Computers

In the Sputnik/Explorer era spacecraft had only a single mode of operation; e.g., take data and transmit it real time until something fails. In the 1960s and early 1970s it was common for spacecraft to have several modes; the equipment that controlled these modes worked much like the sequencer in a dishwasher and were essentially timers for relay functions. As missions became more complicated the sequencers developed into computers. JPL and the planetary missions led the conversion because the communication delays over planetary distances placed a high premium on spacecraft autonomy. For example, consider a spacecraft near Mars. If a failure occurred, it would take 15 min for the data to arrive at Earth. Even a simple failure with clear symptoms would take at least 1 h to respond to, including 30 min in transmission delay. Not many failures are this patient.

In addition, the planetary missions were becoming more and more complex. The Viking mission to Mars in the 1970s required a lander as well as an orbiter, and the command controller, without question, had become a computer. The Viking Command Computer was single-fault-tolerant and dual-redundant throughout. Either computer could do the complete mission in the event of a failure. The machine used 4K of plated wire memory, the development of which became a major schedule risk. The memory was divided into four equal parts. The first three could be set as read only, write, write protected, or read/write. The last 1K was always read/write. The computer used 18 bit words, 6 for operating codes and 12 for addresses. The clock issued interrupt signals every hour, second, and 10 ms, similar to the clocks on the Mariners. With the computer came the first flight-quality software development program for a spacecraft.

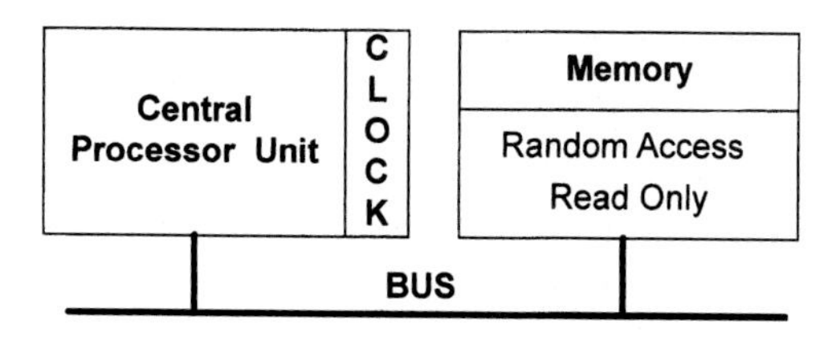

Fig. 8.13 Computer block diagram.

8.5.2 Central vs Distributed Computers

Up to this point, we have emphasized the computer role in the command system; however, there is an additional and different computer job onboard a spacecraft: that of data handling. Spacecraft data is composed of two distinct types: engineering data that describe the performance of the spacecraft subsystems as a function of time, and the science or payload data that are the very purpose for the existence of the spacecraft. The data control functions are analog-to-digital conversion, encoding, formatting, and storing. All of these functions are ideal for computer work.

The ideal computer for data control would feature very large memory and data handling characteristics; modest processor speeds would be acceptable. The ideal computer for attitude control functions would feature very high speeds, matrix manipulation features, and modest throughput. A single computer cannot be optimized for both tasks.

In the 1970s, the central main frame architecture was used, and memory was very limited. Software was usually written in machine language to conserve memory. The Voyager spacecraft, typical of that era, had redundant computers with 4K, 16-bit, memories. It lifted off in 1978 with approximately 30 memory locations unused. In a modern spacecraft, distributed computers on a network are used, memory is extensive, ever larger throughput is desired, and there is an array of high-order languages in use. The extent and criticality of software has risen to the point that software is often considered a separate subsystem. The Galileo spacecraft, launched in 1989, has 18 microcomputers each with 36K, 8-bit memories.

*8.5.3 Anatomy of a Computer**

All computers have four common elements, a central processing unit (CPU), memory, clock, and bus, as shown in Fig. 8.13.

8.5.3.1 CPU. The CPU is the brain of the computer. It contains the circuitry to access memory locations, interpret instructions, implement instructions, and perform logic functions. A major element of the CPU is the arithmetic logic unit (ALU), which retrieves digital numbers from memory locations for calculations and sends the results to memory. A fraction of 1 ms in the life of a CPU might look like this:

*In the preparation of this section, the author made frequent reference to the *Computer Desktop Encyclopedia*,[9] an excellent source for anyone working around these machines.

Get the number in memory 1; put it in the register
Divide the number in the register by the number in memory 2; put result in the register
Determine if the register is larger than the number in memory 3
If yes, send the register number to memory 3
If no, erase the number; set the register to all zeros

8.5.3.2 Floating point processor. The arithmetic unit may be either a floating point processor or a fixed point processor. Floating point is a processor that can handle additional digits describing the decimal point location. (The number 2.204×10^{12} is a floating point number.) Floating point allows a computer to handle more complicated mathematics than a fixed point processor. Attitude control calculations require a floating point machine. A floating point processor is generally slower than a fixed point processor.

8.5.3.3 Instruction set. An instruction set is the collection of machine language commands that the computer recognizes; this smallest executable element of a machine language. These instructions are digital words not readily recognized by humans. High-order languages (BASIC, COBAL, HAL-S) have commands in English words. When a high-order language is *compiled*, it is converted to machine language.

8.5.3.4 Reduced instruction set computer (RISC). A RISC machine is optimized for processing speed. Reducing the instruction set is one of the methods used to increase speed. The antithesis of a RISC machine is a complete instruction set (CISC) machine. A machine that has not reduced the instruction set for speed is called a non-RISC or CISC, complex instruction set computer. A CISC machine can handle complicated processes with less code at the price of reduced speed.

8.5.3.5 Bus. The bus is a high-speed parallel connection between the operating elements of a computer. The number of conductors in a bus is 8, 16, 32, or 64. The parallel paths in a bus normally correspond to the register size and word size and hence, a bus transfers all the bits in a word simultaneously. A CPU usually has at least two buses, an address bus and a data bus. Thus, when the CPU places an address on the address bus it is instantly presented with the memory contents on the data bus. An internal bus is called the *local* bus. Specialized buses are also used between the CPU and peripherals.

8.5.3.6 Memory. The *byte* is the universal unit of data storage; it is the smallest addressable unit of storage. A byte contains eight digital bits, ones and zeros. One byte can describe 256 different characters, letters, numbers, or punctuation. In some cases a byte may contain an additional parity bit for error checking.

Computers are rated by the number of bytes of each type of memory they contain. For example, the Magellan ACS computer contained RAM space for 32 thousand bytes of digital data, 32K. (Note that 1K of computer memory is 1024 bytes, not 1000.) One Space Shuttle computer contains RAM space for 192 million bytes of data, often stated as "192 meg."

There are numerous types of memory in a modern computer. Read only memory (ROM) and random access memory (RAM) are the oldest types; however, numerous subtypes with various useful properties have evolved with time. The new RHPPC603e computer, from Honeywell, has three different memory types: 128 kbytes of DRAM, 4 Mbytes EEPROM, and 128K PROM.

RAM can be read or overwritten by the CPU any number of times. RAM is divided into sections the size of a computer byte. Each byte location has an address. The CPU can read the byte in any RAM address or write a new byte to any address. RAM is volatile; that is, all memory goes to zeros and is lost with a power outage.

ROM is "hard wired," cannot be overwritten, and is not destroyed by power loss (nonvolatile). Information is permanently installed in memory when the chip is manufactured. The primary spacecraft use of ROM is to contain the spacecraft restart code for use after a power interruption or major failure. The main difficulty with spacecraft ROM is that it is a long lead item. The restart steps must be specified in great detail before the spacecraft design is well understood. These restart steps cannot be changed after the spacecraft design is developed. For that reason, new types of memory chips are now used for spacecraft nonvolatile memory. *PROM* and *EPROM* devices have the same properties in use as ROM; however, data are initially placed in memory by different methods.

DRAM (D-RAM or dynamic RAM) is the most common type of RAM, usually consisting of a transistor and a capacitor. It must be refreshed hundreds of times a second to retain information.[9] This type of memory loses all information with a power interruption. *SDRAM* or *SyncDRAM* is a new type of dynamic RAM with sophisticated features that make it fast enough to be synchronized with the CPU clock.

EEPROM (electrically erasable programmable read only memory) is a relatively new memory type that can be reprogrammed in flight and is nonvolatile for substantial time periods in the event of power loss. EEPROM can be used for bulk memory as well as the restart functions that once required ROM. Writing to EEPROM is slower than writing to RAM and requires a higher voltage.

Flash memory is an emerging memory technology that is nonvolatile and rewriteable. It can be used as bulk memory similar to EEPROM, but can be packed more densely than EEPROM. Flash memory cannot be overwritten at the byte level; entire blocks of memory must be overwritten at a time. A block can be 512 or more bytes.

A computer is rated by the number of bits it can hold in its register at one time, which is also the number of bits it can process at one time; this is called its *word size*. The Viking Orbiter computer (circa 1970) was an 8-bit machine. The RAD6000 computer (circa 2000) is a 32-bit machine. For a given clock rate a 32-bit machine can process data at four times the rate of an 8-bit machine.

8.5.3.7 Speed. The CPU contains a quartz crystal clock that sends a stream of timed pulses, also called the heartbeat, to the computer. The *clock rate* (or clock speed) is the interval of these time pulses, usually measured in hertz. At a clock rate of 100 MHz, 1 s is divided into 100 million equally spaced intervals, each marked by a pulse being sent to the CPU. The clock rate for the Viking Orbiter CPU was 100 Hz. The clock rate for a Pentium 4 CPU is 1500 MHz. A *super scalar* computer can perform multiple operations during a single pulse interval.

Table 8.12 Functions of flight software

System	Software function
CDS	Command verification Command distribution Data collection Data formatting Data encoding
Attitude control	Sensor data processing Data filtering Ephemeris propagation Attitude determination Reaction wheel control Thruster control Orbit propagation Antenna pointing Solar array pointing Instrument pointing
Power	Battery charge control Load control
Fault protection	Redundancy management Anomaly response
Operating system	Executive tasks Device drivers Run-time kernel

8.5.6 Software

The typical functions of the flight software are shown in Table 8.12. The attitude control system has the largest share. The software determines the spacecraft attitude (best fit) based on input from each sensor and prepares a corrective response output to the spacecraft actuators. The software also selects alternative modes of operation. Magellan used a reaction wheel and thruster modes of attitude correction; the mode was chosen by software based on preset criteria. There was a special maneuver mode for conducting a star scan that provided the input for attitude update. There was also a software-controlled wheel unloading maneuver using thrusters.

8.5.6.1 Operating system. The operating system software is the master control for the computer. The system contains input/output *device drivers*, *executive* software and a *run-time kernel*. The kernel is a permanent resident of the computer hard wired in ROM; it is the first software to run when the computer is turned on. Some high-order languages require a run-time kernel to perform repetitive utility functions. The executive controls the scheduling of tasks, interrupt handling, diagnostics, and memory error detect/correct. There are specialized device drivers for each type of equipment that interfaces with the computer; for example, the gyros and the reaction wheels. (There are device drivers in a personal computer to control the mouse, printer, and keyboard.) Device drivers accept and interpret input from a device and format commands for output to the device. Typically operating system software uses 20% of the memory and timing resource. DOS and Windows are familiar examples of operating system software for personal computers. There are two spacecraft operating systems in common use: 1) UNIX and 2) VMS. The operating system is not normally developed by the spacecraft project; it is a separate software normally provided with the computer.

8.5.6.2 Fault protection. The fault protection software can vary widely in complexity. In its simplest form it detects a failure by monitoring instrumentation parameters, then it puts the spacecraft in a safe mode and waits for ground instructions. In its most sophisticated form it is an expert system employing multiple levels of protective action without ground operations assistance. Fault protection software is particularly important to planetary spacecraft that have large round-trip command times.

8.5.6.3 Software development. As a result of the critical nature of flight software and the high cost of late changes, the development process has been formalized in a manner analogous to hardware, as shown in Fig. 8.14. The major phases of development are 1) requirements analysis, 2) design, 3) coding, 4) qualification, 5) system level testing, and 6) operational maintenance. Flight software is *verified* and *validated* (V&V) as well as qualified.

Verification is an internal assessment of a software product that assures that it meets its specification and complies with conventions and standards. Verification is largely independent of functional behavior. Validation is an independent, external assessment of a system of software products that assures that the system meets its functional requirements within its intended environment. Validation is largely independent of internal structure. V&V is often performed by an organization different from the organization that wrote the code.

DESIGN

Software Planning
Software Requirements
System Requirements Analysis
Software System Allocation
Requirement Definition
S/W Design
Preliminary Design
Detailed Design
Software Code and Checkout
SRR
SDR
PDR
CDR

TEST & OPERATIONS

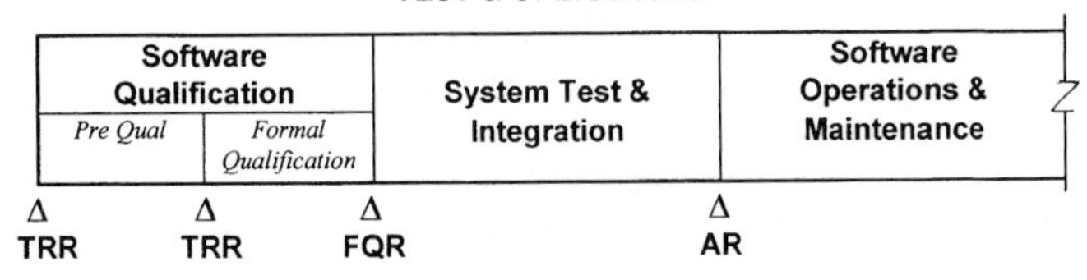

AR = Acceptance Review
CDR = Critical Design Review
FQR = Formal Qualification Review
PDR = Preliminary Design Review
SDR = System Design Review
SRR = System Requirements Review
TRR = Test Readiness Review

Fig. 8.14 Flight software design process.

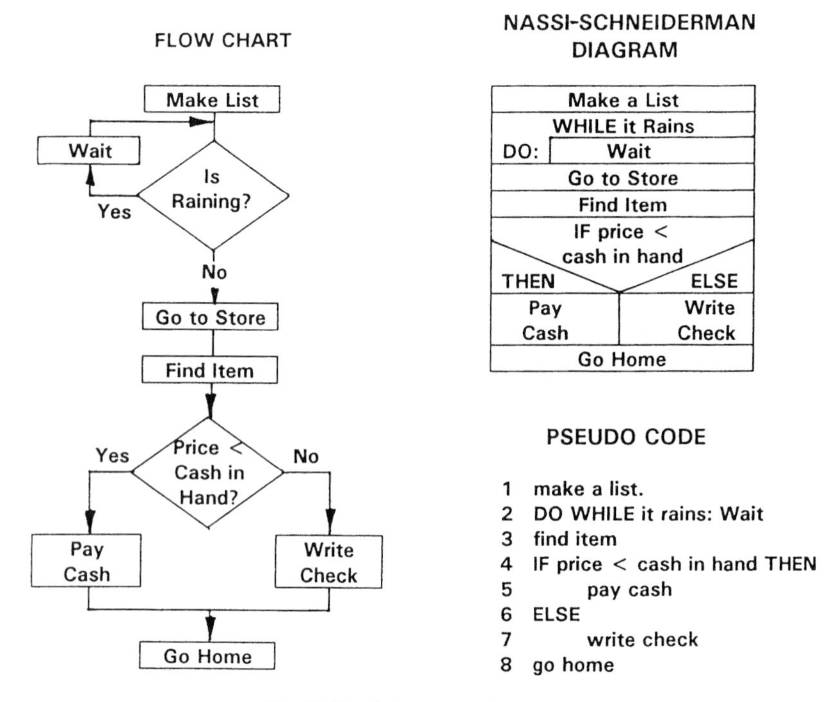

Fig. 8.15 Software design tools.

It is difficult and important to test the software in all possible modes, for example, on each side of each branching statement. One implication of the software test program is the necessity of an adequate test bed. In all probability the flight spacecraft schedule will not accept all of the testing required and a *spacecraft simulator* will be necessary. A good spacecraft simulator is very expensive; an inadequate simulator will miss real problems and generate unreal problems.

When software is *designed*, as opposed to coded, it is written in pseudo code or diagrammed by flow chart or by Nassi–Schneiderman[12] diagrams. Figure 8.15 shows the plan for a shopping trip in pseudo code, flow chart, and Nassi–Shneiderman diagram.

The flow chart is the oldest and most common design method of the three. Nassi–Schneiderman diagrams are particularly useful in ensuring that all cases downstream of a decision statement are handled. These diagrams are also helpful in dividing a software into independent modules. Pseudo code is a design tool that portrays the functions to be coded in English sentences. Pseudo code reveals the code structure and is useful in thinking through the software module; it can be written faster than code because the details of syntax are not important. It is also useful for estimating lines of code and throughput. Notice that the step from pseudo code to Pascal, C, or Quick Basic is a small one indeed.

In the preliminary design phases, requirements are allocated to the software, the size of the software is estimated in lines of code, and a timing analysis is performed. These subjects are discussed in the following sections.

8.5.7 *Timing Analysis*

Timing analysis is used to determine if a time-critical software operation can be completed in a given computer within the time available. The analysis is also used in the selection of a computer. Later in the design, timing analysis is used to track the timing margin and ultimately to determine if the software or computer must be redesigned. The steps in the process are as follows: 1) Enumerate the arithmetic and logic functions in the software module. 2) Determine the computer speed for each function. 3) Calculate the elapsed time for all functions. 4) Compare calculated and required times.

Example 8.3 Digital Filter Timing Analysis

Calculate the time required for a digital filter calculation and the timing margin if the central processor is required to make this computation 20 times per second. The digital filter software performs the following functions each time the filter runs:

		Number of instructions			
Function	Frequency	Add/Sub	Multiply	Divide	Trig
Digital filter	20/s	360	460	0	50

For the computer being used the computation speed is as follows:

	Add/Sub	Multiply	Divide
Fixed point instructions, μs	0.9	2.7	5.7
Floating point instructions, μs	2.3	3.9	9.6

Trigonometric functions can be approximated by

$$\begin{aligned} \text{Trig} &= 30\,\text{Add} + 8\,\text{Mpy} + 2\,\text{Div} \\ &= 30(2.3) + 8(3.9) + 2(9.6) \\ &= 119.4\ \mu\text{s}\,(\text{floating point}) \end{aligned}$$

Calculate the amount of computer time used by the digital filter in 1 s if trigonometric functions are floating point and the remaining instructions are 50% fixed point and 50% floating point:

$$\begin{aligned} &\text{Fixed point add} &&= 360 \times 0.5 \times 0.9 \times 20 = 3240\ \mu\text{s} \\ &\text{Fixed point multiply} &&= 460 \times 0.5 \times 2.7 \times 20 = 12420\ \mu\text{s} \\ &\text{Floating point add} &&= 360 \times 0.5 \times 2.3 \times 20 = 8260\ \mu\text{s} \\ &\text{Floating point multiply} &&= 460 \times 0.5 \times 3.9 \times 20 = 17940\ \mu\text{s} \\ &\text{Floating point trig} &&= 50 \times 119.4 \qquad\qquad\quad = 5970\ \mu\text{s} \\ & && \qquad\qquad\qquad\qquad\qquad\quad\ \ \overline{47830\ \mu\text{s}} \end{aligned}$$

How much timing margin remains in each second? Normal engineering calculation would show that the timing margin is 95.2%; however, computer margins are calculated in an unusual way. When the unused capability is equal to the used capability that is considered 100% margin:

$$\text{Margin} = 100\left(\frac{\text{Allocation} - \text{Used}}{\text{Used}}\right)$$

$$\text{Margin} = 100\left(\frac{1 - 0.04783}{0.04783}\right) = 1991\%$$

The operation of the example digital filter requires 4.8% of the candidate computer throughput. The analysis of a complete spacecraft requires an analysis, similar to the example, for each of the software functions (see Table 8.12 for functions). The sum of the time required for all functions plus overhead would then be compared with the total computer throughput.

Example 8.4 Spacecraft Computer Timing Analysis

Each function in the complete spacecraft software has been analyzed; the resulting instruction mix is shown in Table 8.13. The computer is the same as that used in the prior example.

In the early phases of the design, the time required for all of the computer tasks should be less than half of the time available or a timing margin of 100%. Software timing and size margin are required for four reasons, two of which are peculiar to software: 1) software development problems, 2) uncertainties in the estimate, 3) in the late phases of development it is much less expensive to solve problems in software than hardware, and 4) software needs margin after liftoff for the solution of in-flight problems. Table 9.13 shows a 113% timing margin, a

Table 8.13 Timing analysis example

Instruction	No. required per second	Computation speed, μs	Time required, μs/s
Add/subtract	2421	2.3	5568.3
Multiply	3104	3.9	12105.6
Divide	2133	5.7	12158.1
Trig function	763	119.4	91102.2
Square root	141	89.3	12591.3
Vector add	476	224.3	106766.8
Vector cross	437	153.3	66992.1
Vector dot	646	123.3	79651.8
Test and branch	141	13.4	1889.4
Store and load	112	18.8	2105.6
Subtotal			390931.2
Operating System (20%)			78186.2
Total			469117.4

desirable condition:

$$\text{Margin} = 100\left(\frac{1 - 0.469117}{0.469117}\right) = 113\%$$

An *instruction mix* is a tabulation of the number of instructions of each type (floating point add/subtract, fixed point multiply, etc.) used by a given spacecraft in a given period of time, usually one second. There are standard instruction mixes, e.g., Dhrystone, Gibson mix, and Whetstone, with which computers are rated; thus the computers in Table 9.11 each have a throughput that was determined with a standard mix of instructions. If you assume that your spacecraft software will have an instruction mix identical to the standard mix you can analyze throughput by a simpler technique, as shown in Table 8.14.

The software load listed in Table 8.14 would fit in any computer with a throughput greater than 87 thousand instructions per second and would have adequate margin in any computer with a throughput greater than 180 kips, assuming a reasonable match between the standard and actual instruction mix.

8.5.8 Sizing

Sizing analysis compares the expected software load size to the RAM size in the computer under consideration. The procedure is to 1) estimate the lines of code in the software load, 2) convert lines of code (LOC) to digital words, 3) estimate the words of data that need to be stored in RAM, and 4) compare the software size, in words, plus the data storage needed, to the RAM available. The LOC estimate is also useful in estimating the cost of developing the software.

After the software is coded, the LOC estimate is a simple count of lines. However, the LOC estimate is most valuable when the code is least understood, before the coding is done. There are three common methods of estimating software LOC.

Similarity. Compare software functions, at the lowest level possible, to the known size of similar, previously developed, software. For example, estimate the Kalman filter software to be the size it was on a previous project, taking into account the differences between the projects.

Bottoms up. Divide the software into the smallest possible modules, estimate each, and sum the estimates. Sometimes pseudo code is used to define the modules.

Table 8.14 Functions and throughput of ACS flight software

Function	Instructions	Frequency, Hz	Throughput, kips
Sensor data processing	2220	10	22.2
Kalman filter	70000	0.1	7
Attitude determination	2700	10	27
Reaction wheel control	2500	2	5
Thruster control	600	2	1.2
Orbit propagation	20000	1	20
Ephemeris propagation	4000	1	4
Total			86.4

Table 8.15 Magellan software size

Software	Language	Lines of code
Attitude control	HAL-S	4000
	ATAC-16 Machine language	750
Command and data	1802 assembly language	12000
Fault Protection—ACS	HAL-S	1200
Fault Protection—CDS	1802 assembly language	1400

Top down. Make a top-level comparison of the current project with some past project.

Table 8.15 lists the Magellan flight software code size.

Table 8.16 shows timing and sizing for a typical spacecraft. Data storage requirements are normally about 25% of the code words. The software listed in Table 8.16 requires 330 + 62.4 = 392.4K words of RAM before margin. The conventional wisdom is that a 100% memory margin is required for the same reasons that a 100% timing margin is required; therefore, a computer with 800K word RAM is

Table 8.16 Timing and throughput for a spacecraft

System	LOC ADA	Memory, code words	Memory, data words	Time, ms	Frequency, Hz	Total time, ms/s
CDS						
Telemetry processing	1350	13500	3850	3.00	40.00	120
Command processing	2400	24000	2800	4.50	5.00	22.5
Polling/multiplexing	1200	12000	2100	4.00	10.00	40
Formatting	600	6000	700	2.00	2.50	5
Configuration table	975	9750	2100	3.00	5.00	15
Telecommunication	0	0	0	0	0	0
Uplink processing	900	9000	6000	3.00	5.00	15
Downlink processing	600	6000	1050	2.00	10.00	20
Attitude control	0	0	0	0	0	0
Attitude determination	1500	15000	3500	2.50	40.00	100
Attitude control	2400	24000	4200	3.50	10.00	35
Ephemeris processing	975	9750	1750	3.00	5.00	15
Articulation	0	0	0	0	0	0
Solar array control	900	9000	1050	1.80	2.50	4.5
High gain antenna	1200	12000	1400	2.00	40.00	80
Fault protection	0	0	0	0	0	0
Safing	1500	15000	3500	5.00	10.00	50
CDS fault protection	1800	18000	2100	5.00	10.00	50
ACS fault protection	11500	115000	1300	5.00	10.00	50
Operating system	1000	10000	25000	2.00	90.00	180
Utilities	2200	22000				0
Totals	33000	330000	62400			802

indicated. The timing margin is

$$\text{Timing margin} = 100\left(\frac{1 - 0.802}{0.802}\right) = 25\%$$

Note that you could split the computational chores between an attitude control computer and a CDS computer.

Modern computers with good speed and ample memory have greatly reduced the criticality of timing and throughput analysis. They still must be done but it is no longer such a critical issue.

8.5.9 Languages

A computer language is a collection of commands, functions, and statements that are the building blocks for a body of software in that language. There are two basic types of computer languages: 1) *assembly language* and 2) *high-order languages* (HOLs). An assembly language instructs the computer at the most fundamental level. Assembly language instructions correspond one to one with computer or machine language instructions. An assembly language is unique to a given computer. Programming in assembly language requires a detailed knowledge of the particular computer and of the specific commands of the machine.

The advantages of assembly language are as follows: 1) The resulting code is very compact (fewest possible lines of code). 2) The fastest throughput is achieved. 3) There is no compiling step required. The disadvantages of assembly language are as follows: 1) Coding is difficult, error prone, time consuming and expensive. 2) The code is difficult to read, therefore, difficult to debug and maintain.

HOLs use sophisticated commands and functions called by English language names. An experienced person finds them easy to read and debug. They require a compiler, which is a software package compatible with the computer and the language being used. The compiler translates source code prepared in a high-order language into equivalent machine language code. Table 8.17 lists some common HOLs.

Table 8.17 High-order languages

Language	Application
FORTRAN	Particularly suited for engineering calculations, in use for decades
Jovial	DOD standard language of the 1980s
ADA	New DOD standard language
Pascal	ADA precursor; university and commercial use
C	Extensive use for commercial programming, particularly for Windows environment
LISP	Designed specifically for artificial intelligence
Visual Basic	Object oriented language (commercial) for Windows environment
COBOL	Designed specifically for business data processing; commercial use

Table 8.18 Estimating CDS mass and power

Equipment	Mass, kg	Power, W
Equipment relatively independent of data rates	30	20
Telecommunication interface		
Data flow control		
Signal conditioning		
Command processing		
Spacecraft clock		
Computer	2	10
Science data processor (not req'd for Comsats)	15 kg	2W + 1 W/instrument
Engineering data processor	10	5
Data storage (SSR)	0.25 kg/Gbit	1 W/Gbit

8.6 Estimating Subsystem Mass and Power

Table 8.18 shows mass and power estimating relationships that can be used in the early phases of a project before more accurate data can be acquired.

References

[1]*Recommendations for Space Data System Standards, Packet Telemetry* (Blue Book), Consultative Committee for Space Data Systems Standards, CCSDS Secretariat, NASA, Washington, DC, 1992; also available on the internet at www.ccsds.org.

[2]Wertz, J. R., and Larson, W. J., *Space Mission Analysis and Design*, 2nd Ed., Kluwer Academic Publishers, Dordrecht, The Netherlands, 1992.

[3]"RHPPC, Single Board Computer," Honeywell International, Inc., Clearwater, FL, Jan. 2002.

[4]Sklar, B., *Digital Communications, Fundamentals and Applications*, PTR Prentice–Hall, Englewood Cliffs, NJ, 1988.

[5]Piscane, V. L., and Moore, R. C. (eds.), *Fundamentals of Space Systems*, Oxford Univ. Press, New York, 1994.

[6]Edelson, R. E. (ed.), *Telecommunications System Design Techniques Handbook*, Jet Propulsion Lab., California Inst. of Technology, TM 33-571, Pasadena, CA, July 1972.

[7]Shannon, C. E., and Weaver, W., *The Mathematical Theory of Communication*, Univ. of Illinois Press, Urbana, IL, 1949.

[8]Webb, E., Cunningham, M., and Leach, T. T, "The WARP: Wideband Advanced Recorder Processer for the New Millennium Program, EO-1," IEEE Aerospace Conference, Snowmass, CO, 1997.

[9]Freedman, A., *Computer Desktop Encyclopedia*, 9th Ed., Osborn/McGraw-Hill, New York, 2001.

[10]Stakem, P. H., "Flight Linux Project Target Archetecture Technical Report," QSS Group, Inc., Lanham, MD, Dec. 2001.

[11]Gaona, J. I., "A Radiation Hardened Computer for Satellite Applications," Sandia National Lab., Albuquerque, NM.
[12]Nassi, I., and Schneiderman, B., *ACM SIGPLAN Notices*, Vol. 8. No. 8, Aug. 1973, pp. 12–26.

Problems

8.1 Given the types of computer operating systems shown in the first table, and given the C&DH applications described in the second table, which operating systems would be acceptable? All applications must be run once a second.

Operating system	Size, kbytes	Context switching time,[a] μs
1	32	10
2	256	100
3	1024	10,000
4	70,000	50,000–200,000[b]

[a]Time required to switch between application programs.
[b]Time varies randomly for same event.

C&DH applications	Size, kbytes	No. of instructions
1	10	625
2	100	10250
3	5	300
4	1	50
5	500	6000
6	20	1200

Solution: Operating system no. 1 is the only acceptable one.

8.2 Two types of solid state data recorders are available: Type A, which holds 6×10^9 bits and Type B, which holds 4.5×10^8 bits. What is the minimum number and type of SSDR that will accommodate the following requirements:
(a) The orbit period is 100 min.
(b) Data can be downloaded every other orbit.
(c) Engineering data consists of 1024, 8-bit words per major frame.
(d) A major engineering frame is taken every 5 s.
(e) Payload data are recorded for 20 min per orbit at 20,000 bps.
(f) Payload and engineering data must be recorded in separate recorders.
(g) No single failure can cause loss of payload or engineering data.

8.3 For the mission described in problem 8.2, what is the downlink symbol rate if CSSDS-recommended coding is applied to the data prior to transmission? Payload and engineering data are accomplished in the same downlink. The download must be accomplished in 15 min.

8.4 What is the maximum bus traffic rate in bits per second under the following conditions:

(a) 200 8-bit sensors outputting 10 times/s
(b) 100 1-bit discrete readings read once per second
(c) 100 16-bit sensors read 100 times/s
(d) Attitude control outputting 2000 bits, 40 times/s
(e) Payload A outputting 1 Mbps
(f) Payload B outputting 10 Mbps for 6 min/h.

8.5 Given the instrument data rates shown here, determine the maximum and minimum symbol rate for science data with 10% overhead and CCSDS recommended coding added. The engineering data rate is 1200 bps.

Instrument	Max data rate, bps	Min data rate, bps
1	40,042	700
2	80,000	700
3	4992	150
4	618	618
Total	125,652	2168

Determine (a) the required solid state recorder storage capacity and (b) rate capability, if downlink empties the recorders every 24 h. Store data plus overhead; assume that coding is applied to the bit stream as it leaves the recorders for transmission.

c) What is the communication system downlink symbol rate required to empty the recorders once per day in 12 h?

9
Telecommunication*

9.1 Fundamentals

Communication with a spacecraft takes place over two types of links as shown in Fig. 9.1. An *uplink* carries commands from a ground station to the spacecraft. Commands are specially formatted digital streams that tell the spacecraft what to do. They can be immediately executed or, more frequently, stored for execution at a specified time. Usually a command upload contains instructions for a whole timed sequence of events. For example, a normal upload for Magellan provided the spacecraft with four days of instructions, enough for 30 orbits and 120 maneuvers. A new upload was sent up on the third day.

A *downlink* carries payload (science) data, the data the spacecraft was sent to get, and the engineering or telemetry data, which show the status of the spacecraft subsystems. The engineering and payload data can be in the same or separate data streams. The payload data are usually a much higher data rate than the engineering or command links.

When a ground station is only receiving the downlink data from a spacecraft, the communication is called *one-way*. When the ground system is receiving downlink data and the spacecraft is receiving uplink data, the link is called *two-way*.

The communication links are also used for tracking the spacecraft by measuring the angular position of the downlink signal, by Doppler measurements and by ranging, or two-way Doppler.

In the first section of this chapter, the basic physics of radio communication will be discussed. In Sections 9.2 and 9.3 the communication link is described along with link calculations. In Section 9.4 the use of the communication link for navigation and tracking is reviewed, Section 9.5 describes some of the aspects of system design and performance estimation, and finally Section 9.6 describes the ground stations that support the spacecraft.

9.1.1 Electromagnetic Waves

Radio-frequency signals are electromagnetic waves, as are x rays, gamma rays, and light. All electromagnetic waves are generated when electric charges are accelerated, that is, when an electric current is changed. Although a constant current produces an electric field, electromagnetic radiation only occurs when current is changed, increased, or decreased. A current that oscillates in simple harmonic motion with frequency f radiates an electromagnetic wave of the same frequency. As the name implies, all electric fields are paired with a matching magnetic field. The

---

*The author wishes to acknowledge, with gratitude, the contributions made by Charles H. Green to this chapter.

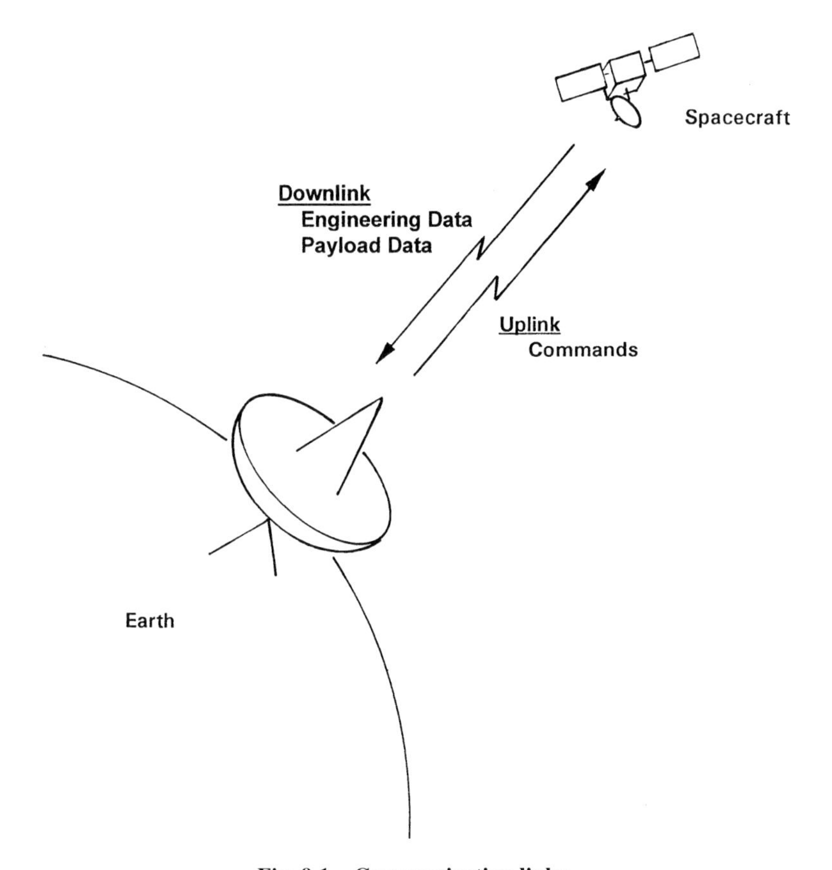

Fig. 9.1 Communication links.

electric and magnetic field vectors are in phase and perpendicular to each other,[1] as shown in Fig. 9.2.

A radio *transmitter* generates an oscillating current, which in turn generates an electromagnetic wave. Conversely, the presence of electromagnetic radiation generates an alternating current in a conductor at the radio receiver. The alternating current in a house oscillates at 60 Hz. It is a simple matter to detect induced 60-Hz current in any isolated conductor in the house. The attendant 60-Hz electromagnetic radiation can also be readily detected as an audible hum on an AM radio.

A complete description of the interrelation of electric current, magnetic fields, and electromagnetic waves is provided by Maxwell's equations[2]: a set of four differential equations that form a unified theory of electromagnetic motion. The beauty and power of Maxwell's work, the first unified theory, created a sensation that persists to today. These four equations incorporated all that was known at the time, 1857, about electricity and magnetism. They simplify to Ampere's theorem, Coulomb's law, Biot–Servart law, and Gauss's law. In addition, the existence of electromagnetic radiation was predicted; the existence of which was later

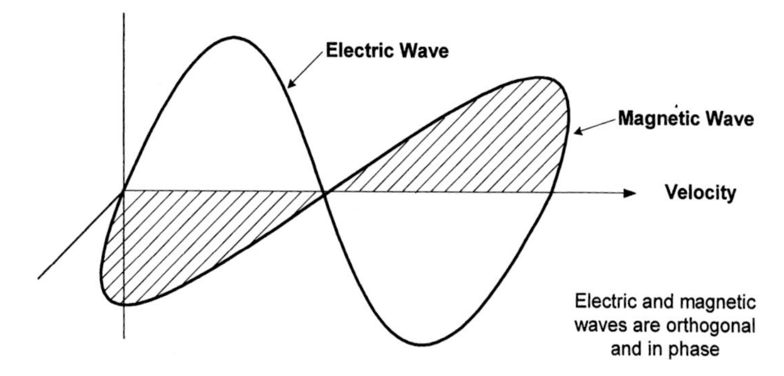

Fig. 9.2 Electromagnetic waves.

demonstrated by Heinrich Hertz. The theory also predicted that electromagnetic radiation should travel at the speed of light, also later demonstrated. (It is interesting that Maxwell's work has also been shown to be perfectly consistent with the special theory of relativity, introduced by Einstein many years later.) Guglielmo Marconi, in 1901, extended Hertz's work into a system that transmitted a message across the Atlantic from Cornwall to Newfoundland—all from four differential equations.

For all of the importance of Maxwell's equations, solutions to them do not come easily. For our purpose it is sufficient to 1) consider only the electrical component of the wave pair, 2) to assume steady-state sinusoidal variation, and 3) assume vacuum conditions. The resulting wave motion is shown in Fig. 9.3.

In Fig. 9.3 the waveform is the projection of vector $\boldsymbol{E}$ rotating around a central point at a phase angle ϕ. The fundamental relationships describing the wave motion are

$$e = E \sin \omega t \tag{9.1}$$

$$\omega = 2\pi f \tag{9.2}$$

$$\phi = \omega t \tag{9.3}$$

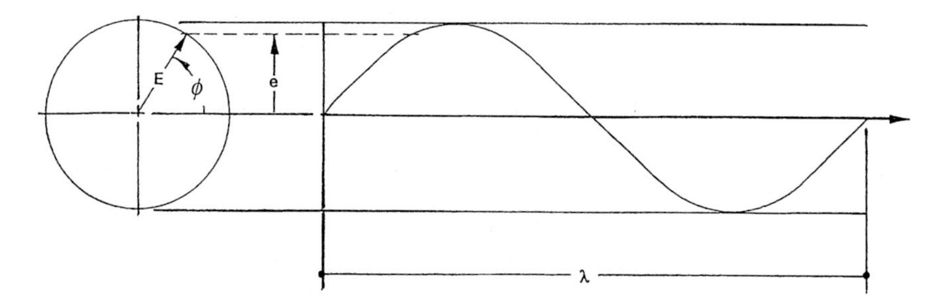

Fig. 9.3 Electromagnetic wave motion.

$$\lambda = \frac{v}{f} = \frac{c}{f} \tag{9.4}$$

$$P = \frac{1}{f} = \frac{2\pi}{\omega} \tag{9.5}$$

where

E = magnitude of an electric field vector
e = instantaneous amplitude of an electric field
ω = angular rotation rate of the $\boldsymbol{E}$ vector, rad/s
ϕ = phase angle of vector $\boldsymbol{E}$
t = time, s
f = frequency, the number of cycles per unit time, Hz
l = wavelength, linear distance between corresponding points on a wave, m
v = wave propagation velocity, usually the speed of light
c = speed of light, 186,280 miles/s or 2.99798×10^8 m/s
P = period, time required for each cycle

It has been shown by numerous methods that electromagnetic waves travel at the speed of light, c, a vacuum or dry air.[3]

The electromagnetic spectrum, shown in Fig. 9.4, is astonishingly broad. Wavelengths can be as long as the diameter of the Earth or as short as 10^{-12} cm. The spectrum is continuous; however, various frequency and wavelength ranges have been named as shown in Fig. 9.4. Of the entire spectrum only the very narrow wavelength range between 0.000016 to 0.000028 in. is visible to the human eye. The radio spectrum extends from 10 kHz to 300 GHz by international agreement.

In an Atlantic City Conference in 1947, the oldest subdivision scheme, shown in Table 9.1, was established.[4]

Spacecraft communication takes place in several frequency bands, which are designated by letter as shown in Table 9.2. The letter designations date from WWII during which they were used to conceal classified radar frequencies.

The most commonly used bands for planetary spacecraft are *S* and *X* band. Earth orbiters, notably TDRSS, also use K and Ku bands (Ka and Ku are subset of K band). S andXbands are divided into channels.Agiven spacecraft will be assigned one or more of these channels by NASA or the U.S. Air Force.

Table 9.1 Early radio-band designations

Band	Name	Frequency
VLF	Very low	3–30 kHz
LF	Low	30–300 kHz
MF	Medium	300–3000 kHz
HF	High	3–30 MHz
VHF	Very high	30–300 MHz
UHF	Ultra high	300–3000 MHz
SHF	Super high	3–30 GHz
EHF	Extremely high	30–300 GHz

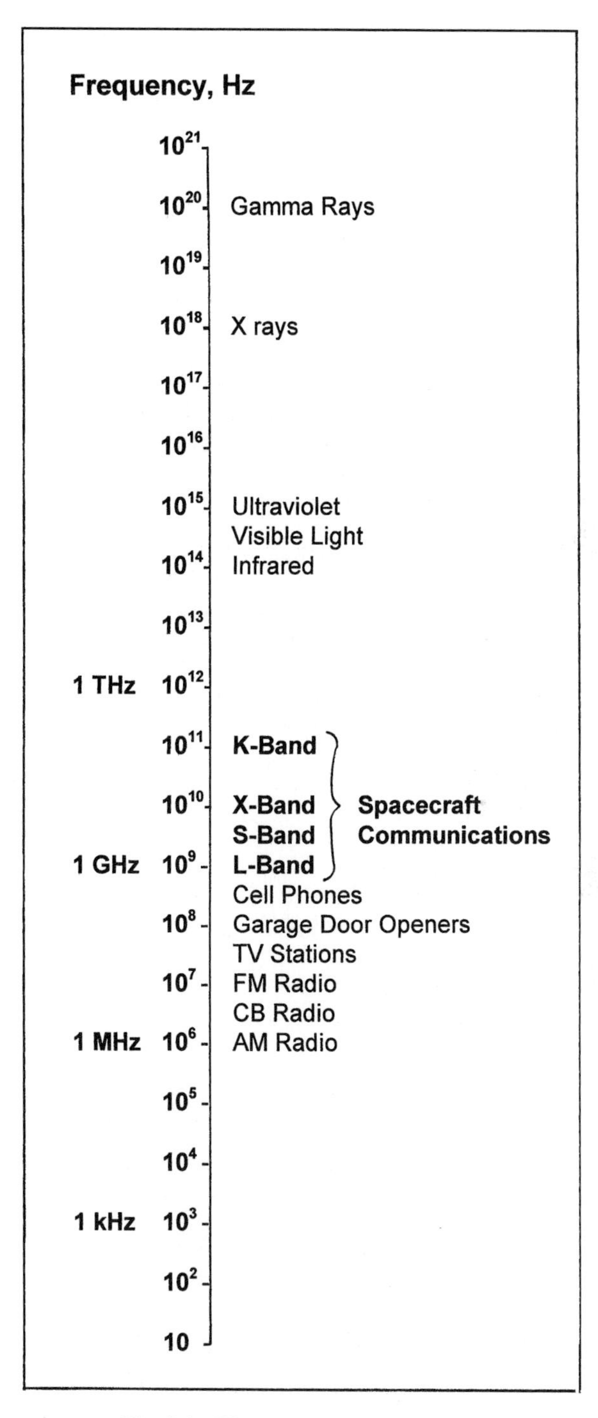

Fig. 9.4 Electromagnetic spectrum.

Table 9.2 Letter-designated frequency bands

Band	Freq. range, GHz
P	0.225–0.39
J	0.35–0.53
L	0.39–1.55
S	1.55–3.9
C	3.9–6.2
X	6.2–10.9
K	10.9–36.0
Ku	10.9–18
Ka	18–31
Q	36.0–46.0
V	46.0–56.0
W	56.0–100.0

9.1.2 Phase

The phase angle of an electromagnetic wave is the angle of rotation of the ***E*** vector (see Fig. 9.3). Two waves can be described in part by the phase difference. Figure 9.5 shows three waves of equal frequency and amplitude and different phases. Phase difference can be expressed in terms of time, angle, or distance,

$$\delta = \phi_2 - \phi_1 \tag{9.6}$$

$$\delta = 2\pi \frac{\Delta x}{\lambda} \tag{9.7}$$

$$\Delta t = \frac{\Delta x}{v} \tag{9.8}$$

where δ = phase shift or phase difference between waves, rad; Δx = difference in distance between the waves measured at corresponding points, and Δt = time delay between waves measured at corresponding points.

9.1.3 Doppler Effect

When a relative velocity exists between an rf transmitter and receiver, the frequency received is not the same as the frequency transmitted. When they are moving away from each other, the frequency received is less than the transmitted

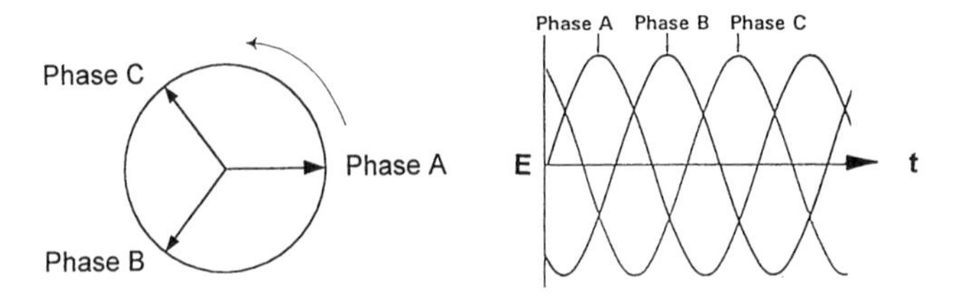

Fig. 9.5 Three waves differing in phase.

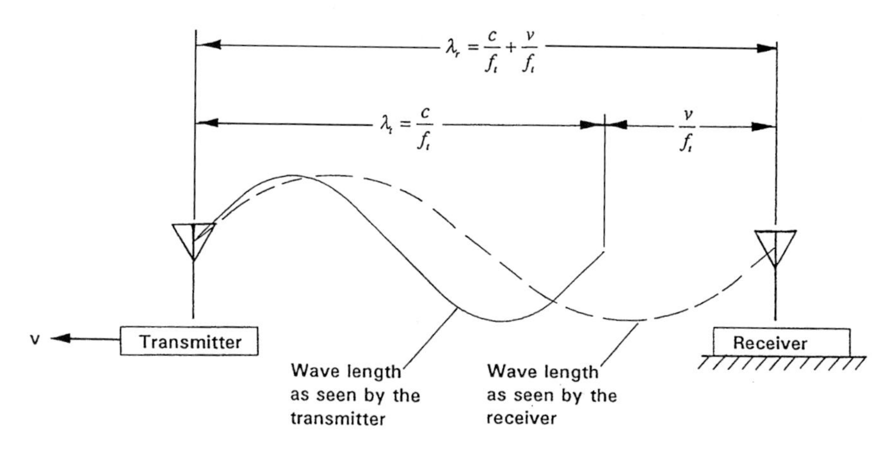

Fig. 9.6 Doppler effect.

frequency and vice versa. This frequency change is called the Doppler effect or Doppler shift (see Fig. 9.6).

The Doppler effect is extensively used to measure spacecraft velocity relative to the receiver. The wavelength received is altered by the relative velocity as follows:

$$\lambda_r = \frac{c}{f_t} + \frac{v}{f_t} \tag{9.9}$$

$$\frac{c}{f_r} = \frac{c}{f_t} + \frac{v}{f_t} \tag{9.10}$$

where λ_r = the wavelength received, f_t = the frequency transmitted, f_r = the frequency received, and v = relative velocity between the transmitter and receiver.

Because c is substantially greater than v,

$$f_r = f_t\left(1 \pm \frac{v}{c}\right) \tag{9.11}$$

The sign in Eq. (9.11) is negative when the distance between transmitter and receiver is increasing and is positive when the distance is decreasing.

The Doppler frequency f_d is

$$f_d = f_t\frac{v}{c} \tag{9.12}$$

Doppler measurements of spacecraft signals yield spacecraft radial velocity from the ground station and the integral gives range. These measurements are best made with an unmodulated carrier (a simple sine wave).

9.1.4 Polarization

Polarization is a property of electromagnetic radiation describing the spatial orientation of the field vectors. An electromagnetic wave is *linearly polarized* when the electric field lies in a plane perpendicular to the direction of propagation.

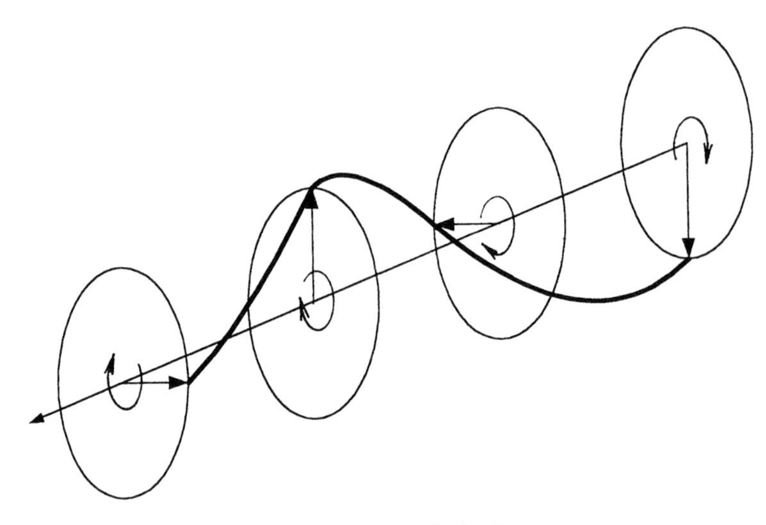

Fig. 9.7 Circular polarized wave.

The wave shown in Fig. 9.3 is linearly polarized in the x–y plane. The general expression for linearly polarized wave motion is

$$y = y_0 \sin(kx - \omega t) \tag{9.13}$$

There are other types of polarization. If the end of a string is moved in a circle at a constant speed, a wave propagates along the string such that each element in the string moves in a circle. Such a wave is *circularly polarized.* Circular polarization is shown in Fig. 9.7.

It takes two equations to describe circularly polarized motion, the first being (9.13) and the second one being

$$z = z_0 \sin(kx - \omega t + \delta) \tag{9.14}$$

where δ is the phase difference between the y and z components. If the phase difference is zero, the wave is linearly polarized. If the phase difference is $\pi/2$, we can write[1]

$$\frac{y^2}{y_0^2} + \frac{z^2}{z_0^2} = \sin^2(kx - \omega t) + \cos^2(kx - \omega t) = 1 \tag{9.15}$$

This is the equation of an ellipse; it describes *elliptical polarization*, of which circular polarization is a special case. Waves can have right- or left-handed circular or elliptical polarization. Figure 9.7 shows right circular polarization.

The transfer of power between two antennas is dependent on the polarization of the antennas; when the polarization of the two antennas is not perfectly matched, there are polarization losses in the transfer of radiated power.

9.1.5 Decibels

The language of communications engineers is "dB-ese." Decibels (dB) are power ratios that work like this:

$$P(\mathrm{dB}) = 10\log\left(\frac{P}{P_{\mathrm{ref}}}\right)$$
$$\frac{P}{P_{\mathrm{ref}}} = 10^{\left[\frac{P(dB)}{10}\right]} \tag{9.16}$$

The power reference is 1 W or 1 mW, hence, dB_w and dB_m.

$$P(\mathrm{dB}_w) = 10\log\left(\frac{P}{1\,\mathrm{W}}\right)$$
$$P\,(\mathrm{dB}_m) = 10\log\left(\frac{P}{1\,\mathrm{mW}}\right) = 10\log\left(\frac{P}{1\,\mathrm{W}}\right) + 30 \tag{9.17}$$

The dB notation is 10 times the log to the base 10 of a ratio of two power measures. Some examples are shown here:

$$1\,\mathrm{W} = 0\,\mathrm{dB}_w = 30\,\mathrm{dB}_m$$
$$10\,\mathrm{W} = 10\,\mathrm{dB}_w = 40\,\mathrm{dB}_m$$
$$1000\,\mathrm{W} = 1\,\mathrm{kW} = 30\,\mathrm{dB}_w = 60\,\mathrm{dB}_m$$

Table 9.3 shows some power ratios and equivalent dBs to give you a feel for how they work.

The use of dBs dates back to the time when link calculations were made with logarithms or mechanical desk calculators. In that era, accurate multiplication was substantially more difficult than addition or subtraction. By using dBs, multiplication becomes addition of the equivalent dBs. This advantage disappeared when hand-held computers arrived, but the use of dBs is an entrenched custom.

Table 9.3 Power ratios and dB equivalents

Power ratio	dB	Power ratio	dB
1	0	1.0	0
2	3	0.9	−0.5
3	4.8	0.8	−1.0
4	6	0.7	−1.55
5	7	0.6	−2.2
6	7.78	0.5	−3.0
7	8.45	0.4	−4.0
8	9	0.3	−5.32
9	9.54	0.2	−7.0
10	10	0.1	−10

You can use dBs to express voltage ratios if you are careful. When dBs are used for voltage, remember that dBs are 20 times the log of a voltage ratio.

$$V(\mathrm{dB}) = 20 \log \left(\frac{V}{V_{\mathrm{ref}}} \right) \tag{9.18}$$

The factor of 2 enters the voltage relation because power is proportional to the square of voltage; therefore, a power ratio is twice the log of a voltage ratio. Some examples are shown here:

$$1\ \mathrm{V} = 0\ \mathrm{dB}$$

$$10\ \mathrm{V} = 20\ \mathrm{dB}$$

$$100\ \mathrm{V} = 40\ \mathrm{dB}$$

Gain G of antennas are represented as power ratios or dBs. The reference is an isotropic source (to be explained later):

$$G_{\mathrm{dBi}} = 10 \log \frac{\text{Power Gain of Antenna}}{\text{Power from Isotropic Source}} \tag{9.19}$$

Gains of amplifiers are power ratios:

$$G_{\mathrm{amp}} = 10 \log \frac{\text{Power out of Amplifier}}{\text{Power into Amplifier}} \tag{9.20}$$

Losses L are represented in dBs; for example, coax insertion loss,

$$L_{\mathrm{coax}} = 10 \log \frac{P_{\mathrm{out}}}{P_{\mathrm{in}}} \tag{9.21}$$

9.1.6 Modulation

Modulation is a process by which a signal is used to vary a characteristic of the carrier wave. The signal contains the data to be transmitted; it is usually a relatively low frequency or digital. Consider the carrier as a pure tone of say, 3 GHz, for example. If you were to quickly turn this off and on at a rate of a thousand times a second, we would say the carrier is being *modulated* with a frequency of 1 kHz. Turning the carrier off and on would be disastrous to a spacecraft, but changing, or modulating, the amplitude or phase would be acceptable. Spacecraft carrier signals are modulated by shifting the carrier phase slightly at a given rate. One scheme is to modulate the carrier with a frequency, for example, near 1 MHz. This 1-MHz modulation is called a *subcarrier*. The subcarrier is in turn modulated to carry individual phase shifts that are designated to represent groups of binary ones and zeros—the spacecraft's telemetry or command data. The amount of phase shift used in modulating data onto the subcarrier is referred to as the modulation index and is measured in degrees.

Demodulation is the process of detecting the subcarrier and processing it separately from the carrier, detecting the individual binary phase shifts, and decoding them into digital data for further processing. The same processes of modulation and demodulation are used commonly with Earth-based computer systems and fax machines transmitting data back and forth over a telephone line. The device used for this is called a *modem*, short for modulator-demodulator. Modems use an audio frequency carrier that the telephone system can readily handle.

The carrier frequency is a sinusoidal wave of a frequency selected for good transmission. For example, an FM radio transmitter would use a carrier frequency of about 100 MHz, and it would transmit signals in the range of 20 to 20,000 Hz (music). The 100-MHz signal is modulated at 20 to 20,000 Hz. Why modulate? The aperture of an effective antenna should be at least as large as the wavelength of the signal. The wave length is $\lambda = c/f$, where c is the speed of light or about 3×10^8 m/s. An antenna to receive 20 Hz would be 15,000,000 m, or 9000 miles, in diameter; for 20,000 Hz the diameter would be 15,000 m or 9 miles! In other words, out of the question. If you modulate a 100-MHz carrier, the antenna size comes down to the 3-m diameter.

The sinusoidal carrier wave C is characterized by amplitude, phase, and frequency as follows:

$$C = A\cos(ft + \phi) \tag{9.22}$$

where A is the amplitude, f is the frequency, t is time, and ϕ is the phase. Frequency, phase, and amplitude can be varied as a function of time in proportion to amplitude of the signal.

9.1.6.1 Analog data. Figure 9.8 shows the two most common methods of modulating an analog chanel. In amplitude modulation (AM) the amplitude of the carrier wave is proportional to the amplitude of the signal. In frequency or phase modulation (FM or PM) the frequency or phase of the carrier is proportional to the amplitude of the signal. Spacecraft use FM or PM because the transmitter can be operated at maximum power and highest power efficiency. FM and PM systems can also operate at a lower signal-to-noise ratio (SNR) than AM systems; however, these systems have a nonlinear response to weak signals.[5] Performance is relatively independent of signal strength until a threshold value is reached below which performance drops rapidly. This effect can be observed as you drive away from an FM station; radio performance is reasonably independent of distance until the link is suddenly lost altogether. AM systems do not exhibit this behavior; they require a greater SNR and degrade gracefully as SNR is reduced.

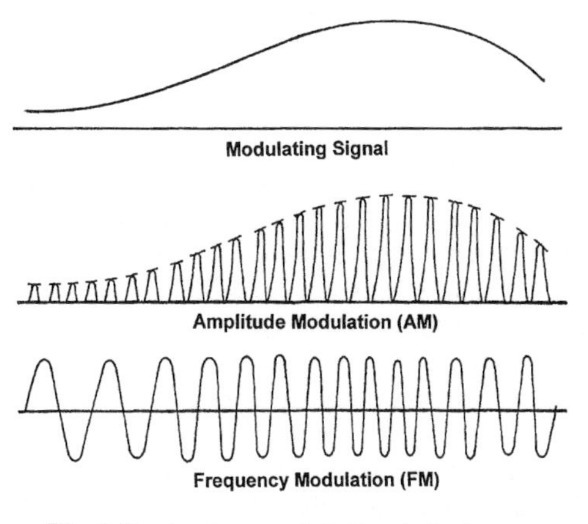

Fig. 9.8 Analog modulation techniques.

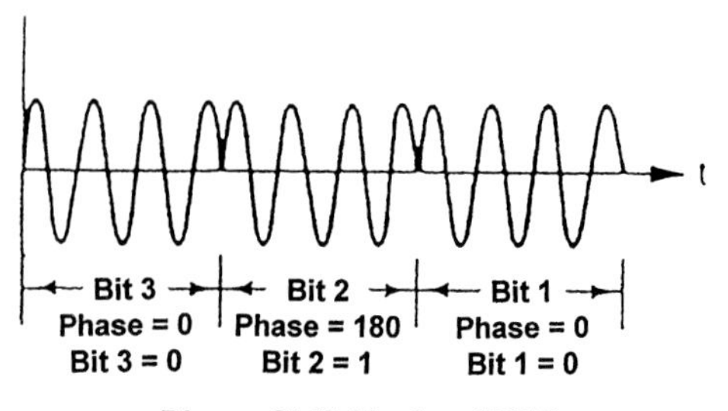

Phase Shift Keying (PSK)

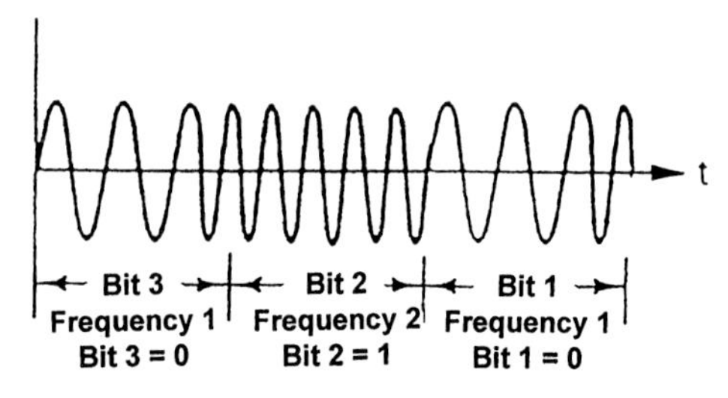

Frequency Shift Keying (FSK)

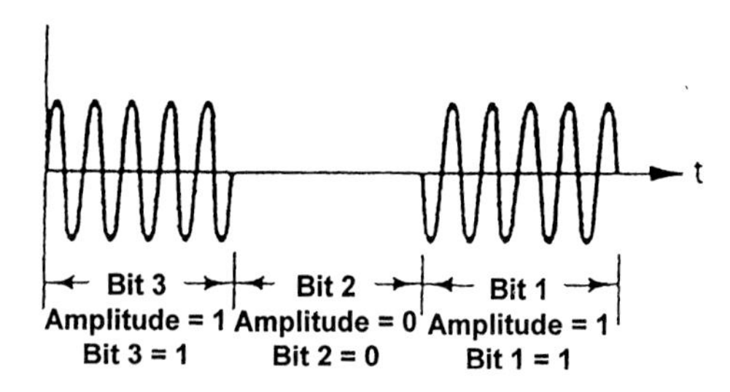

Amplitude Shift Keying (ASK)

Fig. 9.9 Pulse code (digital) modulation.

9.1.6.2 Digital data. Amplitude, frequency, and phase modulation techniques can also be used to transmit digital data. Since about 1970, spacecraft have used digital modulation or *pulse code modulation* (PCM) exclusively. A major advantage of digital data and digital modulation is that the bit error rate can be substantially reduced by coding techniques. The use of digital computers for data management makes digitizing the data essential in any case. Figure 9.9 shows the most common digital modulation methods.

Binary phase shift keying (BPSK) is done by setting the carrier phase to 0 deg to transmit a binary zero and shifting the phase to 180 deg to transmit a binary one as shown in Fig. 9.9.

Binary frequency shift keying (BFSK) switches between two frequencies distinct from the carrier frequency.

Binary amplitude shift keying (BASK) shifts between two amplitudes; however, it is seldom used in spacecraft digital communication. It is interesting to note that Morse code is a form of BASK.

The most common method of modulating a spacecraft communication carrier frequency is phase-shift keying. Shifting between two phases, 0 and 180 deg, is shown in Fig. 9.9; however, phase can be shifted in any number of steps (M-ary PSK). With quaternary phase shift keying (QPSK) there are four phase states to which the carrier can be set, representing the symbols 11, 10, 01, and 00. The advantage of QPSK over BPSK is that, for a given bit error rate it requires only half the bandwidth. Higher orders of shift keying such as 8PSK are also used infrequently. The required bandwidth of 8PSK is only one-half of that for QPSK; however, the phase states are closer together and harder for the receiver to distinguish, resulting in a higher bit error rate for a given power level.

9.2 Communication Links

A communication link consists of a set of transmitting equipment, receiving equipment, and the rf link between them. In this section we will discuss the elements of a link starting with the spacecraft transmitter, which first modulates the subcarrier then modulates the carrier. Then we will discuss the elements of the ground station receiver.

9.2.1 Transmitter Subcarrier Modulation

The first stage of the transmitter uses the binary data bits to modulate the subcarrier; Fig. 9.10 shows the steps in this process. As shown in Fig. 9.10, the input to the subcarrier modulator is a digital bit stream, and the output is a modulated subcarrier, which, in the frequency domain, has a spectrum of the sin x/x type. Data are gathered from all over the spacecraft, from payload sensors and from engineering

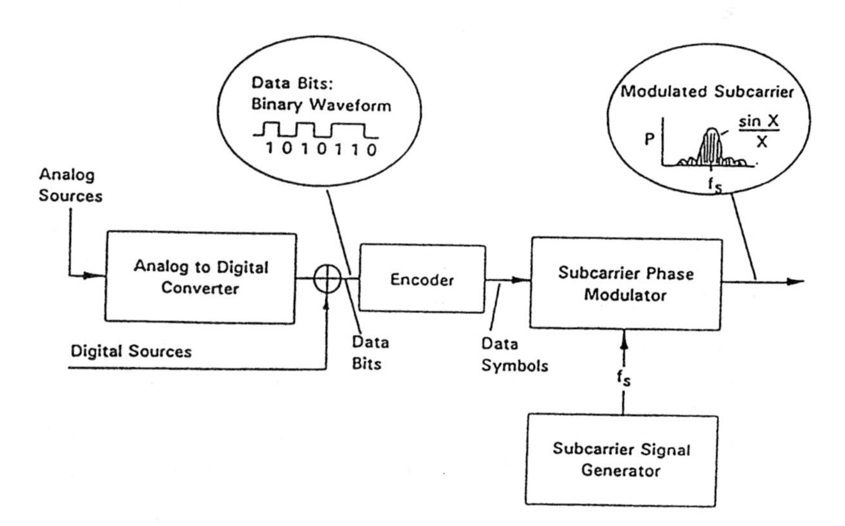

Fig. 9.10 Modulating the subcarrier.

instruments. Analog data are converted to a digital data interleaved into the digital data stream. The data bits are encoded and sent to the subcarrier modulator. A digital phase modulator is shown. The subcarrier is a fixed frequency sine wave, generated by the signal generator and modulated in response to the signal stream.

Encoded data symbols are used to modulate the subcarrier. The subcarrier frequency and wave form are set by the subcarrier signal generator. The subcarrier frequency is controlled by any of several types of oscillator. The wave form can be either square wave or sinusoidal.

9.2.2 Transmitter Carrier Modulation

The second stage of the transmitter uses a subcarrier to modulate the carrier; Fig. 9.11 shows the steps in this process.

9.2.2.1 Carrier signal generation. The carrier is a pure sine wave generated at a precise frequency. Most spacecraft and all planetary spacecraft generate this frequency in two ways: 1) by reference to the spacecraft oscillator or 2) by generating a downlink that is phase coherent with an uplink signal being received. Phase coherence means that the downlink carrier phase is synchronous with the received phase of the uplink carrier. This process is called *two-way coherent mode.* In this mode the downlink frequency is controlled at a fixed ratio to the uplink frequency being received. The equipment that does this is called a *transponder.*

Some of the transponder turnaround ratios required to communicate with three major ground station networks are shown in Table 9.4.

The two-way coherent mode allows the ground station to measure two-way Doppler and to track the spacecraft frequency more precisely. The ground station may insert a pseudorandom PN code or tones or both into the uplink for ranging. From the turnaround time of the ranging code, the MOS team can determine spacecraft range very accurately.

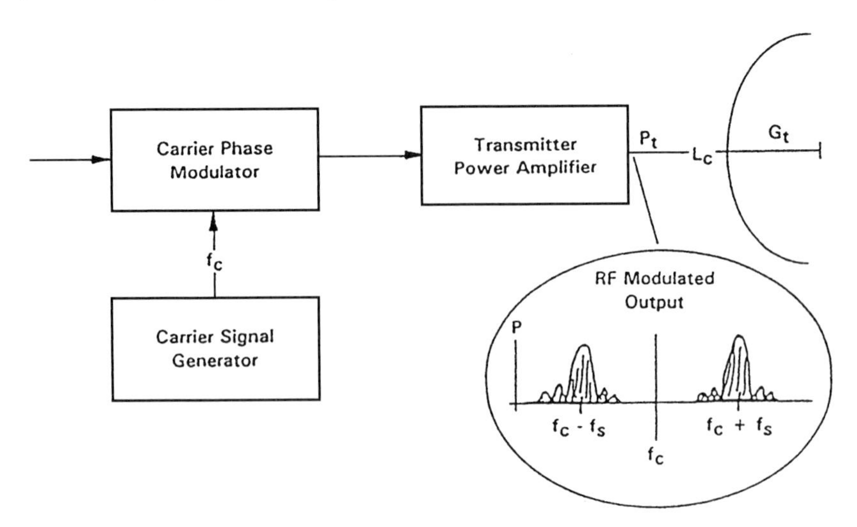

Fig. 9.11 Modulating the carrier.

Table 9.4 Transponder turnaround ratios

Ground stations	Band, up/down	Turnaround ratio
STDN	S/S	240/221
TDRSS	S/S	240/221
	K/K	1600/1469
DSN	S/S	240/221
	S/X	880/221
	X/X	880/749
	X/S	240/749

The output power spectrum of a carrier phase modulator is shown in Fig. 9.11. The center frequency is the carrier frequency; it is a pure sine wave, which produces a spectral line. The phase modulation process has produced two $\sin x/x$ spectra spaced at $\pm f_s$, about the carrier frequency, where f_s is the subcarrier frequency. The amount of energy in the carrier spectral line can be controlled by the transmitter. It is desirable to retain transmitted energy in the carrier in order to facilitate signal acquisition and Doppler measurements.

9.2.2.2 Power amplifier. The signal from the carrier phase modulator is too weak for transmission; after the wave is shaped, power amplification is required. There are two basic types of power amplifiers: solid-state and traveling-wave tube amplifiers (TWTA). TWTAs have been in use longer and have less technology risk; however, their design and construction is still somewhat empirical. In addition, they produce more rf power output at higher efficiency. Solid-state amplifiers tend to be smaller, lighter, and require less power.

The output of the power amplifier is filtered to remove frequencies outside the bandwidth of interest and passed to the antenna for transmission.

9.2.2.3 Antennas. Antennas are generally categorized as omnidirectional ("omni") or directional. Directional antennas focus the radiated power into a beamwidth of a few tenths of a degree to several tens of degrees and come in a variety of types. The beamwidth ϕ is the angle between the half power points (3 dB down) relative to the boresight power. Antenna pattern measurement parameters are shown in Fig. 9.12.

An ideal isotropic antenna would distribute radiated energy evenly over the surface of a sphere of 4π steradians (square radians) or 41,253 square degrees. The product of gain and field of view for a perfect antenna is always 41,253:

$$G\theta = 41{,}253 \tag{9.23}$$

Real antennas are not this efficient; a good rule of thumb for broad-beam antennas is

$$G = \frac{30{,}000}{\theta\phi} \tag{9.24}$$

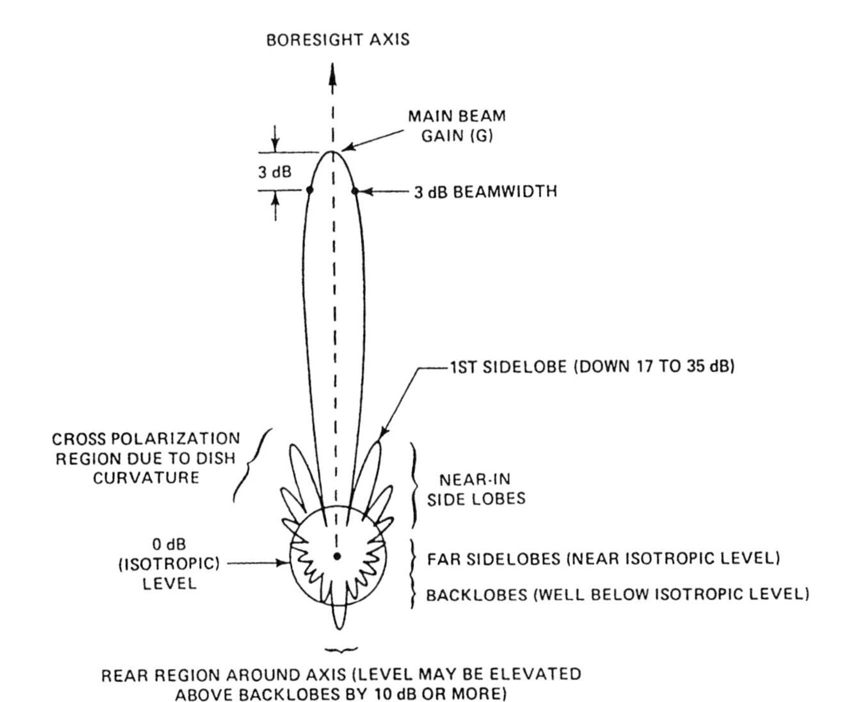

Fig. 9.12 Antenna pattern. (Copyright AIAA, reproduced with permission; Ref. 5, p. 439.)

where θ and ϕ are the height and width of the beam in degrees. Expressing Eq. (24) in decibels yields

$$G = 44.77 - 10\log(\theta) - 10\log(\phi) \tag{9.25}$$

The gain of an antenna is a function of aperture area and antenna efficiency η

$$G = \eta \frac{4\pi f^2 A}{c^2} \tag{9.26}$$

where A = area of antenna aperture and η = antenna efficiency.

Antenna efficiency is a function of feed blockage, spill-over loss, aperture illumination, and transmission line loss. The product ηA is called the effective area A_e. For a parabolic antenna with an efficiency of 55%, Eq. (26) can be reduced to a more convenient dB form:

$$G = 17.8 + 20\log(\bar{f}) + 20\log(D) \tag{9.27}$$

where $\bar{f}$ is in gigahertz and D is in meters.

Table 9.5 shows the characteristics of a number of antenna types; in Table 9.5 frequency is in gigahertz, and diameter and wave length are in meters. The common feed systems for a parabolic antenna are shown in Fig. 9.13.

Table 9.5 Antenna characteristics[5,6]

Configuration	Peak gain, dBi	Beam width, deg	Pattern
Half-wave dipole	1.64	—	
Planar array	$10 \log\left(\frac{A}{\lambda^2}\right) + 8$	—	
Turnstile	0.6	—	—
Horn	$20 \log\left(\frac{D}{\lambda}\right) + 7$ (Typically 5 to 20 dBi)	$\frac{72\lambda}{D}$	—
Bi-cone	$5 \log\left(\frac{D}{\lambda}\right) + 3.5$ (Typically 5 dBi)	Typically 45×360	
Helix	$10 \log\left(\frac{D^2 L}{\lambda^3}\right) + 20.2$ (Typically 5 to 20 dBi)	$\frac{16.6}{\sqrt{D^2 L/\lambda^3}}$	
Parabola	$20 \log(\bar{f}) + 20 \log(D) + 17.8$ (Typically 10 to 65 dBi)	$\frac{65.3\lambda}{D}$	
Yagi	$\approx 12 dBi$	—	

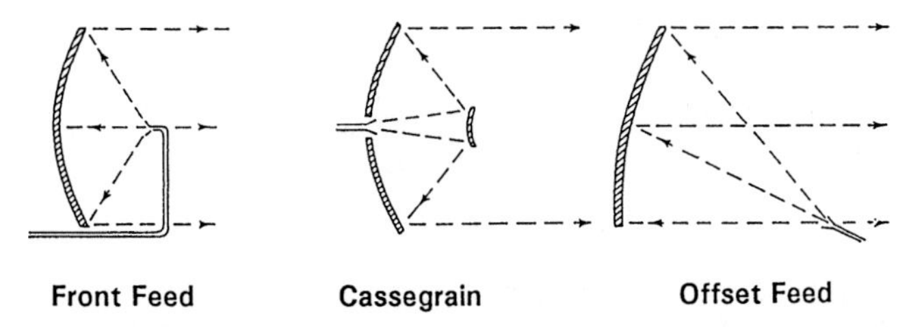

Fig. 9.13 Parabolic antenna feed systems. (*Spacecraft Systems Engineering*, Fortescue and Stark, copyright John Wiley and Sons, Ltd., 1995, reproduced with permission.)

Table 9.6 Polarization loss

	Wave			
Antenna	Vertical	Horizontal	Right circular	Left circular
Vertical	0 dB	Infinite	3 dB	3 dB
Horizontal	Infinite	0 dB	3 dB	3 dB
Right circular	3 dB	3 dB	0 dB	Infinite
Left circular	3 dB	3 dB	Infinite	0 dB

Parabolic antennas are the best selection when a narrow beam is required. The usual feed arrangement is a wave guide running through or around the reflector to the focus of the parabola. In a Cassegrain feed design a horn feeds the subreflector, which is shaped to spread the beam over the face of the parabolic reflector with the resulting high gain on the boresight. The subreflector and support truss block some of the beam and scatter power into the side lobes; this loss is on the order of 1.5 dB (Ref. 7). This loss can be avoided by an offset feed.

Horn antennas can provide full Earth coverage for frequencies above C-band. In its simplest form this system is fed by a waveguide, which is flared at the end to supply the desired aperture.

Helical antennas or end-fire antennas are preferred for S-band medium-gain antennas. For example, NAVSTAR uses helix antennas at 1.5 GHz to provide full Earth coverage from an altitude of 16,000 km.[8] They are used for rather wide beams and gains about 14 dB or less. Antenna output must be circular polarized. If the polarization of the transmitter and receiver do not match exactly, there will be a polarization loss; Table 9.6 shows polarization losses associated with a polarization mismatch.

Phased arrays provide gain by exciting many separate radiating elements, each of which has no directive properties. The combination can, however, have a very narrow beam when the elemental beams interact constructively and destructively. The beam can be steered by controlling the phase of each element. The advantages of this arrangement are that one array can produce a large number of beams simultaneously, and these can be steered electronically over wide angles without mechanical equipment. In addition the array can be mounted rigidly to structure. The major disadvantage is the complexity and number of amplifier units. The number of elements to be controlled can be several hundred.[8]

Polarization antenna output can be polarized. Most space applications use circular polarization. If polarization of transmitter and receiver do not match exactly, there will be a polarization loss, which can be large (see Table 9.6).

Antennas with like polarization have a small polarization loss, about 0.05 dB, as a result of imperfect matching. It is desirable to maintain antenna figure control, the figure departure from ideal, to within 1/20th of a wavelength. For large antennas and high frequencies this requirement can be very demanding.

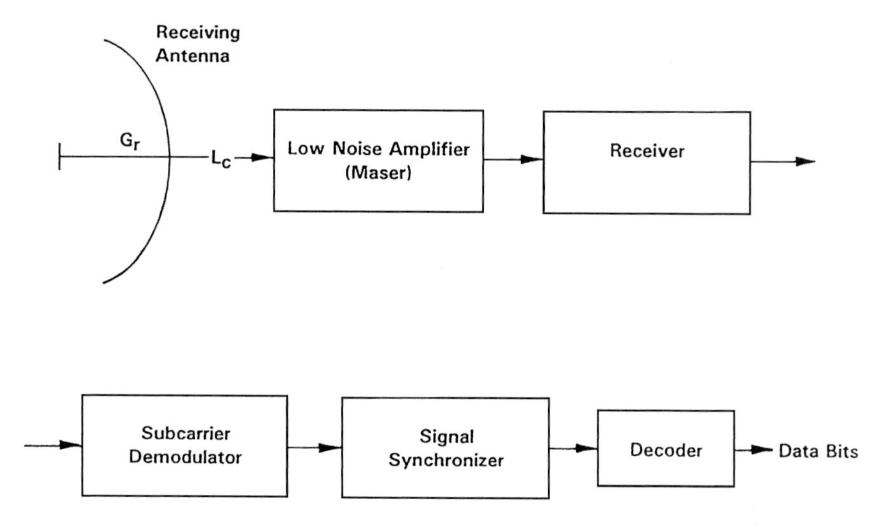

Fig. 9.14 Typical ground receiver.

9.2.3 Ground Receiver

The ground station receiver is not particularly constrained by power and weight considerations. As a result, the ground stations can be depended upon to provide most of the gain in a link as we will discuss later. Figure 9.14 shows the block diagram of a ground station; it is based on a Deep Space Network (DSN) station, but the functions are generally applicable.

9.2.3.1 Phase lock. A receiver, either spacecraft or ground, locks onto a signal using a phase lock loop circuit, which combines a voltage oscillator and a phase comparator designed so that the oscillator tracks the phase of the incoming signal. When this is achieved, it is common to refer to the receiver as "locked" or "in lock." When a receiver is in lock, the communication link is complete, and data can be transferred. When lock is lost, the data flow is interrupted, and bits are lost. The lost bits are not normally recoverable.

9.2.3.2 Maser. The rf wave is brought from the antenna, over as short a span as possible, to the low noise maser. The maser has the lowest noise figure of any type of preamplifier. (Maser stands for microwave amplification by stimulated emission radiation.) Its operation is similar to that of a laser except it operates at microwave frequencies. The maser is maintained at liquid helium (or liquid nitrogen) temperatures. The output energy comes from unstable electrons at high energy level stimulated by the signal to release energy in transit to a stable level.

9.2.3.3 Oscillators. The carrier frequency is provided by an oscillator. Spacecraft timing standard is typically a crystal oscillator, which consists of a quartz crystal coupled to an active amplifier. The vibrating crystal is coupled to the circuit by the piezoelectric effect, whereby stress in a crystal creates charges on its faces. The Applied Physics Laboratory of Johns Hopkins University, Laurel, Maryland, has developed a spacecraft ultra-stable oscillator (USO) with an output frequency of 19.14 MHz, which can be multiplied up to the frequencies the spacecraft requires. The frequency stability is better than 2×10^{-13} for integration times in the range of 1000 s. After a 25-day temperature stabilizing period the USO is expected to have a daily aging rate of 2×10^{-10} and a long-term frequency change of less than 1×10^{-6} over five years.

The ground station reference is typically a hydrogen maser standard with better stability than the USO. At the DSN stations this frequency is provided by a temperature-controlled hydrogen-maser–based frequency standard. The stability of the DSN hydrogen masers is ± 1 s in 30 million years. The masers also control the DSN master clocks, which distribute universal time.

9.3 Link Design

The object of a link design is to determine if the transmitted power of a given design is adequate to successfully transfer the desired data rate. The link is evaluated by systematic tabulation of gains and losses to arrive at the transmitted power for a given symbol rate, range, and losses. The steps in this process are as follows: 1) estimate the data rate, 2) select the maximum bit error rate (usually specified by customer), 3) select frequency (usually specified by customer), 4) select modulation and coding (usually specified by customer), 5) estimate the symbol rate, 6) set transmitter power, 7) set antenna gains, 8) estimate system gains and losses, 9) calculate receiver noise, and 10) prepare a link table for each link. In this section each of these steps will be discussed along with examples.

9.3.1 Effective Isotopic Radiated Power

The power per unit area or power density at a given distance from a transmitting source is

$$P_0 = \frac{P_t G_t}{4\pi R^2} \tag{9.28}$$

where

P_t = power input to transmitting antenna, W
P_0 = power density, power per unit area on the surface of a sphere of radius R centered at the transmitting source, W/m^2
G_t = gain of the transmitting antenna (gain is a unitless power ratio relating transmitted power to an ideal isotopic antenna)
$4\pi R^2$ = surface area of a sphere of radius R, m^2
R = distance between transmitter and any point of interest, usually a receiver, m

Note that the power density at a receiver is proportional to the product $P_t G_t$. This dependence leads to one of the traditional trades between antenna gain and transmitter power. The receiver responds to power density and is indifferent to how it is achieved. The product is called effective isotopic radiated power (EIRP):

$$\text{EIRP} = P_t G_t \tag{9.29}$$

Example 9.1 EIRP Optimization

The purpose of this example is to illustrate the trade between antenna size and power required, based on weight. Other factors, such as attitude control interactions, are ignored for the present purposes.

Consider a spacecraft transmitting system operating at a frequency of 8.4 GHz. The antenna system and the part of the power supply related to the telecommunication system is allocated a total weight of 150 lb.

It is desired to achieve a maximum EIRP within this weight limit. The spacecraft antenna is a parabolic reflector. An analysis of the weight W_a of this type of antenna yields the following relation:

$$W_a = C_a + 2D^2$$

where C_a is a constant, D is the diameter of the antenna in feet, and W_a is the antenna weight in pounds. Similarly, the power supply weight W_p can be approximated as

$$W_p = C_p + 0.5P_t$$

where C_p is a constant and P_t is transmitted power in watts. The sum of C_a and C_p is 40 lb, and the efficiency of the antenna is 65%. (In practice, simple analytic expressions for weight can be inadequate. Indeed the design may require choices among existing components, leading to discontinuous relationships.)

Find the maximum EIRP and the corresponding antenna diameter and transmitted power.

The design constraint is

$$W_a + W_p = C_a + 2D^2 + C_p + 0.5P_t = 150\,\text{lb}$$

$$2D^2 + 0.5P_t = 110\,\text{lb}$$

$$P_t = 220 - 4D^2$$

The problem is to maximize EIRP under this constraint. From Eq. (9.24) the antenna gain is

$$G_t = \eta(\pi D/\lambda)^2$$

$$\lambda = c/f = 3 \times 10^8/(8.4 \times 10^9) = 0.0357\,\text{m} = 0.117\,\text{ft}$$

$$\text{EIRP} = P_t G_t = (0.65)\,(220 - 4D^2)\,(\pi D/0.117)^2$$

Setting the derivative of EIRP with respect to D equal to zero, the value of D for maximum EIRP is

$$D = 5.244\,\text{ft}$$

The corresponding value of transmitted power is found to be

$$P_t = 220 - 4D^2 = 110\,\text{W}$$

and the maximum EIRP is

$$\mathrm{EIRP_{max}} = 110(0.65)(5.244\pi/0.117) = 1.42 \times 10^6\ \mathrm{W}$$

or

$$10\ \log(1.42 \times 10^6) + 30 = 91.52\ \mathrm{dB_m}$$

9.3.2 Free Space Path Loss

The power at a receiver is

$$P_r = P_0 A_r \tag{9.30}$$

$$P_r = \frac{P_t G_t A_r}{4\pi R^2} \tag{9.31}$$

where P_r = received power at the antenna terminals, W; A_r = effective area of the receiving antenna, commonly called the capture area, m^2; and G_r = gain of the receiving antenna, and

$$P_r = \frac{P_t G_t G_r \lambda^2}{(4\pi)^2 R^2} \tag{9.32}$$

The power loss reflected in Eq. (9.32) is caused entirely by the distance between the two antennas. This loss is called the *free space path loss* or *just path loss*. It is not caused by absorption; it is caused by increasing spherical surface area as radius increases. Path loss L_p is

$$L_p = \frac{(4\pi)^2 R^2}{\lambda^2} \tag{9.33}$$

And because $\lambda = c/f$, then

$$L_p = \frac{(4\pi)^2 R^2 f^2}{c^2} \tag{9.34}$$

Path loss is more conveniently expressed in decibels:

$$L_P = 20\log(4\pi) + 20\log(10^3) + 20\log(R) + 20\log(10^9)$$
$$+ 20\log(\bar{f}) - 20\log(3 \times 10^8) \tag{9.35}$$

$$L_p = 92.44 + 20\log(R) + 20\log(\bar{f}) \tag{9.36}$$

where L_p = path loss, dB; R = radius or range, km; and $\bar{f}$ = frequency, GHz; or

$$L_p = 36.6 + 20\log\left(\bar{f}\right) + 20\log(R) \tag{9.37}$$

where R = range, statute miles and $\bar{f}$ = frequency, MHz. Path loss is clearly a major driver in a planetary link design and is significant in geosynchronous link design. Note the units you must use with Eq. (9.36) and (9.37). Power received is

$$P_r = \frac{P_t G_t G_r}{L_p} \tag{9.38}$$

9.3.3 Noise

Signal power is always received along with unwanted noise. Noise degrades the ability to correctly interpret the signal. The smaller the SNR, the more difficult it is to segregate the signal. Thus the SNR is the key measure of the quality of a communication link.

The principle sources of noise are 1) cosmic noise, which comes from all of the bodies in the universe; 2) atmospheric noise from rain, fog, and lightening; 3) man-made noise of all kinds from electric motors, spark plugs, neon lights, and the like; and 4) thermal noise in all electronic devices, called Johnson noise.

9.3.3.1 Effective noise temperature. Most noise is generated by the thermal state of matter; all bodies at a temperature higher than 0 K emit thermal noise for which the noise power is given by

$$P_n = kT_eB \tag{9.39}$$

where

P_n = noise power, W
k = Boltzman constant, 1.38×10^{-23} W-s/K
B = noise bandwidth interval, Hz
T_e = effective noise temperature, deg K

[Note that Eq. (9.39) cannot be used to calculate noise power in optical links; thermal noise is secondary in optical systems.]

If the noise source is thermal, the effective temperature will be the thermodynamic temperature of the body. However, regardless of the noise source, it can be treated as thermally induced by calculation of an effective noise temperature. If thermal noise produced in different devices is uncorrelated, it can be summed on a power basis. For example, the effective thermal noise of an antenna can be added directly to the effective noise of a receiver. A noisy amplifier can be represented by an ideal noiseless amplifier and an effective noise temperature input (see Fig. 9.15).

If the device is purely passive, such as a transmission line, the effective input temperature T_{ei} will be a function of the thermodynamic temperature T_t of the device and the loss in the device L. Assuming the thermodynamic temperature of the device is 290 K, a common assumption,

$$T_{ei} = T_t(L - 1) = 290\,(L - 1) \tag{9.40}$$

where T_{ei} is measured at the input of the device. The equivalent temperature of the device measured at the output T_{eo} is

$$T_{eo} = T_t\frac{L - 1}{L} \tag{9.41}$$

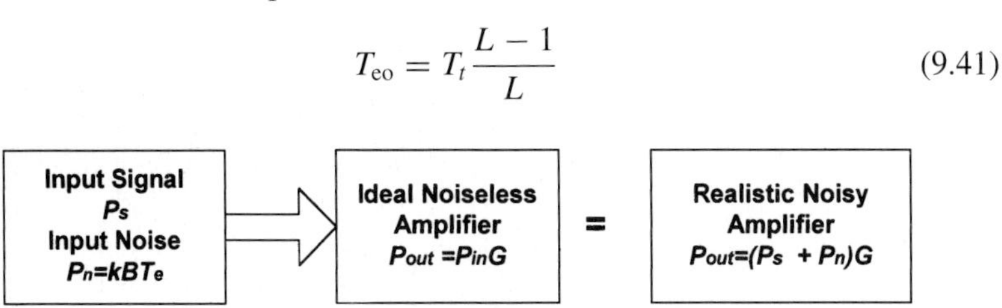

Fig. 9.15 Representation of noisy amplifier.

9.3.3.2 Noise figure. The noise figure is used to measure the noise induced by a real electronic device compared with an ideal device. For a linear electronic device (a two-port device), the noise figure F is defined as

$$F = \frac{(\mathrm{SNR})_{\mathrm{in}}}{(\mathrm{SNR})_{\mathrm{out}}} \tag{9.42}$$

For an ideal noiseless device the SNR will be unchanged by the device. Any noise present at the input would be amplified by the gain, G, as would the signal, and the ratio would not change. The noise figure for an ideal device is 1. Remembering that $P_{\mathrm{ni}} = kTB$ and that $S_o/S_i = G$, Eq. (9.42) can be rewritten as

$$F = \frac{P_{\mathrm{no}}}{P_{\mathrm{ni}}G} = \frac{P_{\mathrm{no}}}{kT_0BG} \tag{9.43}$$

where

F = noise figure
P_{no} = noise power out, W
G = circuit gain
P_{ni} = noise power in, W
T_0 = standard reference temperature, 290 K

If we introduce ΔP_n, which is the noise generated by the device itself, then

$$F = 1 + \frac{\Delta P_n}{kT_0B_nG} \tag{9.44}$$

$$F = 1 + \frac{T_e}{T_0} \tag{9.45}$$

Equation (9.45) gives the relationship between effective temperature and noise figure for a two-port device.

9.3.4 System Noise Temperature

The system noise temperature of a receiver–antenna system, as shown in Fig. 9.16, is the sum of the effective antenna noise temperature, the effective

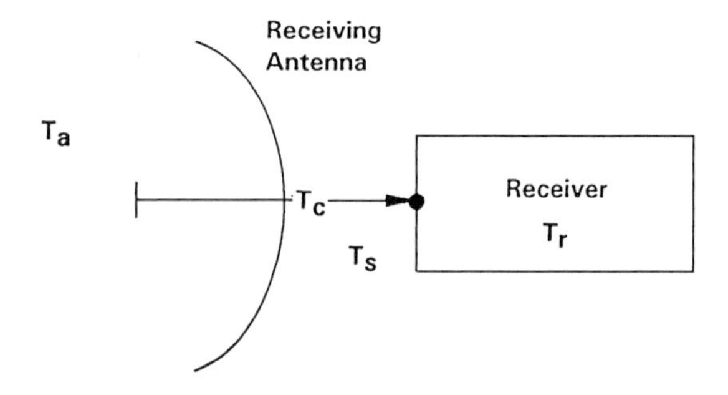

Fig. 9.16 System noise temperature.

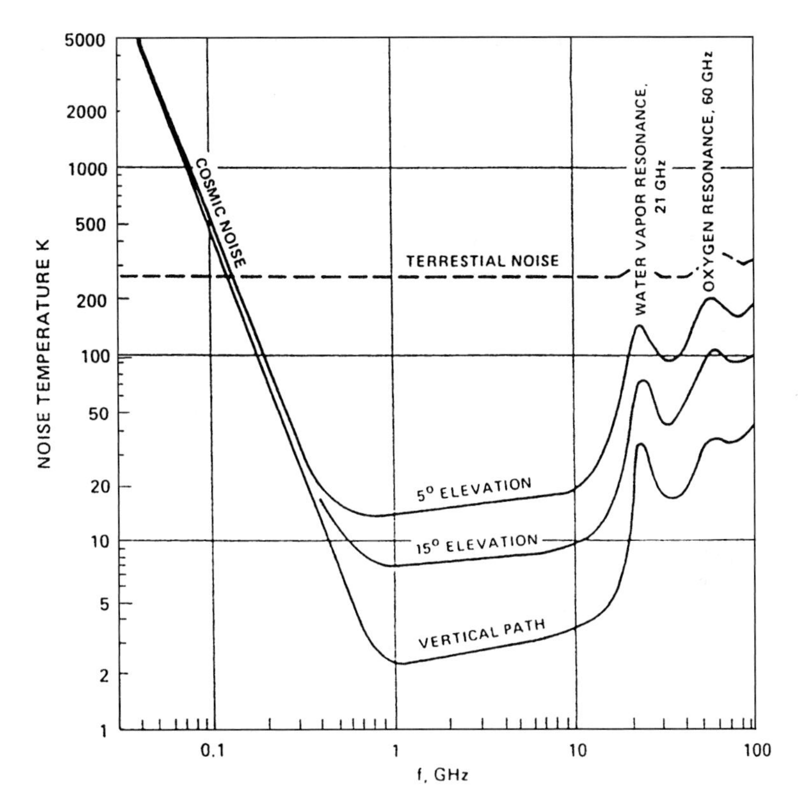

Fig. 9.17 Effective antenna temperature. (Copyright AIAA, reproduced with permission; Ref. 5, p. 448.)

transmission cable noise temperature, and the effective receiver noise temperature:

$$T_s = T_{(\text{antenna})} + T_{(\text{cable})} + T_{(\text{receiver})} \tag{9.46}$$

The effective cable temperature includes the contribution of any passive components between the antenna and the receiver amplifier, typically transmission cable, and a bandpass filter. The effective cable temperature is computed as a simple resistor.

The effective antenna temperature includes contributions of cosmic noise, weather noise, Earth noise if the Earth is in view, man-made noise, and warm-body emission. In short, the antenna noise includes all external effects plus thermal noise internal to the antenna. Figure 9.17 shows antenna temperatures for ground antennas pointed upward at various angles.

The effective temperature of a downward-looking Earth orbiter antenna is the effective temperature of the Earth, 290 K, as shown by the dotted line in Fig. 9.17.

The receiver effective temperature is computed from its noise figure, measured at the input,

$$T_e = T_0 (F - 1) \tag{9.47}$$

Summing the antenna, cable, and receiver effective temperatures as measured at the receiver input yields the receiver system noise temperature:

$$T_s = \frac{T_a}{L_c} + T_t \frac{L_c - 1}{L_c} + T_0 (F - 1) \tag{9.48}$$

where

T_s = effective system temperature, K
T_a = effective antenna temperature including all external noise, K
T_t = thermodynamic temperature of the cable, K
L_c = cable loss
F = noise figure for the receiver defined at the terminals

System noise temperatures are often referenced to the antenna terminals; different reference points lead to equivalent but numerically different effective temperatures.

Example 9.2 System Noise Temperature

What is the system noise temperature of a spacecraft receiver with the following characteristics:

$$\text{Antenna temperature } T_a = 90\,\text{K}$$

$$\text{Cable loss } L_c = 0.5\,\text{dB}$$

$$\text{Cable temperature } T_c = 290\,\text{K}$$

$$\text{Receiver noise figure } F = 3.16$$

Converting the cable loss to a ratio yields $L_c = 1.12$. From Eq. (9.48) the system noise temperature is

$$T_s = \frac{100}{1.12} + 290\left(\frac{0.12}{1.12}\right) + 290(2.16) = 746.77\,\text{K}$$

9.3.5 Signal-to-Noise Ratio

It is traditional to speak of the communications link performance in terms of SNR, which is defined as the ratio of received signal power to noise power. System noise power is

$$P_n = kT_sB \tag{9.49}$$

where B = bandwidth, Hz; and k = Boltzmann's constant, 1.381×10^{-23} J/K, and

$$N_0 = kT_s \tag{9.50}$$

$$\text{SNR} = \frac{P_r}{P_n} \tag{9.51}$$

$$\text{SNR} = \frac{P_t G_t G_r \lambda^2}{(4\pi)^2 R^2 k T_s B L_c} \tag{9.52}$$

and substituting path loss

$$\text{SNR} = \frac{P_t G_t G_r}{L_p k T_s B L_c} \tag{9.53}$$

9.3.6 Probability of Bit Error

Shannon's law states that a certain minimum energy per bit E_b is required for error-free communication. It has also been shown for Gaussian noise distribution and phase shift keying that

$$P_e \propto \sqrt{\frac{E_b}{N_0}} \tag{9.54}$$

where P_e = probability of error or bit error rate; N_0 = noise spectral density, W/Hz, and E_b = energy per bit, W-s or J.

This relation has been solved for various coding techniques as described in Chapter 8. If we set bandwidth equal to symbol rate R_s, we can then say

$$\text{SNR} = \frac{P_r}{P_n} = \frac{E_b R_s}{N_0 B} = \frac{E_b}{N_0} \tag{9.55}$$

Now combining Eq. (55), the link equation becomes

$$\frac{E_b}{N_0} = \frac{P_t G_t G_r}{L_p k T_s L_c R_s} \tag{9.56}$$

The probability of a bit error is a function of the efficiency of coding. This subject is covered in Chapter 8, in the coding section.

9.3.7 Modulation Index

After modulation some of the transmitted power is present in the carrier and some in the data; the power distribution is shown in Fig. 9.18 in the frequency domain.

The carrier is transmitted as a single frequency fc, the center frequency in Fig. 9.18. The data are transmitted in two spectra, or lobes, centered about the carrier frequency. The center of the data spectra is at the carrier frequency $\pm$ the subcarrier frequency fs. The power distribution between the carrier and the

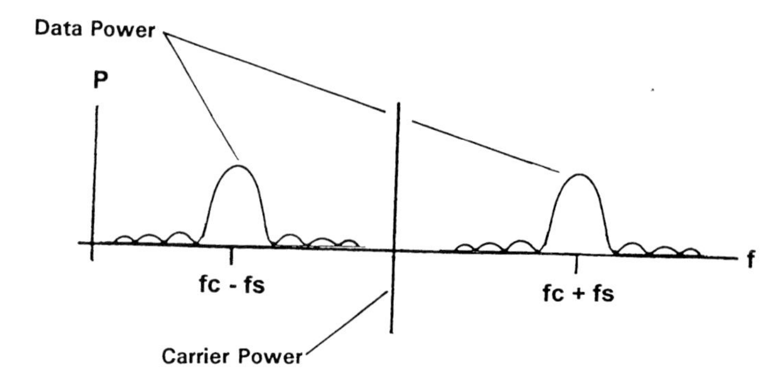

Fig. 9.18 Power distribution.

side lobes, which contain the data, is a variable in the design of the transponder. The modulation index is the parameter by which this distribution is described. The carrier represents a power loss to the data channel and vice versa. For a channel employing a sinusoidal subcarrier, the power ratios are

$$\frac{P_c}{P_t} = 20\log[J_0(\phi)] \tag{9.57}$$

$$\frac{P_d}{P_t} = 20\log[\sqrt{2}J_1(\phi)] \tag{9.58}$$

where

P_c/P_t = carrier power to total power, dB
P_d/P_t = data power to total power, dB
J_0 = zero-order Bessel function
J_1 = first-order Bessel function
ϕ = modulation index, radians

For a channel employing a square-wave subcarrier, the power ratios are

$$\frac{P_c}{P_t} = 20\log(\cos\phi) \tag{9.59}$$

$$\frac{P_d}{P_t} = 20\log(\sin\phi) \tag{9.60}$$

The solution to these equations is plotted in terms of modulation loss in Fig. 9.19, for sinusoidal and square-wave subcarriers.

The power in the carrier assists the receiving station in locking onto the channel. Lock-up is particularly demanding for a planetary link because of the distance and relative velocities involved. In a link analysis it is necessary to provide adequate data channel margin and adequate carrier margin.

9.3.8 Pointing Loss

Calculated and tabulated antenna gain is always maximum gain, that is, gain along the boresight. The attitude control system has the task of pointing spacecraft

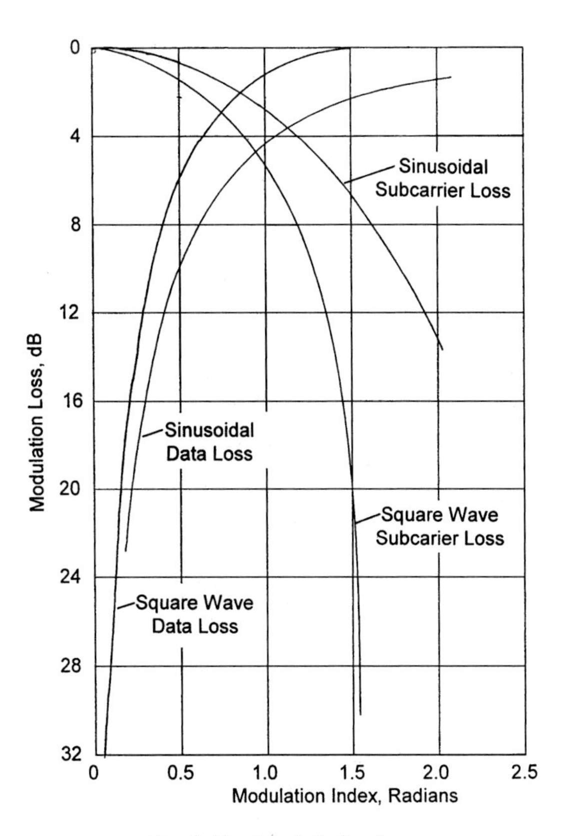

Fig. 9.19 Modulation loss.

antennas, and there is error in the process. Equation (9.61), courtesy Wertz and Larson,[6] offers a method of estimating the pointing loss:

$$L_\theta = -12\left(\frac{e}{\theta}\right)^2 \tag{9.61}$$

where L_θ = loss from peak gain caused by pointing error, dB; e = pointing error, deg; and θ = half-power beam width, deg.

9.3.9 Atmospheric Attenuation

Figure 9.20 shows the one-way vertical attenuation of the atmosphere without the effect of rain or fog.

Atmospheric attenuation is essentially zero from 0.5 to 5 GHz, which explains the popularity of this frequency range (S-band). Figure 9.20 is for vertical transmission. The losses increase at lower angles as a result of the increase in atmosphere traversed; the atmospheric losses at 10 deg are about 3 dB higher than zenith.

At frequencies above 3 GHz, the presence of rain can substantially increase atmospheric attenuation.

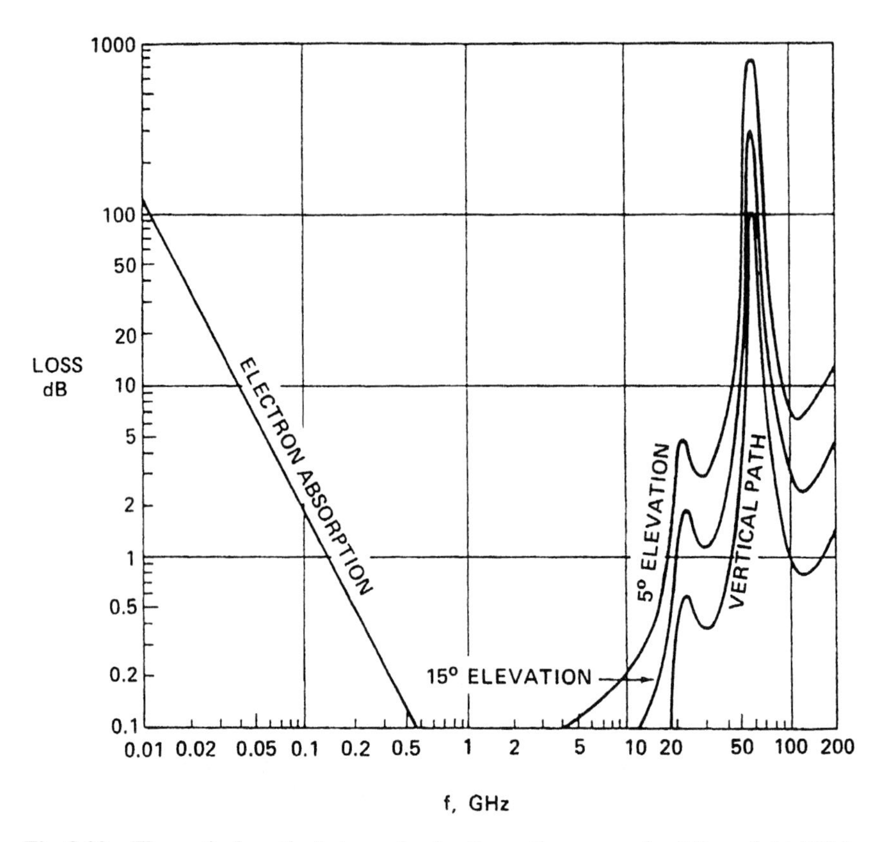

Fig. 9.20 Theoretical vertical atmospheric attenuation—no rain. (Copyright AIAA, reproduced with permission; Ref. 5, p. 443.)

Rain, sufficiently heavy, can overwhelm any communication link above about 5 GHz (see Gordon,[10] p. 179). One approach is to design the link with adequate margin, without rain, and to accept data loss if the atmospheric water exceeds the available margin.

9.3.10 Link Tables

A link table is a systematic method of recording the gain, loss, and tolerance data for a link. To build a link table, it is convenient to convert the link equation to decibels:

$$\frac{E_b}{N_0} = P_t + \sum G - \sum L + 228.6 - 10\log(T_s) - 10\log(R_s) \qquad (9.62)$$

where 228.6 dB is the Boltzmann constant and T_s = system noise temperature, K, and R_s = symbol rate, sps.

Example 9.3 Link Table

In the following example we will construct the high-rate downlink table for the Magellan spacecraft. This is one of 13 telecommunication links that Magellan used. More data were transmitted over this link than the data from all prior planetary programs combined. The link connects the Magellan high-gain antenna and any one of the DSN 70-m ground antennas. The link parameters are shown in Table 9.7.

The detailed steps in creating a link table follow the link from the transmitter through each gain and each loss down the link through the receiver. We will take each of these steps for the Magellan-high rate telemetry link.

1) Set frequency, GHz (given): 8.4259
2) Set bit error rate (given): 10^{-5}
3) Set range, km: 258×10^6
4) Set symbol rate, ksps (268.8 kbps coded): 537.6
5) Transmitter power is 21.5 W (given): +13.32 dB
6) Calculate cable loss, transmitter to antenna (measured): −0.06 dB
   (Magellan cable loss is unusually low because the transmitter is only a few inches from the feed.)
7) Calculate the transmitting antenna gain [diameter = 3.658 m (given), and peak gain using Eq. (9.27)]:

$$G_t = 17.8 + 20\log(3.658) + 20\log(8.425) = 47.57\,\text{dB}$$

   (Actual measured gain was 48.8 dB): +48.8 dB
8) Calculate EIRP = 5 + 6 + 7: 62.06 dB
9) Calculate the free space path loss, from Eq. (9.36):

$$L_p = 92.44 + 20\log(R) + 20\log(\bar{f})$$

$$= 92.44 + 20\log(2.58\text{ E }8) + 20\log(8.4259) = \qquad -279.18\,\text{dB}$$

10) Estimate atmospheric attenuation at 10 deg from Fig. 9.20: −0.15 dB

Table 9.7 Magellan high-rate telemetry channel

Parameter	Value
Frequency	8.4259 GHz ($\lambda = 0.03558$ m)
Data rate	268.8 kbps
Coding	Convolutional, constraint length 7, rate $\frac{1}{2}$
Polarization	RCP or LCP
Subcarrier frequency	960 kHz
Max range	2.58 E 8 km
Transmitter power	21.5 W
Cable loss	0.06 dB
High-gain antenna, parabolic, diameter	3.658 m
Max gain	48.8 dB

11) Estimate polarization loss: −0.12 dB
12) Give DSN receiver gain, 70-m antenna (given in Table 9.15): +74.00 dB_i
13) Calculate pointing loss. Beam width from Table 9.5 is

$$\theta = \frac{65.3\lambda}{D} = \frac{(65.3)(0.03561)}{3.658} = 0.636 \text{ deg}$$

Calculate pointing loss, assume a pointing accuracy of ±0.1 deg, from Eq. (9.61):

$$L_\theta = -12\left(\frac{e}{\theta}\right)^2 = -12\left(\frac{0.1}{0.636}\right)^2 = -0.30 \text{ dB} \qquad -0.30 \text{ dB}$$

14) Estimate receiver cable loss: 0.0 dB
15) Calculate total received power = 8 + 9 + 10 + 11 + 12 + 13 + 14: −143.69 dB
16) Receiver system noise temperature at zenith is given in Table 9.15 as 21 K (in practice a near horizon value would be used): 21 K
17) Calculate system noise density:

$$-228.6 + 10\log(\#16) = -215.38 \qquad -215.38 \text{ dB/Hz}$$

It is now necessary to calculate the link margin for the carrier *and* for the side lobes, which contain the data. The modulation index is a design variable that is used to control the power distribution between the carrier and the data.

Carrier Link Performance

18) Use a modulation index of 78 deg (given), and calculate carrier power to total power ratio from Eq. (9.59):

$$\frac{P_c}{P_t} = 20\log(\cos 78) \qquad -13.64 \text{ dB}$$

19) Calculate received carrier power = 15 + 18 = −157.33 dB
20) Calculate carrier noise bandwidth, 30 Hz (given): 14.77 dB-Hz
21) Calculate carrier signal to noise, 19-17-20: 43.27 dB
22) Calculate carrier signal to noise required by DSN: 10.00 dB
23) Calculate carrier link margin, 21–22: 33.27 dB

Data Link Performance

24) Calculate data power/total power from Eq. (9.60):

$$\frac{P_d}{P_t} = 20\log(\sin 78) \qquad -0.19 \text{ dB}$$

25) Calculate data power received, 15 + 24 = −143.88 dB
26) Calculate data symbol rate, −10 log(#4) − 30 = −57.30 dB-Hz
27) E_b/N_0 achieved = 25 + 26 − 17: +14.19 dB
28) E_b/N_0 required, from Chapter 8, Fig. 8.6: +4.10 dB
29) Calculate data link margin = 27–28: +10.09 dB

The preceding link table is for nominal conditions. In practice there would be a worst-case and a best-case table for each link.

Link tables are ideal subjects for computer spreadsheets. The Magellan link table for one of the S-band uplinks is shown in Table 9.8.

9.3.11 Conjunctions

Planetary conjunctions cause a unique communication hazard. A conjunction occurs when the Earth, sun, and a planet are aligned as shown in Fig. 9.21.

At superior conjunction the Earth and planet are diametrically opposed; the maximum communication distance occurs at superior conjunction. At inferior conjunction the planets are at their minimum range. Communication is impossible within about 3 deg of alignment for a superior conjunction with an outer planet or any conjunction with an inner planet. The link margin for Magellan at the superior conjunction of 1990 is shown in Fig. 9.22.

A planetary spacecraft must be entirely self-sufficient during a conjunction. Communication of any kind is impossible. Engineering data are stored at a low rate during the period and played back after the conjunction.

9.3.12 Optical Systems

Optical transmission powered by narrow-beamed lasers has major potential for communications. The amount of data that can be modulated onto a carrier is directly proportional to carrier frequency. Because of the extremely high frequencies generated by lasers, they have enormous potential capacity to carry information. They have not been used for spacecraft communication because of high absorption by water vapor in the atmosphere. Recent advances in satellite constellations (Iridium et al.) have introduced a need for intersatellite links (ISLs).[16] For ISLs operating at symbol rates greater than 100 msps, laser links are lighter than conventional

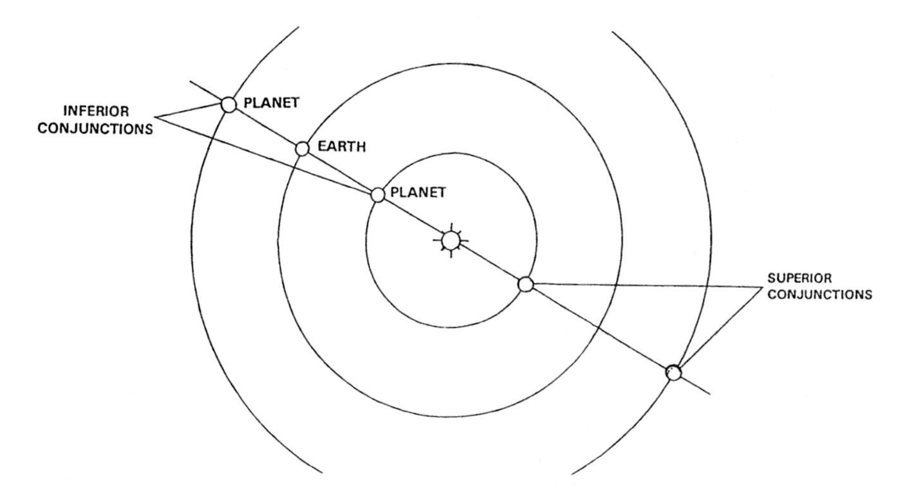

Fig. 9.21 Planetary conjunctions.

Table 9.8 Magellan uplink design table (200 million km)

Parameter	Value	Comments
1 Frequency, GHz	2.1160	MGN S-band
2 Bit error rate,	1.0E − 05	Specified
3 Range, km	2.0E + 08	—
4 Symbol rate, ksps	0.1250	62.5 bps (conv. rate 1/2)
5 Transmitter power, dB	43.00	20 kW emergency
6 Cable loss, dB	0.00	—
7 Antenna gain, dBi	67.00	34 m HE (high efficiency)
8 EIRP, dB	110.00	5 + 6 + 7
9 Free space path loss, dB	−264.93	92.44 + 20 log(#1) + 20 log(#3)
10 Atmospheric attenuation, dB	−0.17	10-deg elevation
11 Polarization loss, dB	−1.00	—
12 S/C antenna gain, dBi	5.30	—
13 Pointing loss, dB	−5.00	—
14 Receiver cable loss, dB	−1.95	—
15 Total received power, dB	−157.75	8 + 9 + 10 + 11 +12 + 13 + 14
16 Receiver noise temperature, K	125.70	—
17 S/C antenna temperature, K	100.00	—
18 System noise temperature, K	225.70	16 + 17
19 System noise density, dB/Hz	−205.06	−228.6 + 10 log(#18)
Carrier performance		
20 Carrier power/total power, dB	−5.35	Eq. (9.59); 1 rad
21 Carrier power received, dB	−163.10	15 + 20
22 Carrier noise band width, dB-Hz	13.00	20 Hz
23 Carrier/noise ratio received, dB	28.96	21-19-22
24 Carrier/noise ratio required, dB	10.00	DSN requirement
25 Carrier margin, dB	18.96	23–24
Command performance		
26 Command power/total power, dB	−1.50	Eq. (9.60); 1 rad
27 Command power received, dB	−159.25	15 + 26
28 Command symbol rate, dB-Hz	−20.97	−10*log(#4) − 30
29 E_b/N_0 achieved, dB	24.84	27 + 28 − 19
30 E_b/N_0 required, dB	4.20	Chapter 8, Fig. 8.6
31 Command link margin, dB	20.64	29–30

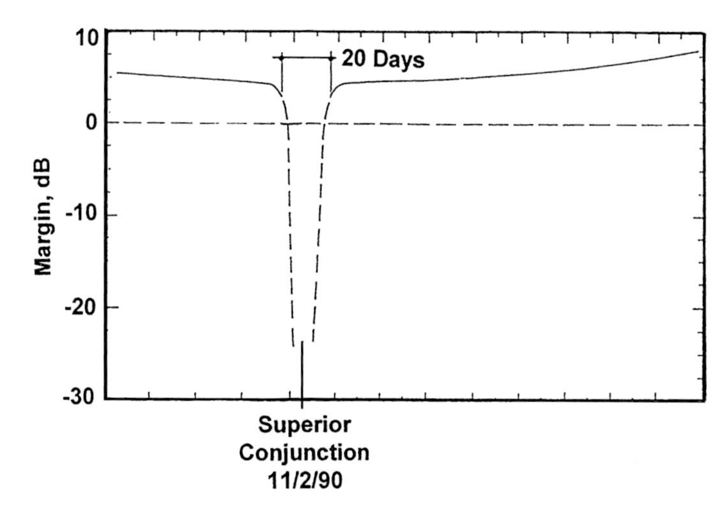

Fig. 9.22 Magellan link margin near superior conjunction.

links[11] and have been used at symbol rates above 300 msps. They have the additional advantage that frequencies are not yet controlled and allocated. A laser communication link is shown schematically in Fig. 9.23.

The first civil high-data-rate optical communications system developed by ESA was launched in 1998 on the SPOT-4 Earth resources satellite. This system, called SILEX (Semiconductor Intersatellite Link Experiment), will transmit image data from Spot-4 to the French ARTEMIS geosynchronous satellite (launch 2002) then down to a ground-based data-processing center near Toulouse. The advantage of this double hop is that ARTEMIS, in geosynchronous orbit (GEO), is in view of the SPOT-4, in low Earth orbit (LEO), for a much longer period than any Earth station. In turn ARTEMIS is always in view of Toulouse. The result is an almost continuous data stream from a spacecraft in LEO to a single ground station. SILEX will also be used for experiments with the Japanese OICETS satellite.

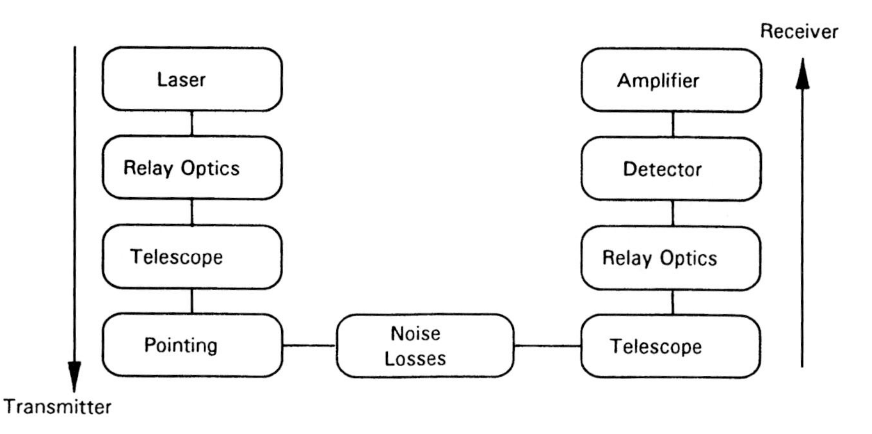

Fig. 9.23 Laser communication link.

The SILEX transmits 50 Mbps at a wavelength of 800 to 860 nm. The light source is a solid-state GaAlAs laser diode operated typically at 60 mW optical power. The modulation scheme is on/off switching. The optical data detector is a photodiode. To concentrate the optical beam in the direction of the partner satellite, the light is transmitted via a 25-cm aperture telescope providing a beamwidth of only 10 microrad (0.000057 deg). In the SPOT-4 configuration where the partner spacecraft is the order of 40,000 km away, the light travel time is about 0.17 s. During this period, the partner moves about 1000 m, whereas the beamwidth is only about 300 m; consequently, extremely tight beam pointing is required.[16]

9.4 Communication System Design

9.4.1 Magellan Communication System Example

The Magellan communication system is unusual in two ways: 1) the system operates at S-band and X-band, and 2) the high-gain antenna is time shared with the radar system. During a portion of each mapping orbit, the antenna is pointed downward and is used for radar data taking. During most of the remainder of the orbit, the antenna is pointed at Earth for data downlink. Two frequency bands were required to accommodate the S-band radar and a high data rate on X-band. The spacecraft has three major communication modes: cruise mode, mapping mode, and emergency mode. The links and data rates are shown in Fig. 9.24 and Table 9.9.

The science data, with engineering data interleaved, can be transmitted 268.2 kbps over the high-gain antenna (HGA) to the 70-m DSN stations. At this data rate an entire mapping pass is transmitted to ground every orbit. Normal commanding was done at S-band via the medium-gain antenna. In an emergency situation where

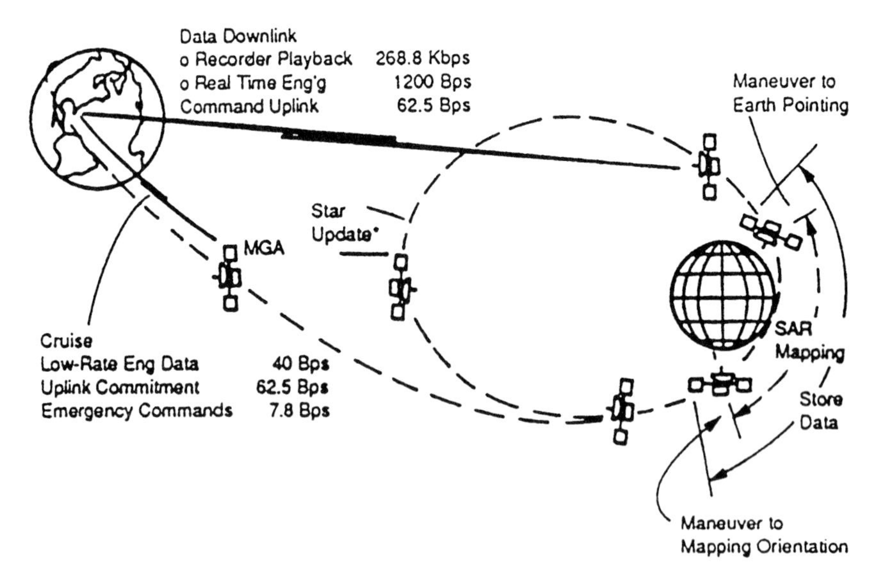

Fig. 9.24 Magellan communication links. (Reproduced with permission of Lockheed Martin; Ref. 12.)

Table 9.9a Magellan downlinks

Phase	Data type	Data rate	Band	Antenna	DSN net, m
Cruise	Recorded engineering	115.2 kbps	X	HGA	34
Cruise	Real-time engineering	40 bps	S	MGA	34
Orbit	Recorded science and engineering	268.8 kbps	X	HGA	70
Orbit	Recorded science and engineering	115.2 kbps	X	HGA	34
Orbit	Real-time engineering	1200 bps	S	HGA	70
Orbit	Emergency engineering	40 bps	S	MGA	70

attitude control is lost, commands can be uplinked to Magellan via the low-gain antenna at rates as low as 7.8 bps. Emergency engineering is downlinked over the MGA at 40 bps. Commands are normally uplinked at 62.5 bps, over X- or S-band. If commands are received at X-band, they are downconverted to S-band before they are forwarded to the command and data system The medium-gain antenna (MGA) is a conical body fixed horn, which transmits engineering data and receives commands at S-band. Spacecraft maneuvering is not required to point the antenna during the primary mission.

The Magellan communications system block diagram is shown in Fig. 9.25. The system is completely redundant and cross strapped. It can receive modulated or unmodulated uplink carrier over the S/X band high gain antenna.

The NASA standard deep-space transponders can use S-band uplink carrier to coherently excite the downlink at either S- or X-band. The transmit/receive ratios are 240/221 and 880/221 (two-way tracking). The X-band uplink can be coherently down converted to downlink at S- or X-band; the transmit/receive ratios for this mode are 240/749 and 880/749 (two-way coherent). With no uplink, the downlink frequency is derived from the spacecraft oscillator. The transponder uses redundant 22 w TWTAs in X-band and 5-W solid-state amplifiers at S-band.

The transmitter downlink path is as follows. Working from point **A** in Fig. 9.25, two digital bit streams enter the control unit and are forwarded to the transponder. One enters from data storage or real time from the payload. The other stream comes from the command and data handling system with telemetry on engineering status. The data streams are modulated onto subcarriers, which are in turn modulated onto the carrier output. If the transponder is in the two-way coherent mode, the downlink carrier frequency is controlled at a fixed ratio to the frequency of the uplink carrier

Table 9.9b Magellan uplinks

Phase	Data type	Data rate, bps	Band	Antenna	DSN net, m
Cruise	Normal commands	62.5	S	MGA	34
Cruise	Emergency commands	7.8	S	LGA	70
Orbit	Normal commands	62.5	S	HGA	34
Orbit	Emergency commands	7.8	S	LGA	70

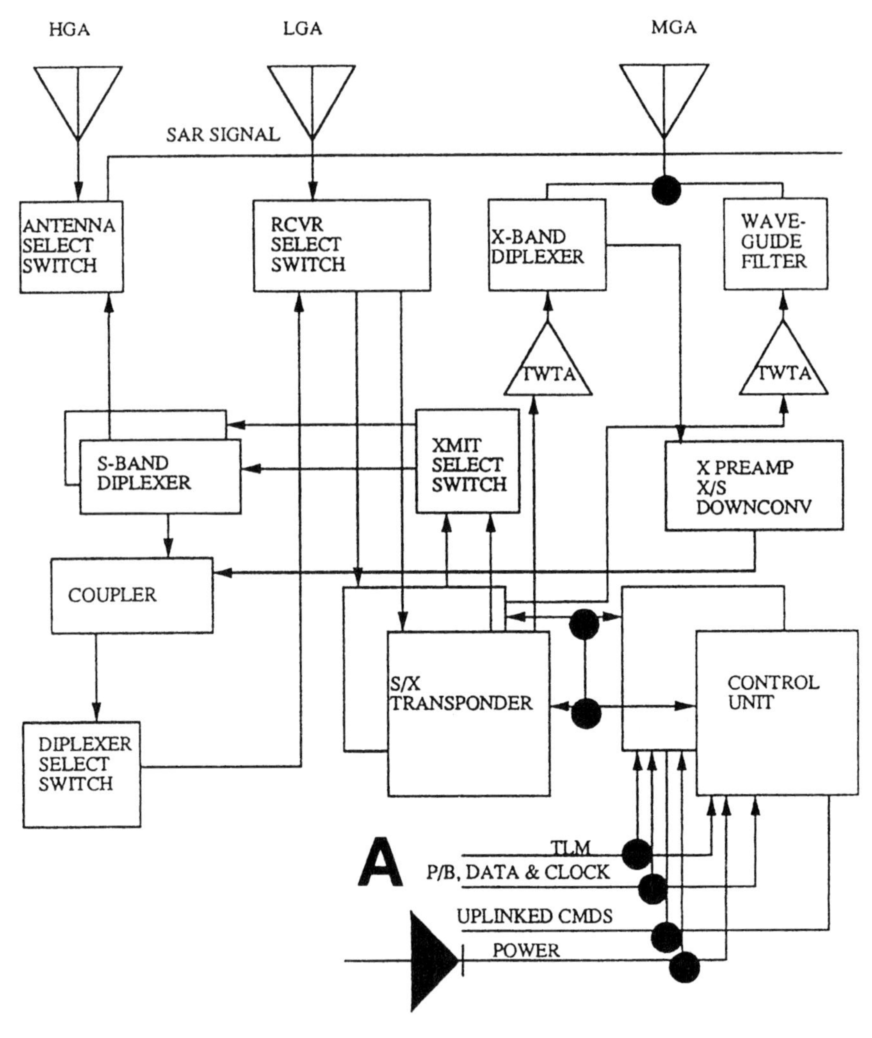

Fig. 9.25 Magellan communication system. (Reproduced with permission of Lockheed Martin; Ref. 12.)

being received. In noncoherent mode the downlink frequency is controlled by reference to the spacecraft oscillator. The composite signal is then routed out of the transponder through a transmit select switch. If transmission is to occur over X-band, the signal is sent to the TWTA for amplification; if the transmission is to occur over S-band, it is sent a solid-state amplifier. The signal is then filtered to reduce second- and higher-order harmonics from the signal. From the filter the composite signal travels through a diplexer and finally to the selected antenna where it radiates to a ground station. The diplexer allows a transmitter and receiver to share the same antenna. It also isolates the transmitter from the receiver port at the receiver's frequency so that the transmitter does not lock, jam, or damage the receiver.

The receiver's uplink path is as follows. The ground station modulates a digital command onto a subcarrier and further modulates the subcarrier onto the uplink carrier frequency to form a composite uplink signal. The composite signal enters the subsystem through the antenna to the receiver select switch, which then selects transponder A or B. The composite signal travels through the transponders low-pass filter, which rejects unwanted transmitter harmonics and frequency spurs that might exist above the diplexer's stop band. The signal then moves into the transponder's receiver, which demodulates it and forwards the data plus a receiver-in-lock indication to the command, and data subsystem, which validates and processes the command. There was no physical way to turn off the command receiver. In two-way coherent mode, ranging tones, or code, are demodulated in the receiver and sent to the transmitter for turnaround and modulation into the downlink signal.

The system accepts recorded data from CDS at 268.8 or 115.2 bps. The data is $K = 6$, rate $\frac{1}{2}$, convolutionally coded, and biphase modulated onto the subcarrier. The subcarrier signal phase modulates either X- or S-band carrier. The maximum bit error rate is 1×10^{-3} for downlink and 1×10^{-5} for command uplink.

9.4.2 Subsystem Interfaces

The communications system interfaces with every subsystem on the spacecraft except propulsion. The power system interface is influenced by the trade studies made between transmitter power and antenna gain. The attitude control system will limit the stability the HGA can expect. If greater accuracy is required, closed-loop tracking can be incorporated in the communication system. There are two kinds in use: *monopulse* and *conical scan* tracking. With either selection antenna alignment is improved at the expense of a more complicated communication system. The power system will be the recipient of requirements for power and special power conditioning. The rf power selected will be a significant impact to the power system. The structural arrangement will need to provide unobstructed field of view and accurate alignment. The CDS system interacts with data rate, data storage, and coding options. CDS and communications systems are so tightly related that for the first 20 years of the space program they were considered to be one system.

9.4.3 Traveling-Wave Tube Amplifiers

One of the difficult areas in a communication system design is rf power amplification. The older method is the use of a TWTA (see Fig. 9.26). In a traveling wave tube an electron beam interacts continually with an RF wave circulating around the beam for a distance long in comparison to wavelength. The interaction is such that rf power is increased substantially at the expense of electron beam energy. Radio frequency output can increase 20 to 60 dB.

TWTAs require voltages that are not available from the spacecraft power system, −1000 V, +5000 V, etc. As a result, each TWTA comes with a power supply designed to match what the TWTA needs. (TWTA normally means the tube and

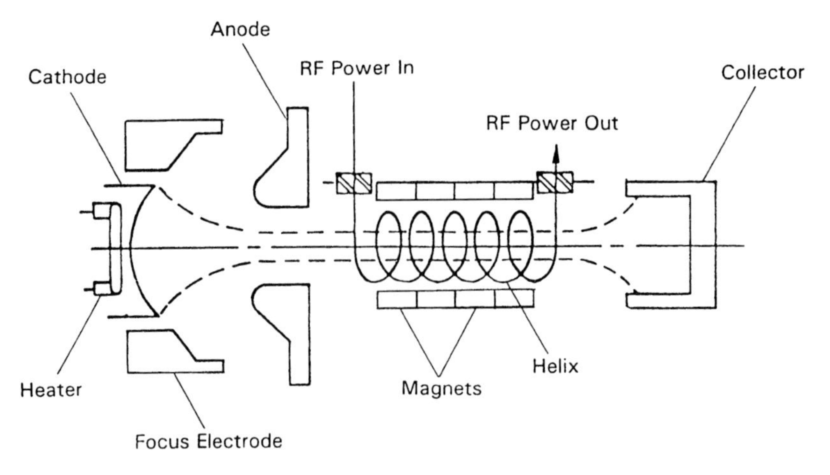

Fig. 9.26 Traveling-wave tube amplifier.

power supply; TWT normally means the tube only.) The overall rf to dc power efficiency of a TWTA (including power supply) is usually between 25 and 35%; it is less for solid-state devices. The overall power efficiency of the Magellan TWTA was 33%. Figure 9.27 shows TWTA input power and mass as a function of output power based on nine spacecraft TWTAs. The mass includes the amplifier as well as the traveling-wave tube.

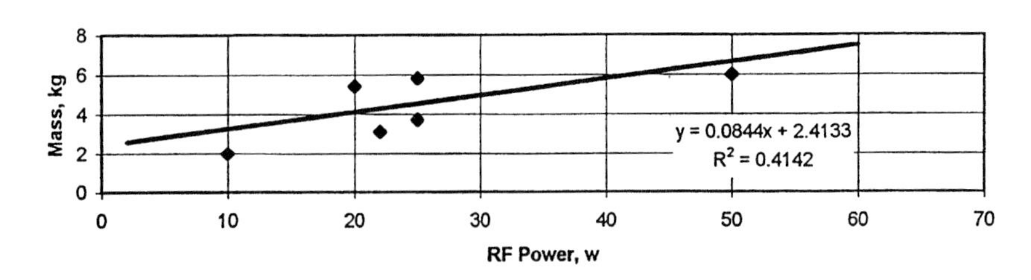

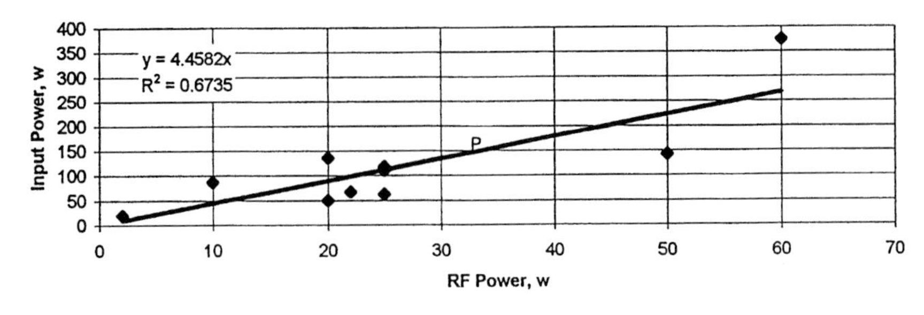

Fig. 9.27 TWTA power and mass.

Table 9.10 Characteristics of some SSPAs

Frequency	Output power, W	Efficiency, %	Weight, kg	Spacecraft
L-band	20	—	—	INMARSAT
L-band	10	40	—	—
S-band, 2.3 GHz	120	53	4.5	—
C-band, 4 GHz	10	31	—	DFH2
C-band, 4 GHz	50	42	1.6	—
X-band	13	25	1.7	Pathfinder
X-band	15	—	—	Stardust
—	40	33	5.1	DSCS
Ku-band, 12 GHz	16	29	1.75	—
Ka-band	2.2	13	0.66	DS-1
K-band, 20 GHz	3	14	—	—

9.4.4 Solid-State Power Amplifiers

A newer, potentially more reliable and lighter method of rf amplification is emerging in the solid-state power amplifier (SSPA), which accomplishes the same rf amplification as a TWTA. Solid-state power amplifiers have fewer failure modes, lower weight, and high voltage is not required; thus, the high voltage power supply is eliminated.

SSPAs have been developed from L-band through K-band. The performance of some SSPAs is shown in Table 9.10.

9.4.5 Transponders

A transponder is a form of repeater that generates a downlink carrier, which is 1) in phase with an uplink signal and 2) at a frequency, which is an exact multiplier of the uplink signal. For example, the Magellan transponder can use S-band uplink carrier to excite the downlink coherently at either S- or X-band. The transmit/receive ratios are 240/221 and 880/221 (two-way tracking). The X-band uplink can be coherently downconverted to downlink at S- or X-band; the transmit/receive ratios for this mode are 240/749 and 880/749 (two-way coherent). With no uplink the downlink frequency is derived from the spacecraft oscillator. The transponder used redundant 22-W TWTAs in X-band and 5-W solid-state amplifiers at S-band.

The newer Small Deep Space Transponder,[13] which flew on Deep Space-1, is an all digital unit that provides Ka-band or X-band downlink from an X-band uplink. It also incorporates the command detection unit and telemetry modulation unit into the transponder package. The transponder weighs 3 kg and consumes 12.9 W power.

9.4.6 Antennas

Parabolic antennas are best suited for high-gain antennas with gains over 20 dB and beamwidths less than 15 deg. For medium- and low-gain antennas you may

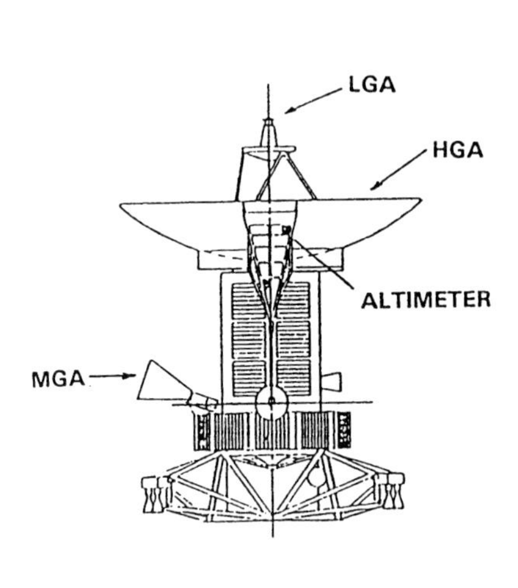

HIGH GAIN ANTENNA: Parabolic
Diameter = 3.658 m
X-Band:
Gain = 48.8 dB
Polarization = RCP or LCP
S-Band
Gain = 36.6 dB
Polarization = Horizontal
Beam Width = 2.4°
Mass = 58.5 kg

MEDIUM GAIN ANTENNA: Horn
Aperture Diameter = 40 cm
S-Band Gain = 18.9 dB
Polarization = RCP
Mass = 2.1 kg

ALTIMETER ANTENNA: Pyramidal Horn
Aperture = 68 x 28 cm
S-Band Gain = 18.5 dB
Polarization = Linear
Beam Width = 30° x 10°
Mass = 7 kg

LOW GAIN ANTENNA: Turnstile

Fig. 9.28 Magellan antenna system. (Reproduced with permission of Lockheed Martin.)

wish to select other types of antennas that are lighter. Helix and horn antennas are good choices for medium-gain applications. A low-gain (hence wide beam) antenna is required for conditions where the vector to the ground station is not predictable, for example, emergency conditions, maneuvers, or launch. A biconical horn is a good choice for low-gain antenna; it produces a toroidal beam with a beamwidth 40 × 360 deg. Figure 9.28 shows the Magellan antenna system.

The Magellan high-gain antenna was used for high-data-rate telemetry (268.6 ksps) at X-band and as the synthetic aperture radar antenna for surface images made at S-band. The medium-gain horn was used for routine command reception and low-rate-data transmission. The low-gain turnstile, mounted on the high-gain feed, was used for emergency command reception.

Figure 9.29 shows mass as a function of diameter for rigid parabolic antennas. The best fit for the antenna data in Fig. 9.29 is

$$W_a = 2.89\, D^2 + 6.11 D - 2.59 \tag{9.63}$$

where W_a = mass of parabolic antenna and feed, kg, and D = diameter of antenna, m. Figure 9.29 is for rigid antennas. There is a new class of furlable mesh parabolic antennas that provide lower mass and allow diameters greater than launch vehicle shrouds. These antennas are more expensive and require an unfurling mechanism.

Table 9.11, which shows the mass of a variety of medium- and low-gain antennas, can be used for mass estimation by analogy. The Magellan communication system weight statement and power summary, shown in Table 9.12, is useful for estimating X- or S-band communication system mass and power by analogy.

Table 9.11 Antenna mass

Type	Frequency band	Mass, kg	Size, m	Program
Horn	S	2.1	0.4	Magellan
Pyramid horn	S	7.0	0.68 × 0.28	Magellan
Turnstile	S	0.3	——	Magellan
Horn	C	3.1	0.3 × 0.65	Intelsat V
Quad helix	L	1.8	0.4 × 0.4 × 0.5	Intelsat V
Conical spiral	S	1.2	——	FLTSATCOM

Table 9.12 Magellan communication system mass and power summary (reproduced with permission of Lockheed Martin)

Component	Mass, kg
Transponder (2)	7.6
Control unit (2)	10.9
TWTA (2)	6.2
RFS components	8.0
RFS subsystem	**32.7**
High-gain antenna	58.5
Medium-gain antenna	2.1
Coax cable	7.8
Total system	**101.1**

Component	Power, W
Control unit	12.2
X-Exciter	1.4
S-Transmitter	24.2
Receiver	6.8
TWTA	67.0
X/S downconverter	2.1
Orbital average	**92.56**

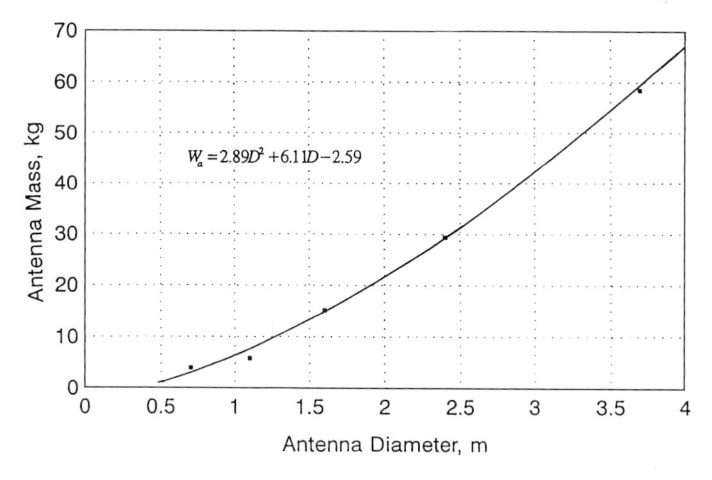

Fig. 9.29 Parabolic antenna mass.

9.5 Ground Stations

Tracking stations have three functions: 1) receiving downlink data and transmitting it to the operations team; 2) unlinking spacecraft commands; and 3) determining range, range rate, azimuth and elevation of the spacecraft. The processing of downlink data is not normally done at the ground station nor is the operations team located at the station. The U.S. Government operates seven different spacecraft communication networks[5]; these four are the most commonly used: 1) Space Tracking and Data Network (STDN) managed by NASA Goddard, Space Flight Center, Greenbelt, Maryland; 2) Tracking and Data Relay Satellite System (TDRSS) managed by NASA Goddard; 3) Space Ground Link System (SGLA) managed by the U.S. Air Force, and 4) Deep Space Network (DSN) managed by the Jet Propulsion Laboratory, Pasadena, California, for NASA.

9.5.1 Space Tracking and Data Net

STDN is the oldest and original NASA satellite communication and tracking system. The system is controlled and maintained by NASA Goddard Space Flight Center. It is composed of the ground terminal at White Sands, New Mexico, and ground stations around the world, as shown in Table 9.13.

Included in the STDN network are telephone, microwave, radio, submarine cables are various computer systems. The major switching centers are at Goddard Space Flight Center, Jet Propulsion Laboratory, and Cape Canaveral. The computer system controls tracking antennas, and commands and processes data for transmission to Johnson Space Center (JSC) and Goddard Space Flight Center (GSFC) control centers. Shuttle data from all tracking stations are funneled into the main switching computers at GSFC and rerouted to JSC, without delay, through domestic communication satellites.

The tracking station at Ponce de Leon Inlet, Florida, near New Smyrna Beach, is used during launch to provide a data link that is free of solid motor exhaust interference. This is primarily a shuttle service. Also supporting STDN are several instrumented U.S. Air Force aircraft that are situated, upon request, at almost any location in the world. This service is primarily to support the space Shuttle. The global coverage of the STDN stations is shown in Fig. 9.30.

Table 9.13 STDN ground stations

Location	Call	Service
Ascension Island	ACN	S-band, UHF
Bermuda	BDA	S-band, C-band, UHF
Canberra, Australia	CAN	S-band
Dakar, India	——	UHF
Guam,	GWM	S-band
Kauai, HI	HAW	S-band, UHF
Merritt Island, FL	MIL	S-band, UHF
Ponce de Leon, FL	PDL	S-band
Santiago, Chili	AGO	S-band
Wallops, VA	WFF	S-band

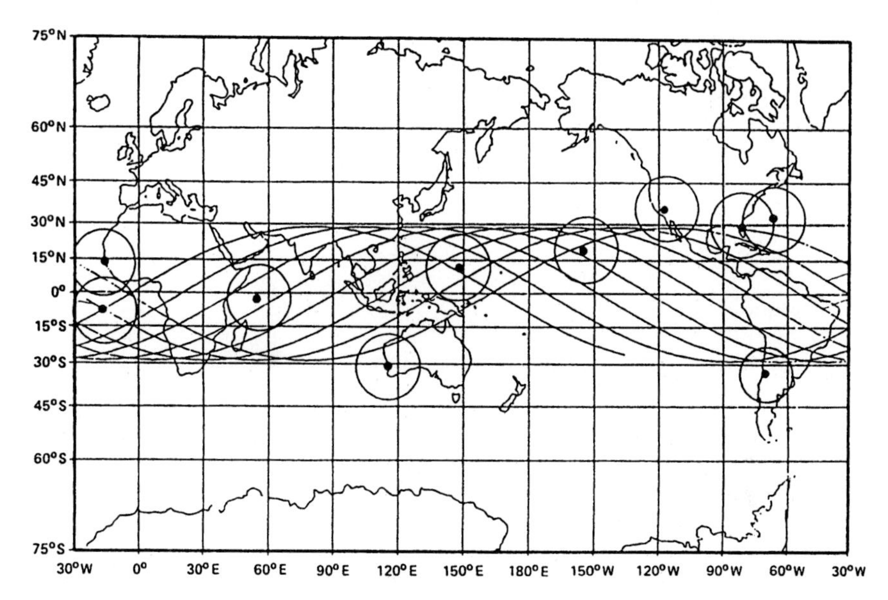

Fig. 9.30 STDN coverage. (Copyright AIAA, reproduced with permission; Ref. 5, p. 454.)

Note that STDN requires numerous handoffs and has unavoidable gaps in coverage. Gaps are much worse for orbits with inclinations above 30 deg. The STDN system is being phased out in favor of TDRSS.

9.5.2 Tracking and Data Relay Satellite

TDRSS, shown in Fig. 9.31, uses a constellation of (currently eight) spacecraft in geosynchronous orbit. The system serves as a radio relay carrying voice, analog, TV, or digital data. The constellation can provide continuous communication with

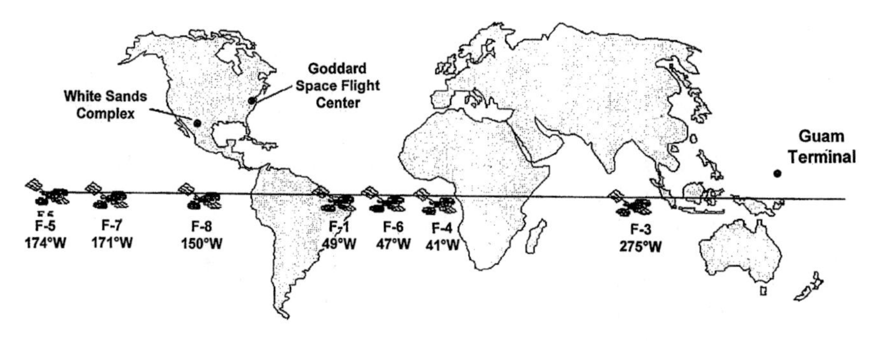

Fig. 9.31 TDRSS constellation. (Reproduced courtesy of NASA Goddard Space Flight Center.)

any Earth-orbiting spacecraft at latitudes of less than about 80 deg. In downlink, a spacecraft communicates with the closest TDRSS; the data are relayed to the ground station at White Sands, New Mexico, or Guam, where they can be relayed to the appropriate spacecraft operations center by ground line or radio link. In uplink, commands are sent to White Sands and uplinked to the spacecraft via one or more of the TDRSS spacecraft.

The system offers S-band, C-band, and high-capacity Ku- and Ka-band with single- or multiple-access channels. The C-band transponders work at 12 to 14 GHz, the Ku-band at 14 GHz. Both single-access and multiple-access channels are offered.

The spacecraft in the constellation are in three states of readiness: 1) operational; 2) stored spare, available for immediate callup; and 3) powered-down standby, which requires about a week to bring to operational state.

The original seven TDRSS spacecraft were designed and built by TRW's Defense and Space Systems Group. When launched in 1982, TDRS 1, at 2300 kg, was one of the largest of the communication satellites. The three-axis-stabilized, nadir-pointing spacecraft is designed with three distinct and separable modules: the equipment module, the communication payload module, and the antenna module.

The new generation TDRS H, I, and J competition was won by Boeing Satellite Systems, formerly Hughes Space. TDRS H, I, and J are scheduled to be launched by 2002. TDRS H, I, and J replicate the services of TDRS 1 through 7, and in addition, Ka band service, enhanced multiple access, forward data rates of 25 Mbps and return data rates of 800 Mbps are provided. The services supplied by spacecraft 1 through 7 and the new H, I, and J are summarized in Table 9.14.

Operational control of the TDRSS spacecraft is from the White Sands complex near Las Cruces, New Mexico. The TDRSS project is directed by Goddard Space Flight Center, Greenbelt, Maryland.

9.5.3 Deep Space Net

Planetary spacecraft, deep space probes (for example, NEAR), and Earth-orbiting spacecraft above about 5000 km use the Deep Space Network (DSN), which is managed by the Jet Propulsion Laboratory (JPL), Pasadena, California.

Table 9.14 Services of TDRSS

Access	Service	TDRS 1 thru 7	TDRS H, I, J
Single access	S-band forward	300 kbps	300 kbps
	S-band return	6 Mbps	6 Mbps
	Ku-band forward	25 Mbps	25 Mbps
	Ku-band return	300 Mbps	300 Mbps
	Ka-band forward	N/A	25 Mbps
	Ka-band return	N/A	800 Mbps
	Links/spacecraft	2S, 2 Ku	2S, 2 Ku, 2 Ka
Multiple access	Forward links/spacecraft	1@10 kbps	1@300 kbps
	Return links/spacecraft	5@100 kbps	5@3 Mbps

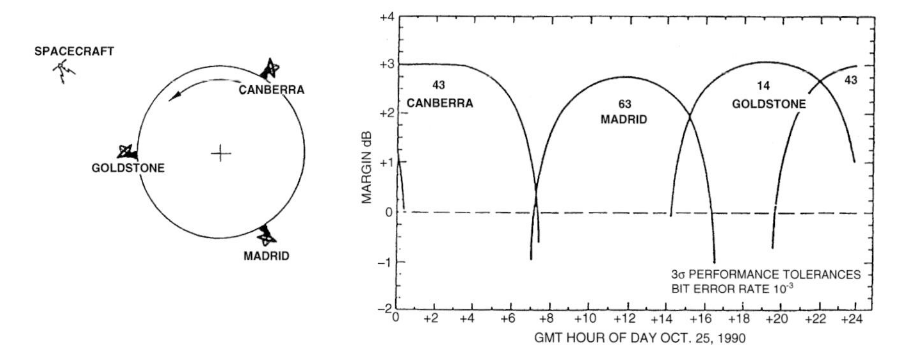

Fig. 9.32 A day in communication with the DSN.

The DSN consists of three ground stations spaced around the globe such that a planetary spacecraft in, or near, the ecliptic plane always has at least one in view. There are stations at Goldstone, California; Canberra, Australia; and Madrid, Spain; the daily activities of a planetary spacecraft involve all three stations (see Fig. 9.32). There is a DSN support and test station at Kennedy Space Center (KSC) on Merritt Island, Florida (MILA). MILA is very likely the first DSN station a planetary spacecraft will link to; this will occur during prelaunch testing at KSC.

9.5.3.1 70-m subnet. The antenna systems of a given size at each station are called a *subnet*. For example, the 70-m subnet includes a 70-m-diam parabolic antenna at Goldstone, Madrid, and Canberra. The 70-m antennas were originally built as 64-m antennas in 1964 in support of the Mariner 4 mission to Mars. All three were expanded to 70-m diam in 1982–1988 to support Voyager 2's Neptune encounter. These big antennas are the backbone of planetary communication; the Goldstone 70-m system is also used for radio astronomy.

9.5.3.2 34-m high-efficiency (HEF) subnet. This subnet was installed to replace the older 34-m standard subnet, which had a polar axis design and was originally built with 26-m reflectors. The latter upgraded to 34 m required repositioning and lifting the entire system onto concrete footings so that the reflector would not strike the ground at low pointing angles. The 34-m STD subnet has been dismantled at Madrid and Canberra. The 34-m STD at Goldstone was converted to an educational resource and renamed the Goldstone-Apple Valley Radio Telescope.

The modern 34-m HEF was designed with more efficient azimuth-elevation mounting and a 34-m precision reflector from the start. The subnet is used primarily for deep space missions, but may occasionally support a mission in high Earth orbit. The subnet can support X-band uplink and downlink and S-band downlink.

9.5.3.3 34-m beam-wave-guide (BWG) subnet. These, newest of the subnets, can be recognized by a hole in the center of the reflector where the feed horn cone full of microwave equipment would normally be. The BWG design does not require sensitive equipment mounted in the space around the feed horn where it is difficult to access and maintain. Instead, the beam wave guide directs the microwave beam through waveguides and by precision microwave mirrors down into the basement where equipment can be stably and precisely mounted and where access for maintenance is not a problem. This subnet generally supports X-band and S-band uplink and downlink. At Goldstone, Ka-band uplink and downlink is also supported.

9.5.3.4 26-m subnet. This subnet is used for rapid tracking of Earth-orbiting spacecraft usually in high Earth orbits. It was originally built in support of the Apollo mission in 1967–1972.

DSN time is a precious commodity because it is shared by all planetary programs including the few that are done by ESA. The 70-m antennas are particularly scarce because there are fewer of them, and all key events (encounter, orbit insertion, etc.) use the big antennas. Designs that minimize DSN time, particularly 70-m antenna time, should be recognized in your trade studies. The DSN station characteristics are listed in Table 9.15.

9.5.4 Acquisition of Signal

All ground stations require handoffs and frequent acquisition of signal (AOS) as the spacecraft passes from station to station. Acquisition of signal by the receiving station is a multidimensional problem. First, the ground antenna must be pointed at the location of the a spacecraft in direction and azimuth angle. (This can be done by the DSN to a few thousandths of a degree.) Next, the carrier must be located by a search over a narrowband of the frequency spectrum. Next, a phase acquisition is performed; the phase is then locked on by a phase-locked loop, which is an electronic servocontrol loop. Symbol synchronization, which is similar to phase synchronization, is next. The receiver must be able to produce a square wave that passes through zero at the same time as the received signal; this is called *symbol lock*. The final step is *frame synchronization*. Frame synchronization is required when the data are organized in blocks, frames, or packets. When direction, azimuth,

Table 9.15 DSN characteristics—X-band (Ref. 14)

	34-m high-efficiency station[a]	70-m station
Receiving parameters		
Nominal frequency	8.4–8.5 GHz	8.4–8.5 GHz
Antenna gain	66.2 dBi	74.0 dBi
System noise temp., zenith	18 K	21 K
Transmitting parameters		
Nominal frequency	7.15–7.24 GHz	—
Transmitter power	0.2–20 kW	—
Antenna gain	66 dBi	—

[a]34-m parameters are for diplexed operation.

frequency, phase, symbol, and frame synchronization are achieved by the receiver, the system is said to be *in lock*.

9.6 Space Navigation and Tracking*

Spacecraft navigation is intimately associated with the communication system, particularly on a deep-space mission. Navigating a spacecraft using a radio link requires the following information with respect to a receiving station: 1) radial distance to spacecraft, 2) radial velocity of spacecraft, 3) velocity of spacecraft in the plane of the sky, and 4) angular direction to spacecraft.

All of these data are with respect to an antenna mounted on Earth, which rotates on its axis and rotates about the sun. In addition, a planetary spacecraft is rotating about the sun or rotating about a planet, which rotates about the sun. These planetary motions are well understood and can be combined with the preceding four sets of spacecraft/antenna relative data to produce a three-dimensional model of spacecraft motion in inertial coordinates. The model then can reproduce the history of spacecraft position and predict the future spacecraft positions.

Future position predictions (called "predicts") are used to determine if trajectory correction maneuvers (TCMs) are required. The history of spacecraft positions is an essential part of the reconstruction of its observations and images of the planet it is studying. The Magellan primary data were a collection of digital data sets, each representing a strip of Venus surface approximately 25 km wide and 16,000 km long. Minutely accurate position history was necessary to convert the strips into a map.

The following paragraphs will describe how spacecraft position and velocity data are obtained.

9.6.1 Velocity Measurement

Doppler measurements of the spacecraft downlink carrier frequency are the primary measurement of spacecraft velocity perpendicular to the plane of the sky. For this purpose, an extremely stable downlink frequency is required. It is necessary to measure Doppler shifts on the order of 1 Hz in a carrier frequency of many gigahertz. It is currently impossible to maintain such stability with spacecraft onboard equipment. The ingenious solution is to have the spacecraft generate a downlink carrier frequency, which is phase coherent to the extremely accurate uplink frequency it receives.

9.6.1.1 Two-way coherent mode. In coherent mode the ground station controls the frequency of the downlink. This is done by transmitting an uplink with a very accurate and stable frequency. At the DSN stations this frequency is provided by a temperature-controlled hydrogen-maser-based frequency standard. The stability of the DSN hydrogen masers is ± 1 s in 30 million years. The masers also control the DSN master clocks, which distribute universal time.

---

*The information in this section draws heavily on the Jet Propulsion Laboratory publication, "Basics of Spaceflight,"[17] an educational document also available online at www.jpl.nasa.gov/basics.

On the spacecraft the uplink frequency is multiplied by a preset factor, for example 240/210, to provide the same extraordinary accuracy on a downlink frequency. (The uplink and downlink frequencies must differ, of course, to prevent direct signal interference.) The DSN calls the coherent mode of communication two-way coherent or just two way.

The two-way coherent mode permits the measurement of velocity-induced Doppler shift to within 1 Hz. The resulting Doppler shift is directly proportional to the radial component of the spacecraft's velocity, relative to the DSN antenna. This relative velocity can be corrected for the rates of movement of the Earth in its revolution about the sun and its axial rotation, to produce the spacecraft velocity perpendicular to the plane of the sky, in inertial coordinates.

9.6.1.2 Transmission modes. Most spacecraft can also invoke a noncoherent mode that does not use the uplink frequency as a downlink reference. Instead, the spacecraft uses its onboard oscillator as a reference for generating its downlink frequency. This mode is known as *two-way noncoherent* (the mode is called "twink" by the DSN). When a spacecraft is receiving an uplink from one station and its coherent downlink is being received by two stations, the downlink is said to be *three-way coherent.*

9.6.1.3 Data loss in mode change. A data loss can occur when the communication mode is switched. Consider a ground station in lock, on a noncoherent downlink from a spacecraft. (The DSN calls this situation one-way transmission.) When the ground station sends uplink to the spacecraft, it will lock on to the uplink and switch to coherent mode using the uplink frequency. The new downlink frequency will be different, more stable but different. The ground station will see a loss of signal on the old frequency and lose lock. The ground station can anticipate this event; the timing is precisely known and can react rapidly, but a small loss of data is common. It is wise not to plan critical data transfer during a mode change or a ground station handoff.

9.6.2 Range Measurement (Ranging)

A uniquely coded ranging pulse is added to the uplink to a spacecraft, and its transmission time is recorded. When the spacecraft receives the ranging pulse, it returns the pulse on its downlink. The time it takes the spacecraft to turn the pulse around within its electronics is known from prelaunch testing. When the pulse is received at the DSN, its true elapsed time is determined, and the spacecraft's distance can then computed to centimeter accuracy. Distance can also be determined, as well as its angular position, using VLBI as described in Section 9.6.3.1.

9.6.3 Angular Measurement

The angles at which the DSN antennas point are recorded with high accuracy (0.001 deg), but even more precise angular measurements, can be provided by VLBI and by differenced Doppler.

9.6.3.1 VLBI. A VLBI observation of a spacecraft begins when two DSN stations on separate continents, separated by a very long baseline, track a single

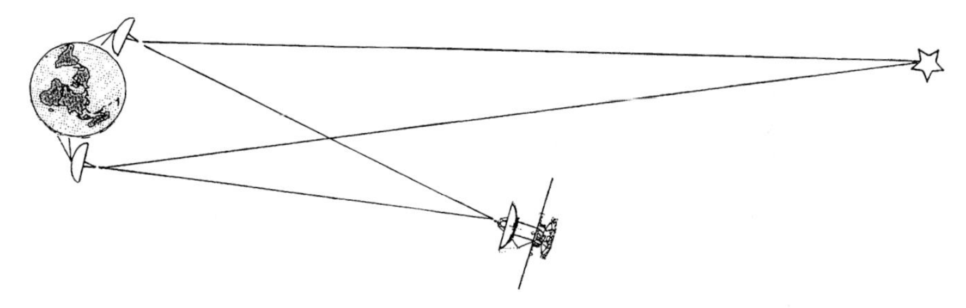

Fig. 9.33 VLBI observations.

spacecraft simultaneously as shown in Fig. 9.33. High-rate recordings are made of the downlink's wave fronts by each station, together with precise timing data. DSN antenna pointing angles are also recorded.

After a few minutes, and while still recording, both DSN antennas slew directly to the position of a quasar, which is an extragalactic object whose position is known with high accuracy. Then they slew back to the spacecraft and end recording a few minutes later. Correlation and analysis of the recorded data yield a very precise triangulation from which both angular position and radial distance can be determined. This process requires knowledge of each station's location with respect to the location of Earth's axis with very high precision. Currently, these locations are known to within 3 cm. Their locations must be determined repeatedly because the location of the Earth's rotational axis varies several meters over a period of a decade and continental plate motion on the order of centimeters also occurs.

9.6.3.2 Differenced Doppler. Differenced Doppler can provide a measure of a spacecraft's changing three-dimensional position. To visualize this, consider a spacecraft orbiting a planet. If the orbit is in a vertical plane edge on to you, as shown in Fig. 9.34, you would observe the downlink to take a higher frequency as it travels toward you. As it recedes away from you, and behind the planet, you

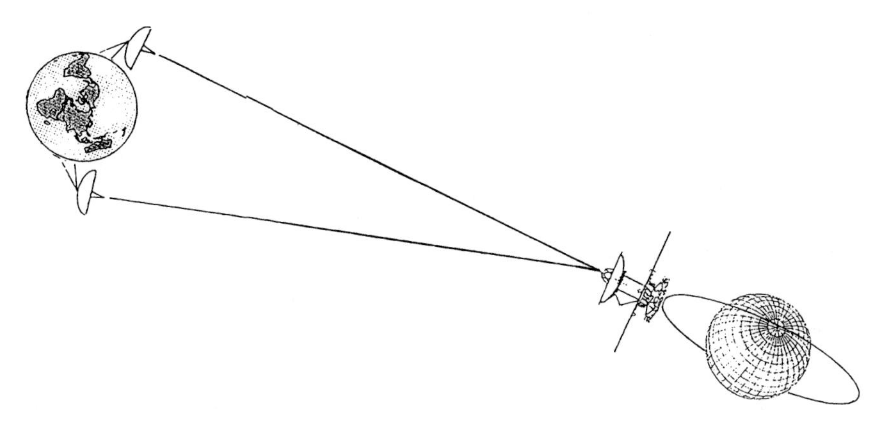

Fig. 9.34 Differenced Doppler measurements.

notice a lower frequency. Now, imagine a second observer halfway across the Earth. Because the orbit plane is not exactly edge-on as that observer sees it, the other observer will record a slightly different Doppler signature. If you and the other observer compare notes and difference your data sets, you would have enough information to determine both the spacecraft's changing velocity and position in three-dimensional space. Two DSNs separated by a large baseline do exactly this. One DSN provides an uplink to the spacecraft so it can generate a stable downlink, which is received by both stations. The differenced data sets, high-precision knowledge of DSN station positions, as well as a precise characterization of atmospheric refraction, make it possible for DSN to measure spacecraft velocities accurate to within hundredths of a millimeter per second, and angular position to within 9 nanoradians.

9.6.4 Optical Navigation

Spacecraft that are equipped with imaging instruments can use them to observe the spacecraft's destination planet against a known background star field. These images are called OPNAV images. Interpretation of them provides a very precise data set useful for refining knowledge of a spacecraft's trajectory.

9.6.5 Orbit Determination

Orbit determination computer solutions use the laws orbital mechanics, along with the navigation data just described, and the mass of the sun and planets, their ephemeris and barycentric movement. The effects of the solar wind, a detailed planetary gravity field model, attitude management thruster firings, and atmospheric friction are also used in the computations. Orbit determination is an iterative process, building upon the results of earlier solutions.

A fascinating quote from "Basics of Space Flight" available online at http://www.jpl.nasa.gov/basics shows how far this science has come:

> The highly automated and refined process of orbit determination is taken for granted today. During the effort to launch America's first satellites, the JPL Explorer 1 and 2, a room-sized IBM computer was employed to compute the satellite's trajectory using Doppler data from Cape Canaveral and a few other tracking sites. The late Caltech physics professor Richard Feynman was asked to come to the Lab and assist with difficulties encountered in processing the data. He accomplished all of the calculations by hand, revealing the fact that Explorer 2 had failed to achieve orbit, and had come down in the Atlantic Ocean. The JPL mainframe was eventually coaxed to reach the same result, hours after Professor Feynman had departed for the weekend.

Once a planetary spacecraft's solar orbital parameters are known, they are compared to those desired. To correct any discrepancy, a TCM may be planned and executed. This involves computing the direction and magnitude of the vector required to correct to the desired trajectory. Usually TCM maneuvers are performed just after insertion before launch errors build and again near the target planet to refine arrival targeting. The design of these maneuvers was discussed in Chapter 4.

References

[1]Tipler, P. A., *Physics*, Worth Publishers, New York, 1976.

[2]Maxwell, Clark, *A Treatise on Electricity and Magnetism*, unabridged 3rd ed. Dover, New York, 1954.

[3]Pearson, S. I., and Maler, G. J., *Introductory Circuit Analysis*, Krieger, Huntington, NY, 1977.

[4]Koelle, H. H. (ed.), *Handbook of Astronautical Engineering*, McGraw–Hill, New York, 1961.

[5]Griffin, M. G., and French, J. R., *Space Vehicle Design*, AIAA Education Series, AIAA, Washington, DC, 1991.

[6]Wertz, J. R., and Larson, W. J. (eds.), *Space Mission Analysis and Design*, 2nd ed., Kluwer Academic, Dordrecht, The Netherlands, 1992.

[7]Evans and Britain, *Antennas*, 1998.

[8]Fortescue, P., and Stark, J., *Spacecraft Systems Engineering*, Wiley, New York, 1996.

[9]Nyquist, H., "Certain Topics on Telegraph Transmission Theory," *Transactions of American Institute of Electrical Engineers*, 1928.

[10]Gordon, G. D., and Morgan, W. L., *Principles of Communications Satellites*, Wiley, New York, 1993.

[11]Chan, V. W. S., "Intersatellite Optical Heterodyne Communications Systems," *Lincoln Laboratory Journal*, 1988, pp. 169–183.

[12]"Magellan Spacecraft Final Report," Lockheed Martin, MGN-MA-011, Denver, CO, Jan. 1995.

[13]Chen, C.-C., Shambayati, S., Makovsky, A., Taylor, F. H., Herman, M. I., Zingales, A., Nuckolls, C., and Seimsen, K., "Small Deep Space Transponder DS-1 Technology Report," Jet Propulsion Lab., Pasadena, CA (no date); also available on-line at http://nmp.jpl.nasa.gov/ds1.

[14]Berman, A. L., Elliano, L. W., Hampton, E. C., Osoro, J. R., Short, A. B., and Tate, T. N., *Deep Space Network Preparation Plan, Magellan Project*, Jet Propulsion Lab., JPL D-4852, Rev. A, Pasadena, CA, March 1989.

[15]Edleson, R. E. (ed.), *Telecommunication Systems Design Techniques Handbook*, Jet Propulsion Lab., TM 33-571, Pasadena, CA, 1972.

[16]Oppenhauser, G., and Nielsen, T. T., *Aerospace America*, April 2002, pp. 20–22.

[17]Doody, D., and Stephan, G., *Basics of Space Flight Learner's Workbook*, Jet Propulsion Lab., JPL D-9774, Rev. A, Pasadena, CA, Dec. 1995.

Problems

9.1 A spacecraft uses a parabolic high-gain antenna 12 ft in diameter with an efficiency of 42%. The X-band downlink frequency is 8402 MHz. The desired EIRP is 92 dB_m.

(a) Estimate the antenna gain.
(b) What is the required transmitter power in dB_m?
(c) Estimate the half-power beamwidth, deg.

9.2 A high-gain parabolic antenna is used for both S-band and X-band transmissions to Earth at planetary ranges. The diameter of the antenna is 12 ft. The frequency at S-band is 2.2 GHz and at X-band is 8.4 GHz. It is necessary to limit antenna pointing loss to less than 3 dB.

(a) Within what angle limits must the attitude control system maintain the antenna boresight in order to meet the pointing loss requirement?

(b) At which frequency is the pointing requirement most restrictive?

9.3 A digital spacecraft to Earth communication system operating at 2.256 GHz and using a 32:6 biorthoganal block code is designed to operate at a bit error rate (BER) of 10^{-3} when the receiving antenna is pointed at the spacecraft near the horizon. The cable loss is 0.5 dB, and the receiver noise figure is 1.5 dB. The antenna noise temperature is 90 K near the horizon and 1.5 K near zenith. During operation, as the spacecraft moves and the receiving antenna is pointed near the zenith, what BER can be achieved?

Solution: BER = 2.2 E-5.

9.4 A spacecraft is in orbit around Venus and transmits an acquisition carrier signal to the DSN. The radial velocity between spacecraft and DSN station varies as a result of the axial rotation of the planets and the orbital motion of the spacecraft about Venus. The radial velocity of the spacecraft with respect to Earth ranges between +3.6576 km/s to −3.6576 km/s. Concurrently and in an uncorrelated manner, the radial velocity of the DSN station ranges from 9.144 km/s to 6.096 km/s. If the spacecraft transmitter frequency is 8.5 GHz, what are the maximum and minimum recieved frequencies that the DSN must accomodate?

9.5 The required EIRP of a spacecraft transmitting system is 70 dB_m at a frequency of 2.3 GHz. It is desired to achieve the EIRP with minimum weight. An analysis of the weight of parabolic antennas in this size range yields the following approximation:

$$W_a = C_a + 2D^2$$

where *Ca* is a constant, the antenna diameter *D* is in feet and weight, and *Wa* is in pounds. Similarly, the power supply weight *Wp* is

$$W_p = C_p + 0.5P_t$$

where *Cp* is a constant, *Pt* is power in watts, and W_p is in pounds.

Find the antenna diameter and the transmitted power for a minimum sum of the antenna and power supply weights.

Solution: $D = 0.965$ ft and $P = 45.46$ dBm.

9.6 The system noise temperature T_s is calculated at the input to the receiver of a ground station. It has contributions from the antenna T_{ant}, from the cabling losses from antenna to receiver T_c, and from the receiver itself T_r. The frequency received at the ground antenna pointed at the horizon is 2297 MHz. The antenna noise temperature T_a is 90 K. The cable losses are 1 dB. The receiver noise figure F is 3 dB. The reference temperature T_0 is 290 K. Determine the percentages of system noise temperature contributed by T_{ant}, T_c, and T_r.

9.7 The DSN station 70-m parabolic reflector antenna system is used to receive spacecraft S-band carrier at 2298 MHz. For acquisition at a range of 100 million km, it is desired to have a signal-to-noise ratio of 20 dB in a 10-Hz noise bandwidth.

The receiving antenna has an efficiency of 65.6% and a noise temperature of 100 K. Cable loss from the antenna to the receiver is 0.1 dB. The receiver noise figure is 0.168 dB.

(a) What is the system noise temperature?
(b) What is the receiving antenna gain in decibels?
(c) What is the free space path loss?
(d) What is the required spacecraft effective radiated power?

Solution (d): 7.58E-7 W.

9.8 It is desired to receive the rf carrier signals of an Earth-orbiting spacecraft at a ground station just as it rises over the horizon at a range of 3000 km. The minimum required signal-to-noise power ratio is 30 dB in a 5-kHz bandwidth. The satellite transmitter/antenna is linearly polarized, and the effective radiated power is 33 dBm. The carrier frequency is 4 GHz. The cable loss from the antenna to receiver is 1.0 dB, and the receiver noise figure is 3 dB.

Dertemine the minimum required diameter of a parabolic antenna having an efficiency of 60% and operating with right circular polarization. The reference temperature at the receiver is 290 K.

10
Structures*
Alfred Herzl

10.1 Introduction

Structure is the stuff that holds things together. It provides support for all load environments from prelaunch through launch and includes on-orbit loads. In this chapter we include mechanisms with structure to encompass all spacecraft mechanical support systems.

Structure is often what you see when you look at a spacecraft, but a spacecraft designer's goal is to minimize structure as much as possible. Just think, if you could save structural mass you could add more fuel to extend life, or more transponders to generate income on a communication satellite, or more instruments on a scientific investigation. As long as it meets the functional requirements of the rest of the mission system, structure is best when minimized. Section 10.2 will show how this is done.

To be successful, the structure must survive all environments without detrimental deformation. Environments are loads on the structure. Loads are converted to stresses, then stresses are converted to strains, and then strains are converted to deformation. In this way environments cause deformation and potentially failure; the job of the structures designer is to prevent this failure, as described in Section 10.3.

Section 10.2 explains why mechanisms are required—to stow structures in payload fairings and to operate equipment in flight. Section 10.4 describes some of the key rules in designing and analyzing mechanisms.

Sections 10.5 and 10.6 detail some of the analyses and tests required to verify the functionality and performance of mechanical equipment. Because there are an infinite number of options for *how* to verify hardware, these chapters just explain what the analyses and tests are and why they are used.

10.1.1 Structures Development Planning

When you are planning the development of a spacecraft structure, you need to know what the tasks are, what is required to perform these tasks in the form of resources, and how long they will take. That philosophy is great for building the 100,000th car, but most spacecraft are unique in design. Why does that matter? The first time you do anything complex like building a spacecraft things always go wrong. Yes, it helps to detail out the planning steps, but things *always* go wrong. A vendor will ship bolts that are different than those ordered; delivered materials properties are not as expected; someone dropped a wrench on the brittle material; a mistake was made in calculation of strength, and the hardware broke in test; parts

---

*Chapter 10, "Structures," is included in this text courtesy of Lockheed Martin Corporation.

do not fit, either as a result of tolerances or machining errors; loads are determined to increase six months before flight. It is a scary world out there.

Fortunately, many organizations have developed plans either to avoid these problems or to be able to recover from them. First, an overall program flow of tasks is developed for the structure, such as that shown in Fig. 10.1. This flow is repetitive in nature; as programs mature and problems occur, developers need to build in a chance to get things right before it is too late. Therefore, most programs will perform a complete load and stress analysis of the spacecraft design in the first few months of a program, prior to PDR. Obviously, this approach will not have the confidence to be ready to launch, but all of the "big picture" problems are uncovered at this level. Next a more detailed design and analysis is performed, where an attempt is made to uncover *all* potential problems, and plans are made to avoid them, prior to CDR. This is where material is ordered and fabrication started, and some problems, which will require workarounds, are discovered. The third column concentrates on fabrication, assembly, and testing. This is where the program *wants* to uncover problems—no surprises after launch! These problems will require workarounds, which cause shedding of blood, sweat, and sometimes tears. But all of that will be well worth it when the final verification analysis report is complete prior to launch, which documents all of the work done, how all of the problems were resolved, and how the spacecraft is ready for launch.

A summary of the typical structural design and verification approach contained in this chapter is presented in Fig. 10.1. The general flowchart of Fig. 10.1 is subdivided into detailed task plans for the structures subsystem, as some complex projects can have hundreds and even thousands of task entries. Although this might seem difficult to manage, greater detail usually means more problems are avoided.

10.1.2 Coordinate Systems, Sign Conventions, and the Metric System

Coordinate systems generally use one of three options: airplane, rocket, or spacecraft. Airplanes use a coordinate system with x originating at the nose and increasing to the rear. Rotation about this axis is roll. The axis out the wings is y, with rotation pitch, and the z axis is yaw. Launch vehicles use a similar coordinate system, but the x axis originates at ground level in the liftoff configuration. y is still out the sides and perpendicular to the plane of trajectory curvature. Note, however, that sign convention requires the use of the right-hand rule (if your right-hand fingers are pointed in the x direction then rotated towards y, your thumb points in the z direction). Therefore, if y is starboard on an airplane, either y or z has to be flipped 180 deg on a rocket.

Spacecraft occasionally use airplane or rocket sign convention. Usually it is important to align the spacecraft with the launch vehicle coordinate system to avoid errors. Once in orbit, the spacecraft often takes on a different coordinate system. For instance, instruments might want to point nadir (towards a planet) all of the time; the side of the spacecraft on which the instruments are mounted might be identified as nadir, and the opposite as antinadir or zenith. Sometimes the velocity vector is important, because of reentry fluxes in the atmosphere as with a reentry capsule, and velocity and antivelocity are used. And sometimes the solar flux on a spacecraft is the most important factor in design and operation of a

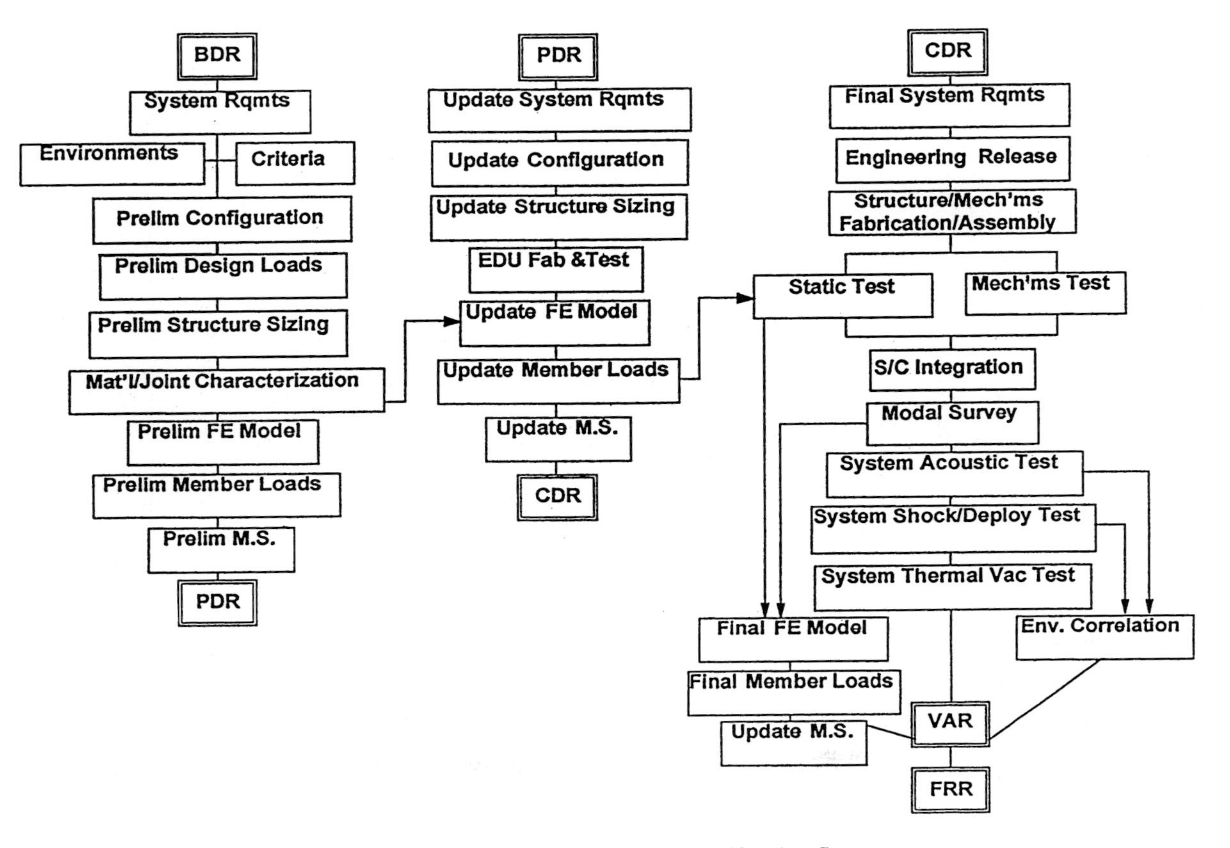

Fig. 10.1 Structural design and verification flow.

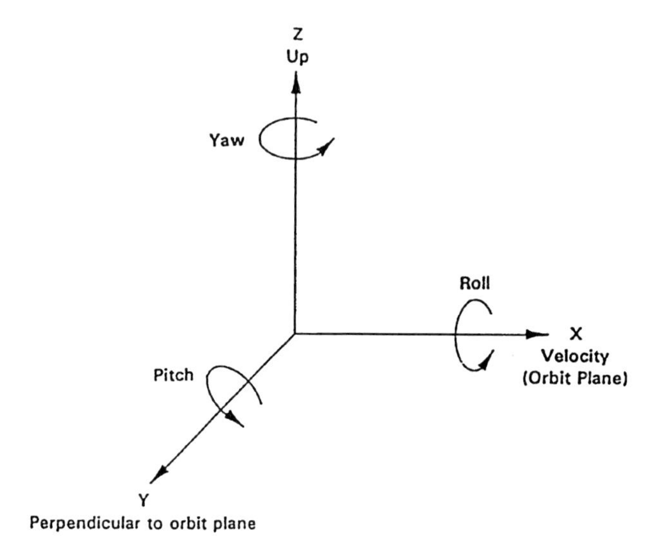

The X axis is in the orbit plane; positive X is in the spacecraft velocity direction. The X axis is also the roll axis. The Z axis points up, or opposed to the gravity vector. The Z axis also aligns with the launch vehicle long axis, which is pointed up on the launch stand. The Y axis is orthogonal to the orbit plane.

Fig. 10.2 Common structural axis system.

spacecraft that maintains a constant attitude to the sun, and solar/antisolar is used. Figure 10.2 shows the most common spacecraft axis system.

Design and analysis uses U.S. customary or the International System of Units (SI). Typically, customers request that the metric system be used to avoid potentially catastrophic problems as with NASA's Mars '98 Orbiter. But much hardware in the United States is unavailable in metric sizes. Can you imagine going to a hardware store and buying lumber with dimensions 2 cm by 4 cm by 2 m? It will be a while. Similarly, hardware available to spacecraft is mostly in nonmetric units. Designers simply need to work with this, usually by indicating on drawings a primary dimension (metric) and a reference dimension (U.S.) in parentheses. Other metric units are easier to implement. For instance, kilograms are actually easier to use than slugs, and pounds are a unit of force, not mass. Until procured raw materials can be dimensioned in metric units, including tolerances, full adoption of metric-only dimensions is not realistic.

10.2 Spacecraft Configuration Design

Initial sketch drawings of the configuration are one of the first steps in a spacecraft design. An initial attempt is made to incorporate all known requirements and constraints into a reasonable general arrangement. Two general arrangements are normally required: 1) a launch configuration that will fit within the launch vehicle shroud and 2) a mission configuration with antennas, booms, and science platforms deployed. The spacecraft mechanisms list starts with the mechanisms required for these deployments.

The general arrangement is an iterative process, as are all of the subsystem designs. As soon as an arrangement is finished, a list of potential problems and potential improvements is developed from which trade studies are conducted and another configuration developed. As the design progresses, the changes in the configuration become less global, and the general arrangement has been fixed. Changes continue at a lower level.

The structural support concept, the general thermal design, the electronics layouts, the attitude control layout, the propulsion layout, and the mechanisms list all flow from the general arrangement.

10.2.1 Constraints

Unlike an airplane, the form of a spacecraft has no function; any shape will do. However, there are numerous constraints on the arrangement. General arrangement is a compromise between the requirements for 1) field of view for instruments, antennas, radiators, solar panels, rocket motors; 2) thermal control requirements; 3) center of mass; 4) ratio of moments of inertia (for spinners); 5) propulsion lever arm; 6) constraints of launch vehicle shroud; and 7) constraints on sun vector and Earth vector.

10.2.1.1 Field of view. Much of the equipment onboard a spacecraft requires a certain field of view. Most scientific instruments—all antennas, all rocket motors, all solar panels, and most electronics, and all radiators—require a view of space. Voyager, shown in Fig. 10.3, is a classic Jet Propulsion Laboratory (JPL) spacecraft arrangement and illustrates a design complying with field-of-view constraints.

The Voyager mission was a multiplanet flyby. The science instruments, all but one, required an unobstructed view of each target planet with the planet at any

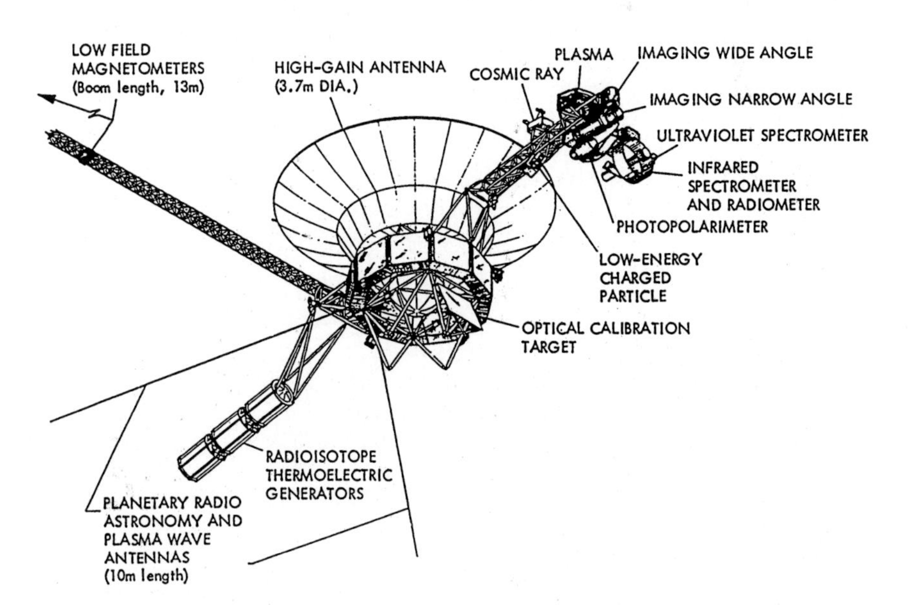

Fig. 10.3 Voyager spacecraft. (Reproduced courtesy of NASA/JPL/Caltech; Ref. 10, p. 17.)

position with respect to the spacecraft. This stiff requirement was met with an articulated instrument platform shown at the right of Fig. 10.3.

The high-gain antenna was fixed to the top of the bus and pointed at Earth by maneuvering the spacecraft. Direct mounting to the bus provided maximum pointing stability for the antenna. An antenna requires a clear field of view in the entire hemisphere above the aperture plane. Magellan, Galileo, TELESTAR, and numerous spacecraft dedicate an entire hemisphere to a high-gain antenna.

The radioisotope thermoelectric generators (RTGs), lower left in Fig. 10.4, require a maximum view of space to reject excess heat. Placing them on a deployable platform met this requirement. Although there are axial obstructions to the field of view, the cylindrical slice of sky centered on the long axis of the RTGs is the most important and is completely clear. The RTGs also emit radiation harmful

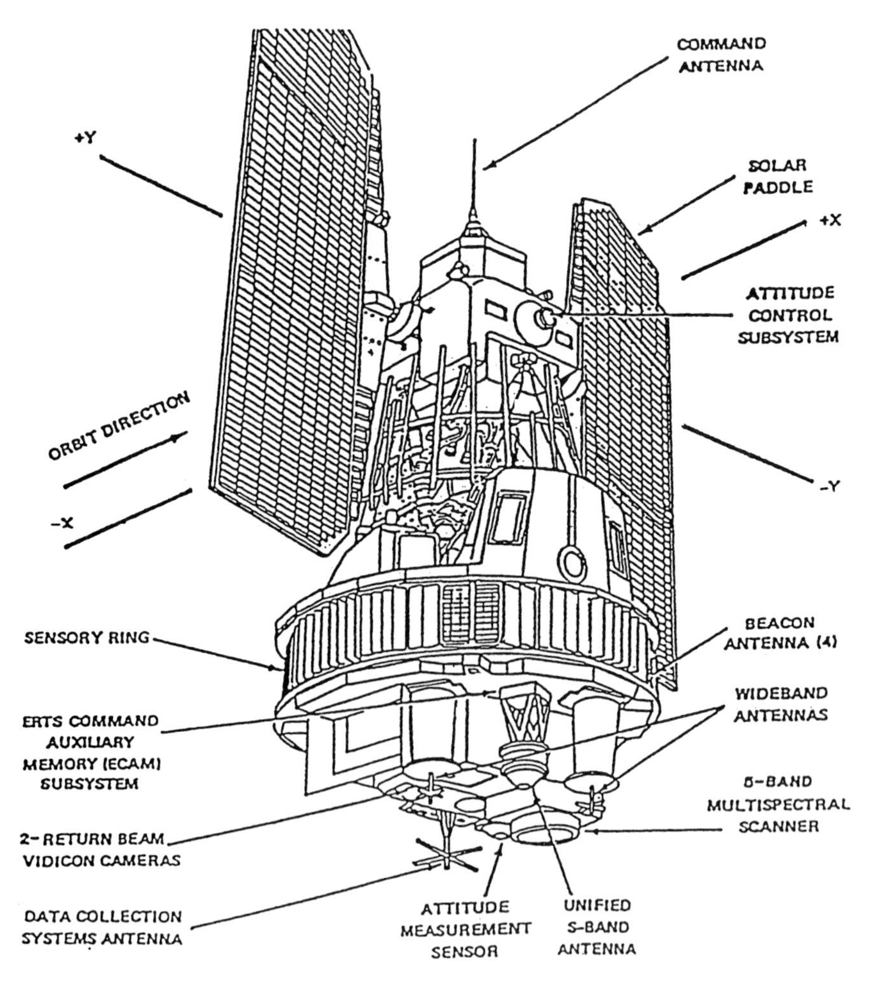

Fig. 10.4 LANDSAT C. (Reproduced courtesy of NASA Goddard; Ref. 13.)

to electronics. The deployment structure also provides some separation distance, which moderates this problem.

The 10-sided bus body is an excellent arrangement from a number of viewpoints. The electronics is installed in each of the 10 boxes. One side of each has a clear view of space for heat rejection, which is controlled by louvers. More about this polygon configuration type, which was first devised by JPL, is given later in this chapter.

10.2.1.2 Nadir pointing. Nadir pointing is a special, and common, field-of-view requirement, which dictates that the spacecraft z axis be pointed at the planet center at all times. LANDSAT C, Fig. 10.4, is a typical nadir-pointing spacecraft. The entire nadir face was dedicated to the mounting of Earth-sensing instruments. The spacecraft bus equipment was mounted above the instrument ring and between two rotating solar panels. The panels were folded back like elephant ears, to fit within the launch vehicle shroud.

10.2.1.3 Moments of inertia. Spinners must constrain the moments of inertia around the spacecraft axes for stability. A spinner must have its mass distributed like a fly wheel with the spin axis coincident with the axis of maximum moment of inertia. For dynamic stability the moment of inertia about the spin axis must be greater than the moment of inertia about any other axis (see Fig. 10.5). If this criterion is not met, the spacecraft will change spin axis with time to arrive at a stable spin around the axis of greatest moment of inertia. Explorer I, the first U.S. spacecraft, demonstrated this principle in flight. This spacecraft was shaped like a bullet and was launched spinning around the long axis, the axis of least moment of inertia. After a few days the MOS team detected that the rotation had switched to the axis of maximum moment of inertia. A rigid spacecraft, without damping, is statically and stably spinning around the axis of minimum moment of inertia; however, this is not practical in the real world of damped, slightly flexible structures.

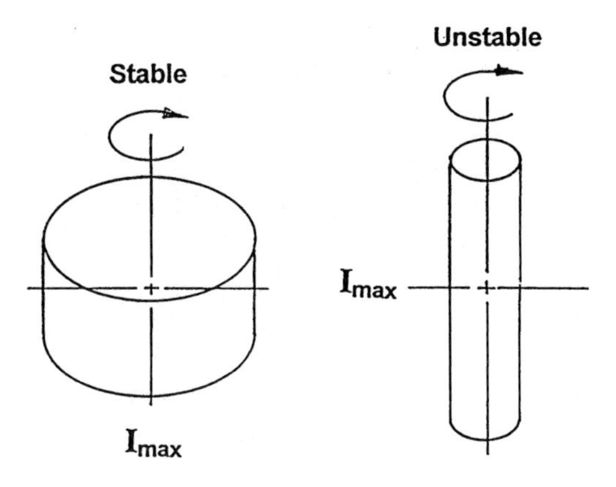

Fig. 10.5 Spin stability.

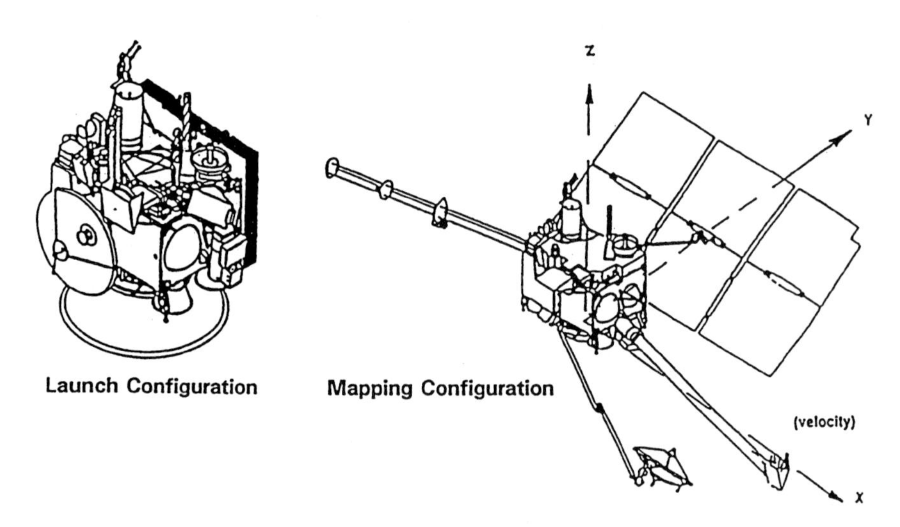

Fig. 10.6 Mars Orbiter configurations. (Reproduced with permission of Lockheed Martin; Ref. 14.)

10.2.1.4 Launch vehicle. The launch vehicle shroud limits the spacecraft to a dynamic envelope diameter about the z axis and to a length in the z direction. The diameter is usually the most severe restriction. Shroud diameters are typically 8, 10, 12, and 15 ft; the shuttle bay envelope is 15 ft. For a given launch vehicle, larger shroud diameters mean heavier shrouds and lower allowable spacecraft mass. For almost all spacecraft the shroud size causes a launch configuration that is different from the deployed configuration with attendant deployment mechanisms. The Mars Observer stowed and deployed configuration is shown in Fig. 10.6.

The launch vehicle adapter makes the mechanical transition from the launch vehicle interface to the spacecraft structure; the adapter is the responsibility of the spacecraft team, and the adapter mass is charged to the spacecraft. The *International Reference Guide to Space Launch Systems*[1] contains data on launch vehicle shrouds and mechanical interfaces for all active launch vehicles.

10.2.1.5 Propulsion system. The propulsion system requires the longest lever arm obtainable, which is normally limited by the shroud diameter. All rocket motors require a clear field of view aft of the nozzle exit plane. If a clear field is not provided, the exhaust gases will be deflected as will the thrust vector; the net result is a greater than anticipated fuel consumption and reduced mission life. Problems with plume impingement are not easy to find on the ground but are immediately obvious in flight. It is also important to consider the possibility of propulsion exhaust contaminating thermal surfaces and optics. If solid rocket motors are used, the thrust line must pass through the center of mass. The most common way to achieve this alignment is to center the solid motor on the $-z$ axis.

Propellant tanks are heavy components and are normally centered on the z axis of three-axis-stabilized spacecraft. Spinner designs normally employ two or four tanks mounted on the outer perimeter equidistant from the spin axis for high moment of inertia.

10.2.1.6 Sun vector. The solar panel axis must be perpendicular to the sun vector if two-degree-of-freedom articulation is to be avoided. The sun vector is strictly prohibited from the field of view of light-sensitive instruments such as cameras, star scanners, horizon sensors, and spectrometers. The instruments can be designed to protect against sunlight (with cover mechanisms for example), or the attitude control system (ACS) computer can protect the instruments via software. The preferred approach is to designate sun-free faces and mount the instruments accordingly; even in this case backup protection should be considered.

It is highly desirable to keep the sun off of louvers and other heat-rejection faces. The typical heat-rejection faces are the electronics mount plates and the shunt radiator for the power system. In addition, some instruments have cooling radiators. It is possible to design the thermal control system to withstand sun on the heat rejection faces as Magellan did, but it is desirable to avoid this situation if possible.

10.2.1.7 Togetherness. There are several equipment items that need to be mounted together:

1) It is desirable for the telecommunications system to be installed as close to the antennas as possible to minimize cable losses.

2) Batteries are another heavy component. It would be tempting to place batteries across from each other, equidistant from the center of mass for balance reasons; however, multiple batteries must be installed side by side and with the power control equipment in order to save cable mass and complexity.

3) The attitude control sensors and gyros need to be attached to a common, very rigid structure. On Magellan a beryllium optical bench was used to mount the gyros and star scanner. The attitude control optical bench, as well as telecommunications mounting, is shown in Fig. 10.7.

4) It is desirable to mount liquid propulsion equipment on or near the tanks and to configure the system as a separable module so that hazardous operations can be accomplished remotely. The Magellan modular propulsion system is shown in Fig. 10.8. Solid motors are inherently modular; however, special mounting arrangements are required to bring loads into solid rocket motor (SRM) cases, especially point loads.

5) If the command and data handling (C&DH) system and the ACS share a central computer, it is important that they be located together. Short communication lines to a computer are important.

10.2.1.8 Staging planes. It is usually better to leave the launch vehicle adapter on the spent launch vehicle, thus minimizing spacecraft mass. With this arrangement the staging plane is at the upper face of the adapter, and the launch vehicle interface is at the lower face of the adapter.

If the spacecraft carries a solid rocket motor, it is desirable to stage the spent motor and support structure, thereby reducing the mass load on the attitude control system. With this arrangement the spacecraft will have two staging planes.

10.2.1.9 Pyrotechnic shock. Electronic equipment is particularly susceptible to damage from pyrotechnic shock. It is highly desirable to separate electronics from ordnance devices. Because the shock energy is transmitted through

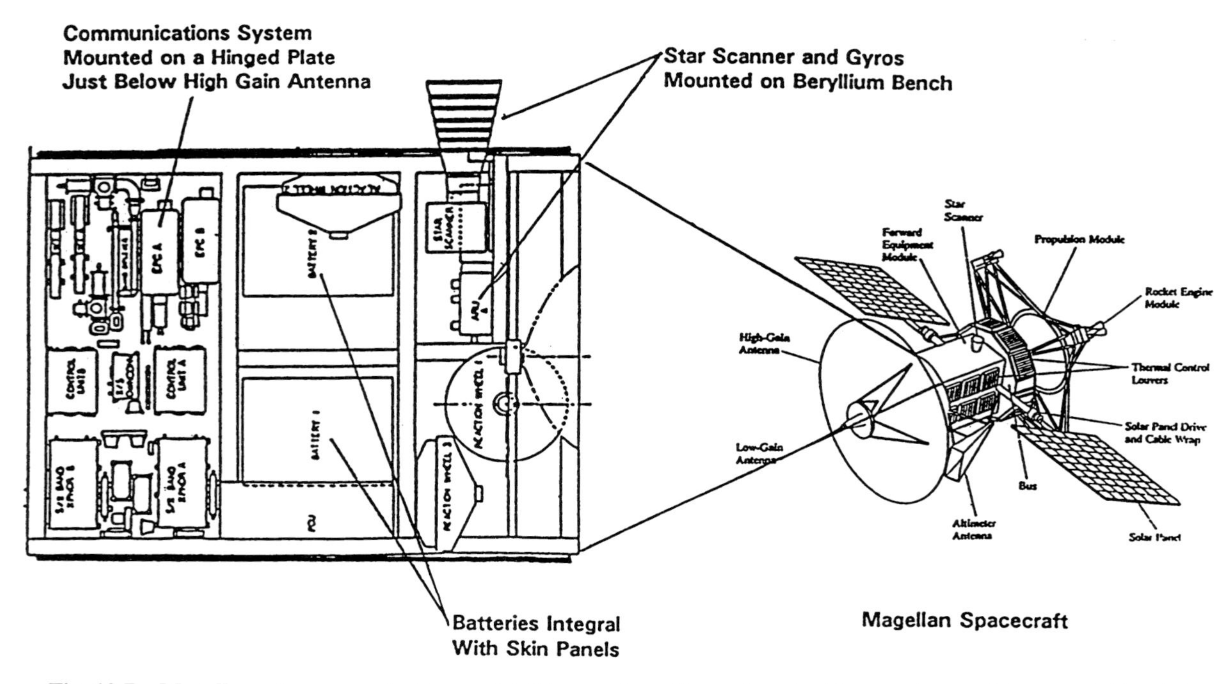

Fig. 10.7 Magellan telecommunication and ACS mounting. (Reproduced with permission of Lockheed Martin.)

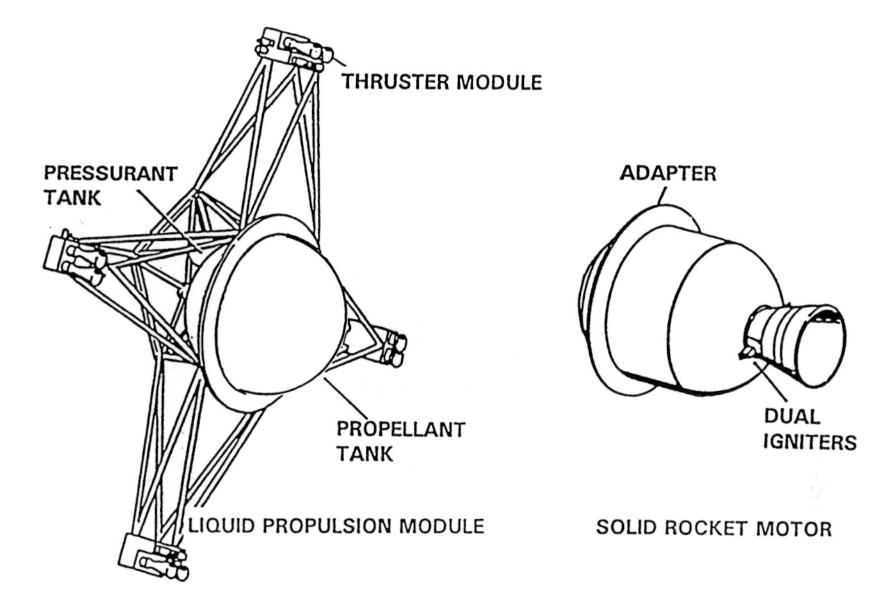

Fig. 10.8 Magellan propulsion system. (Reproduced with permission of Lockheed Martin.)

the structure, separation in this case means a long structural path from the shock source to the electronics. The propulsion system, staging planes, and deployment devices are the usual ordnance sources. Mechanical joints attenuate shock well. Low shock ordnance, which is designed to minimize shock energy, is also available. It is particularly important to consider pyrotechnic shock in early layouts because testing cannot occur until late in a program when high-fidelity structure and ordnance are available. By this time design options are limited and expensive.

10.2.2 General Arrangement Types

10.2.2.1 JPL polygon. The JPL polygon (Fig. 10.9) was first used on Mariner 1 and 2 in 1962 and was the dominant design for three-axis-stabilized planetary spacecraft for 40 years, being used on all Mariners, Viking Orbiter, Voyager, Magellan, Galileo, and some non-JPL spacecraft, for example, FLTSATCOM, HEO.

As shown in Fig. 10.10, the electronic boards are mounted directly to the front shear plate, and a louver is installed on the opposite side. Cables are connected to a ring harness that runs around the inside of the bus. There are no electronics boxes. The electronics are physically protected by the load-bearing bus structure. All electronics box weight is saved. The negative side of the integral electronics design is that the electronics must be packaged on custom boards that fit the bus. Also, special fixtures are required to remove boards when the structure is carrying loads. Note that the open center of the bus is an excellent location for propellant tanks on the z axis near the c.g.

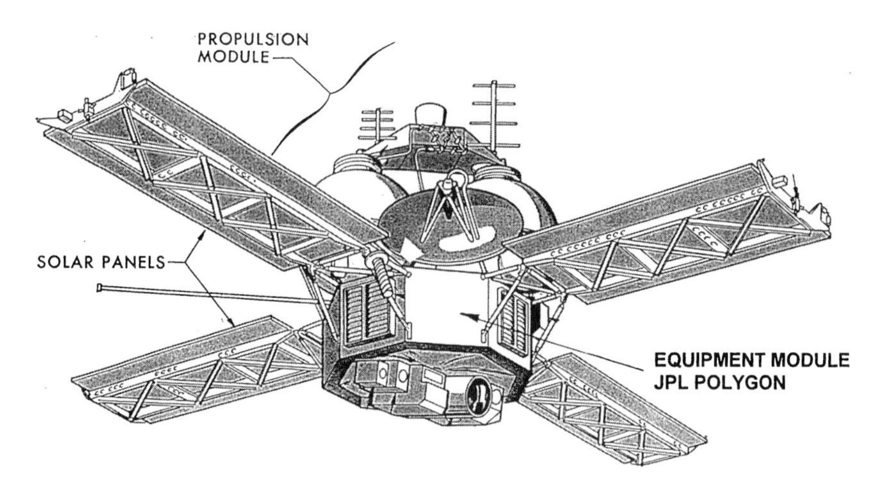

Fig. 10.9 Mariner 9 showing JPL polygon electronics packaging. (Reproduced with permission of NASA/JPL/Caltech; Ref. 19, p. 23.)

Each bay has a clear hemispherical view of space. This is an excellent arrangement for instrument view and for thermal control. The number of bays can be selected to serve the equipment needing view or good heat rejection. Eight- and ten-bay configurations have flown. The center opening inside the polygon is an excellent location for cables and for propulsion tanks or solid motors. Access to the bays is obtained by removal of the inner and outer shear plates. The advantages of polygon bay configuration are 1) very efficient use of weight and volume; 2) inherently stiff, strong structure; 3) excellent thermal control arrangement; 4) access to the electronics and cabling is good; 5) simple, direct, protected path around the inside of the polygon, for power and data cabling; and 6) excellent hemispherical field of view from each bay.

The disadvantages of polygon bay configuration are 1) off-the-shelf electronics requires repackaging and 2) in heavier spacecraft bay removal might require special tooling.

10.2.2.2 Hinged panel. FLTSATCOM also used polygon buses, two bolted together, as shown in Fig. 10.11. Note from Fig. 10.11 that the bay shear plates are hinged for easy equipment access and that electronic boxes are mounted on the shear plates rather than integral with them. This arrangement allows existing electronics to be used in black boxes without repackaging. Magellan also used the hinged panel arrangement to mount some equipment.

The advantages of hinged panel construction are listed here: 1) very efficient use of weight and volume; 2) inherently stiff, strong structure; 3) excellent thermal control arrangement; 4) excellent access to the electronics; 5) black box electronics can be used without repackaging; 6) simple, direct, protected path for power and data cabling (in the center of the polygon); and 7) excellent hemispherical field of view from each bay.

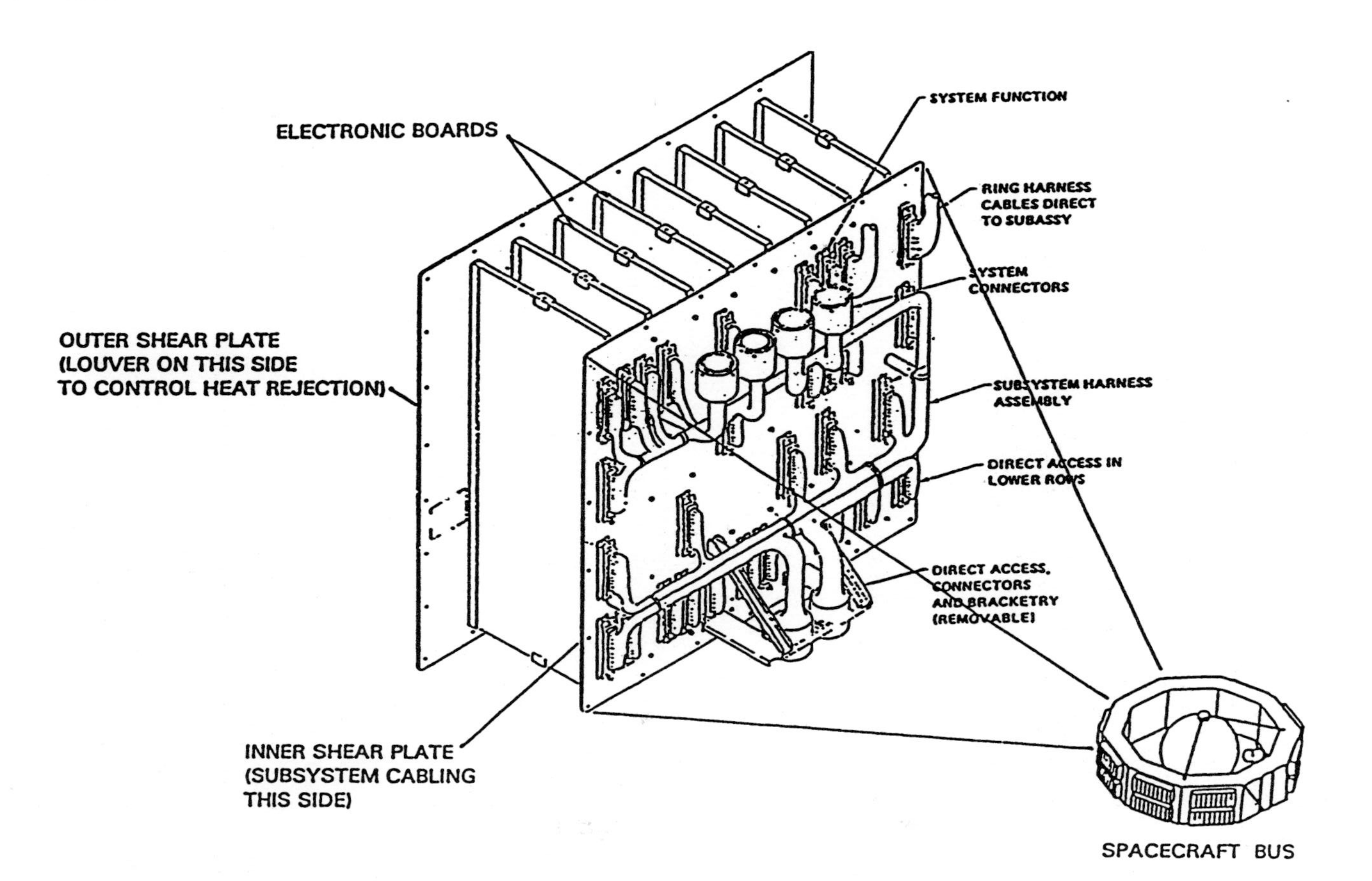

Fig. 10.10 JPL dual shear plate packaging.[20]

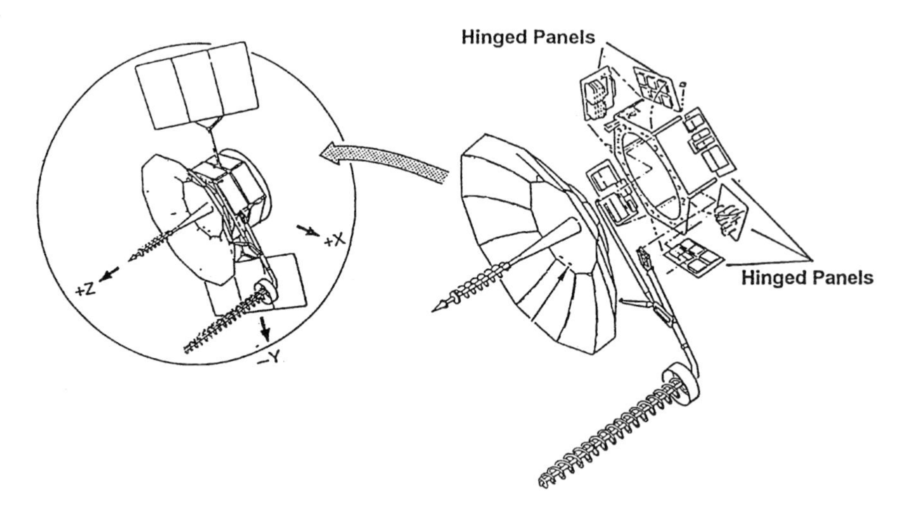

Fig. 10.11 FLTSATCOM hinged panels. (Reproduced courtesy of TRW; Ref. 2, pp. 2–6.)

In heavier spacecraft the hinged panel construction may require bay removal equipment.

10.2.2.3 Spinning cylinder/internal shelf. Figure 10.12 shows the classic cylindrical spinner design that originally was developed by Hughes Aircraft Company (now Boeing Satellite Systems) and used for a large number of spacecraft. The spacecraft shown is TELSTAR 3; however, the same general arrangement was used for SBS, ANIK, PALAPA-B, WESTAR, and others. All were dual spinners. In the launch configuration, shown on the left, the high-gain antenna is folded down, and the aft solar panel cylinder is retracted.

The equipment shelf, on which the high-gain antenna and telecommunication equipment is installed, is inertially fixed, and the remainder of the spacecraft spins, including the apogee kick motor, the external cylindrical solar panels, and an internal equipment shelf. The shelf is adjacent to a polished heat-rejection area in the cylindrical panel cylinder. The electronics is attached to the top of the spun shelf, and the propulsion system is attached to the bottom of the shelf. The Landsat C (Fig. 10.2) is an example of a three-axis-stabilized, nadir-pointing spacecraft that uses shelf-mounted equipment. The nadir-pointing instruments are mounted on the bottom of the shelf, and the electronics is mounted on the top, internal to the spacecraft. The Magellan spacecraft (Fig. 10.5) used shelf mounting for some equipment.

The advantages of internal shelf mounting are 1) integrates well with a cylindrical spinner configuration and 2) can use existing black boxes without repackaging.

The disadvantages of internal shelf mounting are 1) access is substantially more difficult; 2) view of space is limited to equipment near the outer walls; 3) less efficient mass and volume usage. (For a spinner the cylinder diameter and height is normally set by solar-panel area requirements; therefore, efficient use of the shelf area is not an issue.)

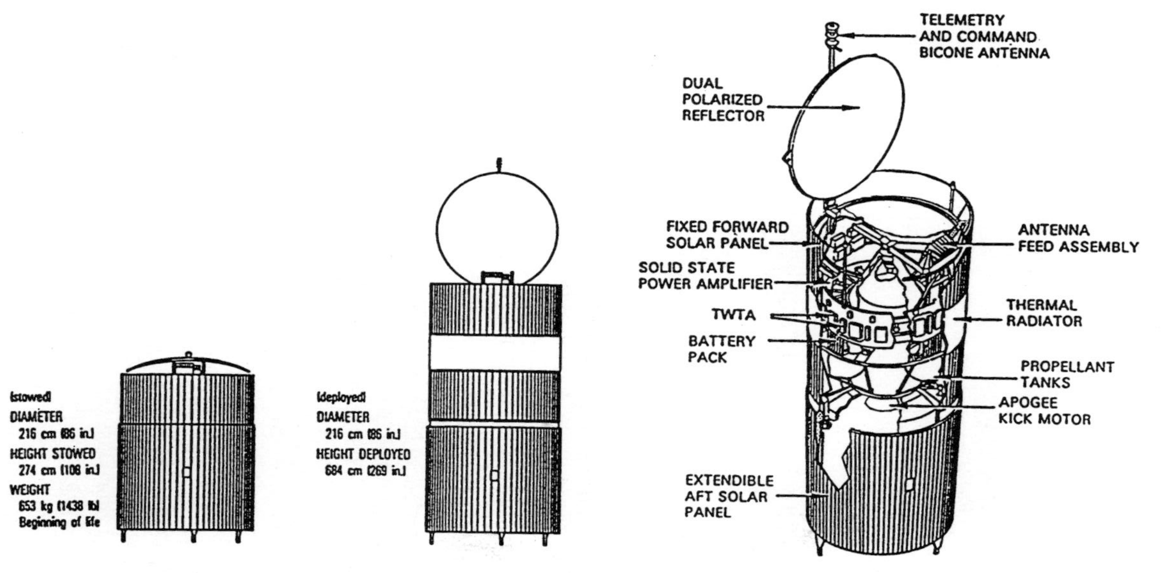

Fig. 10.12 TELSTAR 3. (Reproduced with permission of Boeing Satellite Systems.)

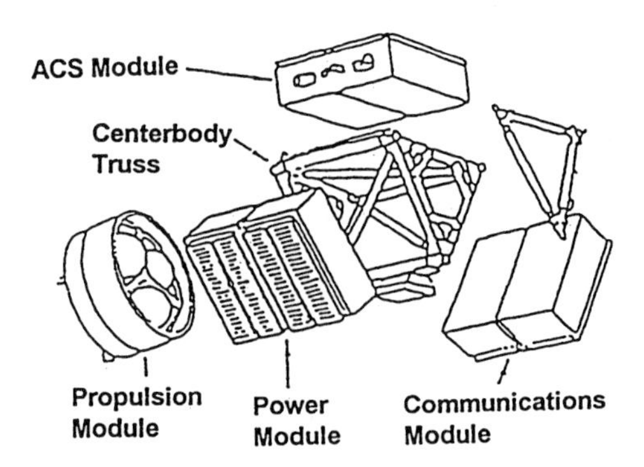

Fig. 10.13 Landsat 4 centerbody structure.[3]

10.2.2.4 Centerbody truss. Landsat 4, shown in, Fig. 10.13, was one of the first spacecraft to use a centerbody truss and attach subsystem modules to the truss. In the Landsat 4 design entire subsystems, power, attitude control, communications, and propulsion were packaged as a single box. Each box was attached to a triangular central truss. The design provides each subsystem excellent access to space, and each subsystem is splendidly accessible. The design set the trend for all future Landsats.

On the downside, subsystems must be designed to a given size and shape. Use of similar equipment from another program is not possible without repackaging. (This problem could be avoided by oversizing the center truss or by designing the center truss late in the design release.) The complexity of the exterior shape makes thermal analysis complicated and less accurate. In addition, the shape makes thermal cavities, hot spots, possible. The advantages of centerbody mounting are 1) very efficient use of weight and volume; 2) light, simple structure; 3) excellent hemispherical field of view from each module; 4) excellent access to the modules; and 5) simple, direct, protected path for power and data cabling.

The disadvantages of centerbody mounting are 1) subsystems must be designed as a module of a given shape, and 2) the complex exterior shape makes thermal hot spots a concern.

10.2.2.5 Cubical box. Many modern spacecraft configurations are a simple, nearly cubical, box. This configuration is driven, in part, by the fact that corner joints are heavy, so the fewer the better. This is particularly true of graphite epoxy structure. The Mars Observer, Mars Global Surveyor, NEAR, global positioning system (GPS), and Star Dust exemplify this trend. Figure 10.14 shows the GPS cubical box gener arrangement.

10.2.3 Example-Preliminary General Arrangement

To make even the most preliminary equipment arrangement, you need the following: 1) payload instruments field of view and size, 2) electronic equipment mass (volume is also desirable but can be estimated), 3) solar-panel area, 4) high-gain antenna size, 5) solid rocket motor size, 6) propulsion tanks sizes,

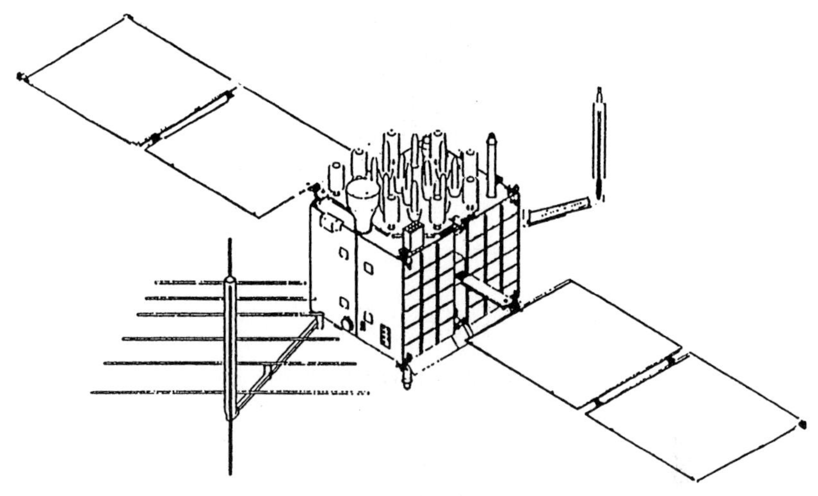

GPS Cubical Box Structure

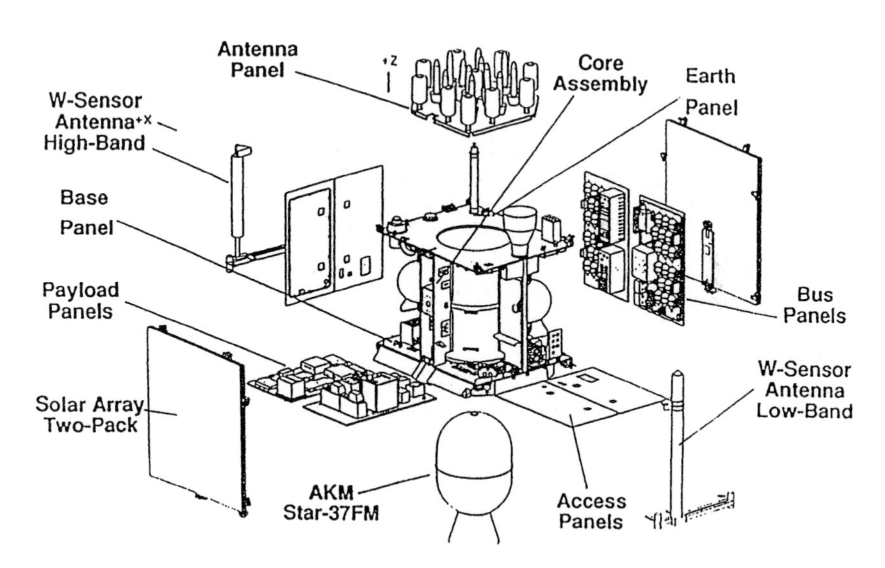

Exploded View

Fig. 10.14 Global positioning system spacecraft. (Reproduced with permission of Lockheed Martin; Ref. 18.)

7) attitude control equipment list and field of view, 8) location of the sun vector and target planet vector, 9) launch vehicle mechanical interface, and 10) launch vehicle shroud dynamic envelope. For example, establish a preliminary general arrangement to meet the following requirements:

1) Launch vehicle interface is 18 equally spaced bolts on a 57.3-in.-diam bolt circle.

2) dynamic envelope 80-in.-diam.

3) The electronics system weighs 600 lb: a) assume subsystem packing density of 20 lb/ft^3, and b) assume subsystem packaging envelope height of 10 in.

4) An 18-in.-diam cold-gas sphere is required for spacecraft attitude control.

5) A 25-in.-diam spherical SRM kick stage is to be used to obtain orbit and then is jettisoned; weight = 500 lb.

6) A 3-ft-diam rigid (not furlable) parabolic antenna is required to have hemispherical field of view on the nadir ($-z$) side of the spacecraft; weight = 20 lb.

7) Solar-array area = 50 ft^2; weight = 20 lb.

8) The payload is a fixed Earth-pointing camera 12 in. in diameter, which has a 5-deg half-angle field of view; weight = 30 lb.

9) A star scanner is used for gyro update.

10) The nadir-pointing spacecraft is to be placed in sun-synchronous Earth orbit.

Estimating the footprint of the electronics compartment,

$$\frac{600}{(20)(0.83)} = 36\,\text{ft}^2$$

Arbitrarily assume a rectangular box with sides 5 ft high; using two sides for electronics produces an equipment plate width of

$$\frac{(36)}{(2)(5)}12 = 43.2\,\text{in.}$$

Thus the two rectangular equipment plates would be 60×44 in. The rectangular equipment box would be $60 \times 44 \times W$, where W can be made as big as the dynamic envelope will allow. The emerging general arrangement is shown in Fig. 10.15 in the stowed configuration. From the requirements it is clear that the nadir face ($-z$) must be occupied by the high-gain antenna and the camera. The solid rocket motor will occupy the $+z$ face. The solar panels can occupy the x or y faces. The solar-panel area can be supplied by two solar panels 2.5×10 ft, which are folded once into 2.5×5 ft packages. The gas sphere can also be installed on the z axis. The star scanner needs a narrow, unobstructed field of view and positive protection from seeing the sun.

Cylindrical structure was arbitrarily chosen to mount the spherical SRM and the attitude control gas sphere. A conical launch vehicle adapter is a common choice. Trusses could be used in these locations. Trusses can be compared to shell structure in a later trade study. Gyros and star scanner are mounted to a common plate, and the ACS and C&DH systems are collocated. The circular launch vehicle interface is shown by a dotted line below the launch vehicle adapter. Two separation planes are planned; one at the upper end of the launch vehicle adapter and another at the SRM girth for use after the SRM burn.

The separation ordinance and propulsion ordnance is well separated from the electronics. If ordnance is chosen for solar-panel deployment, the layout will have to be modified to provide additional separation. Figure 10.16 shows the general arrangement in the deployed configuration.

The solar arrays are mounted on the x axis. Because the orbit is sun synchronous, the sun will be in the y-axis hemisphere; the spacecraft always can bring the sun perpendicular to the solar panels by a single-axis yaw maneuver. The nadir plane

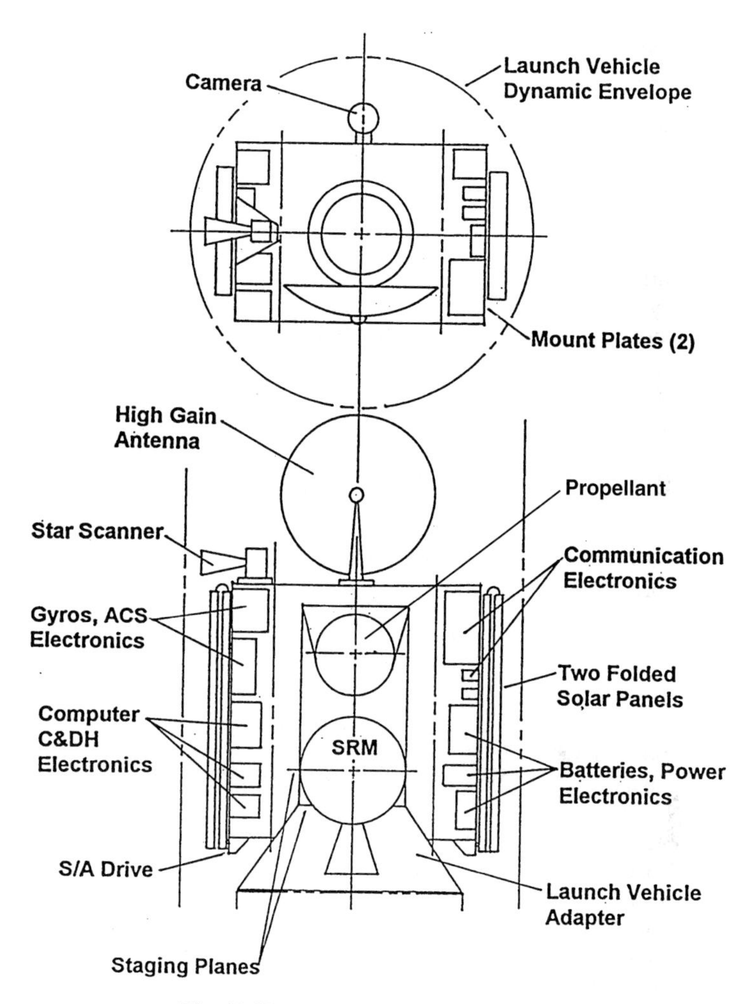

Fig. 10.15 Example spacecraft stowed.

contains the camera and the high-gain antenna. The sun cannot normally shine on the nadir plane, and so the shunt radiator is mounted here.

Here is a quick checklist for evaluating configurations:

1) Science instrument field of view (FOV)
2) Antennas FOV
3) Rocket engine impingement; clear aft of exit plane
4) Ordnance shock path
5) Electronics heat rejection
   a) Free view of space
   b) Sun free face
6) Shunt radiator heat rejection
   a) Free view of space
   b) Sun free face

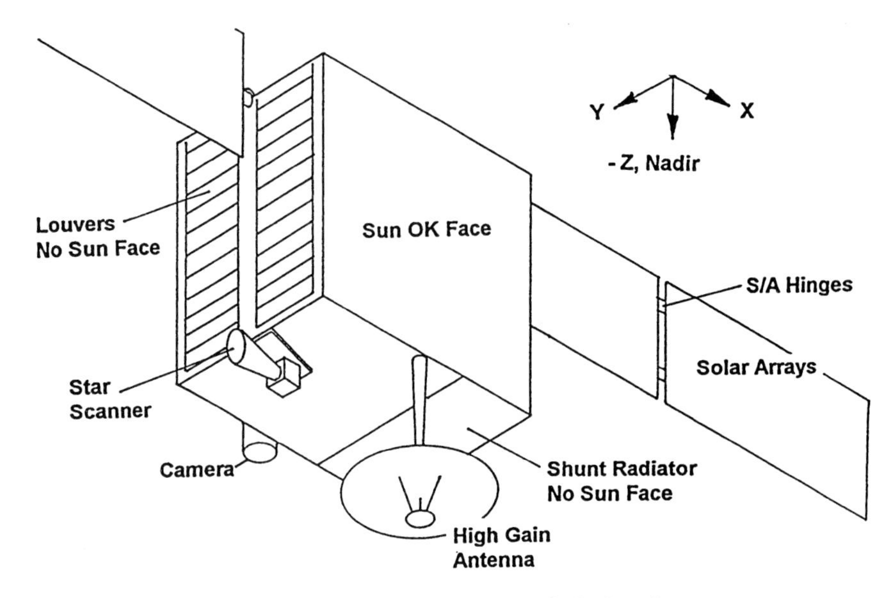

Fig. 10.16 Example spacecraft deployed.

7) ACS sensor FOV
8) ACS sensors and gyros on common plate
9) Propulsion equipment mounted with tanks, modular design possible
10) Maximum moment arm for ACS thrusters
11) Propellant tanks balanced with respect to c.g., at any propellant load
12) Telecommunication electronics near antennas, especially high-gain antenna
13) Sun OK faces picked, sun vector located
14) Sun free faces picked
15) Light sensitive instruments on sun free faces (cameras, star scanners, horizon sensors, spectrometers)
16) Center of mass near geometric center
17) Moments of inertia acceptable (especially spinners)
18) Structure types picked, load paths reasonable
19) Batteries together and near the control electronics
20) C&DH near ACS if common computer used

10.2.4 Preliminary Mass Estimation

Structure and mechanism mass can be estimated accurately after members are analyzed in detail and sized. In the early phases before structural analysis has even started, it is necessary to estimate structural mass by using relationships based on prior programs. For early mass estimates use the initial general arrangement drawing. Divide the structure into major assemblies. Use the following mass estimating relationships to estimate the mass of each:

10.2.4.1 Primary load shell structure. Use 5 to 15 kg/m^2 (4 to 11 kg/m^2 for composites) depending on the type of construction, material, factor of safety, static or dynamic loading, and difficulty of fabrication.

10.2.4.2 Lightly loaded fairing. Use 1 to 4 kg/m^2 (1 to 3 kg/m^2 for composites) depending on material minimum gauge, vibration loading, and attachment difficulty.

10.2.4.3 Equipment support structure. Use 15% of equipment mass.

10.2.4.4 Launch vehicle adapter. Use the relation developed in Chapter 2.

10.2.4.5 Solar-array substrate. See Chapter 6.
Total the mass of structure assemblies, and multiply by a factor of 1.2 to allow for tolerances, joints, thermal coatings, and the like. Note that 1 $lb/ft^2 = 4.88$ kg/m^2. The structure mass is typically 20–30% of the spacecraft dry mass.

10.2.5 Structure Types

The structural concept, and preliminary selection of structure type, flows from the general arrangement. Different structure types may be chosen for each major element of the spacecraft. Figure 10.17 shows some of the common types.

Skin-stringer construction is very frequently used for the central bus structure; graphite epoxy composites are an emerging technology that offers lighter weight for this application. Trusses are used in point load situations, typically propulsion system mounting and launch vehicle interface structure. Honeycomb is used for stiff, large surface area situations like solar panels and large antennas.

10.3 Structure Design

Spacecraft are subjected to various environments during their fabrication, transportation and handling, testing, and service life. This section defines these various environments and the criteria for combination of the loads induced by those environments.

10.3.1 Environments

10.3.1.1 Preliminary loads. For conceptual and preliminary design to locate structural members and approximately size them, the applied loads to the spacecraft must be estimated. Launch vehicles and upper stages have developed structural capabilities for payload launch weight or "throw" weight. Knowing the propulsive capability at different stages and knowing the vehicle weight at launch gives a measure of the acceleration environment to which the spacecraft will be exposed, as

$$\text{Steady-state acceleration} = \text{Propulsive force/Mass} \tag{10.1}$$

This acceleration is most commonly expressed in g, where $1g$ = Earth's gravitational acceleration = 32.2 ft/s^2. This g loading is commonly referred to as a load

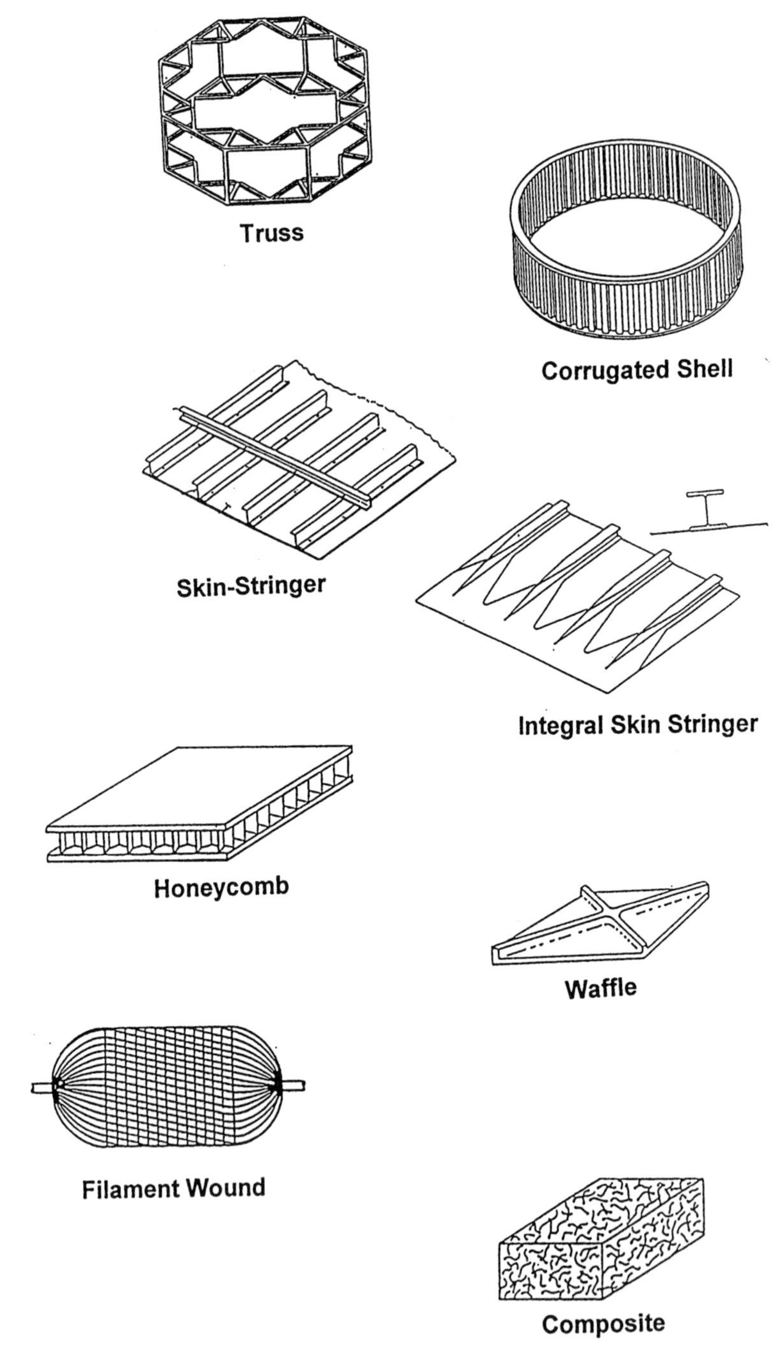

Fig. 10.17 Common types of spacecraft structure.

factor = −acceleration. Thus a 1,000,000-lb launch thrust acting on a 800,000-lb launch vehicle will give a steady-state acceleration of $1.25g$. At stage 3 ignition the 200,000-lb thrust will act on the remaining 100,000-lb vehicle with a steady-state acceleration of $2g$.

In addition to the steady-state components, the transient loads exciting the lower vehicle modes will cause large spacecraft loads. Transients can also be expressed as load factors. Transient load estimates in the early stages of design are derived from test data and previous flight data. The third stage ignition just mentioned may show a transient peak of $3g$ at the motor-to-vehicle interface. Although structural damping could lower the acceleration response of this transient on the vehicle mass, the transient could also excite certain vehicle modes, depending on how close the frequency of the vehicle modes are to the frequencies of the motor and the transient forcing function. In some cases this could result in a significant amplification as the vehicle resonates.

These load amplifications must be calculated, and the best way to determine this effect is to calculate the vehicle modal frequencies (eigenvalues) as well as the shape of the vehicle as it deforms in the modes (eigenvectors). Although it is possible to perform some analyses without computers, the availability of main-frames to solve many simultaneous equations makes them ideal for calculating structural responses. This is because of the advent of numerical structural analysis in the form of finite elements or finite differences. Finite difference methods work with the equilibrium equations in differential form, and finite differences are substituted for the derivatives that appear. Finite element analysis works with an integral form usually based on the principle of minimum potential energy. Finite element analysis is the most common approach and the best suited for complex shapes. Only finite element analysis will be discussed here.

Any structure can be mathematically divided into smaller pieces that lend themselves to straightforward solution. These pieces are called finite elements. Each finite element consists of mass and stiffness terms that attach to grid points or nodes. The mass and stiffness terms represent the mass and stiffness of the real structure, which can be modeled using standard finite elements such as beams and plates, or by more advanced terms. The equations of statics or dynamics are written at each of these nodes in each direction; however, many springs are attached to those nodes. Generally there are six degrees of freedom at each node, three translation and three rotation. At some of the nodes, degrees of freedom may be eliminated, such as when a piece of structure is attached to ground and is no longer allowed to move. All of the equations are then assembled in a matrix and solved, the solution giving magnitudes of deflection at each node, load or stress in each spring, interface loads, etc.

In general, the smaller the pieces the more accurate the analysis is. The stiffness and mass matrices can also be solved for their roots, which are the eigenvalues.

Statics:

$$F = kx \tag{10.2}$$

Dynamics:

$$F(t) = ma + cv + kx \tag{10.3}$$

Eigenvalues:

$$\text{Roots of } (k - \lambda m) \tag{10.4}$$

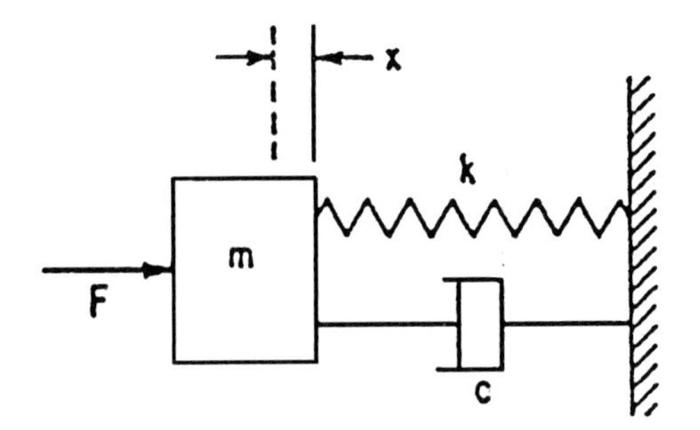

Fig. 10.18 Vibrating structure.

10.3.1.2 Vibration. All structure can be converted into the terms k, c, and m (see Fig. 10.18). The damping term c is usually given a constant value for all structure. This leaves only k and m as variables. If $c = 0$, the frequency of a single-mass single-spring system is

$$f_n = \frac{1}{2\pi}\left(\frac{k}{m}\right)^{0.5} \tag{10.5}$$

The frequencies of many other systems can be calculated exactly, and some are shown on the following pages. Without damping, the system would vibrate forever under any external force input. Damping can vary widely depending on material type and method of attachment. Normal spacecraft structures use 1–2% of critical damping. This is viscous damping and is a function of velocity. Another type of damping is coulomb damping, which is a function of friction.

For most structural elements the primary contribution to the critical design load will be from steady-state and transient loads. Structure that supports other hardware such as electronic boxes will often not have modes less than 50 Hz. In this case random vibration is usually the highest load. The hardware can be analyzed by driving its interface with a random vibration acceleration input, or an equivalent analytical method can be developed using power spectral density (PSD) as a function of response frequency. The random vibration power input to most spacecraft peaks in the 100- to 500-Hz range, as shown in Fig. 10.19.

By using Mile's equation, an equivalent static load factor can be developed:

$$LF = \sigma\left[\left(\frac{\pi}{2}\right)(PSD)\,Q\,f_n\right]^{0.5} \tag{10.6}$$

where

LF = equivalent static load factor
σ = three standard deviations
PSD = power spectral density, G^2/Hz level at the natural frequency of the part
Q = amplification factor = $\frac{1}{2}\zeta$
ζ = % critical damping
f_n = natural frequency.

Sound transmitted energy, or acoustics, acts on all structure that has surface area. The more surface area and the lighter the structure, the more it will respond to

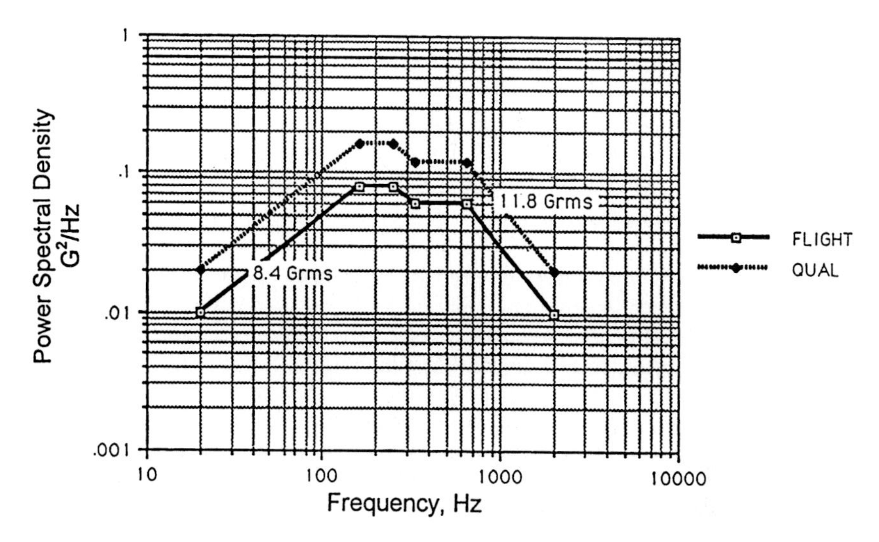

Fig. 10.19 Vibration power input.

acoustic input. Examples are solar arrays, antennas, and large sheet metal surfaces. Analytical methods of calculating stresses in structure caused by acoustics are less well understood than transients, but some calculations can be made assuming the air pressure takes the shape of the structural modes.

10.3.1.3 Load verification. As the design matures, the models also mature. After the design is complete and most flight structure is manufactured, the spacecraft is tested to determine the eigenvalues, eigenvectors, and damping. This is performed in a modal survey test. The structure is excited at various locations with low-level random or sine vibration input. The structural response is measured at many locations with accelerometers. When the data are recorded, the accelerometers will respond most when the frequency of the input vibration reaches a structural mode (resonance frequency). The mode shapes can be determined by mapping the response of the different accelerometers. Once the modal test is complete, the finite element model can be updated, and the loads on the structure are verified to be safe for flight.

Transient loads are verified to be safe for flight by performing load cycle analyses. For most of the structure, transients will be the biggest contributor to total load. A description of forces with respect to time is available for all boosters. When these forcing functions are applied to the spacecraft model (particularly when the spacecraft model is mathematically coupled to the rest of the flight vehicle), loads and stresses in finite elements can be obtained for each increment of time. This analysis is called a *loads analysis cycle*. The final verification that these loads are safe for flight is to perform a static-proof load test.

Random vibration testing is considered essential for all of flight hardware electronics boxes and mechanisms. Testing verifies not only that the hardware will survive flight, but is often increased in intensity to verify workmanship of the hardware.

Virtually all spacecraft undergo an acoustic test with all of the flight hardware installed. Acoustic tests are controlled by computers and monitored by the loads analyst using microphones and accelerometers. Acoustic energy is logarithmic and doubles with each 6 dB. Acoustic tests are usually run with an overall energy level of 142 to 148 dB.

10.3.1.4 Fabrication environment. Almost all structural parts experience some loading conditions during the manufacturing and assembly process. Knowledge of the fabrication process is required to understand if these loading conditions are significant. The fabrication environment for flight hardware will consider the combined effects of the fabrication process together with all other loading environments existing simultaneous with the fabrication event. Some typical fabrication environment examples and recommended practices are listed in Table 10.1.

If stress relief is not provided, known residual stresses must be accounted for in the analysis of subsequent loading conditions. Aerospace structures are nominally very thin, and it is not unusual for panels, dishes, or beams to require added support to preclude their collapse under their own weight, especially at intermediate stages of their assembly.

Table 10.1 Fabrication environment

Structure	Manufacturing load or problem	Recommended practice
Thin metallic panels	Cracks formed when machining thin webs or pockets	Choose ductile materials, design for proper edge distances, design for recommended taper and fillet limits
Thick metallic panels	Warping with the release of internal residual stresses	Choose ductile materials, design for proper edge distance, add stress relief
Thick metallic parts	Nonuniform heat treatment	Less than maximum strength properties must be considered in analysis
Sheet metal	Material yield during forming process	Design for allowable bend radii or stamping criteria for given material
Metal tubing	Material yield during bending process	Design for allowable bend radii for given material
Weldments	Residual stress	Develop weld schedule early, give proper attention to weld tooling, add stress relief
Weldments	Reduced strength in heat affected zones	Provide postweld heat treat, account for reduced properties in analysis (in some cases, like welding tubes of certain alloys, the heat affected zone may extend the entire length of the tube)
Panels, webs	Cracks formed when drilling or tapping holes	Choose ductile materials, select compatible drill bit, design for proper edge distance

10.3.1.5 Handling and transport environment. Spacecraft handling can occur at any level of spacecraft assembly from component to payload. Sources of ground handling loads are hoisting and lifting, jacking, towing and moving, braking, and ground, air or sea transport.

Handling and transport loads for flight hardware must consider all loading existing simultaneous with the handling or transport event. The specific handling configuration must be considered as well as mass of protective covers, instrumentation, and ancillary equipment. It is good practice to make certain that handling and transport loads do not constitute the critical design condition.

10.3.1.6 Liftoff/ascent environment. The maximum predicted launch acceleration is determined from the combined effects of quasi-static acceleration and the transient response of the vehicle as a result of engine ignition, air loads, engine burnout, and stage separation. Theses loads are unique to the ascent system and the dynamic interaction between the ascent vehicle and the spacecraft.

Quasi-static limit loads and accelerations are specified in the payload/launch vehicle interface agreements. These launch vehicle load factors (LLF) are typically defined in the longitudinal (thrust axis) and lateral directions. These load factors must be verified by coupled loads analysis.

Portions of the liftoff/ascent phase might also subject the spacecraft to a spin environment. For components lying away from the spin axis, the radial loads from spinning might be larger than the lateral loads.

10.3.1.7 Vibration environments. The random vibration environment is caused by mechanically transmitted vibration from impingement of rocket engine and aerodynamically generated acoustic fields on the launch vehicle as well as mechanically transmitted structure-borne rocket engine vibrations. The maximum predicted random vibration environment is specified as an acceleration spectral density (ASD; often called a power spectral density), based on a frequency resolution of one-third octave or narrowband analysis over a frequency range of 20 to 2000 Hz.

The sine vibration environment is caused by mechanically transmitted vibration from launch vehicle structure ringing. This environment is usually an acceleration input at a single frequency. The maximum predicted sine vibration environment is specified as g over the frequency range of the input.

10.3.1.8 Acoustic environment. The maximum predicted acoustic environment is the extreme value of fluctuating pressure occurring, with varying degree, on all surfaces of launch and space vehicles. This environment occurs during liftoff, powered flight, and reentry. The maximum predicted acoustic environment is specified as a sound pressure level (SPL) typically based on one-third octave bands over a frequency range of 31.5 to 10,000 Hz. An example of the acoustic level inside the fairing of a typical launch vehicle is shown in Fig. 10.20.

For structures that have significant high-frequency structural resonance and are acoustically responsive (that is, solar arrays and large antennas), it may be necessary to combine acoustically induced loads with structure-borne random vibration loads to produce the total vibroacoustic loads.

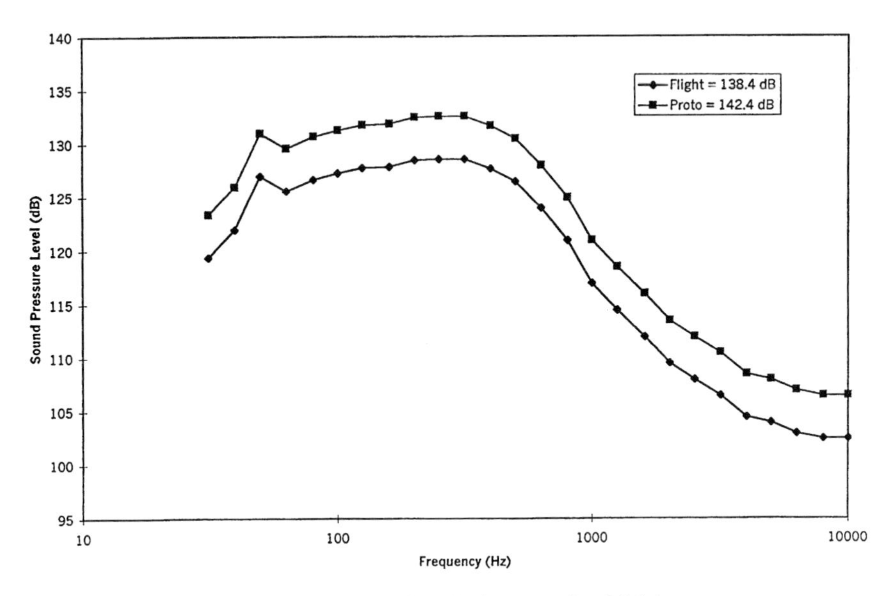

Fig. 10.20 Typical SPL inside a payload fairing.

10.3.1.9 Pyro shock. The pyrotechnic shock environment imposed on the flight structure and components is caused by structural response when the spacecraft or launch vehicle electro-explosive devices are activated. Resultant structural response accelerations resemble the form of superimposed complex decaying sinusoids that decay to a few percent of their maximum acceleration in 5 to 15 ms. The maximum predicted pyrotechnic shock environment is specified as the maximax absolute shock response spectrum as determined by the response of a number of lightly damped single-degree-of-freedom systems using a dynamic amplification factor Q of 10 at the resonant frequency. This shock response spectrum is determined at frequency intervals of one-sixth octave band or less over a frequency range of 100 to 10,000 Hz. A typical shock response is shown in Fig. 10.21.

10.3.1.10 Ambient temperature. In-flight temperatures are discussed in Chapter 7. In general, ground ambient temperatures are controlled as necessary to ensure that they do not drive the design of space hardware. Typical ground handling temperature limits are as follows: storage, 50 to 80°F; factory, 67 to 77°F; transportation, 60 to 80°F; launch site, 25 to 100°F.

10.3.2 Materials

Based on the specific requirements for the structural element under design consideration, one material type may have advantages over another. By far the most common material used on spacecraft is aluminum. It is found in many alloys and tempers, it is lightweight, easy to machine, weld, and handle, and offers significant structural advantages over other materials. Alloy steels, stainless steels, nickel-based alloys, magnesium, beryllium, titanium, reinforced plastics, and composites are almost always found on spacecraft as structural elements.

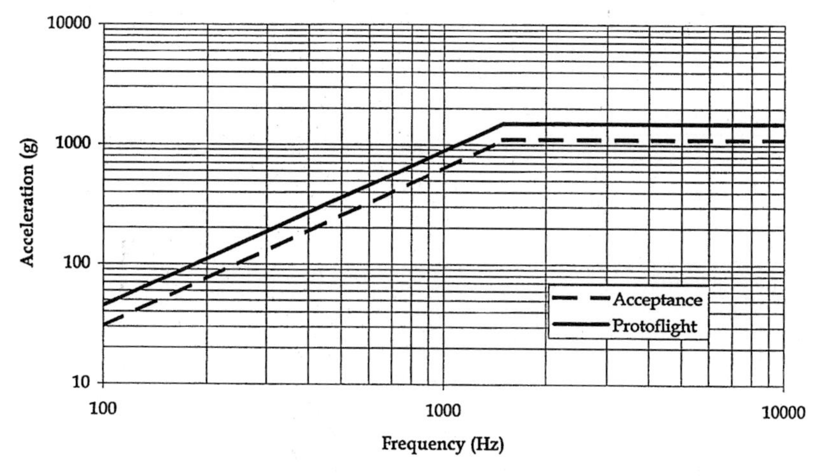

Fig. 10.21 Typical pyro-shock spectrum.

Material properties of interest to the designer are listed here:

F_{tu} = ultimate tensile strength
F_{ty} = yield tensile strength (yield is usually determined using the 0.002 in./in. (0.2%) offset method)
F_{cy} = yield compressive strength
F_{bru} = ultimate bearing strength, a uniaxial allowable derived from two-dimensional test data, also called F_{bry}
F_{bu} = ultimate bending strength, takes advantage of the nonlinear nature of stress vs strain curves and is cross-section dependent, also called F_{by}
F_s = ultimate shear strength
ε = strain at failure, gives a measure of ductility (toughness)
α = coefficient of thermal expansion
C = coefficient of thermal conductivity
E = modulus of elasticity or Young's modulus, average ratio of stress to strain below the elastic limit
G = shear modulus or modulus of rigidity
μ, ν = Poisson's ratio
ρ = density
K = Bulk modulus, ratio of normal stress, applied to all six sides of a cube, to the change in volume.
K_{ic} = fracture toughness

Other properties to consider are machinability, weldability, corrosion resistance, and hardness. Material properties can be obtained from Military Specifications, primarily MIL-HDBK-5, *Metallic Materials and Elements for Aerospace Vehicle Structures,*[4] or from industry standards. Composite material properties are given in MIL-HDBK-17, *Advanced Composites Design Guide.*[5] Table 10.2 summarizes the properties of the most common materials.

Table 10.2 shows representative properties from MIL-HDBK-5 for educational purposes; for design consult the latest revision of MIL-HDBK-5[4] for the material

Table 10.2 Structural material properties

Material	Form	Density, lb/in.^{3}	F_{tu}, ksi	F_{ty}, ksi	E, 10^3 ksi	G, 10^3 ksi
Aluminum sheet						
	2014-T6	0.101	65	58	10.5	4.0
	2024-T3	0.101	64	47	10.5	4.0
	6061-T6	0.093	42	36	9.9	3.8
	7075-T6	0.101	76	67	10.3	3.9
Beryllium						
	Sheet	0.067	65	42	42.5	20
	Hot Pressed	0.066	40	30	42	20
Magnesium						
Sheet	AZ31B-O	0.0639	32	18	6.5	2.4
Extrusion	AZ31B-F	0.0639	35	21	6.5	2.4
Stainless steel						
	17-7PH	0.276	177	150	29	11.5
Titanium						
Forged	6Al-4V	0.160	135	125	16	6.2
Sheet	6Al-4V	0.160	134	126	16	6.2

specification, treatment, and thickness you are using. The sheet properties shown are for the thinner gauges.

Materials are most often selected based on their physical properties with respect to weight. In particular, the ratios E/ρ (stiffness) and F_{tu}/ρ (strength) are the most common. Figure 10.22 compares common structural materials on the basis of strength/weight and stiffness/weight.

The metals steel, titanium, aluminum, and magnesium all have approximately the same stiffness ratio. If the structural element's stiffness is linear with thickness, as in an axial strut member, none of the four materials provides an advantage over the other. If the stiffness is linear with thickness squared or cubed, as in a beam element, lighter materials, such as aluminum, offer significant weight savings.

Because of its corrosion susceptibility, magnesium is rarely used. Other materials such as beryllium and composites offer E/ρ ratios well in excess of most other metals, but may be eliminated because of cost or toxicity problems. The F_{tu}/ρ ratio can vary a lot, not only from material to material, but also as a function of alloy elements, heat treat, and cold work. Few steels are three times as strong as aluminum, although aluminum is only one-third the density of steel. Titanium, on the other hand, can be three times as strong as aluminum with only 1.6 times its density.

Composites can be specifically tailored to obtain the particular mechanical properties required. Table 10.3 summarizes the properties of two typical composite materials.

Preliminary loads and stress analysis are performed in the preliminary design phase. Rough calculations of mass properties and member sizes are required.

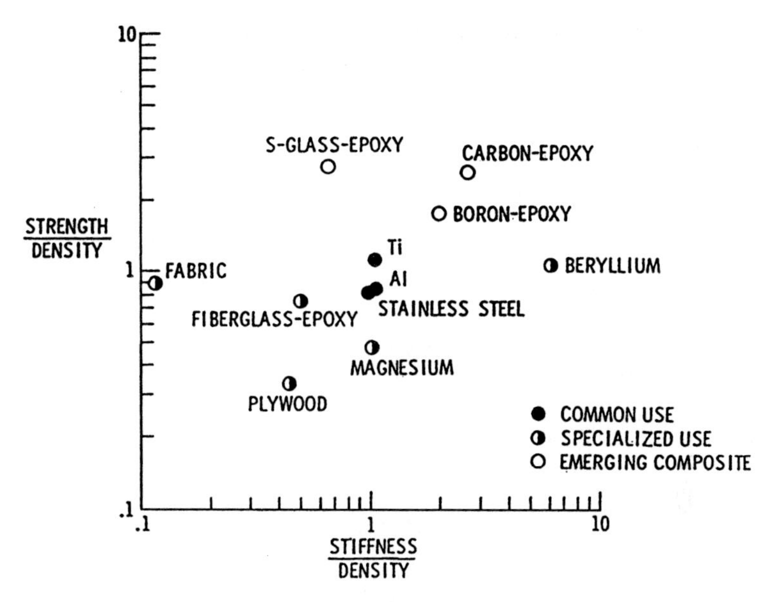

Fig. 10.22 Strength and stiffness comparison. (Copyright AIAA, reproduced with permission; Ref. 6.)

Table 10.3 Typical composite materials

Material property	M55J	K13C2U
Elastic modulus (MSI)—E_{11}	34.00	60.0
Elastic modulus (MSI)—E_{22}	0.735	0.618
Elastic modulus (MSI)—E_{33}	0.735	0.618
Poison ratio—ν_{12}	0.35	0.34
Poison ratio—ν_{23}	0.001	0.001
Tension		
Ultimate strength (KSI)—UTS_{11}	250	254
Ultimate strength (KSI)—UTS_{22}	2.8	2.4
Ultimate strength (KSI)—UTS_{33}	2.8	2.4
Compression		
Ultimate strength (KSI)—UTS_{11}	88.2	44.8
Ultimate strength (KSI)—UTS_{22}	12.6	16.8
Ultimate strength (KSI)—UTS_{33}	12.6	16.8
Shear		
Elastic modulus (MSI)—G_{12}	0.517	0.390
Ultimate strength (KSI)—USS_{12}	8.9	4.8
CTE		
CTE_x	−1.0 μin./in./°C	−0.69 μin./in./°C
CTE_y*	29 μin./in.°C	29 μin./in./°C
CTE_y*	29 μin./in.°C	29 μin./in./°C

Some preliminary testing of hardware can be performed to verify its structural or functional adequacy, for example, fracture toughness testing of thick plates and structural bond strength testing of composites.

10.3.3 Loads Analysis

Limit loads and pressures, which are the maximums expected to act on a structure throughout its service life, are determined for 1) the basis of structural design and 2) as the basis for the loads to be applied in structural proof, ultimate, and life strength testing. Limit loads are determined for all structural elements and for every potentially critical loading condition likely to be encountered in the life of the structure. This section defines the loads combination process for determining structural limit loads.

Limit loads are defined by load environment, loadpath, mass properties, and load uncertainty factors. This interrelation is shown in Fig. 10.23.

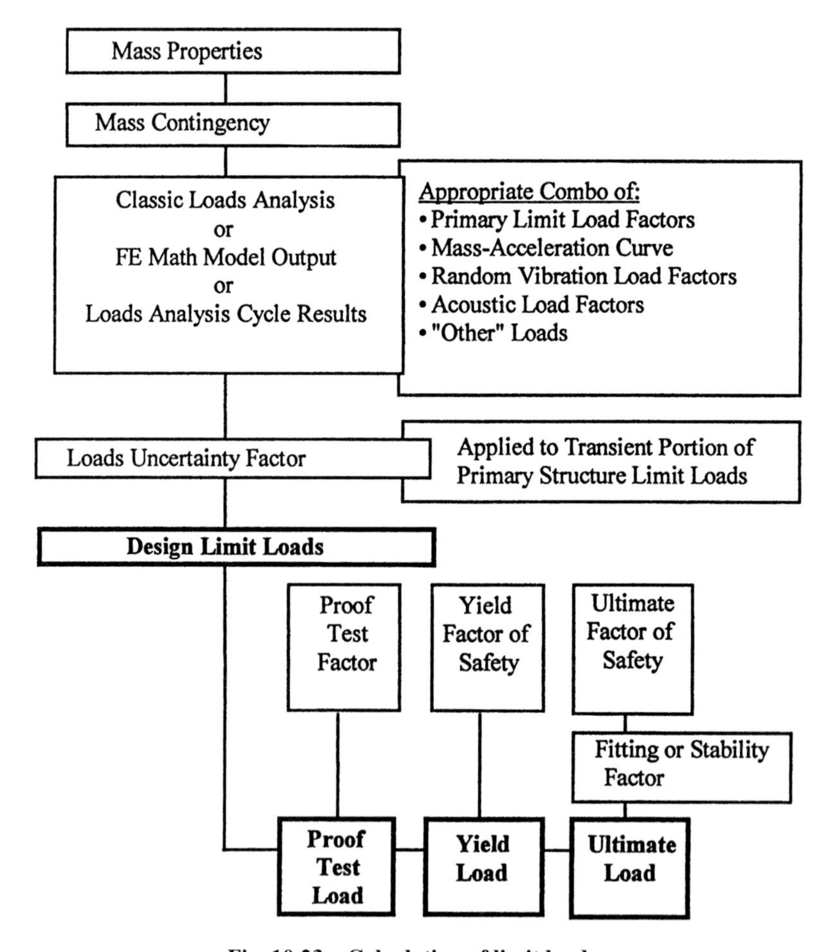

Fig. 10.23 Calculation of limit loads.

Table 10.4 Recommended loads uncertainty factors

Factor	Starting point	Ending point
1.40	Start of program	Sufficient development testing or CDR level design maturity
1.25	Development testing or CDR level design maturity	Correlation of finite element model with flight hardware static or modal test
1.10	Correlation of finite element model	Launch

10.3.3.1 Loads uncertainty factor. A loads uncertainty factor is applied to the primary structure's low-frequency transient loads to compensate for unknowns pertaining to the coupling effects of a launch vehicle and payload. These factors are defined in Table 10.4.

10.3.3.2 Mass properties. The application of the worst-case combination of load factors to the design mass properties constitutes the predicted load condition. Design mass properties should include contingency mass and unallocated margin in a worst-case manner for loads and or frequency.

10.3.3.3 Primary structure loads. Primary structure is defined as any structure, the failure of which would result in the general failure of the spacecraft. For example, the launch vehicle adapter is a primary structure. Primary structure is analyzed for liftoff/ascent environment plus the loads from supported mass. The lateral loads are applied simultaneously with the axial loads in all possible sign combinations. Lateral loads are applied in the worst-case direction.

10.3.3.4 Secondary structure loads. Secondary structure is defined structure, the failure of which would not result in failure of the mission nor in substantial limit of objectives. Secondary structure is loaded mainly by vibro-acoustic loads. In preliminary design secondary structure can be analyzed in two major ways as shown in Fig. 10.24.

10.3.3.5 Option 1—Mass acceleration curve. A mass acceleration curve (MAC) represents the upper bound on acceleration for a given component/structure mass. The MAC curve is presented as g as a function of item mass (physical MAC) or modal mass (modal MAC) and is typically developed from load cycle data and models for a particular launch vehicle/upper-stage combination and a payload "similar" to the payload being designed.

The physical MAC load factor is generally applied in a single worst-case axis of the item being analyzed. The launch vehicle thrust is added to the MAC load factor. The physical MAC load factor should also be combined with any other loads (pressure, thermal, misalignment, etc.) present. When using the physical MAC, the support structure should be checked at its own weight plus the supported weight at the predicted acceleration. The modal MAC (Fig. 10.25) is generally

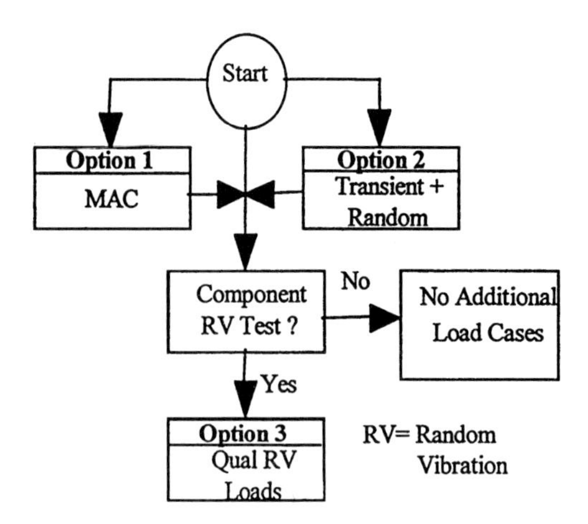

Fig. 10.24 Preliminary structural analysis options.

applied to a spacecraft or component modal model in each of the three orthogonal axes and root sum squared with the modal portion. That result is added to the steady-state loads in the launch vehicle thrust (axial) direction. The modal MAC loads should also be combined with other loads (pressure, thermal, misalignment, etc.).

Figure 10.25 also shows that the MAC is usually conservative compared to the verification load cycle (VLAC).

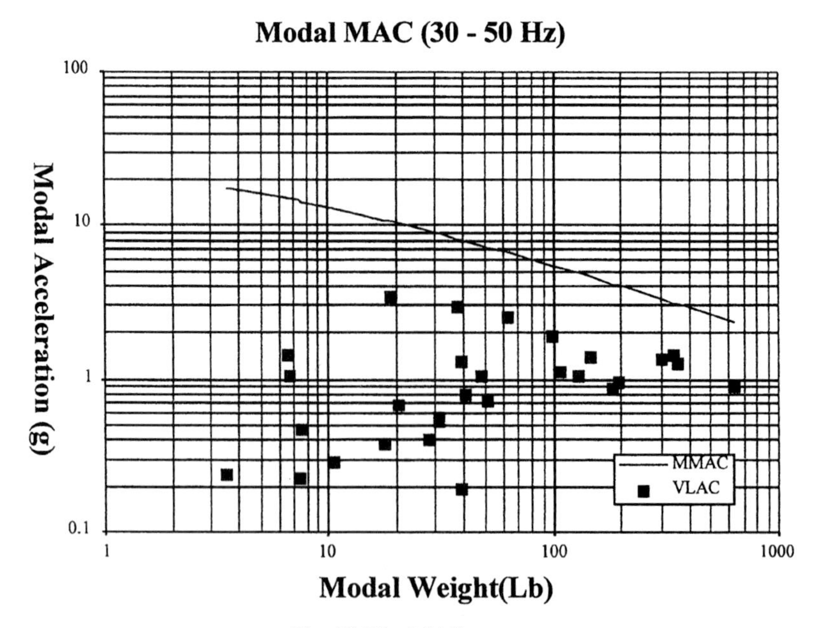

Fig. 10.25 MAC curve.

Table 10.5 Recommended loads uncertainty factors

Case	Axial	Lateral (axis 1)	Lateral (axis 2)
1	$\pm\sqrt{(S_{Ax} \pm T_{Ax})^2 + R_{Ax}^2}$ $\pm O_{Ax}$	$\pm S_{Lat1} \pm T_{Lat1} \pm O_{Lat1}$	$\pm S_{Lat2} \pm T_{Lat2} \pm O_{Lat2}$
2	$\pm S_{Ax} \pm T_{Ax} \pm O_{Ax}$	$\pm\sqrt{(S_{Lat1} \pm T_{Lat1})^2 + R_{Lat1}^2}$ $\pm O_{Lat1}$	$\pm S_{Lat2} \pm T_{Lat2} \pm O_{Lat2}$
3	$\pm S_{Ax} \pm T_{Ax} \pm O_{Ax}$	$\pm S_{Lat1} \pm T_{Lat1} \pm O_{Lat1}$	$\pm\sqrt{(S_{Lat2} \pm T_{Lat2})^2 + R_{Lat2}^2}$ $\pm O_{Lat2}$

10.3.3.6 Option 2—Transient plus random vibration loads. Acceptance random vibration loads or load factors can be combined with steady-state loads or load factors in a rss fashion, as shown in Table 10.5. To completely analyze the hardware, all three cases in Table 10.5 must be analyzed. These loads need to be applied in all possible combinations with the lateral loads to generate the worst-case stresses. Depending on model fidelity, the portion of the load from the launch vehicle steady state and transient environment can be superseded by load cycle results. For secondary structure a maximum cutoff of three times the overall spectrum G_{RMS} is acceptable for computing the random vibration loads or load factors.

Where S_i is the launch steady-state load or load factor; T_i the launch transient load or load factor; R_i the acceptance level random vibration load or load factor from Mile's equation or equivalent (may be limited to three times the overall spectrum G_{RMS}); and O_i the other load or load factor (thermal, pressure, misalignment, etc.).

10.3.3.7 Tertiary structure loads. Tertiary structure is defined as cantilevered components and the support structure (if any) for these components. Electronics boards are also classified as tertiary structure. Tertiary structure is loaded mainly by vibroacoustic loads. In a manner similar to secondary structure, tertiary structure will be analyzed per one of the options shown in Fig. 10.24. The main difference in the analysis of secondary and tertiary structure is in the treatment of random vibration loading. For tertiary structure no cutoff value is permitted for computing random vibration loads in option 2.

10.3.3.8 Acoustically excited structure. Large surface area, lightweight structures are susceptible to excitation as a result of acoustic energy impingement. For preliminary analysis of these structures, launch loads are combined with acoustic loads as shown in Table 10.6.

These load factors need to be applied in all possible sign combinations with the lateral loads applied in the direction to generate the worst-case stresses. Depending on model fidelity, the portion of the load from the launch vehicle steady state and transient environment can be superseded by load cycle results.

Where S_i is the launch steady-state load or load factor; T_i the launch transient load or load factor; A_i the acceptance level acoustic load or load factor; and O_i the other load or load factor (thermal, pressure, misalignment, etc.).

Table 10.6 Acoustically excited structure

Case	Axial	Lateral (axis 1)	Lateral (axis 2)
1	$\pm\sqrt{(S_{\mathrm{Ax}} \pm T_{\mathrm{Ax}})^2 + A_{\mathrm{Ax}}^2}$ $\pm O_{\mathrm{Ax}}$	$\pm S_{\mathrm{Lat1}} \pm T_{\mathrm{Lat1}} \pm O_{\mathrm{Lat1}}$	$\pm S_{\mathrm{Lat2}} \pm T_{\mathrm{Lat2}} \pm O_{\mathrm{Lat2}}$
2	$\pm S_{\mathrm{Ax}} \pm T_{\mathrm{Ax}} \pm O_{\mathrm{Ax}}$	$\pm\sqrt{(S_{\mathrm{Lat1}} \pm T_{\mathrm{Lat1}})^2 + A_{\mathrm{Lat1}}^2}$ $\pm O_{\mathrm{Lat1}}$	$\pm S_{\mathrm{Lat2}} \pm T_{\mathrm{Lat2}} \pm O_{\mathrm{Lat2}}$
3	$\pm S_{\mathrm{Ax}} \pm T_{\mathrm{Ax}} \pm O_{\mathrm{Ax}}$	$\pm S_{\mathrm{Lat1}} \pm T_{\mathrm{Lat1}} \pm O_{\mathrm{Lat1}}$	$\pm\sqrt{(S_{\mathrm{Lat2}} \pm T_{\mathrm{Lat2}})^2 + A_{\mathrm{Lat2}}^2}$ $\pm O_{\mathrm{Lat2}}$

10.3.3.9 Internally pressurized structure. An internally pressurized structure must sustain the combination of pressure and load shown in Table 10.7.

10.3.3.10 Pressurized vessels. Internally pressurized vessels will sustain the combination of pressure and load shown in Table 10.8.

10.3.3.11 Pressurized components and lines. Internally pressurized components and lines must sustain the combination of pressure and load shown in Table 10.9.

10.3.4 Stress Analysis

Virtually all failure criteria for structures can be thought of in terms of deformation. For instance, crushing, buckling, crack growth, stretching, bending, twisting, and bursting can all be thought of as geometric deformations that exceed some acceptable limit. For example, stretching beyond a materials elastic limit might occur so that it remains plastically deformed. If this is considered unacceptable to the mission, the structure will have failed and must be redesigned. Sometimes elastic deformations can be excessive. Most modern buildings and bridges are designed to limit elastic deflection. Displacement is the critical failure condition for

Table 10.7 Internally pressurized structure

Case	Applied pressure	Applied load	Accept test	Proto/qual test	Comment
1	Burst	None	——	Burst press	Dedicated qual unit or qual by similarity Burst test includes cycle requirement
2	Proof	Yield	Proof pressure	Proof press and load	——
3	Max pressure	Ultimate	——	——	Pressure consistent with load environment; may be multiple cases

Table 10.8 Pressurized vessels

Case	Applied pressure	Applied load	Accept test	Proto/qual test	Comment
1	Burst	None	——	Burst press	Dedicated qual unit or qual by similarity Burst test shall include cycle requirement
2	Proof	Yield	Proof pressure	Proof press and load	——
3	Max pressure	Ultimate	——	——	Pressure consistent with load environment; may be multiple cases

floors, which cannot bend too much before the occupants "feel" uncomfortable, and for tall buildings, which dynamically sway in the wind and can make occupants nauseous. Similarly, spacecraft structures must limit elastic deformation so that parts do not bump together either as a result of mechanical force or thermal deformation. The most critical condition for many modern spacecraft designs is to make sure structural frequencies are above a certain threshold. Below this value structural resonances can occur, which can cause extremely large loads and destruction of hardware. Stiff designs prevent resonances with launch vehicle forcing functions, but stiff structures get heavy. Another option is load isolation systems are designed to have one fundamental frequency, which is designed to be separate from all load frequencies. This can dramatically reduce loads, but also increases deflections.

Once the geometric deformation criteria are understood, the applied loads are related to the failure criteria. This cannot be done directly except by test. By analysis, the deformation is converted to strain in the materials using the part geometries. For instance, in a column, the deflection on a column of length l causes a strain d/l. Similarly, in a beam, the deflection is measured as curvature or ρ, and the strain is y/ρ, where y is the original depth of the beam (independent of length). But in a beam it is often easier to calculate the stretch or contraction in the extreme fibers (top or bottom) and compare that to the original length to calculate strain. This approach works for pressure vessels and other structures as well.

By analysis, strains can be converted into stresses using the elastic/plastic material properties such as Young's modulus. For most materials, $stress = E^{*}strain$, where E = Young's modulus, which is the slope of the linearly elastic portion of a material's stress/strain curve. For advanced composite materials that are brittle,

Table 10.9 Pressurized components and lines

Case	Applied pressure	Applied load	Accept test	Proto/qual test	Comment
1	Burst	Ultimate	——	——	Analysis only
2	Proof	Yield	Proof pressure	——	——

that is, have no plasticity, stress is all we care about. For most metals that have significant plasticity, it is sometimes acceptable to take advantage of the tremendous energy absorbing capability beyond the elastic limit. For instance, some steels have more than 50% strain at failure and up to 100 times the energy absorbing capability as compared to the elastic region only. This makes most metals much safer to work with—there is much more strain available than is normally accounted for. Unfortunately, once a material is strained into the plastic deformation region, by definition, it does not return to its original state. And "bent up" structure is usually not acceptable. Thus, plastic materials are treated the same as brittle materials in most analyses. Exceptions include susceptibility to handling damage and cyclic load resistance, as explained later in this chapter.

Once loads are understood, stress analysis converts loads into stress by understanding 1) load magnitudes, 2) load frequencies, 3) load distributions on the structure, 4) how the load is transmitted through the structure (analysis of load paths), and 5) geometry of the structure. Often complex mathematical models are used to perform this series of tasks. Many books have been written on methods of performing this analysis, and many aerospace companies have developed rules on how to perform this operation. These steps will be addressed later in this chapter.

To better understand, manufacture, and analyze the structure, it is decomposed into structural elements that are easily understood, such as beams (tubes, bars, I-beams, or other cross sections), struts, plates, columns, shear panels, pressure vessels, stiffeners, stringers, ring frames, boxes, lands, bosses, lugs, intercostals, and many other structural member types.

10.3.4.1 Simplifying assumptions. The common simplifying assumptions in stress analysis are 1) small deflections, 2) linear stress vs strain relationship, 3) structure idealized as either beam or plate, 4) symmetry taken advantage of, 5) lightweight secondary structure ignored, and 6) structure idealized as statically determinant.

10.3.4.2 Force resolution. Forces and moments can be represented by vectors: position indicates direction and point of origin; length represents magnitude. Moments are represented by a vector perpendicular to the plane in which they are acting.

Parallel forces acting in opposite directions separated by a distance d produce a moment $M = Fd$. The moment M is represented by a vector $\boldsymbol{M}$, which can be spatially located with the right-hand rule. Vector mechanics applies to forces and moments:

1) Any number of concurrent forces can be resolved into a single resultant force.

2) Any number of concurrent moments can be resolved into a single resultant moment.

3) Any number of nonconcurrent forces and moments can be resolved into a single force and a single moment.

Example 10.1 Resolution of Forces

1) Given: Force shown in Fig. 10.26.

Find: x and y components of the force.

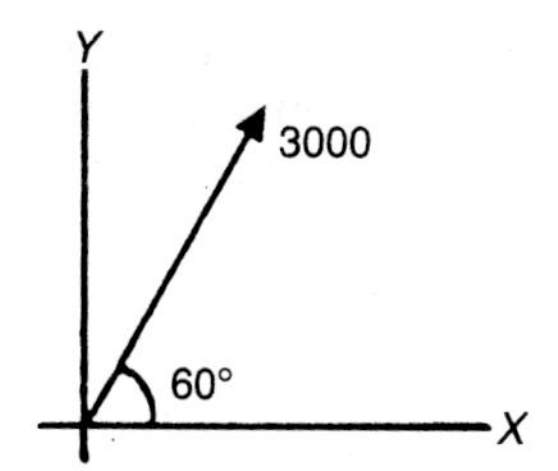

Fig. 10.26 Example 1.

Solution:

$$F_x = (3000)\cos(60) = 1500\,\text{lb}$$

$$F_y = (3000)\sin(60) = 2598\,\text{lb}$$

2) Given: Force $F = 5000$ lb with line of action shown in Fig. 10.27.
Find: F_x, F_y, F_z.
Solution:

$$\text{Length, } bh = \sqrt{(20)^2 + (30)^2 + (40)^2} = 53.85$$

$$F_x = 5000\left(\frac{40}{53.85}\right) = 3714$$

$$F_y = 5000\left(\frac{30}{53.85}\right) = 2786$$

$$F_z = 5000\left(\frac{20}{53.85}\right) = 1857$$

3) If F_x in the preceding example were 9285 lb, what would the force F be?
Solution:

$$F = 9285\left(\frac{53.85}{40}\right) = 12{,}500$$

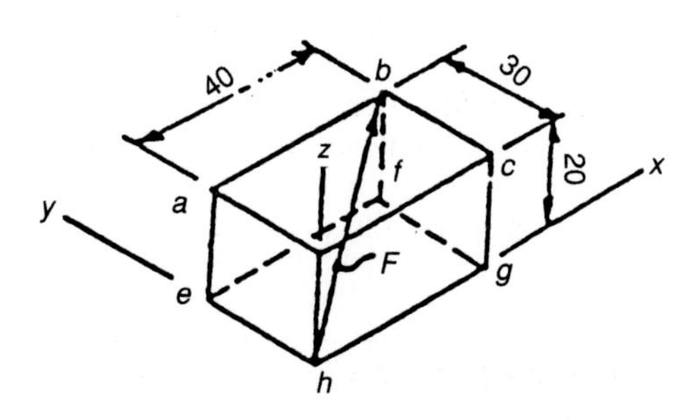

Fig. 10.27 Example 2.

10.3.4.3 Force equilibrium. All six degrees of freedom of a body, or a point, must be in dynamic equilibrium at all times. All forces must balance in accordance with the following rules:

$$\sum F_{x\text{-dir}} = F_{X1} + F_{X2} + \cdots + F_{Xn} = ma_X \quad (10.7)$$

and for static equilibrium

$$ma_X = 0 \quad (10.8)$$

where $F_{x\text{-dir}}$ is the sum of the forces in the x direction, m the mass, and a_x the acceleration in the x direction.

Clearly Eqs. (10.7) and (10.8) can be written for the y direction and the z direction as well. For moments

$$\sum M_{X\text{-axis}} = M_{X1} + M_{X2} + \cdots + M_{Xn} = I\alpha_X \quad (10.9)$$

or

$$\begin{aligned}\sum M_{X\text{-axis}} &= F_{Y1}Z_1 + F_{Y2}Z_2 + \cdots + F_{Yn}Z_n + F_{Z1}Y_1 + F_{Z2}Y_2 + \cdots + F_{Zn}Y_n \\ &= I\alpha_X \end{aligned} \quad (10.10)$$

and for static equilibrium

$$\sum M_{X\text{-axis}} = 0 \quad (10.11)$$

The equations for moments about the y and z axes are analogous. For the coplanar case only two force summations and one moment summation are required.

10.3.4.4 Free-body diagrams. When analyzing a piece of structure, it is convenient to isolate that piece of structure and to resolve all of the forces acting on it to demonstrate static or dynamic equilibrium. Figure 10.28 shows examples.

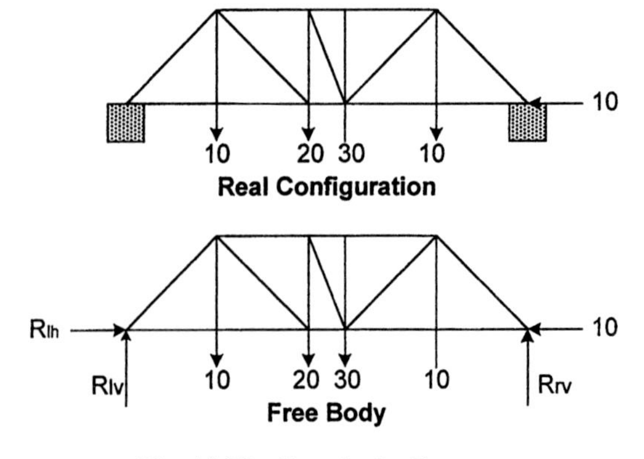

Fig. 10.28 Free-body diagram.

10.3.4.5 Determinant. A structure is said to be *determinant* if the structure is stable and the reactions on the free body can be calculated using geometry only; structural stiffness is not required. A structure is indeterminate if more reactions are available on the free-body diagram than the minimum that are required to calculate their magnitudes. Structural stiffness must be employed to solve for the reactions.

10.3.4.6 Superposition. If all of the analysis input is linear, the effects of one applied load can be added (superposed) to another. The result is the same as if both loads were simultaneously applied.

10.3.4.7 Section properties. Many inertial properties are used to describe engineering elements, including areas, masses, etc. Those of particular interest to the stress analyst describe the cross-sectional properties of a beam or bar. The first and second area moments are the most important for shear and bending analysis. The first moment of area is

$$M = \sum y_i A_i \tag{10.12}$$

$$M = A y_0 \tag{10.13}$$

where

M = first moment of area
y_i = linear distance from the reference axis to the centroid of area A_i
y_0 = linear distance from the reference axis to the centroid of the total area A
A_i = an elemental area
A = total area of the section

The second moment of area, or area moment of inertia I, is

$$I = \sum y_i^2 A_i \tag{10.14}$$

$$I = A y_0^2 \tag{10.15}$$

These moments are calculated about the neutral axis, the axis around which the bending stress is zero; the neutral axis of a section subject to bending only passes through the centroid. The neutral axis location is

$$y_{na} = \frac{\sum A_i y_i}{\sum A_i} \tag{10.16}$$

For a rectangular cross section of height h in the y direction and width b in the x direction, the moment of inertia about its neutral axis is

$$I_{xx} = \frac{bh^3}{12} \tag{10.17}$$

The parallel axis theorem allows calculation of I_2 about any axis knowing I_1 about any parallel axis and the distance d between them:

$$I_2 = I_1 \pm Ad^2 \tag{10.18}$$

Equation (10.18) is useful for calculating area moment of inertia for complex sections. The sign in Eq. (10.18) is negative if axis 2 is closer to the neutral axis than axis 1.

The section properties of a hollow figure are obtained by subtracting the moment of inertia of the hole from the moment of inertia of the complete figure. For example, a hollow rectangular shape h_1, b_1 with a hole measuring h_2, b_2 has a moment of inertia of

$$I = \frac{b_1 h_1^3 - b_2 h_2^3}{12} \tag{10.19}$$

with respect to an x–x axis located as shown in Fig 10.22.

Example 10.2 Beam Section Properties

Consider the simplified I-beam cross section shown in Fig. 10.29.

The steps in solving for the beam section properties are as follows:

1) Divide the beam into three rectangular sections A, B, and C.
2) Calculate the area of each rectangle.
3) Calculate the moment of inertia for each rectangle from $I_{CG} = bh^3/12$.
4) Calculate elemental I_{xx} from $I_{xx} = I_{CG} + A_i y_i^2$.
5) Calculate beam neutral axis location from $y_{na} = \Sigma A_i y_i / \Sigma A_i$.
6) Calculate section I_{na} from $I_{na} = I_{xx} - A y_{na}^2$.

The calculations are shown in Table 10.10.

The location of the neutral axis is

$$y_{na} = \frac{143}{22} = 6.5\,\text{cm}$$

and the moment of inertia about the neutral axis is

$$I_{na} = 1327.33 - (22)(42.25) = 397.83\,\text{cm}^4$$

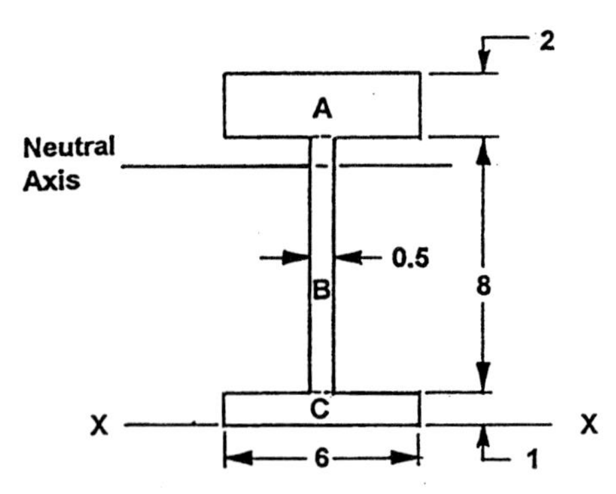

Fig. 10.29 Beam cross section.

Table 10.10 Calculating section properties

Section	b	h	A	y_i	Ay	Ay^2	I_{cg}	I_{xx}
A	6	2	12	10	120	1200	10.00	1204.00
B	0.5	8	4	5	20	100	21.33	121.33
C	6	1	6	0.5	3	1.5	0.50	2.00
Total	—	—	22	—	143	—	—	1327.33

Figure 10.30 lists area moments of inertia and other properties of common sections. Structural beam section properties are published by the American Institute of Steel Construction.

10.3.4.8 Beam analysis. Consider a beam with a distributed load $W_{(x)}$ and two simple supports as shown in Fig. 10.31. The sum of the vertical shear at any section is equal to the sum of all of the vertical shear on one side of the section:

$$V_{(a)} = \int_0^a W_{(x)}\, d(x) \tag{10.20}$$

where $V_{(a)}$ is the vertical shear at any section. The shear is positive when the vertical forces on one side of the section tends to move upward. A section of a beam is always normal to the long axis. In calculating shear, the reactions must be considered as applied loads.

The bending moment $M_{(a)}$ at any section is equal to the sum of the moments caused by all of the forces on one side of the section:

$$M_{(a)} = \int_0^a V_{(x)}\, d(x) \tag{10.21}$$

In calculating moments, the reactions must be considered as applied forces.

Beam curvature $e_{(a)}$ is the reciprocal of radius of curvature and is

$$e_{(a)} = \frac{M_{(a)}}{EI} \tag{10.22}$$

The slope of curvature at any section of a beam $\theta_{(a)}$ is

$$\theta_{(a)} = \int_0^a e_{(a)}\, dx \tag{10.23}$$

The beam deflection at any section $\Delta_{(a)}$ is

$$\Delta_{(a)} = \int_0^a \theta_{(a)}\, dx \tag{10.24}$$

Stress in a beam is primarily caused by moment and

$$\sigma = \frac{M}{z} \tag{10.25}$$

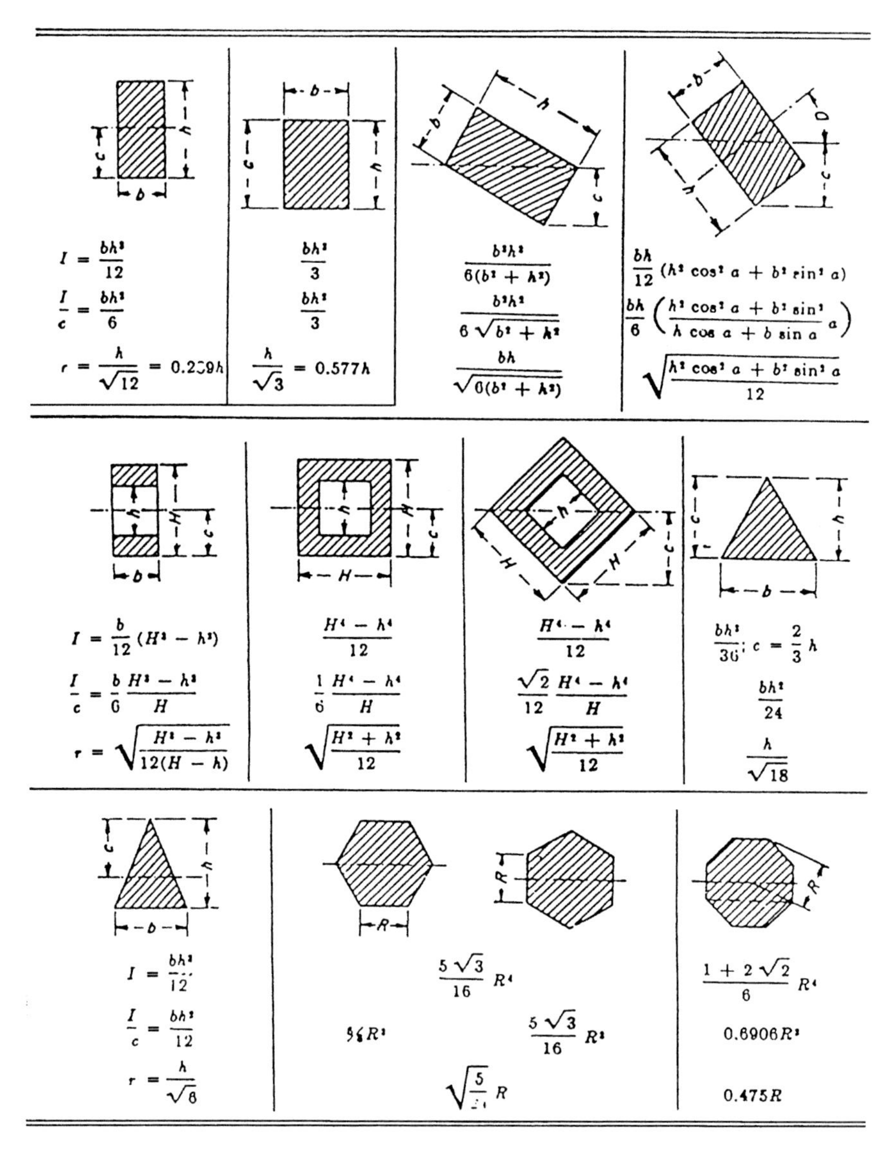

Fig. 10.30 Properties of various sections. (Reproduced with permission of Research and Education Association; Ref. 7, p. 869.) I = moment of inertia; I/c = section modulus; $r = \sqrt{I/A}$ = radius of gyration.

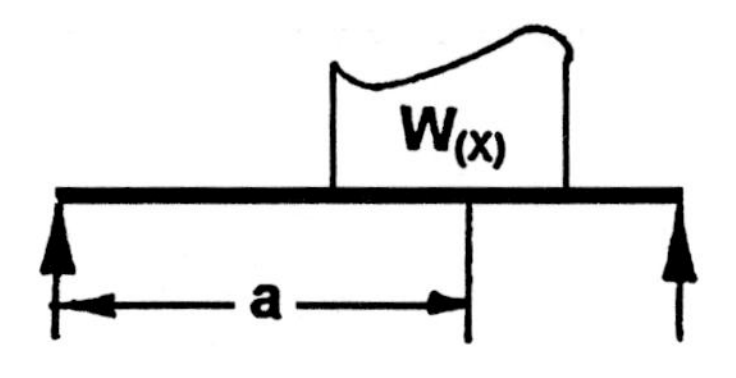

Fig. 10.31 Beam.

or

$$\sigma = \frac{Mc}{I} \tag{10.26}$$

where σ is the stress in a beam at position c, psi; c the distance from the centroid of a beam section, in.; M the applied moment, in.-lb; and $z = I/c$ the section modulus, in.^{3}

Example 10.3 Beam Analysis

Calculate the moment vertical shear and end point reactions for the uniformly loaded beam shown in Fig. 10.32. Given a uniform load $W_{(x)}$ of 10 lb/in. and a beam length of 60 in.

To obtain the right-side reaction force, sum the moments about the left support:

$$\sum M_L = 0$$

$$0 = (10)(60)(30) - (60)R_R$$

$$R_R = 300 \text{ lb}$$

The left-side reaction force can be obtained from symmetry, sum of the moments about the left reaction, or sum of the forces:

$$\sum F = 0$$

$$0 = (10)(60) - 300 - R_L$$

$$R_L = 300 \text{ lb}$$

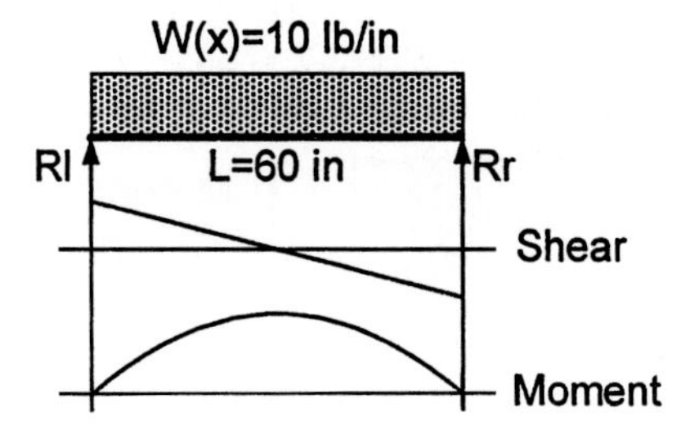

Fig. 10.32 Example 10.3 beam.

Calculating vertical shear from Eq. (10.20),

$$V_{(x)} = \int_0^x -10\,\mathrm{d}x = -10x + C_1$$

where C_1 is an integration constant. When $x = 0$, $V_x = 300$ lb; therefore, $C_1 =$ 300 lb, and

$$V_{(x)} = 300 - 10x$$

From Eq. (10.21)

$$M_{(x)} = \int_0^x V_x\,\mathrm{d}x = \int_0^x (300 - 10x)\,\mathrm{d}x$$

$$M_{(x)} = 300x - 5x^2 + C_2$$

where C_2 is an integration constant. When $x = 0$, $M(x) = 0$, and

$$300(0) - 5(0)^2 + C_2 = 0$$

$$C_2 = 0$$

$$M_{(x)} = 300x - 5x^2$$

Find maximum $M_{(x)}$:
Maximum moment occurs when $\mathrm{d}M/\mathrm{d}x = V = 0$, and

$$V_{(x)} = 0 = 300 - 10x$$

$$x = 30 \text{ in.}$$

Therefore, maximum moment is

$$M_{\max} = 300(30) - 5(30)^2$$

$$M_{\max} = 4500 \text{ in.-lb}$$

This result can be checked by forming a free-body diagram of the left half of the beam as shown in Fig. 10.33.

For this free body

$$\sum M_{\text{center}} = 0 \; M_{\text{center}} + 300(30) - 10(30)(15) = 0$$

$$M_{\text{center}} = -4500 \text{ in.-lb}$$

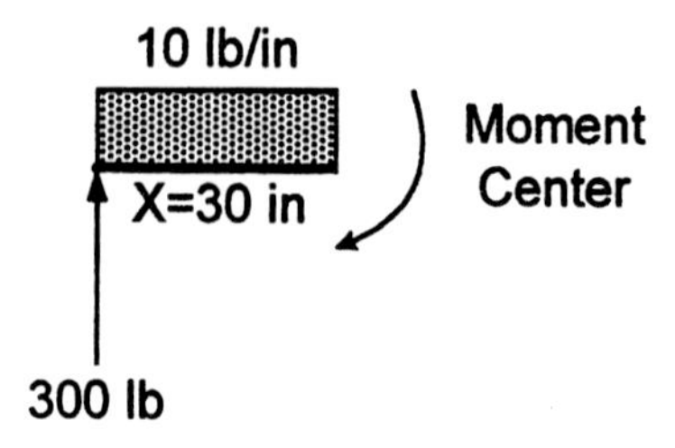

Fig. 10.33 Left-half free body.

Clockwise moments are positive. The free-body diagram checks the sum of the moments calculation.

If the beam cross section has the following properties, find the maximum stress caused by moments.

$$I_Y = 12{,}431 \text{ in.}^4$$

$$c = 2.5 \text{ in.}$$

From Eq. (26)

$$\sigma = \frac{4500(2.5)}{12431} = 905 \text{ psi}$$

If the beam is made of 7075-T7351 aluminum plate with a tensile ultimate of 50,000 psi and a factor of safety of 1.25 is used, the margin of safety based on ultimate is

$$M.S. = \frac{F_{\text{tu}}}{F.S.\,\sigma} - 1 = \frac{50000}{(1.25)(905)} - 1 = 43.2$$

The beam will have compression stresses on one side and tension stresses on the other. It might require two analyses to determine the lowest margin.

In the preceding example we assumed that stress governed the design; however, stiffness requirements (natural frequency or allowable deflection) may govern.

Example 10.4 Beam Analysis

Consider the beam and loading shown in Fig. 10.34.

The distributed load is 10 lb/in. acting perpendicular to the beam. The requirements are F.S.$_u$ = 1.4, F.S.$_y$ = 1.1, and deflection $\Delta \leq 1.0$ in.

Assume the beam is an aluminum tube with the following properties: F_{tu} is the ultimate tensile stress = 70 ksi; F_{ty} is the yield stress in tension = 60 ksi; F_{su} is the ultimate shear stress = 45 ksi; and E is the Young's modulus = 10^7 psi.

Taking the moments about a:

$$M_a = 10(100)(50) = 50{,}000 \text{ in.-lb}$$

The allowable stress as a result of bending is the lesser of

$$S_b = \frac{70000}{1.4} = 50{,}000$$

$$S_b = \frac{60{,}000}{1.1} = 54{,}500$$

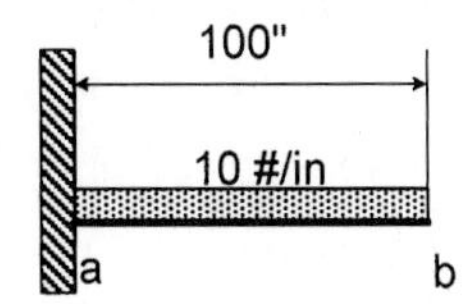

Fig. 10.34 Example 10.4 beam.

Thus, the maximum allowable bending stress is 50 ksi, and the section modulus must be greater than

$$z = \frac{50{,}000}{50{,}000} = 1 \text{ in.}^3$$

Try a 3-in. o.d. $\times$ 0.188 thick tube with $z = 1.0969$, $I = 1.6454$. Check beam deflection

$$\Delta = \frac{Wl^3}{8EI} \tag{10.27}$$

where W is the total load, uniformly distributed, lb; l the beam length, in.; and I the section moment of inertia, in.^{4}

And

$$\Delta = \frac{1000(100)^3}{8(10^7)(1.6454)} = 7.597 \text{ in.} \qquad \text{(too flexible)}$$

The beam is too flexible with the chosen tube. Try 7-in. o.d. $\times$ 0.109t, $I = 13.910$ in.^{4}:

$$\Delta = \frac{10^9}{8(10)^7 13.91} = 0.899 \text{ in.}$$

Now check shear strength. The maximum shear stress always occurs at the centroid of the beam section and is equal to

$$S_s = \frac{VQ}{Ib}$$

where V is the vertical shear force acting on the section, lb; Q the static moment, taken about the neutral axis, of that area of the section that is above the neutral axis $= A\overline{y}$, in.^{3}; and b the beam width at neutral plane, in. And

$$Q = (\pi r t)\left(\frac{2r}{\pi}\right) = 3.5(0.109)(2)(3.5) = 2.67 \text{ in.}^3$$

$$S_s = \frac{1000(2.67)}{13.91(2)(0.109)} = 880.4 \text{ psi}$$

10.3.4.9 Forces in truss members. For purposes of analysis, it is assumed that trusses consist of straight bars connected at their ends with frictionless pins and that all loads enter the truss at pin joints. Under these idealized conditions the stress in the members is either tension or compression, without bending or torsion. The tension and compression in truss members derived from analysis are called *primary stresses*. The stresses caused by riveted end joints and nonideal pin joints are called *secondary stresses*. The following discussion deals only with primary stresses.

The general procedure for analysis of an ideal truss is as follows:

1) Define all loads and reactions on the truss.

2) Determine the stresses in any member by the following:

a) Take a section, an imaginary cut through the truss, including the member being analyzed, which divides the truss into two pieces.

b) Isolate either of these pieces.

c) Replace each cut member with a force representing the stress in the member or the force existing in the piece that was removed.

d) Use summation of forces and moments to evaluate the stress in the cut bar or bars.

Assume either tension or compression in analyzing a member; if the force comes out negative, reverse the assumption.

Example 10.5 Analysis of a Planar Truss

Given the planar truss and loads shown in Fig. 10.35, find the forces in members *AB*, *BH*, *AH*, *BC*, and *CG*.

Step 1: Find reaction at *A*. The free-body diagram is shown in Fig. 10.36.

$$\sum F_H = 0 = A_H - 1000$$

$$A_H = 1000\,\text{lb}$$

$$\sum M_E = 0$$

$$0 = 4(30)A_v - 3(30)(2000) - 3(30)(5000) - 30(2000) - 40(1000)$$

$$A_v = 6083\,\text{lb}$$

Step 2: Find forces in members. Cut out a free body (see Fig. 10.37), which cuts member *AB* and *AH* to determine F_{AB} and F_{AH}

$$\sum F_y = 0$$

$$0 = 6083 - \frac{40}{50}\,F_{\text{AB}}$$

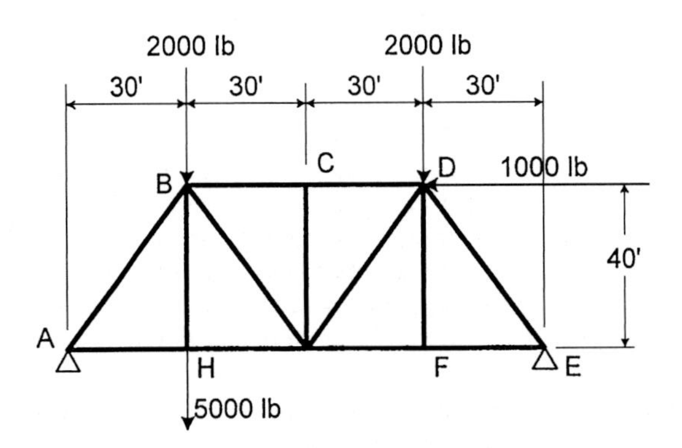

Fig. 10.35 Example planar truss.

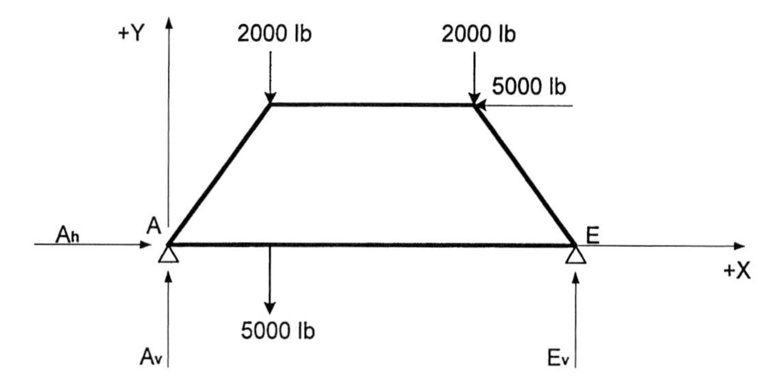

Fig. 10.36 Free-body diagram.

$$F_{\mathrm{AB}} = 7604 \text{ lb } (C)$$

$$\sum F_x = 0$$

$$0 = 1000 - 7604\left(\frac{30}{50}\right) + F_{\mathrm{AH}}$$

$$F_{\mathrm{AH}} = 3562 \text{ lb } (T)$$

Making a free body of the members joined at point H (see Fig. 10.38) will produce F_{BH}.

$$\sum F_y = 0$$

$$0 = F_{\mathrm{BH}} - 5000$$

$$F_{\mathrm{BH}} = 5000\ (T)$$

Similarly, F_{CG} can be shown to be zero, and $F_{\mathrm{BC}} = 3875$ (C).

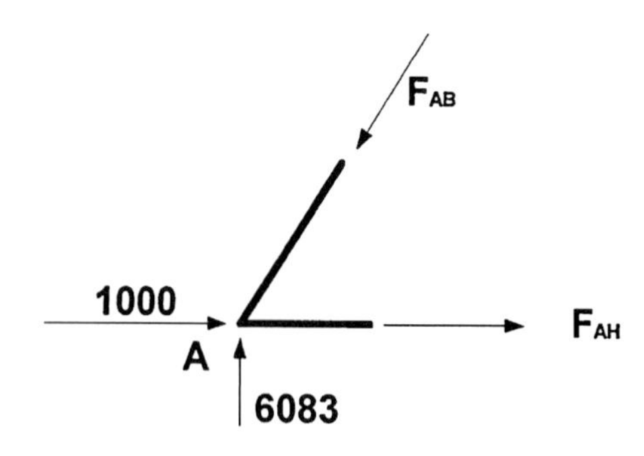

Fig. 10.37 Free-body diagram.

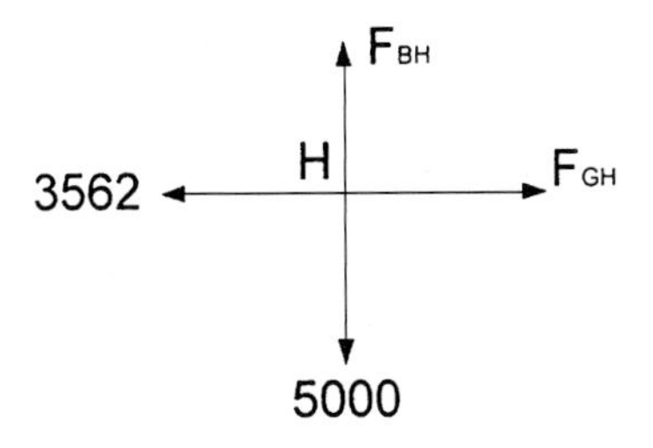

Fig. 10.38 Free body.

10.3.4.10 Instability. Structure with a relatively low stiffness-to-strength ratio is subject to failure by buckling or crippling. Examples are column buckling, web shear buckling, and thin-walled tube collapse.

10.3.4.11 Energy analysis. To calculate deflections and to analyze statically indeterminate structures, the stiffness of the hardware must be known and used in analysis. Thus, not only forces but also deflections must be calculated. When both of these are used, the strain energy must be minimized to solve the problem. These methods are also called work methods because work is twice the strain energy:

$$W = FD \tag{10.28}$$

and

$$\text{Strain Energy} = \int \sigma \varepsilon \, \mathrm{d}V \tag{10.29}$$

$$= \frac{FD}{2} \tag{10.30}$$

10.3.4.12 Equivalent axial load. When analyzing large diameter thin-walled tubes, the moment plus axial load effect can be expressed as a single term—the equivalent axial load P_e. The equivalent axial load is calculated by converting the stress from the moment into an axial stress and then into an axial load:

$$P_e = fA = \frac{Mc}{I} A \tag{10.31}$$

$$P_e = \frac{Mr(2\pi rt)}{\pi r^3 t} \tag{10.32}$$

$$P_e = 2\frac{M}{r} \tag{10.33}$$

where

A = cross-sectional area of a thin walled tube
f = maximum stress in the tube wall from the moment
M = moment on the tube

Table 10.11 Typical safety factors

Expendable launch vehicles	1.25 ultimate
Shuttle payloads	1.40 ultimate
Pressure vessels	2.00 ultimate
Ground handling	2.00 yield
Equipment	3.00 ultimate

I = area moment of inertia of the tube section
r = radius of the tube
t = thickness of the tube
c = greatest distance from the outer fiber of the tube to the center

10.3.4.13 Factor of safety. Factor of safety relates limit load to the strength of the part. Factor of safety accounts for uncertainty in the loads, analysis methods, and materials strength. Higher safety factors give greater conservatism. Factor of safety can be given with respect to the yield strength or to the ultimate strength. Typical examples are given in Table 10.11.

Government organizations and aerospace companies all have their own standards for what factor of safety will be used under any given condition.

10.3.4.14 Margin of safety. Margin of safety is a measure of the structural capability of a part remaining after applying limit load.

$$\text{MS} = \frac{\text{Allowable load (or stress)}}{\text{(Factor of safety) [Limit load (or stress)]}} \tag{10.34}$$

Structural integrity of flight structures can be verified by analysis alone or by a combination of analysis and test.

10.3.4.15 Material strength properties. Conventional material strengths and a wealth of other mechanical and physical properties are contained in MIL-HDBK-5.[4] Use of material properties from this source is normally a contract requirement. Where values for desired materials are not available, they must be determined by test. Test procedures are also set forth in MIL-HDBK-5.

Material strength in MIL-HDBK-5 is designated as A-basis, B-basis, or S-basis. A-basis and B-basis values are based on statistical evaluation of controlled and reported test results. Values, which are based on undocumented data, are classified as S-basis. It is clearly desirable to use A-basis allowables, particularly in flight-critical structure.

10.3.4.16 Strength properties of advanced structural materials. Many composite materials have not been characterized sufficiently to establish A- or B-basis properties. Extensive testing and higher safety factors are needed to use these materials.

10.3.4.17 Fracture mechanics. Fracture mechanics is a method of analyzing structures to verify that they are safe for the entire mission life. All materials are assumed to be flawed, and the analysis verifies that the largest flaw that cannot be detected by inspection or proof test will not grow to an unstable size and cause failure of the hardware.

10.3.4.18 Margin of safety. The structural margin of safety is based on theoretical analysis and substantiated by test. The margins so determined are used as final indicators of available strength or adequate stiffness. Margin of safety must be greater than or equal to zero and reflect the most critical mode of failure and whether the structure is yield, ultimate, or deflection critical.

10.3.4.19 Stress concentrations. The distribution of stress across the effective cross-sectional area of a structural element is nominally uniform. Any irregularities, however, such as abrupt changes in section, notches, or holes, cause stresses to increase locally at those points. These increases can be several times the nominal stress. Similarly, application of loading is localized in many cases, and stress patterns are disturbed by eccentricities or discontinuities. Stress concentration factors are applied in these cases.

10.3.4.20 Fasteners. The allowable strengths for fasteners are also obtained from MIL-HDBK-5.[4] Where special fasteners are required, allowable strengths must be obtained from vendor data or from tests. Fasteners with an ultimate tensile strength of 160 ksi are preferred for their thread life characteristics over both lower and higher strength fasteners. Cold rolled threads with the MIL-S-8879 UNJ thread form are preferred for thread life for all structural fasteners. A positive locking feature other than preload induced friction is required on all flight hardware fasteners.

10.4 Mechanisms

Detailed design of mechanisms is a highly specialized field and is well beyond the scope of this book. In this section the types of mechanisms that are used on spacecraft are discussed along with examples. In the early design phases it is adequate to isolate the functions that mechanisms will be required to perform, to select a general type of mechanism, and tabulate mass and power requirements. Table 10.12 lists the functions of spacecraft mechanisms from several spacecraft programs.

Table 10.12 shows that the requirement for mechanisms are derived from the following: 1) the actions necessary to convert the stowed, or launch, configuration into the cruise configuration; 2) the articulation required for solar panels, antennas, and science instruments; 3) the actions necessary to transfer power from articulating solar panels to fixed body; 4) the staging processes; and 5) dual spinning, for example, spin-bearing assemblies. Note that the functions in Table 10.12 can be divided into one-time functions, such as stage separation, and continuous functions, such as scan platform pointing. This distinction makes a significant difference in the mechanism design approach.

Table 10.12 Functions of spacecraft mechanisms

Functions	MGN[a]	MO[b]	Voy[c]	GLL[d]	I-6[e]	PVO[f]
Solar-panel deployment	1	2	—	—	1	—
Solar-panel unfolding	—	2	—	—	—	—
Solar-panel articulation, axes	1	2	—	—	—	—
Power transfer-slip rings	—	—	—	—	—	—
Power transfer-cable wrap	2	2	—	—	—	—
Antenna deployment	—	1	—	1	10	1
Antenna articulation, axes	—	2	—	—	—	—
Scan platform deployment	—	—	1	—	—	—
Scan platform articulation, axes	—	—	2	—	—	—
Booms	—	2	1	1	—	—
Launch locks	—	—	—	Y	4	—
Spin bearing	—	—	—	1	1	1
Separation mechanisms	2	1	—	2	1	2
ACS type:						
Three axis	X	X	X	—	—	—
Dual spin	—	—	—	X	X	X

[a]MGN = Magellan. [b]MO = Mars Orbiter. [c]Voy = Voyager. [d]GLL = Galileo.
[e]I-6 = Intelsat VI. [f]PVO = Pioneer Venus Orbiter.

As shown in Table 10.12, the mechanism requirements for a spinning spacecraft are different than for a three-axis-stabilized machine. Dual spinners require a multifunction spin-bearing assembly between the fixed and spinning segments. Spinners normally do not require solar-panel deployment, articulation, or power transfer. One of the major advantages of spinners is that scanning instruments can be accommodated without mechanisms.

The severity of the launch vehicle shroud constraints determine the complexity of the deployment process.

INTELSAT VI has nested deployment of the antennas and feeds on the upper deck, which entails sequenced deployments each of which must be successful before the next can be started.[8] Spacecraft with a kick stage typically have two separation functions: one from the launch vehicle and one from the spent kick stage. Magellan, Voyager, and PVO are examples of this situation.

Mechanisms are a major source of in-flight spacecraft failures because of the following:

1) The design of mechanisms is largely empirical.

2) Mechanisms are normally custom designed for each spacecraft and do not have a flight history.

3) It is difficult and imprecise to test space mechanisms in a laboratory environment with gravity and the atmosphere.

A major objective of the early design work on a spacecraft is to minimize the mechanisms required.

10.4.1 Common Mechanisms

The design of mechanisms is a specialty field in itself. An excellent detailed engineering text on spacecraft mechanism design is *NASA Space Mechanism Handbook*[9], edited by R. L. Fusaro. The design, analysis, and test of mechanisms use MIL-A-83577B as a guide. The following paragraphs are a general survey of some common spacecraft mechanisms.

10.4.1.1 Squibs. The most elemental ordnance device is a *squib* or an *initiator*, which is used to activate a mechanism such as an explosive nut or an ordnance valve. In a squib electrical energy is used to heat bridge wires (some have dual bridge wires) that ignite an explosive train, which in turn releases a burst of gas at high temperature. NASA now has a standard initiator (NSI) for this purpose.

10.4.1.2 Ordnance devices. Ordnance devices use chemical energy from an initiator to actuate a one-time-use mechanism. Table 10.13 lists common ordnance devices.

Explosive bolts and explosive nuts are functionally interchangeable; either destroys the tension carrying capability of a nut/bolt combination. Explosive nuts generally cause less shock to adjacent equipment.

Ordnance devices are empirical designs; high design margins and statistical testing are used as a result. It is common practice to use redundant devices, each with redundant squibs, and design each device to operate properly with less than half the gas pressure from one squib.

10.4.1.3 Solar-panel deployment. Solar-panel deployment is usually accomplished by use of a spring-loaded hinge, which is released by an ordnance

Table 10.13 Common ordnance devices

Device	Function
Explosive nuts	Activation of the squib releases the nut from the stud.
Explosive bolt	Activating the squib breaks the bolt.
Pin puller	Activating the squib retracts a pin.
Cable cutter	Activation of the squib cuts a cable.
Ordnance valve, normally open	Activation of the squib closes the valve permanently (see propulsion chapter).
Ordnance valve, normally closed	Activation of the squib opens the valve permanently.
Igniter	Activation of the squib starts an explosive train that ignites a rocket motor. (See propulsion chapter; also safe and arm devices.)
Super Zip™	Used for staging. Activation of the squib starts an explosive train that fractures structure.
Separation clamps	Used for staging. Activation of the squib releases a nut or bolt holding a staging clamp closed.

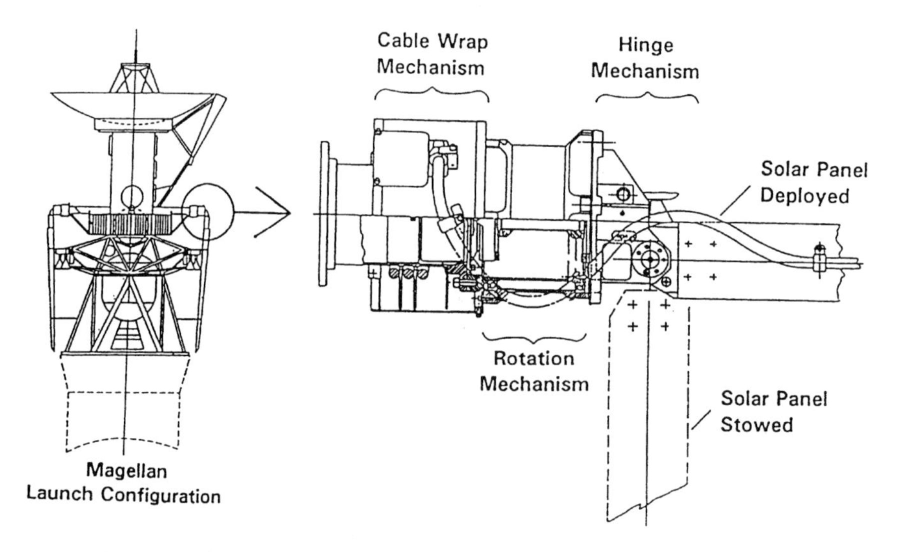

Fig. 10.39 Magellan solar-panel deployment/rotation mechanism. (Reproduced with permission of Lockheed Martin.)

device. The Magellan solar-panel deployment method is typical for three-axis-stabilized spacecraft (see Fig. 10.39). For launch the solar panels were stowed folded downward and latched in position with explosive pins at the midpoint. There are spring-loaded hinges located at the bus attachment point; the hinge mechanism is shown in Fig. 10.39. When the panel was released by the pin pullers, the springs raised the panels into the flight position where they were locked in place. There is a microswitch in each mechanism to verify the deployed and latched condition. When the panels were deployed in flight, there was no initial indication of panels latched. When the IUS stage I ignited, both panels latched.

10.4.1.4 Solar-panel articulation. Unless the spacecraft is sun-oriented, the solar panels must be articulated in one or two axes to keep the panels normal to the sun. The Magellan mission was designed such that one axis of rotation would be adequate; spacecraft rotation provided the second axis. The mechanism has two components: 1) a rotation mechanism, which was a motor/bearing set, and 2) a cable wrap system, which controlled the wrap and unwrap of the power cable as the panel was rotated. The cable wrap assembly contained 55 in. of cable and provided for 350 deg of rotation. As an alternative to the cable wrap mechanism, a slip ring mechanism could have performed the same function. The stepper drive motor has an integral potentiometer to provide position information. The Magellan deployment and rotation mechanism weighed 13.8 kg (each).

Magellan was designed for a single mechanism to provide deployment and one-axis rotation; clearly two separate mechanisms could have been used.

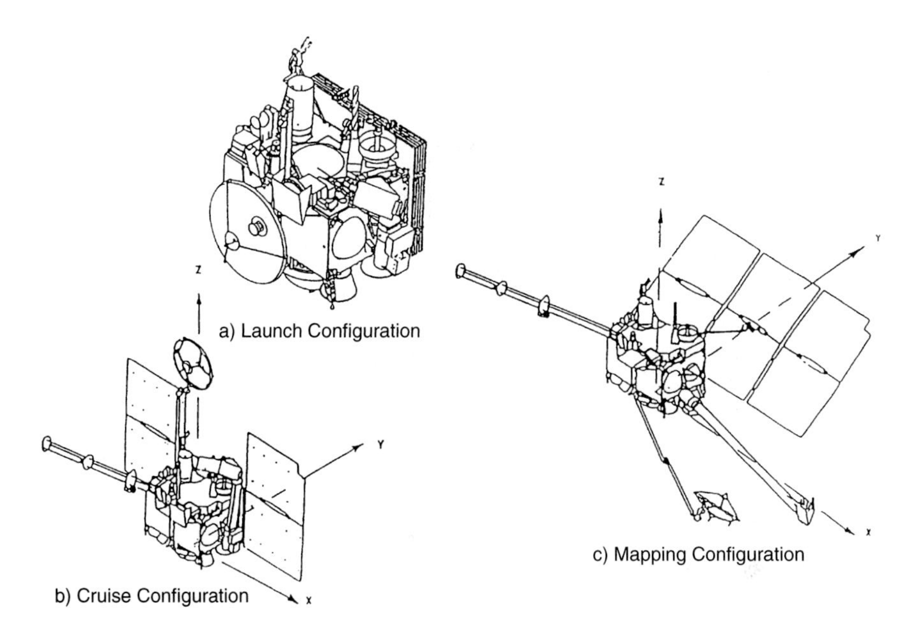

Fig. 10.40 Mars Observer solar-panel deployment. (Reproduced with permission of Lockheed Martin; Ref. 14.)

10.4.1.5 Folded solar panels. The general arrangement of some spacecraft requires solar-panel folding to accommodate the launch vehicle shroud. Mars Observer, shown in Fig. 10.40, is an example of a folded-panel configuration.

Figure 10.40a shows the spacecraft in the stowed configuration with the six-segment solar array completely folded. Figure 10.40b shows the spacecraft cruise configuration with the solar panel partially unfolded. Figure 10.40c shows the mapping configuration in Mars orbit with the panel completely deployed.

The six panels and a two-section solar-panel boom were folded into the launch position and restrained by cables. At each of the two deployments, the appropriate panel segments and boom segments were released by redundant cable cutters. Spring-loaded hinges drove the segments into the deployed positions. The velocity of the deploying segments was controlled by passive dampers.

10.4.1.6 Spring-driven hinges. Spring-driven hinges are common mechanisms as evidenced by the Magellan, Mars Observer, and INTELSAT VI. Figure 10.41 shows the INTELSAT VI viscous-damped, spring-driven hinge, which is used in five different mechanisms on the spacecraft.

A minimum motor torque three times the maximum operating torque (200% margin) is good design practice and a requirement on INTELSAT vehicles. With high torque, damping is often required to reduce the latch-up velocity of the mechanism. The drive shaft is pinned to a fixed structure; the housing is attached to the driven structure. The deployment spring imparts a constant torque to the housing, The housing rotates relative to the shaft, forcing fluid (Dow Corning 200) through the main gap at a controlled rate. Near the end of the stroke, the main gap decreases, forcing a reduction in rotational velocity. Because fluid viscosity is a function of temperature, the temperature of the mechanism must be within temperature limits prior to initiating deployment. This hinge design incorporates redundant microswitches to verify successful deployment.[8] Redundant microswitches are highly desirable in order to differentiate between a mechanism failure and a microswitch failure.

10.4.1.7 Launch locks. Dual-spin spacecraft require launch locks to fix the spun and despun sections during the launch process. Releasing the launch locks is part of the deployment process. INTELSAT VI, for example, had four launch locks installed at 90-deg intervals of the despun shelf. The latches were released by explosive bolt despun segments.[8]

10.4.1.8 Spin-bearing mechanism. Dual-spin spacecraft require a multifunction spin-bearing assembly between the spun and despun sections. The functions of the assembly are to 1) maintain the proper rotation rate between the two sections, 2) transfer power, and 3) transfer signals. Rotation rate is maintained by electric motors. Slip rings transfer power and signals across the interface. INTELSAT VI used redundant 24-pole dc motors each developing 4.8 N-m torque. If required, both motors can be operated simultaneously, doubling the torque. The slip ring assembly contained 44 signal slip rings and 4 power slip rings. The power transfer capacity is 2000 W.[8]

10.4.1.9 Staging mechanisms. There are three basic approaches to staging separation:

1) Hold the staging interface together with either explosive nuts or bolts. To stage, blow the ordnance. This is a simple approach with much flight history. It is best suited for a structure designed for point loads, for example, a truss.

2) Hold the staging interface together with a two- or three-segment clamp with the segments held together by explosive nuts. This method is well suited for an interface with distributed loads, for example, a shell structure or a solid rocket motor case.

3) Use Super-Zip™ technique, also best for shell structure. In this technique a linear-shaped charge is built into the interface structure at the staging plane. When the ordnance is fired, the structure fractures all of the way around the shell at the interface plane (like a zipper). After the interstage structure is split, by whatever means, balanced, calibrated springs are used to force a small positive separation velocity with minimal overturning moment.

10.4.1.10 Scan platforms. Scan platforms are required when pointing requirements exceed (or conflict with) the maneuvering capability of the spacecraft. Supporting a narrow-angle camera requires a scan platform. The Voyager spacecraft, which produced such impressive images of the outer planets, is a good illustration of the successful use of a scan platform. During each encounter, while the spacecraft points the large parabolic high-gain antenna at Earth, a set of sensors must be accurately aimed at a succession of objects of scientific interest. To accomplish this, the cameras as well as the sensors for infrared radiation, polarimetry, and UV spectroscopy were mounted on a scan platform, which is shown on the right in Fig. 10.41.

The scan platform can be precisely pointed around two axes. The instruments are mounted on the platform with a common boresight. The total gimbaled mass is 102.5 kg (226 lb). Driver circuits for the scan motors, one for each axis, are located in the ACS electronics and commanded by the ACS computer. The angular range is 360 deg in azimuth and 210 deg in elevation. The platform is slewed one axis at a time with selectable slew rates, which are 0.083 deg/s and 0.0052 deg/s. Camera line of sight is controlled to within 2.5 mrad.

The scan platform was stowed for launch with a set of two, one-axis deployment hinge mechanisms visible in the platform truss near the high-gain antenna in Fig. 10.41.

10.4.1.11 Booms. A number of instruments, notably magnetometers, require booms in order for the measurements to be free from the influence of the spacecraft. The booms extend to the deployed position after spacecraft separation; some require retraction as well. The three common types of booms are shown in Table 10.14.

Figure 10.41 shows the 13-m deployable Astromast (Astro Aerospace, Carpinteria, California). It is a common deployable boom used on Voyager, Galileo, and Cassini. This boom is stowed inside a canister only 23 cm in diameter and 66 cm long. The Astromast is a triangular truss with fiberglass longitudinal members held in place by fiberglass triangles that are spaced 14 cm apart. The truss is stiffened by tensioned, collapsible diagonal filaments. Releasing the canister lid allows the boom to deploy like a jack-in-the-box.

The boom is stowed by twisting the entire structure so that the diagonal filaments and triangles are nearly in contact with each other. This stores a considerable elastic energy in the assembly. When the canister is opened by an ordnance device, a lanyard is played out, which controls the velocity of extension.[10]

The Viking soil sample boom is a *de Havilland* type boom, which is frequently used for deployment of antennas or instruments. The de Havilland boom is made from a strip of beryllium copper that is heat treated in a tubular form. The strip is then flattened and rolled into a coil. When it is deployed, the strip is unwound through a flat-to-tubular guide where the original tubular shape is recovered. A fairly rigid sensor support or whip antenna results. (The action is similar to that of a carpenter's steel tape measure.) The Viking lander used a variant on the de Havilland boom to extend and retract the soil sample collector. The boom, when extended, was a spring steel tube with two extended flanges (see Fig. 10.42). The flanges contained a series of sprocket holes that were used by the spool in the extend/retract mechanism. During the process, the tube was pressed flat, thereby losing its stiffness.

The boom mechanism provided three-axis articulation as well as extension and retraction for the soil sampler. The boom could be extended 10 ft and rotated 108 deg, covering about 130 ft^2 of the surface in front of the lander.

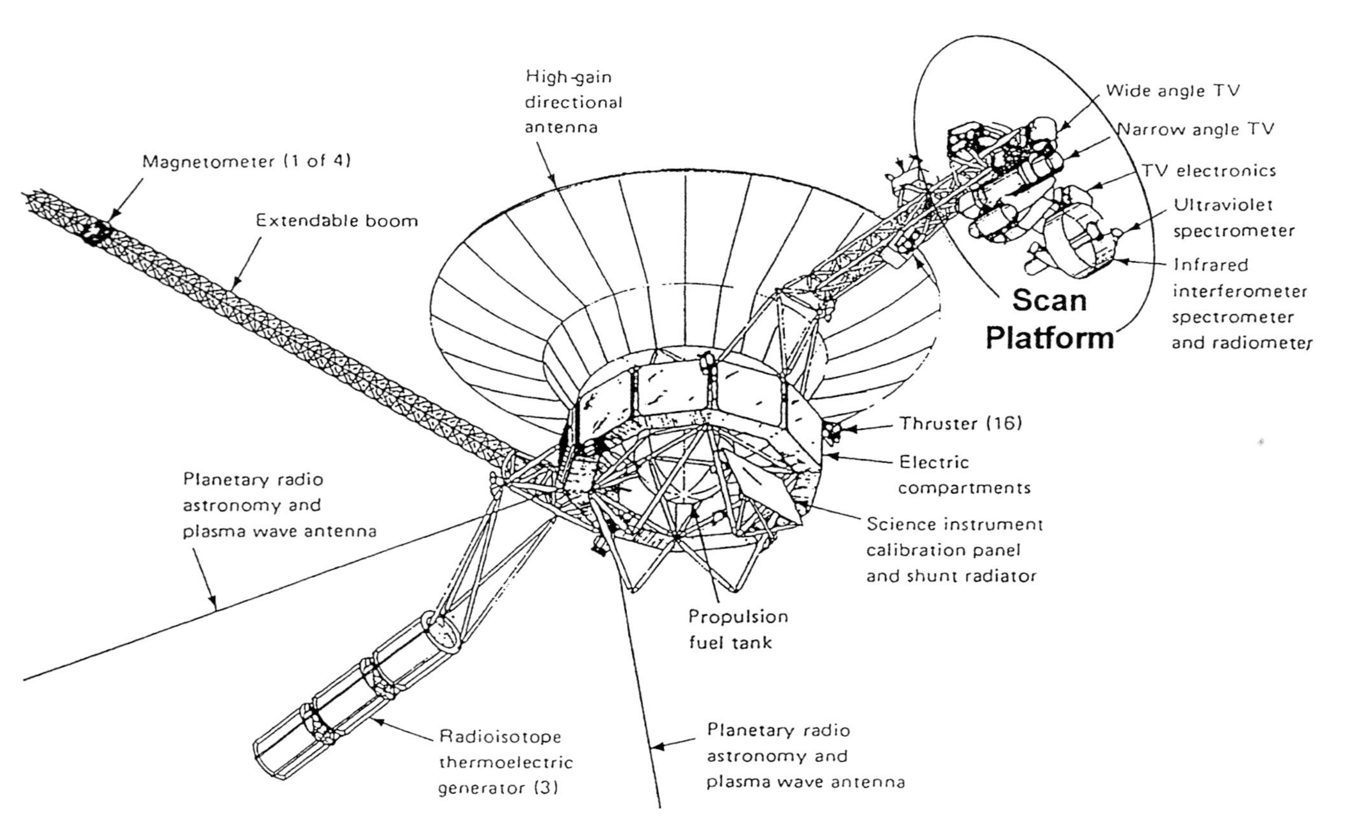

Fig. 10.41 Voyager spacecraft. (Reproduced courtesy of NASA/JPL/Caltech; Ref. 10, p. 17.)

Table 10.14 Spacecraft booms

Type	Construction/function
Hinged tube	Hinged tube spring loaded to deployed position, released by pinpuller, locked down Simplest Least expensive Example: Voyager RTG truss (Fig. 10.41), Magellan Solar Panel (Fig. 10.39)
de Havilland boom Stowed Deployed	Spring steel tube is flattened into a ribbon and rolled onto a reel Deployed by unrolling ribbon, which springs into a stiff tube when uncoiled, similar to carpenter's steel tape measure Can be restowed Length is limited by loads Example: Viking Lander soil sampler boom (Fig. 10.42)
Astromast	Deployable truss: Coiled longerons, flexible intracostals 50:1 Deployed to stowed lengths attainable Stored energy snaps truss into deployed shape Rigid deployed Can be restowed Example: Voyager magnetometer boom (Fig. 10.41)

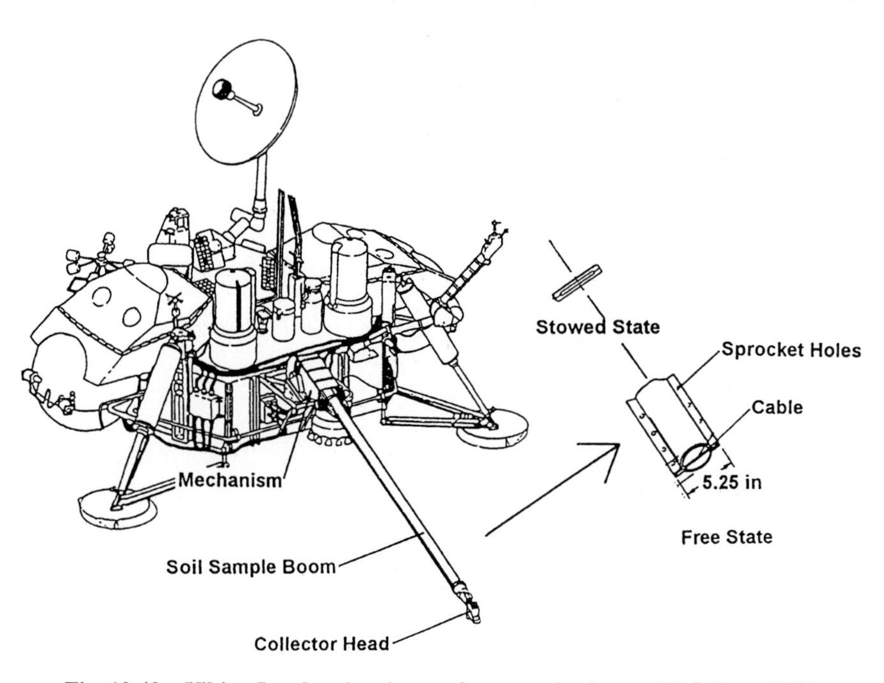

Fig. 10.42 Viking Lander showing surface sampler boom. (Ref. 12, p. 175.)

References

[1]Isakowitz, S. J., *International Reference Guide to Space Launch Systems*, AIAA, Washington, DC, 1991.

[2]Reeves, E. I., *FLTSATCOM, Case Study in Spacecraft Design*, Professional Study Series, AIAA, Washington, DC, Aug. 1979.

[3]Bachofer, B. T., *Landsat D, Case Study in Spacecraft Design*, Professional Study Series, AIAA, Washington, DC, Aug. 1979.

[4]"*Metallic Materials and Elements for Aerospace Vehicle Structures*," Air Force Materials Lab., MIL-HDBK-5, Wright Patterson AFB, OH, 1987.

[5]"*Advanced Composites Design Guide*," Air Force Materials Lab., MIL-HDBK-17, Wright Patterson AFB, OH.

[6]*Proceedings of the AIAA/ASME 7th Annual Structures and Materials Conference*, AIAA, New York, 1966.

[7]Fogiel, M. (ed.), *Handbook of Mathematical, Scientific, and Engineering Formulas, Graphs, Transforms*, Research and Education Association, Piscataway, NJ, 1984.

[8]Dest, L. R., Bouchechez, J. P., Serafini, V. R., Schavietello, M., and Volkert, K. J., "INTELSAT VI Spacecraft Bus Design," *COMSAT Technical Review*, Vol. 21, No. 1, Spring 1991.

[9]Fusaro, R. L. (ed.), *NASA Space Mechanisms Handbook*, NASA, Glenn Research Center, Cleveland, OH, July 1999.

[10]"*Voyager to Jupiter and Saturn*," NASA Science and Technical Information Office, NASA SP-420, Washington, DC, June 1977.

[11]Corless, W. R., "*Scientific Satellites*," NASA SP-133, Washington, DC, 1967.

[12]Holmberg, N. A., Faust, R. P., and Holt, H. M., *Viking 75 Spacecraft Design and Test Summary*, Vol. I, Ref. Publ. 1027, NASA, Washington, DC, Nov. 1980.

[13]"Landsat-C, Improved Landsat to Give Better View of Earth Resources," NASA Press Kit 78-22, Washington, DC, 1978.

[14]Komro, F., and Kaskiewicz, "Mars Observer Spacecraft Overview," General Electric Space Systems Div. (currently Lockheed Martin), Nov. 1991.

[15]Jones, C. P., and Risa, T. H., "The Voyager Spacecraft System Design," AIAA Paper 81-0911, May 1981.

[16]"Magellan Mission to Venus, Press Information," Martin Marietta (currently Lockheed Martin), Feb. 1989.

[17]"Magellan Spacecraft Final Report," Martin Marietta (currently Lockheed Martin), Jan. 1995.

[18]"NAVSTAR Global Positioning System (GPS) Phase II, System Review," Martin Marietta Astro Space (currently Lockheed Martin), Feb. 1994.

[19]"Mariner Mars 1971," NASA/JPL/CalTech, JPL TM 33-449, Pasadena, CA, Sept. 1969.

[20]Griffin, M. D., and French, J. R., *Space Vehicle Design*, AIAA, Washington, DC, 1991.

Problems

10.1 Calculate the 3 sigma load factor on a spacecraft component box, shown in Fig. 10.43. Use the flight level PSD vs frequency data from Fig. 10.19 at the box natural frequency of 200 Hz, and a damping factor of 2%.

Solution: 75.2 *g*.

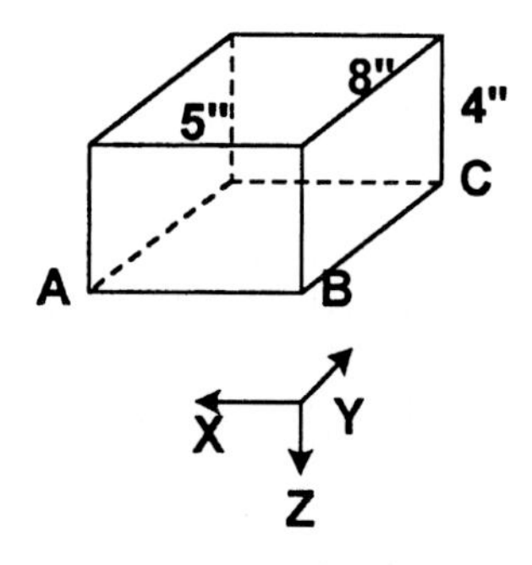

Fig. 10.43 Component box.

10.2 If the frequency in problem 10.1 is increased to 400 Hz, what is the maximum limit on Q to maintain the same load factor.

10.3 If the box center of gravity is 0.5 in. above the geometric center and the box weighs 10 lb, what are the reaction forces on the box at *A*, *B*, *C*, and *D* due to the load factor applied in the directon?

Partial solution: $Ax = 188$ lb.

10.4 Find the neutral axis of the cross section shown in Fig. 10.44. Find the moments of inertia, I_x and I_y.

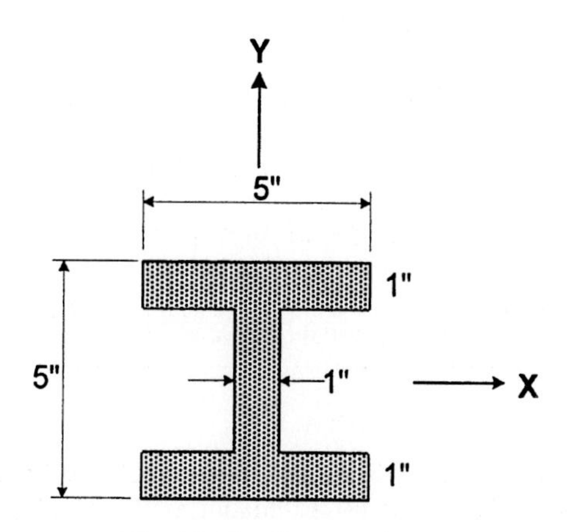

Fig. 10.44 Cross section 1.

10.5 Find the neutral axis of the cross section shown in Fig. 10.45. Find the moments of inertia, I_x and I_y.

Partial solution: $I_x = 43.08$ in.^{4}

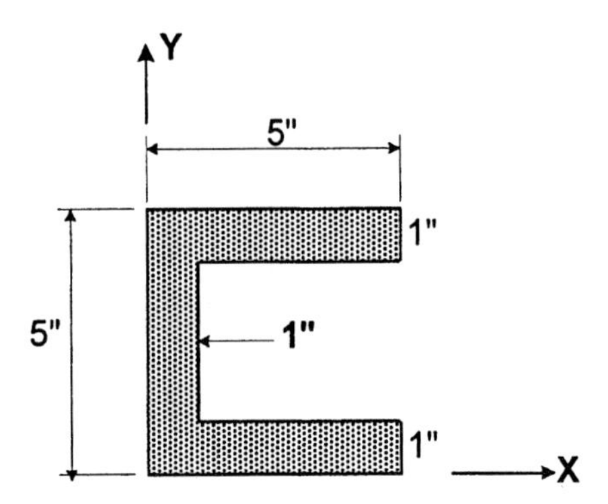

Fig. 10.45 Cross section 2.

10.6 A beam supports two component boxes as shown in Fig. 10.46. Box A weighs 120 lb and its center of gravity is 2.5 in. above the beam. Box B weighs 240 lb and its center of gravity is 1.5 in. above the beam.

(a) Draw the free body diagram.
(b) Draw the shear diagram.
(c) Draw the moment diagram.
(d) What is the maximum moment?

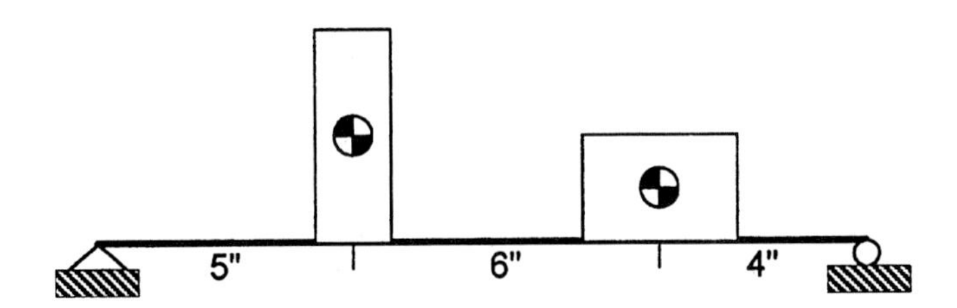

Fig. 10.46 Beam with component boxes.

10.7 In Fig. 10.46, add an additional 2 g lateral load in the axis of the beam. Draw free body, shear and moment diagrams, as above. What is the maximum moment?

10.8 Loads applied from the spacecraft to the launch vehicle adapter are shown Fig. 10.46, at $x = 0$, $y = 0$, $z = 60$ in. In addition the moment about the z-axis is 100,000 in. lb.

(a) Draw the free body diagram of the adapter.
(b) What are the loads applied by the adapter to the launch vehicle?

10.9 In Fig. 10.47, there are four attachment points to the launch vehicle at a 100 in. diam on the axes, $+x$, $+y$, $-x$ and $-y$. Assume an infinitely stiff launch vehicle. What are the loads on the launch vehicle at the four points? Note that bolts react only tension, compression, or tangential shear.

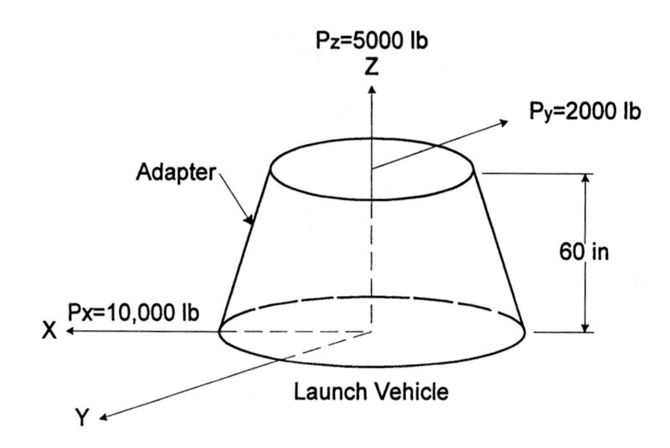

Fig. 10.47 Launch vehicle adapter.

10.10 Given the truss structure loaded as shown in Fig. 10.48, what is the reaction *Ra*, the load in member *AB* and the load in member *CG*?

Partial solution: Load $CG = 8000$ lb (compression).

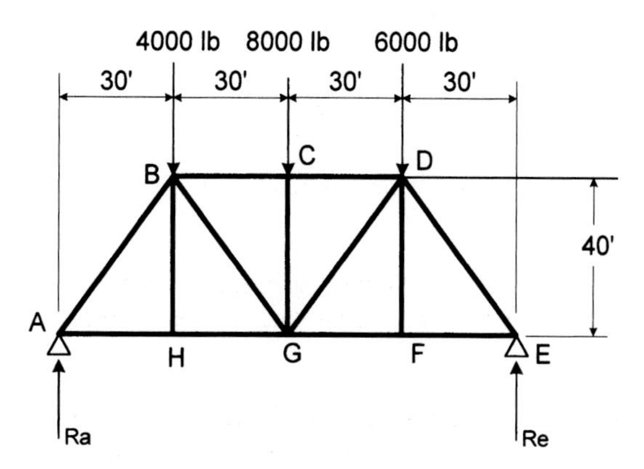

Fig. 10.48 Truss structure.

10.11 What is the maximum load P that the beam shown in Fig. 10.49 can carry given:

(a) Bending stress governs, ignore shear.
(b) Ultimate tensile strength = 60,000 psi.
(c) Ultimate factor of safety = 1.5, yield factor of safety = 1.2.
(d) Bending stress = Mc/I.
(e) Moment of inertia = $bh^3/12$.

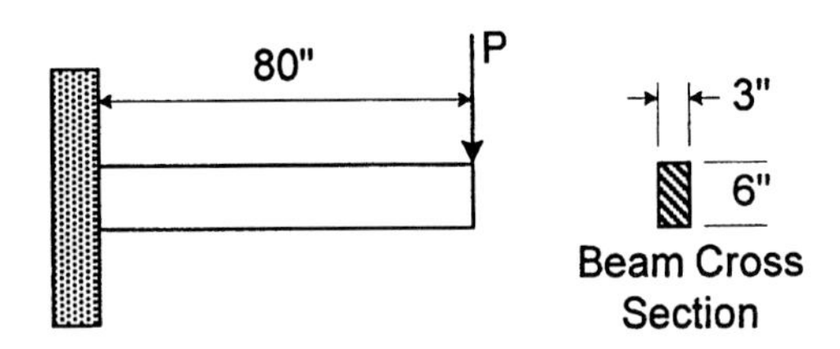

Fig. 10.49 Beam.

Appendix A
Acronyms and Abbreviations

The aerospace business is replete with acronyms and abbreviations. As much as they are maligned, they save a lot of time and paper. The real downside of acronyms is they can obstruct understanding. This Appendix is designed to help alleviate that problem.

A	ampere
Å	angstrom
AAS	American Astronautical Society
AACS	attitude and articulation control
ABM	apogee boost motor
ac	alternating current
ACE	attitude control electronics (Also used by JPL to designate mission control manager)
ACS	attitude control system
ACWP	actual cost of work performed
AEC	Atomic Energy Commission
A/D	analog to digital
AFRO	Air Force Representatives Office
AGC	automatic gain control
AGE	aerospace ground equipment
AGP	additional general provision
Ah, A-h	ampere-hour(s)
AIAA	American Institute of Aeronautics and Astronautics
AKM	apogee kick motor
Al	aluminum
alt	altitude or altimetry
ALU	arithmetic logic unit
AM	amplitude modulation
AM0	air mass zero
AMD	angular momentum dumping (reaction wheels or momentum wheels)
AO	announcement of opportunity
AOS	acquisition of signal
AP	ammonium perchlorate
APL	approved parts list
APS	application program set
ARC	Ames Research Center
arc-s	arc second(s)
ARR	ATLO readiness review
ARU	attitude reference unit (gyro package)
As	arsenate

ASD	acceleration spectral density
ASIC	application-specific integrated circuit
ASK	amplitude shift keying
ATL	acceptance test level
ATLO	assembly, test, and launch operations
ATP	authority to proceed
AU	astronomical unit, mean distance Earth to sun
Az	azimuth
BASK	binary amplitude shift keying
BAT	bench acceptance test
BCA	battery charge assembly
BCE	bench checkout equipment or best current estimate
BCH	Bose–Chadhuri–Hocquehem (a type of block coding)
BCR	battery charge regulator
BCWP	budgeted cost of work performed
BDR	battery discharge regulator
BER	bit error rate
BFSK	binary frequency shift keying
BIU	bus interface unit
BMD	Ballistic Missile Defense Organization
BOL	beginning of life
BOM	bill of material
bps	bits per second
BPSK	binary phase shift keying
BWG	beam wave guide
c	speed of light
C	celsius, battery capacity, or coulomb
CAD	computer aided design
CAS	cost accounting standard
CBE	current best estimate
CCAFS	Cape Canaveral Air Force Station
CCSDS	Consultative Committee for Space Data Systems
CCB	change control board
CCD	charge coupled device
CCIR	International Consultative Committee on Radio Communications
Cd	cadmium
cd	candela (SI unit)
C&D	command and data
CDB	central data base
CDS or C&DH	command and data system or handling
CDMA	code-division multiple access
CDR	critical design review
CDRL	contract data requirements list
CDT	central daylight time
CDU	command detector unit
Ce	cesium
CEI	contract end item

CER cost estimating relationship
cg, c.g. center of gravity
CISC complete or complex instruction set computer
CLTU command link transmission unit
cm centimeter(s)
CM, c.m. center of mass
CM configuration management
CMD command
CMG control moment gyro
CMOS complementary metal oxide semiconductor
CoDR conceptual design review
COM communications or cost of money
COMSAT communication satellite
COMSEC communications security
COSPAR Committee on Space Research
COTS commercial off-the-shelf
CPC computer program component
CPT channel parameter table
CPU central processor unit
CPV common pressure vessel (battery)
CRAD contractual research and development
CRES corrosion resistant steel (stainless steel)
CRT cathode ray tube
CST central standard time
CSCI computer software configuration item
CTPB carboxyterminated polybutadiene
CU control unit
CV command verification
CW continuous wave
CWBS contract work breakdown structure
CY calendar year

DARPA Defense Research Project Organization
DAS data acquisition system
DB double base (solid propellant)
dB decibel(s)
dBi decibel(s) referenced to an isotropic antenna
dBm decibel(s) referenced to a milliwatt
DC design complete
dc direct current
DCM direction cosine matrix
DCS digital communication system
DDT&E design, development, test, and evaluation
DEMUX demultiplexer
DET direct energy transfer (power system)
DIU data interface unit
DMA direct memory access
DMS data management system
DMU data management unit

DMSP	Defense Meteorological Satellite Program
DOD	Department of Defense or depth of discharge
DOE	Department of Energy
DOF	degrees of freedom
DOR	doppler one-way ranging
DPA	destructive physical analysis
DRM	drafting room manual
DSCS	Defense Satellite Communications System
DSN	Deep Space Network
DSP	defense support program
DTC	design to cost
DT&E	design, test, and evaluation
DTM	design test model
DTR	digital tape recorder
E	east
EAC	estimate at completion (cost)
EBA	electronics bay assembly
EBS	electron bombarded silicon
ECA	estimated cost analysis
ECP	engineering change proposal
ECR	engineering change request
EDAC	error detection and correction
EDF	engineering data formatter
EDL	entry descent and landing
EDR	experiment data record
EDT	eastern daylight time
EDU	engineering development unit
EEC	extendible exit cone
EEO	elliptical Earth orbit
EGSE	electrical ground support equipment
EHF	extremely high frequency
EIRP	effective isotropic radiated power
ELV	expendable launch vehicle
EM	electromagnetic or engineering model
EMC	electromagnetic compatibility
EMF	electromagnetic force
EMD	engineering and manufacturing development
EMI	electromagnetic interference
EMR	electromagnetic radiation
EOL	end of life
EOM	end of mission
EOR	end of record
EOS	Earth observing system
EPROM	erasable programmable read-only memory
EPS	electrical power system
ERS	Earth resources system
ESA	European Space Agency
ESD	electrostatic discharge

ESMC	Eastern Space and Missile Center
EST	eastern standard time
ET	ephemeris time (mean solar time)
ETA	explosive transfer assembly (ordnance, similar to a fuse)
ETC	estimate to complete
ETR	Eastern Test Range
ev	electron volt
EVA	extravehicle activity (space walk)
EW	electronic warfare
F	fahrenheit
FAA	Federal Aviation Agency
FAR	federal acquisition regulation (procurement regulation)
F-B-C	faster-better-cheaper (programs)
FCC	Federal Communication Commission
FDM	frequency division multiplexing
FEA	finite element analysis
FEM	finite element module
FET	field effects transistor
FFP	firm fixed price
FHST	fixed head star tracker
FLTSATCOM	Fleet Satellite Communication
FM	frequency modulation
FMEA	failure modes and effects analysis
FMECA	failure modes effects and criticality analysis
FMF	free molecular flow
FOC	faint object camera (Hubble)
FOR	frame of reference
FOS	faint object spectrograph
FOV	field of view
FP	fault protection
FPGA	field programmable gate array
FR	failure report
FRR	flight readiness review
FRUSA	flexible roll-up solar array
FS	factor of safety
FSW	flight software
FSK	frequency shift keying
ft	foot or feet
FY	fiscal year
G	gain (in communications)
g	gram
g	acceleration of gravity
Ga	gallium
GaAs	gallium arsenide (solar cell type)
G&A	general and administrative (expense)
Gb	gigabit(s)

GCF	ground communications facility
GDS	ground data system or goldstone
Ge	germanium
G/E, G-E	graphite epoxy
GEO	geosynchronous Earth orbit
GFE	government furnished equipment
GFP	government furnished property
GHz	gigahertz
GIDEP	governmentindustry data exchange program
GIGO	garbage in, garbage out
GMT	Greenwich mean time
GN&C	guidance, navigation, and control
GOES	geostationary environmental satellite
GPS	global positioning system
GSE	ground support equipment
GSFC	Goddard Space Flight Center
GSTDN	ground spaceflight tracking and data network
GRO	Gamma Ray Observatory
GTO	geosynchronous transfer orbit
Gy	gray (SI unit)
H	hydrogen
h	hour(s)
He	helium
HEAO	High Energy Astronomical Observatory
HEF	high efficiency (DSN Station)
HEO	high Earth orbit or highly elliptical orbit
HEPA	high efficiency particulate air (filter type)
HGA	high gain antenna
HIG	hermetic integrating gyro
HPF	hazardous processing facility (KSC)
HOL	high-order language
HQ	headquarters
HRCR	hardware requirements certification review
HRG	hemispherical resonating gyro
HRS	high resolution spectrograph
HSA	horizon sensor assembly
HST	Hubble Space Telescope
HTPB	hydroxy-terminated polybutadiene
H/W	hardware
Hz	hertz (cycle per second)
I&A	integration and assembly
IAR	independent annual review
IA&T	integration assembly and test
I&C	instrument and communication subsystem
IC	integrate circuit
ICA	independent cost analysis
ICBM	intercontinental ballistic missile

ICD	interface control document
IF	intermediate frequency
I/F	interface
IFOV	instrument field of view or instantaneous field of view
IFRB	International Frequency Regulation Board
IFSS	instrumentation and flight safety system
IM	integration meeting
IMU	inertial measurement unit
in.	inch(s)
I/O	input/output
IOC	initial operational capability (usually a date)
IPS	inertial positioning system or instructions per second
IPV	individual pressure vessel (battery)
IR	infrared
IR&D	independent research and development
IRAS	infrared astronomy satellite
IRBM	intermediate range ballistic missile
IRR	independent readiness review
IRU	inertial reference unit
ISA	instruction set architecture
ISL	intersatellite links
ISO	International Standards Organization
ISOE	integrated sequence of events
IST	integrated system test
I&T	integration and test
ITAR	international traffic in arms regulation
ITL	integration, test, and launch
ITLO	integration, test, and launch operations
ITU	International Telecommunications Union
IUE	International Ultraviolet Explorer
IUS	inertial upper stage (for Shuttle)
IV&V	independent verification and validation
J	joule
JIS	joint integration simulation
JPL	Jet Propulsion Laboratory, a division of California Institute of Technology
JSC	Johnson Space Center
k	kilo, a multiplier of a thousand
K	kelvin
kb	kilobit(s)
K-band	range of frequencies of about 12 to 40 GHz
kbps	kilobit(s) per second
kg	kilogram(s)
kHz	kilohertz
kips	thousands of instructions per second
km	kilometer(s)

krad	kiloradian(s)
KSC	Kennedy Space Center (also Cape Canaveral)
ksps	thousands of samples per second
kW	kilowatt(s)
L	loss (communications)
l	liter
LAN	local area network
LAM	liquid apogee motor
LaRC	Langley Research Center
Lat, lat	latitude
L-band	range of frequencies of about 1 to 2 gHz
lb	pound(s)
LBB	leak before burst (tanks)
LCP	left-hand circular polarization
LDEF	Long Duration Exposure Facility
LED	light emitting diode
LEO	low Earth orbit
LET	linear energy transfer
LeRc	Lewis Research Center
LGA	low gain antenna
LH2	liquid hydrogen
LICS	line item cost summary
LLF	launch vehicle load factor
LMC	link monitor and control
LNA	low noise amplifier
LO2	liquid oxygen
LOC	line(s) of code
LOE	level of effort
Lon, lon	Longitude
Loos	launch operations and orbital support
LOS	loss of signal or line of sight
LOX	liquid oxygen (also LO2)
LPF	low pass filter
LPM	liquid propulsion module
LRU	line replacement unit
LTF	leak test facility
LV	launch vehicle
LVA	launch vehicle adapter
lx	lux (SI unit)
M	million(s) or mega
m	meter(s)
MA	multiple access
MAC	mass acceleration curve
MAF	Michaud Assembly Facility
M&R	maintenance and repair
MAT	multiple access transponder
Mb	megabit(s)

MBA	multiple beam antenna
MCC	mission control center
MCT	mission control team
MDA	multiple docking adapter
MDT	mission design team or mountain daylight time
MEA	main engine assembly (Shuttle)
MEL	master equipment list
MEO	medium Earth orbit
MEP	main electrical plug (medium altitude)
Metsat	meteorology satellite
MGA	medium-gain antenna
MGN	Magellan spacecraft
MGS	Mars Global Surveyor
MGSE	mechanical ground equipment
MHz	megahertz
mi	mile
MILA	Merritt Island
MIL-STD	military standard
min	minute(s)
MIPS, mips	millions of instructions per second
MIRF	mmneuver implementation/reconstruction file
ml	milliliter(s)
MLI	multilayer insulation
mm	millimeter(s)
MMH	monomethylhydrazine
MMIC	monolithic microwave integrated circuit
MMS	multimission modular spacecraft
MMU	manned maneuvering unit
MO	Mars Observer
MON	mixed oxides of nitrogen
MOI	Mars orbit insertion
MOS	mission operations system or metal oxide semiconductor
MOSFET	metal oxide semiconductor field effects transistor
M&P	materials and processes
MPDF	maneuver performance data file
Mrad	milliradians
MRB	material review board
ms	millisecond(s)
MSFC	Marshall Space Flight Center
MST	mountain standard time
MTBF	mean time between failure
MU	memory unit
MUX	multiplex or multiplexer
MW	momentum wheel
N	newton(s) or north
N/A	not applicable
NACA	National Advisory Committee for Aeronautics (NASA predecessor)

NAR	non-advocate review
NASA	National Aeronautics and Space Administration
NASCOM	NASA Communications Network
NATO	North Atlantic Treaty Organization
NBFM	narrow band frequency modulation
NDE	nondestructive evaluation
NDI	nondestructive inspection
NDT	nondestructive test
NE	near encounter phase in a fly-by mission
Ni	nickel
NiCd	nickel–cadmium
NiH2	nickel–hydrogen
N-m	newton-meter(s)
nm	nautical mile or nano meter (10^{-6} meter)
nmi	nautical mile
N-MOS	N-type metal oxide semiconductor
NOAA	National Oceanic and Atmospheric Administration
NPSH	net positive suction head
NRE	nonrotating Earth
NRL	Naval Research Laboratory
NRZ	non-return to zero
NSI	NASA standard initiator (ordnance device)
NSF	National Science Foundation
NSPAR	nonstandard parts approval list
NSSC-1	NASA Standard Computer, model 1
nT	nano tesla
NTO	nitrogen tetroxide
NVR	nonvolatile residue
OAO	Orbiting Astronomy Observatory
OBC	onboard computer
OBDH	onboard data handling
OBS	onboard software
OCC	operations control center
OMB	Office of Management and Budget
OMS	orbital maneuvering system (Shuttle)
OR	operations research
OSC	orbital sciences corporation
OSD	Office of Secretary of Defense
OSI	open system interconnect protocol
OSO	orbiting solar observatory
OSR	optical solar reflector
OT&E	operational test and evaluation
OTM	orbital trim maneuver
OTV	orbital transfer vehicle
OWLT	one way light time (Earth to spacecraft)
Pa	pascal
P-MOS	P-type metal oxide semiconductor

PA	product assurance
PAM	payload assist module
PAM-A	payload assist module-Atlas size
PAM-D	payload assist module-Delta size
PBAA	polybutadiene-acrylic acid polymer
PBAN	polybutadiene-acrylic acid-acrylonitrile terpolymer
PCA	propellant control assembly
PCB	parts control board
PCM	pulse code modulation or phase change material
PCU	power control unit
PDA	percent defective allowable
PDMA	phase division multiple access
PDT	Pacific daylight time
PDP	procurement data package
PDR	preliminary design review
PDT	Pacific daylight time
PDS	payload data system
PDU	power distribution unit
PE	post encounter
PFR	problem/failure report
Phz	peta hertz (10^{15})Hz
PI	principal investigator
PIP	payload integration plan
	project implementation plan
PIWG	payload integration working group
PLL	phase locked loop
PM	program manager or phase modulation
PMD	propellant management device
PMP	parts, materials, and processes
PMS	performance measuring system
PMSR	preliminary mission and system review
PN	pseudo-random noise
POC	payload operations control
POR	power-on reset
PPL	preferred parts list
PPT	peak power tracker (power system)
PQ	position quaternion
PR	procurement requisition
PRB	program review board
PRD	program pequirement document
PROM	programmable read-only memory
PROP	propulsion
PRK	phase-reverse shift keying
PRR	preship readiness review
PS	polysulfide
PSD	power spectral density
PST	Pacific standard time
PSU	pyro switching unit
PSK	phase shift keying

PTFE polytetrafluororthylene
Pu plutonium
PVC polyvinyl chloride
PWR power

QA quality assurance
QC quality control
QCM quartz crystal microbalance
QPL qualified parts list
QPSK quaternary phase shift keying
QTL quality test level

R/T real time
RA reliability assurance or right ascention
rad radian(s)
RAM random access memory
RBV return beam vidicon
RCP right circular polorized
RCS reaction control system
RCV receive
REA rocket engine assembly
REM rocket engine module
rf radio frequency
RFI radio frequency interference
RFP request for proposal
RFS radio frequency system
RFSE radio frequency support equipment
RGA rate gyro assembly
RHU radioisotope heater unit
RISC reduced instruction set computer
RIU remote interface unit
RLG ring laser gyro
RMS root mean square
ROM read-only memory or rough order of magnitude
RISC reduced instruction set computer
RP-1 rocket proplellant 1 (kerosene)
R-S Reed–Solomon (data code)
RSO range safety officer
RTI real time interrupt
RTLT round trip light time
RTG radioisotope thermoelectric generator
RTS remote tracking station
RTU remote terminal unit
RW reaction wheel
RWA reaction wheel assembly

S siemens (an SI unit)
s seconds

SA	solar array or single access
S&A	safe and arm unit (ordnance)
SAA	South Atlantic anomaly
SAD	solar array drive
SADE	solar array drive electronics
SADM	solar array drive mechanism
SAE	Society of Automotive Engineers
SAT	single access transponder
SAR	synthetic aperture radar, software acceptance review, or system acceptance review
SAW	surface acoustic wave
SB	small business
S-Band	range of frequencies of about 2 to 4 GHz
SBC	single board computer
SBE	S-band exciter
S/C	spacecraft
SCC	stress corrosion cracking
SCET	spacecraft event time
SCF	satellite control facility
SCLK	spacecraft clock
SCP	standard control processor
SCRAMJET	supersonic combustion RAMJET
SCT	spacecraft team
SCU	signal conditioning unit
SD	science data
SDF	science data formater
SDI	Strategic Defense Initiative
SDRL	supplier data requirements list
SE	support equipment
SEB	source evaluation board
SEC	single error correction
sec	seconds
SEE	single event effects (high energy particle events, usually electronics effects)
SEF	sequence of events file
SEL	single event latch up
SEMP	system engineering management plan
SEU	single event upset
SFOF	space flight operations facility
SGL	space to ground link
SI	international system of units
Si	silicon
SIPS	spacecraft interface performance specification
SIS	software interface specification
SLC	spacecraft launch complex
SLOC	source lines of code
SLR	satellite laser ranging
SMP	software management plan
SMM	solar max mission

SNAP system for nuclear auxiliary power (RTG)
SNR signal to noise ratio
SOC state of charge
SOE sequence of events
SOW statement of work
SPC stored program command
SPF single point failure
SPL sound pressure level
SPV single pressure vessel
SPS satellite power system
S3R sequential switching shunt regulation
SRCR software review certification record
SRD system requirements document
SRM solid rocket motor
SRP solar radiation pressure
SR&QA safety, reliability, and quality assurance
SRR system requirements review
SRS software requirements specification
SS subsystem
SSA sun sensor assembly or solid state amplifier or Space Station Alpha
SSB single side band
SSD Space Systems Division (Air Force)
SSDR solid state data recorder
SSE space support equipment
SSM single surface mirror
SSMA spread-spectrum multiple access
SSO sun synchronous orbit
SSPA solid state power amplifier
SSR software requirements specification or solid state recorder or sequential shunt regulator
STD standard
STDN Spacecraft Tracking and Data Network
STR or STRU structure and mechanisms
STS Space Transportation System (Shuttle)
STScI Space Telescope Science Institute
SUG software users guide
SSME Space Shuttle main engine
Sv sievent (SI unit)
S/W software
SWG science working group

T tetra = 10^{12} or tesla
TacSat tactical communication satellite
TAG technical advisory group
TAU thousand AU mission
TBD to be determined
TBS to be supplied

TC	thermal control
TCD	trajectory characteristics document
TCM	trajectory correction maneuver
TCS	thermal control system
TCV	thrust chamber valve
TDL	telemetry display language
TDM	time division multiplexer or technical direction memorandum
TDMA	time division multiple access
TDR	test discrepancy document
TDRS	Tracking and Data Relay Satellite
TDRSS	Tracking and Data Relay Satellite System
T&H	transportation and handling
TID	total ionizing dose
TIM	technical interchange meeting
TIROS	television infrared observation satellite
TLM	telemetry
TML	total mass loss
TNT	trinitrotoluene
TOF	time of flight
TOS	transfer orbit stage
TPA	telemetry processor assembly
TPM	technical performance measurement
TQM	total quality management
TRL	technical readiness level
TRR	test readiness review
TS	telecommunication subsystem
TTL	transistor-transistor logic
TT&C	telemetry, tracking, and command
TVC	thrust vector control
TWT	traveling wave tube
TWTA	traveling wave tube amplifier
TX	transmission or transmitter
TYP	typical

UARS	Upper Atmospheric Research Satellite
UCD	user control directive
UDMH	unsymmetrical dimethyl hydrazine
UHF	ultra high frequency
UQPSK	unbalanced quadrature phase shift keying
USA	upper stage adapter
USB	upper side band or unified S-band
USAF	United States Air Force
USI	user interface
USNO	United States Naval Observatory
USO	ultra-stable oscillator
UT	universal time
UTC	universal time, coordinated
UV	ultraviolet

V	volt(s)
Vac	volt(s) alternating current
VAB	vehicle assembly building (Shuttle)
VAFB	Vandenberg Air Force Base
VC	virtual channel
VCDU	virtual channel data unit
VCHP	variable conductance heat pipe
VCO	voltage controlled oscillator
Vdc	voltage direct current
VHF	very high frequency
VHSIC	very high speed integrated circuit
VLAC	verification load cycle
VLBI	very long baseline interferometry
VLSI	very large scale integrated circuit
VOI	Venus orbit insertion
VOIR	Venus orbiting imaging radar
VRM	Venus radar mapper
V&V	verification and validation
W	watt(s)
WARC	World Administrative Radio Conference
Wb	weber
WBFM	wide band frequency modulation
WBS	work breakdown structure
WCA	worst-case analysis
W-h	watt hour(s)
WI	Wallops Island
WSMC	Western Space and Missile Center
WSTF	White Sands Test Facility
WRT	with respect to
WS	work station
WSPG	White Sands Proving Ground
Wt	weight
WTR	WesternTtest Range
X	experimental
XPD	cross polar discrimination
XSU	cross-strap unit
YSM	yaw steering mode
ZOE	zone of exclusion
ΔV	Delta V (velocity change)
μm	micrometer, 10^{-6} m
2-DOF	two degrees of freedom
3D, 3-D	three dimensional
3-DOF	three degrees of freedom

Appendix B
Spacecraft Design Data

Appendix B is the professional reference appendix. Once you understand how to make the calculations described in the preceding chapters, this is the place to look up all of those numbers you are going to need.

B.1 Constants and Conversions

The most common constants and unit conversions are listed in the following three tables. For a comprehensive set of conversions to the SI system see NIST *Guide for the Use of the International System of Units (SI).*[1]

Table B.1 Important constants

Parameter	English unit	SI unit
Astronomical unit	92,955,620 miles	149,597,870 km
Density of water (4°C)	62.4266 lb/ft^3	999.978 kg/m^3
e	2.71626	same
Gravitational acceleration-Earth	32.1740 ft/s^2	9.08665 m/s^2
Julian century	36,525 days	same
Light year	5.879×10^{12} miles	9.4607×10^{12} km
Mechanical equivalent heat, J	778 ft-lb/Btu	4190 calories/joule
Pi	3.1415926535	same
Solar constant at Earth	434.9 ± 1.6 Btu/ft^2-h	1371 ± 5 W/m^2 (NASA)
Speed of light (vacuum)	9.8357×10^8 fps	299,792.5 km/s
Standard atmosphere	14.696 psia	1.01325×10^5 pascal
Universal gas constant, R_u	1545 ft-lb$_f$/lb^m-mole °R	8314 N-m/kg-mole-°K

Table B.2 Prefixes for exponents 10^n

n	Prefix	n	Prefix
1	deka	−1	deci
2	hecto	−2	centi
3	kilo	−3	milli
6	mega	−6	micro
9	giga	−9	nano
12	tetra	−12	pico

Table B.3 Common unit conversions

Multiply		By	To get
Acceleration	ft/s^2	0.3048	m/s^2
Angles	deg	0.01745329252	rad
	radians	57.2957795131	deg
Area	ft^2	0.09290304	m^2
	$in.^2$	6.4516	cm^2
Density	$lb/in.^3$	27679.905	kg/m^3
	$lb/in.^3$	2.7679905×10^{-2}	kg/cc
	lb/ft^3	16.018463	kg/m^3
	lb/ft^3	0.01601846	gm/cc
Force	lb_f	4.448221615	N
Impulse, specific	s	9.80665	N-s/kg
Impulse, total	lb-s	4.448221615	N-s
I_{sp}	$lb_f/s\text{-}lb_m$	9.8066516	N-s/kg
Length	ft	0.3048	m (exact)
	in.	0.0254	m (exact)
	nm	1.852	km (exact)
	stat mile	1.60934	km
	km	0.62137	stat mile
Mass	lb_m	0.45359237	kg (exact)
	kg	2.2046	lb
	slug	14.594	kg
Moment of inertia	$slug\text{-}ft^2$	1.355818	$kg\text{-}m^2$
Momentum	$lb_f\text{-}ft\text{-}s$	1.355818	N-m-s
Pressure	psi	6894.7572	pascal, Pa
	psf	47.880258	pascal, Pa
Speed	km/s	3280.84	ft/s
	ft/s	0.3048	m/s (exact)
	rpm	0.1047198	rad/s
Torque	ft-lb	1.355818	N-m
Thrust	lbf	4.448222	N
Temperature	°F	°F + 459.7	°R
	°C	°C + 273.2	°K
Velocity	ft/s	0.3048	m/s (exact)
	km/s	3280.84	ft/s
Volume	ft^3	0.0283168	m^3
	$in.^3$	1.6387×10^5	m^3

Table B.4 SI contrived units

Quantity	SI name	Symbol	Fundamental units
Frequency	Hertz	Hz	1/s
Force	Newton	N	kg-m/s^2
Pressure, stress	Pascal	Pa	N/m^2
Energy, work, heat	Joule	J	N-m
Power	Watt	W	N-m/s
Electrical charge	Coulomb	C	A-s
Capacitance	Farad	F	A-s/V
Electrical resistance	Ohm	Ω	V/A
Conductance	Siemens	S	A/V
Magnetic flux	Weber	Wb	V-s
Magnetic flux density	Tesla	T	V-s/m^2, N/A-m
Inductance	Henry	H	V-s/A
Absorbed dose	Gray	Gy	J/kg
Radioactive source activity	Becquerel	Bq	1/s

B.2 Mass Estimating and Budgeting

This section contains weight-estimating relationships that can be used for very early estimates of on-orbit dry mass. These relationships are based on statistics from prior spacecraft. Note that estimating relationships based on history, by definition, do not reflect the latest state of the art. Figure B.1 shows the on-orbit dry mass of seven geosynchronous communication satellites as a function of payload mass.

A JPL study of 46 Earth-orbiting navigation, communications, meteorology, Earth resources, and astronomy spacecraft showed that the on-orbit dry mass was bounded by the limits of 3 and 7 times the payload mass with the average ratio being 4.8; Fig. B.2.

Figure B.3 shows planetary spacecraft on-orbit dry mass vs payload mass for 11 planetary spacecraft.

The classes of design listed in Table B.5 are class 1: a new spacecraft (one of a kind, first generation); class 2: the next generation spacecraft based on a previously developed family, which expands in complexity or capability within an established design envelope; and class 3: a production-level development based on an existing design for which multiple units are planned and where a significant amount of standardization exists.

In Table B.6, for communication satellites, telecommunication is not shown separately; it is included in the CDS mass. The subsystem weights are shown as percentages of spacecraft dry weight.

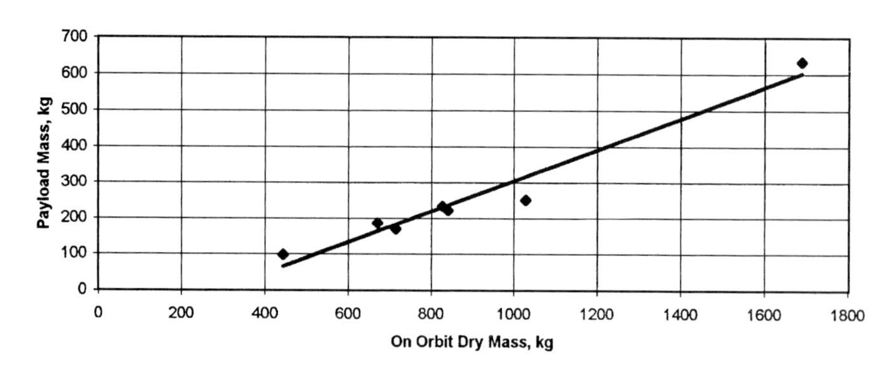

Fig. B.1 Communication spacecraft mass.

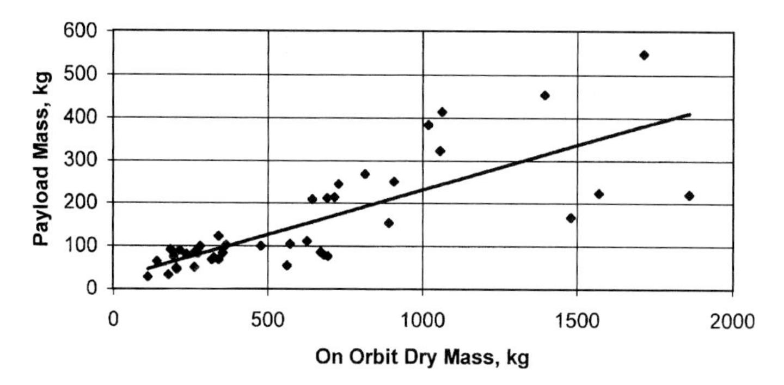

Fig. B.2 Earth orbiting spacecraft.

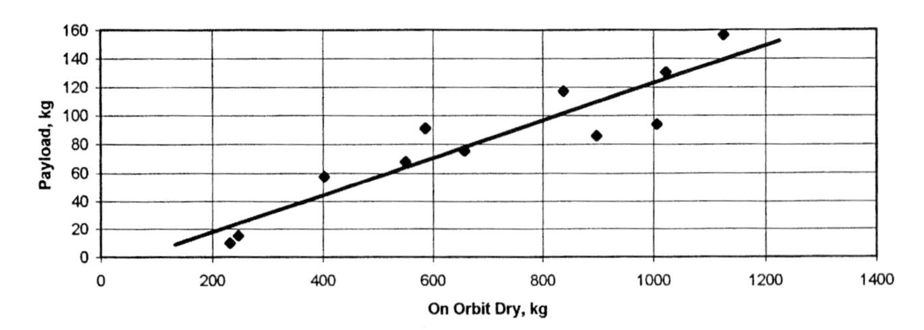

Fig. B.3 Planetary spacecraft mass.

Table B.5 AIAA recommended mass contingencies[a]

	Minimum standard weight contingencies, %														
	Proposal stage			Design development stage											
	Bid Class			*CoDR* Class			*PDR* Class			*CDR* Class			*PRR* Class		
Description/ categories	1	2	3	1	2	3	1	2	3	1	2	3	1	2	3
Category AW, 0–50 kg 0–110 lb	50	30	4	35	25	3	25	20	2	15	12	1	0	0	0
Category BW, 50–500 kg 110–1102 lb	35	25	4	30	20	3	20	15	2	10	10	1	0	0	0
Category CW, 500–2500 kg 1102–5511 lb	30	20	2	25	15	1	20	10	0.8	10	5	0.5	0	0	0
Category DW, 2500 kg and up	28	18	1	22	12	0.8	15	10	0.6	10	5	0.5	0	0	0

[a]Data copyright AIAA, reproduced with permission; Ref. 2.

Table B.6 Subsystem on-orbit dry mass allocation guide

	Comsats[a]		Metsats[b]		Planetary		Other	
Subsystem	with P/L[c]	GFE P/L	with P/L	GFE P/L	with P/L	GFE P/L	with P/L	GFE P/L
Structure, %	21	29	20	29	26	29	21	30
Thermal, %	4	6	3	4	3	3	3	4
ACS, %	7	10	9	13	9	10	8	11
Power, %	26	35	16	23	19	21	21	29
Cabling, %	3	4	8	12	7	8	5	7
Propulsion, %	7	10	5	7	13	15	5	7
Telecom, %	—	—	4	6	6	7	4	6
CDS, %	4	6	4	6	6	7	4	6
Payload, %	28	—	31	—	11	—	29	—

[a]Comsat = communication satellite. [b]Metsat = meteorology or weather satellite. [c]P/L = payload.

B.3 Power Estimating and Budgeting

Table B.7 Total spacecraft power estimating relationships

Spacecraft mission	Power estimating relationship
Communications	$P_t = 1.1568 P_{pl} + 55.497$
Meteorology	$P_t = 602.18 \ln(P_{pl}) - 2761.4$
Planetary	$P_t = 332.93 \ln(P_{pl}) - 1046.6$
Other missions	$P_t = 210 + 1.3 P_{pl}$

Table B.8 Subsystem power allocation guide

	Percentage of subsystem total			
Subsystem	Comsats	Metsats	Planetary	Other
Thermal control	30	48	28	33
Attitude control	28	19	20	11
Power	16	5	10	2
CDS	19	13	17	15
Communications	0	15	23	30
Propulsion	7	0	1	4
Mechanisms	0	0	1	5

Table B.9 AIAA recommended power contingencies[a]

	Minimum standard power contingencies, %														
	Proposal stage			Design development stage											
	Bid Class			*CoDR* Class			*PDR* Class			*CDR* Class			*PRR* Class		
Description/ categories	1	2	3	1	2	3	1	2	3	1	2	3	1	2	3
Category AP, 0–500 W	90	40	13	75	25	12	45	20	9	20	15	7	5	5	5
Category BP, 500–1500 W	80	35	13	65	22	12	40	15	9	15	10	7	5	5	5
Category CP, 1500–5000 W	70	30	13	60	20	12	30	15	9	15	10	7	5	5	5
Category DP, 5000 W and up	40	25	13	35	20	11	20	15	9	10	7	7	5	5	5

[a]Data copyright AIAA, reproduced with permission; Ref. 2.

Design maturity is expressed in Table B.9 as a function of the major reviews conducted during a design: the bid (proposal or bid stage), CoDR (conceptual design review), PDR (preliminary design review), CDR (critical design review), PRR (preship readiness review), and FRR (flight readiness review).

The classes of design listed in Table B.9 are class 1: a new spacecraft (one of a kind first generation); class 2: the next generation spacecraft based on a previously developed family, which expands in complexity or capability within an established design envelope; and class 3: a production-level development based on an existing design for which multiple units are planned and where a significant amount of standardization exists.

B.4 Launch Vehicle Interface Checklist

Payload accommodation data of the following type are available for essentially any launch vehicle in the *International Reference Guide to Space Launch Systems*.[3]

Integration issue	Accommodation
Payload Compartment	
Maximum payload diameter	
Maximum length	
Payload Integration	
Payload adapter diameter	
Integration start	
Last countdown hold w/o recycling	
On-pad storage capability	
Last access to payload	
Environment	
Maximum axial load	
Maximum lateral load	
Minimum lateral/longitudinal frequency	
Maximum acoustic level	
Overall sound pressure	
Maximum flight shock	
Maximum dynamic pressure on fairing	
Maximum aeroheating rate at fairing separation	
Maximum pressure change rate in fairing	
Cleanliness level in fairing	
Payload Delivery	
Orbit injection accuracy (3 sigma)	
Attitude accuracy (3 sigma)	
Nominal payload separation rate	
Deployment rotation rate available	
Loiter duration in orbit	
Maneuvers available (thermal, collision avoidance)	
Multiple payloads possible	

The launch vehicle adapter is an assembly that adapts the launch vehicle structure to the spacecraft structure and provides for spacecraft separation; the adapter mass is a strong function of the spacecraft mass, as shown in Fig. B.4.

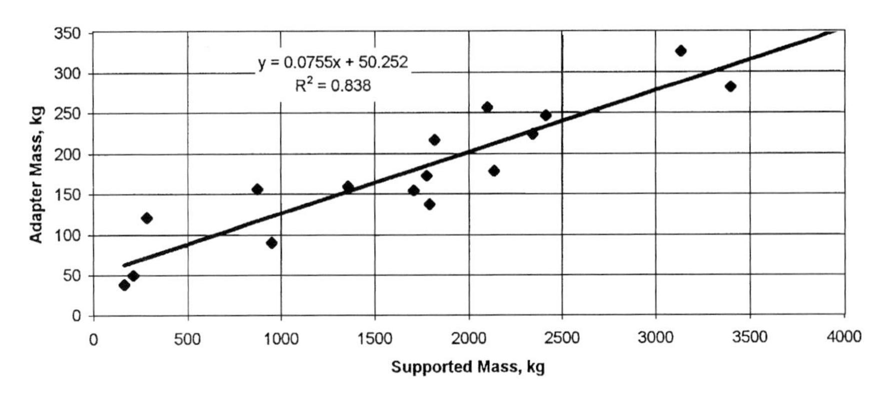

Fig. B.4 Launch vehicle adapter mass.

The equation in Fig. B.4 can be rounded to

$$LVA = 0.0755LM + 50 \tag{B.1}$$

where

LVA = launch vehicle adapter mass, kg
LM = spacecraft launch mass, kg

B.5 Orbital Mechanics Data

The parameters in Table B.10 are the most frequently used planetary constants. The gravitational parameters are from Lyons and Dallas.[4] The mean equatorial radii and zonal coefficients are from *The Astronautical Almanac*.[5] Table B.11 lists the physical characteristics of the planets, sun, and moon extracted from Refs. 4–6.

Table B.10 Frequently used planetary constants

Body	μ, km^3/s^2	R_0, km	A_r, deg/s	J_2	Mean solar distance, km $\times 10^6$
Mercury	22,032.1	2439.7	0.0000711		57.9
Venus	324,858.8	6051.8	−0.0000171	0.000027	108.2
Earth	**398,600.4**	**6378.14**	**0.0041781**	**0.00108263**	**149.6**
Mars	42,828.3	3397	0.0040613	0.001964	228.0
Jupiter	126,711,995.4	71,492	0.0100756	0.01475	778.4
Saturn	37,939,519.7	60,268	0.0093843	0.01645	1433
Uranus	5,780,158.5	25,559	−0.0058005	0.012	2883
Neptune	6,871,307.8	24,764	0.0062073	0.004	4517
Pluto	1020.9	1195	−0.0006524		5820
Moon	4902.8	1737.4	0.0001525	0.0002027	
Sun	132,712,439,935.5	696,000	0.0001642		

Table B.11 Physical characteristics of the planets, sun, and moon

Symbol		☿	♀	⊕	♂	♃	♄	♅	♆	♇	⊙	☾
Planet		Mercury	Venus	Earth	Mars	Jupiter	Saturn	Uranus	Neptune	Pluto	Sun	Moon
Natural satellites		0	0	1	2	16	18	15	8	1	9	0
Equatorial radius,	⊕ = 1	0.383	0.949	1.00	0.533	11.20	9.45	4.01	3.88	0.19	109.12	0.272
R_0	km	2439.7	6051.8	6378.14	3397.0	71492	60268	25559	24764	1195	696000	1737.4
	miles	1516.0	3760.4	3963.19	2110.9	44423.1	37449.0	15881.9	15387.6	742	432474	1080
Oblateness	J2	—	2.7E-5	0.00108	0.00196	0.01475	0.01645	0.012	0.004	—	—	202.7E-6
density	⊙ = 1	3.85	3.72	3.91	2.80	0.94	0.50	0.92	1.25	0.78	1.0	2.37
	⊕ = 1	0.985	0.950	1.00	0.71	0.241	0.127	0.236	0.319	0.2	0.255	0.606
	gm/cc	5.43	5.24	5.515	3.94	1.33	0.70	1.30	1.76	1.1	1.409	3.34
	lb/ft^3	339	327	344.29	246	83	43.7	81.2	109.9	69	87.96	208.8
Mass[a]	⊙ = 1	1.66E-7	2.45E-6	3.00E-6	3.23E-7	9.55E-4	2.858E-4	4.36E-5	5.17E-5	8E-9	1.00	3.7E-8
	⊕ = 1	0.0552	0.815	1.00	0.1074	317.83	95.16	14.50	17.20	0.0025	3.329E-5	0.0123
	$kg \times 10^{24}$	0.33022	4.869	5.9742	0.64191	1898.8	568.5	86.625	102.78	0.015	1.9891E-6	0.073483
Surface gravity	⊕ = 1	0.377	0.904	1.00	0.378	2.351	0.914	0.792	1.12	0.72	27.94	0.165
(Equatorial)	cm/s^2	370	887	980.665	371	2312	896	777	1100	720	27398	162
	ft/s^2	12.130	29.085	32.1740	12.162	75.641	29.407	25.482	36.035	23.165	898.942	5.309
Inclination of equator[b]	deg	0.01	177.36	23.45	25.19	3.13	26.73	97.77	28.32	122.53	—	6.68
Period of rotation[c]	days	58.6462	−243.019	0.99727	1.02596	0.41354	0.44401	−0.71833	0.67125	−6.387.2	25.38	27.32166
Rotation rate, A_r	deg/s $\times 10^{-4}$	0.7111	−0.171	41.781	40.613	100.756	93.843	−58.005	62.073	−6.524	1.6417	1.525
Albedo[d]		0.106	0.65	0.367	0.150	0.52	0.47	0.51	0.41	0.30	—	0.12
Surface	°F	322	854	59	−67	−140	−290	−360	−362	−370	9944	−72
Temperature	°K	440	730	288	218	129	97	58	56	50	5780	215[e]
Pressure	bar	2E-15	90	1	0.007	—	—	—	—	1.2E-5	—	—

[a]Excludes satellites; includes atmosphere.
[b]Inclination of the equator to the orbital plane of the planet.
[c]Period of rotation is the rotation of the equator with respect to a fixed reference frame in sidereal days. A negative sign indicates a retrograde rotation with respect to the pole in the northern hemisphere. The period is measured in days of 86,400 sidereal seconds.
[d]Albebo listed is the ratio of illumination of the planet at zero phase angle to the illumination produced by a plane, absolutely white Lambert surface at the same radius and position as the planet.
[e]Lunar surface temperature is for the sunny side; dark side = 123°K.

Table B.12 Orbital elements of the planets[a,b]

Symbol		☿	♀	⊕	♂	♃	♄	♅	♆	♇
Planet		Mercury	Venus	Earth	Mars	Jupiter	Saturn	Uranus	Neptune	Pluto
Semimajor axis, a	AU	0.387098	0.723327	0.99998	1.52372	5.2033	9.58078	19.2709	30.1927	39.3782
	Mil km	57.90904	108.2082	149.5979	227.9423	778.4026	1433.2643	2882.8856	4516.7636	5890.8948
Perihelion radius	Mil km	46.0015	107.474	147.204	206.673	740.580	1355.061	2758.835	4467.034	4438.990
Aphelion radius	Mil km	69.817	108.942	152.092	249.212	816.225	1511.467	3006.936	4566.493	7342.800
Sphere of influence	Mil km	0.111	0.616	0.924	0.577	48.157	54.796	51.954	80.196	3.400
Eccentricity, e		0.205625	0.006785	0.01667	0.09331	0.04859	0.054563	0.043030	0.01101	0.246466
Inclination, i	deg	7.0050	3.3946	0.0	1.8498	1.3047	2.4853	0.7730	1.7677	17.1365
Period	days	87.969	224.70	365.26	686.986	4335.28	10831.77	30899.498	60597.03	90257.27
	$\oplus = 1$	0.240	0.6699	1.0	1.881	11.869	29.654	84.59	165.897	247.097
Synodic period	days	115.88	583.92	—	779.94	398.88	378.09	369.66	367.49	366.73
Perihelion velocity	km/s	58.976	35.259	30.285	26.496	13.708	10.163	7.083	5.481	6.105
Aphelion velocity	km/s	38.859	34.784	29.292	21.974	12.4376	9.111	6.499	5.361	3.690
Mean motion	deg/day	4.0923	1.6021	0.9856	0.5240	0.08308	0.0536	0.04303	0.01101	0.2464
Longitude:										
Ascending node, Ω	°	48.318	76.671	—	49.550	100.468	113.632	74.048	131.785	110.321
Perihelion, ϖ	°	77.438	131.25	103.059	336.011	15.627	88.626	172.887	22.04	224.462

[a]Elements for 13 December 1998, referred to the mean ecliptic of date. Elements change slowly with time.
[b]Data for Earth are actually for the moon–Earth barycenter.

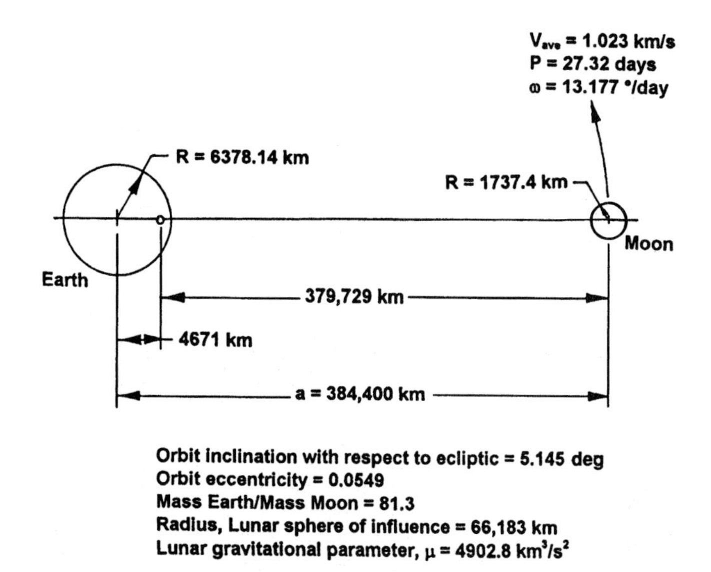

Fig. B.5 Characteristics of the Earth–moon pair.

Table B.12 lists the orbital elements of the planets for 13 December 1998 (JD 2451160.5) referred to as the mean ecliptic of date. Semimajor axis, eccentricity, and longitude are from *The Astronautical Almanac*[5]; the remaining elements are calculated assuming two-body motion. Orbital elements change with time; for precise work see *The Astronautical Almanac*. Figure B.5 shows the important characteristics of the Earth/moon pair.

B.6 Propellant Properties

Table B.13 Properties of propellants

Propellant	Symbol	Molecular weight	Freezing point, °F	Boiling point, °F	Density @ 68°F	Vapor pressure psia	Vapor pressure °F
Chlorine triflouride	ClF_3	92.46	−105.4	53.15	1.825	20.8	110
Fluorine	F_2	38	−363	−307	1.51[a]	5.0	−322
Hydrazine	N_2H_4	32.05	35.6	236.3	1.008	0.2	68
Hydrogen	H_2	2.02	−435	−423	0.071[a]	1.02	−435
MMH	$CH_3N_2H_3$	46.08	−62.1	188.2	0.8765	0.70	68
Nitric acid	HNO_3	63.02	−42.9	185.5	1.513	0.93	68
Nitrogen tetroxide	N_2O_4	92.02	11.8	70.1	1.447	13.92	68
Oxygen	O_2	32	−361.8	−297.6	1.14[a]	7.35	−308
RP-1	$CH_{1.9-2.0}$	175	−48	360 to 500	0.806	0.02	68
UDMH	$(CH_3)_2N_2H_2$	60.10	−71	146	0.793	2.38	68

[a]At normal boiling point.

B.7 Tsiolkowski Equations

For zero g, no drag conditions:

$$\Delta V = g_c I_{sp} \ln\left(\frac{W_i}{W_f}\right) \tag{B.2}$$

$$W_p = W_i\left[1 - \exp\left(-\frac{\Delta V}{g_c I_{sp}}\right)\right] \tag{B.3}$$

$$W_p = W_f\left[\exp\left(\frac{\Delta V}{g_c I_{sp}}\right) - 1\right] \tag{B.4}$$

W_i = initial vehicle weight, lb
W_f = final vehicle weight, lb
W_p = propellant weight required to produce the given ΔV, lb
ΔV = velocity increase of the vehicle, ft/s
g_c = gravitational constant, 32.1740 ft/s^2

Density of nitrogen tetroxide (from 60° to 160°F):

$$\rho = 91.060 - 0.0909(T - 60)\ \text{lb/ft}^3$$

where T = liquid temperature °F.

Density of hydrazine (from 0 to 60°C):

$$\rho = 65.0010 - 0.0298(°\text{F}) - 8.7023\text{E-6}(°\text{F})^2,\ \text{lb/ft}^3$$

$$\rho = 1025.817 - 0.8742(°\text{C}) - 0.0005(°\text{C})^2,\ \text{kg/m}^3$$

Table B.14 Specific gravity of common propellants

Propellant	Temp., °F	Specific gravity
Hydrazine	67	1.008
	120	0.984
Liquid hydrogen	−423	0.071
Monomethylhydrazine	67	0.8788
	100	0.8627
Nitrogen tetroxide	67	1.447
	120	1.37
Liquid oxygen	−320	1.23
	−297	1.14
RP-1	76	0.807
	300	0.58

Table B.15 Liquid propellant combustion gas properties

Propellant	Property	Value
Hydrazine		
	Molecular weight	13.04
	Ratio of specific heats	1.27
	Combustion temperature, °R	2210
	Density, lb/ft^3	62.67
N_2O_4/MMH		
	Molecular weight	21.52
	Ratio of specific heats	1.25
	Combustion temperature, °R	6145
	Bulk density, lb/ft^3	71.67
	Mixture ratio	1.50

Table B.16 Solid propellant properties

Propellant	Property	Value
PBAA/AP/AL		
	Molecular weight	22
	Ratio of specific heats	1.26
	Combustion temperature, °R	6260
	Density, lb/in.^{3}	0.064
	Burning rate, in./s	0.32
	Pressure exponent	0.35
PBAN/AP/AL		
	Molecular weight	22
	Ratio of specific heats	1.26
	Combustion temperature, °R	6260
	Density, lb/in.^{3}	0.064
	Burning rate, in./s	0.55
	Pressure exponent	0.33
CTPB/AP/AL		
	Molecular weight	22
	Ratio of specific heat	1.26
	Combustion temperature, °R	6160
	Density, lb/in.^{3}	0.064
	Burning rate, in./s	0.45
	Pressure exponent	0.40
HTPB/AP/AL		
	Molecular weight	22
	Ratio of specific heats	1.26
	Combustion temperature, °R	6160
	Density, lb/in.^{3}	0.065
	Burning rate, in./s	0.282
	Pressure exponent	0.30

(continued)

Table B.16 Solid propellant properties (continued)

Propellant	Property	Value
PS/AP/AL		
	Molecular weight	22
	Ratio of specific heats	1.67
	Combustion temperature, °R	5460
	Density, lb/in.^{3}	0.062
	Burning rate, in./s	0.31
	Pressure exponent	0.33
PVC/AP/AL		
	Molecular weight	22
	Ratio of specific heats	1.26
	Combustion temperature, °R	6260
	Density, lb/in.^{3}	0.064
	Burning rate, in./s	0.45
	Pressure exponent	0.35

B.8 Gas and Material Properties

Table B.17 Constants for gases

Gas	M	cp	cv	k	R
Air	29	0.24	0.1715	1.4	53.3
Argon, A	39.9	0.125	0.0749	1.668	38.7
Hydrogen	2.016	3.421	2.4354	1.405	767
Nitric Oxide, NO	30	0.2378	0.1717	1.384	51.4
Nitrogen	28	0.2484	0.1776	1.4	55.1
Oxygen	32	0.2193	0.1573	1.394	48.25
Helium	4	1.25	0.754	1.659	386
Propane	44.1	0.404	0.360	1.12	34.1
Freon 14	88			1.3	17.5

Table B.18 Material density

Material	Density, kg/m^3	Density, lb/in.^{3}
Elastomer	997	0.036
Teflon®	2130	0.077
Aluminum	2710	0.098
Titanium	4430	0.160

B.9 General Arrangement Checklist

1) Science instrument FOV
2) Antennas FOV
3) Rocket engine impingement; clear aft of exit plane
4) Ordnance shock path
5) Electronics heat rejection
   - free view of space
   - sun-free face
6) Shunt radiator heat rejection
   - free view of space
   - sun-free face
7) ACS sensor FOV
8) ACS sensors and gyros on common plate
9) Propulsion equipment mounted with tanks, modular design possible
10) Maximum moment arm for ACS thrusters
11) Propellant tanks balanced with respect to cg, at any propellant load
12) Telecommunication electronics near antennas, especially HGA
13) Sun OK faces picked, sun vector located
14) Sun-free faces picked
15) Light sensitive instruments on sun-free faces (cameras, star scanners, horizon sensors, spectrometers)
16) Center of mass near geometric center
17) Moments of inertia acceptable (especially spinners)
18) Structure types picked, load paths reasonable
19) Batteries together and near the control electronics
20) C&DH near ACS if common computer used

References

[1]Taylor, B. N., *Guide for the Use of the International System of Units (SI)*, U.S. Dept. of Commerce Special Publ. 811, Washington, DC, April 1995.

[2]*Guide for Estimating and Budgeting Weight and Power Contingencies*, ANSI/AIAA-G-020, April 1992.

[3]Isakowitz, S. J., Hopkins, J. P., and Hopkins, J. B., *International Reference Guide to Space Launch Systems*, 3rd ed., AIAA, Washington, DC, 1991.

[4]Lyons, D. T., and Dallas, S. S., *Magellan Constants and Models Document*, Jet Propulsion Lab., PD 630-79, Rev. C, California Inst. of Technology, Pasadena, CA, 1988.

[5]*The Astronautical Almanac for the Year 1998*, U.S. Naval Observatory and Royal Greenwich Observatory, U.S. Government Printing Office, Washington, DC, 1998.

[6]Lang, R. K., *Astrophysical Data: Planets and Stars*, Springer-Verlag, New York, 1992.

Index

TEXTS PUBLISHED IN THE AIAA EDUCATION SERIES

Elements of Spacecraft Design
Charles D. Brown 2002
ISBN 1-56347-524-3

Dynamics, Control, and Flying Qualities of V/STOL Aircraft
James A. Franklin 2002
ISBN 1-56347-575-8

Orbital Mechanics, Third Edition
Vladimir A. Chobotov 2002
ISBN 1-56347-537-5

Basic Helicopter Aerodynamics, Second Edition
John Seddon and Simon Newman 2001
ISBN 1-56347-510-3

Aircraft Systems: Mechanical, Electrical, and Avionics Subsystems Integration
Ian Moir and Allan Seabridge 2001
ISBN 1-56347-506-5

Design Methodologies for Space Transportation Systems
Walter E. Hammond 2001
ISBN 1-56347-472-7

Tactical Missile Design
Eugene L. Fleeman 2001
ISBN 1-56347-494-8

Flight Vehicle Performance and Aerodynamic Control
Frederick O. Smetana 2001
ISBN 1-56347-463-8

Modeling and Simulation of Aerospace Vehicle Dynamics
Peter H. Zipfel 2000
ISBN 1-56347-456-6

Applied Mathematics in Integrated Navigation Systems
Robert M. Rogers 2000
ISBN 1-56347-445-X

Mathematical Methods in Defense Analyses, Third Edition
J. S. Przemieniecki 2000
ISBN 1-56347-396-6

Finite Element Multidisciplinary Analysis
Kajal K. Gupta and John L. Meek 2000
ISBN 1-56347-393-3

Aircraft Performance: Theory and Practice
M. E. Eshelby 1999
ISBN 1-56347-398-4

Space Transportation: A Systems Approach to Analysis and Design
Walter E. Hammond 1999
ISBN 1-56347-032-2

Civil Jet Aircraft Design
Lloyd R. Jenkinson, Paul Simpkin, and Darren Rhodes 1999
ISBN 1-56347-350-X

Structural Dynamics in Aeronautical Engineering
Maher N. Bismarck–Nasr 1999
ISBN 1-56347-323-2

Intake Aerodynamics, Second Edition
E. L. Goldsmith and J. Seddon 1999
ISBN 1-56347-361-5

Integrated Navigation and Guidance Systems
Daniel J. Biezad 1999
ISBN 1-56347-291-0

Aircraft Handling Qualities
John Hodgkinson 1999
ISBN 1-56347-331-3

Performance, Stability, Dynamics, and Control of Airplanes
Bandu N. Pamadi *1998*
ISBN 1-56347-222-8

Spacecraft Mission Design, Second Edition
Charles D. Brown *1998*
ISBN 1-56347-262-7

Computational Flight Dynamics
Malcolm J. Abzug *1998*
ISBN 1-56347-259-7

Space Vehicle Dynamics and Control
Bong Wie *1998*
ISBN 1-56347-261-9

Introduction to Aircraft Flight Dynamics
Louis V. Schmidt *1998*
ISBN 1-56347-226-0

Aerothermodynamics of Gas Turbine and Rocket Propulsion, Third Edition
Gordon C. Oates *1997*
ISBN 1-56347-241-4

Advanced Dynamics
Shuh-Jing Ying *1997*
ISBN 1-56347-224-4

Introduction to Aeronautics: A Design Perspective
Steven A. Brandt, Randall J. Stiles, John J. Bertin, and Ray Whitford *1997*
ISBN 1-56347-250-3

Introductory Aerodynamics and Hydrodynamics of Wings and Bodies: A Software-Based Approach
Frederick O. Smetana *1997*
ISBN 1-56347-242-2

An Introduction to Aircraft Performance
Mario Asselin *1997*
ISBN 1-56347-221-X

Orbital Mechanics, Second Edition
Vladimir A. Chobotov, Editor *1996*
ISBN 1-56347-179-5

Thermal Structures for Aerospace Applications
Earl A. Thornton *1996*
ISBN 1-56347-190-6

Structural Loads Analysis for Commercial Transport Aircraft: Theory and Practice
Ted L. Lomax *1996*
ISBN 1-56347-114-0

Spacecraft Propulsion
Charles D. Brown *1996*
ISBN 1-56347-128-0

Helicopter Flight Dynamics: The Theory and Application of Flying Qualities and Simulation Modeling
Gareth D. Padfield *1996*
ISBN 1-56347-205-8

Flying Qualities and Flight Testing of the Airplane
Darrol Stinton *1996*
ISBN 1-56347-117-5

Flight Performance of Aircraft
S. K. Ojha *1995*
ISBN 1-56347-113-2

Operations Research Analysis in Test and Evaluation
Donald L. Giadrosich *1995*
ISBN 1-56347-112-4

Radar and Laser Cross Section Engineering
David C. Jenn *1995*
ISBN 1-56347-105-1

Introduction to the Control of Dynamic Systems
Frederick O. Smetana *1994*
ISBN 1-56347-083-7

Tailless Aircraft in Theory and Practice
Karl Nickel and Michael Wohlfahrt *1994*
ISBN 1-56347-094-2

Mathematical Methods in Defense Analyses, Second Edition
J. S. Przemieniecki *1994*
ISBN 1-56347-092-6

Hypersonic Aerothermodynamics
John J. Bertin 1994
ISBN 1-56347-036-5

Hypersonic Airbreathing Propulsion
William H. Heiser and David T. Pratt 1994
ISBN 1-56347-035-7

Practical Intake Aerodynamic Design
E. L. Goldsmith and J. Seddon 1993
ISBN 1-56347-064-0

Acquisition of Defense Systems
J. S. Przemieniecki, Editor 1993
ISBN 1-56347-069-1

Dynamics of Atmospheric Re-Entry
Frank J. Regan and Satya M. Anandakrishnan 1993
ISBN 1-56347-048-9

Introduction to Dynamics and Control of Flexible Structures
John L. Junkins and Youdan Kim 1993
ISBN 1-56347-054-3

Spacecraft Mission Design
Charles D. Brown 1992
ISBN 1-56347-041-1

Rotary Wing Structural Dynamics and Aeroelasticity
Richard L. Bielawa 1992
ISBN 1-56347-031-4

Aircraft Design: A Conceptual Approach, Second Edition
Daniel P. Raymer 1992
ISBN 0-930403-51-7

Optimization of Observation and Control Processes
Veniamin V. Malyshev, Mihkail N. Krasilshikov, and Valeri I. Karlov 1992
ISBN 1-56347-040-3

Nonlinear Analysis of Shell Structures
Anthony N. Palazotto and Scott T. Dennis 1992
ISBN 1-56347-033-0

Orbital Mechanics
Vladimir A. Chobotov, Editor 1991
ISBN 1-56347-007-1

Critical Technologies for National Defense
Air Force Institute of Technology 1991
ISBN 1-56347-009-8

Space Vehicle Design
Michael D. Griffin and James R. French 1991
ISBN 0-930403-90-8

Defense Analyses Software
J. S. Przemieniecki 1990
ISBN 0-930403-91-6

Inlets for Supersonic Missiles
John J. Mahoney 1990
ISBN 0-930403-79-7

Introduction to Mathematical Methods in Defense Analyses
J. S. Przemieniecki 1990
ISBN 0-930403-71-1

Basic Helicopter Aerodynamics
J. Seddon 1990
ISBN 0-930403-67-3

Aircraft Propulsion Systems Technology and Design
Gordon C. Oates, Editor 1989
ISBN 0-930403-24-X

Boundary Layers
A. D. Young 1989
ISBN 0-930403-57-6

Aircraft Design: A Conceptual Approach
Daniel P. Raymer 1989
ISBN 0-930403-51-7

Gust Loads on Aircraft: Concepts and Applications
Frederic M. Hoblit 1988
ISBN 0-930403-45-2

Aircraft Landing Gear Design: Principles and Practices
Norman S. Currey 1988
ISBN 0-930403-41-X

Mechanical Reliability: Theory, Models and Applications
B. S. Dhillon *1988*
ISBN 0-930403-38-X

Re-Entry Aerodynamics
Wilbur L. Hankey *1988*
ISBN 0-930403-33-9

Aerothermodynamics of Gas Turbine and Rocket Propulsion, Revised and Enlarged
Gordon C. Oates *1988*
ISBN 0-930403-34-7

Advanced Classical Thermodynamics
George Emanuel *1987*
ISBN 0-930403-28-2

Radar Electronic Warfare
August Golden Jr. *1987*
ISBN 0-930403-22-3

An Introduction to the Mathematics and Methods of Astrodynamics
Richard H. Battin *1987*
ISBN 0-930403-25-8

Aircraft Engine Design
Jack D. Mattingly, William H. Heiser, and Daniel H. Daley *1987*
ISBN 0-930403-23-1

Gasdynamics: Theory and Applications
George Emanuel *1986*
ISBN 0-930403-12-6

Composite Materials for Aircraft Structures
Brian C. Hoskin and Alan A. Baker, Editors *1986*
ISBN 0-930403-11-8

Intake Aerodynamics
J. Seddon and E. L. Goldsmith *1985*
ISBN 0-930403-03-7

The Fundamentals of Aircraft Combat Survivability Analysis and Design
Robert E. Ball *1985*
ISBN 0-930403-02-9

Aerothermodynamics of Aircraft Engine Components
Gordon C. Oates, Editor *1985*
ISBN 0-915928-97-3

Aerothermodynamics of Gas Turbine and Rocket Propulsion
Gordon C. Oates *1984*
ISBN 0-915928-87-6

Re-Entry Vehicle Dynamics
Frank J. Regan *1984*
ISBN 0-915928-78-7